Celebrating J.D. Murray's contributions to Mathematical Biology

Philip K. Maini · Mark A. J. Chaplain ·
Mark Lewis · Jonathan A. Sherratt
Editors

Celebrating J.D. Murray's contributions to Mathematical Biology

Spin-off from the journal *Bulletin of Mathematical Biology* in the Topical Collection: Celebrating J.D. Murray's contributions to Mathematical Biology

Springer

Editors
Philip K. Maini
Mathematical Institute, Radcliffe
Observatory Quarter
University of Oxford
Oxford, UK

Mark Lewis
Department of Mathematical
and Statistical Sciences
University of Alberta
Edmonton, AB, Canada

Mark A. J. Chaplain
School of Mathematics and Statistics
University of St. Andrews
St. Andrews, UK

Jonathan A. Sherratt
Department of Mathematics
Heriot-Watt University
Edinburgh, UK

ISBN 978-1-0716-2406-7

This Springer imprint is published by the registered company Springer Science+Business Media, LLC part of Springer Nature.
The registered company address is: 1 New York Plaza, New York, NY 10004, U.S.A.

Contents

Bulletin of Mathematical Biology (2022) 84:13
https://doi.org/10.1007/s11538-021-00955-8

Special Collection: Celebrating J.D. Murray's Contributions to Mathematical Biology

Philip K. Maini[1] · Mark A. J. Chaplain[2] · Mark A. Lewis[3] · Jonathan A. Sherratt[4]

Published online: 4 December 2021

In the mid-1960s, after a distinguished early career in fluid dynamics, Jim Murray realised how important it was for mathematicians to become involved in biology, where the term is used in its broadest sense to also include medicine, epidemiology and ecology. Thus, he became one of the founders of modern mathematical biology, tackling an astonishingly diverse range of subjects with vision, originality and creativity. Here, we select a few examples that illustrate his great versatility.

In 1983, Murray proposed a new model for self-organisation in biological pattern formation. The prevailing model at the time was the coupled system of reaction–diffusion equations proposed by Turing who hypothesized that spatial pattern arose due to cells responding to chemical pre-patterns, set up by the phenomenon of diffusion-driven instability (Turing 1952). It was known that mesenchymal cells generate large contractile forces that deform the extracellular environment. This observation formed the new model proposed by Murray and colleagues (Oster et al. 1983; Murray et al. 1983) in which it was hypothesized that these forces led to cells aggregating into self-organised patterns. This was a completely new way of looking at the problem of self-organisation in tissues and they showed how the resultant system of highly nonlinear partial differential equations led to patterns that were consistent with those observed in two well-studied examples, namely, skin-organ primordia (specifically feather germs) and skeletal patterning in the vertebrate limb. Murray was one of the very first people to combine mechanics with biochemistry, developing what is now known as the mechanochemical theory of pattern formation. This theory found

✉ Philip K. Maini
philip.maini@maths.ox.ac.uk

1 Wolfson Centre for Mathematical Biology, Mathematical Institute, University of Oxford, Oxford OX2 6GG, UK

2 School of Mathematics and Statistics, Mathematical Institute, University of St Andrews, St Andrews KY16 9SS, UK

3 Department of Mathematical and Statistical Sciences, CAB 545B, University of Alberta, Edmonton, AB T6G 2G1, Canada

4 Department of Mathematics, Heriot-Watt University, Edinburgh EH14 4AS, UK

application in a range of areas, including the propagation of post-fertilization waves on eggs (Lane et al. 1987).

In parallel with this new approach to pattern formation, Murray extended the classical pre-pattern reaction–diffusion theory to study the effects of domain size and growth. A particularly striking application of this theory is to animal coat markings (Murray 1981, 1988). He further developed this approach to address the issue of how patterns evolved in the context of evolution (Oster et al. 1988). Using a chemotaxis model, he showed that the sex-dependent stripe patterns exhibited by the alligator *Alligator mississippiensis* could be explained by incubation temperature, and did not necessarily have to be genetically sex-linked (Murray et al. 1990). A similar modelling framework was proposed to account for the diverse pigmentation patterning observed on snakes (Murray and Myerscough 1991).

Both mechanochemical and reaction–diffusion mechanisms were then combined to study the initiation and sequential positioning of teeth in *Alligator mississippiensis* (Sneyd et al. 1993). This model also accounts for jaw growth, age structure of epithelial cells, and age-dependent production of cell adhesion molecules, and shows that these added complexities are necessary to account for the observed patterning.

Murray also explored cell movement in a different context, namely that of epidermal wound healing, in a series of papers that proposed that the biochemical regulation of mitosis is key to the healing of wounds (Sherratt and Murray 1990). The model was then used to make predictions on the effects on wound closure of regulating mitosis and wound shape (Sherratt and Murray 1992). While the aforementioned papers addressed adult epithelial wound healing, Murray also investigated epithelial wound healing in embryos, where the mechanism is quite different, involving the contraction of actin cables at the wound edge (Sherratt et al. 1992). Further work involved deriving a mechanical model for dermal wound healing which allowed for a study of how mechanical effects on, and remodelling of, extracellular matrix could affect the extent of scar tissue formation (Tracqui et al. 1995a, b).

Murray proposed a number of models in epidemiology to investigate disease spread. For example, Murray et al. (1986) uses a simple model for the transmission of rabies among foxes to quantify its spread should the disease be introduced to England. Parameter values were determined from the literature (and calculated from data) and the model used to investigate the effectiveness of different control strategies (for example, vaccination versus culling). Bentil and Murray (1993) used age-structured and non-age-structured models to investigate the spread of bovine tuberculosis infection in badgers. The model was analysed using a logical parameter search method to determine the values the model parameters must take to exhibit key types of observed behaviours. The values are then shown to be in good agreement with those from the literature. Nelson et al. (2000) presents a model for HIV AIDS which includes a delay in virus production when a cell is infected, as well as accounting for the fact that drugs are not 100% efficent. A detailed analysis of this model improves upon previous estimates of infected cell loss rates.

In ecology, the papers of Lewis and Murray (1993) and White et al. (1996a, b) developed and analysed novel models for wolf pack territory formation that took into account movement, scent marking, and predator–prey interactions. Analyses of these models revealed, amongst other results, steady states that contained buffer zones devoid of large numbers of wolves. Not only did this agree with field observations, it also suggested a mechanism for how prey, such as deer and moose, may survive in areas next to their main predator. This work has implications more generally and, in particular, for the social organization of humans. Indeed, in Volume II of his book, *Mathematical Biology*, Murray extends these ideas in the context of intertribal warfare (Murry 2003).

In the mid-1990s Murray began work on modelling the growth of brain tumours (glioma) in a series of highly influential papers upon which, some 25 years on, much ongoing research is still building. Tracqui et al. (1995a, b) considered a simple model of glioma cell proliferation and infiltration. Fitting model simulations to computerized tomography (CT) scans using optimization techniques allowed the model to be parametrised and then used to predict the effect of chemotherapy on the spread of the tumour, accounting for sensitive and resistant cell populations. Woodward et al. (1996) used a similar model to explore different surgical resection strategies. While these papers investigated the model on two-dimensional cross-sections of the brain, Burgess et al. (1997) extended the modelling to three spatial dimensions. An important model prediction here is that cell diffusion was actually a more important component of glioma growth than proliferation rate.

A further significant extension involved taking into account for the first time, to our knowledge, spatial heterogeneity within the brain. Specifically, Swanson et al. (2000) extended the model to incorporate the enhanced rate of cell motility in white matter as compared to that in grey matter. This paper used magnetic resonance imaging (MRI) and CT imaging data to simulate tumour growth on an anatomically accurate brain domain. The inclusion of grey and white matter then allowed the model to predict pathways of probable tumour infiltration, thus identifying areas of the brain that may be more susceptible to invasion and should therefore be targeted for treatment. Moreover, the model was used to predict how much of an expanding tumour could be missed due to the limitations of image detection methods and how this then allows us to begin to determine the true extent of invasion, which has important surgical implications. The importance of using mathematical models as virtual tumours to complement and enhance information gained from imaging and inform therapy is highlighted in Swanson et al. (2002) and the review article Swanson et al. (2003).

In the social sciences, as well as addressing intertribal conflict (as mentioned above), Murray has also proposed a mathematical model for the dynamics of marital interaction. Couples were observed and their interactions described via a Rapid Couples Interaction Scoring System (RCISS). A mathematical model, consisting of a coupled system of nonlinear discrete (in time) equations, was then developed to account for how various interactions contributed to each partner's RCISS score (Cook et al. 1995). Analysis of the model revealed a number of steady states, each of which describes a certain type of marriage. By parametrising the model from data on a particular couple's interactions, the model was found to be able to predict the probability of divorce, in a fixed period of time, to a very high degree of accuracy. The model was also used to predict how changing specific aspects of a couple's behaviour would affect their relationship (Gottman et al. 2002, Murray 2003).

The above is just a small sample of Murray's ground-breaking research and, while his personal research has been hugely influential on the development of mathematical biology, it is only one part of his contribution. Right from the start of his work on biological applications, he collaborated with experimental biologists. In doing so, he helped to establish a yardstick for genuinely applied mathematics in the life sciences that now permeates the thinking and policy-making in the field. Murray has also been extremely successful in attracting people from other areas of applied mathematics into mathematical biology. His infectious enthusiasm for the subject is an important part of this, coupled with the wide range of exciting biological problems he has at hand, and his incredible ability to get to the core of the problem, seek out the key question, and find a way to answer it.

In 1983 Murray established the Oxford Centre for Mathematical Biology, through which he organised a wide-ranging visitor programme. A high proportion of today's senior figures in mathematical biology spent time at the Centre in the 1980s, and were strongly influenced by its interdisciplinary ethos and collaborative spirit. Furthermore, he has trained and mentored a generation of mathematical biologists as graduate students and postdocs, inspiring creativity and originality in their work. Many of these trainees now hold influential positions in academic departments around the world.

Murray's book *Mathematical Biology*, first published in 1989 and then updated to the present version comprising two volumes (Murry 2002, 2003), is one of the most influential books published in the field. It forms the essential text for most high level mathematical biology courses taught worldwide and has been translated into several different languages. It is one of the reasons why mathematical biology has been transformed from a niche subject to an established part of most university mathematics courses.

Murray has received many deserved accolades during his career. He is a Fellow of the Royal Society, a Fellow of the Royal Society of Edinburgh, and a Foreign Member of the French Academy. He was awarded the London Mathematical Society (LMS) Naylor Prize (1988–1990), the Society of Mathematical Biology (SMB) Akira Okubo prize (2005), the Royal Society Bakerian Medal and Prize Lecture (2009), the Institute of Mathematics and its Applications (IMA) Gold Medal (2009), the European Academy of Sciences Leonardo da Vinci Medal (2011) and the William Benter Prize in Applied Mathematics (2012). He has also been awarded honorary doctorates by several universities, and, in 2006, the University of Washington created the James D, Murray Chair of Neuropathology, a donor endowed chair in perpetuity.

This Special Collection celebrates the diversity of Jim Murray's contribution to the field of mathematical biology. It contains 17 invited articles and, here, we briefly outline what is in each article.

While Murray's work investigated certain areas of tumour growth and therapy, mathematical modelling is now being used to address many different aspects of cancer treatment. Cassidy et al. (2020) addresses the problem of the toxic side-effects of cytotoxic chemotherapy treatment for cancer. Building on previous work that proposed strategies to reduce chemotherapy-induced neutropenia (lack of circulating neutrophils), this paper extends the model to include monocyte production. Using clinical data to determine model parameters, it is shown that monocytopenia precedes neutropenia and hence that monocytropenia can be used as a clinical marker to facilitate the delivery of an optimal dosing strategy to reduce, or completely eliminate, neutropenia.

Murray's ground-breaking work on glioma has been extended in many different directions as advances in experimental technology reveal new observations. Curtin et al. (2020) extends a previous model of glioblastoma (GBM) to investigate clinical observations that suggest that distal tumour recurrence was more likely to occur in cases of ischemia following surgery. A nutrient-transport equation is added to the original model, and is parametrized using glucose uptake rates in GBM. The model is then used to determine which mechanisms are most likely to result in distal recurrence and it is found that tumours with faster migration and slower proliferation rates are more likely to recur in response to ischemia. A detailed simulation study is carried out to investigate different recurrence scenarios.

Murray's work on spatial disease spread and how it can be controlled, most famously on rabies in foxes, has inspired many researchers to ask similar questions for different disease systems. Anita et al. (2021) ask how bacterial disease spreads spatially in olive trees via insects and how this can be controlled via management strategies such as weed cut, treated nets and resistant cultivars. They find that a "containment band", where control measures are judiciously applied, is sufficient to prevent overall spread of the disease. This is in some ways similar to the earlier discovery by Murray and colleagues that a vaccination band ahead of the rabies front could be sufficient to contain its spread.

Murray's work on pattern formation highlighted how important domain geometry is for pattern selection. While his work in this area was primarily at the tissue level, Seirin-Lee (2021) shows that geometry may also play a role at the cell level. This paper develops a partial differential equation model for cytoplasmic polarity within a cell in the context of asymmetric cell division. The model describes the interaction of key membrane proteins with those in the bulk of the cell. A detailed numerical simulation study illustrates how cell geometry can affect the observed behaviour of the model and shows that to understand how cell polarity is established in asymmetric cell division, geometry must be taken into account.

Kulesa et al. (2021) highlights how Murray's approach to mathematical modelling "offers a practical template for constructing clear, logical, direct and verifiable models that help to explain complex cell behaviors and direct new experiments" by presenting a brief review of how this approach, used by the authors, has led to new insights into neural crest cell biology and cancer. The paper shows how mathematical models have been used to test and generate hypotheses on the mechanisms that underly collective cell movement in neural crest. In particular, it illustrates how a cell-based mechanistic model identified a number of phenotypic traits that must be possessed by cells, and how these were then validated using gene sequencing and bioinformatics. Furthermore, it is shown how understanding normal development, in the context of neural crest cell migration, can help us understand key aspects of cancer cell metastasis.

While Murray's work on HIV AIDS investigated aspects of within-host dynamics, the paper by Levy et al. (2021) focuses on population spread and investigates the effect of stigma. A mechanistic model for stigma is proposed and incorporated into a model for infection dynamics. The models are parameterised by fitting to data from Kenya and then the full model is used to make predictions on how different components of stigma affect the spread of the disease. The study highlights the importance of gathering data on sociological processes when investigating the spread of infectious diseases.

Murray has worked extensively on the biochemical regulation of cell behaviour in a wide range of physiological contexts. Sneyd et al. (2021) reviews work of this type for the specific case of saliva secretion by acinar cells in the salivary glands. The paper paints a picture of a continual interplay between experiments and modelling, highlighting the effectiveness of models in modifying and enhancing our understanding of saliva generation. The paper concludes with a discussion of the new generation of models currently being developed to utilise the most recent empirical data.

Woolley et al. (2021) addresses the issue of complex spatial pattern formation by building not only on Murray's work but, literally, on Murray himself (see Figure 10 and Figure 11). This paper shows how a two-morphogen reaction–diffusion Turing model can be built so that the parameter space in which patterning occurs, the morphogen phases, and the resulting pattern (up to spatial dimension of two) can all be determined. Moreover, by incorporating spatial and temporal heterogeneities, it is shown how mixed mode patterns can be generated, allowing for arbitrarily complex patterns to be produced. The implications of these results are then discussed in the light of linking theory and experiment.

Much of Murray's pioneering work features multi-stable systems, where nonlinear dynamics can drive multiple possible outcomes. Koch et al. (2021) analyse a classical multi-stable system, mountain pine beetle populations, which can jump between endemic and outbreak states. However, this work develops a new perspective on such systems by incorporating spatial autocorrelation in dispersal dynamics for the beetles. The resulting dynamical system is then fit to detailed outbreak data, shedding light on how outbreaks can sporadically arise across the landscape.

The hallmark of Murray's work on disease dynamics, such as HIV, has been to gain insight from realistic models, tailored to the specific details that can so often affect outcomes. Britton and White (2021) show how the inclusion of multiple infection classes in a honey-bee-mite-virus system is key to our understanding of spontaneous transitions between disease states as bees and mites move from covertly to overtly infected states. These distinct infection classes then play a key role in governing disease outcomes. Nick Britton very sadly passed away in December 2020, while this paper was still in the review stage. On his request, Jane White, his long-time colleague and friend at the University of Bath, revised and finalised the manuscript in response to reviewer comments.

A unifying feature of Murray's varied research work is the use of simple models to explain complex biological phenomena. He showed that low order systems of ordinary and partial differential equations can capture the essence of processes such as embryonic development, tumour growth and disease spread. Kempes et al. (2021) applies this philosophy to the field of astrobiology. The paper uses a simple ordinary differential equation model coupled with previously established scaling relationships of the main macromolecular components of cells. This approach enables prediction of potential biosignatures of life, via variation of elemental ratios in a range of terrestrial biological environments.

Villa et al. (2021) re-examines a key aspect of the Murray–Oster mechanochemical theory of pattern formation, namely that of the role of the constitutive equation. In the original theory the main assumption of the modelling regarding the biomechanical properties of tissue was that it was a linear viscoelastic material governed by the so-called Kelvin-Voigt model. This linear stress-strain relationship permits an equation to be derived for the displacement of the tissue (cell-extracellular matrix system) which, in turn, is used to derive a dispersion relation, whence a prediction of pattern formation. The results of the current paper show that pattern formation is critically dependent upon the constitutive equation chosen and that fluid-like constitutive models have a higher potential for generating patterns than solid-like constitutive models.

As mentioned above, Murray's work explored the idea that, in the context of developmental pattern formation, evolution can be captured by movement, on a vastly different timescale, through parameter space. Of course, at the macroscale level, parameters are largely a model construct as they really describe processes occurring on different length and/or timescales. How to link processes across different scales is now a major change in the field of mathematical biology. Stotsky et al. (2021) use multiscale continuous time random walks and generalised master equations to go from the microscale to the macroscale in the context of transport processes in tissues. It is shown, using signalling in the *Drosophila* wing disc as a case study, how the framework developed can be employed to directly link experimentally observable macroscale properties, such as transport coefficients, to the detailed underlying microscale level tissue properties.

A very important weapon in the armoury of the applied mathematician is asymptotic analysis, a subject which formed the basis for Murray's first book (Murray 1974) and which also plays a role in Efendiev et al. (2021). This paper investigates high intensity focussed ultrasound (HIFU), which is being increasingly used as a promising cancer treatment, without the side effects of more standard treatments. This paper investigates the long-time dynamics of a previously proposed partial differential equation system that aims to model energy deposition in biological tissue due to HIFU. Specifically, a number of theoretical results are derived on the attractors shown to exist for this system.

Pattern formation has been a central component of Murray's work. He primarily studied patterns in embryonic development, but his ideas have subsequently been exported to a wide range of other biological contexts. Sherratt et al. (2021) considers pattern formation in intertidal mussel beds, where the patterning occurs at the scale of the whole ecosystem. The paper makes a detailed comparison between two different hypothesised mechanisms for patterning, and demonstrates important differences between the resulting patterns for both biomass distribution and resilience to disturbances.

Lui and Myerscough (2021) address one of the few research areas in which Murray has not worked, namely, heart disease. This paper proposes a model for the composition of atherosclerotic plaques, which can cause heart attacks and strokes. More specifically, it develops an ordinary differential equation model for time evolution of macrophage cells and lipids. It is shown that the model is amenable to a multiple timescale analysis and the contribution of key cell processes to plaque lipid accumulation is explored in detail.

Fowler (2021) extends the aforementioned work of Murray et al. (1986) on rabid foxes to address the issue of extinction, which is a general problem faced by continuous models of population dynamics, whose applicability is called into question at low numbers of individuals. A number of ways to address/resolve these problems is presented in this paper, and illustrated through application to foxes approaching extinction; oscillatory dynamics with extremely small minima, explored in the context of immune dynamics and microbial growth models; and finally frogspawn, where an age-structured model is proposed and analysed.

The articles in this Special Collection illustrate the synergy between the fields of mathematics and biology (in its broadest sense) that J.D. Murray has demonstrated countless times thoroughout his amazing career: biology inspires new mathematics, while mathematics leads to new biology.

References

Anita S, Capasso V, Scacchi S (2021) Controlling the spatial spread of a Xylella epidemic. Bull Math Biol 83:22

Bentil DE, Murray JD (1993) Modelling bovine tuberculosis in badgers. J Anim Ecol 62:239–250

Britton NF, White KAJ (2021) The effect of covert and overt infections on disease dynamics in honey-bee colonies. Bull Math Biol 83:67

Burgess PK, Kulesa PM, Murray JD, Alvord EC Jr (1997) The interaction of growth rates and diffusion coefficients in a three-dimensional mathematical model of gliomas. J Neuropathol Exp Neurol 56:704–713

Cassidy T, Humphris AR, Craig M, Mackey MC (2020) Characterizing chemotherapy-induced neutropenia and monocytopenia through mathematical modelling. Bull Math Biol 82:104

Cook J, Tyson R, White KAJ, Rushe R, Gottman J, Murray JD (1995) Mathematics of marital conflict: qualitative dynamic mathematical modeling of marital interaction. J Fam Psycol 9(2):110–130

Curtin L, Hawkins-Daarud A, Porter AB, van der Zee KG, Owen MR, Swanson KR (2020) A mechanistic investigation into ischemia-driven distal recurrence of glioblastoma. Bull Math Biol 82:143

Fowler AC (2021) Atto-foxes and other minutiae. Bull Math Biol 83:104

Gottman JM, Murray JD, Swanson CC, Tyson R, Swanson KR (2002) The mathematics of marriage: dynamic nonlinear models. MIT Press, Cambridge

Kempes CP, Follows MJ, Smith H, Graham H, House CH, Levin SA (2021) Generalised stoichiometry and biogeochemistry for astrobiological applications. Bull Math Biol 83:73

Koch D, Lewis MA, Lele S (2021) The signature of endemic populations in the spread of mountain pine beetle outbreaks. Bull Math Biol 83:65

Kulesa PM, Kasemeier-Kulesa JC, Morrison JA, McLennan R, McKinney MC, Bailey C (2021) Modelling cell invasion: a review of what JD Murray and the embryo can teach us. Bull Math Biol 83:26

Lane DC, Murray JD, Manoranjan VS (1987) Analysis of wave phenomena in a morphogenetic mechanochemical model and an application to post-fertilization waves on eggs. IMA J Math Appl Med Biol 4:309–331

Levy B, Correia HE, Chirove F, Ronoh M, Abebe A, Kgosimore M, Chimbola O, Machingauta MH, Lenhart S, White KAJ (2021) Modeling the effect of HIV/AIDS stigma on HIV infection dynamics in Kenya. Bull Math Biol 83:55

Lewis MA, Murray JD (1993) Modelling territoriality and wolf–deer interactions. Nature 366(6457):738–740

Lui G, Myerscough MR (2021) Modelling preferential phagocytosis in atherosclerosis: delineating timescales in plaque development. Bull Math Biol 83:96

Murray JD (1974) Asymptotic analysis. Clarendon Press, Oxford (further material added in asymptotic analysis, applied mathematical sciences, vol 48. Springer-Verlag, New York, 1984, 2nd printing, 1996)

Murray JD (1981) A pre-pattern formation mechanism for animal coat markings. J Theor Biol 88:161–199

Murray JD (1988) Mammalian coat patterns: how the leopard gets its spots. Sci Am 256:80–87

Murray JD (2002) Mathematical biology: I. An introduction 2002 (3rd Edition). Interdisciplinary applied mathematics, vol 17. Springer

Murray JD (2003) Mathematical biology: II. Spatial models and biomedical applications (3rd Edition). Interdisciplinary applied mathematics, vol 18. Springer
Murray JD (2004) Mathematical biology, 3rd edition in 2 volumes: mathematical biology: I. An introduction 2002 (2nd printing 2004) interdisciplinary applied mathematics, vol 17. Springer
Murray JD (2008) Mathematical biology: II. Spatial models and biomedical applications 2003 (2nd printing 2004, 3rd printing 2008) interdisciplinary applied mathematics, vol 18. Springer
Murray JD, Deeming DC, Ferguson MWJ (1990) Size dependent pigmentation pattern formation in embryos of alligator mississippiensis: time of initiation of pattern generation mechanism. Proc Roy Soc Lond B 239:279–293
Murray JD, Myerscough MR (1991) Pigmentation pattern formation on snakes. J Theor Biol 149:339–360
Murray JD, Oster GF, Harris AK (1983) A mechanical model for mesenchymal morphogenesis. J Math Biol 17:125–129
Murray JD, Stanley EA, Brown DL (1986) On the spatial spread of rabies among foxes. Proc Roy Soc Lond B 229:111–150
Nelson P, Perelson AS, Murray JD (2000) Delay model for the dynamics of HIV infection. Math Biosci 163:201–215
Oster GF, Murray JD, Harris AK (1983) Mechanical aspects of mesenchymal morphogenesis. J Embryol Exp Morph 78:83–125
Oster GF, Shubin N, Murray JD, Alberch P (1988) Evolution and morphogenetic rules: the shape of the vertebrate limb in ontogeny and phylogeny. Evolution 42:862–884
Seirin-Lee S (2021) The role of cytoplasmic MEX-5/6 polarity in asymmetric cell division. Bull Math Biol 83:29
Sherratt JA, Ling Q-X, van de Koppel J (2021) A comparison of the "reduced losses" and "increased production" models for mussel bed dynamics. Bull Math Biol 83:99
Sherratt JA, Martin P, Murray JD, Lewis J (1992) Mathematical models of wound healing in embryonic and adult epidermis. IMA J Math Appl Med Biol 9:177–196
Sherratt JA, Murray JD (1990) Models of epidermal wound healing. Proc Roy Soc Lond B 241:29–36
Sherratt JA, Murray JD (1992) Epidermal wound healing: the clinical implications of a simple mathematical model. Cell Transpl 1:365–371
Sneyd J, Atri A, Ferguson MWJ, Lewis MA, Seward W, Murray JD (1993) A model for the spatial patterning of teeth primordia in the alligator: initiation of the dental determinant. J Theor Biol 165:633–658
Sneyd J, Vera-Sigüenza E, Rugis J, Pages N, Yule DI (2021) Calcium dynamics and water transport in salivary acinar cells. Bull Math Biol 83:31
Stotsky JA, Gou J, Othmer HG (2021) A random walk approach to transport in tissues and complex media: from microscale descriptions to macroscale models. Bull Math Biol 83:92
Swanson KR, Alvord EC Jr, Murray JD (2000) A quantitative model for differential motility of gliomas in grey and white matter. Cell Prolif 33:317–329
Swanson KR, Alvord EC Jr, Murray JD (2002) Virtual brain tumors (gliomas) enhance the reality of medical imaging and highlight inadequacies of current therapy. Br J Cancer 86:14–18
Swanson KR, Bridge C, Murray JD, Alvord EC Jr (2003) Virtual and real brain tumors: using mathematical modeling to quantify glioma growth and invasion. J Neurol Sci 216(1):1–10
Tracqui P, Cruywagen GC, Woodward DE, Bartoo GT, Murray JD, Alvord EC Jr (1995a) A mathematical model of glioma growth: the effect of chemotherapy on spatial-temporal growth. Cell Prolif 28:17–31
Tracqui P, Woodward DE, Cruywagen GC, Cook J, Murray JD (1995b) A mechanical model for fibroblast-driven wound healing. J Biol Syst 3:1075–1085
Turing AM (1952) The chemical basis of morphogenesis. Philos Trans Roy Soc B 237:37–72
Villa C, Chaplain MAJ, Gerisch A, Lorenzi T (2021) Mechanical models of pattern and form in biological tissues: the role of stress–strain constitutive equations. Bull Math Biol 83:8
White KAJ, Lewis MA, Murray JD (1996a) A model for wolf-pack territory formation and maintenance. J Theor Biol 178:29–43
White KAJ, Murray JD, Lewis MA (1996b) Wolf–deer interactions: a mathematical model. Proc R Soc Lond B 263:299–305
Woodward DE, Cook J, Tracqui P, Cruywagen GC, Murray JD, Alvord EC Jr (1996) A mathematical model of glioma growth: the effect of extent of surgical resection. Cell Prolif 29:269–288
Woolley TE, Krause AL, Gaffney EA (2021) Bespoke Turing systems. Bull Math Biol 83:41

Bulletin of Mathematical Biology (2022) 84:12
https://doi.org/10.1007/s11538-021-00956-7

Special Tributes

James D. Murray[1,2]

Published online: 3 December 2021

It was with great sadness that I learned of the death of Nick Britton during the preparation of this Special Collection. Nickc was one of my first PhD students in mathematical biology at Oxford, graduating five years before I established the Centre for Mathematical Biology. Nick went on to a long and distinguished academic career, spent mostly at the University of Bath, which founded its own Centre for Mathematical Biology in 2004, with Nick as Co-Director. His research work was wide ranging, and he made important contributions to mathematics applied in many areas of the life sciences, including ecology, epidemiology, evolution, cancer, artificial kidneys and pain—the last of these in collaboration with his wife Suzanne Skevington. In addition to his many research papers, Nick wrote two highly influential books—"Reaction-diffusion Equations and their Applications to Biology" and "Essential Mathematical Biology"—which are key reference texts and teaching tools in our community.
Nick was a longtime member of the Society for Mathematical Biology and served on its Board of Directors from July 2006 until July 2010, and was Chair of the Future Meetings Committee from July 2008—July 2010. But Nick's contribution to mathematical biology was much greater than this. He was an inspirational figure, a dedicated teacher and a great listener, who quietly but effectively influenced the academic trajectory of many students and colleagues. We will miss him.

James D. Murray, January 2021

✉ James D. Murray
jdstmurray@gmail.com

1 Wolfson Centre of Mathematical Biology, University of Oxford, Oxford, UK

2 Department of Applied Mathematics, University of Washington, Seattle, USA

Very sadly, Masayasu Mimura, known to his friends as Mayan, who had accepted the invitation to contribute to this Special Collection, passed away on April 8th, 2021. He came to Oxford in 1976 and was my first postdoc and, while working with me, his interests extended from understanding the qualitative aspects of reaction-diffusion systems to exploring their behaviour in the context of mathematical biology, and we published a number of papers together. He then moved back to Japan and started a research group in mathematical biology at Hiroshima University. Mayan went on to become a very influential figure in Japan. He held many leadership positions, including Director of the Institute for Nonlinear Sciences and Applied Mathematics (INSAM), Faculty of Science, Hiroshima University (1998–2004), overlapping with his time as Chairman of the Department of Mathematical and Life Sciences, Graduate School of Science, Hiroshima University (2000–2001). He then went on to Meiji University, where he was Director of the Meiji Institute of Advanced Study of Mathematical Sciences (2007–2015). He was President of the Japanese Society of Industrial and Applied Mathematics (JSIAM), President of the Japanese Society of Mathematical Biology, and he held many visiting positions.

Mayan had a truly original mind and he advanced the field of reaction-diffusion theory in many different directions. For example, he developed a global bifurcation theory for reaction-diffusion systems in the context of biological and ecological applications, proposed the singular limit procedure in the context of species competition models, worked on free boundary problems in ecology and developed novel models for patterning in bacterial colonies. He had a warm personality, an infectious enthusiasm for his science and was very supportive and encouraging to young researchers. He will be sorely missed.

James D. Murray, May 2021

Publisher's Note Springer Nature remains neutral with regard to jurisdictional claims in published maps and institutional affiliations.

Bulletin of Mathematical Biology (2021) 83:104
https://doi.org/10.1007/s11538-021-00936-x

Society for Mathematical Biology

SPECIAL ISSUE: CELEBRATING J. D. MURRAY

Atto-Foxes and Other Minutiae

A. C. Fowler[1,2]

Received: 29 March 2020 / Accepted: 13 August 2021 / Published online: 31 August 2021

Abstract
This paper addresses the problem of extinction in continuous models of population dynamics associated with small numbers of individuals. We begin with an extended discussion of extinction in the particular case of a stochastic logistic model, and how it relates to the corresponding continuous model. Two examples of 'small number dynamics' are then considered. The first is what Mollison calls the 'atto-fox' problem (in a model of fox rabies), referring to the problematic theoretical occurrence of a predicted rabid fox density of 10^{-18} (*atto-*) per square kilometre. The second is how the production of large numbers of eggs by an individual can reliably lead to the eventual survival of a handful of adults, as it would seem that extinction then becomes a likely possibility. We describe the occurrence of the atto-fox problem in other contexts, such as the microbial 'yocto-cell' problem, and we suggest that the modelling resolution is to allow for the existence of a reservoir for the extinctively challenged individuals. This is functionally similar to the concept of a 'refuge' in predator–prey systems and represents a state for the individuals in which they are immune from destruction. For what I call the 'frogspawn' problem, where only a few individuals survive to adulthood from a large number of eggs, we provide a simple explanation based on a Holling type 3 response and elaborate it by means of a suitable nonlinear age-structured model.

Keywords Atto-foxes · Boom-and-bust · Extinction · Stochastic logistic model · Frogspawn

Preamble

It is a privilege to present this paper in a special issue of the journal in honour of Jim Murray's 90th birthday in January 2021. Jim is of course a legend in the field of mathematical biology. He was also my undergraduate tutor at Corpus Christi Col-

✉ A. C. Fowler
andrew.fowler@ul.ie

1 MACSI, University of Limerick, Limerick, Ireland

2 OCIAM, University of Oxford, Oxford, UK

lege, Oxford, whom I first encountered by candlelight on a murky December evening 50 years ago last year (2020). I was asked recently by his contemporary Fellow Brian Harrison whether Jim had been the inspiration behind my own development as an applied mathematician. I suspect the answer is no, it would have happened anyway, but it is one of those unanswerable questions. What is undoubtedly true is that I have followed his inspirational mantra in aiming to be a genuinely applied mathematician. Jim is a marvellously interesting man, as anyone who reads his enthralling memoir (Murray 2018) will know. Jim and I each grew a beard over the same summer, and of our many overlapping interests—old furniture, home improvement, mediaeval literature, carpentry, red wine—it is perhaps the only thing we have in common where I could lay claim to parity.

1 Introduction

In a recent fascinating conversation with the ecologist Yvonne Buckley, of Trinity College, Dublin, we touched on a number of issues concerning population dynamics which have puzzled me for some time. In this paper I want to draw these conundrums together and offer some palliative solutions.

The central theme is the use of continuous population models in circumstances where the population levels become extremely low. A continuous population model relies on the assumption that the population size varies continuously in time. This requires, for two reasons, that the population size be large. Firstly, because the actual discrete changes in integer numbers need to be viewed as infinitesimal changes, and secondly, because actual finite time gestation periods can only be interpreted as continuous in time if the large population can be taken as a realisation of an evolving probability density. Continuous population models are in particular subject to criticism when they indicate very low population levels, and in this circumstance discrete and/or stochastic models may be preferable (Durrett and Levin 1994).

The issue is simply illustrated by a linear birth-death process (Bartlett 1960) in which, if the population is of size n, individuals have probability $\lambda_n\,\delta t$ of giving birth in a time interval δt and equivalent probability of dying of $\mu_n\,\delta t$. If $p_n(t)$ is the probability that the population is of size n at time t, then it is straightforward to show that

$$\dot{p}_n = (n-1)\lambda_{n-1}p_{n-1} - n(\lambda_n+\mu_n)p_n + (n+1)\mu_{n+1}p_{n+1}; \tag{1.1}$$

this applies for $n \geq 0$, providing we define $p_{-1}=0$. Note that p_0 is the probability of extinction at time t. If $\lambda_n=\lambda$ and $\mu_n=\mu$ are constant, the equation is easily solved with a generating function

$$G(s,t) = \sum_0^\infty p_n s^n, \tag{1.2}$$

and one finds that

$$G_t + (\lambda s - \mu)(1-s)G_s = 0, \tag{1.3}$$

and the solution starting with m individuals ($G = s^m$ at $t = 0$) is

$$G = \left[\frac{\lambda s - \mu + \mu(1-s)e^{(\lambda-\mu)t}}{\lambda s - \mu + \lambda(1-s)e^{(\lambda-\mu)t}} \right]^m. \tag{1.4}$$

The mean of the population is

$$N = \left. \frac{\partial G}{\partial s} \right|_{s=1}, \tag{1.5}$$

and it is simple to show directly from (1.3) that N satisfies the mean field equation

$$\dot{N} = (\lambda - \mu)N. \tag{1.6}$$

The probability of extinction, $p_0 = G(0, t)$, is given by

$$p_0 = \left[\frac{\mu \left\{ e^{(\lambda-\mu)t} - 1 \right\}}{\lambda e^{(\lambda-\mu)t} - \mu} \right]^m = \left[\frac{\mu \left\{ 1 - e^{-(\mu-\lambda)t} \right\}}{\mu - \lambda e^{-(\mu-\lambda)t}} \right]^m, \tag{1.7}$$

and we see that

$$p_0(\infty) = \left(\frac{\mu}{\lambda} \right)^m, \quad \lambda > \mu; \qquad p_0(\infty) = 1, \quad \lambda < \mu. \tag{1.8}$$

If the population is growing ($\lambda > \mu$) and the initial population size is of reasonable size ($m \gg 1$), then the likelihood of extinction is negligible; on the other hand a decaying population ($\lambda < \mu$) will eventually become extinct. Essentially the same conclusion is true if λ and μ are functions of t.

1.1 Nonlinear Stochastic Models

The problem with such discrete models is that their extension to nonlinear processes becomes less tractable. As the simplest example, suppose that the birth rate λ is constant, but that the specific death rate is $\mu_n = \frac{\lambda n}{K}$. In this case we would expect that the mean of the population would satisfy the Verhulst (1845) logistic equation, with carrying capacity K. There is a large literature dealing with such problems; see for example Bartlett et al. (1960), Nåsell (2001), Ovaskainen and Meerson (2010) and Doering et al. (2005). The paper by Ovaskainen and Meerson, in particular, contains many further references. We summarise some of the results here, though perhaps in a slightly different guise.

The equation (1.1) takes the form

$$\dot{p}_n = -\Delta_-(np_n) + \frac{1}{K}\Delta_+(n^2 p_n),$$

$$\Delta_- q_n = q_n - q_{n-1}, \quad \Delta_+ q_n = q_{n+1} - q_n, \tag{1.9}$$

and the generating function defined in (1.2) now satisfies

$$G_\tau + s(1-s)G_s = \frac{(1-s)}{K}(sG_s)_s\,, \tag{1.10}$$

where we have scaled time by writing $\tau = \lambda t$. As is commonly done, the subscripts τ and s in (1.10) denote partial derivatives (but the subscripts n, etc. in (1.9) are indices).

1.1.1 Generating Function

The carrying capacity K is an integer, and if it is not large, we would expect extinction to occur fairly rapidly. In fact, eventual extinction is certain for any finite K, in the sense that $G \to 1$ as $\tau \to \infty$ (thus $p_0(\infty) = 1$). On the face of it, this seems to indicate that a continuum model is doomed. If we consider (1.9) for $n = 0, 1, \ldots$, it is not difficult to show, since p_0 is bounded above and thus $\dot{p}_0 \to 0$ as $\tau \to \infty$, that $p_n \to 0$ for all $n \geq 1$. The two possibilities are that the mean of the population grows unboundedly, or that extinction occurs. The effect of the diffusion term in (1.10) appears to imply the second of these conclusions. If we write $G = 1 + g$ (so g also satisfies (1.10), but its steady state is $g = 0$), then we find

$$\tfrac{1}{2}K\frac{d}{d\tau}\int_0^1 \frac{e^{-Ks}g^2\,ds}{1-s} = -\int_0^1 sg_s^2 e^{-Ks}\,ds, \tag{1.11}$$

and thus indeed $g \to 0$ (since $g = 0$ at $s = 1$). The rate of approach can be estimated by writing $g = \phi(s)e^{\sigma\tau}$, and then conversion of (1.10) to Sturm–Liouville form

$$\left\{se^{-Ks}\phi_s\right\}_s = \frac{\sigma K e^{-Ks}}{1-s}\phi, \quad \phi(1) = 0, \tag{1.12}$$

shows that

$$-\sigma = \inf\left[\frac{\displaystyle\int_0^1 se^{-Ks}\phi'^2\,ds}{K\displaystyle\int_0^1 \frac{e^{-Ks}\phi^2\,ds}{1-s}}\right], \tag{1.13}$$

where the admissible functions ϕ are piecewise smooth functions with $\phi(1) = 0$. A simple estimate (which also betrays the boundary layer structure of the eigenfunctions) is to take

$$\phi = \begin{cases} 1, & s < 1 - \dfrac{\lambda}{K}, \\[2ex] \dfrac{K(1-s)}{\lambda}, & s > 1 - \dfrac{\lambda}{K}, \end{cases} \tag{1.14}$$

whence we find for large K that $-\sigma \lesssim 1.54Ke^{-K}$ (1.54 is the minimum of $\dfrac{e^\lambda - 1}{\lambda^2}$).

From this we see that for large K, extinction takes an exponentially large time to occur. This is similar to the Ehrenfest urn problem and is not a matter for concern when

K is large, but for $O(1)$ values of K, extinction will occur on times $\tau \sim O(1)$. In practice, the distribution approaches a quasi-steady state (Bartlett et al. 1960), which may be determined as follows. In a steady state, (1.9) implies

$$p_n = \frac{(n-1)K}{n^2} p_{n-1}, \quad n \geq 1, \tag{1.15}$$

and obviously $p_0 = 1$, $p_n = 0$ for $n > 1$. The quasi-steady-state assumption is that p_n for $n \geq 1$ approaches a (quasi-)steady state while $p_0 < 1$; the idea is that p_0 varies on a slow time scale. If this is the case then we can solve (1.15) to obtain

$$p_n = \frac{AK^n}{n\,n!}, \quad n \geq 1, \tag{1.16}$$

where A is slowly varying. Applying $G = 1$ at $s = 1$, this gives

$$G = p_0 + \frac{(1-p_0)\displaystyle\int_0^s \frac{(e^{Kx}-1)\,\mathrm{d}x}{x}}{\displaystyle\int_0^1 \frac{(e^{Kx}-1)\,\mathrm{d}x}{x}}; \tag{1.17}$$

we can then use this to calculate

$$\dot{p}_0 = \frac{(1-p_0)}{\displaystyle\int_0^1 \frac{(e^{Kx}-1)\,\mathrm{d}x}{x}} \approx Ke^{-K}(1-p_0), \tag{1.18}$$

using Laplace's method, which confirms the quasi-steady-state hypothesis and is also consistent with the earlier estimate of decay rate. The next question is to provide this quasi-stationary profile. This can be obtained from (1.16) using Stirling's approximation, but it is more illuminating to use a continuum approximation for p_n in order to show that it is obtained on a timescale of $\tau \sim O(1)$.

1.1.2 Continuum Approximation

Keeping K large (when we therefore might expect the continuous model to apply), we revert to (1.1), and then writing

$$n = K\xi, \quad p_n = \frac{p(\xi,\tau)}{K}, \quad \Delta\xi = \frac{1}{K}, \tag{1.19}$$

(1.1) is a discrete approximation to

$$p_\tau + [\xi(1-\xi)p]_\xi = 0, \quad \int_0^\infty p\,\mathrm{d}\xi = 1, \tag{1.20}$$

which can be solved using the method of characteristics, for example with an initial condition $p = \delta(\xi - \xi_0)$, where $K\xi_0$ is the initial population size. It can then be shown that the continuous Verhulst model for the mean population is regained. A nice way to demonstrate this uses the fact that with a delta function as initial condition, the solution must in fact be

$$p = \delta[\xi - N(t)]. \tag{1.21}$$

Substituting this in to (1.20), we find, using the language of generalised functions,

$$[-\dot{N} + \xi(1-\xi)]\delta'(\xi - N) + (1 - 2\xi)\delta(\xi - N) = 0; \tag{1.22}$$

now we multiply by $(\xi - N)$ and use the fact that $x\delta(x) = 0$ and thus $x\delta'(x) = -\delta(x)$ to obtain $[\dot{N} - \xi(1-\xi)]\delta(\xi - N) = 0$, whence we derive the logistic equation

$$\dot{N} = N(1 - N) \tag{1.23}$$

on integrating over $0 < \xi < \infty$.

An extension to this for large K is to use the next term in the expansion of (1.1), which leads to the Fokker–Planck equation

$$p_\tau + \{\xi(1-\xi)p\}_\xi = \frac{1}{2K}\{\xi(1+\xi)p\}_{\xi\xi}, \tag{1.24}$$

and to solve this, we write

$$\xi = N(\tau) + \frac{\eta}{\sqrt{K}}, \quad p = \sqrt{K}\phi(\eta, \tau), \quad \int_{-\infty}^{\infty} \phi \, d\eta \approx 1, \tag{1.25}$$

and if we choose N to satisfy (1.23), then

$$\phi_\tau + (1 - 2N)(\eta\phi)_\eta \approx \tfrac{1}{2}N(1+N)\phi_{\eta\eta}. \tag{1.26}$$

As $\tau \to \infty$, $N \to 1$, and (1.26) has a quasi-steady solution

$$\phi = \frac{1}{\sqrt{2\pi}} e^{-\frac{1}{2}\eta^2}. \tag{1.27}$$

There is a caveat to this result. This is because the approximation in (1.27) is invalid for $\eta \sim \sqrt{K}$. To deal with this, we revert to p and ξ, and since the far-field expression in (1.27) is

$$p \approx \sqrt{\frac{K}{2\pi}} e^{-\frac{1}{2}K(\xi-1)^2}, \tag{1.28}$$

we define

$$p = A(\tau)\, e^{K\psi}, \tag{1.29}$$

where we assume that $\dot{A} \lesssim O(1)$, so that (1.27) corresponds to $\psi \approx -\frac{1}{2}(\xi - 1)^2$, $A = \sqrt{\dfrac{K}{2\pi}}$, for $\xi \approx 1$. We use the language of the operational calculus ($f(\xi + h) = e^{hD} f(\xi)$, where $D = \dfrac{\partial}{\partial \xi}$), to write (1.9) in the form

$$p_\tau = K\left[\left\{\exp\left(-\frac{D}{K}\right) - 1\right\}(\xi p) + \left\{\exp\left(\frac{D}{K}\right) - 1\right\}(\xi^2 p)\right]. \tag{1.30}$$

Using the fact that $\dfrac{D}{K} f(\xi)e^{K\psi} \approx \psi_\xi f(\xi)e^{K\psi}$, we then find that (1.30) takes the approximate form

$$\psi_\tau = Q = \xi(e^{-P} - 1) + \xi^2(e^P - 1), \quad P = \psi_\xi. \tag{1.31}$$

The initial condition we choose for (1.31) should correspond approximately to an initial delta function, which we take to be centred at the steady value $\xi = 1$. Note that an arbitrary additive constant for ψ can be absorbed into A. To represent the initial condition, we will consider the family of functions

$$\psi = -\tfrac{1}{2}a(\xi - 1)^2 \text{ at } \tau = 0, \tag{1.32}$$

where the limit of $a \to \infty$ corresponds to a delta function.

The equation (1.31) is a nonlinear hyperbolic equation of the form

$$F(\xi, P, Q) = Q - \xi(e^{-P} - 1) - \xi^2(e^P - 1) = 0, \tag{1.33}$$

which can be solved by writing it in characteristic form using Charpit's equations. The initial condition (1.32) can be written in parametric form as

$$\xi = \sigma, \quad \psi = \psi_0(\sigma) = -\tfrac{1}{2}a(\sigma - 1)^2, \quad P = P_0(\sigma) = -a(\sigma - 1),$$

$$Q = Q_0(\sigma) = \sigma[e^{a(\sigma-1)} - 1] + \sigma^2[e^{-a(\sigma-1)} - 1] \text{ at } \tau = 0, \tag{1.34}$$

and Charpit's equations reduce to

$$\begin{aligned}
&Q = Q_0(\sigma),\\
&P = \ln\left\{\frac{1}{2\xi^2}\left[(\xi^2 + \xi + Q) \pm \left\{(\xi^2 + \xi + Q)^2 - 4\xi^3\right\}^{1/2}\right]\right\} \equiv P(\xi, \sigma),\\
&\dot{\xi} = \xi e^{-P} - \xi^2 e^P = \mp\left\{(\xi^2 + \xi + Q)^2 - 4\xi^3\right\}^{1/2},\\
&\dot{\psi} = Q + P\dot{\xi}, \quad \Rightarrow \psi = \psi_0(\sigma) + Q\tau + \int_\sigma^\xi P(\eta, \sigma)\,\mathrm{d}\eta,
\end{aligned} \tag{1.35}$$

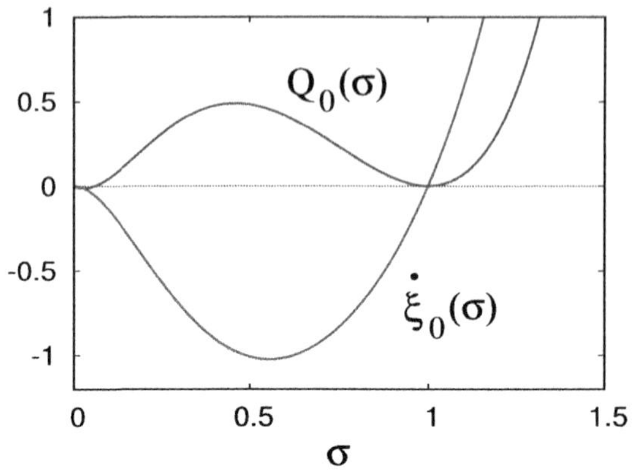

Fig. 1 The functions $\dot{\xi}_0(\sigma)$ defined by (1.36) and $Q_0(\sigma)$ defined in (1.34), using a value $a = 3$

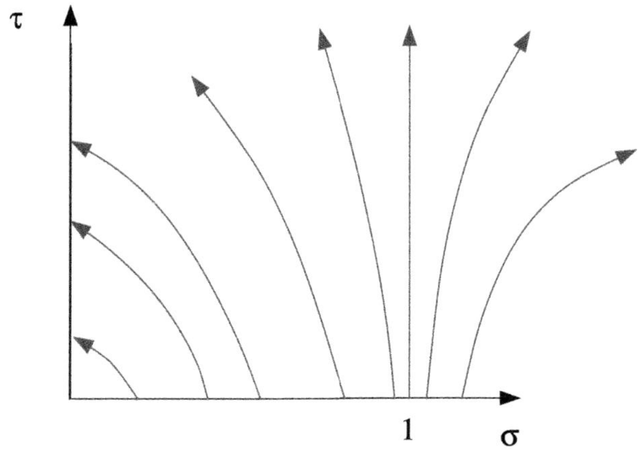

Fig. 2 Schematic characteristic diagram for (1.35)

where the overdots refer to differentiation with respect to τ. The expression for P comes from solving the quadratic equation for e^P given by $F = 0$. The upper or lower signs in (1.35) are chosen so that

$$\dot{\xi}_0(\sigma) = \dot{\xi}\big|_{\tau=0} = \sigma e^{a(\sigma-1)} - \sigma^2 e^{-a(\sigma-1)}. \tag{1.36}$$

This function is plotted in Fig. 1, along with $Q_0(\sigma)$. Evidently the upper sign is chosen for $0 \lesssim \xi < 1$ and the lower one for $\xi > 1$. Thus for small τ, the characteristics move to the left for $\xi < 1$ and to the right for $\xi > 1$. This will remain true unless $\dot{\xi} = 0$, which occurs when $Q = Q_\pm(\xi)$, where

$$Q_\pm(\xi) = -\xi(1 \pm \xi^{1/2})^2 \tag{1.37}$$

are the roots of $\dot{\xi} = 0$. As suggested by Fig. 1, this does not occur, since $Q_0(\sigma) > 0$ (except near $\sigma = 0$, discussed below).

The form of the characteristic diagram is then shown in Fig. 2. For $\xi < 1$, $\dot{\xi} < 0$ and for $\xi > 1$, $\dot{\xi} > 0$. Therefore at large τ, all the characteristics come from the vicinity of $\sigma = 1$, where $Q \approx 0$, and so from (1.35),

$$\dot{\xi} \approx \xi^2 - \xi, \quad P \approx -\ln \xi, \quad \psi \approx \xi \left(\ln \frac{1}{\xi} + 1 \right) - 1, \tag{1.38}$$

which provides the uniformly valid quasi-steady solution for ψ; note that $\psi \sim -\frac{1}{2}(\xi - 1)^2$ as $\xi \to 1$.

A comment is necessary concerning the behaviour at $\xi = 0$. With the precise choice in (1.32), it is clear from (1.34) that for any finite value of a, $Q_0 < 0$ for sufficiently small σ, and also $\dot{\xi} > 0$ there. Thus for large but finite a, a shock will form near $\xi = 0$. This would slightly confuse matters, but in fact this issue is associated with the consequence at finite a that the probability density $p > 0$ at $\xi = 0$. In reality, a better initial condition would have $\psi \to -\infty$ at $\sigma = 0$, so that this region of $\dot{\xi} > 0$ disappears, but the limit is a non-uniform one, since $\xi = 0$ remains a characteristic. It seems to be that the resultant shock is the cause of the necessity of treating p_0 separately.

1.1.3 Long-Time Evolution

The key to extending the result above is to realise that the correct way to formulate a 'continuous' distribution model is to allow p to have delta function behaviour at $\xi = 0$. In keeping with what the discrete model actually implies, we adopt (1.29) and thus (1.30) for ξ strictly positive, and let the probability $p_0(\tau)$ at $\xi = 0$ be finite. Thus the distribution is a Stieltjes one, and we have

$$p_0 + \int_{0+}^{\infty} p(\xi, \tau)\, d\xi = 1, \tag{1.39}$$

where the lower limit 0+ indicates that it is in fact slightly positive ($= \frac{1}{K}$). Using (1.38) (which shows that $\psi \approx -\frac{1}{2}(\xi - 1)^2$ near $\xi = 1$) together with the use of Laplace's method for the integral, we find

$$A \approx \sqrt{\frac{K}{2\pi}}(1 - p_0), \tag{1.40}$$

where A is as in (1.29).

The approximation in (1.31) is essentially that used in the geometric optics approximation of WKB theory (Bender and Orszag 1978), but to obtain a result equivalent to (1.18), we need the next term of the approximation. To find this, we return to (1.30), but now written in the form

$$\psi_\tau = \xi_- e^{K\psi(\xi_-)} - \xi e^{K\psi} + \xi_+^2 e^{K\psi(\xi_+)} - \xi^2 e^{K\psi}, \quad \xi_\pm = \xi \pm \frac{1}{K}, \tag{1.41}$$

and we write $\psi = \Psi_0 + \frac{1}{K}\Psi_1 + \ldots$; at leading order we regain (1.31); selecting the steady solution $\Psi_0 = \xi\left(\ln\frac{1}{\xi} + 1\right) - 1$, the equation for Ψ_1 is easily solved to find $\Psi_1 = -\frac{3}{2}\ln\xi$, so that the physical optics approximation for p is

$$p \approx A\xi^{-3/2}\exp\left[K\left\{\xi\left(\ln\frac{1}{\xi} + 1\right) - 1\right\}\right]. \tag{1.42}$$

This of course looks suspicious at low ξ, but it seems to be all right provided we take $\xi \geq \frac{1}{K}$. In fact, we now derive the equation for p_0 by taking

$$\dot{p}_0 = \frac{p_1}{K} = \frac{1}{K^2}p\left(\frac{1}{K}, \tau\right) \approx \frac{e}{\sqrt{2\pi}}Ke^{-K}(1 - p_0). \tag{1.43}$$

This can be favourably compared to (1.18), since $\frac{e}{\sqrt{2\pi}} \approx 1.084$. Actually, we can see what is happening here, since $\frac{e}{\sqrt{2\pi}}$ is just Stirling's (rather good) approximation to $\frac{1}{n!}$ in (1.16) when $n = 1$. Thus the continuous probability density function approximation to p_n does rather well all the way down to $n = 1$.

The implication of this discussion is that when a continuous model indicates a very small population being maintained for a significant time, then in practice the population will become extinct. The importance of this lies in the fact that it is not uncommon for population models which exhibit oscillations to have precisely this property, and suggests that where the oscillations are the point of focus, the continuous models are in essence incorrect.

A second conundrum which relates continuous models to low population densities is what I will call the frogspawn problem. Frogs produce thousands of eggs (e. g., Beattie 1987), but only a few of these survive to become adult frogs (Calef 1973). If we want to write a continuous model for a frog population which describes the production of thousands of eggs and their maturation as tadpoles to adult frogs, we need to find a way in which a small number of the original eggs can survive. There need to be very few, but importantly these few must not go extinct; how can that be? In a later section, I will describe one possible answer to this query.

2 Atto-Foxes

An enduring issue in population dynamics is what has been called 'the atto-fox problem' (Lobry and Sari 2015). The origin of this term lies in a model suggested by Anderson et al. (1981) to account for the fact that rabies outbreaks tend to recur, and they explained this by showing that oscillations can occur in the model. The work was extended by Murray et al. (1986) to account both for the recurrence and also the spread of the disease by including a term allowing for the 'diffusion' of wandering rabid foxes. An account of the model is given in his book (Murray 1989, chapter 20), since re-published in second and third editions. The epidemiology of fox rabies is described by Bacon (1985) and Toma and Andral (1977) for example. The virus is expressed in the saliva and transmitted by biting. Models of rabies dynamics continue to attract attention (e. g., Liu et al. 2017). However, the simple early models of Anderson and May and of Murray suffer from a defect, which is that the minima of the infected population reach levels which are so low as to imply extinction of the virus. Mollison (1991, p. 31) severely criticised ('this is incredible') the continuous model on two counts, the second of which is the inability of a continuous model to allow populations to become extinct, despite reaching values (in the rabies case) of one atto-fox (10^{-18} of a fox) per square kilometre.

Mollison's advice was that it is essential to use a stochastic model instead of a continuous one, with the consequence of extinction. One way in which local extinction can be circumvented is through spatial heterogeneity. Most simply, if a population is distributed between relatively isolated regions with some contact, then its eradication in one region can be overcome by leakage from a neighbouring patch. For example, measles in the UK in pre-vaccination times (before 1965) was endemic in London but was transmitted to smaller communities (below the critical size necessary to maintain the endemic state) by spatial transmission (Grenfell et al. 2001; Korevaar et al. 2020), in much the same way as contraction of the heart muscle is enabled by spread of the pacemaker activity of the sino-atrial node cells to the excitable cardiac tissue cells.

The simplest way to think about spatial transmission is by the addition of a diffusion term. Of course, this only describes nearest neighbour contacts and is not suitable to describe local outbreaks caused by long-range transmission of 'sparks' (Grenfell et al. 2001), for which a spatial convolution integral would be more appropriate, but it will serve for the present discussion. It is well known that the addition of diffusion to an oscillatory system leads to periodic travelling waves. However, if the minima of the oscillations are so low as to promote extinction, then, as for measles or the heart, a local outbreak will cause the propagation of a solitary travelling wave

Here I want to explore a different possibility, which is that there is something fundamentally lacking in such models, and that is the concept of a reservoir. Let me illustrate this by reference to a simple (dimensionless) population model

$$\varepsilon \dot{f} = sf - f, \tag{2.1}$$

where $\varepsilon \ll 1$ represents the idea that the timescales for growth and decay of the population f are much smaller than that of the slowly varying nutrient source s. We shall see examples built around (2.1) in the following section. If $s < 1$, then the population plummets and will become extinct in practice. A reservoir allows for a second state g and the transfer $f \to g$ is enabled at low values of s; basically, the population hibernates until s increases above one, at which point the awakening $g \to f$ is enabled. This is somewhat similar to the concept of a refuge (Sih 1987), but it is not quite the same.

In the case of rabies, I suggested (Fowler 2000) that a resolution of the issue could be found in the distribution of infection times. SIR-type models (with a rate of recovery of the infected population I of rI) are equivalent to assuming an exponential distribution

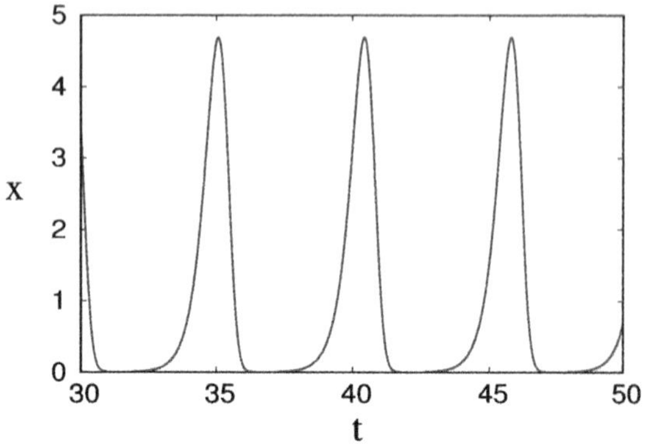

Fig. 3 Solution of the delayed logistic equation (3.1) at $\alpha = 2.5$. The asymptotic limit of large α is indicated by the flat minimum phase (the minimum of x is approximately 0.00158)

of recovery times $R(a) = 1 - e^{-ra}$, where a is the 'age' of the infection, so that $R'(a) = re^{-ra}$ is the recovery time probability density (Fowler and Hollingsworth 2015). But such an assumption is not commonly realistic; if one instead assumes a fixed disease period, then the SIR model reduces to a differential-delay equation (Soper 1929). A more general assumption is that of a gamma distribution (Fowler and Hollingsworth 2015) of the form

$$R'(a) = \frac{r^{\gamma} a^{\gamma-1} e^{-ra}}{\Gamma(\gamma)}, \tag{2.2}$$

which provides a gradual change from the exponential to fixed period distributions as γ increases from 1 to ∞. In fact, experimental work in the 1960s (Parker and Wilsnack 1966; Steck and Wandeler 1980) indicated that though most rabid foxes would have incubation periods of about a month, around 10% of cases would survive for up to six months; in effect these more resistant foxes act as a reservoir for the virus. The point is that it is the small ratio of incubation time to population growth time which causes the exponentially small minima to occur.

3 Boom-and-Bust Dynamics

Oscillations in continuous population models which have extremely low minima occur in many other situations. One simple but remarkable example is the delayed logistic equation (Hutchinson 1948), which can be written in the form

$$\dot{x} = \alpha x(1 - x_1), \quad x_1 \equiv x(t-1). \tag{3.1}$$

In this dimensionless form, α is the ratio of the delay to the specific growth rate, and large values cause periodic solutions to occur with extremely small minima. What is remarkable is that oscillations only occur for $\alpha > 1.57$, but already for $\alpha = 2.5$ the asymptotic limit is visible. This is shown in Fig. 3. The minimum of the oscillations is approximately

$$x_{\min} \approx \alpha \exp(-e^{\alpha} + 2\alpha - 1) \tag{3.2}$$

(Fowler 1982) and is already of order 10^{-3} when $\alpha = 2.5$.

While the model itself is now largely only of academic interest, it is closely related to a simple model of the immune response to an infecting antigen presented by Dibrov et al. (1977a, b), which was analysed by Fowler (1981). In its simplest dimensionless form, the model is given by

$$\dot{g} = \alpha g(1 - a),$$
$$\dot{a} = \beta[g_1 - a\{1 - \kappa + \kappa g\}], \quad g_1 = g(t - 1), \tag{3.3}$$

where a and g denote antibody and antigen densities, respectively. The interpretation of the model is fairly straightforward: the infecting antigen grows but is removed by the antibodies, produced here through stimulation of the humoral immune response, the maturation time of which produces the delay in the response. If

$$\kappa > \kappa_c = \alpha e^{1-\alpha}, \tag{3.4}$$

then the antigen grows unboundedly, otherwise oscillations occur, and these are severe if the delay is large. Specifically, if α and β are large (and thus necessarily κ is small), then we have $a \approx g_1$, and we regain the logistic delay equation. The behaviour of the variables is similar to that shown in Fig. 3.

The immune response time is of the order of days, and much larger than a common infective (e. g., viral) growth time, and it is because of this in the model that the minima in the oscillations are attometric in scale. The immune system is a good deal more complicated than indicated in (3.3), but can be modelled in similar fashion, usually with a continuous model (Perelson 2002), and sometimes with delays (Lee et al. 2009; Rundell et al. 1998) or without (Yan et al. 2016). Commonly extinction occurs, in the sense that viral populations become very low, with the assumption that stochastic elimination occurs at these levels (Yan et al. 2016), but this is not always the case: HIV is one viral disease where an endemic state is maintained for a long period (Perelson 2002), and the herpes virus establishes itself in the body by entering a dormant state which allows for further outbreaks (Nicoll et al. 2012).

3.1 Microbial Growth

Microbial growth models are another source of oscillatory dynamics. A particularly simple example is a model due to Omta et al. (2013), who were interested in oscillations in oceanic calcifiers as a possible cause of periodic ice ages. Their model can be written as

$$\dot{B} = kYCB - dB,$$
$$\dot{C} = I - kCB; \tag{3.5}$$

here B represents biomass, and C a (limiting) nutrient (in Omta *et al.*'s case the nutrient is carbon and the biomass consists of calcifiers such as coccolithophores).

Fig. 4 The solution for b in (3.7) using a value of $\varepsilon = 0.001$ and initial conditions $b = 10$, $c = 1$. The large initial value of b causes the minima to be exponentially small; here the minimum at $t \approx 90.75$ is $b \approx 0.00074$, for example

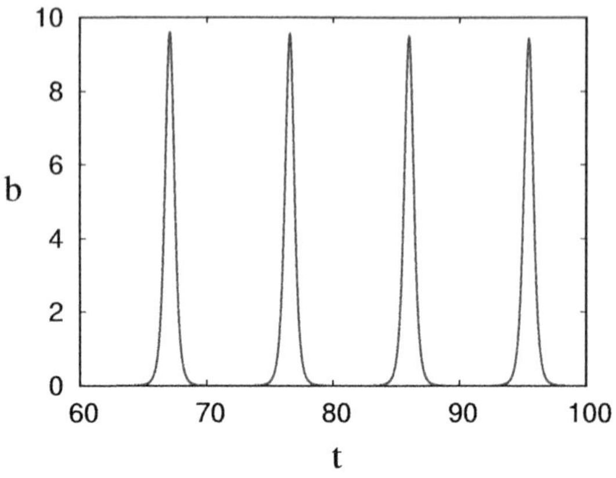

Very similar types of model have been used in describing oscillations in glycolysis (Goldbeter 1996), and plankton blooms (Huppert et al. 2005; Mahadevan et al. 2012). I represents a rate of input of nutrient to the system. It can be noted that if $I = 0$, (3.5) is just an SIR model, in which nutrient represents susceptibles, and the biomass masquerades as the infected. In the present case, the non-zero supply allows recovery, much as the increasing fox population allows oscillation in the rabies model.

By defining the non-dimensional variables and parameter

$$B = \frac{IY}{d}b, \quad C = \frac{d}{kY}c, \quad t \sim \frac{1}{\sqrt{IkY}}, \quad \varepsilon = \frac{\sqrt{IkY}}{d}, \tag{3.6}$$

the model can be written in the dimensionless form (cf. Fowler 2013)

$$\begin{aligned} \varepsilon\dot{b} &= (c-1)b, \\ \dot{c} &= \varepsilon(1-bc), \end{aligned} \tag{3.7}$$

where the small parameter ε is a measure of the ratio of the nutrient supply rate to the biomass growth rate, and the nature of the model is easily understood by writing $b = e^{\theta}$, whence we have

$$\ddot{\theta} + V'(\theta) = -\varepsilon e^{\theta}\dot{\theta}, \tag{3.8}$$

where the potential V is defined by

$$V(\theta) = e^{\theta} - \theta. \tag{3.9}$$

When $\varepsilon \ll 1$, this is a slowly decaying nonlinear oscillator, and if the 'energy' $E = \frac{1}{2}\dot{\theta}^2 + V(\theta)$ is large, then the biomass b exhibits typically spiky 'boom-and-bust' oscillations. The form of these is shown in Fig. 4.

Fowler (2014) extended this model to describe competing microbial populations (heterotrophs and fermenters). Denoting the heterotroph and fermenter populations

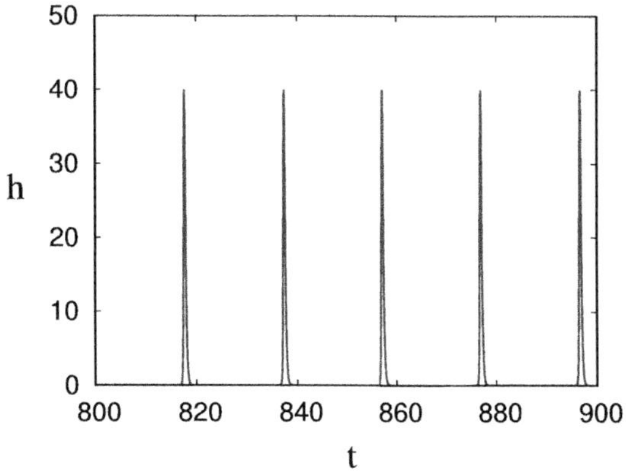

Fig. 5 The solution for h in (3.10) using values $\varepsilon = 0.1$ and $\delta = 0.5$. The minimum of the population is $\sim 10^{-15}$

as h and f, his model is, in dimensionless form,

$$\begin{aligned}
\varepsilon\dot{h} &= \delta hs + hc - h,\\
\varepsilon\dot{f} &= sf - f,\\
\dot{s} &= \varepsilon(1 - sf - \delta hs),\\
\dot{c} &= \varepsilon(sf - hc);
\end{aligned} \tag{3.10}$$

here s and c are two different forms of organic carbon. The heterotrophs can utilise both, whereas the fermenters can only use the s-form, but produce the c-form, which is preferentially used by the heterotrophs. The system thus has the form of an activator-inhibitor system, and, as shown in Fig. 5, boom-and-bust oscillations occur in conditions of starvation i.e., for small ε. For $\varepsilon = 0.2$, the minimum of h is $\approx 1.9 \times 10^{-4}$, but for $\varepsilon = 0.1$, as shown in the figure, it is $\approx 1.8 \times 10^{-15}$. When it is reduced slightly further to the practically estimated value of $\varepsilon = 0.07$ (Fowler et al. 2014), the microbial dimensionless minima are of the order of 10^{-32}, and extinction beckons. Given a common bacterial loading of 10^9 cells g^{-1} (Kirchman 2012, p. 9), this last value corresponds to levels of 1 yocto-cell per gram, thus transcending Mollison's atto-fox per km^2.

In all these examples, extinction looms, and the oscillations are suspect. I want to suggest that extinction can be avoided in reality by the presence of a reservoir. This concept resembles but is distinct from that of a refuge for prey in predator–prey models (e. g., Sih 1987; Haque et al. 2014; Balaban-Feld et al. 2019), although the introduction of that concept was aimed at providing stability for the otherwise structurally unstable Lotka-Volterra model. The purpose and function of a reservoir in the present case is quite different. In the case of the rabies virus, a reservoir could take the form of an endemic host, which could even be the long-lived foxes themselves. But, as pointed out by Sterner and Smith (2006), there are actually many different reservoirs amongst mammals for the rabies virus and its variants. From the point of view of a continuous model, all that is necessary is that there is at least one reservoir where viral extinction does not occur.

In the case of microbes, the existence of bacteria in a dormant state is well recognised (Kaprelyants et al. 1993; Lennon and Jones 2011; Hoehler and Jørgensen 2013). Fowler and Winstanley (2018) suggested a simple model to describe the switch to dormancy; their model can be written in dimensionless form as

$$\begin{aligned} \varepsilon\dot{b} &= (c-1)b - qb + pa, \\ \dot{c} &= \varepsilon(F - bc), \\ \varepsilon\dot{a} &= qb - pa, \end{aligned} \tag{3.11}$$

which generalises (3.7). Here a represents the dormant state, and p and q are switching functions which switch on (q) and off (p) at low nutrient (c) or active biomass (b) levels. This model allows self-sustained oscillations when ε is small, again of boom-and-bust type, and although generally they have less extreme minima than the yocto-cell example, effective extinction of the active biomass can occur; the difference is that the dormant bacterial population remains viable when that happens, providing a nursery for bacterial growth when the environmental stress is reduced. *The reduction of b to tiny levels does not matter.* This suggestion of a latent reservoir is one that can come to the rescue of continuous models when they indicate extinction. One circumstance where a reservoir may not exist is in a viral infection. Many viral infections may be completely eliminated (Rundell et al. 1998; Yan et al. 2016), although there are some where an endemic or latent state is established (Perelson 2002; Nicoll et al. 2012).

4 Frogspawn

Finally I want to consider another problem of small numbers and whether it can be catered for in a continuous model. Many species of plants and animals reproduce by means of the production of thousands, or even millions, of eggs or seeds. Fish provide one example (e. g., Pope et al. 2010), and trees another (Greene and Johnson 1994); helminths, discussed in the conclusions, provide another. In such cases, very few of the offspring survive to adulthood, and the question is: why? I will refer to this as the frogspawn problem, as frogs provide another example of such extreme fecundity.

The population biology of frogs (it should be noted that there are many different species) has been studied by many authors (Berven 1990; Friedl and Klump 1997; Heyer et al. 1975; Smith 1983; Travis et al. 1985). Frogs produce thousands of eggs (Gibbons and McCarthy 1986), some of which later hatch to tadpoles, and still fewer of these make it to adulthood. Roughly speaking, a single adult frog in a stable population ought to produce a single offspring. How can this be?

The reason I find this perplexing is that the normal predation rate of a population F would be $\propto F$ (assuming plenty of predators) so that if the population becomes very low, we arrive at the previous conundrum: why does extinction not occur?

There is in fact a simple possible answer to this problem. Let us consider a population of adult frogs F, and suppose that F_n is the frog population measured at intervals of $\Delta T = 1$ year, thus $F_n = F(n\Delta T)$. If each female produces N eggs per year ($N \sim 1000$ year^{-1}), and the adult death rate is μ (year^{-1}), then the year on year change in the population would be

$$F_{n+1} - F_n = \Delta F = M_n \Delta T - \mu F_n \Delta T, \tag{4.1}$$

where M_n (frog year^{-1}) is the total number of eggs which survive predation and metamorphose to adults. In the absence of predation, $M_n = \frac{1}{2} N F_n$.

The death rate is enhanced for eggs and tadpoles by predation (Waller 1973)); the reason so few tadpoles survive to adulthood is that they are consumed (even by themselves). So the principal loss term is not natural death but juvenile predation. And here is the idea: when the population levels become low, the predators do not find the prey so easily. We can for example model egg and tadpole predation during the year as a term $-kM^2$, where $M = \frac{1}{2} N F_n$ initially, and $\frac{dM}{dt} = -kM^2$, whence we find

$$M_n = \frac{\frac{1}{2} N F_n}{1 + \frac{1}{2} k N F_n \Delta T}. \tag{4.2}$$

There is in fact experimental evidence that this description may be reasonably accurate (see Brockelman 1969, figure 4). A suggestive continuous version of (4.1) is

$$\dot{F} = \mu F \left[\frac{F_0}{\delta F_0 + F} - 1 \right], \tag{4.3}$$

where the overdot denotes differentiation with respect to time, and

$$F_0 = \frac{1}{\mu k \Delta T}, \quad \delta = \frac{2\mu}{N}. \tag{4.4}$$

A typical value of μ is 1 year^{-1} (Berven 1990). With $N \sim 10^3$ year^{-1}, evidently $\delta \ll 1$ (this is the frogspawn problem), but infant predation leads to a stable population $F \approx F_0$, whose size is actually nothing to do with the egg production rate.

This of course is hardly a new idea and is the basis for Holling's type 3 predator response to prey density (Holling 1959, 1961, 1973), in which the individual predator's consumption rate as a function of prey density is S-shaped, and (for example) quadratic at low prey densities. This same idea was used in the spruce budworm model of Ludwig et al. (1978), where the birds' predation of budworm larvae is described thus by Murray (1989, p. 5): 'For small population densities ..., the birds tend to seek food elsewhere...'. The motivation leading to (4.3) has the same effect as the saturating term in the logistic or Verhulst equation:

$$\dot{F} = rF \left(1 - \frac{F}{F_0} \right), \tag{4.5}$$

whose right-hand side has the same unimodal shape as (4.3).

It should be noted that Verhulst's (1845) suggestion of the logistic equation[1] was not based on any process description, but on a wish to describe the weakening (*affaiblement*) of reproduction in the presence of limited resources. The present suggestion of a saturational model is based on quite different considerations.

The resolution of the frogspawn conundrum is simply due to the nonlinear egg predation rate, which allows an equilibrium to occur no matter how many eggs are produced, and whose size depends on the predation coefficient k. The surprising thing is that one normally teaches the Verhulst equation as a response to limited resources: the specific fecundity rate decreases with population size, but here the same effect is due to a quite different mechanism.

4.1 An Age-Structured Model

Of course, (4.3), while suggestive, is rather crude. A more subtle approach is to consider an age-structured model in which the amphibian density $f(t, a)$ depends on both time t and age a. For small a, f represents eggs, then tadpoles, and for $t > T$, say, adult frogs. The total amphibian density is then

$$F = \int_0^\infty f \, da. \tag{4.6}$$

The units of f are taken to be Am psf^{-1} year^{-1}, where Am means amphibians, y is years, and psf is pond square foot. This last unit of area is by analogy with the Jones site model for spruce budworm outbreaks, where the larval density was measured as individuals tsf^{-1}, where tsf means ten square foot of susceptible branch surface area (Jones 1979; Hassell et al. 1999).

We pose the following model for f, in which the subscripts denote partial derivatives:

$$f_t + f_a = -r(a) f^2, \tag{4.7}$$

by analogy with (4.3). The boundary and initial conditions are taken to be

$$f(0, a) = f_i(a), \quad f(t, 0) = f_0(t), \tag{4.8}$$

where the renewal equation for f_0 is taken to be

$$f_0(t) = N \int_T^\infty f(t, a) \, da, \tag{4.9}$$

indicating that mature adult frogs lay N eggs per year. The time T is the age of sexual maturity, commonly 1–2 years (e. g., Friedl and Klump 1997; Berven 1990). The units of r are Am^{-1} psf, and N has units Am Am^{-1} y^{-1} = y^{-1}, or eggs per frog per year.

[1] There was an earlier note from 1838, and this has been provided in translation by Vogels et al. (1975).

If we define

$$R(a)=\int_0^a r(a')\,\mathrm{d}a', \tag{4.10}$$

then the solution is

$$f=\begin{cases}\dfrac{f_i(a-t)}{1+\{R(a)-R(a-t)\}f_i(a-t)}, & t<a,\\[2ex] \dfrac{f_0(t-a)}{1+R(a)f_0(t-a)}, & t>a,\end{cases} \tag{4.11}$$

and for $t>T$ the renewal equation gives the integral delay equation for f_0:

$$f_0(t)=N\left[\int_0^{t-T}\frac{f_0(s)\,\mathrm{d}s}{1+R(t-s)f_0(s)}+\int_0^{\infty}\frac{f_i(\xi)\,\mathrm{d}\xi}{1+\{R(t+\xi)-R(\xi)\}f_i(\xi)}\right]. \tag{4.12}$$

Let us consider the form of $R(a)$. The predation rate r is a rapidly decreasing function of a for $a<T$, so that R monotonically increases to a plateau at $R(T)=\bar{R}$, say. Thereafter R will increase slowly, and if we suppose all frogs die of senescence by age A, say, then $R\to\infty$ as $a\to A$. In this case the second term in the square bracket is zero for $t>A$.

A natural scale for R is thus $\bar{R}$, and it is now convenient to scale the variables as

$$R\sim\bar{R},\quad t,a\sim T,\quad f=\frac{\Phi}{\bar{R}},\quad f_0=\frac{\phi}{\bar{R}},\quad A=T\alpha, \tag{4.13}$$

and this leads to the equation for ϕ,

$$\phi(t)=\Lambda\int_0^{t-1}\frac{\phi(s)\,\mathrm{d}s}{1+R(t-s)\phi(s)},\quad t>\alpha, \tag{4.14}$$

where

$$\Lambda=NT\gg 1, \tag{4.15}$$

and the age structure is given by

$$\Phi(t,a)=\frac{\phi(t-a)}{1+R(a)\phi(t-a)}. \tag{4.16}$$

Because of the large value of Λ, it is relatively easy to solve (4.14). We begin with an example, and consider first the steady state. R is an increasing function, with $R(1)=1$ and $R\to\infty$ as $t\to\alpha$. As an illustration, suppose

$$R=\frac{\alpha-1}{\alpha-a},\quad a>1; \tag{4.17}$$

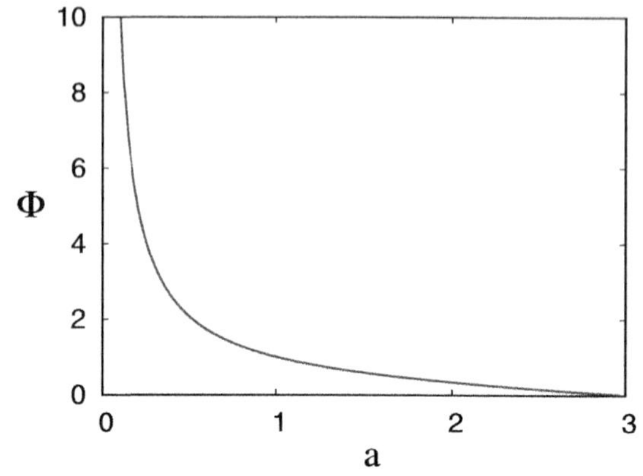

Fig. 6 The dimensionless age distribution Φ given by (4.21), with $\Lambda = 2000$ and $R(a) = \dfrac{\alpha(\alpha-1)a}{\alpha(\alpha-1)-a(a-1)}$, $\alpha = 3$ (corresponding to $N = 1000\ \text{y}^{-1}$, $T = 2$ y, $A = 6$ y). These are typical values for Irish frogs (Gibbons and McCarthy 1984). For visibility the range of Φ is not shown, but at $a = 0$, $\Phi \approx 1530$

the steady solution of (4.14) is then given uniquely by

$$1 = \Lambda(\alpha - 1)\left[1 + \phi \ln\left(\frac{\phi}{1+\phi}\right)\right], \tag{4.18}$$

and for large Λ this is approximately

$$\phi \approx \tfrac{1}{2}\Lambda(\alpha - 1). \tag{4.19}$$

More generally, and even if time-dependent, the observation that $\phi \sim \Lambda$ leads to the approximate result

$$\phi \approx \Lambda \int_1^{\alpha} \frac{da}{R(a)}, \tag{4.20}$$

and a uniform approximation for the age distribution is

$$\Phi = \frac{\displaystyle\int_1^{\alpha} \frac{da}{R(a)}}{\displaystyle\frac{1}{\Lambda} + R(a)\int_1^{\alpha} \frac{da}{R(a)}}, \tag{4.21}$$

which shows that the distribution descends sharply from $O(\Lambda)$ when $a \ll 1$ to $O(1)$ when $a \sim 1$. An illustration of the resulting dimensionless age distribution is shown in Fig. 6.

5 Conclusions

As regards atto-foxes and yocto-cells, we suggest that the resolution of this long-standing issue may be that in practice, vanishing populations seek refuge in a safe haven, whether it be in dormancy or as an endemic remnant in another host reservoir. One example is the ability of bacteria to remain viable in the most inhospitable places (for example deep in the Earth) for an extremely long time (Hoehler and Jørgensen

2013). Plant seeds are another example: metabolic processes essentially shut down until they are stimulated to re-emerge (Bewley 1997; FitzGerald and Keener 2021).

The suggested resolution of the frogspawn problem, whether it be the logistic-type equation (4.3) or the mildly more interesting age-dependent model, seems straightforward, but it raises another issue. If the reduction of the large numbers of eggs is due to an effectively quadratic predation rate, what then is the point of the large number? We can see from (4.21) that it does not really matter how large N and thus Λ is. It is possible that this is controlled by actual space limitation, but also, production of a small number of eggs presumably requires parental care, which is not so easily available for the predator-prone frog, and the large number simply indicates this (Davis and Roberts 2005).

There is a related issue in continuous modelling which arises in a classical model for human infection of the roundworm *Ascaris lumbricoides*. Roundworm infection is endemic in low-income populations with poor sanitation in tropical countries and is one of a number of 'neglected tropical diseases' which are a focus of much international interest (Holland 2013; Hollingsworth et al. 2015). The classic model to describe the infection was put forward by Anderson and May (1985, 1991) and takes the essential form

$$\dot{L} = rM - \mu_2 L,$$
$$\dot{M} = \nu_0 L - \mu_1 M. \tag{5.1}$$

Here M is the adult worm burden in the human small intestine, typically of the order of 10–20 per human, while L is the number of mature eggs in the environment. The lifetimes of eggs and adults are respectively taken to be $\mu_2^{-1} \sim$ 28–84 days, and $\mu_1^{-1} \sim$ 1–2 years. Actually there is an issue here already, because viable *Ascaris* eggs in latrines have been found with ages of the order of up to 15 years (WIN-SA 2011, Chris Buckley, private communication). Leaving that aside, (5.1) is of course linear, but nonlinearity is introduced by a unimodal dependence of the recruitment rate ν_0 on M: at low M this is due to the increasing probability of having male and female worms present in the human, and at high M because of the reduction in fecundity due to crowding. This nonlinearity allows for a stable endemic population.

Aside from the issue of the egg lifetime, there is an issue concerning the recruitment rate ν_0. Anderson and May (1991) effectively avoided estimating this by using measured values of the basic reproduction rate

$$R_0 = \frac{\nu_0 r}{\mu_1 \mu_2} \tag{5.2}$$

in the range 1–5. Adult worms produce up to 2×10^5 eggs per day, so that in a village community of 100 people, we might have $r \sim 10^7\ \mathrm{d}^{-1}$. It then turns out that the transmission coefficient (i. e., rate of uptake of mature eggs in the environment) is $\sim 10^{-10}\ \mathrm{d}^{-1}$, something less than a nano-egg per human per day (Fowler and Hollingsworth 2016). We are back with the atto-fox problem, and the problem is worsened if we select a lower value of μ_2. Actually this is more of a frogspawn-type problem: huge numbers of eggs only result in a small number of adults. While the latter can be understood by crowding effects in the small intestine, it does not explain why the basic reproduction rate is so low. We do not offer a resolution of this conundrum here, but the frogspawn discussion suggests that a more detailed examination of egg survival (which is assumed as a linear decay rate in deriving (5.1)) might be worthwhile.

Acknowledgements Many thanks to Yvonne Buckley for stimulating conversation. I am indebted to David Stirzaker, who pointed me in the direction of the birth-death process literature. This publication has emanated from research conducted with the financial support of Science Foundation Ireland under grant number SFI/13/IA/1923.

References

Anderson RM, Jackson HC, May RM, Smith AM (1981) Population dynamics of fox rabies in Europe. Nature 289:765–771

Anderson RM, May RM (1985) Helminth infections of humans: mathematical models, population dynamics, and control. Adv Parasitol 24:1–101

Anderson RM, May RM (1991) Infectious diseases of humans: dynamics and control. OUP, Oxford

Bacon PJ (ed) (1985) Population dynamics of rabies in wildlife. Academic Press, New York

Balaban-Feld J, Mitchell WA, Kotler BP, Vijayan S, Tov Elem LT, Rosenzweig ML, Abramsky Z (2019) Individual willingness to leave a safe refuge and the trade-off between food and safety: a test with social fish. Proc R Soc B 286:20190826

Bartlett MS (1960) Stochastic population models in ecology and epidemiology. Methuen, London

Bartlett MS, Gower JC, Leslie PH (1960) A comparison of theoretical and empirical results for some stochastic population models. Biometrika 47:1–11

Beattie RC (1987) The reproductive biology of common frog (Rana temporaria) populations from different altitudes in northern England. J Zool 211:387–398

Bender CM, Orszag SA (1978) Advanced mathematical methods for scientists and engineers. McGraw-Hill, New York

Berven KA (1990) Factors affecting population fluctuations in larval and adult stages of the wood frog (Rana Sylvatica). Ecology 71(4):1599–1608

Bewley JD (1997) Seed germination and dormancy. Plant Cell 9:1055–1066

Brockelman WY (1969) An analysis of density effects and predation in Bufo Americanus tadpoles. Ecology 50(4):632–644

Calef GW (1973) Natural mortality of tadpoles in a population of Rana Aurora. Ecology 54(4):741–758

Davis RA, Roberts JD (2005) Embryonic survival and egg numbers in small and large populations of the frog Heleioporus albopunctatus in Western Australia. J Herpetol 39(1):133–138

Dibrov BF, Livshits MA, Volkenstein MV (1977a) Mathematical model of immune processes. J Theor Biol 65:609–631

Dibrov BF, Livshits MA, Volkenstein MV (1977b) Mathematical model of immune processes. II. Kinetic features of antigen-antibody interaction. J Theor Biol 69:23–39

Doering CR, Sargsyan KV, Sander LM (2005) Extinction times for birth-death processes: exact results, continuum asymptotics, and the failure of the Fokker-Planck approximation. Multiscale Model Simul 3(2):283–299

Durrett R, Levin SA (1994) The importance of being discrete (and spatial). Theor. Pop. Biol. 46:363–394

FitzGerald C, Keener J (2021) Red light and the dormancy-germination decision in Arabidopsis seeds. Bull Math Biol 83(3):17
Fowler AC (1981) Approximate solution of a model of biological immune responses incorporating delay. J Math Biol 13:23–45
Fowler AC (1982) An asymptotic analysis of the logistic delay equation when the delay is large. IMA J Appl Math 28:41–49
Fowler AC (2000) The effect of incubation time distribution on the extinction characteristics of a rabies epizootic. Bull Math Biol 62:633–660
Fowler AC (2013) Note on a paper by Omta et al on sawtooth oscillations. SeMA J 62:1–13
Fowler AC (2014) Starvation kinetics of oscillating microbial populations. Math Proc R Irish Acad 114(2):173–189
Fowler AC, Hollingsworth TD (2015) Simple approximation methods for epidemics with exponential and fixed infectious periods. Bull Math Biol 77:1539–1555
Fowler AC, Hollingsworth TD (2016) The dynamics of Ascaris lumbricoides infections. Bull Math Biol 78:815–833
Fowler AC, Winstanley HF (2018) Microbial dormancy and boom-and-bust population dynamics under starvation stress. Theor Popul Biol 120:114–120
Fowler AC, Winstanley HF, McGuinness MJ, Cribbin LB (2014) Oscillations in soil bacterial redox reactions. J Theor Biol 342:33–38
Friedl TWP, Klump GM (1997) Some aspects of population biology in the European tree frog Hyla arborea. Herpetologica 53(3):321–330
Gibbons MM, McCarthy TK (1984) Growth, maturation and survival of frogs Rana temporaria L. Holarct Ecol 7:419–427
Gibbons MM, McCarthy TK (1986) The reproductive output of frogs Rana temporaria (L.) with particular reference to body size and age. J Zool A 209:579–593
Goldbeter A (1996) Biochemical oscillations and cellular rhythms. CUP, Cambridge
Greene DF, Johnson EA (1994) Estimating the mean annual seed production of trees. Ecology 75(3):642–647
Grenfell BT, Bjørnstad ON, Kappey J (2001) Travelling waves and spatial hierarchies in measles epidemics. Nature 414:716–723
Haque M, Rahman MS, Venturino E, Li B-L (2014) Effect of a functional response-dependent prey refuge in a predator? Prey model. Ecol Complex 20:248–256
Hassell DC, Allwright DJ, Fowler AC (1999) A mathematical analysis of Jones site model for spruce budworm infestations. J Math Biol 38:377–421
Heyer WR, McDiarmid RW, Weigmann DL (1975) Predation and pond habits in the tropics. Biotropica 7(2):100–111
Hoehler TM, Jørgensen BB (2013) Microbial life under extreme energy limitation. Nat Rev Microbiol 11(2):83–94
Holland C (ed) (2013) Ascaris: the neglected parasite. Elsevier, Amsterdam
Holling CS (1959) The components of predation as revealed by a study of small mammal predation on the European pine sawfly. Can Entomol 91:293–320
Holling CS (1961) Principles of insect predation. Ann Rev Entomol 6:163–182
Holling CS (1973) Resilience and stability of ecological systems. Ann Rev Ecol Evol Syst 4:1–23
Hollingsworth TD, Pulliam JRC, Funk S, Truscott JE, Isham V, Lloyd AL (2015) Seven challenges for modelling indirect transmission: vector-borne diseases, macroparasites and neglected tropical diseases. Epidemics 10:16–20
Huppert A, Blasius B, Olinky R, Stone L (2005) A model for seasonal phytoplankton blooms. J Theor Biol 236(3):276–290
Hutchinson GE (1948) Circular causal systems in ecology. Ann N Y Acad Sci 50:221–240
Jones DD (1979) The budworm site model. In: Norton CA, Holling CS (eds) Pest management, proceedings of an international conference, Pergamon Press, Oxford, pp 91–155
Kaprelyants AS, Gottschal JC, Kell DB (1993) Dormancy in non-sporulating bacteria. FEMS Microbiol Rev 104:271–286
Kirchman DL (2012) Processes in microbial ecology. OUP, Oxford
Korevaar H, Metcalf CJ, Grenfell BT (2020) Structure, space and size: competing drivers of variation in urban and rural measles transmission. J R Soc Interface 17:20200010

Lee HY, Topham DJ, Park SY, Hollenbaugh J, Treanor J, Mosmann TR, Jin X, Ward BM, Miao H, Holden-Wiltse J, Perelson AS, Zand M, Wu H (2009) Simulation and prediction of the adaptive immune response to influenza A virus infection. J Virol 83(14):7151–7165
Lennon JT, Jones SE (2011) Microbial seed banks: the ecological and evolutionary implications of dormancy. Nate Rev Microbiol 9:119–130
Liu J, Jia Y, Zhang T (2017) Analysis of a rabies transmission model with population dispersal. Nonlinear Anal Real World Appl 35:229–249
Lobry C, Sari T (2015) Migrations in the Rosenzweig-MacArthur model and the atto-fox problem. ARIMA J 20:95–125
Ludwig D, Jones DD, Holling CS (1978) Qualitative analysis of insect outbreak systems: the spruce budworm and forest. J Anim Ecol 47:315–332
Mahadevan A, Dasaro E, Perry M-J, Lee C (2012) Eddy-driven stratification initiates North Atlantic Spring phytoplankton blooms. Science 337(6090):54–58
Mollison D (1991) Dependence of epidemic and population velocities on basic parameters. Math Biosci 107:255–287
Mollison D, Levin SA (1995) Spatial dynamics of parasitism. In: Grenfell BT, Dobson AP (eds) Ecology of infectious diseases in natural populations. CUP, Cambridge, pp 384–398
Murray JD (1989) Mathematical biology. Springer, Berlin
Murray JD (2018) My gift of polio: ~ an unexpected life ~ from Scotland's rustic hills to Oxford's hallowed halls and beyond (independently published)
Murray JD, Stanley EA, Brown DL (1986) On the spatial spread of rabies among foxes. Proc R Soc Lond B 229:111–150
Nåsell I (2001) Extinction and quasi-stationarity in the Verhulst logistic model. J Theor Biol 211:11–27
Nicoll MP, Proença JT, Efstathiou S (2012) The molecular basis of herpes simplex virus latency. FEMS Microbiol Rev 36:684–705
Omta AW, van Voorn GAK, Rickaby REM, Follows MJ (2013) On the potential role of marine calcifiers in glacial-interglacial dynamics. Global Biogeochem Cycles 27:692–704
Ovaskainen O, Meerson B (2010) Stochastic models of population extinction. Trends Ecol Evol 25(11):643–651
Parker RL, Wilsnack RE (1966) Pathogenesis of skunk rabies virus: quantitation in skunks and foxes. Am J Vet Res 27:33–38
Perelson AS (2002) Modelling viral and immune system dynamics. Nat Rev Immunol 2(1):28–36
Pope EC, Hays GC, Thys TM, Doyle TK, Sims DW, Queiroz N, Hobson VJ, Kubicek L, Houghton JDR (2010) The biology and ecology of the ocean sunfish Mola mola: a review of current knowledge and future research perspectives. Rev Fish Biol Fish 20:471–487
Rundell A, DeCarlo R, HogenEsch H, Doerschuk P (1998) The humoral immune response to Haemophilus influenzae type b: a mathematical model based on T-zone and germinal center B-cell dynamics. J Theor Biol 194:341–381
Sih A (1987) Prey refuges and predator-prey stability. Theor Popul Biol 31:1–12
Smith DC (1983) Factors controlling tadpole populations of the chorus frog (Pseudacris triseriata) on Isle Royale. Mich Ecol 64(3):501–510
Steck F, Wandeler A (1980) The epidemiology of fox rabies in Europe. Epidemiol Rev 2:71–96
Sterner RC, Smith GC (2006) Modelling wildlife rabies: transmission, economics, and conservation. Biol Conserv 131:163–179
Soper HE (1929) The interpretation of periodicity in disease prevalence. J R Stat Soc 92:34–73
Toma B, Andral L (1977) Epidemiology of fox rabies. Adv Virus Res 21:1–36
Travis J, Keen WH, Juilianna J (1985) The role of relative body size in a predator-prey relationship between dragonfly naiads and larval anurans. Oikos 45:59–65
Verhulst P-F (1845) Recherches mathématiques sur la loi daccroissement de la population. Mémoires de Académie R de Bruxelles 18:1–38
Vogels M, Zoeckler R, Stasiw DM, Cerny LC (1975) PF Verhulst notice sur la loi que la populations suit dans son accroissement from Correspondence Mathématique et Physique. Ghent. J Biol Phys 3(4):183–192
Waller GC (1973) Natural mortality of tadpoles in a population of Rana aurora. Ecology 54(4):741–758
WIN-SA (2011) What happens when the pit is full? Developments in on-site faecal sludge management (FSM). FSM Seminar, 14–15 March 2011, Durban, South Africa, 43 pp. WIN-SA (Water Information Network South Africa). Stockholm Environmental Institute, www.sei.org

Yan AWC, Caoamd P, McCaw JM (2016) On the extinction probability in models of within-host infection: the role of latency and immunity. J Math Biol 73:787–813

Publisher's Note Springer Nature remains neutral with regard to jurisdictional claims in published maps and institutional affiliations.

Bulletin of Mathematical Biology (2021) 83:99
https://doi.org/10.1007/s11538-021-00932-1

Society for Mathematical Biology

SPECIAL ISSUE: CELEBRATING J. D. MURRAY

A Comparison of the "Reduced Losses" and "Increased Production" Models for Mussel Bed Dynamics

Jonathan A. Sherratt[1] · **Quan-Xing Liu**[2] · **Johan van de Koppel**[3]

Received: 4 September 2020 / Accepted: 20 July 2021 / Published online: 24 August 2021

Abstract

Self-organised regular pattern formation is one of the foremost examples of the development of complexity in ecosystems. Despite the wide array of mechanistic models that have been proposed to understand pattern formation, there is limited general understanding of the feedback processes causing pattern formation in ecosystems, and how these affect ecosystem patterning and functioning. Here we propose a generalised model for pattern formation that integrates two types of within-patch feedback: amplification of growth and reduction of losses. Both of these mechanisms have been proposed as causing pattern formation in mussel beds in intertidal regions, where dense clusters of mussels form, separated by regions of bare sediment. We investigate how a relative change from one feedback to the other affects the stability of uniform steady states and the existence of spatial patterns. We conclude that there are important differences between the patterns generated by the two mechanisms, concerning both biomass distribution in the patterns and the resilience of the ecosystems to disturbances.

Keywords Pattern formation · Mathematical model · Mussels · Reaction–diffusion–advection

✉ Jonathan A. Sherratt
j.a.sherratt@hw.ac.uk

Quan-Xing Liu
liuqx315@gmail.com

Johan van de Koppel
Johan.van.de.Koppel@nioz.nl

1 Department of Mathematics and Maxwell Institute for Mathematical Sciences, Heriot-Watt University, Edinburgh EH14 4AS, UK

2 State Key Laboratory of Estuarine and Coastal Research, School of Ecological and Environmental Sciences, East China Normal University, Shanghai 200241, People's Republic of China

3 Department of Estuarine and Delta Systems, Royal Netherlands Institute for Sea Research and Utrecht University, PO Box 140, 4400 AC Yerseke, The Netherlands

1 Introduction

Pattern formation at the landscape scale is an established feature of many ecosystems across the world. Examples include alternating patches of vegetation and bare ground in arid environments (Bastiaansen et al. 2018; Gandhi et al. 2019); "fairy circles" in Namibian grasslands (Zelnik et al. 2015); patterns of open-water pools in peatlands (Belyea 2007; Eppinga et al. 2009); tussock patterns in freshwater marshes (van de Koppel and Crain 2006; Yu 2010); and labyrinthine patterns in mussel beds (van de Koppel et al. 2005, 2008; Liu et al. 2014). Mathematical modelling has played an important role in the study of this type of pattern formation. Most commonly, models have been used to show that a hypothesised ecological mechanism can indeed generate spatial patterns. A crucial commonality for these underlying processes is that there is a local positive feedback, where organisms improve the conditions in which they grow, but at the same time lower growth conditions at distance, often by means of competition for resources. This scale-dependent alternation between local positive feedback and larger-scale negative feedback has been proposed for many patterned ecosystems, providing a general principle for regular spatial patterns for ecology and beyond. Despite this common ground, a wide variety of specific mechanisms have been proposed as potential generators of regular patterns. Soft-bottomed mussel beds provided an example of a system in which multiple hypothesised mechanisms have been shown to be potential generators of patterns. These large aggregations of mussels form in intertidal regions, most notably in the Dutch Wadden Sea, and are often patterned, with dense clusters of mussels separated by regions of bare sediment. Two possible mechanisms for this patterning have been proposed: (i) "reduced losses"—high mussel density reduces the dislodgement and predation of mussels, because of their adherence to one another; (ii) "increased production"—a high density of mussels increases their growth rate, since the faeces and pseudofaeces they produce raise them towards the algae-rich upper water layers. Mathematical modelling confirms that both of these mechanisms can generate spatial patterns (van de Koppel et al. 2005; Liu et al. 2012). How then can one determine which mechanism is the real driver of the observed pattern formation?

A first step in answering this question was taken by Liu and co-workers (Liu et al. 2012) who calculated bifurcation diagrams for models based on each of the two mechanisms, demonstrating significant differences between the patterns in the two models. However, it is unclear to what extent these differences depend on the formulation of the models and the parameter values, rather than being due to the alternative underlying patterning mechanisms. The objective of this paper is to address this issue in a comprehensive way. We will do this by developing a hybrid model that includes both the "reduced losses" and "increased production" mechanisms, with their relative importance controlled by a single tuning parameter. Our hybrid model is based on the "reduced losses" model proposed by van de Koppel et al. (2005) and the "increased production" model proposed by Liu et al. (2012).

We begin by describing the two models separately, starting with the "reduced losses" model (RLM). Denoting the mussel density by $\widetilde{m}(\widetilde{x}, \widetilde{t})$ and the density of algae (the main mussel food source Dolmer 2000; Øie et al. 2002) by $\widetilde{a}(\widetilde{x}, \widetilde{t})$, the model equations are:

$$\frac{\partial \widetilde{a}}{\partial \widetilde{t}} = \overbrace{\widetilde{F}\cdot(\widetilde{A}-\widetilde{a})}^{\substack{\text{transfer to/}\\ \text{from upper}\\ \text{water layers}}} - \overbrace{\widetilde{B}\,\widetilde{a}\,\widetilde{m}}^{\substack{\text{consumption}\\ \text{by mussels}}} + \overbrace{\widetilde{V}\,\partial\widetilde{a}/\partial\widetilde{x}}^{\substack{\text{advection}\\ \text{by tide}}} \tag{1a}$$

$$\partial \widetilde{m}/\partial \widetilde{t} = \underbrace{\widetilde{C}\,\widetilde{a}\,\widetilde{m}}_{\text{birth}} - \underbrace{\widetilde{E}\,\widetilde{m}/(\widetilde{K}+\widetilde{m})}_{\substack{\text{dislodgement}\\ \text{by waves}}} + \underbrace{\widetilde{D}\,\partial^2\widetilde{m}/\partial\widetilde{x}^2}_{\substack{\text{random}\\ \text{movement}}}. \tag{1b}$$

Here $\widetilde{t}$ is time and $\widetilde{x}$ is distance away from the shore; $\widetilde{A}$, $\widetilde{B}$, $\widetilde{C}$, $\widetilde{D}$, $\widetilde{E}$, $\widetilde{F}$, $\widetilde{K}$ and $\widetilde{V}$ are positive parameters. Since its initial formulation in 2005 (van de Koppel et al. 2005), model (1) and minor extensions have been studied in a number of papers, from both applications and mathematical viewpoints (Wang et al. 2009; Liu et al. 2012; Sherratt 2013; Ghazaryan and Manukian 2015; Cangelosi et al. 2014; Sherratt and Mackenzie 2016).

The alternative "increased production" model (IPM) centres around the build-up of faeces and pseudofaeces (van Broekhoven et al. 2015) under mussel beds, which contribute to the underlying sediment. Since the mussels' food source (algae) mainly resides in upper water layer, this sediment deposition raises the mussels towards their food source and thus further promotes their growth (Liu et al. 2012, 2014). In this case, we denote the mussel and algal densities by $\widehat{m}(\widehat{x},\widehat{t})$ and $\widehat{a}(\widehat{x},\widehat{t})$, respectively, and we use $\widehat{s}(\widehat{x},\widehat{t})$ to represent the amount of accumulated sediment. Here $\widehat{t}$ and $\widehat{x}$ are time and distance away from the shore, respectively. Then, the model of Liu et al. (2012) is:

$$\partial \widehat{a}/\partial \widehat{t} = \overbrace{\widehat{F}\cdot(\widehat{A}-\widehat{a})}^{\substack{\text{transfer to/}\\ \text{from upper}\\ \text{water layers}}} - \overbrace{\widehat{B}\,\widehat{a}\,\widehat{m}(\widehat{s}+\widehat{\eta}\widehat{S}_0)/(\widehat{s}+\widehat{S}_0)}^{\substack{\text{depth-dependent}\\ \text{consumption}\\ \text{by mussels}}} + \overbrace{\widehat{V}\,\partial\widehat{a}/\partial\widehat{x}}^{\substack{\text{advection}\\ \text{by tide}}} \tag{2a}$$

$$\partial \widehat{m}/\partial \widehat{t} = \overbrace{\widehat{C}\,\widehat{a}\,\widehat{m}(\widehat{s}+\widehat{\eta}\widehat{S}_0)/(\widehat{s}+\widehat{S}_0)}^{\text{birth}} - \overbrace{\widehat{E}\,\widehat{m}}^{\text{death}} + \overbrace{\widehat{D}_m\,\partial^2\widehat{m}/\partial\widehat{x}^2}^{\substack{\text{random}\\ \text{movement}}} \tag{2b}$$

$$\partial \widehat{s}/\partial \widehat{t} = \underbrace{\widehat{P}\,\widehat{m}}_{\text{production}} - \underbrace{\widehat{Q}\,\widehat{s}}_{\text{erosion}} + \underbrace{\widehat{D}_s\partial^2\widehat{s}/\partial\widehat{x}^2}_{\text{dispersal}} \tag{2c}$$

where $\widehat{A}$, $\widehat{B}$, $\widehat{\eta}\,(<1)$, $\widehat{S}_0$, $\widehat{C}$, $\widehat{D}_m$, $\widehat{D}_s$, $\widehat{E}$, $\widehat{F}$, $\widehat{P}$, $\widehat{Q}$ and $\widehat{V}$ are positive parameters. The function $(\widehat{s}+\widehat{\eta}\widehat{S}_0)/(\widehat{s}+\widehat{S}_0)$ has the property of increasing from a nonzero value ($\widehat{\eta}$) when $\widehat{s}=0$ towards the saturation level 1 as $\widehat{s}\to\infty$; this reflects the increase in the growth rate of the mussel population with increased proximity to the (algal-rich) upper water layers. Beyond these properties, the precise functional form is arbitrary.

Liu et al. (2012) calculated bifurcation diagrams for models (1) and (2) as a way of comparing their predictions for pattern formation. One of their observations was a distinct difference between the mussel densities in the patterns predicted by the two models. They reported that the average mussel density within patterns in the reduced losses model (1) is considerably greater than that in the spatially uniform steady state, but that in the increased production model (2) the two densities are similar. In this paper, we will present a more detailed study of this difference between the models with the aim of clarifying whether it is due to different model formulations and parameters, or whether it can genuinely be attributed to the different underling mechanisms. The difference is important because it provides a potential avenue for testing which model applies in a particular ecological context; mere observation of patterns is not sufficient for this since both models predict pattern formation for a wide range of ecologically plausible parameters. One issue that necessitates a detailed study is that (as we will show) there are multiple patterned solutions for any given set of ecological parameters.

As a first step, we consider a two-equation analogue of (2), which will facilitate direct comparison between the two model frameworks. Our aim is not to obtain a formal two-equation approximation to (2); rather, we seek a prototypical model of the increased production mechanism, and we use (2) as a starting point for formulating this. In this spirit, we will apply two simplifications to (2), neither of which is a good quantitative approximation, but both of which maintain the key qualitative features of the model. Firstly, we make a quasi-steady-state assumption on the kinetics of $\widehat{s}$, setting $\widehat{s} = \left(\widehat{P}/\widehat{Q}\right)\widehat{m}$ in (2a,b). In order for this to be a good quantitative approximation, it would be necessary that $\widehat{P}$ and $\widehat{Q}$ are significantly larger than comparable rate parameters in other equations. This is not actually the case: the timescale of sediment production and erosion is broadly similar to that of mussel birth and death, and for example the values estimated by Liu et al. (2012) for $\widehat{P}$ and $\widehat{Q}$ are the same. However, $\widehat{s}$ does not play a central role in the pattern formation process, and the only effects of the quasi-steady-state assumption are on quantitative details. Secondly, to simplify the two-equation model we replace $\left(\widehat{s}+\widehat{\eta}\widehat{S}_0\right)/\left(\widehat{s}+\widehat{S}_0\right)$ by $\widehat{s}/\widehat{S}_0$. Quantitative validity of this approximation would require that $\widehat{S}_0 \gg \widehat{s} \gg \widehat{\eta}\widehat{S}_0$, which holds for many of the patterns predicted by (2) using the parameter estimates in Liu et al. (2012), but not all. Note that (Liu et al. 2012) does not actually show any plots of the sediment profile in the patterned solutions of (2); we show an example in Fig. 1. In fact, the choice of functional form for the dependence of mussel birth on sediment level $\widehat{s}$ in Liu et al. (2012) was somewhat arbitrary, and the prediction of pattern formation is not sensitive to this functional form. In particular, the nonzero value of $\left(\widehat{s}+\widehat{\eta}\widehat{S}_0\right)/\left(\widehat{s}+\widehat{S}_0\right)$ when $\widehat{s} = 0$ is not important, for the following reason. In Fig. 1, it is clear that the mussel and sediment patterns are approximately in phase; this is a typical feature of simulations, and it has a mathematical basis from linear stability analysis that we outline in "Appendix". It follows that if $\widehat{s}$ is small then $\widehat{m}$ will also be small, implying that the growth rate will necessarily be small.

These two simplifications give the model

$$\partial\widehat{a}/\partial\widehat{t} = \widehat{F}\cdot\left(\widehat{A}-\widehat{a}\right) - \left(\widehat{B}\,\widehat{P}/\widehat{Q}\widehat{S}_0\right)\widehat{a}\,\widehat{m}^2 + \widehat{V}\,\partial\widehat{a}/\partial\widehat{x} \tag{3a}$$

$$\partial\widehat{m}/\partial\widehat{t} = \left(\widehat{C}\,\widehat{P}/\widehat{Q}\widehat{S}_0\right)\widehat{a}\,\widehat{m}^2 - \widehat{E}\,\widehat{m} + \widehat{D}_m\,\partial^2\widehat{m}/\partial\widehat{x}^2. \tag{3b}$$

Numerical studies show that the qualitative features of pattern formation in (3) are broadly similar to those in (2); an example comparison is shown in Fig. 1. However, this similarity is not a key foundation of our study. Rather, our approach is to regard (3) as a prototypical model for the "increased production" mechanism, which can easily compared to the prototypical model (1) for the "reduced losses" mechanism.

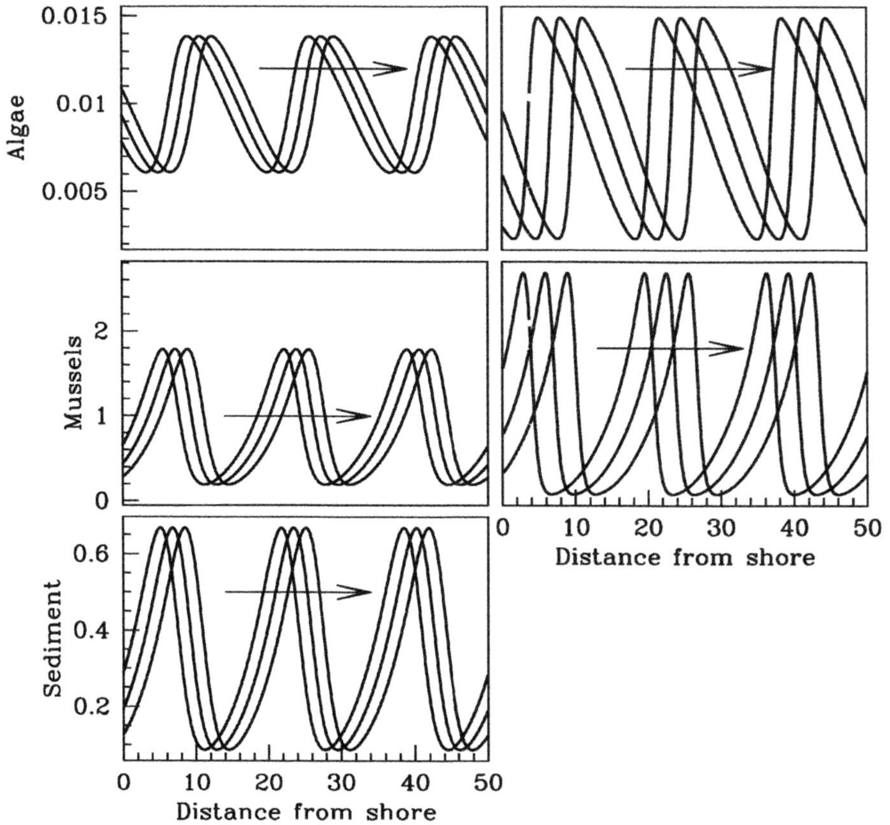

Fig. 1 A comparison of pattern solutions of the full increased production model (2) and the reduced model (3). Before solving, we nondimensionalised the models. Rescalings (6) are not convenient for (3), and so we used a different set of rescalings taken from Liu et al. (2012); details are given in "Appendix". This transforms (2) and (3) to (A.1) and (A.2), respectively. We used the domain $0 < \check{x} < 50$ with periodic boundary conditions, and with parameter values $\check{\alpha} = 50$, $\check{\beta} = 200$, $\check{\eta} = 0.1$, $\check{\nu} = 360$, $\check{D} = 1$, $\check{\delta} = 320$, $\check{\theta} = 2.5$ and $\check{\delta} = 1$. (All notations are defined in "Appendix"). We began by solving (A.1), with the variables $(\check{a}, \check{m}, \check{s})$ set to the spatially uniform steady state $(\check{a}_s, \check{m}_s, \check{s}_s)$ plus the small mixed-mode perturbation $0.01(\check{a}_s, \check{m}_s, \check{s}_s)\sum_{i=1}^{i=10}\cos(\pi\, i\, \check{x}/50)$. The solutions were plotted at times $\check{t} = 600,\ 602,\ 604$. We then used the solutions for $\check{a}$ and $\check{m}$ at $\check{t} = 604$ as initial conditions for (A.2), again solving from $\check{t} = 0$ and plotting the solutions at times $\check{t} = 600,\ 602,\ 604$, although we translated the solutions in the negative $\check{x}$ direction by 3.75 space units in order to facilitate comparison. (Since we are using periodic boundary conditions, the solutions are invariant to translation.)

Our next step is to nondimensionalise the two two-equation models. Previous papers on (1) and (2) have reduced them to dimensionless forms (van de Koppel et al. 2005; Wang et al. 2009; Liu et al. 2012); however if one applies these rescalings to (1) and (3), respectively, then the resulting pairs of equations are not easily comparable. We use different rescalings that are constructed specifically to facilitate comparison between the two dimensionless models. We give the rescalings in full because our subsequent work depends crucially on the way in which the (dimensional) algal supply rate affects the dimensionless parameters in the two models. For (1), we substitute

$$\begin{array}{llll}\widetilde{a}^* = (\widetilde{C}/\widetilde{F})\widetilde{a} & \widetilde{m}^* = (\widetilde{B}/\widetilde{F})\widetilde{m} & \widetilde{x}^* = (\widetilde{F}/\widetilde{D})^{1/2}\widetilde{x} & \widetilde{t}^* = \widetilde{F}\widetilde{t} \\ \widetilde{\alpha}^* = \widetilde{A}\widetilde{C}/\widetilde{F} & \widetilde{\beta}^* = \widetilde{E}/(\widetilde{F}\widetilde{K}) & \widetilde{\xi}^* = \widetilde{F}/(\widetilde{B}\widetilde{K}) & \widetilde{\nu}^* = \widetilde{V}/(\widetilde{D}^{1/2}\widetilde{F}^{1/2})\end{array} \tag{4}$$

which gives

$$\partial\widetilde{a}^*/\partial\widetilde{t}^* = \widetilde{\alpha}^* - \widetilde{a}^* - \widetilde{a}^*\widetilde{m}^* + \widetilde{\nu}^*\,\partial\widetilde{a}^*/\partial\widetilde{x}^* \tag{5a}$$

$$\partial\widetilde{m}^*/\partial\widetilde{t}^* = \widetilde{a}^*\widetilde{m}^* - \widetilde{\beta}^*\widetilde{m}^*/(1+\widetilde{\xi}^*\widetilde{m}^*) + \partial^2\widetilde{m}^*/\partial\widetilde{x}^{*2}. \tag{5b}$$

Here the asterisks denote dimensionless variables and parameters. The dimensionless parameters $\widetilde{\alpha}^*$, $\widetilde{\nu}^*$, $\widetilde{\xi}^*$ and $\widetilde{\beta}^*$ are defined by a combination of dimensional constants, but readers may find it useful to interpret intuitively $\widetilde{\alpha}^*$ as corresponding to the algal supply from upper water layers, $\widetilde{\nu}^*$ as corresponding to the tide strength, $\widetilde{\xi}^*$ as corresponding to the ability of intermussel bonds to reduce dislodgement by waves, and $\widetilde{\beta}^*$ as corresponding to the per capita dislodgement rate for isolated mussels.

For (3), we substitute

$$\begin{array}{ll}\widehat{a}^* = \dfrac{\widehat{C}\widehat{P}^{1/2}}{\widehat{B}^{1/2}\widehat{F}^{1/2}\widehat{Q}^{1/2}\widehat{S}_0^{1/2}}\widehat{a} & \widehat{m}^* = \left(\dfrac{\widehat{B}\widehat{P}}{\widehat{F}\widehat{Q}\widehat{S}_0}\right)^{1/2}\widehat{m} \\ \widehat{x}^* = (\widehat{F}/\widehat{D}_m)^{1/2}\widehat{x} \quad \widehat{t}^* = \widehat{F}\widehat{t} & \\ \widehat{\alpha} = \dfrac{\widehat{A}\widehat{C}\widehat{P}^{1/2}}{\widehat{B}^{1/2}\widehat{F}^{1/2}\widehat{Q}^{1/2}\widehat{S}_0^{1/2}} & \widehat{\beta} = \widehat{E}/\widehat{F} \quad \widehat{\nu} = \widehat{V}/(\widehat{D}_m\widehat{F})^{1/2}\end{array} \tag{6}$$

which gives

$$\partial\widehat{a}^*/\partial\widehat{t}^* = \widehat{\alpha}^* - \widehat{a}^* - \widehat{a}^*\widehat{m}^{*2} + \widehat{\nu}^*\,\partial\widehat{a}^*/\partial\widehat{x}^* \tag{7a}$$

$$\partial\widehat{m}^*/\partial\widehat{t}^* = \widehat{a}^*\widehat{m}^{*2} - \widehat{\beta}^*\widehat{m}^* + \partial^2\widehat{m}^*/\partial\widehat{x}^{*2}. \tag{7b}$$

Again the asterisks denote dimensionless variables and parameters. Equations (7) are the Klausmeier model, which has been well studied as a model of semi-arid vegetation (Klausmeier 1999; Sherratt and Lord 2007; van der Stelt et al. 2013), and whose kinetics are the same as those of the Gray–Scott model, which has important applications to chemistry (Chen and Ward 2009; Doelman et al. 1997; Gray and Scott 1984). The dimensionless parameters $\widehat{\alpha}$, $\widehat{\nu}$ and $\widehat{\beta}$ are defined by a combination of dimensional constants, but readers may find it useful to interpret intuitively $\widehat{\alpha}$ as corresponding to the algal supply from upper water layers, $\widehat{\nu}$ as corresponding to the tide strength, and $\widehat{\beta}$ as corresponding to the per capita dislodgement rate of mussels by waves.

The basic approach of this paper is to construct a hybrid model, which includes both the reduced losses and increased production feedback mechanisms. The model will include a new (dimensionless) parameter λ, which takes values between 0 and 1 and which controls the relative importance of the two mechanisms, with the equations

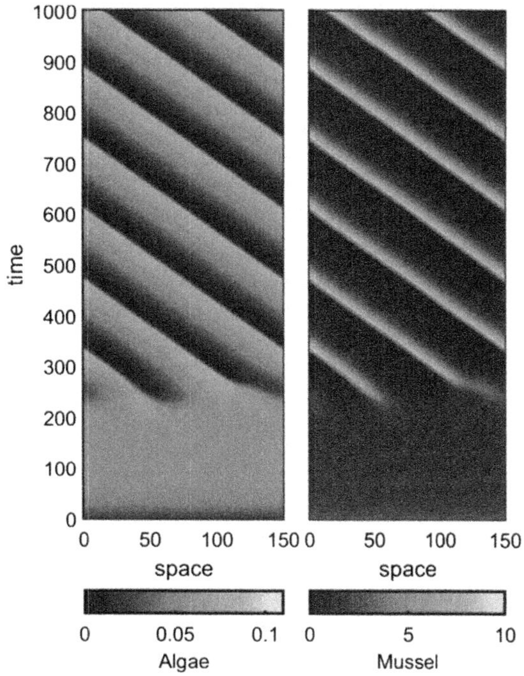

Fig. 2 A typical pattern solution of the hybrid model (8). We show the evolution of a pattern following a small random perturbation of a uniform state. The parameter values are $\alpha = 0.2$, $\beta = 0.1$, $\nu = 100$, $\xi = 0.5$, $\lambda = 0.25$

reducing to (5) when $\lambda = 1$ and to (7) when $\lambda = 0$. This approach is made possible by the close similarity between the two pairs of dimensionless equations (5) and (7), and our hybrid model is:

$$\partial a/\partial t = \alpha - a - am\,[\lambda + (1-\lambda)m] + \nu\,\partial a/\partial x \tag{8a}$$
$$\partial m/\partial t = am\,[\lambda + (1-\lambda)m] - \beta m\left[1 - \lambda + \lambda/\left(1+\xi m\right)\right] + \partial^2 m/\partial x^2. \tag{8b}$$

Note that a dimensional model can be recovered from (8) by using either (4) or (6) as rescalings.

Figure 2 shows a typical example of a spatially patterned solution of (8). A central question that we will address in this paper is how the mussel density in such patterns changes as the supply rate of algae is varied. In view of this, it is important to note that the algal supply parameters $\widetilde{A}$ in (1) and $\widehat{A}$ in (2) do not appear in the rescalings for $\widetilde{m}^*$ or $\widehat{m}^*$ and that the only dimensionless parameters that they affect are $\widetilde{\alpha}^*$ and $\widehat{\alpha}^*$. Therefore, we can address our question by considering how the (dimensionless) mussel density m in patterned solutions of (8) changes with the (dimensionless) parameter α.

2 Uniform Steady States of the Hybrid Model

For all parameter values, model (8) has a mussel-free steady state $a = \alpha$, $m = 0$. Spatially uniform steady states with $m \neq 0$ must satisfy

$$a = \alpha / \left[1 + \lambda m + (1-\lambda) m^2\right] \tag{9}$$

$$0 = \alpha\,[\lambda + (1-\lambda)m]\cdot(1+\xi m) - \beta\,[\lambda + (1-\lambda)\xi m]\cdot\left[1 + \lambda m + (1-\lambda)m^2\right]. \tag{10}$$

Equation (10) is a cubic polynomial whose roots are very complicated algebraically, and we investigated them using numerical continuation. Starting points for this are provided by the extreme cases $\lambda = 0$ (increased production model) and $\lambda = 1$ (reduced losses model), for which the spatially uniform steady states have a much simpler form; moreover, their stability can be determined quite easily; here and throughout this section, the "stability" that we consider is to spatially uniform perturbations. For $\lambda = 0$, $(\alpha, 0)$ is stable for all parameter values, and if $\alpha > 2\beta$ there are two other steady states:

$$a = \frac{2\beta^2}{\alpha - \sqrt{\alpha^2 - 4\beta^2}} \qquad m = \frac{\alpha - \sqrt{\alpha^2 - 4\beta^2}}{2\beta} \tag{11}$$

which is unstable and

$$a = \frac{2\beta^2}{\alpha + \sqrt{\alpha^2 - 4\beta^2}} \qquad m = \frac{\alpha + \sqrt{\alpha^2 - 4\beta^2}}{2\beta}. \tag{12}$$

The stability of (12) depends on parameters, but it is always stable if $\beta < 2$ which holds for realistic parameter estimates (Liu et al. 2012). For $\lambda = 1$, $(\alpha, 0)$ is stable if and only if $\alpha < \beta$. When (α/β) lies between 1 and $(1/\xi)$, there is one other nonnegative steady state

$$a = (\beta - \xi\alpha)/(1-\xi) \qquad m = (\alpha - \beta)/(\beta - \xi\alpha) \tag{13}$$

which is stable if $\xi < 1$.

With these expressions for the steady states when $\lambda = 0$ and $\lambda = 1$ as starting points, we used the software package AUTO (Doedel 1981; Doedel et al. 1991, 2006) to track the steady states and their stability as λ is varied. The results are significantly different when ξ is above and below 1, and Figs. 3 and 4 illustrate the two cases. We will discuss first the case $\xi < 1$ (Fig. 3) and then $\xi > 1$ (Fig. 4), before using our results to construct a diagram (Fig. 5) showing the existence of a positive stable-steady state in a parameter plane. However, a few general comments about the stability of the steady states are a useful preliminary. The form of (10) shows that the steady states depend on the parameters α and β only through the ratio α/β. In general, their stability does depend on the individual values of α and β, but we will show later that when $\beta < 2$, the stability also depends only on the ratio α/β. Realistic parameter estimates are consistent with $\beta < 2$ (Liu et al. 2012), and therefore, the stability shown in Figs. 3 and 4 applies under this assumption. In the figures, we plot the mussel density m at the steady states against λ, for a series of values of the parameter ratio α/β. Recall that α and β are most usefully interpreted as representing the rates of algal supply and

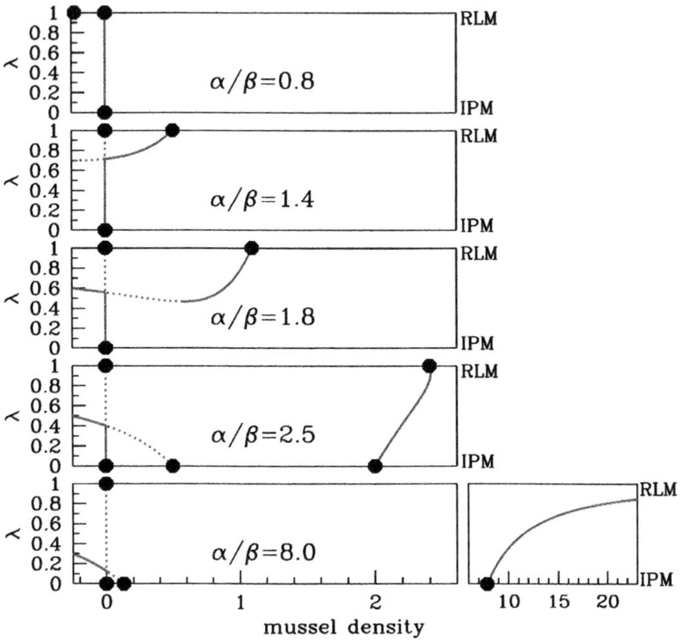

Fig. 3 Steady states for $\xi < 1$. We plot the steady-state solutions of (8) when $\xi = 0.15$, which is typical of behaviour when $\xi < 1$. We plot mussel density m against the parameter λ. Solid / dashed lines indicate stable / unstable steady states, calculated by numerical continuation using AUTO (Doedel 1981; Doedel et al. 1991, 2006), with λ as continuation parameter. Dots indicate the steady states for the reduced losses ($\lambda = 1$) and increased production ($\lambda = 0$) models; these are $(a, m) = (\alpha, 0)$ plus (11), (12) and (13). The five different cases shown in the figure are separated by four critical values of α/β: 1, $1/\lambda^0_{\text{min}} \approx 1.55$, 2, and $1/\xi \approx 6.67$; a full description is given in the main text. To improve clarity, the plot for $\alpha/\beta = 8.0$ is shown in two parts with different horizontal scalings. The stability of the steady states shown in the figure is valid, provided that $\beta < 2$, which applies for realistic parameter estimates (Liu et al. 2012)

mussel loss, respectively, although of course they are dimensionless parameter ratios that involve a combination of ecological quantities (see (4) and (6)).

Considering first $\xi < 1$, five different cases are shown in Fig. 3. When α/β is small, the corresponding algal supply is insufficient to maintain a mussel population for any value of λ, and the only steady state is the mussel-free state, which is always stable. As α/β increases through 1, (13) becomes positive, and there is a stable steady state for sufficiently large values of λ. This solution branch actually has a local minimum, at $\lambda = \lambda_{\text{min}}$ say. The value of m giving this local minimum is negative when α/β is just above 1, but it increases with α/β and passes through zero when (10) has a double root at $m = 0$. A straightforward calculation shows that this occurs when $\alpha/\beta = 1/\lambda^0_{\text{min}}$, where λ^0_{min} is the corresponding value of λ and is given by

$$\lambda^0_{\text{min}} = \frac{\sqrt{5 - 4\xi} - 1}{2(1 - \xi)}.$$

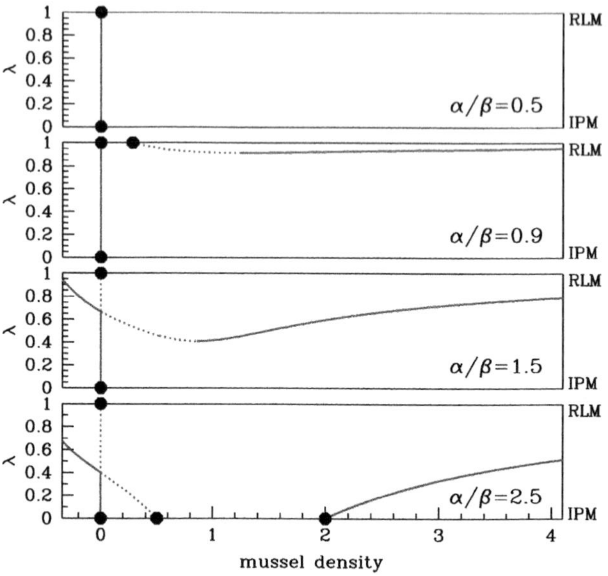

Fig. 4 Steady states for $\xi > 1$. We plot the steady-state solutions of (8) when $\xi = 1.5$, which is typical of behaviour when $\xi > 1$. We plot mussel density m against the parameter λ. Solid / dashed lines indicate stable / unstable steady states, calculated by numerical continuation using AUTO (Doedel 1981; Doedel et al. 1991, 2006), with λ as the continuation parameter. Dots indicate the steady states for the reduced losses ($\lambda = 1$) and increased production ($\lambda = 0$) models; these are $(a, m) = (\alpha, 0)$ plus (11), (12) and (13). The four different cases shown in the figure are separated by three critical values of α/β: $1/\xi \approx 0.67$, 1 and 2; a full description is given in the main text. The steady-state solution branches depend on α and β only through their ratio α/β; their stability does depend on the separate values, but the stability of the steady states shown in the figure is valid whenever $0 < \beta < 2$, which applies for realistic parameter estimates (Liu et al. 2012). Figure 7 shows the corresponding bifurcation diagram for a larger value of β

For the value of ξ used in Fig. 3, $\lambda_{\min}^0 \approx 0.646 \Rightarrow 1/\lambda_{\min}^0 \approx 1.55$. The numerical results indicate a change in stability at the local minimum, as one would expect from general bifurcation theory: $\lambda = \lambda_{\min}$ is a saddle–node bifurcation point (Kuznetsov 2004, Ch. 3). Therefore, when (α/β) is just above $1/\lambda_{\min}^0$, there are two positive steady states for $\lambda > \lambda_{\min}$, one stable and one unstable. Figure 5a shows that $\lambda_{\min}$ decreases as (α/β) increases, reaching 0 at $\alpha/\beta = 2$. This heralds the appearance of nonzero steady states for $\lambda = 0$ (the increased production model case): see (11) and (12). The final change in qualitative behaviour occurs when $\alpha/\beta = 1/\xi$, when the value of m at the nonzero steady state (13) for $\lambda = 1$ passes through infinity and becomes negative. For $\alpha/\beta > 1/\xi$, the solution branch starting at (12) when $\lambda = 0$ does not intersect $\lambda = 1$; instead, $m \to \infty$ as $\lambda \to 1^-$. This is illustrated in the right-hand subpanel of Fig. 3 for $\alpha/\beta = 8$.

The onset of patterning in (8) occurs when a spatially uniform steady state becomes unstable to inhomogeneous perturbations—this is a Turing or Turing–Hopf bifurcation. A prerequisite for this is a positive steady state that is stable to homogeneous perturbations, and the above results enable us to determine when such a steady state exists (for $\xi < 1$). Firstly, we note that there is never more than one positive stable

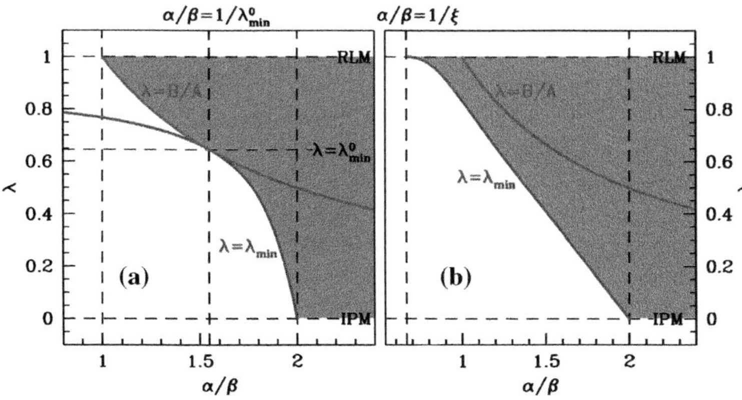

Fig. 5 Existence of a positive stable steady state. The shading shows typical examples of the parameter region for which model (8) has a positive steady state that is stable to homogeneous perturbations, for **a** $\xi < 1$ and **b** $\xi > 1$. The blue curve is the locus of λ_{min}, calculated by a two-parameter numerical continuation using AUTO (Doedel 1981; Doedel et al. 1991, 2006), with λ and α as continuation parameters. The red curve is $\alpha/\beta = 1/\lambda$: the mussel-free steady state $a = \alpha$, $m = 0$ is stable below this curve and unstable above it. The parameter values used for the plots were $\beta = 0.1$ and **a** $\xi = 0.15$, **b** $\xi = 1.5$

steady state.[1] When $\alpha/\beta > 2$, there is a stable steady for all λ, with the exception of $\lambda = 1$ when $\alpha/\beta > 1/\xi$. For $2 > \alpha/\beta > 1/\lambda^0_{\text{min}}$, the condition for a positive steady state is $\lambda > \lambda_{\text{min}}$, while for $1/\lambda^0_{\text{min}} > \alpha/\beta > 1$ the condition is $\lambda > (\beta/\alpha)$, since $m = 0$ is a solution of (10) when $\alpha/\beta = 1/\lambda$. Finally when $\alpha/\beta < 1$, there are no positive stable steady states. Figure 5a illustrates the parameter region giving a positive stable steady state for one value of ξ between 0 and 1.

We now consider the case of $\xi > 1$, which is a little simpler (Fig. 4) . Again, when α/β is small, the only steady state is the mussel-free state, which is always stable. As α/β increases through $1/\xi$, a branch of positive steady states appears. From the viewpoint of bifurcation theory, this solution branch arises from a transcritical bifurcation at $(a, m) = (0, \infty)$ when $\lambda = 1$; for fixed $\lambda < 1$, the appearance of the steady states corresponds to a saddle–node bifurcation as α/β is increased. As for $\xi < 1$, the solution branch has a local minimum, at $\lambda = \lambda_{\text{min}}$ say, which separates stable and unstable parts of the solution branch. At $\alpha/\beta = 1$, the mussel density at the nontrivial steady state (13) for $\lambda = 1$ changes sign, and for $\alpha/\beta > 1$ the mussel-free steady state is unstable for values of λ above the intersection point of the two solution branches. In the language of bifurcation theory, there is another transcritical bifurcation, this time at $(a, m) = (\alpha, 0)$ when $\lambda = 1$. The value of λ_{min} decreases as α/β increases, and it reaches zero at $\alpha/\beta = 2$, heralding the appearance of positive steady states for $\lambda = 0$. Figure 5b shows a typical plot of λ_{min} against α/β for $\xi > 1$. This curve bounds the parameter region giving a positive steady state, which is shaded in the figure. Note that for $\alpha/\beta > 1$ there is a positive steady state for an interval of

[1] This statement is based on evidence from our numerical continuation study, but is not proven. In particular, there could in principle be a closed branch of spatially uniform steady states that exists for a range of values of λ in $(0, 1)$ but does not extend to either $\lambda = 0$ or $\lambda = 1$. However, we have not found any evidence of such a closed branch.

λ values up to 1 (including $\lambda = 0$), but not for $\lambda = 1$ itself. Figure 5b illustrates the parameter region giving a positive stable steady state for one value of ξ greater than 1.

To conclude our discussion of steady states, we return to the issue of their stability; specifically, we will justify our assertion that provided $\beta < 2$, stability depends on the parameters α and β only through their ratio α/β. For a system of two coupled ODEs the stability of a steady state depends on the trace and determinant of the Jacobian (a.k.a. stability or community) matrix. For a steady state of (8) with $m \neq 0$, the determinant of the Jacobian matrix depends on α and β only through their ratio α/β, while the trace has the form $\mathcal{T} = \mathcal{F}_1\beta - \mathcal{F}_2$. Here $\mathcal{F}_1$ and $\mathcal{F}_2$ are both strictly positive and depend on λ, ξ and the ratio α/β; their algebraic forms are complicated[2]. However, since $\mathcal{F}_2 > 0$ it follows that for any set of values of λ, ξ and α/β, $\mathcal{T} < 0$ for sufficiently small β. Then, the steady state is stable / unstable when the determinant of the Jacobian matrix is positive / negative, which depends on α and β only through their ratio α/β. In the latter case, the steady state is unstable for all β, but when the determinant is positive, the steady state will only be stable for $\beta < \beta_{\text{crit}} = \mathcal{F}_2/\mathcal{F}_1$. Thus for $\beta > \beta_{\text{crit}}$ there are no stable steady states with positive mussel density.

Our method of numerical continuation gives the value of m at the (unique) positive steady state with positive Jacobian determinant at any point in the (α/β)–λ plane for which this steady state exists; these are the shaded regions in Fig. 5. Using this, we calculated β_{crit}, and the results are shown as colour maps in Fig. 6. For both $\xi < 1$ and $\xi > 1$, the minimum value of β_{crit} is 2, occuring at $\lambda = 0$ and $\alpha/\beta = 2$; here, the value of $m = 1$. Hence whenever $\beta < 2$, $\mathcal{T} < 0$ and stability depends only on the ratio α/β. To illustrate the difference in steady-state stability that can occur for larger values of β, Fig. 7 shows the same bifurcation diagrams as in Fig. 3 but for $\beta = 6$. The solution branch curves are the same, but for $\alpha/\beta = 1.8$ and 2.5 the stability is clearly different.

Our objective is to study pattern formation in (8) as λ varies between 0 and 1. Therefore, we require that there is a positive stable steady state for all values of λ, including the extreme case $\lambda = 1$ (reduced losses model). This only occurs when $\xi < 1$ and $2 < \alpha/\beta < 1/\xi$ (fourth panel in Fig. 3), and we restrict attention to parameter values satisfying these constraints, and also $\beta < 2$.

3 Spatial Pattern Formation

When the parameters α, β, ξ and λ are such that (8) has a positive steady state that is stable to spatially uniform perturbations, there is the possibility of pattern formation in the spatial model (8), because the steady state can be destabilised by the advection and diffusion terms. This is sometimes known as "differential flow instability" (Rovinsky and Menzinger 1992). Conditions for patterning can be found via a standard procedure. We linearise equations (8) about the steady state (a_s, m_s) and substitute $(a - a_s, m - m_s) = (a_0, m_0)\exp(ikx + \mu t)$. The requirement that the constants a_0 and m_0 are not both zero gives a dispersion relation, i.e. an equation for μ in terms of k. The steady

[2] Denoting by m_s the steady-state value for m, $\mathcal{F}_2 = 1 + \lambda m_s + (1-\lambda)m_s^2$ and $\mathcal{F}_1 = \left[(1-\lambda)(1+\xi m_s)(1+(1-\lambda)\xi m_s) + \left(\lambda + (1-\lambda)m_s\right)\lambda\xi\right]m_s / \left[(\lambda + (1-\lambda)m_s)(1+\xi m_s)^2\right]$.

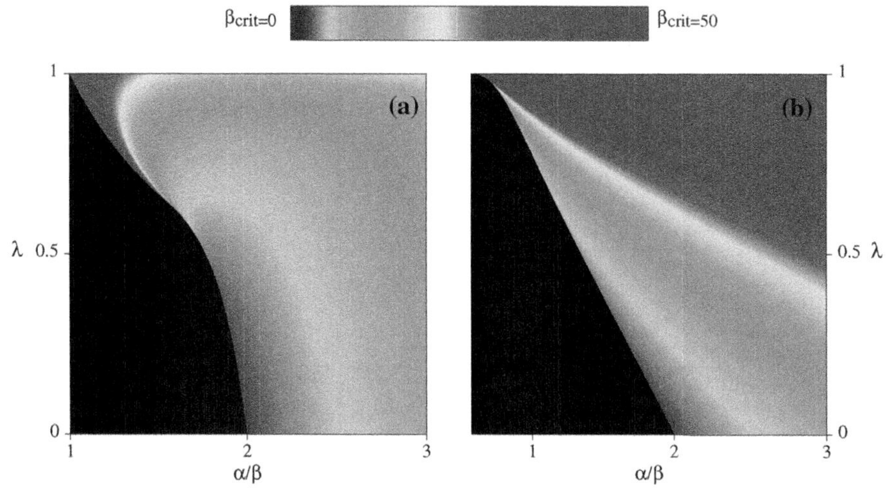

Fig. 6 β_{crit} for (a) $\xi = 0.15$, (b) $\xi = 1.5$. For values of β above β_{crit}, there are no stable steady states with positive mussel density. The colour indicates β_{crit} according to the scalebar, with black indicating that there is no positive steady state. Thus, the black region in these plots corresponds to the unshaded region in Figs. 3 and 4. The key message from these plots is that in both cases the minimum value of β_{crit} occurs at $\alpha/\beta = 2$ and $\lambda = 0$, at which point $\beta_{\text{crit}} = 2$

state is stable if Re $\mu < 0$ for all real k and unstable if Re $\mu > 0$ for some real k. For a more detailed explanation, see Murray (2003), Perumpanani et al. (1995), Sherratt (2005), Siteur et al. (2014). Using this approach, we calculated the curves in the α–ν plane on which the steady state loses stability, leading to patterns. This is illustrated in Fig. 8, which shows that the α–ν region giving patterns shrinks as λ is increased.

The curves plotted in Fig. 8 give an upper limit on the algal supply parameter α for spatial patterning, for given values of the other parameters. Intuitively, when α exceeds this upper limit the food supply to the mussels is large enough to maintain a spatially uniform mussel population. There is also a minimum value of α below which patterns do not occur, because the food supply is insufficient to maintain even a spatially patterned mussel population; this is discussed in more detail below. Between these two threshold values of α there are spatial patterns. Although the patterns have a constant shape, they are not stationary, but rather they move in the positive x direction (away from the shore) because of the directed movement of the algae (towards the shore). Mathematically, such movement is a standard feature of patterns in reaction–diffusion–advection equations (Rovinsky and Menzinger 1992; Perumpanani et al. 1995; Malchow 2000), and it is reflected by the imaginary part of the growth rate μ being nonzero. Intuitively, the model predicts that algal density is higher on the off-shore side of a mussel band compared to the on-shore side, because of consumption in the band. This causes a net growth of mussels on the off-shore side and a net loss on the on-shore side, and this causes a gradual net off-shore migration of the band (Song et al. 2017). Although such movement is intrinsic to patterned solutions of (8), it is not observed in real mussel beds, presumably because the (deliberately) simplistic modelling omits some stabilising feature(s) of the real system—one possibility for this

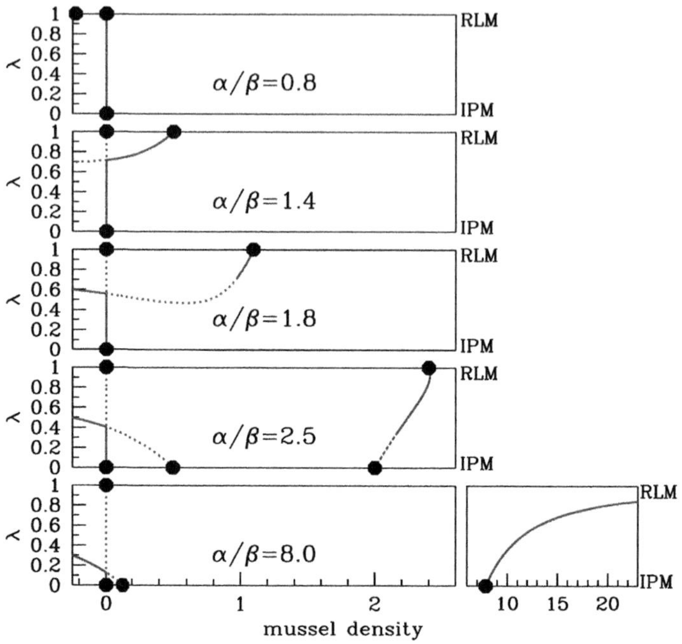

Fig. 7 Steady states for $\xi < 1$ and $\beta > 2$. We plot mussel density m against the parameter λ for the steady-state solutions of (8) when $\xi = 0.15$ and $\beta = 6$. Solid / dashed lines indicate stable / unstable steady states, calculated by numerical continuation using AUTO (Doedel 1981; Doedel et al. 1991, 2006), with λ as continuation parameter. Dots indicate the steady states for the reduced losses ($\lambda = 1$) and increased production ($\lambda = 0$) models; these are $(a, m) = (\alpha, 0)$ plus (11), (12) and (13). The five different cases shown in the figure are separated by four critical values of α/β: 1, $1/\lambda^0_{\text{min}} \approx 1.55$, 2, and $1/\xi \approx 6.67$; a full description is given in the main text. To improve clarity, the plot for $\alpha/\beta = 8.0$ is shown in two parts with different horizontal magnifications. These plots should be compared with Fig. 3, which shows the corresponding plots when $\beta < 2$: the steady states are the same but there are differences in stability for $\alpha/\beta = 1.8$ and 2.5

is the oscillatory nature of tidal flow, which has recently been incorporated into the reduced losses model (1) and which does lead to patterns without large-scale migration (Sherratt and Mackenzie 2016).

Being periodic solutions moving with constant shape and speed, the patterns are "periodic travelling waves", and general theory for this type of solution implies that for a given value of the algal supply α (and of the other parameters) there is a family of patterns, with the migration speed and wavelength varying within the family (Sherratt and Smith 2008; Sherratt 2012; van der Stelt et al. 2013; Rademacher and Scheel 2007). A convenient way of illustrating the wave family is to plot the patterning region in the α—speed plane, and some examples are shown in Fig. 9 with calculations done using the software package WAVETRAIN (Sherratt 2012). In these plots, the patterning region is shaded; in keeping with Fig. 8 the region shrinks as λ is increased. The right-hand boundary of the patterning region is a curve on which patterns with a particular speed are initiated: mathematically this is a locus of Hopf bifurcations in the travelling wave ODEs. The left-hand boundary is for the most part a curve along which the pattern

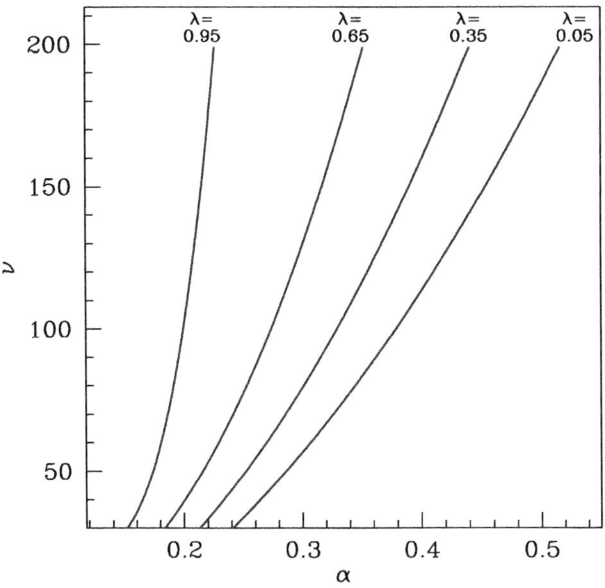

Fig. 8 Pattern onset. We plot the patterning onset curves for (8) in the α–ν plane for $\beta = 0.1$ and $\xi = 0.5$ fixed, for 4 values of the parameter λ. As α and ν are varied from lower right to upper left in the parameter plane, the locally stable positive steady state loses stability as the patterning onset curve is crossed. Note that in contrast to much of our work on steady states, the results in this and subsequent figures depend on the values of both α and β, rather than just their ratio. The curves were calculated by determining the dispersion relation as described in the main text and then solving numerically the condition for this dispersion relation to have a double root. The case $\lambda = 0$ provides a starting point for this solution, since (8) then reduces to the Klausmeier model for which patterning conditions have been studied in detail in previous work (e.g. Sherratt 2005; Siteur et al. 2014; Sherratt 2015). From this starting point, we gradually increased λ, solving the double root conditions numerically using the solution at the previous value of λ as an initial guess. Note that to enable direct comparison, the values of β and ξ used in this figure are the same as in Figs. 9, 10, 11, 12 and 13

wavelength becomes infinitely long: mathematically this is a locus of homoclinic solutions. For one of the cases shown in Fig. 9 ($\lambda = 0.95$) a small part of the left-hand boundary consists instead of a locus of folds; this implies multiple pattern solutions in a small region of parameter space, but this has no practical significance for realistic parameters.

4 Mussel Density in Spatial Patterns

Our basic objective is to compare the average mussel density in spatial patterns with the density in the spatially uniform steady state from which the patterns have developed. Therefore, we tracked the form of the pattern along solution branches of (8) as α is varied with the other parameters fixed and compared this with the steady state for the same value of α. Again we used the software package WAVETRAIN (Sherratt 2012) for this calculation. Since there is a range of different patterned solutions for any value of α within the patterning range (see Fig. 9), it is necessary to impose some

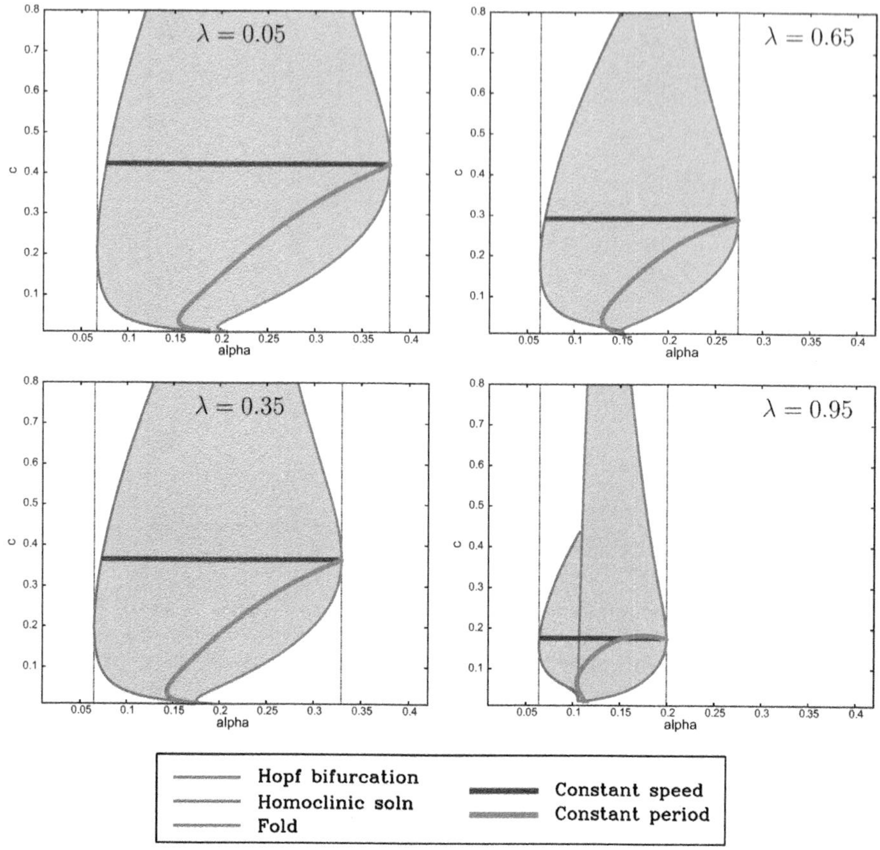

Fig. 9 Pattern existence. We show examples of the regions (shaded) of the α-migration speed plane in which patterns exist, for $\beta = 0.1$, $\xi = 0.5$ and $\nu = 100$. The regions are bounded by loci of Hopf bifurcations, homoclinic solutions and solution branch folds in the travelling wave ODEs. The vertical dotted lines show the values of α between which there are pattern solutions of some speed. Thus, the right-hand vertical line corresponds to pattern onset: mathematically there is a Turing–Hopf bifurcation in (8) at this value of α. We also show curves of constant migration speed and of constant wavelength (period) passing through the Turing–Hopf point. The various computations and plots were done using the software package WAVETRAIN (Sherratt 2012). The shaded region includes patterns that are both stable and unstable as solutions of (8). We include unstable patterns in the plots because they can be ecologically relevant in some situations (see, for example, Sherratt et al. 2009). Note that to enable direct comparison, the same parameters are used in Figs. 9, 10, 11, 12 and 13 . There is no special significance to the four values of λ that we have chosen, and in particular our use of $\lambda = 0.05$ and 0.95 rather than 0 and 1 is arbitrary

form of constraint in order to specify the solution branch to be followed. A number of simulation-based studies on semi-arid vegetation patterns (Sherratt 2013; Siteur et al. 2014; Siero et al. 2015) have found that as rainfall is varied slowly, the biomass within a vegetation pattern changes, but its wavelength remains the same—except for occasional large and abrupt changes in wavelength that occur when patterns of a given wavelength lose stability. Previous studies by one of us (Sherratt 2013, 2016) found a corresponding result in both the reduced losses model (1) and the increased production model (2) for mussel bed patterns. Therefore, in the present work we considered patterns of constant wavelength as α was decreased; we started our calculations at the pattern onset (Turing–Hopf bifurcation) point. However, our results do not depend critically on this assumption of constant wavelength, and to emphasise this we also considered patterns of constant speed (and therefore varying wavelength) as α was decreased. Both of the resulting solution branches are shown in Fig. 9. Figures 10 and 11 show typical results from these calculations. In both the constant speed and constant wavelength cases, there is an increasing separation between the pattern and steady-state solution branches as λ is increased between 0 and 1. Recall that Liu et al. (2012) reported that the average mussel density in the spatial patterns was much greater than that in the steady state for the "reduced losses" model, while there was relatively little difference in the two densities for the "increased production" model. In our hybrid model (8), there is a gradual transition from the increased production to the reduced losses feedback mechanism as λ is increased at either constant speed or constant wavelength. Therefore, the results in Figs. 10 and 11 are consistent with the findings of Liu et al. (2012).

The vertical dashed lines in Figs. 10 and 11 are at values of α that are 20% below the pattern-onset values. In Figs. 12 and 13, we show the corresponding patterns, plotting mussel density against space. The steady-state mussel density is also indicated in these figures. This shows that in contrast to the increased production mechanism, feedback of decreased losses type (λ near 1) causes high mussel densities in the peaks of the patterns; it is this that leads to the high level of mean mussel density, compared to the steady state.

To quantify the difference between the pattern and steady-state mussel densities, we calculated a single numerical measure of this difference. The starting point for our calculation is the variation with α of the mean mussel density in patterns and of the steady-state mussel density; typical examples of this are illustrated in Figs. 10 and 11. We then calculated the average over α of the difference between the two densities and divided this by the average of the steady-state density. We restricted our averaging to values of α for which the pattern and the steady state both exist, and for which the steady state is positive. In some cases, there is a fold in the pattern solution branch (for example the $\lambda = 0.95$ case in Fig. 11), and then, we restricted attention to the part of the solution branch between the fold and the pattern onset (Turing–Hopf bifurcation) point. This prevents the results from being skewed by patterns with very low densities that are (typically) unstable as solutions of (8). We repeated this procedure for a sequence of values of α, and for three values of ξ, for both the constant speed and constant wavelength solution branches. The results are shown in Fig. 14. This figure illustrates the broad generality of two conclusions. First, the (average) mussel density in the patterns is greater than that in the steady state—this follows from all of the differences plotted in Fig. 14 being positive. And secondly there is a gradual increase in the average difference in mussel density as λ is increased from 0 to 1, i.e. as the feedback mechanism gradually shifts from increased production to decreased losses.

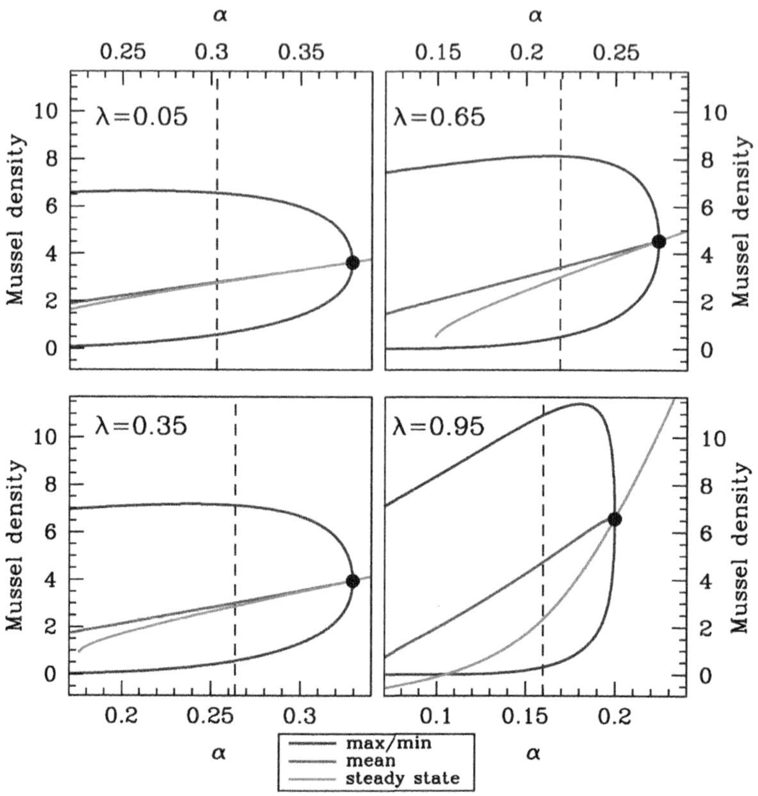

Fig. 10 Density versus α at constant speed. We show comparisons of the average mussel density in spatial patterns with the steady-state mussel density, when the migration speed is held constant at its value at pattern onset (the Turing–Hopf bifurcation point, indicated by ●). Using the software package WAVETRAIN (Sherratt 2012), we tracked the form of the pattern along solution branches of (8) as α is varied with the other parameters fixed. We plot the maximum, minimum and mean (average) mussel density in the patterns and also the steady-state mussel density. The figure shows an increasing separation between the pattern and the steady-state solution as λ is increased between 0 (increased production model) and 1 (decreased losses model). The dashed vertical lines indicate the values of α used in Fig.12. Note that to enable direct comparison, the same parameters are used in Figs. 9, 10, 11, 12 and 13

5 Recovery Time

Previous modelling predicts that self-organisation into spatial patterns significantly increases the resilience of mussel beds to disturbances, relative to the resilience of spatially uniform mussel populations (van de Koppel et al. 2005; Liu et al. 2012). Moreover, in their study comparing models based on the reduced losses and increased production mechanisms, Liu et al. (2012) reported that this increased resilience occurs to a greater extent in the reduced losses model than in the model based on increased production. However, it is again unclear to what extent this difference depends on the formulation of the models and the parameter values, rather than being due to the alternative underlying patterning mechanisms. To clarify this, we investigated resilience in our hybrid model (8), by removing a fixed proportion of the total mussel population

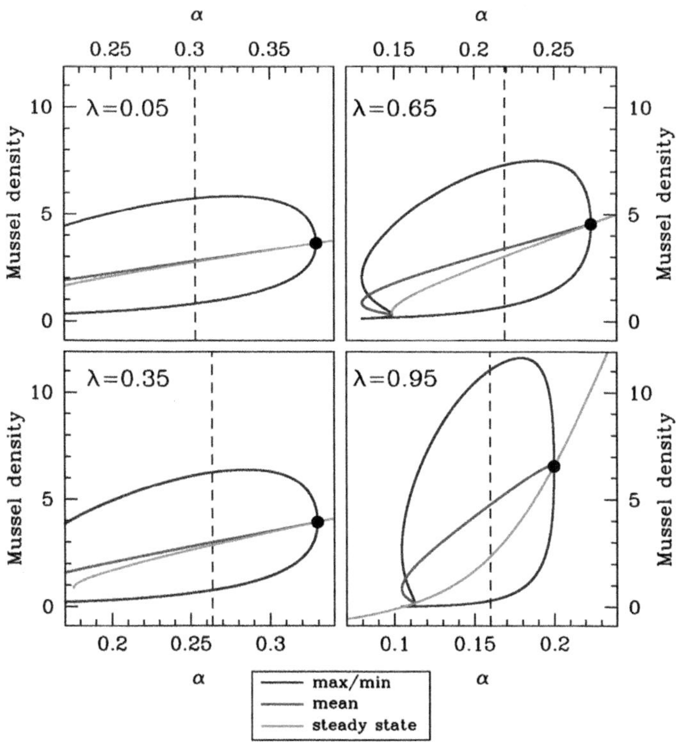

Fig. 11 Density versus α at constant wavelength. We show comparisons of the average mussel density in spatial patterns with the steady-state mussel density, when the wavelength is held constant at its value at pattern onset (the Turing–Hopf bifurcation point, indicated by ●). Using the software package WAVETRAIN (Sherratt 2012), we tracked the form of the pattern along solution branches of (8) as α is varied with the other parameters fixed. We plot the maximum, minimum and mean (average) mussel density in the patterns and also the steady-state mussel density. As in Fig. 10, this figure shows an increasing separation between the pattern and the steady-state solution as λ is increased between 0 (increased production model) and 1 (decreased losses model). The dashed vertical lines indicate the values of α used in Fig. 13. Note that to enable direct comparison, the same parameters are used in Figs. 9, 10, 11, 12 and 13

from a patterned solution, via a spatially varying perturbation, and monitoring the time taken to return to the original pattern. A typical result is shown in Fig. 15, confirming that the recovery time increases gradually as one varies the parameter λ from $\lambda = 0$ (increased production) to $\lambda = 1$ (reduced losses). This shows that the difference in resilience can indeed be attributed to the underlying pattern mechanism, rather than being a function of parameter values or other model details.

6 Discussion

The concept of self-organisation has been particularly successful in providing a framework for the emergence of spatial patterns in a wide range of systems. Despite its generality, a wide array of possible self-organisation models have now emerged for an

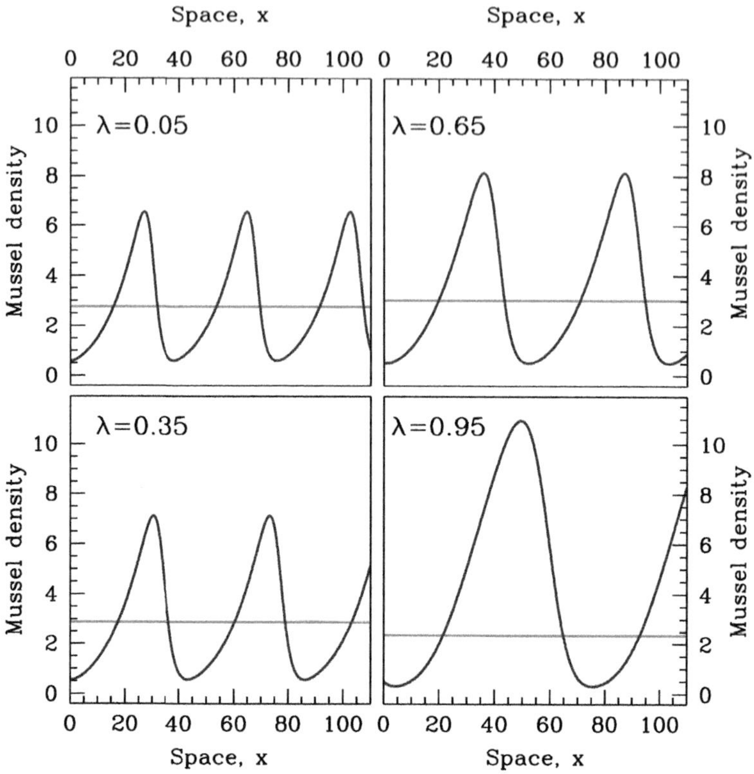

Fig. 12 Patterns for the same speed as at the pattern onset point. We show examples of mussel density patterns given by model (8). We plot patterns with the same migration speed as at the pattern onset (Turing–Hopf bifurcation) point, with the value of α set at 20% below that at the pattern onset point. These values of α are indicated by dashed vertical lines in Fig.10. In each panel, we also show the steady-state mussel density (horizontal line). The patterns were calculated by numerical continuation of the travelling wave ODEs starting at a Hopf bifurcation point, using the software package WAVETRAIN (Sherratt 2012). Note that to enable direct comparison, the same parameters are used in Figs. 9, 10, 11, 12 and 13

equally large array of example ecosystems, and attempts for unification have been limited. This limits our understanding of the functional differences between self-organised ecosystems in a comprehensive way. Here we analysed a generalised model of pattern formation in ecosystems using a model that combines two general mechanisms for pattern formation: "increased production" and "reduced losses". We showed that both mechanisms predict Turing-type regular patterns to develop, but that there are clear differences between the two models.

In reality, the reduced losses and increased production mechanisms are not alternatives—both will apply to some extent, and they act in concert to generate spatial patterns. Our hybrid model (8) provides a means of including both mechanisms, with a single new parameter, and our detailed study of pattern generation provides a comprehensive framework for understanding the solutions of this model. A key take-home message of our work concerns sensitivity to variations in the balance between the reduced losses and increased production mechanisms. Figures 14 and 15 show that

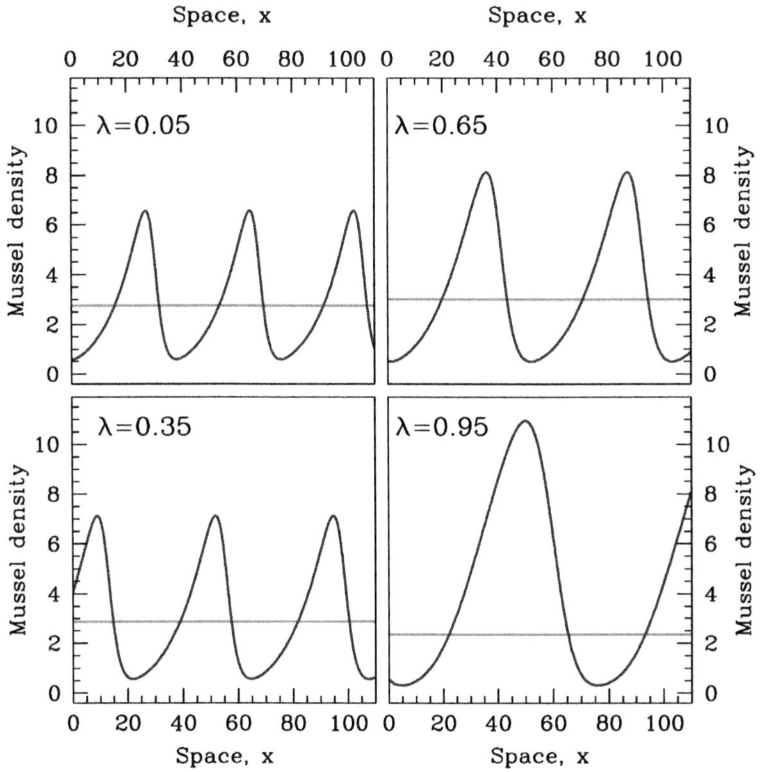

Fig. 13 Patterns for the same wavelength as at the pattern-onset point. We show examples of mussel density patterns given by model (8). We plot patterns with the wavelength as at the pattern-onset (Turing–Hopf bifurcation) point, with the value of α set at 20% below that at the pattern onset point. These values of α are indicated by dashed vertical lines in Fig. 11. In each panel, we also show the steady-state mussel density (horizontal line). The patterns were calculated by numerical continuation of the travelling wave ODEs starting at a Hopf bifurcation point, using the software package WAVETRAIN (Sherratt 2012). Note that to enable direct comparison, the same parameters are used in Figs. 9, 10, 11, 12 and 13

a change in this balance has a much greater effect in a system that is dominated by the reduced losses mechanism than for a system dominated by the increased production mechanism. This highlights that when the reduced losses mechanism is dominant, it is particularly important to investigate the possibility and extent of additional feedback mechanisms.

Our work has been set in the specific context of spatially patterned mussels beds. However, one or other of the "decreased losses" and "increased production" mechanisms lie at the heart of many other patterned ecosystems. Vegetation patterns are a characteristic feature of semi-arid ecosystems, consisting of alternating patches of vegetation and bare ground (Bastiaansen et al. 2018; Gandhi et al. 2019). It is widely accepted that a mechanism of increased production type plays a key role in the generation of these patterns; specifically, higher vegetation levels increase the rate at which rain water infiltrates into the soil, leading to higher plant growth rates (Klausmeier 1999; Rietkerk et al. 2002; Meron 2012). Indeed, as we have already remarked, the

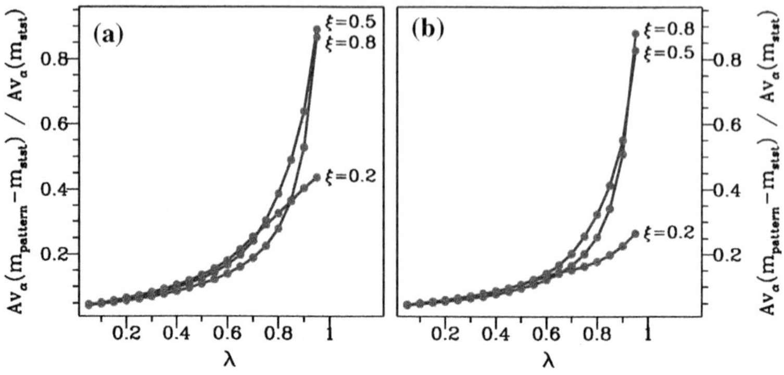

Fig. 14 Quantitative details of the comparison between the average mussel density in spatial patterns and the steady-state mussel density. In (**a**, **b**) the migration speed and wavelength, respectively, are held constant at their value at pattern onset (the Turing–Hopf bifurcation point); examples of the two corresponding solution branches are shown in Fig. 9. For each pair of λ and ξ values, we used the software package WAVETRAIN (Sherratt 2012) to track the form of the pattern along these solution branches as α is varied with the other parameters fixed. We then calculated the average over α of the difference between the mean mussel density and the steady-state mussel density and divided this by the average of the steady-state mussel density. This gives a single number comparing the mussel density in spatial patterns and in the steady state, and we plot this as a function of λ for $\xi = 0.2$, 0.5 and 0.8. The plots show an increasing separation between the pattern and steady-state solution branches as λ is increased between 0 (increased production model) and 1 (decreased losses model). However, there is no clear trend in the way in which the difference in mussel densities varies with the parameter ξ. The other parameters are $\beta = 0.1$ and $\nu = 100$

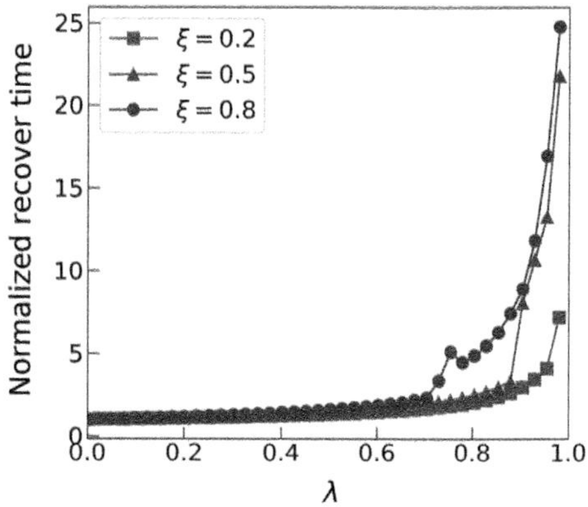

Fig. 15 Recovery time versus **λ**. Quantitative comparison of recovery time from directional numerical simulation on model (8). A spatial pattern is generated in model (8); then, a single pattern period was extracted and a perturbed version of this was used as an initial condition on a spatial domain of length equal to one wavelength, with periodic boundary conditions. The perturbation consisted of a removal of 20% of the total biomass, varying randomly in space. The "recovery time" is a measure of the time taken for the solution to return to the patterned state; it is defined as the time by which solution is within 99% of the pre-perturbation pattern. We found that the details of the perturbation had a negligible effect on the recovery time. There is a clear trend for this recovery time to increase with the parameter λ, that is as the feedback mechanism gradually shifts from increased production to decreased losses. The parameter values are $\alpha = 0.2$, $\beta = 0.1$, $\nu = 100$

$\lambda = 0$ limit of (8) gives the widely used Klausmeier model for vegetation patterning (Klausmeier 1999; Sherratt and Lord 2007; van der Stelt et al. 2013). Another ecosystem in which a mechanism of "increased production" type drives pattern formation is the sequence of ridges and hollows found in many peat bogs (Morris et al. 2011). In fact, two different mechanisms combine to generate these patterns, with each mechanism being of "increased production" type. The elevated ridges are drier than neighbouring hollows, which causes an increased production rate of peat and an increase in the thickness of the surface acrotelm layer. This further amplifies the height difference between the ridges and hollows. In addition, the higher evapotranspiration rates by the vascular plants on the ridges lead to increased nutrient accumulation, which generates faster plant growth and thus a thicker acrotelm layer. Mathematical models based on these mechanisms confirm pattern formation (Morris et al. 2011; Eppinga et al. 2009) and have been verified by comparison with field data (Eppinga et al. 2008).

An example of an ecosystem in which pattern formation is driven by a mechanism of "reduced losses" type is provided by patterns of hummocks and hollows on intertidal mudflats. Diatoms accumulate on the top of hummocks, forming a biofilm that is strengthened by extracellular secretions, and this promotes sedimentation; in the hollows, water accumulation inhibits the corresponding processes. This was first proposed as a pattern generation mechanism by Rietkerk and van de Koppel (2008) and has subsequently been tested via both modelling and field data (Weerman et al. 2010).

The occurrence of the two alternative mechanisms in a wide range of ecosystems highlights the importance of understanding the similarities and differences in the patterns that they generate. In particular, the combination of modelling studies and field work has the potential to distinguish the two mechanisms, and our results open up a number of different avenues for such future cross-disciplinary research.

Appendix

In Fig. 1, we show a comparison between two- and three-equation models for the increased production mechanism. As a prelude to these simulations, we derived dimensionless versions of the models that can easily be compared. Rescalings (6) do not extend to the three-equation model (2) in a natural way; therefore, we use instead the rescalings proposed by Liu et al. (2012):

$$\check{t} = \widehat{E}\widehat{t} \quad \check{x} = \left(\frac{\widehat{E}}{\widehat{D}_m}\right)^{1/2}\widehat{x} \quad \check{a} = \left(\frac{\widehat{E}}{\widehat{F}\widehat{A}}\right)\widehat{a} \quad \check{m} = \left(\frac{\widehat{P}}{\widehat{E}\widehat{S}_0}\right)\widehat{m} \quad \check{s} = \widehat{s}/\widehat{S}_0$$
$$\check{\alpha} = \widehat{F}/\widehat{E} \quad \check{\beta} = \widehat{B}\widehat{S}_0/\widehat{P} \quad \check{\nu} = \left(\widehat{E}\widehat{D}_m\right)^{-1/2}\widehat{\nu} \quad \check{\eta} = \widehat{\eta}$$
$$\check{D} = \widehat{D}_s/\widehat{D}_m \quad \check{\theta} = \widehat{Q}/\widehat{E} \quad \check{\delta} = \widehat{C}\widehat{F}\widehat{A}/\widehat{E}^2.$$

This transforms (2) to the system

$$\partial\check{a}/\partial\check{t} = 1 - \check{\alpha}\check{a} - \check{\beta}\check{a}\check{m}(\check{s}+\check{\eta})/(\check{s}+1) + \check{\nu}\,\partial\check{a}/\partial\check{x} \tag{A.1a}$$
$$\partial\check{m}/\partial\check{t} = \check{\delta}\check{a}\check{m}(\check{s}+\check{\eta})/(\check{s}+1) - \check{m} + \partial^2\check{m}/\partial\check{x}^2 \tag{A.1b}$$
$$\partial\check{s}/\partial\check{t} = \check{m} - \check{\theta}\check{s} + \check{D}\,\partial^2\check{s}/\partial\check{x}^2. \tag{A.1c}$$

Applying the same rescalings to (3) gives

$$\partial\check{a}/\partial\check{t} = 1 - \check{\alpha}\check{a} - \left(\check{\beta}/\check{\theta}\right)\check{a}\check{m}^2 + \check{\nu}\,\partial\check{a}/\partial\check{x} \tag{A.2a}$$
$$\partial\check{m}/\partial\check{t} = \left(\check{\delta}/\check{\theta}\right)\check{a}\check{m}^2 - \check{m} + \partial^2\check{m}/\partial\check{x}^2. \tag{A.2b}$$

In Fig. 1 we compare solutions of (A.1) and (A.2) for parameter values taken from Liu et al. (2012). The uniform steady state $(\check{a}_s, \check{m}_s, \check{s}_s)$, which is used in the initial conditions for the simulations in Fig. 1, is given by

$$\check{s}_s = \frac{\check{\delta} - \check{\alpha} - \check{\beta}\check{\theta}\check{\eta} + \sqrt{\left(\check{\delta} - \check{\alpha} - \check{\beta}\check{\theta}\check{\eta}\right)^2 - 4\check{\beta}\check{\theta}\left(\check{\alpha} - \check{\delta}\check{\eta}\right)}}{2\check{\beta}\check{\theta}}$$
$$\check{m}_s = \check{\theta}\check{s}_s$$
$$\check{a}_s = \frac{1+\check{s}_s}{\check{\delta}\left(\check{s}_s + \check{\eta}\right)}$$

(see Liu et al. 2012 for details).

One notable feature of Fig. 1 is that in the solution of the three-equation model, the patterns for $\check{m}$ and $\check{s}$ are approximately in phase. Motivation for this small phase difference comes from linear stability analysis of the spatially uniform steady state from which patterns bifurcate. Linearising (A.1) about this steady state and substituting $(\check{a} - \check{a}_s, \check{m} - \check{m}_s, \check{s} - \check{s}_s) = (\check{a}_0, \check{m}_0, \check{s}_0)\exp(ik\check{x} + \mu\check{t})$ gives

$$\left(\lambda + \check{\theta} + \check{D}k^2\right)\check{s}_0 = \check{m}_0.$$

Therefore, the phase difference between the $\check{m}$ and $\check{s}$ patterns at the Turing–Hopf bifurcation point is

$$\arg\left[\left(\mathrm{Re}\,\lambda + \check{\theta} + \check{D}k^2\right) + i\,\mathrm{Im}\,\lambda\right].$$

Since $|\mathrm{Im}\,\lambda/k|$ is the pattern migration speed, which is small, the phase difference is also small. There is no *a priori* reason for the phase difference not to increase, potentially significantly, as the pattern amplitude increases, but our numerical simulations suggest that this does not happen.

References

Bastiaansen R, Jaïbi O, Deblauwe V, Eppinga MB, Siteur K, Siero E, Mermoz S, Bouvet A, Doelman A, Rietkerk M (2018) Multistability of model and real dryland ecosystems through spatial self-organization. Proc Natl Acad Sci USA 115:11256–11261

Belyea LR (2007) Climatic and topographic limits to the abundance of bog pools. Hydrol Process 21:675–687

Cangelosi RA, Wollkind DJ, Kealy-Dichone BJ, Chaiya I (2014) Nonlinear stability analyses of Turing patterns for a mussel-algae model. J Math Biol 70:1249–1294

Chen W, Ward MJ (2009) Oscillatory instabilities and dynamics of multispike patterns for the one-dimensional Gray-Scott model. Eur J Appl Math 20:187–214

Doedel EJ (1981) Auto, a program for the automatic bifurcation analysis of autonomous systems. Cong Numer 30:265–384

Doedel EJ, Keller HB, Kernévez JP (1991) Numerical analysis and control of bifurcation problems: (I) bifurcation in finite dimensions. Int J Bifurc Chaos 1:493–520

Doedel EJ, Govaerts W, Kuznetsov YA, Dhooge A (2006) Numerical continuation of branch points of equilibria and periodic orbits. In: Doedel EJ, Domokos G, Kevrekidis IG (eds) Modelling and computations in dynamical systems. World Scientific, Singapore, pp 145–164

Doelman A, Kaper TJ, Zegeling P (1997) Pattern formation in the one dimensional Gray–Scott model. Nonlinearity 10:523–563

Dolmer P (2000) Algal concentration profiles above mussel beds. J Sea Res 43:113–119

Eppinga MB, Rietkerk M, Borren W, Lapshina ED, Bleuten W, Wassen MJ (2008) Regular surface patterning of peatlands: confronting theory with field data. Ecosystems 11:520–536

Eppinga MB, De Ruiter PC, Wassen MJ, Rietkerk M (2009) Nutrients and hydrology indicate the driving mechanisms of peatland surface patterning. Am Nat 173:803–818

Gandhi P, Iams S, Bonetti S, Silber M (2019) Vegetation pattern formation in drylands. In: D'Odorico P, Porporato A, Runyan C (eds) Dryland ecohydrology. Springer, New York

Ghazaryan A, Manukian V (2015) Coherent structures in a population model for mussel-algae interaction. SIAM J Appl Dyn Syst 14:893–913

Gray P, Scott SK (1984) Autocatalytic reactions in the isothermal, continuous stirred tank reactor: oscillations and instabilities in the system A+2B→3B; B→C. Chem Eng Sci 39:1087–1097

Klausmeier CA (1999) Regular and irregular patterns in semiarid vegetation. Science 284:1826–1828

Kuznetsov YA (2004) Elements of applied bifurcation theory. Springer, New York

Liu Q-X, Weerman EJ, Herman PM, Olff H, van de Koppel J (2012) Alternative mechanisms alter the emergent properties of self-organization in mussel beds. Proc R Soc Lond B 14:20120157

Liu Q-X, Weerman EJ, Gupta R, Herman PM, Olff H, van de Koppel J (2014) Biogenic gradients in algal density affect the emergent properties of spatially self-organized mussel beds. J R Soc Interface 11:20140089

Malchow H (2000) Motional instabilities in predator-prey systems. J Theor Biol 204:639–647

Meron E (2012) Pattern formation approach to modelling spatially extended ecosystems. Ecol Model 234:70–82

Morris PJ, Belyea LR, Baird AJ (2011) Ecohydrological feedbacks in peatland development: a theoretical modelling study. J Ecol 99:1190–1201

Murray JD (2003) Mathematical biology II: spatial models and biomedical applications. Springer, New York

Øie G, Reitan KI, Vadstein O, Reinertsen H (2002) Effect of nutrient supply on growth of blue mussels (Mytilus edulis) in a landlocked bay. In: Vadstein O, Olsen Y (eds) Sustainable increase of marine harvesting: fundamental mechanisms and new concepts. Kluwer Academic Publishers, Dordrecht, pp 99–109

Perumpanani AJ, Sherratt JA, Maini PK (1995) Phase differences in reaction-diffusion-advection systems and applications to morphogenesis. IMA J Appl Math 55:19–33

Rademacher JDM, Scheel A (2007) Instabilities of wave trains and Turing patterns in large domains. Int J Bifurc Chaos 17:2679–2691

Rietkerk M, van de Koppel J (2008) Regular pattern formation in real ecosystems. Trends Ecol Evol 23:169–175

Rietkerk M, Boerlijst MC, van Langevelde F, HilleRisLambers R, van de Koppel J, Prins HHT, de Roos A (2002) Self-organisation of vegetation in arid ecosystems. Am Nat 160:524–530

Rovinsky AB, Menzinger M (1992) Chemical instability induced by a differential flow. Phys Rev Lett 69:1193–1196

Sherratt JA (2005) An analysis of vegetation stripe formation in semi-arid landscapes. J Math Biol 51:183–197

Sherratt JA (2012) Numerical continuation methods for studying periodic travelling wave (wavetrain) solutions of partial differential equations. Appl Math Comput 218:4684–4694

Sherratt JA (2013) History-dependent patterns of whole ecosystems. Ecol Complex 14:8–20

Sherratt JA (2015) Using wavelength and slope to infer the historical origin of semi-arid vegetation bands. Proc Natl Acad Sci USA 112:4202–4207

Sherratt JA (2016) Using numerical bifurcation analysis to study pattern formation in mussel beds. Math Model Nat Phenom 11:86–102

Sherratt JA, Lord GJ (2007) Nonlinear dynamics and pattern bifurcations in a model for vegetation stripes in semi-arid environments. Theor Pop Biol 71:1–11

Sherratt JA, Mackenzie JJ (2016) How does tidal flow affect pattern formation in mussel beds? J Theor Biol 406:83–92

Sherratt JA, Smith MJ (2008) Periodic travelling waves in cyclic populations: field studies and reaction-diffusion models. J R Soc Interface 5:483–505

Sherratt JA, Smith MJ, Rademacher JDM (2009) Locating the transition from periodic oscillations to spatiotemporal chaos in the wake of invasion. Proc Natl Acad Sci USA 106:10890–10895

Siero E, Doelman A, Eppinga MB, Rademacher JD, Rietkerk M, Siteur K (2015) Striped pattern selection by advective reaction-diffusion systems: resilience of banded vegetation on slopes. Chaos 25:036411

Siteur K, Siero E, Eppinga MB, Rademacher J, Doelman A, Rietkerk M (2014) Beyond Turing: the response of patterned ecosystems to environmental change. Ecol Complex 20:81–96

Song Y, Jiang H, Liu Q-X, Yuan Y (2017) Spatiotemporal dynamics of the diffusive mussel-algae model near Turing–Hopf bifurcation. SIAM J Appl Dyn Syst 16:2030–2062

van Broekhoven W, Jansen H, Verdegem M, Struyf E, Troost K, Lindeboom H, Smaal A (2015) Nutrient regeneration from feces and pseudofeces of mussel Mytilus edulis spat. Mar Ecol Progress (Ser) 27) 534:107–120

van de Koppel J, Crain CM (2006) Scale-dependent inhibition drives regular tussock spacing in a freshwater marsh. Am Nat 168:E136–E147

van de Koppel J, Rietkerk M, Dankers N, Herman PM (2005) Scale-dependent feedback and regular spatial patterns in young mussel beds. Am Nat 165:E66-77

van de Koppel J, Gascoigne JC, Theraulaz G, Rietkerk M, Mooij WM, Herman PMJ (2008) Experimental evidence for spatial self-organization and its emergent effects in mussel bed ecosystems. Science 322:739–742

van der Stelt S, Doelman A, Hek G, Rademacher JDM (2013) Rise and fall of periodic patterns for a generalized Klausmeier–Gray–Scott model. J Nonlinear Sci 23:39–95

Wang RH, Liu Q-X, Sun GQ, Jin Z, van de Koppel J (2009) Nonlinear dynamic and pattern bifurcations in a model for spatial patterns in young mussel beds. J R Soc Interface 6:705–718

Weerman EJ, Van de Koppel J, Eppinga MB, Montserrat F, Liu QX, Herman PM (2010) Spatial self-organization on intertidal mudflats through biophysical stress divergence. Am Nat 176:E15–E32

Yu BG (2010) Dynamic behaviour of a plant-wrack model with spatial diffusion. Commun Nonlinear Sci Numer Simul 15:2201–2205

Zelnik YR, Meron E, Bel G (2015) Gradual regime shifts in fairy circles. Proc Natl Acad Sci USA 112:12327–12331

Publisher's Note Springer Nature remains neutral with regard to jurisdictional claims in published maps and institutional affiliations.

Bulletin of Mathematical Biology (2021) 83:96
https://doi.org/10.1007/s11538-021-00926-z

Society for Mathematical Biology

SPECIAL ISSUE: CELEBRATING J. D. MURRAY

Modelling Preferential Phagocytosis in Atherosclerosis: Delineating Timescales in Plaque Development

Gigi Lui[1] · **Mary R. Myerscough**[1]

Received: 19 October 2020 / Accepted: 20 July 2021 / Published online: 14 August 2021

Abstract

Atherosclerotic plaques develop over a long time and can cause heart attacks and strokes. There are no simple mathematical models that capture the different timescales of rapid macrophage and lipid dynamics and slow plaque growth. We propose a simple ODE model for lipid dynamics that includes macrophage preference for ingesting apoptotic material and modified low-density lipoproteins (modLDL) over ingesting necrotic material. We use multiple timescale analysis to show that if the necrosis rate is small then the necrotic core in the model plaque may continue to develop slowly even when the lipid levels in plaque macrophages, apoptotic material and modLDL appear to have reached equilibrium. We use the model to explore the effect of macrophage emigration, apoptotic cell necrosis, total rate of macrophage phagocytosis and modLDL influx into the plaque on plaque lipid accumulation.

Keywords Efferocytosis · Cholesterol · Asymptotic analysis

1 Introduction

Atherosclerotic plaques may form in artery walls throughout the body but are particularly likely to occur in areas where there is low shear stress on the blood vessel walls, such as at junctions and curves (Wentzel et al. 2012). A plaque starts as a simple injury to the endothelium, a layer of cells that lines blood vessel walls. This may lead to the development of a fatty streak in the artery wall. If the fatty streak does not resolve, then a plaque forms. Over many years, this plaque develops a necrotic core of free lipids and cellular debris, usually covered by a cap of smooth muscle cells and collagen. If this cap ruptures, the plaque releases necrotic material into the blood stream, which

MRM acknowledges funding from two Discovery Grants, DP160104685 and DP200102071, from the Australian Research Council.

✉ Mary R. Myerscough
mary.myerscough@sydney.edu.au

[1] School of Mathematics and Statistics, The University of Sydney, NSW 2006, Australia

promotes the formation of blood clots. If these blood clots block an artery in the heart or the brain, then a heart attack or stroke ensues (Libby 2002; Tabas 2010).

Following endothelial injury, low-density lipoproteins (LDL) inside the vessel walls become modified due to the release of free radicals by the damaged endothelium. This modified LDL (modLDL) stimulates an immune response which draws immune cells, principally macrophages, into the artery wall (Libby 2002). Macrophages ingest the modLDL particles and also dead cells that have undergone controlled cell death (apoptosis), including other macrophages that have become apoptotic (Moore et al. 2003). If all goes well, macrophages remove modLDL and damaged cells, and repair damage to the artery wall so that the initial injury resolves (Bäck et al. 2019). When an injury does not resolve, the artery wall remains in a state of chronic sterile inflammation where the population of macrophages is elevated and lipid continues to enter the artery wall, both on LDL particles and as lipid in the cell membranes of the macrophages themselves (Swirski et al. 2006). Establishment of chronic inflammation occurs over a period of days or weeks, but the growth of the lipid-laden necrotic core occurs over a period of months or years in humans. In this paper, we use mathematical modelling to consider the role of preferential phagocytosis by macrophages in producing these two distinct timescales.

Macrophages ingest and remove pathogenic material inside the body via the process known as phagocytosis. In particular, in atherosclerotic plaques, they consume and remove modified LDL, apoptotic material which includes whole apoptotic cells and apoptotic blebs that form as apoptotic cells start to break apart, and necrotic material which includes cellular debris and free lipids (Moore et al. 2003). Cells that are undergoing apoptosis send out chemokines to attract macrophages (so-called "find me" signals) and then express "eat me" receptors on their cell membranes to facilitate ingestion by live macrophages (Kojima et al. 2017). Likewise modLDL is readily ingested via scavenger receptors on the macrophages' surface. Necrotic material, however, has fewer specific binding sites to facilitate phagocytosis and may also be toxic to macrophages. Hence, macrophages are most likely to ingest apoptotic material and least likely to ingest material in the necrotic core (Kojima et al. 2017).

In this paper, we present a population model for cells and lipids in an atherosclerotic plaque. This model tracks the number of macrophages, the lipids that these cells contain, together with lipid in modLDL, apoptotic materials and in the necrotic core. We show that the model predicts two distinct timescales so that the necrotic core continues to grow on a long timescale, even once cell numbers and other lipid quantities have reached equilibrium. We examine the consequences of different phagocytosis rates and other cellular behaviour, and finally put these results in the context of observed plaque biology.

The modelling in this paper is based on the ODE models of Ford et al. (2019a). Our model also explicitly represents the dynamics of modLDL in the plaque so that it focuses strongly on lipid dynamics rather than cell dynamics. The models of Ford et al. (2019a, b) and Meunier and Muller (2019) were the first to explicitly track lipid loads in plaque macrophages. This model extends those ideas with the specific purpose of examining the effects of macrophage preferential phagocytosis on plaque fate.

2 Formulating the Model

The model is formulated as a set of ordinary differential equations for the population of macrophages $M(t)$, the amount of lipid, including both endogenous and ingested lipid, that is contained in macrophages $A_M(t)$, the lipid in apoptotic material $A_P(t)$, the lipid in modLDL $\ell(t)$ and the lipid in the necrotic core $N(t)$. Here t represents time in suitable units.

"Lipid" here includes cholesterol attached to the protein scaffold of modLDL particles, lipids in cellular membranes and intracellular free cholesterol and cholesterol esters. All of these are ingested by cells via phagocytosis and efferocytosis. We do not include LDL and cellular components other than lipid in this model, although these are also ingested by cells, as our aim here is to examine lipid accumulation. The implicit assumption is that lipid accumulation is not strongly influenced by the ingestion of other components of apoptotic cells or necrotic material (Ford et al. 2019b). In terms of experimental measurement, lipid consumption rates, for example, could best be thought of in terms of cholesterol molecules per unit time.

Macrophages in atherosclerotic plaques consume lipid via phagocytosis in the form of modLDL, from apoptotic bodies and cells, and from the necrotic core (Kojima et al. 2017). We assume that there is a maximum rate at which each macrophage can ingest lipids, which is likely to be determined by the surface area of the macrophage and its ability to internally process ingested material (Schrijvers et al. 2007). For this reason, we model the total rate of lipid ingestion as a saturating function of the lipids available for ingestion. We assume that macrophages ingest lipid from modLDL, apoptotic material and necrotic material, respectively, at different rates due to their different abilities to engage with these lipids (Khan et al. 2003; Schrijvers et al. 2007). Apoptotic material is typically ingested so quickly that it is rare to see an apoptotic cell under the microscope (Kojima et al. 2017). This rapid ingestion is due to the presence on apoptotic cells of receptors and chemokines that promote their phagocytosis (Tabas 2010). Necrotic material will be ingested most slowly due to the toxicity of this material to cells in its vicinity and its lack of receptors and other cues for phagocytosis (Kojima et al. 2017). ModLDL is ingested by macrophages via scavenger receptor of their surface, but phagocytosis of modLDL does not seem to be as rapid as phagocytosis of apoptotic material (Thon et al. 2018). We assume that macrophages have access to all these types of lipids when they are present in the artery wall but will preferentially ingest apoptotic material, then lipid in modLDL and only ingest necrotic material slowly as the last preference. We further assume that these lipids compete for macrophage phagocytic capacity (Khan et al. 2003). Using these assumptions, we model the rate of lipid ingestion per macrophage as

$$f(A_P, \ell, N) = \theta \frac{\zeta_P A_P + \zeta_\ell \ell + \zeta_N N}{\mu + (\zeta_P A_P + \zeta_\ell \ell + \zeta_N N)}. \tag{1}$$

Here θ governs the overall rate that a macrophage can phagocytose lipid, μ is the half saturation constant and ζ_P, ζ_ℓ and ζ_N with $\zeta_P \geq \zeta_\ell \geq \zeta_N$ determine the rates that different types of lipids are consumed. This function, which represents preferential phagocytosis, is a key innovation in this model.

2.1 Macrophage Population

The rate of change of the macrophage population is given by

$$\frac{\mathrm{d}M}{\mathrm{d}t} = \alpha \frac{A_M}{\kappa_\alpha + A_M} - \beta M - \gamma M \tag{2}$$

where the first term on the right-hand side models macrophage recruitment from the blood stream into the artery wall, $-\beta M$ represents the rate that macrophages become apoptotic and $-\gamma M$ represents the rate that macrophages leave the population by emigrating from the artery wall. The parameters β and γ are both positive constants. Macrophage recruitment is stimulated by inflammatory cytokines that are produced by macrophages when they consume lipid. Accumulated intracellular cholesterol promotes the production of these inflammatory cytokines (Zhu et al. 2008). We therefore assume that cytokine production and hence macrophage recruitment are both functions of A_M the total lipid contained in all macrophages. The macrophage recruitment function must necessarily be saturating as there is a finite number of monocytes (macrophage precursors) that are available to be recruited from the blood stream. The half saturation constant is κ_α and the positive constant α governs the maximum rate of macrophage recruitment. Equation (2) follows closely the equation for the macrophage population in the ODE model of Ford et al. (2019a).

2.2 Lipid Contained in Macrophages

We assume that each macrophage contains a quantity of endogenous lipid a_0 in its cell membranes in addition to any lipid that it has ingested. When a macrophage undergoes apoptosis, this endogenous lipid becomes apoptotic lipid along with the accumulated ingested lipid. Under this assumption, the equation for the change in lipid carried by macrophages is

$$\frac{\mathrm{d}A_M}{\mathrm{d}t} = a_0 \alpha \frac{A_M}{\kappa_\alpha + A_M} - \beta A_M - \gamma A_M + \theta M \frac{\zeta_P A_P + \zeta_\ell \ell + \zeta_N N}{\mu + (\zeta_P A_P + \zeta_\ell \ell + \zeta_N N)}. \tag{3}$$

Here the first term on the right-hand side represents the rate that lipid enters the plaque in the form of endogenous lipid in macrophages that are recruited from the bloodstream, $-\beta A_M$ and $-\gamma A_M$, and model the rate at which lipid is lost from the macrophage population when macrophages, respectively, become apoptotic or emigrate. The final term describes the rate at which the macrophage population ingests lipid through phagocytosis.

2.3 Apoptotic Material

The rate of change of lipid in apoptotic material is given by

$$\frac{\mathrm{d}A_P}{\mathrm{d}t} = \beta A_M - \nu g(A_P) A_P - \theta M \frac{\zeta_P A_P}{\mu + (\zeta_P A_P + \zeta_\ell \ell + \zeta_N N)} \tag{4}$$

where βA_M models the rate at which lipid in apoptotic material accumulates following macrophage apoptosis and $\nu g(A_P)A_P$ models the rate at which apoptotic cells become necrotic. Here ν controls the maximum linear rate of lipid necrosis. The last term on the right-hand side is the rate at which apoptotic lipid is ingested by macrophages. We set $g(A_P) = A_P/(\kappa_\nu + A_P)$, where κ_ν is the half saturation constant. This function is intended to represent the increase in the rate of necrosis if we assume that low levels of apoptotic lipid A_P produce a nonlinearly increasing rate of necrosis that saturates to a linear rate as A_P becomes large, independently of other dynamics. In this paper, we will often assume that $g(A_P) \equiv 1$ in simplified versions of the model.

2.4 ModLDL Lipid

The rate of change of modLDL lipid, ℓ is modelled by

$$\frac{\mathrm{d}\ell}{\mathrm{d}t} = \xi - \theta M \frac{\zeta_\ell \ell}{\mu + (\zeta_P A_P + \zeta_\ell \ell + \zeta_N N)} \tag{5}$$

where ξ, assumed constant in this model, is the rate at which modLDL lipid enters the artery wall and the second term is the rate that lipid on modLDL is ingested by macrophages.

2.5 Necrotic Core Lipid

The rate of change for lipid in the necrotic core is given by

$$\frac{\mathrm{d}N}{\mathrm{d}t} = \nu g(A_P)A_P - \theta M \frac{\zeta_N N}{\mu + (\zeta_P A_P + \zeta_\ell \ell + \zeta_N N)} \tag{6}$$

where the first term on the right-hand side represents the rate at which apoptotic cells become necrotic with the resultant transfer of lipid from apoptotic material to the necrotic core, and the second term models the rate at which macrophages ingest lipid and remove it from the necrotic core.

The artery wall contains endogenous macrophages at the onset of plaque formation so we assume the following initial conditions:

$$M(0) = M_0, \quad A_M(0) = a_0 M_0, \quad A_P(0) = 0, \quad \ell(0) = 0, \quad N(0) = 0. \tag{7}$$

We nondimensionalise Eqs. (2) to (6) by setting:

$$\widetilde{t} = \beta t, \qquad \widetilde{a} = \frac{\beta}{a_0 \alpha} a, \qquad \widetilde{M}(\widetilde{t}) = \frac{\beta + \gamma}{\alpha} M(t), \tag{8}$$

$$\widetilde{A_M}(\widetilde{t}) = \frac{\beta}{a_0 \alpha} A_M(t), \quad \widetilde{A_P}(\widetilde{t}) = \frac{\beta}{a_0 \alpha} A_P(t), \quad \widetilde{\ell}(\widetilde{t}) = \frac{\beta}{a_0 \alpha} \ell, \quad \widetilde{N}(\widetilde{t}) = \frac{\beta}{a_0 \alpha} N(t). \tag{9}$$

This produces the following dimensionless parameters:

$$\Omega = \frac{\beta+\gamma}{\beta}, \quad \Phi = \frac{\nu}{\beta}, \quad \Theta = \frac{\theta}{a_0(\beta+\gamma)}, \quad \Xi = \frac{\xi}{a_0\alpha}, \tag{10}$$

$$\widetilde{\zeta_\ell} = \frac{\zeta_\ell}{\zeta_P}, \quad \widetilde{\zeta_N} = \frac{\zeta_N}{\zeta_P}, \tag{11}$$

$$\widetilde{\kappa_\alpha} = \frac{\beta}{a_0\alpha}\kappa_\alpha, \quad \widetilde{\kappa_\nu} = \frac{\beta}{a_0\alpha}\kappa_\nu. \tag{12}$$

Dropping the tildes, we have the nondimensional set of equations:

$$\frac{1}{\Omega}\frac{\mathrm{d}M}{\mathrm{d}t} = \frac{A_M}{\kappa_\alpha + A_M} - M \tag{13}$$

$$\frac{\mathrm{d}A_M}{\mathrm{d}t} = \frac{A_M}{\kappa_\alpha + A_M} - \Omega A_M + \Theta M \frac{A_P + \zeta_\ell \ell + \zeta_N N}{\mu + (A_P + \zeta_\ell \ell + \zeta_N N)} \tag{14}$$

$$\frac{\mathrm{d}A_P}{\mathrm{d}t} = A_M - \Phi \frac{A_P}{\kappa_\nu + A_P} A_P - \Theta M \frac{A_P}{\mu + (A_P + \zeta_\ell \ell + \zeta_N N)} \tag{15}$$

$$\frac{\mathrm{d}\ell}{\mathrm{d}t} = \Xi - \Theta M \frac{\zeta_\ell \ell}{\mu + (A_P + \zeta_\ell \ell + \zeta_N N)} \tag{16}$$

$$\frac{\mathrm{d}N}{\mathrm{d}t} = \Phi \frac{A_P}{\kappa_\nu + A_P} A_P - \Theta M \frac{\zeta_N N}{\mu + (A_P + \zeta_\ell \ell + \zeta_N N)}. \tag{17}$$

3 The Four-Equation Model

The set of five Eqs. (13) to (17) is difficult to analyse. To produce an analytically tractable model, we neglect modLDL lipid. This is a reasonable assumption, given that apoptosis and defective efferocytosis drive late-stage plaque dynamics and not the ingestion of modLDL (Thorp and Tabas 2009), even though the presence of modLDL in the artery wall is the stimulus that initiates plaque growth. We also assume that macrophage recruitment occurs at a constant rate; that is, the number of monocytes in the blood stream is sufficiently high that it is not significantly impacted by plaque macrophage recruitment. We further assume that $g(A_P) \equiv 1$, so that the necrosis rate is linear, or in other words that the number of apoptotic cells A_P is sufficiently large that $g(A_P)$ is close to saturation. These last two assumptions are essentially equivalent to taking the limit as $\kappa_\alpha, \kappa_\nu \to 0$. The model then becomes:

$$\frac{1}{\Omega}\frac{\mathrm{d}M}{\mathrm{d}t} = 1 - M \tag{18}$$

$$\frac{\mathrm{d}A_M}{\mathrm{d}t} = 1 - \Omega A_M + \Theta M \frac{A_P + \zeta N}{\mu + (A_P + \zeta N)} \tag{19}$$

$$\frac{\mathrm{d}A_P}{\mathrm{d}t} = A_M - \Phi A_P - \Theta M \frac{A_P}{\mu + (A_P + \zeta N)} \tag{20}$$

$$\frac{\mathrm{d}N}{\mathrm{d}t} = \Phi A_P - \Theta M \frac{\zeta N}{\mu + (A_P + \zeta N)}, \tag{21}$$

where $\zeta = \zeta_N$ for notational ease, and with initial conditions

$$M = M_0, \quad A_M = A_{M_0}, \quad A_P = 0, \quad N = 0. \tag{22}$$

These equations have a unique steady state at

$$M^* = 1, \quad A_M^* = \frac{1}{\Omega - 1}, \quad A_P^* = \frac{\mu}{\Theta(\Omega-1)+\mu\Phi(\Omega-1)-1}, \tag{23}$$

$$N^* = \frac{\mu^2\Phi(\Omega-1)}{\zeta(\Theta(\Omega-1)-1)(\Theta(\Omega-1)+\mu\Phi(\Omega-1)-1)}. \tag{24}$$

Since $\Omega > 1$ if $\gamma \neq 0$, A_M^* is always positive, and A_P^* and N^* are positive if $\Theta(\Omega - 1) > 1$.

Linear analysis about this steady state is not revealing as the signs of the eigenvalues of the Jacobian cannot be determined analytically. However, numerical exploration suggests that the steady state given in equations (23) and (24) is an attractor when the steady state exists and is positive, that is, when $\Theta(\Omega - 1) > 1$.

We assume that the rate of necrosis of apoptotic cells is very small, that is, $\Phi \ll 1$. In healthy tissue, almost no apoptotic cells become necrotic. In early-stage plaque, this is also true, although as atherosclerotic plaques develop, the necrosis rate increases (Tabas 2005). Since we are considering plaques essentially in the weeks after initial plaque onset, it is reasonable to take Φ as small. In a later section, we consider what happens when Φ is larger in the five-equation model. We also assume that $\zeta \ll 1$; that is, that macrophages consume necrotic lipid at a much slower rate than they consume apoptotic material. This is likely as necrotic material does not contain the receptors or generate the signals that facilitate the rapid consumption of apoptotic cells. It is also toxic to macrophages and hence is likely to be more slowly phagocytosed (Kojima et al. 2017). Numerical solutions (see Fig. 1a) suggest that this produces a solution where the lipids in macrophages and apoptotic material rapidly reach equilibrium while the quantity of necrotic lipid develops much more slowly. We therefore analyse the system of Eqs. (13) to (17) using a multiple timescale approach.

4 Multiple Timescale Analysis of the Four-Equation Model

In this analysis, we assume that the plaque is generally healthy with low rates of necrosis; that is, apoptotic cells and other material are rapidly ingested by macrophages so that the rate that apoptotic cells become necrotic is very small. We also assume that any necrotic material is removed only very slowly by macrophages. Therefore, we set

$$\Phi = \epsilon\bar{\Phi}, \quad \text{and} \quad \zeta = \epsilon\bar{\zeta}, \tag{25}$$

where $\epsilon \ll 1$ and $\bar{\Phi}$ and $\bar{\zeta}$ are both of order 1. We substitute these expressions for Φ and ζ into Eqs. (13) to (17), assuming that all other parameters are of order 1, and use

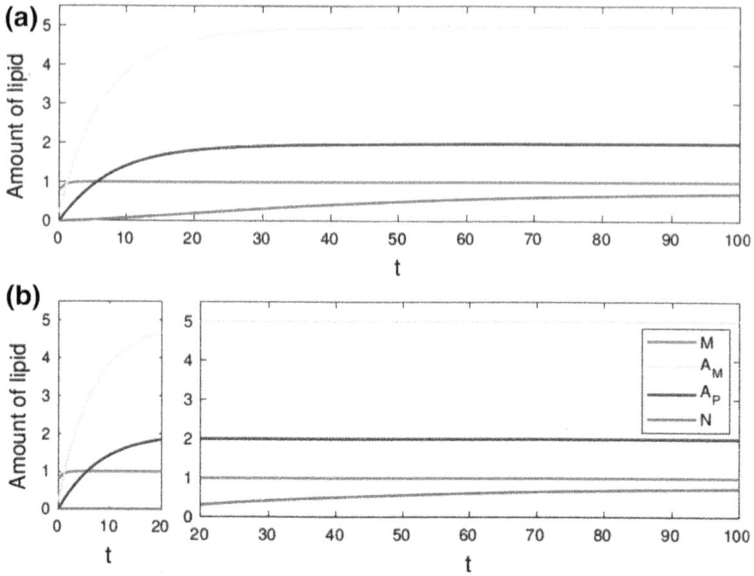

Fig. 1 Four-equation model (18)–(21) and its two-timescale approximation. Plot **a** shows the numerical solution of the model. Plots in **b** show numerical solutions to the fast system equations (26)–(29) in the left panel and the analytic solution to the slow system (35)–(38) in the right panel. Parameter values for both plots are $\Omega = 1.2$, $\Theta = 30$ and $\mu = 10$ with initial conditions $M_0 = 0.8$ and $A_{M_0} = 0.2$. For **a** $\Phi = 0.01$ and $\zeta = 0.01$. For **b** $\epsilon = 0.01$ and $\bar{\Phi} = \bar{\zeta} = 1$. Hence, the parameter values in plots **a** and **b** are equivalent

the binomial expansion to expand rational expressions where ζ is in the denominator. This gives

$$\frac{1}{\Omega}\frac{\mathrm{d}M}{\mathrm{d}t} = 1 - M \tag{26}$$

$$\frac{\mathrm{d}A_M}{\mathrm{d}t} = 1 - \Omega A_M + \Theta M \frac{A_P + \epsilon\bar{\zeta}N}{\mu + A_P}\left(1 - \epsilon\frac{\bar{\zeta}N}{\mu + A_P} + \ldots\right) \tag{27}$$

$$\frac{\mathrm{d}A_P}{\mathrm{d}t} = A_M - \epsilon\bar{\Phi}A_P - \Theta M \frac{A_P}{\mu + A_P}\left(1 - \epsilon\frac{\bar{\zeta}N}{\mu + A_P} + \ldots\right) \tag{28}$$

$$\frac{\mathrm{d}N}{\mathrm{d}t} = \epsilon\left(\bar{\Phi}A_P - \Theta M \frac{\bar{\zeta}N}{\mu + A_P} + \ldots\right). \tag{29}$$

This suggests that t is a fast variable, M, A_M and A_P are all fast variables and N is a slow variable.

The limiting case, when $\epsilon \to 0$, gives an approximation for the fast system:

$$\frac{1}{\Omega}\frac{\mathrm{d}M}{\mathrm{d}t} = 1 - M \tag{30}$$

$$\frac{\mathrm{d}A_M}{\mathrm{d}t} = 1 - \Omega A_M + \Theta M \frac{A_P}{\mu + A_P} \tag{31}$$

$$\frac{\mathrm{d}A_P}{\mathrm{d}t} = A_M - \Theta M \frac{A_P}{\mu + A_P} \tag{32}$$

$$\frac{\mathrm{d}N}{\mathrm{d}t} = 0. \tag{33}$$

A set of solutions for this system is illustrated in the left panel in Fig. 1b. Note that on this timescale $N \approx 0$ using the initial conditions given in (22). The steady state in the fast variables is

$$M^* = 1, \quad A_M^* = \frac{1}{\Omega - 1}, \quad A_P^* = \frac{\mu}{\Theta(\Omega - 1) - 1}, \tag{34}$$

where the asterisk specifies the steady state in the fast timescale.

Let $t = \tau/\epsilon$ where τ is the slow timescale. Substituting this into Eqs. (26) to (29) gives the equations for the evolution on the slow timescale:

$$\epsilon \frac{1}{\Omega} \frac{\mathrm{d}M}{\mathrm{d}\tau} = 1 - M \tag{35}$$

$$\epsilon \frac{\mathrm{d}A_M}{\mathrm{d}\tau} = 1 - \Omega A_M + \Theta M \frac{A_P + \epsilon \bar{\zeta} N}{\mu + A_P} \left(1 - \epsilon \frac{\zeta \bar{N}}{\mu + A_P} + \dots\right) \tag{36}$$

$$\epsilon \frac{\mathrm{d}A_P}{\mathrm{d}\tau} = A_M - \epsilon \bar{\Phi} A_P - \Theta M \frac{A_P}{\mu + A_P} \left(1 - \epsilon \frac{\zeta \bar{N}}{\mu + A_P} + \dots\right) \tag{37}$$

$$\frac{\mathrm{d}N}{\mathrm{d}\tau} = \bar{\Phi} A_P - \Theta M \left(\frac{\bar{\zeta} N}{\mu + A_P} + \dots\right). \tag{38}$$

The solution to the limit of these equations as $\epsilon \to 0$ yields the steady state values for M, A_M and A_P, given in Eq. (34). Substituting these values into Eq. (38) gives a first-order equation in N only:

$$\frac{\mathrm{d}N}{\mathrm{d}\tau} + \frac{\bar{\zeta}(1 + \Theta(1 - \Omega))}{\mu(1 - \Omega)} N = -\frac{\bar{\Phi}\mu}{1 + \Theta(1 - \Omega)}. \tag{39}$$

This equation, together with the matching condition $N = 0$ at $\tau = 0$, can be solved explicitly to give an expression for the growth of the necrotic core over the long timescale:

$$N(\tau) = -\frac{\mu^2 \bar{\Phi}(1 - \Omega)}{\bar{\zeta}(1 + \Theta(1 - \Omega))^2} \left(1 - \exp\left(-\frac{\bar{\zeta}(1 + \Theta(1 - \Omega))}{\mu(1 - \Omega)} \tau\right)\right). \tag{40}$$

As $\tau \to 0$, $N \to 0$, and as $\tau \to \infty$, N approaches a finite limit. The solution for $N(\tau)$ is illustrated in the second panel of Fig. 1b.

This comparatively simple two-timescale solution is biologically and conceptually important. It demonstrates that in a plaque that does not appear to be growing, a very low level of necrosis, together with macrophage preferential phagocytosis, can lead to the slow growth of a necrotic core which will, in time, lead to the onset of more significant pathologies beyond chronic inflammation. Numerical solutions of

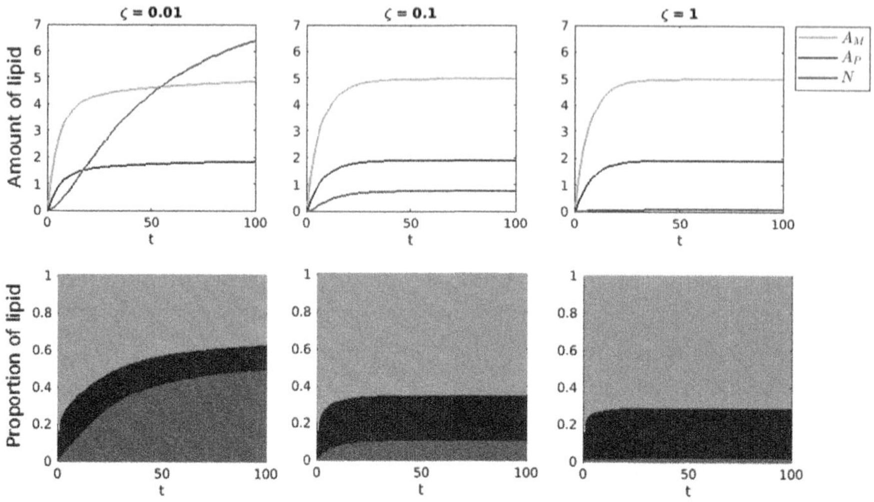

Fig. 2 Solutions for lipid loads in the four-equation model (18)–(21) for different values of ζ as shown. The parameter values are $\Omega = 1.2$, $\Phi = 0.1$, $\Theta = 30$, $\mu = 10$ with initial conditions $M_0 = 0.8$, $A_{M_0} = 0.2$. The top row shows the change of the quantity of each lipid type with time. The bottom row shows how the proportion of each lipid type in the plaque changes with time

the four-equation model (Fig. 2) show that the order of magnitude of ζ has very little impact on the amount of lipid contained in macrophages and in apoptotic material. However, ζ has a significant impact on both the size and the rate of development of the necrotic core, and hence, ζ impacts both the total amount of lipid in the plaque and the proportion of that lipid that is in the necrotic core.

5 Numerical Solutions of the Full Five-Equation Model

The four-equation model enabled analysis that showed explicitly that a very small rate of necrosis of apoptotic cells produced a two-timescale effect on plaque growth and composition, where initial inflammation (indicated by M, A_M and A_P) rapidly reached equilibrium, while the necrotic core continued to grow. We return now to the five-equation model to examine the effect of including modLDL in the model, the effect of saturation of the macrophage recruitment rate and the effect of the rate that apoptotic cells can be ingested.

We solved the five-equation model (13)–(17) with initial conditions $M(0) = M_0$, $A_M(0) = A_{M_0}$, $\ell(0) = A_P(0) = N(0) = 0$, using the parameter set in Table 1 as the default. We explored the effect of different macrophage behaviour on the growth and make-up of the plaque. We used the MATLAB routine ode45. Despite the two timescale behaviour, the equations appeared to be sufficiently well behaved to be solved using methods for standard ODEs.

Figure 3 shows the effects of changing the orders of magnitude of ζ_ℓ and ζ_N. Recall that in the nondimensional system, ζ_ℓ and ζ_N specify macrophage preference for phagocytosis of modLDL and of necrotic core lipids, respectively, relative to their

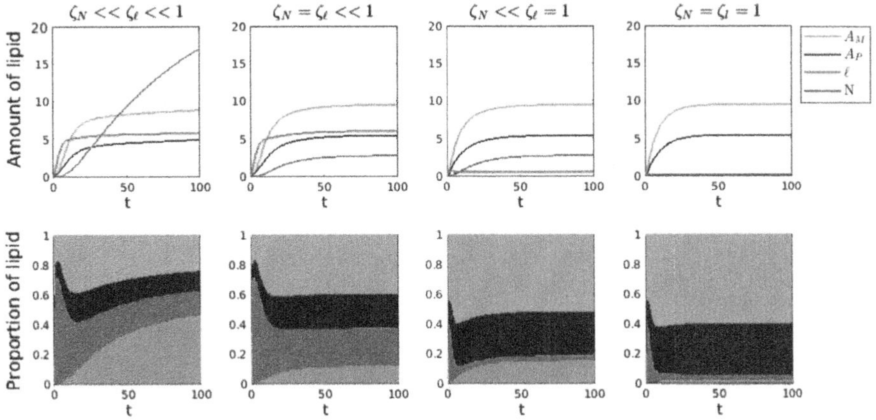

Fig. 3 Solutions for lipid loads in the full, five-equation model (13)–(17) for different combinations of ζ_N and ζ_ℓ as shown. The parameter values are $\Omega = 1.2$, $\Phi = 0.1$, $\Theta = 30$, $\Xi = 1$, $\mu = 10$, $\kappa_\alpha = \kappa_\nu = 1$ with initial conditions $M_0 = 0.8$, $A_{M_0} = 0.2$. The combinations of ζ_N and ζ_ℓ, respectively, from left to right, are $(0.01, 0.1)$, $(0.1, 0.1)$, $(0.1, 1)$ and $(1, 1)$. The top row shows the evolution of lipid quantities over time and the bottom row shows plaque composition over time

preference for ingesting apoptotic material. The biology underlying these assumptions is explained in Sect. 2. This figure shows four different cases.

1. $\zeta_N \ll \zeta_\ell \ll 1$: The plaque slowly develops a very large lipid necrotic core and a substantial amount of modLDL lipid, specified by ℓ, remains in the artery wall and is not ingested by macrophages. This case is comparable to case where $\zeta = 0.01$ in the four-equation model, shown in the left-hand column of diagrams in Fig. 2. The behaviour of A_M and A_P in the five-equation model in this case is very similar to in the four-equation model but the necrotic core material accumulates more rapidly in the five-equation model, possibly due to the extra quantity of lipid that is introduced by the inclusion of modLDL.
2. $\zeta_N = \zeta_\ell \ll 1$: The lipid content of the necrotic core reaches equilibrium at about the same time as the macrophage population, the apoptotic material and modLDL. The lipid core in this case is an order of magnitude smaller than when $\zeta_N \ll \zeta_\ell \ll 1$. ModLDL is still present in the artery wall in similar amounts to case 1, but the necrotic core is much smaller.
3. $\zeta_N \ll \zeta_\ell = 1$: Here modLDL is ingested at the same rate as the apoptotic material. Only a small amount modLDL remains in the artery wall compared to cases 1 and 2, but the amount of lipid in the necrotic core is similar to case 2. Cases 2 and 3 in the five-equation model are both comparable to the case where $\zeta = 0.1$ in the four-equation model (middle column in Fig. 2). The evolution of A_M, A_P and N is qualitatively similar across these three cases, although the amount of lipid in each of these classes is higher in the five-equation model than in the four-equation model.
4. $\zeta_N = \zeta_\ell = 1$: In this case, macrophages consume lipid in the necrotic core, modLDL and lipid in apoptotic matter at equal rates, with no preference for any of these over the others. The amount of both modLDL lipid and necrotic lipid in the

Table 1 Parameter values used in numerical solutions of the five-equation model

Symbol	Value	Description
Ω	1.2	Ratio of total macrophages that leave the macrophage population to macrophages that undergo apoptosis
Φ	0.1	Rate of necrosis of apoptotic material
Θ	30	Controls total rate of macrophage ingestion across all lipids
Ξ	1	Rate of modLDL influx
ζ_ℓ	0.1	Controls rate of macrophage ingestion of modLDL
ζ_N	0.01	Controls rate of macrophage ingestion of lipid from necrotic core
μ	10	Half saturation constant in macrophage ingestion function
κ_α	1	Half saturation constant in macrophage recruitment function
κ_υ	1	Half saturation constant in necrosis function
M_0	0.8	Initial condition for M
A_{M_0}	0.2	Initial condition for A_M

plaque is very low. In this case, almost all the lipid is inside the macrophages or in apoptotic material and there is an effective process of recycling, where lipid in apoptotic material is rapidly recycled through macrophage ingestion (Ford et al. 2019b). This case is comparable to the case where $\zeta = 0.1$ in the four-equation model (right-hand column of Fig. 2). In cases 2 and 3, the evolution of A_M, A_P and N is qualitatively similar between the two models but with higher amounts of lipid in the five-equation model, possibly due to the inclusion of modLDL.

Other parameters will also affect plaque composition. We explored the effects of

- Ω, the ratio of total numbers of macrophages leaving the plaque (both via apoptosis and via emigration) to those that leave via apoptosis,
- Φ the rate at which apoptotic matter becomes necrotic,
- Θ which governs the rate of macrophage phagocytosis and
- Ξ, the rate of modLDL entry into the artery wall.

Equations (13)–(17) are solved using the parameters in Table 1 and the results at $t = 10000$ (effectively steady state) were plotted for different values of the parameter under investigation.

5.1 Effect of Varying Ω (Emigration vs. Apoptosis Rate)

When $\Omega = 1$, all macrophages leave the plaque macrophage population via apoptosis. As Ω increases from 1, an increasing proportion of macrophages leave the population via emigration out of the plaque (Fig. 4). When emigrating macrophages leave the plaque, they take with them the lipid that they have accumulated. When a macrophage does not emigrate, the lipid that it contains is recycled within the plaque when it

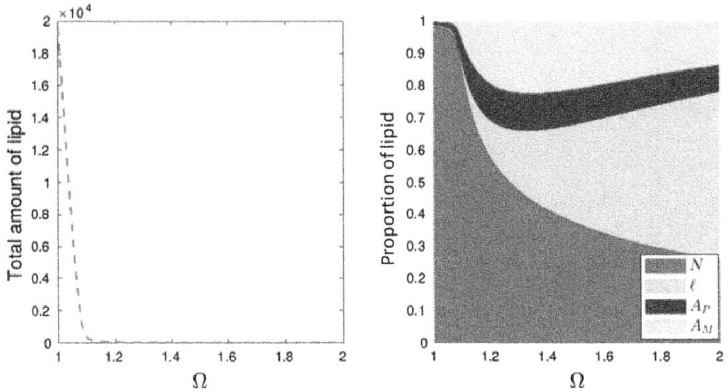

Fig. 4 Effect of parameter Ω on the total amount of lipid (left) and the plaque composition (right) at $t = 10000$. In the left-hand panel, the total amount of lipid plotted is given by $A_M + A_P + \ell + N$. Parameter values, except Ω, are those given in Table 1

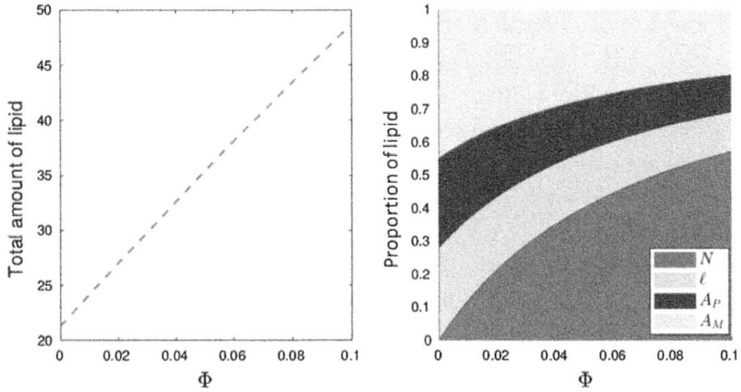

Fig. 5 Effect of parameter Φ on the total amount of lipid (left) and the plaque composition (right) at $t = 10000$. In the left-hand panel, the total amount of lipid plotted is given by $A_M + A_P + \ell + N$. Parameter values, except Φ, are those given in Table 1

undergoes apoptosis. High numbers of apoptotic cells will also increase the linear rate of necrosis, due to the saturating function $g(A_P)$. Hence, decreasing Ω so that it is less than about 1.1 rapidly increases plaque lipid load. However, once Ω is increased beyond about 1.2, macrophages do not stay in the plaque long enough to ingest modLDL effectively. The amount of modLDL becomes a higher proportion of plaque lipid than lipid in macrophages and apoptotic material combined for $\Omega \gtrsim 1.5$.

5.2 Effect of Varying Φ (Necrosis Rate)

As Φ increases, a higher proportion of apoptotic material becomes necrotic and is not ingested as apoptotic material by macrophages. This leads to increased amounts of lipid in the necrotic core (Fig. 5). Since this necrotic core material is ingested by macrophages at a much lower rate than apoptotic material, this increases the total lipid in the plaque.

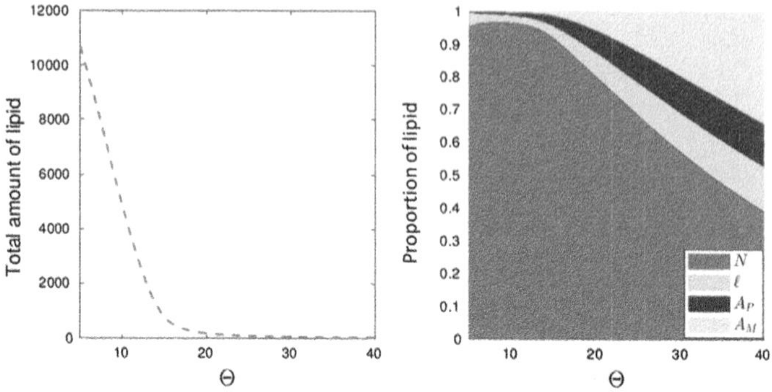

Fig. 6 Effect of parameter Θ on the total amount of lipid (left) and the plaque composition (right) at $t = 10000$. In the left-hand panel, the total amount of lipid plotted is given by $A_M + A_P + \ell + N$. Parameter values, except Θ, are those given in Table 1

5.3 Effect of Varying Θ (Overall Macrophage Lipid Uptake Rate)

For very low Θ, plaque lipids loads are high (Fig. 6), up to three orders of magnitude greater than those in Fig. 5, although comparable with those in Fig. 4. When Θ is very low, almost no phagocytosis or efferocytosis takes place. This has two consequences. Without effective efferocytosis, a high proportion of apoptotic cells become necrotic and contribute to the necrotic core, and emigrating macrophages contain only very small quantities of nonendogeneous lipid so macrophage emigration does not remove significant amounts of lipid from the plaque. This means that almost all the lipid that enters is recycled within the plaque, producing a large necrotic core. Additionally, for very low Θ, modLDL lipid will continuously accumulate at an approximately constant rate Ξ, and ℓ will become large over time. As Θ increases, there is more capacity for macrophages to ingest lipids from the necrotic core and both the proportion of lipid in the core and the total amount of lipid decrease. This strongly suggests that total macrophage phagocytic capacity is important in reducing the overall size of the necrotic core.

5.4 Effect of Varying Ξ (modLDL Influx into Plaque)

The rate of influx of modLDL into plaque, represented by Ξ in this model, is believed to be correlated with the lipid load on LDL particles in the blood stream. This cholesterol on LDL particles in the blood stream is colloquially known as "bad cholesterol". In clinical practice, patients with high levels of blood LDL cholesterol are regarded as being at high risk of cardiovascular disease.

In this model, when Ξ is low, the plaque lipid load is low with only a small proportion of lipid in the necrotic core (Fig. 7). As Ξ increases, there is a point where macrophages are overwhelmed by the amount of modLDL in the artery wall, and therefore, they do not have the capacity to ingest lipid from the necrotic core. The necrotic core contains a higher proportion of the total plaque lipids as Ξ increases past this point and the overall lipid load of the plaque also increases.

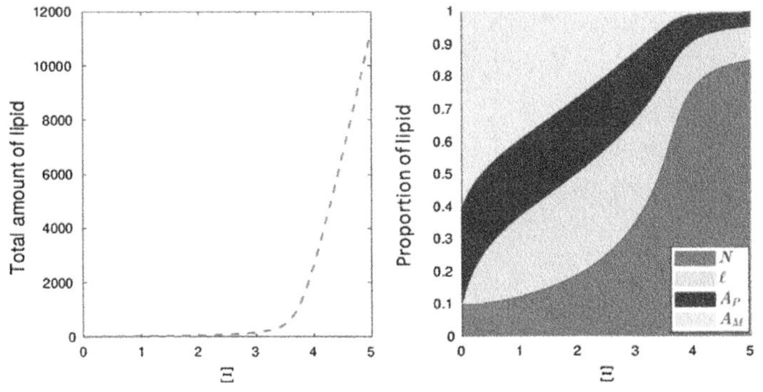

Fig. 7 Effect of parameter Ξ on the total amount of lipid (left) and the plaque composition (right) at $t = 10000$. In the left-hand panel, the total amount of lipid plotted is given by $A_M + A_P + \ell + N$. Parameter values, except Ξ, are those given in Table 1

5.5 Effects of Changing the Phagocytosis Rate ζ_N of Necrotic Core Lipids in Combination with Other Parameters

The five-equation model (13)–(17), using parameters in Table 1, was run until $t = 200$. At this time, the system of ODEs was close to steady state. We then changed the values of ζ_N and one other parameter (either Ω, Φ, Θ or Ξ) and found the values of N, the necrotic lipid load, at $t = 1000$ (that is 800 units of time later). Figure 8 shows the change in the amount of necrotic core lipid as ζ_N along with Θ, Φ, Θ and Ξ, respectively, are changed. In this figure, ζ_N changes over several orders of magnitude. When $\zeta_N = 10$, so that macrophages ingest necrotic material at 10 times the rate that they ingest apoptotic material (a biologically very unlikely scenario) plaques are uniformly healthy with low necrotic lipid loads. As ζ_N decreases, plaques acquire increasing amounts of necrotic lipid, particularly when macrophage emigration rate Ω and total ingestion rate Θ are low and the necrosis rate Φ and rate of influx of modLDL Ξ are high.

6 Discussion

The formation of atherosclerotic plaques in the artery wall is the outcome of processes at many different timescales—the pulse of the blood flow that determines wall shear stress (seconds), phagocytosis of apoptotic cells (minutes), the movement of macrophages into the artery wall (hours), the death of cells (weeks) and the formation of a necrotic core (months or years). Linking these timescales in a mathematically and scientifically consistent way is an important step towards realising a comprehensive computational model for atherosclerotic plaque formation and regression.

The model presented here is very simple. It does not, for example, account for the many cytokines that contribute to plaque growth, nor the phenotypic diversity of plaque macrophages (Bäck et al. 2019). However, by reducing the plaque lipid dynamics to its simplest components in the model we have been able to show that the long

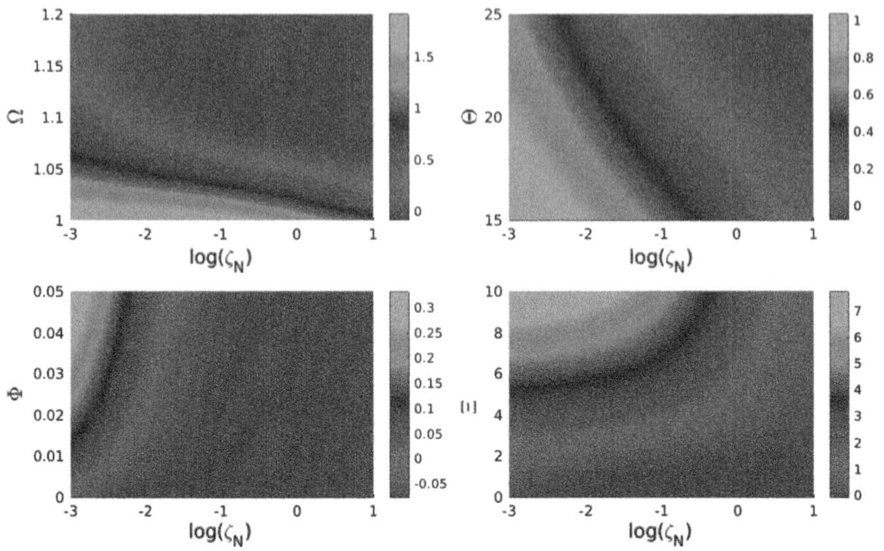

Fig. 8 Change of plaque necrotic lipid load due to varying ζ_N simultaneously with each of the parameters Ω, Φ, Θ and Ξ, respectively, from their values in Table 1. The system was run until $t = 200$ with Table 1 values, and then, ζ_N and one of the other parameters were changed while holding all other parameters constant and the system was then run again until $t = 1000$. The resulting colour plot illustrates the change in necrotic lipid load between $t = 200$ with Table 1 parameters and $t = 1000$ after the equations had been solved where one of the parameters has been altered, in addition to ζ_N

timescale of the development of the necrotic core can arise from the shorter timescale dynamics of lipid processing through phagocytosis of modLDL and of apoptotic cells and their lipid loads. When a small proportion of apoptotic material becomes necrotic and macrophages avoid ingesting this material, then a large necrotic core will slowly grow, even when the other components of the system—macrophages, modLDL and apoptotic material—appear to have reached equilibrium. The model suggests that macrophage preferential phagocytosis, due to cytokine signals and receptors on apoptotic material and the toxicity of necrotic material, may create an environment where the slow accumulation of necrotic material is inevitable.

The accumulation of necrotic material and the growth the necrotic core is promoted by low rates of emigration and phagocytosis, and high rates of necrosis and high levels of modLDL in the artery wall. It is well known, experimentally, that inhibiting macrophage emigration from plaques fosters plaque growth (van Gils et al. 2011; Llodrá et al. 2004). Recent scientific research has identified raising efferocytosis (phagocytosis specifically of dead and apoptotic cellular material) as a promising therapeutic target (Flores et al. 2020). Increasing the rate that macrophages ingest apoptotic material will also reduce the rate of secondary necrosis of apoptotic material that is not rapidly consumed.

Real plaques are extremely complicated (Bäck et al. 2019) and to produce even a simple model plaque that can undergo regression, by reducing the number of macrophages in the plaque and removing necrotic material will require more sophisticated models to be developed. Nevertheless, this model establishes a good foundation for further modelling.

Declarations

Conflict of interest The authors declare that they have no conflict of interest.

References

Bäck M, Yurdagul A, Tabas I, Öörni K, Kovanen P (2019) Inflammation and its resolution in atherosclerosis: mediators and therapeutic opportunities. Nat Rev Cardiol 16:389–406

Flores A, Hosseini-Nassab N, Jarr KU, Ye J, Zhu X, Wirka R, Koh A, Tsantilas P, Wang Y, Nanda V, Kojima Y, Zeng Y, Lotfi M, Sinclair R, Weissman I, Ingelsson E, Smith B, Leeper N (2020) Pro-efferocytic nanoparticles are specifically taken up by lesional macrophages and prevent atherosclerosis. Nat Nanotechnol 15(2):154

Ford H, Byrne H, Myerscough M (2019) A lipid-structured model for macrophage populations in atherosclerotic plaques. J Theor Biol 479:48–63

Ford H, Zeboudj L, Purvis G, ten Bokum A, Zarebski A, Bull J, Byrne H, Myerscough M, Greaves D (2019) Efferocytosis perpetuates substance accumulation inside macrophage populations. Proc R Soc B Biol Sci 286(1904):20190730

Khan M, Pelegaris S, Cooper M, Smith C, Evan G, Betteridge J (2003) Oxidised lipoproteins may promote inflammation through the selective delay of engulfment but not binding of apoptotic cells by macrophages. Atherosclerosis 171:21–29

Kojima Y, Weissman I, Leeper N (2017) The role of efferocytosis in atherosclerosis. Circulation 135:476–489

Libby P (2002) Inflammation in atherosclerosis. Nature 420(6917):868–874

Llodrá J, Angeli V, Liu J, Trogan E, Fisher EA, Randolph GJ (2004) Emigration of monocyte-derived cells from atherosclerotic lesions characterizes regressive, but not progressive plaques. Proc Natl Acad Sci USA 101(32):11779–11784

Meunier N, Muller N (2019) Mathematical study of an inflammatory model for atherosclerosis: a nonlinear renewal equation. Acta Appl Math 161(1):107–126

Moore K, Sheedy F, Fisher E (2003) Macrophages in atherosclerosis: a dynamic balance. Nat Rev Immunol 13:710–721

Schrijvers D, De Meyer G, Herman A (2007) Phagocytosis in atherosclerosis: molecular mechanisms and implications for plaque progression and stability. Cardiovasc Res 73:470–480

Swirski F, Libby P, Aikawa E, Alcaide P, Luscinskas F, Weissleder R, Pittet M (2006) Ly-6C(hi) monocytes dominate hypercholesterolemia-associated monocytosis and give rise to macrophages in atheromata. J Clin Investig 117(1):195–205

Tabas I (2005) Consequences and therapeutic implications of macrophage apoptosis in atherosclerosis. The importance of lesion stage and phagocytic efficiency. Arterioscler Thromb Vasc Biol 25:2255–2264

Tabas I (2010) Macrophage death and defective inflammation resolution in atherosclerosis. Nat Rev Immunol 10:36–46

Thon M, Ford H, Gee M, Myerscough M (2018) A quantitative model of early atherosclerotic plaques parameterized using in vitro experiments. Bull Math Biol 80:175–214

Thorp E, Tabas I (2009) Mechanisms and consequences of efferocytosis in advanced atherosclerosis. J Leukoc Biol 86:1089–1095

van Gils J, Derby M, Fernandes L, Ramkhelawon B, Ray T, Rayner K, Parathath S, Distel E, Feig J, Alvarez-Leite J, Rayner A, McDonald T, O'Brien K, Stuart L, Fisher E, Lacy-Hulbert A, Moore K (2011) The neuroimmune guidance cue netrin-1 promotes atherosclerosis by inhibiting the emigration of macrophages from plaques. Nat Immunol 13(5):136–143

Wentzel J, Chatzizisis Y, Gijsen F, Giannoglou G, Feldman C, Stone P (2012) Endothelial shear stress in the evolution of coronary atherosclerotic plaque and vascular remodelling: current understanding and remaining questions. Cardiovasc Res 96(2):234–243

Zhu X, Lee JY, Timmins J, Brown J, Boudyguina E, Mulya A, Gebre A, Willingham M, Hiltbold E, Mishra N, Maeda N, Parks J (2008) Increased cellular free cholesterol in macrophage-specific Abca1 knock-out mice enhances pro-inflammatory response of macrophages. J Biol Chem 283:22930–22941

Publisher's Note Springer Nature remains neutral with regard to jurisdictional claims in published maps and institutional affiliations.

Bulletin of Mathematical Biology (2021) 83:95
https://doi.org/10.1007/s11538-021-00928-x

Society for Mathematical Biology

SPECIAL ISSUE: CELEBRATING J. D. MURRAY

Dimension Estimate of Uniform Attractor for a Model of High Intensity Focussed Ultrasound-Induced Thermotherapy

M. A. Efendiev[1,2] · J. Murley[3] · S. Sivaloganathan[4,5]

Received: 4 October 2020 / Accepted: 20 July 2021 / Published online: 7 August 2021

Abstract

High intensity focussed ultrasound (HIFU) has emerged as a novel therapeutic modality, for the treatment of various cancers, that is gaining significant traction in clinical oncology. It is a cancer therapy that avoids many of the associated negative side effects of other more well-established therapies (such as surgery, chemotherapy and radiotherapy) and does not lead to the longer recuperation times necessary in these cases. The increasing interest in HIFU from biomedical researchers and clinicians has led to the development of a number of mathematical models to capture the effects of HIFU energy deposition in biological tissue. In this paper, we study the simplest such model that has been utilized by researchers to study temperature evolution under HIFU therapy. Although the model poses significant theoretical challenges, in earlier work, we were able to establish existence and uniqueness of solutions to this system of PDEs (see Efendiev et al. Adv Appl Math Sci 29(1):231–246, 2020). In the current work, we take the next natural step of studying the long-time dynamics of solutions to this model, in the case where the external forcing is quasi-periodic. In this case, we are able to prove the existence of uniform attractors to the corresponding evolutionary processes generated by our model and to estimate the Hausdorff dimension of the attractors, in terms of the physical parameters of the system.

Keywords High intensity focussed ultrasound · Uniform attractors · PDEs · Hausdorff dimension

---

Dedicated to James D. Murray: Pioneer, Teacher, Colleague and Friend, on the occasion of his 90th birthday.

✉ S. Sivaloganathan
ssivalog@uwaterloo.ca

Extended author information available on the last page of the article

1 Introduction

High intensity focussed ultrasound (HIFU) is a minimally invasive surgical technique that can be used to thermally ablate both malignant and benign tumours, as well as to cauterize injured vessels and organs (to staunch internal bleeding), with minimal damage to surrounding tissue. HIFU has been proposed as an alternative to surgery for treatment of cancer and various tumour types, including prostate, breast, brain, and renal cancer, amongst others. Currently, it is also used for palliative care, for example, to alleviate the pain resulting from the metastasis of malignant tumours to bone tissue. The emergence of HIFU as a powerful new therapeutic modality in the treatment of cancers promises to revolutionize cancer care worldwide. The use of mathematical modelling to predict the effects of HIFU for thermal ablation has facilitated the effective implementation of this therapeutic modality for certain disorders such as osteoid osteomas, essential tremors and prostate ablation (see, Dobrakowski et al. 2014; Hudson et al. 2018; Modena et al. 2019; Sagaonkar et al. 2014). Figure 1 illustrates a typical set-up for clinical administration of HIFU to treat uterine fibroids. The further development and generalization of mathematical models for soft tissue lesions, cavitation and disruption of the blood brain barrier suggest significant opportunities for mathematics to contribute to the development of HIFU as the "gold standard" for cancer therapy. Over the past 50 years, high intensity focussed ultrasound has become a subject of increasing interest among medical researchers. HIFU causes selective tissue necrosis in a very well-defined volume, at a variable distance from the transducer, through heating or cavitation. For the past two to three decades, the use of HIFU has been investigated in many clinical settings. The most abundant clinical trial data come from studies on the treatment of prostatic disease, although early research did examine its potential applications in neurosurgery. More recently, horizons have been considerably broadened, and the potential of HIFU as a non-invasive surgical tool has been demonstrated in many settings including the treatment of tumours of the liver, kidney, breast, bone, uterus and pancreas, in addition to its successful use in rectifying conduction defects in the heart, as well as to perform surgical haemostasis, and for the relief of chronic pain of malignant origin. Further clinical evaluation and developments will undoubtedly follow, but recent technological advances suggest that HIFU is likely to play an increasingly significant role in future surgical practice. While the quest continues for a reliable and minimally invasive alternative to open surgery, the endoscopic revolution is well underway and there is much research activity in other relevant fields such as laser, radiofrequency, cryo-, thermo- and brachy-therapies. Lithotripsy is now an established treatment for kidney stones and gallstones, but currently, the only pervasive, non-invasive modalities used in mainstream clinical oncology are chemotherapy and radiotherapy, and while effective in many instances, they both have significant, associated negative side effects. High intensity focussed ultrasound has the potential to provide clinicians with another truly non-invasive, targeted treatment option. Its scope is not, however, limited to the treatment of only malignant cancers, but can also be used in a palliative setting for relief of chronic pain, for haemostasis or even for the treatment of cardiac conduction or congenital anomalies. In this current work, we are primarily focussed on a simple model of energy deposition in biological tissue (and the subsequent evolution of the temperature field) as a result of HIFU interaction with

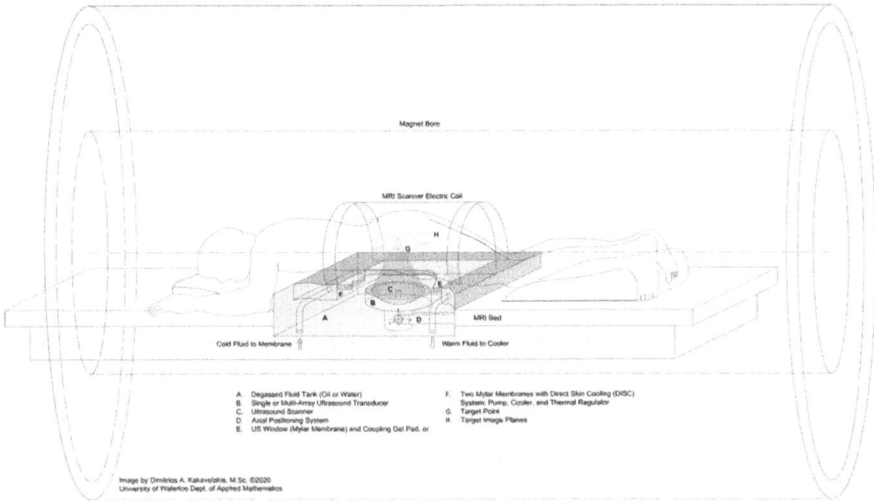

Fig. 1 An MRI-guided HIFU apparatus. (Figure by D. Kakavelakis, PhD candidate, UW.)

biological tissue and not with the actual propagation of ultrasound waves through biological tissue, per se. This model and several others have been presented in the literature to computationally explore the effects of HIFU therapy (see, for example, Hariharan et al. 2007; Hudson et al. 2018; Kaltenbacher 2015; Kreider et al. 2013).

2 Model and problem formulation

The focus of this paper is on establishing the existence of a uniform attractor and dimension estimates of this attractor (and not on the actual modelling of HIFU therapy). In this section, we briefly discuss the model analysed. This is the simplest model that has been widely used in the literature in order to model the evolution of the temperature field, under HIFU therapy (see Bailey et al. 2003 and references therein). The model is comprised of a system of coupled partial differential equations: a linear wave equation for the propagation of ultrasound through tissue, coupled to the Pennes bioheat equation (a linear heat equation with a source term for the the acoustic energy deposition obtained from the solution to the wave equation) and a thermal dose model (to control the acoustic energy deposition of ultrasound exposure). The acoustic pressure derived from the acoustic wave model, $q(t, x)$, is used to determine the heat source term, $h(t, x)$, in the Pennes bioheat equation. The heating caused by focussed ultrasound is a result of the energy absorption from the acoustic wave passing through the tissue. This is reflected in the heat source term by defining an absorption coefficient, α, for each tissue type, to correlate the tissue heat with the acoustic intensity. It is possible to show that

$$\tilde{h}(t, x) = \alpha I = \frac{\alpha q^2(t, x)}{\rho c_0} \tag{1}$$

where I is the acoustic intensity, ρ is the tissue density, and c_0 is the speed of the acoustic wave in the tissue (see Bailey et al. 2003).

The tissue temperature, $T(t, x)$, resulting from HIFU heating is determined from the Pennes bioheat equation, with source term given by (2.1),

$$\rho c_p \frac{\partial T(t, x)}{\partial t} = \nabla \cdot (k \nabla T(t, x)) + \omega_b c_{p,b}(T_a - T(t, x)) + \tilde{h}(t, x) \tag{2}$$

Hence, it is this Eq. (2) that is the focus of the analysis to follow. The third term in (2.2), the blood perfusion term, can be heuristically justified on the basis of Newton's Law of cooling (see (Pennes 1948) for a more detailed discussion of the Pennes bioheat equation). The Pennes bioheat equation has been the subject of much discussion, as it is less justified physically, and we refer the reader to Wissler (1987) and the review paper by Bhowmik et al. (2013), for further discussion of more physically justifiable extensions of this equation. Thus, using a common approximation to treat the soft tissue as an incompressible fluid, we may use a more standard term that takes convective effects into account in the third term of Eq. (2).

Let Ω be a bounded domain in $\mathbb{R}^n$. In our application, $n = 3$, but the following proofs work in general. As a consequence of the assumption that the soft tissue behaves like an incompressible fluid, we have

$$\nabla_x . \mathbf{v}(t, x) = 0 \tag{3}$$

for the fluid velocity of the biological tissue. The velocity and temperature are related by Darcy's law in the following manner:

$$\mathbf{v}(t, x) = \nabla_x P(t, x) - T(t, x) \tag{4}$$

where P is the fluid pressure in the biological tissue and is a constant vector parameter to match the scalar temperature term with the other vector terms. Finally, using this fluid velocity, we arrive at a modified heat equation where the ad hoc blood perfusion term is replaced with a more theoretically sound convection term (Chen and Holmes 1980),

$$\frac{\partial T}{\partial t}(t, x) = \Delta_x T(t, x) - \mathbf{v}(t, x) \cdot \nabla_x T(t, x) + h(t, x) \tag{5}$$

Although the assumption of incompressibility may appear unjustified in this context—and indeed this observation would be valid if our interest was in the propagation of acoustic waves through a biological medium—our focus is on the transmission of heat through biological tissue and the evolution of the temperature field. In this case, the assumption of incompressibility can be justified (see, for example, Lesniewski et al. 2010, chapter one of Wu 2016, or Wu 2018). These papers and the references therein on acoustic streaming provide justification for the assumption of incompressibility under appropriate conditions. We note, however, that our analytical approach for establishing the existence of a uniform attractor and dimension estimates carry through, even in the case of slight compressibility. In general, our governing equation must be coupled with a nonlinear wave equation, the solvability of which in this case requires additional technical requirements and would make the analysis more cumbersome. However, we note that the inclusion of the Westervelt or KZK equation (instead of the linear wave equation) in the model does not affect our analysis, since these nonlinear wave equations only impact Eq. (5) through the source term h(t,x). The system of non-autonomous Eqs. (3–5) defines the fluid velocity in terms of tissue temperature and fluid pressure, and thus, a solution to this system of non-autonomous equations is given by a pair of functions (P, T). More general boundary conditions of Neumann or Robins type are clearly necessary to model more physically realistic scenarios; however, in the current work, we use the simplest Dirichlet boundary conditions. We assume that the initial distribution of these functions in Ω and a set of non-homogeneous boundary conditions are given a priori by

$$P(t=\tau, x) = P_\tau(x); \quad P(t, x')|_{x'\in\partial\Omega} = P_{bd}(x') \tag{6}$$

$$T(t=\tau, x) = T_\tau(x); \quad T(t, x')|_{x'\in\partial\Omega} = T_{bd}(x') \tag{7}$$

The following compatibility conditions must also be imposed:

$$P_\tau(x')|_{x'\in\partial\Omega} = P_{bd}(x') \quad T_\tau(x')|_{x'\in\partial\Omega} = T_{bd}(x') \tag{8}$$

Without loss of generality (Efendiev et al. 2020), we may assume $P_{bd}(x) = 0$.

Note, however, that it is possible to define the fluid pressure P in terms of the temperature T by applying the divergence operator to Eq. (4). Thus, from Eq. 3, we obtain

$$\Delta_x P(t, x) = \cdot \nabla_x T(t, x)$$

This leads to the following full system of non-autonomous equations that constitute our model:

$$\Delta_x P(t, x) = \cdot \nabla_x T(t, x), \qquad x \in \Omega, t \geq \tau \tag{9}$$

$$\mathbf{v}(t, x) = \big(\nabla_x((-\Delta_x)^{-1} \cdot (\nabla_x T) + T)\big), \qquad x \in \Omega, t \geq \tau \tag{10}$$

$$\frac{\partial T}{\partial t}(t, x) = \Delta_x T(t, x) - \mathbf{v}(t, x) \cdot \nabla_x T(t, x) + h(t, x) \tag{11}$$

$$P(t=\tau, x) = P_\tau(x); \quad P(t, x')|_{x'\in\partial\Omega} = P_{bd}(x') \tag{12}$$

$$T(t=\tau, x) = T_\tau(x); \quad T(t, x')|_{x'\in\partial\Omega} = T_{bd}(x') \tag{13}$$

Let $Q_{\tau,\eta} := [\tau, \eta] \times \Omega$ for $\eta \geq t \geq \tau$. We seek a solution of (9–13):

$$\big(T(t, x), P(t, x)\big) \in W^{(1,2),p}(Q_{\tau,\eta}) \times W^{(1,3),p}(Q_{\tau,\eta}) \tag{14}$$

which satisfy the above equations, in the sense of distributions. Note that $u \in W^{(1,2),p}(Q_{\tau,\eta})$, by definition, means that

$$u \in L^p(Q_{\tau,\eta}),\ \partial_t u \in L^p(Q_{\tau,\eta}),\ D_x^\alpha u \in L^p(Q_{\tau,\eta})$$

for $0 \le |\alpha| \le 2$. The value of $p = p(n)$ (where n is the dimension of the underlying domain $\Omega \subset \mathbb{R}^n$) can be chosen "a posteriori" to guarantee compact embedding (Simon 1987).

$W^{(1,2),p}(Q_{\tau,\eta}) \subset C\big([\tau,\eta], C^\epsilon(\Omega)\big)$ for some $\epsilon \in (0, \frac{3}{2}]$. It is known that for sufficiently large $p = p(n) >> 1$, where n is the dimension of $\Omega \in \mathbb{R}^n$, this compact embedding holds (see Simon 1987). From (14), it follows

$$\begin{aligned} T_\tau(x) &\in W^{2(1-\frac{1}{p}),p}(\Omega), \qquad P_\tau(x) \in W^{3-\frac{2}{p},p}(\Omega) \\ T_{bd}(x') &\in W^{2-\frac{1}{p},p}(\partial\Omega), \qquad P_{bd}(x') \in W^{3-\frac{1}{p},p}(\partial\Omega) \end{aligned} \tag{15}$$

where $W^{s,p}(\Omega)$ for non-integer real s is a fractional Sobolev space (Triebel 1978).

As already indicated, we seek a solution

$$\big(T(t,x), P(t,x)\big) \in W^{(1,2),p}(Q_{\tau,\eta}) \times W^{(1,3),p}(Q_{\tau,\eta})$$

that satisfies (9–11), in a weak sense under compatibility conditions (8).

To study the long-time dynamics of the solution of the non-autonomous system of Eqs. (9–13), an indispensable first step is the proof of uniqueness and global existence in time of (9–13), and this was established in our prior paper (Efendiev et al. 2020). For clarity and convenience, we sketch the proof of well-posedness of Eqs. (9–13).

We introduce a pseudodifferential operator of zeroth-order $\Psi(x, D)$ defined by

$$\Psi(x,D)T(t,x) := -\big(+\nabla_x(-\Delta_x)^{-1}\cdot\nabla_x\big)T(t,x) \tag{16}$$

Substituting (10) into (11) results in the following equation:

$$\partial_t T(t,x) - \Delta_x T(t,x) = -\nabla T(t,x)\cdot\Psi(x,D)T(t,x) + h(t,x) \tag{17}$$

The proof of existence of solutions in the appropriate spaces indicated in (14, 15) is based on the Leray–Schauder principle. To this end, we consider a family of equations that depends on $\lambda \in [0, 1]$, such that for $\lambda = 1$, the new equations coincide with Eq. (17) with initial and boundary conditions given by (12), (13). If the solutions of this family of equations are uniformly bounded with respect to $\lambda \in [0, 1]$, the Leray–Schauder technique uses a proof of the existence of a solution to the equation when $\lambda = 0$, to also prove the existence of a solution to the equations when $\lambda = 1$—that is, to Eq. (17), with initial boundary conditions given by (13). Since the solution of Eq. (17) may be used to derive the solutions $(P(t,x), T(t,x))$ of (9–13), this establishes that the system of PDEs and initial/boundary conditions (9–13), that constitute our model, is well-posed.

Our first step is to reduce (17) subject to (13), to Leray–Schauder form, and we begin by replacing $v(t, x)$ with $v(\lambda; t, x)$ in (4), such that

$$\mathbf{v}(\lambda; t, x) = \nabla_x P(\lambda; t, x) - \lambda T(\lambda; t, x)$$

With this definition of v, Eq. (17) takes the form

$$\frac{\partial T}{\partial t}(\lambda; t, x) = \Delta_x T(\lambda; t, x) - \lambda \Psi(t, x) T(\lambda; t, x) \cdot \nabla_x T(\lambda; t, x) + h(t, x) \quad (18)$$

Let $\tilde{T}(t, x)$ be a solution of

$$\begin{aligned} \frac{\partial \tilde{T}}{\partial t}(t, x) &= \Delta_x \tilde{T}(t, x) + h(t, x) \\ \tilde{T}(\tau, x) &= \tilde{T}_\tau(x) \in W^{2(1-\frac{1}{p}),p}(\Omega); \\ \tilde{T}(t, x')|_{x' \in \partial\Omega} &= \tilde{T}_{bd}(x') \in W^{2-\frac{1}{p},p}(\partial\Omega) \end{aligned} \quad (19)$$

The initial boundary value problem (19) has a unique solution in $W^{(1,2),p}(Q_{\tau,\eta})$ (Ladyzhenskaya et al. 1968). Note that $\tilde{T}(t, x) = T(\lambda = 0; t, x)$.

Let $T^*(\lambda; t, x) = T(\lambda; t, x) - \tilde{T}(t, x)$. Then, T^* satisfies the family of equations

$$\begin{aligned} \frac{\partial T^*}{\partial t}(\lambda; t, x) &= \Delta_x T^*(\lambda; t, x) \\ &- \lambda \Psi(x, D)(T^*(\lambda; t, x) + \tilde{T}(t, x)) \cdot \nabla_x (T^*(\lambda; t, x) + \tilde{T}(t, x)) \\ T^*(\lambda; t = \tau, x) &= 0; \qquad T^*(\lambda; t, x)|_{x \in \partial\Omega} = 0 \end{aligned} \quad (20)$$

In our earlier paper, we proved the following "a priori" dissipative estimate uniformly in $\lambda \in [0, 1]$:

$$\begin{aligned} &||T^*(\lambda; t, x)||_{W^{(1,2),p}(Q_{\tau,\eta})} + ||P^*(\lambda(t, x)||_{W^{(1,3),p}(Q_{\tau,\eta})} \\ &\leq Q\big(||T_\tau(x)||_{W^{2(1-\frac{1}{p}),p}(\Omega)}\big) e^{-\alpha(t-\tau)} \\ &\quad + Q\big(||T_{bd}(x)||_{W^{2-\frac{1}{p},p}(\partial\Omega)} + ||P_{bd}(x)||_{W^{3-\frac{1}{p},p}(\partial\Omega)}\big) \end{aligned} \quad (21)$$

where $t \geq \tau, \tau \in \mathbb{R}, \alpha > 0$, and Q is a function that is independent of λ and initial data T_τ, provided that $p = p(n) >> 1$ is sufficiently large to guarantee the compactness of the embedding

$$W^{(1,2),p}(Q_{\tau,\eta}) \subset\subset C\big([\tau, \eta], C^\epsilon(\Omega)\big) \quad (22)$$

for some $\epsilon \in (0, \frac{3}{2}]$.

Inequality (21) may be used to prove that the solutions to $T(\lambda; t, x) = \tilde{T}(t, x) + T^*(\lambda; t, x)$ are uniformly bounded in λ. This uniform estimate allows us to apply the

Leray–Schauder technique to (17) with boundary and initial conditions (13), which leads to the existence of at least one solution $T(1; t, x) = T(t, x)$ in $W^{(1,2),p}(Q_{\tau,\eta})$.

To establish uniqueness of the solution in $W^{(1,2),p}(Q_{\tau,\eta})$, we assume that there are at least two distinct solutions of the system of Eqs. (9–13) in $W^{(1,2),p}(Q_{\tau,\eta}) \times W^{(1,3),p}(Q_{\tau,\eta})$. Let $(T_1(t, x), P_1(t, x))$ and $(T_2(t, x), P_2(t, x))$ be two of these solutions of (9–13) in $W^{(1,2),p}(Q_{\tau,\eta}) \times W^{(1,3),p}(Q_{\tau,\eta})$. Let $w(t, x) = T_1(t, x) - T_2(t, x)$. One can easily see $w(t, x)$ satisfies

$$\begin{cases} \partial_t w(t, x) = \Delta_x w(t, x) + \big(\nabla_x T_1(t, x)\big).\Psi(x, D)w(t, x) \\ \qquad\qquad +\big(\nabla_x w(t, x)\big).\Psi(x, D)T_2(t, x) \\ w(t, x)|_{\partial\Omega} = w(t, x)|_{t=\tau} = 0 \end{cases} \tag{23}$$

Let $p = p(n) >> 1$ be sufficiently large to guarantee compact embedding (22). As a result of this embedding, we have $||\nabla_x T_1(t, x)||_{L^\infty(\Omega)} \leq C_*$ and $||\Psi(x, D)T_2(t, x)|| = ||v_2(t, x)||_{L^\infty(\Omega)} \leq C_*$ for some constant C_*. By multiplying the PDE for $\partial_t w(t, x)$ in the form (17, 13), by $w(t, x)$ and integrating over $x \in \Omega$, we obtain (Efendiev et al. 2020)

$$\begin{aligned} &\frac{1}{2}\partial_t ||w(t, x)||^2_{L^2(\Omega)} + ||\nabla_x w(t, x)||^2_{L^2(\Omega)} \\ &\quad \leq C_1 ||w(t, x)||^2_{L^2(\Omega)} + \frac{1}{2}||\nabla_x w(t, x)||^2_{L^2(\Omega)} \end{aligned} \tag{24}$$

By integrating (24) over $[\tau, t]$ and using Gronwall's inequality (Renardy and Rogers 2006), we obtain $w(t, x) \equiv 0$, which establishes the uniqueness of this solution.

The well-posedness of (9), (13) generates a two-parameter family of maps (*a process*) $U(t, \tau) : W^{2(1-\frac{1}{p}),p}(\Omega) \to W^{2(1-\frac{1}{p}),p}(\Omega)$, mapping the initial condition $T_\tau(x)$ to $T(t, x)$ at a fixed time t.

Our next goal is to prove the existence of uniform attractors of the processes $U(t, \tau)$. To this end, we will recall some basic definitions and theorems of uniform attractors of PDEs.

3 Uniform attractor for HIFU model

In what follows, we use the methods of Chepyzhov and Vishik (1994), Chepyzhov and Efendiev (1998) to prove the existence of a uniform attractor and estimate its Hausdorff dimension. We begin with some preliminaries. Let E be a Banach space. In E, we consider the non-autonomous Cauchy problem

$$\begin{cases} \partial_t u = A(u, t) \\ u|_{t=\tau} = u_\tau, u \in E, \tau \in \mathbb{R}, \ t \geq \tau, \end{cases} \tag{25}$$

where $A(u, t) : E_1 \text{ x } \mathbb{R} \to E_0$ is a family of nonlinear operators and E_1 and E_0 are Banach spaces. We assume that the embeddings $E_1 \subset E \subset E_0$ are everywhere dense.

From our earlier work (Efendiev et al. 2020), we may assume that the initial boundary value problem (9–13) has a unique solution $u(t) \in E \forall t \geq \tau$ and $\forall \tau \in \mathbb{R}$ and $u_\tau \in E$.

Consider the two-parameter family of maps $\{U(t, \tau), t \geq \tau\}$, $U(t, \tau) : E \rightarrow E$, such that

$$U(t, \tau) : u_\tau \rightarrow u(t).$$

Definition 3.1 A family of maps $\{U(t, \tau)\}$ is called a *process* if

- $U(\tau, \tau) = Id$ (identity) and
- $U(t, s) \odot U(s, \tau) = U(t, \tau)$ for all $t \geq s \geq \tau, \tau \in \mathbb{R}$

In this paper, we are mainly interested in processes generated by non-autonomous evolution equations such as (9–13). It is clear that a process is a natural generalization of a semigroup $S_t : E \rightarrow E$, which corresponds to autonomous evolution equations; that is, $A(u, t) \equiv A_0(u)$. Note that in this case, it is easy to see that

$$U(t, \tau) = U_0(t - \tau)$$

Our main goal is to study the large-time asymptotics of a process $\{U(t, \tau)\}$ generated by Eq. (25): that is, the behaviour of trajectories $u(t) = U(t, \tau)u_\tau$ of (25) when $t - \tau$ tends to infinity. As we will see below, the large-time dynamics of a process can be described in terms of attractors. (We give a precise definition of *attractor* later.) As was shown by Chepyzhov and Vishik (1994), an adequate theory of attractors for non-autonomous equations is obtained by considering a family of processes $\{U(t, \tau)\}$, instead of a single process, where $g \in \Sigma$ is a functional parameter. For the convenience of the reader, we recall basic definitions from Chepyzhov and Vishik (1994). Consider the family of Cauchy equations

$$\partial_t u = A_{g(t)}(u), \tag{26}$$

$$u|_{t=\tau} = u_\tau, \qquad u_\tau \in E \tag{27}$$

where for any fixed $t \in \mathbb{R}$, $A_{g(t)}(u)$ is, in general, a nonlinear operator acting from a Banach space E_1 to a Banach space E_0 and $E_1 \subset E \subset E_0$ (everywhere dense). We assume that a functional parameter $g(t)$ belongs to a certain closed set Σ in $C_b(\mathbb{R}, W)$, where $\{C_b(\mathbb{R}, W)$ denotes the space of bounded continuous functions on $\mathbb{R}$ with values in a certain metric space W, with $T(h)\Sigma = \Sigma$, where

$$T(h)g(t) := g(t + h), \qquad h \in \mathbb{R} \tag{28}$$

We suppose that the problem (26)-(27) is well-posed for any symbol $g(t) \in \Sigma$, so that any solution $u(t) \in E$ (we specify in each case what we mean by "solution") can be represented as

$$\begin{aligned} u(t) &= U_g(t, \tau)u_\tau, \\ g &= g(t) \in \Sigma, u_\tau \in E, \tau \in \mathbb{R}, t \geq \tau \end{aligned} \tag{29}$$

Due to the uniqueness theorem for (26)-(27), operators defined by (29) define a process and satisfy the following translation identity:

$$U_{T(h)g}(t,\tau) = U_g(t+h,\tau+h), \qquad \forall h \geq 0,\ t \geq \tau,\ \tau \in \mathbb{R} \tag{30}$$

Let $S(t): E \times \Sigma \to E \times \Sigma$ be the family of operators defined by

$$S(t)(u,g) = (U_g(t,0)u, T(t)g), \qquad t \geq 0,\ (u,g) \in E \times \Sigma \tag{31}$$

It is not difficult to see that the family of operators $\{S(t)\}$ defined by (31) forms a semigroup on the extended phase space $E \times \Sigma$. We use this fact throughout this paper.

Before formulating the main result for a family of processes $\{U_g(t,\tau)\}$ from a dynamical viewpoint, we recall some definitions (see Chepyzhov and Vishik 1994; Hale 1988). Let E be the Banach space as above, and denote by $\beta(E)$ the set of all bounded subsets of E.

Definition 3.2 A set $B_0 \subset E$ is called a *uniformly* (with respect to Σ) *absorbing* (or *attracting*) set for the family of processes $\{U_g(t,\tau)\}$ if for any $\tau \in \mathbb{R}$ and any $B \subset \beta(E)$ there exists $T = T(\tau, B) \geq \tau$ such that

$$\underset{g\in\Sigma}{\cup} U_g(t,\tau)B \subset B_0, \qquad \forall t \geq T \text{ (absorbing property)},$$

and

$$\lim_{t\to+\infty} \sup_{g\in\Sigma} \operatorname{dist}_E\big(U_g(t,\tau)B, P\big) = 0, \qquad \forall \tau \in \mathbb{R},\ B \subset \beta(E) \text{ (attracting property)}.$$

Definition 3.3 A closed set $\mathbb{A}_\Sigma \neq \varnothing$ is called a *uniform* (with respect to Σ) *attractor* of the $\{U_g(t,\tau)\}$ if it is uniformly attracting and is contained in any closed uniformly attracting set $\tilde{\mathbb{A}}$ (minimality property).

Following (Haraux 1991), a family of processes possessing a compact, uniformly absorbing (uniformly attracting) set are called uniformly compact (or uniformly asymptotically compact) processes. Now, we are in a position to formulate the main result on the existence of attractors for a family of processes. Let $\Pi_1: E \times \Sigma \to E$ be the projection operator defined by $\Pi_1(u,g) = u$.

Theorem 3.4 *Let a family of processes $\{U(t,\tau)\}$, $g \in \Sigma$,acting in the Banach space E be uniformly compact and $(E \times \Sigma, E)$ continuous. Then, the semigroup $\{S(t)\}: E \times \Sigma \to E \times \Sigma$ defined by (31) possesses a global attractor $\mathcal{A}$.Moreover, $\mathbb{A} = \Pi_1\mathcal{A}$ is the uniform attractor of the family of processes $\{U_g\}$.*

For a proof, see Chepyzhov and Vishik (1994).

In the following section, we will apply Theorem 3.4 to our hyperthermia model in the case where the external force is quasi-periodic in t. Let $\bar{h}(t,x) \in \mathcal{H}\big(h(t,x)\big)$, where $\mathcal{H}\big(h(t,x)\big)$ denotes the convex hull of a given quasi-periodic function $h(t,x)$ of t. Note that, by definition,

$$\mathcal{H}(h(t,x)) := \overline{\{T(\theta)h(t,x):\theta \in \mathbb{R}\}}^{C_b(\mathbb{R},W)}$$

that is, the closure in $C_b(\mathbb{R}, W)$ of the set of all translations of the given quasi-periodic function h. Depending on the context, we will take $W = L_p(\Omega)$ or $W = L_\infty(\Omega)$. On the other hand, a quasi-periodic (in t) function $h(t,x)$ can be represented as

$$h(t,x) = \tilde{h}(\alpha_1 t, \ldots, \alpha_k t, x)$$

where $\tilde{h}(\omega_1, \ldots, \omega_j + 2\pi, \ldots, \omega_k, x) = \tilde{h}(\omega_1, \ldots, \omega_k, x)$ and the numbers $\alpha_1, \ldots, \alpha_k$ are rationally independent. When $\tilde{h}(\omega_1, \ldots, \omega_k, x)$ is a continuous function on $\mathbb{T}^k$ (k-dimensional torus), one can easily see that the hull $\mathcal{H}(h)$ is a set

$$\begin{aligned}\mathcal{H}\big(h(t,x)\big) = \{&\tilde{h}(\alpha_1 t + \omega_{10}, \ldots, \alpha_k t + \omega_{k0}, x);\\ &\omega_0 = (\omega_{10}, \ldots, \omega_{k0}) \in \mathbb{T}^k\}\end{aligned} \tag{32}$$

Thus, given (32), it is reasonable to consider the torus $\mathbb{T}^k$ as the symbol space through the map

$$\begin{aligned}\mathbb{T}^k \ni \omega_0 \to\ & h(\alpha t + \omega_0, x)\\ &:= \tilde{h}(\alpha_1 t + \omega_{10}, \ldots, \alpha_k t + \omega_{k0}, x) \in C_b(\mathbb{R}, W).\end{aligned} \tag{33}$$

We set

$$T(\theta)\omega_0 = [\omega_0 + \alpha\theta] := \omega_0 + \alpha\theta \ (\mathrm{mod}\ \mathbb{T}^k) \tag{34}$$

Clearly, $T(\theta)\mathbb{T}^k = \mathbb{T}^k$.

We will apply Theorem 3.4 to the family of processes generated by (9), (13).

4 Long-time dynamics of solutions for hyperthermia model

In this section, we prove the existence of a uniform attractor of our model Eqs. (3–7).

$$\begin{cases}\nabla_x \cdot \mathbf{v}(t,x) = 0, & x \in \Omega, t \geq \tau\\ \mathbf{v}(t,x) = \nabla_x P(t,x) - T(t,x), & x \in \Omega, t \geq \tau\\ \frac{\partial T}{\partial t}(t,x) = \Delta_x T(t,x) - \mathbf{v}(t,x).\nabla_x T(t,x) + h(t,x) & x \in \Omega, t \geq \tau\\ P(t,x)|_{x \in \partial\Omega} = P_{bd}(x) & x \in \Omega\\ T(0,x) = T_o(x); \qquad T(t,x')|_{x' \in \partial\Omega} = T_{bd}(x') & x \in \Omega\end{cases}$$

where $h(t, x)$ is a quasi-periodic external source, Ω is a bounded domain in $\mathbb{R}^n$ with a sufficiently smooth boundary, $\big(T(t, x), P(t, x)\big)$ are unknown functions (temperature and pressure, respectively), and $= (\gamma_1, \ldots, \gamma_N)$ is a given constant vector, say $|| = 1$.

As already indicated above, we will apply Theorem 3.4 to the family of Cauchy problems

$$\begin{cases} \nabla_x \cdot \mathbf{v}(t,x) = 0 & x \in \Omega, t \geq \tau \\ \mathbf{v}(t,x) = \nabla_x P(t,x) - T(t,x), & x \in \Omega, t \geq \tau \\ \frac{\partial T}{\partial t}(t,x) = \Delta_x T(t,x) - \mathbf{v}(t,x).\nabla_x T(t,x) + \bar{h}(t,x) & x \in \Omega, t \geq \tau \\ P(t,x)|_{x \in \partial\Omega} = P_{bd}(x), & x \in \Omega \\ T(0,x) = T_o(x); \qquad T(t,x')|_{x' \in \partial\Omega} = T_{Bd}(x') & x \in \Omega \end{cases} \tag{35}$$

with $\bar{h}(t, x) \subset \mathcal{H}\text{ull}(h(t, x))$.

To apply Theorem 3.4 in our case, we have to prove the existence of a uniformly absorbing (uniformly attracting) set. To this end, let us recall that our problem (17), (13) generates a family of processes $\{\mathbb{U}_{\omega_o}(t, \tau)|t \geq \tau\}$, $\omega_o \in \mathbb{T}^k$ in the space $\mathbb{V}_0 = W^{2-\frac{1}{p},p}(\Omega) \cap \{T(t, x)|_{x \in \partial\Omega} = T_{bd}(x)\}$. Specifically,

$$\begin{aligned} \mathbb{U}_{\omega_o}(t,\tau) &: \mathbb{V}_0 \to \mathbb{V}_0 \qquad t \geq \tau \\ \mathbb{U}_{\omega_o}(t,\tau) &: T_\tau(x) \to T(t,x) \end{aligned} \tag{36}$$

where $T(t, x)$ is a solution of (35) where $\tilde{h}$ takes the form $\tilde{h}(\alpha t + \omega_0, x) = \tilde{h}(\alpha_1 t + \omega_{10}, \ldots, \alpha_k t + \omega_{k0}, x)$.

Proposition 4.1 *The family of processes constructed above for our model (35), that is, $\{\mathbb{U}_{\omega_o}(t, \tau) \mid t \geq \tau, \tau \in \mathbb{R}, \omega_o \in \mathbb{T}^k\}$, $\mathbb{U}_{\omega_o}(t, \tau) : \mathbb{V}_0 \to \mathbb{V}_0$ is uniformly bounded and uniformly compact in $\mathbb{V}_0$.*

Proof In our previous paper (Efendiev et al. 2020), we obtained the following dissipative estimate:

$$\begin{aligned} &||T(t,x)||_{W^{(1,2),p}(Q_{\tau,\eta})} + ||P(t,x)||_{W^{(1,3),p}(Q_{\tau,\eta})} \\ &\quad \leq Q\Big(||T_\tau(x)||_{\mathbb{V}_0}\Big)e^{-\alpha(t-\tau)} + Q(||T_{bd}(x')||_{W^{2-\frac{1}{p},p}(\partial\Omega)} \\ &\qquad + ||P_{bd}(x')||_{W^{3-\frac{1}{p},p}(\partial\Omega)} + C||\bar{h}(t,x)||_{C\left(\mathbb{T}^k, L^p(\Omega)\right)}, \alpha > 0 \end{aligned} \tag{37}$$

where constant $C, \alpha > 0$, and the monotonic function Q are independent of $T_\tau(x)$. From this dissipative estimate, it follows that

$$||\mathbb{U}_{\omega_o}(t,\tau)T_\tau(x)||_{\mathbb{V}_0} \leq C_3\big(||T_\tau(x)||_{\mathbb{V}_0}\big) \tag{38}$$

where constant C_3 depends on initial data $||T_\tau(x)||_{\mathbb{V}_0}$. This proves uniform boundedness of the process $\mathbb{U}_{\omega_o}(t,\tau)$ with respect to $\omega_o \in \mathbb{T}^k$. Moreover, the same dissipative estimate also implies that the set

$$\mathbb{B}_1 = \{\psi(x) \in \mathbb{V}_0 \mid ||\psi(x)||_{\mathbb{V}_0} \le C_*\} \tag{39}$$

(where C_* is a sufficiently large constant depending on $||T_{bd}(x)||_{W^{2-\frac{1}{p},p}(\Omega)}$ and $\sup_{\omega_o\in\mathbb{T}^k} ||\tilde{h}([\omega_o + \alpha t], x)||_{L^p(\Omega)}$, but independent of $\omega_o \in \mathbb{T}^k$) is a uniformly absorbing set for the family $\{\mathbb{U}_{\omega_o}\} \in \mathbb{V}_0$.

To apply Theorem 3.4, we need to show the existence of a compact absorbing set in $\mathbb{V}_0$. However, the set $\mathbb{B}_1$ is not compact in $\mathbb{V}_0$. To obtain a compact absorbing set, consider the new set

$$\mathbb{B}_{abs} := \bigcup_{\omega_o\in\mathbb{T}^k} \bigcup_{\tau\in\mathbb{R}} \mathbb{U}_{\omega_o}(\tau+1,\tau)\mathbb{B}_1 \tag{40}$$

□

Proposition 4.2 *The set $\mathbb{B}_{abs} \subset \mathbb{V}_0$ defined by Eq. (40) is a compact set in $\mathbb{V}_0$.*

Proof To show this, we consider

$$\hat{T}(t,x) = (t-\tau)\big[T(t,x) - T^*(x)\big] \tag{41}$$

where $T^*(x)$ is a solution of

$$\begin{cases} \Delta_x T^*(x) = 0, & x \in \Omega \\ T^*(x')|_{x'\in\partial\Omega} = T_{bd}(x') \in W^{2-\frac{1}{p},p}(\partial\Omega) \end{cases} \tag{42}$$

From elliptic regularity theory (Efendiev 2009; Nirenberg 1974; Renardy and Rogers 2006), it follows that $T^*(x) \in W^{2,p}(\Omega)$. Obviously,

$$\begin{cases} \frac{\partial\hat{T}}{\partial t} = \Delta_x\hat{T} + (t-\tau)\nabla_x T(t,x)\cdot\Psi(x,D)T(t,x) - T^*(x) - \bar{h}(t,x) \\ \hat{T}(t=\tau,x) = 0, \qquad \hat{T}(t,x')|_{x'\in\partial\Omega} = 0 \end{cases} \tag{43}$$

Here, we use the fact that $T(t,x)$ is a solution of

$$\begin{cases} \partial_t T(t,x) - \Delta_x T(t,x) = -\nabla T(t,x)\cdot\Psi(x,D)T(t,x) + \bar{h}(t,x) \\ T(t=\tau,x) = T_\tau(x) \in \mathbb{B}_1 \\ T(t,x')|_{x'\in\partial\Omega} = T_{bd}(x') \in W^{2-\frac{1}{p},p}(\partial\Omega) \end{cases} \tag{44}$$

□

Remark 4.3 Here, we impose additional requirements on on the right-hand side of (44). Namely, $\bar{h}(t,x) \in L^\infty(Q_{\tau,\eta})$.

Let $h^*(t,x) := (t-\tau)\nabla_x T(t,x)\Psi(x,D)T(t,x) - T^*(x) - \bar{h}(t,x)$ and $T_\tau(x) \in \mathbb{B}_1$. Then, due to $||\nabla_x T||_{L^\infty(Q_{\tau,\eta})} \le C$, $||v(t,x)||_{L^\infty(Q_{\tau,\eta})} \le C$, we obtain $h^*(t,x) \in L^\infty(Q_{\tau,\eta})$. Consequently, for any $q > p$, since $\bar{h}(t,x) \in L^\infty(Q_{\tau,\eta}) \subset L^q(Q_{\tau,\eta})$, we obtain that $h^*(t,x) \in L^q(Q_{\tau,\eta})$.

$$||\hat{T}(t,x)||_{W^{(1,2),q}(Q_{\tau,\eta})} \le Q_2(||T_{bd}(x')||_{W^{2-\frac{1}{q},q}(\partial\Omega)} + ||P_{bd}(x')||_{W^{3-\frac{1}{q},q}(\partial\Omega)})$$

where Q_2 is a monotonicity function. Consequently, the set $\mathbb{U}_{\omega_o}(t,\tau)\mathbb{B}_1 - T^*(x)$ is bounded in $\mathbb{V}_0$. Since $T^*(x) \in W^{2,p}(\Omega)$ where $p >> 1$ and $q > p$, then $\mathbb{B}_{abs} = \mathbb{U}_{\omega_o}(t,\tau)(t,\tau)\mathbb{B}_1$ is compact in $\mathbb{V}_0$. Here, we use our new assumption $\bar{h}(t,x) \in L^\infty(Q_{\eta,\tau})$ for the first time. Hence, Proposition 4.2 is proved, and as a consequence, we obtain that the family of processes $\{\mathbb{U}_{\omega_o}(t,\tau) | t \ge \tau\}$ possesses a uniform attractor in $\mathbb{V}_0$.

We denote by $\mathbb{A}^{un}_{HT} \subset \mathbb{V}_0$ (hyperthermia) a uniform attractor for a family of processes $\{\mathbb{U}_{\omega_o}(t,\tau), t \ge \tau\}$, $\omega_o \in \mathbb{T}^k$. In the next section, we prove an estimate for the Hausdorff dimension of the uniform attractor $\mathbb{A}^{un}_{HT}$ in $\mathbb{V}_0$.

Definition 4.4 *Let E be a Banach space, let $Y \subset E$ be a compact subset of E, and let $d, \epsilon \in \mathbb{R}_+$. Let the sets $\{B_{r_i}(y_i)$ be all coverings of Y* for $r_i \le \epsilon$ and $y_i \in Y$. *The d -dimensional Hausdorff measure of Y, $\mu(Y,d)$, is defined as*

$$\mu(Y,d) = \lim_{\epsilon\to 0^+} \inf_{\{B_{r_i}(y_i)\}} \sum_i r_i^d = \sup_{\epsilon\to 0^+} \inf_{\{B_{r_i}(y_i)\}} \sum_i r_i^d \tag{45}$$

Then, the number

$$\dim_{\text{Haus}}(Y,E) = \inf\{d \mid \mu(Y,d) = 0\} \tag{46}$$

is called the Hausdorff dimension of Y in E (Temam 2012).

Note that $\mathbb{U}_{\omega_o}(t,\tau) : \mathbb{V}_0 \to \mathbb{V}_0$ generates the semigroup

$$\begin{aligned} &S(t){:}\mathbb{V}_0 \times \mathbb{T}^k \to \mathbb{V}_0 \times \mathbb{T}^k \\ &S(t)(T_o(x),\omega_o) = \left(\mathbb{U}_{\omega_o}(t,0)T_o(x), [\alpha t + \omega_o]\right) \\ &T_o(x) \in \mathbb{V}_0, \omega_o \in \mathbb{T}^k, t \ge 0 \end{aligned} \tag{47}$$

which in turn corresponds to the following autonomous dynamical system

$$\begin{cases} \partial_t T(t,x) = \Delta_x T(t,x) - \nabla T(t,x) \cdot \Psi(x,D)T(t,x) + \bar{h}(t,x) \\ \partial_t \omega(t) = \alpha \\ T(0,x) = T_o(x), \qquad \omega(0) = \omega_o, \qquad T(t,x')|_{x'\in\partial\Omega} = T_{bd}(x') \end{cases} \tag{48}$$

where $T_\tau(x) \in \mathbb{V}_0$, $T_{bd}(x) \in W^{2-\frac{1}{p},p}(\partial\Omega)$, $\omega_o \in \mathbb{T}^k$, and $\Psi(x, D)$ is defined as in (16), or

$$\begin{cases} \partial_t y(t) = My(t), \qquad y(t)|_{t=0} = y_o \\ y(t) := \big(T(t,x), \omega(t)\big) \in \mathbb{V}_0 \times \mathbb{T}^k \\ M(y(t)) := (\Delta_x T(t,x) - \nabla T(t,x) \cdot \Psi(x,D)T(t,x) + \bar{h}(\omega(t),x), \alpha) \end{cases} \tag{49}$$

where $\omega(t) = [\alpha t + \omega_o]$, $\omega_o \in \mathbb{T}^k$ is a transformation of ω_0 on the torus $\mathbb{T}^k$, and $M : \mathbb{V}_0 \times \mathbb{T}^k \to \mathbb{V}_0 \times \mathbb{T}^k$. Since the process $\{\mathbb{U}_{\omega_o}(t,\tau),\ t \geq \tau\}$ defined by (36) is uniformly compact, then due to Theorem 3.4 the semigroup $S(t)$ defined by

$$S(t)(T_o(x), \omega_o) = (\mathbb{U}_{\omega_o}(t,0)T_o(x), [\alpha t + \omega_o])$$

possesses a global attractor $\mathcal{A}_{HT}$ in $\mathbb{V}_0 \times \mathbb{T}^k$. Moreover, the projection $\Pi_1 \mathcal{A}_{HT}$ is the uniform attractor of the processes $\mathbb{U}_{\omega_o}(t,\tau) : \mathbb{V}_0 \to \mathbb{V}_0$ defined by $\mathbb{U}_{\omega_o}(t,\tau)T_\tau(x) = T(t,x)$, where $T(t,x)$ is a solution of (1) and $\Pi_1 : \mathbb{V}_0 \times \mathbb{T}^k \to \mathbb{V}_0$, $\Pi_1\big(\xi(x), \nu\big) = (\xi(x), 0)$. We denote by $\mathbb{A}^{un}_{HT} = \Pi_1 \mathcal{A}_{HT}$.

Obviously, $\dim_{\text{Haus}}(\mathbb{A}^{un}_{HT}, L^2(\Omega)) \leq \dim_{\text{Haus}}(\mathcal{A}_{HT}, L^2(\Omega) \times \mathbb{T}^k)$, where $\dim_{\text{Haus}}(\cdot,\cdot)$ denotes the Hausdorff dimension as defined in Definition 4.5. Hence, to obtain an estimate for $\dim_{\text{Haus}}(\mathbb{A}^{un}_{HT}, L^2(\Omega))$, it is sufficient to obtain an upper bound for $\dim_{\text{Haus}}(\mathcal{A}_{HT}, L^2(\Omega) \times \mathbb{T}^k)$. Note that $\mathcal{A}_{HT} \subset \mathbb{V}_0 \times \mathbb{T}^k = W^{2(1-\frac{1}{p}),p}(\Omega) \times \mathbb{T}^k \subset L^2(\Omega) \times \mathbb{T}^k$. To this end, we will assume that $\tilde{h} \in C^1(\mathbb{T}^k, L^\infty(\Omega))$, and in addition, use a well-known formula of Constantin, Foias and Temam (Temam 2012).

Let $\mathbb{E}_d$ be any d-dimensional subspace in the Hilbert space $L^2(\Omega) \times \mathbb{R}^k$ containing some orthonormal family in $L^2(\Omega) \times \mathbb{R}^k$, $z_j \in \mathbb{E}_d$, $j = 1, 2, \ldots, d$, belonging to $H^2(\Omega) \times \mathbb{R}^k$ such that $\big(z_i(x), z_j(x)\big)_{L^2(\Omega)\times\mathbb{R}^k} = \delta_{ij}, i, j = 1, \ldots, d$, where $z_j(x) = (\theta_j(x), \nu_j) \in L^2(\Omega) \times \mathbb{R}^k$ and $\theta_j(x)|_{\partial\Omega} = 0$. Let $y(t) = \big(T(t,x), \omega(t)\big)$ where $T(t,x) = \mathbb{U}_{\omega_o}(t,0)T_o(x)$ is a solution of (35) and $\omega(t) = [\alpha t + \omega_o]$, $y(0) \in \mathcal{A}_{HT}$. If

$$\lim_{t\to\infty} \frac{1}{t} \int_0^t \sup_{\mathbb{E}_d} \sum_{j=1}^d \Big(M'\big(y(\tau)\big)z_j, z_j\Big) \mathrm{d}\tau < 0 \tag{50}$$

holds, then due to the aforementioned formula of Constantin, Foias and Temam,

$$\dim_{\text{Haus}}(\mathcal{A}_{HT}, L^2(\Omega)) \leq d \tag{51}$$

By $M'\big(y(t)\big)$, we denote a quasi-differential of the mapping M defined by (49)

$$\begin{cases} M'\big(y(t)\big)z(x) = \Delta_x\theta(x) - \nabla_x T(t,x) \cdot \Psi(x,D)\theta(x) \\ \qquad\qquad -\nabla_x\theta(x) \cdot \Psi(x,D)T(t,x) + \tilde{h}'\big(\omega(t),x\big)\nu \\ z(x) = \big(\theta(x), \nu\big) \in L^2(\Omega) \times \mathbb{R}^k \\ \tilde{h}'\big(\omega(t),x\big) := \Big(\frac{\partial \tilde{h}}{\partial \omega_1}, \ldots, \frac{\partial \tilde{h}}{\partial \omega_k}\Big) \end{cases} \tag{52}$$

with $\theta(x)|_{\partial\Omega} = 0$. In (50), we take this quasi-differential of M at the point $y(\tau) \in \mathcal{A}_{HT}$. Hence,

$$
\begin{aligned}
&\Big(M'\big(y(t)\big)z(x), z(x)\Big) \\
&= -\|\nabla_x \theta(x)\|^2 - \big(\nabla_x T(t,x)\cdot \Psi(x,D)\theta(x), \theta(x)\big) - \big(\nabla_x \theta(x)\cdot \Psi(x,D) T(t,x), \theta(x)\big) \\
&\qquad + \int_\Omega \Big(\tilde{h}'\big(\omega(t), x\big), \nu\Big)_{\mathbb{R}^k} \cdot \theta(x)\mathrm{d}x \\
&= -\|\nabla_x \theta(x)\|^2 + \big(T(t,x), \Psi(x,D)\theta(x)\nabla_x\theta(x)\big) + \int_\Omega \Big(\tilde{h}'\big(\omega(t), x\big), \nu\Big)_{\mathbb{R}^k} \cdot \theta(x)\mathrm{d}x \\
&= -\|\nabla_x \theta(x)\|^2 + \big(T(t,x), \nabla_x P(t,x)\cdot\theta(x)\nabla_x\theta(x)\big) - \big(T(t,x), \gamma\theta(x)\nabla_x\theta(x)\big) \\
&\qquad + \int_\Omega \Big(\tilde{h}'\big(\omega(t), x\big), \nu\Big)_{\mathbb{R}^k} \cdot \theta(x)\mathrm{d}x \\
&\le -\frac{1}{2}\|\nabla_x\theta(x)\|^2 \\
&\qquad + C^{**}\|\theta(x)\|^2 + \int_\Omega |\tilde{h}'\big(\omega(t), x\big)|^{\frac{1-\delta}{2}} \cdot |\tilde{h}'\big(\omega(t), x\big)|^{\frac{1+\delta}{2}} \cdot |\nu| \cdot |\theta(x)|\mathrm{d}x \\
&\le -\frac{1}{2}\|\nabla_x\theta(x)\|^2 + C^{**}\|\theta(x)\|^2 + \frac{b}{2}\int_\Omega |\tilde{h}'(\omega(t), x)|^{1-\delta}|\theta(x)|^2\mathrm{d}x \\
&\qquad + \frac{|\nu|^2_{\mathbb{R}^k}}{2b}\int_\Omega |\tilde{h}'(\omega(t), x)|^{1+\delta}\mathrm{d}x
\end{aligned} \tag{53}
$$

where $0 < \delta < 1$, b is an arbitrary positive number, and C^{**} is a constant dependent on $\|T_{bd}(x')\|$ but independent of ω_o.

Let $G_1 := \max_{\omega\in\mathbb{T}^k} \int_\Omega |\tilde{h}'(\omega(t), x)|^{1+\delta}dx$. Hence, for any $z = (\theta(x), \nu) \in L^2(\Omega)\times\mathbb{R}^k$ where $\theta(x)|_{x\in\partial\Omega} = 0$.

$$
\begin{aligned}
&(Az, z) := (A_1\theta,\ \theta) + (A_2\nu,\ \nu) \le \\
&\le -\frac{1}{2}\|\nabla_x\theta(x)\|^2 + C^{**}\|\theta(x)\|^2 + \frac{b}{2}\int_\Omega |\tilde{h}'(\omega(t), x)|^{1-\delta}|\theta(x)|^2 dx + |\nu|^2_{\mathbb{R}^k}\frac{G_1}{2b}
\end{aligned} \tag{54}
$$

where A_j, $j = 1, 2$, are quadratic forms defined by

$$
\begin{cases}
(A_1\theta, \theta) = -\frac{1}{2}\|\nabla_x\theta(x)\|^2 + C^{**}\|\theta(x)\|^2 + \frac{b}{2}\int_\Omega |\tilde{h}'(\omega(t), x)|^{1-\delta}|\theta(x)|^2\mathrm{d}x \\
(A_2\nu, \nu) = \frac{G_1}{2b}(\nu, \nu)
\end{cases} \tag{55}
$$

Let us recall that our basic goal is to estimate the expression

$$
\sum_{j=1}^{d}\Big(M'\big(y(t)\big)z_j, z_j\Big)_{L^2(\Omega)\times\mathbb{R}^k}
$$

where $z_j = (\theta_j(x), \nu_k) \in L^2(\Omega)\times\mathbb{R}^k$ with $(z_i, z_j) = \delta_{ij}$, $i, j = 1, 2, \ldots, d$.

Proposition 4.5 *Let* $z_j = (\theta_j(x), \nu_j)$, $j = 1, 2, \ldots, d$, *be any orthonormal system in* $L^2(\Omega) \times \mathbb{R}^k$. *Then, there exists some integer* k_1 *such that* $0 \leq k_1 \leq k$, *some orthonormal (in* $L^2(\Omega)$*) vectors,* $\bar{\theta}_1, \ldots, \bar{\theta}_{d-k_1}$, *and some orthonormal (in* $\mathbb{R}^k$*) vectors,* $\bar{\nu}_1, \ldots, \bar{\nu}_{k_1}$, *such that*

$$\sum_{j=1}^{d}(Az_j, z_j) \leq \sum_{i=1}^{d-k_1}(A_1\bar{\theta}_i, \bar{\theta}_i) + \sum_{m=1}^{k_1}(A_2\bar{\nu}_m, \bar{\nu}_m) \tag{56}$$

Proof Let $\mathbb{E} \subset \mathbb{L}^2(\Omega) \times R^k$ be the subspace of the form

$$\mathbb{E} = \{\beta_1\theta_1(x) + \cdots + \beta_d\theta_d(x)\} \times \mathbb{R}^k \tag{57}$$

where $\beta_j \in \mathbb{R}^1$, $j = 1, 2, \ldots, d$. In $\mathbb{E}$, there is a scalar product induced from $L^2(\Omega) \times \mathbb{R}^k$. Consider the restriction of (Az, z) to $\mathbb{E}$. Note that $z_j = (\theta_j(x), \nu_j) \in \mathbb{E}$ and is orthonormal in $\mathbb{E}$. Then,

$$\begin{aligned}
&\big(A(\beta_1\theta_1 + \cdots + \beta_d\theta_d, \nu), (\beta_1\theta_1 + \cdots + \beta_d\theta_d, \nu)\big)\\
=&A_1(\beta_1\theta_1 + \cdots + \beta_d\theta_d, \beta_1\theta_1 + \cdots + \beta_d\theta_d) + (A_2\nu, \nu)\\
=&\sum_{i,j=1}^{d} \big(A_1\theta_i(x), \theta_j(x)\big) \cdot \beta_i\beta_j + (A_2\nu, \nu)\\
=&(\mathbb{B}\beta, \beta) + (A_1\theta, \theta)
\end{aligned} \tag{58}$$

From (58), it follows that the operator A is block diagonal, that is,

$$A = \begin{pmatrix} \mathbb{B} & 0 \\ 0 & A_2 \end{pmatrix}$$

and from linear algebra, it is known that A can be transformed into diagonal form

$$A = \begin{pmatrix} \lambda_1 & & & & & \\ & \ddots & & & 0 & \\ & & \lambda_d & & & \\ & & & \nu_1 & & \\ & 0 & & & \ddots & \\ & & & & & \nu_k \end{pmatrix}$$

by orthogonal transformations in $L^2(\Omega)$ and $\mathbb{R}^k$, respectively. Let $\bar{\theta}_1(x), \ldots, \bar{\theta}_d(x)$ and $\bar{\nu}_1, \ldots, \bar{\nu}_k$ be orthonormal in $L^2(\Omega)$ and $\mathbb{R}^k$ eigenvectors of $\mathbb{B}$ and A_2, respectively. Obviously, orthonormal eigenvectors of $A : L^2(\Omega) \times \mathbb{R}^k \to L^2(\Omega) \times \mathbb{R}^k$ have the form $(\bar{\theta}_i, 0)$ and $(0, \bar{\nu}_m)$, $i = 1, \ldots, d$, $m = 1, \ldots, k$. We denote $\zeta_i(x) = (\bar{\theta}_i, 0)$, $\zeta_m = (0, \bar{\nu}_m)$. Then, due to Courant's principle (Chepyzhov and Efendiev 2000), we have

$$\sum_{j=1}^{d}\left(Az_j, z_j\right) \leq \sum_{j=1}^{d}\left(A\zeta_j, \zeta_j\right) \tag{59}$$

where $\{\zeta_j(x)\}$, $j = 1, \ldots, d$ are eigenvectors of the matrix A in $\mathbb{E}$ corresponding to the greater eigenvalues of the block operator A. Without loss of generality, we assume that (due to the block structure A) eigenvectors of A are $(\bar{\theta}_1, 0), \ldots, (\bar{\theta}_{d-k_1}, 0), (0, \bar{v}_1), \ldots, (0, \bar{v}_{k_1})$, where $0 \leq k_1 \leq k$, $k_1 \in \mathbb{N}$. □

Corollary 4.6 *There exist vectors orthonormal in $L^2(\Omega)$ $\bar{\theta}_1, \ldots, \bar{\theta}_{d-k_1}$, $0 \leq k_1 \leq k$, such that*

$$\begin{aligned}
&\sum_{j=1}^{d}(Az_j, z_j) \leq \\
&\leq \sum_{j=1}^{d-k_1}\left(A_1\bar{\theta}_j, \bar{\theta}_j\right) + \sum_{m=1}^{k_1}\left(A_2\bar{v}_m, \bar{v}_m\right) \\
&\leq -\frac{1}{2}\sum_{j=1}^{d-k_1}||\nabla_x\bar{\theta}_j(x)||^2 + C^{**}\sum_{j=1}^{d-k_1}||\bar{\theta}_j(x)||^2 \\
&\qquad + \frac{b}{2}\sum_{j=1}^{d-k_1}\int_\Omega |h'(\omega(t), x)|^{1-\delta}|\theta_j(x)|^2\mathrm{d}x + \frac{G_1}{2b}k_1 \\
&\leq -\frac{1}{2}\sum_{j=1}^{d-k_1}||\nabla_x\bar{\theta}_j(x)||^2 + C^{**}\sum_{j=1}^{d-k_1}||\bar{\theta}_j(x)||^2 \\
&\qquad + \frac{b}{2}\int_\Omega |\bar{h}'(\omega(t), x)|^{1-\delta}\sum_{j=1}^{d-k_1}|\theta_j(x)|^2\mathrm{d}x + \frac{G_1}{2b}k \\
&\leq -\frac{1}{2}\sum_{j=1}^{d-k_1}||\nabla_x\bar{\theta}_j(x)||^2 + C^{**}\sum_{j=1}^{d-k_1}||\bar{\theta}_j(x)||^2 \\
&\qquad + \frac{b}{2}\int_\Omega |\bar{h}'(\omega(t), x)|^{1-\delta}\Big(\sum_{j=1}^{d-k_1}|\theta_j(x)|^2\Big)\mathrm{d}x + \frac{G_1}{2b}k \\
&\leq -\frac{1}{4}\sum_{j=1}^{d-k_1}||\nabla_x\bar{\theta}_j(x)||^2 - \frac{1}{4}\sum_{j=1}^{d-k_1}||\nabla_x\bar{\theta}_j(x)||^2 + C^{**}\sum_{j=1}^{d-k_1}||\bar{\theta}_j(x)||^2 \\
&\qquad + \frac{b}{2}\int_\Omega |\tilde{h}'(\omega(t), x)|^{1-\delta}\rho(x)\mathrm{d}x + \frac{G_1}{2b}k
\end{aligned} \tag{60}$$

where $\rho(x) := \sum_{j=1}^{d-k_1} |\theta_j(x)|^2$. *On the other hand,*

$$\int_\Omega |h'(\omega(t),x)|^{1-\delta}\rho(x)\mathrm{d}x = \int_\Omega \Big(\frac{1}{\epsilon}|h'(\omega(t),x)|^{1-\delta}\Big)\Big(\epsilon\rho(x)\Big)\mathrm{d}x$$

Hence,

$$\begin{aligned}\int_\Omega |\tilde{h}'(\omega(t),x)|^{1-\delta}\rho(x)\mathrm{d}x &\le \frac{n}{n+2}\int_\Omega \Big(\epsilon\rho(x)\Big)^{1+\frac{2}{n}}\mathrm{d}x \\ &+ \frac{2}{n+2}\int_\Omega \Big(\frac{1}{\epsilon}|\tilde{h}'(\omega(t),x)|^{1-\delta}\Big)^{1+\frac{n}{2}}\mathrm{d}x \\ &\le \frac{n}{n+2}\epsilon^{1+\frac{2}{n}}\int_\Omega \big(\rho(x)\big)^{1+\frac{2}{n}}\mathrm{d}x + \frac{2}{n+2}\big(\frac{1}{\epsilon}\big)^{1+\frac{n}{2}}G_2\end{aligned} \tag{1}$$

where $G_2 = \max_{\omega\in\mathbb{T}^k}\int_\Omega |\tilde{h}'(\omega(t),x)|^{(1-\delta)\frac{n+2}{2}}$. *Hence, due to Proposition 4.4,*

$$\begin{aligned}&\sum_{j=1}^{d}\Big(M'\big(y(\tau)\big)z_j, z_j\Big) \le \\ &\le -\frac{1}{4}\sum_{j=1}^{d-k_1}||\nabla_x\theta_j||^2 + \frac{G_1}{2b}k + \int_\Omega C^{**}\rho(x)\mathrm{d}x - \frac{1}{4}C_0\int_\Omega \big(\rho(x)\big)^{1+\frac{2}{n}}\mathrm{d}x \\ &\quad + \frac{bn}{2(n+2)}\int_\Omega \epsilon^{\frac{n+2}{n}}\big(\rho(x)\big)^{1+\frac{2}{n}}\mathrm{d}x + \frac{b}{n+2}\big(\frac{1}{\epsilon}\big)^{\frac{n+2}{2}}G_2 \\ &\le -\frac{1}{4}\sum_{j=1}^{d-k_1}||\nabla_x\theta_j||^2 + \int_\Omega \Big(\big(-\frac{C_0}{4} + \frac{bn}{2(n+2)}\epsilon^{\frac{n+2}{n}}\big)\big(\rho(x)\big)^{1+\frac{2}{n}} + C^{**}\rho(x)\Big)\mathrm{d}x \\ &\quad + \frac{G_1}{2b}k + \frac{b}{n+2}\big(\frac{1}{\epsilon}\big)^{\frac{n+2}{2}}G_2\end{aligned}$$

where C_0 *is a constant from the Lieb–Thirring inequality,*

$$\int_\Omega \sum_{j=1}^{d-k_1}|\nabla_x w_j|^2\mathrm{d}x \ge C_0\int_\Omega \big(\sum_{j=1}^{d-k_1}|w_j|^2\big)^{1+\frac{2}{n}}\mathrm{d}x \tag{61}$$

Choosing $\epsilon << 1$ *such that*

$$\frac{nb}{2(n+2)}\epsilon^{\frac{n+2}{n}} = \frac{C_0}{8} \tag{62}$$

Then,

$$\sum_{j=1}^{d}\Big(M'\big(y(\tau)\big)z_j, z_j\Big) \leq$$
$$\leq -\frac{1}{4}\sum_{j=1}^{d-k_1}||\nabla_x\theta_j||^2 + \int_\Omega\Big(-\frac{C_0}{8}\big(\rho(x)\big)^{1+\frac{2}{n}} + C^{**}\rho(x)\Big)\mathrm{d}x$$
$$+\frac{G_1}{2b}k + \frac{b}{n+2}\big(\frac{1}{\epsilon}\big)^{\frac{n+2}{2}}G_2 \tag{63}$$

By using the extremum test on the variable $\xi = \rho(x) \geq 0$*, it can be shown that the integrand in (63)* $-\frac{C_0}{8}\big(\rho(x)\big)^{1+\frac{2}{n}} + C^{**}\rho(x)$ *obtains its maximum value at*

$$\rho(x) = (2C^{**})^{\frac{n}{2}}\Big(\frac{4n}{C_0(n+2)}\Big)^{\frac{n}{2}} = (2C^{**})^{\frac{n}{2}}C_4 \tag{64}$$

Thus,

$$\sum_{j=1}^{d}\Big(M'\big(y(\tau)\big)z_j, z_j\Big) \leq$$
$$\leq -\frac{1}{4}\sum_{j=1}^{d-k_1}||\nabla_x\theta_j||^2 + \int_\Omega (2C^{**})^{\frac{n}{2}}C_4\mathrm{d}x + \frac{G_1}{2b}k + \frac{b}{n+2}\big(\frac{1}{\epsilon}\big)^{\frac{n+2}{2}}G_2$$
$$\leq -\frac{1}{4}\sum_{j=1}^{d-k_1}||\nabla_x\theta_j||^2 + \frac{G_1}{2b}k + \frac{b}{n+2}\big(\frac{1}{\epsilon}\big)^{\frac{n+2}{2}}G_2 + |\Omega|(2C^{**})^{\frac{n}{2}}C_4$$
$$\leq -\frac{\lambda_1}{4}(d-k_1) + \frac{G_1}{2b}k + \frac{b}{n+2}\big(\frac{1}{\epsilon}\big)^{\frac{n+2}{2}}G_2 + |\Omega|(2C^{**})^{\frac{n}{2}}C_4$$
$$\leq -\frac{\lambda_1}{4}(d-k) + \frac{G_1}{2b}k + \frac{b}{n+2}\big(\frac{1}{\epsilon}\big)^{\frac{n+2}{2}}G_2 + |\Omega|(2C^{**})^{\frac{n}{2}}C_4$$
$$<0$$

This implies that

$$dim_{Haus}\mathcal{A}_{HT} \leq k + \frac{|\Omega|}{\lambda_1}(2C^{**})^{\frac{n}{2}}C_4 + \frac{G_1}{2b\lambda_1}k + \frac{G_2 b}{\lambda_1(n+2)}\big(\frac{1}{\epsilon}\big)^{\frac{n+2}{2}} \tag{65}$$

To simplify this inequality further, we set (62),

$$\epsilon^{\frac{n+2}{n}} = \frac{C_0 2(n+2)}{8nb}$$
$$\epsilon = \Big(\frac{C_0(n+2)}{4n}\Big)^{\frac{n}{n+2}}\Big(\frac{1}{b}\Big)^{\frac{n}{n+2}}$$

$$\frac{1}{\epsilon} = \left(\frac{4n}{C_0(n+2)}\right)^{\frac{n}{n+2}} b^{\frac{n}{n+2}}$$
$$\left(\frac{1}{\epsilon}\right)^{\frac{n+2}{2}} = \left(\frac{4n}{C_0(n+2)}\right)^{\frac{n}{2}} b^{\frac{n}{2}} = C_4 b^{\frac{n}{2}}$$

and introduce the constant

$$C_4^{**} = |\Omega| C_4$$

From (65), it follows that

$$dim_{Haus}(\mathcal{A}_{HT}, L^2(\Omega) \times \mathbb{R}^k) \leq k + \frac{C_4^{**}}{\lambda_1} + \frac{G_1}{2b\lambda_1}k + C_4 \frac{G_2 b^{\frac{n+2}{2}}}{\lambda_1(n+2)} \tag{66}$$

Recall that b is an arbitrary positive constant and so we choose b such that the last two terms in (66) will be equal. This implies that

$$C_4 \frac{G_2 b^{\frac{n+2}{2}}}{\lambda_1(n+2)} = \frac{G_1}{2b\lambda_1}k$$
$$b^{\frac{n+4}{2}} = \frac{G_1(n+2)}{2C_4G_2}k$$
$$b = \left(\frac{G_1(n+2)}{2C_4G_2}k\right)^{\frac{2}{n+4}} \tag{67}$$

Let $C_5 = \left(\frac{G_1}{2}\right)^{\frac{n+2}{n+4}} \left(\frac{C_4G_2}{n+2}\right)^{\frac{2}{n+4}}$. *Substituting (67) into (66), we rewrite (66) as*

$$dim_{Haus}(\mathcal{A}_{HT}, L^2(\Omega) \times \mathbb{R}^k) \leq k + \frac{C_4^{**}}{\lambda_1} + \frac{2}{\lambda_1} C_5 k^{\frac{n+2}{n+4}} \tag{68}$$

Thus, we are able to conclude finally with a formulation of our main result:

Theorem 4.7 *Let* $\mathbb{A}_{HT}^{un}$ *be the uniform attractor of* $\{\mathbb{U}_{\omega_o}(t,\tau),\ t \geq \tau\}$ *defined by (36). Then,*

$$dim_{Haus}(\mathbb{A}_{HT}^{un}, L^2(\Omega)) \leq k + \frac{C_4^{**}}{\lambda_1} + \frac{2}{\lambda_1} C_5 k^{\frac{n+2}{n+4}} \tag{69}$$

5 Conclusion

Understanding the asymptotic behaviour of dynamical systems is a fundamental question in modern applied mathematics. One way to tackle this problem for dissipative dynamical systems is to consider its global attractors. A basic question then is to study the existence of a global attractor, and once this is established, it is natural to study important properties of the global attractor, such as its dimension, dependence on parameters and regularity of the attractor. In this paper, we have answered many of these questions for a simple mathematical model that has been used in the literature to simulate the effect of HIFU to treat bone osteomas (when used to ablate a region of bone tissue at an adjustable focal point of a HIFU transducer). The model takes into consideration both the convective and diffusive transport of heat, together with inhomogeneous initial and boundary conditions. One of the objectives of the current paper has been to build on our earlier work (which established well-posedness of the system of coupled partial differential equations that constitute this model) and prove the existence of a uniform attractor to this non-autonomous system of PDEs. Our proof of the existence and uniqueness of the solution (in earlier work) was based on the Leray–Schauder principle and the use of a priori estimates. In the current work, we study the long-time dynamics of solutions to a simple model (equations (2.3-2.7)) and establish the existence of a non-autonomous attractor. This puts the theoretical basis of this mathematical model on a firm foundation and completes what we consider to be a fundamental first step in a truly applied mathematical approach to "real-world" problems. We emphasize again that our method extends to the case where the Westervelt or KZK equations are used instead of the linear wave equation, and for the case where the biological medium is assumed to be slightly compressible. A second (equally important) step in modelling efforts is the thorough numerical exploration of the HIFU problem and comparison with experimental data, and this has already been carried out to a great extent in several publications, see, for example, (Cavicchi and O'Brien 1984; Etehadtavakol 2020; Makarov and Ochmann 1996, 1997; Nyborg 1981; Roohi et al. 2021; Shen et al. 2005). This, we consider to be an equally important step, since:

(a) it often highlights shortcomings of the model, where crucial underlying mechanisms (driving the underlying problem studied) may have been neglected. This, in turn, often leads to further developments and refinement of mathematical tools and techniques that result in a mathematical framework that (to a great extent) closely mirrors the physical problem it purports to model,

(b) This application of "Ockham's razor" (which is a quintessentially applied mathematical approach to study real-world problems) can often draw attention to shortcomings in existing models and, when appropriately applied, can lead to a model that provides a unique "in silico" approach to studying applications of HIFU (devoid of the ethical questions and challenges that face experimental, biomedical scientists when carrying out "in vivo" experiments).

(c) Finally, a well-developed mathematical model that provides a reliable framework to study a biomedical problem can be used to generate hypotheses which can subsequently be investigated experimentally "in vivo". In this manner, a good mathematical model can throw light on the interplay of various subprocesses and lead to important insights into a particular therapeutic intervention or treatment strategy that can be confirmed by "in vivo" studies.

As evident from the enormous growing interest and the already substantial body of work in HIFU (both experimental and theoretical) over the past two decades, this novel, non-invasive, therapeutic modality has emerged as an important, novel technology, with a myriad of potential applications. HIFU provides a non-invasive, non-ionizing therapy that can be used to thermally ablate tissue at a target location, while minimally affecting the surrounding tissue. These (and other considerations) suggest significant opportunities for mathematics to contribute to the development of HIFU as the "gold standard" for cancer therapy, as well as in treating other medical conditions and diseases.

Acknowledgements S. Sivaloganathan is grateful to NSERC for support of this research through an NSERC Discovery Grant; M.A. Efendiev is grateful to the University of Waterloo, Canada, for the award of the James D. Murray Distinguished Visiting Professorship and to the University of Marmara, Turkey, for the Rector's Distinguished Visiting Professorship, during which time much of this research was carried out.

References

Bailey MR, Khokhlova VA, Sapozhnikov OA, Kargl SG, Crum LA (2003) Physical mechanisms of the therapeutic effect of ultrasound (a review). Acoust Phys 49(4):369–388

Bhowmik A, Singh R, Repaka R, Mishra SC (2013) Conventional and newly developed bioheat transport models in vascularized tissues: a review. J Therm Biol 38(3):107–125

Cavicchi T, OBrien WJ (1984) Heat generation by ultrasound in an absorbing medium. J Acoust Soc Am 76(4):1244–1245

Chen MM, Holmes KR (1980) Microvascular contributions in tissue heat transfer. Ann N Y Acad Sci 335(1):137–150

Chepyzhov VV, Vishik MI (1994) Attractors of nonautonomous dynamical systems and their dimension. J Math Pure Appl 73(3):279–333

Chepyzhov VV, Efendiev MA (2000) Hausdorff dimension estimation for attractors of nonautonomous dynamical systems in unbounded domains: an example. Commun Pure Appl Math 53(5):647–665

Dobrakowski PP, Machowska-Majchrzak AK, Labuz-Roszak B, Majchrzak KG, Kluczewska E, Pierzchała KB (2014) MR guided focused ultrasound: a new generation treatment of Parkinsons disease, essential tremor and neuropathic pain. Interv Neuroradiol 20(3):275–282

Efendiev MA (2009) Fredholm structures, topological invariants and applications. American Institute of Mathematical Sciences

Efendiev MA, Murley J, Sivaloganathan S (2020) A coupled pde model of high intensity ultrasound heating of biological tissue. Adv Appl Math Sci 29(1):231–246

Etehadtavakol M (2020) Survey of numerical bioheat transfer modelling for accurate skin surface measurements. Therm Sci Eng Prog 20:100681

Hale JK (1988) Asymptotic behavior of dissipative systems. American Mathematical Soc

Hariharan P, Myers MR, Banerjee RK (2007) HIFU procedures at moderate intensities - effect of large blood vessels. Phys Med Biol 52:3493–3513

Haraux A (1991) Systèmes dynamiques dissipatifs et applications. Masson, Moulineaux

Hudson TJ, Looi T, Pichardo S, Amaral J, Temple M, Drake JM, Waspe AC (2018) Simulating thermal effects of MR-guided focused ultrasound in cortical bone and its surrounding tissue. Med Phys 45(2):506–519

Kaltenbacher B (2015) Mathematics of nonlinear acoustics. Evol Equ Control Theory 4(4):447–491

Kreider W, Yuldashev PV, Sapozhnikov OA, Farr N, Partanen A, Bailey MR, Khokhlova VA (2013) Characterization of a multi-element clinical HIFU system using acoustic holography and nonlinear modeling. IEEE Trans Ultrason Ferroelectr Freq Control 60(8):1683–1698

Ladyzhenskaya OA, Solonnikov VA, Ural'ceva NN (1968) Linear and quasi-linear equations of parabolic type. American Mathematical Soc

Lesniewski P, Stepin B, Thomas JC (2010) Ultrasonic Streaming in incompressible fluids-modelling and measurements. Proceedings of 20th International Congress on Acoustics

Makarov S, Ochmann M (1996) Nonlinear and thermoviscous phenomena in acoustics, part I. Acta Acust united Ac 82(4):579–606

Makarov S, Ochmann M (1997) Nonlinear and thermoviscous phenomena in acoustics, part II. Acta Acust United Ac 83(2):197–222

Modena D, Bassano D, Elevelt A, Baragona M, Hilbers PAJ, Westenberg MA (2019) HIFUpm: a visual environment to plan and monitor high intensity focused ultrasound treatments. In VCBM: 207-211

Nirenberg L (1974) Topics in nonlinear functional analysis. American Mathematical Soc

Nyborg WL (1981) Heat generation by ultrasound in a relaxing medium. J Acoust Soc Am 70(2):310–312
Pennes HH (1948) Analysis of tissue and arterial blood temperatures in the resting human forearm. J Appl Physiol 1(2):93–122
Renardy M, Rogers RC (2006) An introduction to partial differential equations. Springer Science & Business Media, Berlin
Roohi R, Baroumand S, Hosseinie R, Ahmadi G (2021) Numerical simulation of HIFU with dual transducers: The implementation of dual-phase lag bioheat and non-linear Westervelt equations. Int Commun Heat Mass Transf 120:105002
Salgaonkar VA et al (2014) Model based feasibility assessment and evaluation of prostate hyperthermia with a commercial MR guited endorectal HIFU ablation array. Med Phys 41(3):033301
Shen W, Zhang J, Yang F (2005) Modeling and numberical simulation of bioheat transfer and biomechanics in soft tissue. Math Comput Model 41:1251–1265
Simon J (1987) Compact sets in the space Lp (0, T; B). Ann Mat Pure Appl 4(146):65–96
Temam R (2012) Infinite-dimensional dynamical systems in mechanics and physics, vol 68. Springer Science & Business Media, Berlin
Triebel H (1978) Interpolation theory, function space, differential operators. North-Holland, Amsterdam
Wissler EH (1987) Comments on Weinbaum and Jijis discussion of their proposed bioheat equation. J Biomech Eng 109(4):226–233
Wu JR (2016) Handbook of contemporary acoustics and its applications. World scientific
Wu J (2018) Acoustic streaming and its applications. Fluids 3(4):108

Affiliations

M. A. Efendiev[1,2] · J. Murley[3] · S. Sivaloganathan[4,5]

M. A. Efendiev
messoud.efendiyev@gmail.com ; m.efendiyev@marmara.edu.tr

J. Murley
jmurley@uwaterloo.ca

1 Institute of Computational Biology, Helmholtz Center Munich, 85764 Neuherberg, Germany

2 Present Address: Department of Mathematics, Marmara University, Istanbul, Turkey

3 University of Waterloo, 200 University Avenue West, Waterloo, ON N2L 3G1, Canada

4 Department of Applied Mathematics, University of Waterloo, 200 University Avenue West, Waterloo, ON N2L 3G1, Canada

5 Centre for Math Medicine, The Fields Institute, 222 College Street, Toronto, ON M5T 3J1, Canada

Bulletin of Mathematical Biology (2021) 83:92
https://doi.org/10.1007/s11538-021-00917-0

Society for Mathematical Biology

SPECIAL ISSUE: CELEBRATING J. D. MURRAY

A Random Walk Approach to Transport in Tissues and Complex Media: From Microscale Descriptions to Macroscale Models

Jay A. Stotsky[1] · Jia Gou[2] · Hans G. Othmer[1]

Received: 5 October 2020 / Accepted: 1 June 2021 / Published online: 16 July 2021

Abstract

The biological processes necessary for the development and continued survival of any organism are often strongly influenced by the transport properties of various biologically active species. The transport phenomena involved vary over multiple temporal and spatial scales, from organism-level behaviors such as the search for food, to systemic processes such as the transport of oxygen from the lungs to distant organs, down to microscopic phenomena such as the stochastic movement of proteins in a cell. Each of these processes is influenced by many interrelated factors. Identifying which factors are the most important, and how they interact to determine the overall result is a problem of great importance and interest. Experimental observations are often fit to relatively simple models, but in reality the observations are the output of complicated functions of the physicochemical, topological, and geometrical properties of a given system. Herein we use multistate continuous-time random walks and generalized master equations to model transport processes involving spatial jumps, immobilization at defined sites, and stochastic internal state changes. The underlying spatial models, which are framed as graphs, may have different classes of nodes, and walkers may have internal states that are governed by a Markov process. A general form of the solutions, using Fourier–Laplace transforms and asymptotic analysis, is developed for several spatially infinite regular lattices in one and two spatial dimensions, and the theory is developed for the analysis of transport and internal state changes on general graphs. The goal in each case is to shed light on how experimentally observable macroscale transport coefficients can be explained in terms of microscale properties

Dedicated to James D. Murray on his 90th birthday. Jim was a pioneer in Mathematical Biology, and as a mentor to many his leadership and vision have had an enormous impact on the development of the field.

✉ Hans G. Othmer
othmer@math.umn.edu

1 School of Mathematics, University of Minnesota, 270A Vincent Hall, Minneapolis, USA

2 Department of Mathematics, University of California, 900 University Ave. Skye Hall, Riverside CA 92521, USA

of the underlying processes. This work is motivated by problems arising in transport in biological tissues, but the results are applicable to a broad class of problems that arise in other applications.

Keywords Continuous-time random walk · Multiscale model · Drosophila · Multistate random walk · First passage time

1 Introduction

Within any organism, various spatially distributed networks with widely distinct characteristics exist—the nervous system, the cardiovascular system, and air passages in the lungs of higher organisms are but a few examples. When biochemical processes occur within these biological networks, the network structure can play a significant role in determining the outcome of such processes. For instance, if a biochemical signal secreted by a set of cells has to travel through this network to reach distant cells, transport properties of the network play a crucial role in determining the strength and dynamic behavior of the signal. In order to understand complicated cell-level and tissue-level processes such as morphogenesis, it is crucial to assess the effects of interactions between the complex spatial structures and the properties of associated signal transduction networks that translate chemical signals into cellular responses. This problem is particularly challenging as it is difficult to interpret tissue or organism-scale experiments in terms of microscopic details of reactions and transport, and conversely, it is a challenge to extrapolate local microscale measurements of transport to tissue-level behavior.

A widely-studied example arises in the analysis of transport and transduction of bone morphogenetic protein (BMP) signals during morphogenesis of the wing disk of the fruit fly *Drosophila melanogaster*. Wing disks (Fig. 1a) arise in the embryo at the intersection of a circumferential stripe of the protein wingless (Wg) and an anterior–posterior (AP) stripe of Decapentaplegic (Dpp) (Gou et al. 2020). The disk has two layers of cells separated by a lumen (Fig. 1b), one a layer of columnar epithelial cells

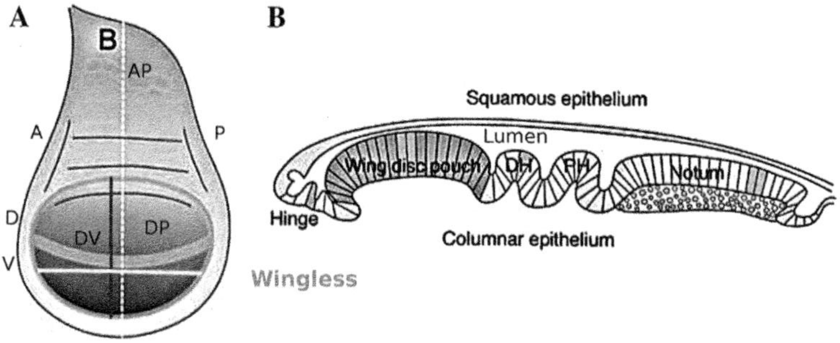

Fig. 1 **A** Wing disk A: anterior, P: posterior, D: dorsal, V: ventral, AP & DV: anterior–posterior and dorsal–ventral boundaries, DP: disk pouch. **B** side view along B in (**A**). Modified from Widmann and Dahmann (2009)

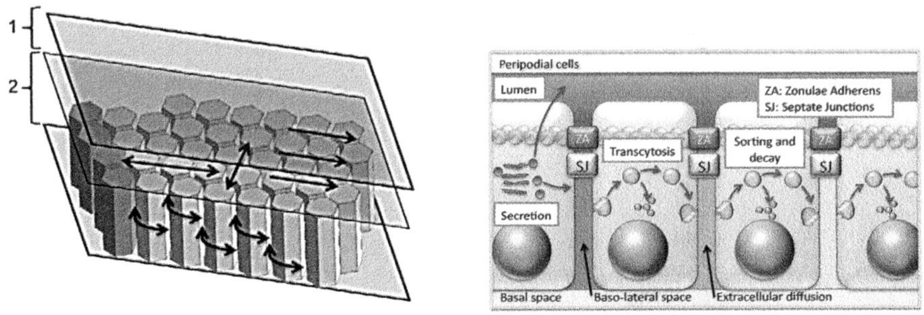

Fig. 2 Left: A schematic of a portion of the wing disk, showing the luminal section (1) at the top, , the cellular array below (2), and the source of Dpp at the left. Right: A cross section of a disk showing the transport processes that affect the morphogen distribution (Othmer et al. 2009). ZA are aka AJ

with the apical side at the lumen and the basal side at the basal lamina, the other a peripodial epithelium overlying the lumen (Fig. 2). The lateral membranes of adjacent columnar cells are connected via adherens junctions (AJs), which are complexes of E-cadherins and adapter proteins, and via septate junctions (SJs) that lie basal to the AJs and separate the extracellular fluid into apical and baso-lateral layers (Gibson and Gibson 2009; Harris and Tepass 2010; Choi 2018).

The wing disk is perhaps the best understood system in which transport of a molecule—in this case the BMP family member Dpp—involves a number of very distinct steps in a geometrically complex tissue. The secretion of Dpp from a stripe of cells along the AP boundary colored red in Fig. 2(left) gives rise to a spatial distribution of Dpp transverse to the AP boundary, but how the distribution is established is still an open question.

Various local processes and paths are involved in the transport of Dpp in the wing disk, and it has been observed that Dpp dispersal occurs both in the lumenal and the baso-lateral space. Dpp is uniformly distributed in the lumen, while in the baso-lateral region a graded Dpp distribution was found (Gibson et al. 2002; Mundt 2013; Harmansa et al. 2017). Extracellular Dpp can bind to membrane receptors, which can restrict its long-range transport by retaining it on cell surfaces (Haerry et al. 1998). It can then be internalized via endocytosis for either lysosomal degradation or recycling to the membrane, as well as for transcellular transport (Entchev et al. 2000; Akiyama et al. 2008). For instance, as was shown for Wg (Yamazaki et al. 2016), Dpp might also undergo apico-basal transport in the columnar cells, which promotes the communication between the apical and baso-lateral regions.

Spatial profiles of Dpp measured via imaging techniques such as FRAP, are usually described with a reaction–diffusion model based on diffusion and first-order decay (Kicheva et al. 2007; Wartlick et al. 2011). Fitting this model to FRAP data from one set of experiments yielded a diffusion coefficient of $0.1\,\mu m^2/s$ (Kicheva et al. 2007). In contrast, Zhou et al. (2012) measured a free diffusion coefficient of $20\,\mu m^2/s$ using fluorescence correlation spectroscopy. The discrepancy between the two values can be understood since the former is a macroscopic parameter, and in reality, the simple diffusion model integrates various local processes, such as receptor-mediated uptake, local degradation and transport in each cell. More recent work shows that

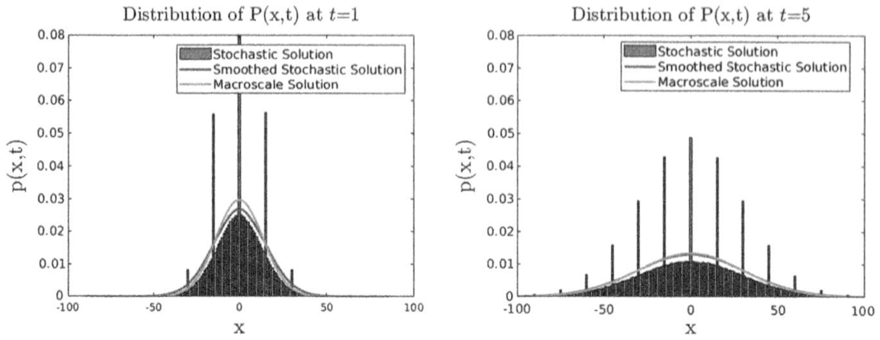

Fig. 3 Comparisons of stochastic simulations and continuum results for two times, $t = \{1, 5\}$ with $n = 15$. The cell-averaged concentration, obtained through smoothing the stochastic data, approaches the continuum macroscale solution as time progresses. The spikes are due to slower jumping rates at the cell membranes as compared with the interior, which leads to local accumulation. The waiting times at interior points are set to be 4.5 times faster than at the junctions and roughly 100 jumps occur per unit time on average

spreading of Dpp in the apical lumen, where free diffusion can occur, plays a minor role in both growth and patterning—most occurs in the baso-lateral space (Mundt 2013; Harmansa et al. 2017). Given the complexity of the geometric structure of the disk and the different local processes that may be involved in Dpp transport, a much more complicated transport model is required to describe the establishment of the Dpp profile in this system.

Herein we describe transport processes as multistate random jump processes in which the walkers are characterized by both their spatial location and their internal state. Spatial states live on lattices and general graphs, chosen so as to represent some aspects of transport in the disk, and as used here, these lattices have a translation-invariant structure that makes the problems amenable to analytical techniques. We analyze a number of examples of Markovian problems to obtain relations for certain transport coefficients in terms of the details of the underlying processes, and develop techniques for use on general graphs and non-Markovian jump processes. While we focus on analytically-tractable problems, numerical approaches can be applied to supplement what can be obtained analytically in more complex applications.

As a motivating example, consider Dpp transport in a 1D domain that is subdivided into a number of cells, each of width L. We overlay a lattice with two types of vertices on the line of cells, $n = L/\Delta x$ vertices within each cell and a distinct vertex at each membrane. We suppose that within each cell a molecule can jump from site to site at a rate μ, and at membrane points a molecule can linger longer, and either return to where it came from or jump into the neighboring cell. A stochastic simulation of the movement of a particle that begins at $x = 0$ leads to the distributions shown at two different times in Fig. 3.

Depending on the waiting-time distributions for a jump at interior points and at the membranes, the time spent at these points can be very different, as shown by the spikes at both times, but when the stochastic results are averaged the result is a normal or Gaussian distribution characteristic of a diffusion process. As we show later, if we define the diffusion coefficient within a cell as $D_m = \mu/\Delta x^2$ and set the mean membrane residence time $\tau = \lambda_0^{-1} = \lambda^{-1}\Delta x$, then the macroscopic diffusion coefficient that characterizes the spread of the molecule's location at long times is $D_M = \frac{\lambda D_m}{\lambda + D_m/L^2}$ under a suitable scaling of the jump rates as the lattice spacing goes to zero. The yellow line in Fig. 3 shows the solution of the macroscale diffusion equation using this diffusion coefficient.

A major objective of this work is to understand in general how parameters describing microscale processes can be "lifted" to parameters for macroscale descriptions of the processes. This has been done in a continuum framework for a line of cells, but for nonlinear reactions the homogenization is not simple (Othmer 1983). This also becomes difficult when describing transport in complex geometries such as the wing disk. We show that the lattice structure often makes it possible to reduce a random process with numerous (even infinite) internal states into random processes with just a few internal states. The effect of this change is to ignore certain local details, by including information about local processes in overall transport coefficients valid at larger length and timescales.

We only deal with linear processes, which leads to linear evolution equations and thus to potentially large spectral problems, and a great deal of earlier work has dealt with similar problems (Othmer and Scriven 1971; Gadgil et al. 2005; Kang et al. 2012; Hu and Othmer 2011; Ciocanel et al. 2020). However, earlier work was focused on simple geometries, and in this article, we expand the scope of these problems to cases where there are multiple types of states that have differing connectivity patterns, and we show how exact macroscale waiting-time distributions can be derived in some cases. The computation of spatial moments for multistate random walks has also been treated earlier (Landman and Shlesinger 1979a, b, 1977; Roerdink and Shuler 1985a, b; Scher and Wu 1981), and while there is some overlap with earlier work, our approach includes both microscale spatial structure, complex geometries, and internal states for the walkers.

An outline of the paper is as follows. In the following section, we provide a brief introduction to the standard theory of continuous-time random walks (CTRWs) in continuum space. We then extend these ideas to lattice walks on graphs in which there are two distinct classes of vertices in the underlying graph, one of which we call primary junctions or vertices, and the other, which we call secondary vertices, live on edges connecting primary junctions. We also consider processes in which the walker has internal states as well, and develop evolution equations for the stochastic evolution of both spatial location and internal state. In Sect. 3, we derive the moment equations for one-dimensional lattices, and in Sect. 4 and 5, we develop different approaches to the "micro-to-macro" analysis in one-dimensional lattices. In Sect. 6, we consider hexagonal lattices and develop the evolution and moment equations for such lattices. In Appendix B, we describe a general formulation by which arbitrary graphs can be treated. For reference, Table 1 contains a list of the abbreviations used throughout.

Table 1 List of abbreviations and recurring notation

Abbreviation	Notation	Meaning
CTRW		Continuous-time random walk
WTD	$\phi(t)$ or $\psi(t)$	Waiting-time distribution
	$\Phi(t)$	Cumulative waiting-time distribution
	$\hat{\Phi}(t)$	Complementary waiting-time distribution
FPT	$f(t)$	First-passage-time density distribution
Junction		A vertex in the primary graph $\mathscr{G}$
SV		Secondary vertices–spatial states on an edge connecting vertices in the primary graph
	$p(\boldsymbol{x}, t\|\boldsymbol{0})$ or $p(\boldsymbol{x}, t)$	Probability of occupying location $\boldsymbol{x}$ at time t after starting at $\boldsymbol{x} = \boldsymbol{0}$ at $t = 0$.
	$q(\boldsymbol{x}, t\|\boldsymbol{0})$ or $q(\boldsymbol{x}, t)$	Probability density of arriving at location $\boldsymbol{x}$ at time t after starting at $\boldsymbol{x} = \boldsymbol{0}$ at $t = 0$.
	$n(\boldsymbol{x}, t)$	Number density of particles at $\boldsymbol{x}$ and t
	$m^{(k)}(t)$	k^{th} spatial moment of $p(\boldsymbol{x}, t)$
	$T(\boldsymbol{x}, \boldsymbol{y})$	Spatial jump distribution operator for a jump process
	$\boldsymbol{K}(t)$	Internal state transition matrix
	$\boldsymbol{\Lambda}(t)$	Diagonal matrix of internal state exit rates
	$W(\boldsymbol{x}, \boldsymbol{y})$	Spatial transition operator for a differential master equation
	$\Gamma(t)$	Memory kernel for a differential master equation
	$\otimes$	Kronecker product operator
	n_k	Number of SVs on an edge
	n_s	Number of internal states of a walker or particle

Blank entries in the notation column apply when the abbreviation has no associated mathematical symbol. Likewise, the abbreviation column is empty for mathematical symbols that are not abbreviated.The dependence on the initial position $\boldsymbol{x} = \boldsymbol{0}$ in p and q is often omitted in the text

2 The Continuous-Time Random Walk

2.1 Single-State Walkers in a Continuum

When the underlying space is a continuum, a continuous-time random walk (CTRW) is a random jump process on R^n in which the particle or walker, which can range from a molecule to an organism, executes a sequence of spatial jumps of negligible duration. In biological contexts, a walker may have many internal states such as the configuration of a molecule that can also change, but to begin we assume that the only change in the jump process is position in space—the case of multiple internal states is treated in the following subsection. We assume that the waiting times between successive jumps at a given position are independent and identically distributed, i.e. if the jumps occur at $T_0, T_1, \ldots$ then the increments $T_i - T_{i-1}$ are independently and identically distributed, and therefore, the jump process is a semi-Markov process (Feller 1968; Karlin and Taylor 1975).

Let $\boldsymbol{\Psi}(\boldsymbol{x}, t|\boldsymbol{y}, 0)$ be the joint conditional probability density that a walker at $\boldsymbol{y}$ at time $t = 0$ remains at $\boldsymbol{y}$ until it jumps at t^- and lands at $\boldsymbol{x}$. The waiting time $\mathscr{T}$ between jumps is governed by the waiting-time distribution (WTD)

$$\phi(\boldsymbol{y}, t) = \int_{R^n} \boldsymbol{\Psi}(\boldsymbol{x}, t|\, \boldsymbol{y}, 0)d\boldsymbol{x} \tag{1}$$

and the spatial distribution of jumps is given by

$$T(\boldsymbol{x}, \boldsymbol{y}) = \int_0^\infty \boldsymbol{\Psi}(\boldsymbol{x}, t|\, \boldsymbol{y}, 0)dt. \tag{2}$$

The cumulative and complementary waiting-time distributions, Φ and $\hat{\Phi}$, respectively, are

$$\Phi(\boldsymbol{y}, t) = \int_0^t \phi(\boldsymbol{y}, s)\, ds = Pr\{\mathscr{T} \leq t\} \tag{3}$$

and

$$\hat{\Phi}(\boldsymbol{y}, t) = \int_t^\infty \phi(\boldsymbol{y}, s)\, ds = 1 - \Phi(t) = Pr\{\mathscr{T} \geq t\}. \tag{4}$$

The fact that we condition on the assumption that the current time at position $\boldsymbol{y}$ can be taken as $t = 0$ means that the jump distribution, $T(\boldsymbol{x}, \boldsymbol{y})$, is independent of the waiting time. In the above we incorporated the current position, $\boldsymbol{y}$, in the waiting-time distribution and this will be used later for lattice walks, but in this section we will omit this. Thus, if the jumps are governed by a Poisson process with parameter λ, the cumulative distribution is $\Phi(t) = 1 - e^{-\lambda t}$ and the waiting-time distribution is $\phi(t) = \lambda e^{-\lambda t}$. This is the only smooth distribution for which the jump process is Markovian (Feller 1968).

CTRWs are particularly useful since experimental observables, such as the moments of the displacement distribution and their dependence on time, can be related to the quantities, ϕ and T that specify the CTRW. In turn, ϕ and T are specified in terms of the microscale properties of the system being studied. To relate observables to ϕ and T we require an evolution equation for the density function $p(\boldsymbol{x}, t|\, 0)$, which is defined such that $p(\boldsymbol{x}, t|\, 0)d\boldsymbol{x}$ is the probability that the position of a jumper which begins at the origin at $t = 0$ lies in the interval $(\boldsymbol{x}, \boldsymbol{x} + d\boldsymbol{x})$ at time t. We derive this equation via equations for some auxiliary quantities.

Let $q_k(\boldsymbol{x}, t)d\boldsymbol{x}\ dt$ be the joint probability that a walker starting at the origin takes its kth step in the interval $(t, t + dt)$ and lands in the interval $(\boldsymbol{x}, \boldsymbol{x} + d\boldsymbol{x})$. Then for any $\boldsymbol{x} \neq \boldsymbol{0}, t > 0$, q_k satisfies the first-order integro-difference equation

$$q_{k+1}(\boldsymbol{x}, t|\boldsymbol{0}) = \int_0^t \int_{R^n} \phi(t-\tau)T(\boldsymbol{x}, \boldsymbol{y})q_k(\boldsymbol{y}, \tau|\boldsymbol{0})\, d\boldsymbol{y}\, d\tau, \qquad k = 1, 2, \ldots \tag{5}$$

The density function for arriving in the interval $(\boldsymbol{x}, \boldsymbol{x} + d\boldsymbol{x})$ in time interval $(t, t+dt)$ after any number of steps is the sum of these, from which we obtain the integral equation

$$q(\boldsymbol{x}, t|\mathbf{0}) = \sum_{k=0}^{\infty} q_k(\boldsymbol{x}, t|\mathbf{0}) = q_0(\boldsymbol{x}, t|\mathbf{0}) + \int_0^t \int_{R^n} \phi(t-\tau)T(\boldsymbol{x}, \boldsymbol{y})q(\boldsymbol{y}, \tau|\mathbf{0})\, d\boldsymbol{y}\, d\tau. \tag{6}$$

Since the walker starts at $\boldsymbol{x} = \mathbf{0}$, $q(\boldsymbol{x}, t|\mathbf{0})$ must satisfy the initial condition $q(\boldsymbol{x}, 0|\mathbf{0}) = \delta(\boldsymbol{x})$ and therefore

$$q(\boldsymbol{x}, t|\mathbf{0}) = \delta(\boldsymbol{x})\delta(t) + \int_0^t \int_{R^n} \phi(t-\tau)T(\boldsymbol{x}, \boldsymbol{y})q(\boldsymbol{y}, \tau|\mathbf{0})\, d\boldsymbol{y}\, d\tau. \tag{7}$$

To obtain the density function $p(\boldsymbol{x}, t|\mathbf{0})$, observe that it is the product of the probability of arriving in $(\boldsymbol{x}, \boldsymbol{x} + d\boldsymbol{x})$ at some time $\tau < t$ and the probability that no transition occurs in the remaining time $t - \tau$. As a result,

$$\begin{aligned} p(\boldsymbol{x}, t|\mathbf{0}) &= \int_0^t \hat{\Phi}(t-\tau)q(\boldsymbol{x}, \tau|\mathbf{0})\, d\tau \\ &= \int_0^t \hat{\Phi}(t-\tau)\{\delta(\boldsymbol{x})\delta(\tau) + \int_0^\tau \int_{R^n} \phi(\tau - s)T(\boldsymbol{x}, \boldsymbol{y})q(\boldsymbol{y}, s|\mathbf{0})\, d\boldsymbol{y}\, ds\}\, d\tau \\ &= \hat{\Phi}(t)\delta(\boldsymbol{x}) + \int_0^t \int_{R^n} \left(\int_s^t \hat{\Phi}(t-\tau)\phi(\tau - s)\, d\tau \right) T(\boldsymbol{x}, \boldsymbol{y})q(\boldsymbol{y}, s|\mathbf{0}) d\boldsymbol{y}\, ds. \end{aligned} \tag{8}$$

From this, one finds that $p(\boldsymbol{x}, t|0)$ satisfies the following renewal equation (Othmer et al. 1988)

$$p(\boldsymbol{x}, t|0) = \hat{\Phi}(t)\delta(\boldsymbol{x}) + \int_0^t \int_{R^n} \phi(t-\tau)T(\boldsymbol{x}, \boldsymbol{y})p(\boldsymbol{y}, \tau|0)\, d\boldsymbol{y}\, d\tau \tag{9}$$

which is sometimes called the *integral master equation* (Mainardi et al. 2000).

Although $p(\boldsymbol{x}, t|\mathbf{0})$ gives the probability density for a single particle to be located at $\boldsymbol{x}$ at time t, in experiments it is more common to observe a concentration distribution associated with many particles. If the initial number density distribution of non-interacting walkers is $f(\boldsymbol{x})$, then at time t, the number density, $n(\boldsymbol{x}, t)$ satisfies

$$n(\boldsymbol{x}, t) = \hat{\Phi}(t)f(\boldsymbol{x}) + \int_0^t \int_{R^n} \phi(t-\tau)T(\boldsymbol{x}, \boldsymbol{y})n(\boldsymbol{y}, \tau)\, d\boldsymbol{y}\, d\tau. \tag{10}$$

The necessary and sufficient condition for conservation of walkers is that the jump distribution satisfies

$$\int_{R^n} T(\boldsymbol{x}, \boldsymbol{y})\, d\boldsymbol{x} = 1.$$

To obtain a differential form of the integral master equation, we define the Laplace transform of an exponentially bounded $f(t)$ as

$$\tilde{f}(s) = \mathscr{L}\{\langle f(t)\rangle\} \equiv \int_0^\infty e^{-st} f(t)\, dt$$

and take the Laplace transform of (10). This leads to

$$\tilde{n}(\boldsymbol{x}, s) = \frac{1 - \tilde{\phi}(s)}{s} f(\boldsymbol{x}) + \tilde{\phi}(s) \int_{R^n} T(\boldsymbol{x}, \boldsymbol{y}) \tilde{n}(\boldsymbol{y}, s)\, d\boldsymbol{y} \tag{11}$$

which can be rearranged to

$$\left\{ \frac{1 - \tilde{\phi}(s)}{s\tilde{\phi}(s)} \right\} \{s\tilde{n}(\boldsymbol{x}, s) - f(\boldsymbol{x})\} = -\tilde{n}(\boldsymbol{x}, s) + \int_{R^n} T(\boldsymbol{x}, \boldsymbol{y}) \tilde{n}(\boldsymbol{y}, s)\, d\boldsymbol{y}. \tag{12}$$

The second bracketed quantity on the left side is the transform of $\dfrac{\partial n(\boldsymbol{x}, t)}{\partial t}$, while the leading factor defines the function

$$\Gamma(t) \equiv \mathscr{L}^{-1} \left\{ \frac{1 - \tilde{\phi}(s)}{s\tilde{\phi}(s)} \right\},$$

which is called the memory function (Mainardi et al. 2000). Thus the inverse transform of (12) is

$$\int_0^\infty \Gamma(t - \tau) \frac{\partial n(\boldsymbol{x}, \tau)}{\partial \tau} d\tau = -n(\boldsymbol{x}, t) + \int_{R^n} T(\boldsymbol{x}, \boldsymbol{y}) \boldsymbol{n}(\boldsymbol{y}, t)\, d\boldsymbol{y}. \tag{13}$$

For an exponential WTD of the form $\phi(t) = \lambda exp(-\lambda t)$, $\Gamma(t) = \delta(t)/\lambda$ and (13) reduces to

$$\frac{\partial n(\boldsymbol{x}, t)}{\partial t} = -\lambda n(\boldsymbol{x}, t) + \lambda \int_{R^n} T(\boldsymbol{x}, \boldsymbol{y}) n(\boldsymbol{y}, t)\, d\boldsymbol{y}. \tag{14}$$

This is the differential form of the master equation (Othmer et al. 1988). A number of examples of choices for $T(\boldsymbol{x}, \boldsymbol{y})$ that lead to either a diffusion equation or a telegrapher's equation are given in Othmer et al. (1988). A choice that will be used later defines a CTRW on a lattice or graph.

In the above, we consider a general jump function $T(\boldsymbol{x}, \boldsymbol{y})$, but if the jumps depend only on the difference $\boldsymbol{\xi} \equiv \boldsymbol{x} - \boldsymbol{y}$ the foregoing equations can be rewritten slightly. For example (14) can be written as

$$\frac{\partial \boldsymbol{n}(\boldsymbol{x}, t)}{\partial t} = \lambda \int_{R^n} T(\boldsymbol{\xi}, \boldsymbol{0}) \{\boldsymbol{n}(\boldsymbol{x} - \boldsymbol{\xi}, t) - \boldsymbol{n}(\boldsymbol{x}, t)\}\, d\boldsymbol{\xi}. \tag{15}$$

Since $T(\boldsymbol{x} - \boldsymbol{y}, 0) = T(\boldsymbol{x}, \boldsymbol{y})$ in this case, we can omit the second argument of T and just write $T(\boldsymbol{x} - \boldsymbol{y})$.

2.2 Walkers with Multiple Internal States and Binding

In biological transport problems, there are often various types of reaction, such as binding, catalysis, and degradation, that occur. One way to include these in the present formulation is to describe a walker in a CTRW as having internal states, each of which describes a distinct internal characteristic of the walker. For example, the internal state may describe the conformation of a protein, or it could describe a molecular type when interconversions occur according to a first-order reaction.

In a later example, binding will only occur in a subset of the spatial locations, thus some properties are attached to the spatial point, not to the internal state of the walker. On the other hand, if there is a probability of spontaneous interconversion between internal states (or death) that is the same for all spatial sites, that is built into the internal dynamics of the walker.

When walkers have a finite number n_s of internal states that evolve according to a Markov chain, the transitions can be included in the foregoing equations by defining the WTDs for internal state changes and the probability of a jump between states. We assume that changes in internal states and spatial position do not occur simultaneously, and thus, the effects of the processes are additive. Suppose that the internal states are $S = (S_1, \cdots, S_{n_s})$ and that they evolve according to a Markov chain with evolution equation

$$\frac{dS}{dt} = RS, \tag{16}$$

where R_{ij} is the transition rate between states j and i. Then the probability of a jump from state j to i is

$$k_{ij} = \frac{R_{ij}}{\sum_{l \neq j} R_{lj}} \tag{17}$$

and the parameter of the Poisson process governing the WTD in state j is the sum of the exit rates from state j, which is

$$\lambda_j = \sum_{l \neq j} R_{lj}. \tag{18}$$

In this case, where the space jumps and internal state changes are themselves Markovian processes, the waiting-time distributions for the composite process may be found explicitly. For the composite process of space jumps and internal state changes, the waiting-time density $\phi_{c_i}(t)$ for a space jump while in the ith state [1], conditioned on no change in the ith state, is the product of the WTD for a pure jump process times the probability of no change in the ith internal state. Thus

$$\phi_{c_i}(t) = \lambda e^{-\lambda t} \times \left(1 - \int_0^t \lambda_i e^{-\lambda_i \tau} d\tau\right) = \lambda e^{-(\lambda+\lambda_i)t}, \tag{19}$$

and the density for a jump in space, given no change in any internal state, is simply the product of these taken over all internal states.

[1] The subscript c indicates the composite process rather than the pure space jump process.

Since the position and state changes cannot occur simultaneously, their effects are additive, and letting $\boldsymbol{q}_k$ be the n_s-vector of states as a function of time and space, the generalization of (5) for arrival at $(\boldsymbol{x}, S)$ at time t is

$$\boldsymbol{q}_{k+1}(\boldsymbol{x},t) = \int_0^t \left\{ \left[\int_{R^n} \boldsymbol{\phi}_c(t-\tau) T(\boldsymbol{x},\boldsymbol{y}) \boldsymbol{q}_k(\boldsymbol{y},\tau) d\boldsymbol{y} \right] + \boldsymbol{K} \Lambda(t-\tau) \boldsymbol{q}_k(\boldsymbol{x},\tau) \right\} d\tau. \tag{20}$$

Here $\boldsymbol{\phi}_c(t) = diag\{\phi_{c_i(t)}\}$ and $\Lambda(t) = diag\{\lambda_i e^{-(\lambda+\lambda_i)t}\}$. The elements of Λ represent the waiting-time densities for the state changes multiplied by the probability that no space jump occurs in $[0, t]$, and the entries of K represent the probabilities of jumps between states, given that one occurs. $\boldsymbol{K}$ is the analog in the set of internal states of the jump kernel T in physical space. Note that for brevity of notation, we have omitted the conditioning on the initial condition in $\boldsymbol{q}$ and $\boldsymbol{p}$.

The vector version of (8) for occupancy in the state $(\boldsymbol{x}, S)$ at time t becomes

$$\begin{aligned}\boldsymbol{p}(\boldsymbol{x},t) = \delta(\boldsymbol{x})\delta(t)S_0 + \int_0^t \Bigg\{ &\left[\int_{R^n} \boldsymbol{\phi}_c(t-\tau) T(\boldsymbol{x},\boldsymbol{y}) \boldsymbol{p}(\boldsymbol{y},\tau) d\boldsymbol{y} \right] \\ &+\boldsymbol{K}\Lambda(t-\tau)\boldsymbol{p}(\boldsymbol{x},\tau) \Bigg\} d\tau,\end{aligned} \tag{21}$$

where S_0 is the vector of the initial internal states. The state of a multiparticle system is now defined by the n_s-vector $\boldsymbol{n}(\boldsymbol{x},t) = (n_1(\boldsymbol{x},t), \cdots, n_{n_s}(\boldsymbol{x},t))$ and the corresponding vector master equation is

$$\frac{\partial \boldsymbol{n}(\boldsymbol{x},t)}{\partial t} = -(\lambda \boldsymbol{I} + \boldsymbol{\Lambda}(0))\boldsymbol{n}(\boldsymbol{x},t) + \lambda \int_{R^n} T(\boldsymbol{x},\boldsymbol{y}) \boldsymbol{n}(\boldsymbol{x}-\boldsymbol{\xi},t)\, d\boldsymbol{\xi} + \boldsymbol{K}\boldsymbol{\Lambda}(0)\boldsymbol{n}(\boldsymbol{x},t). \tag{22}$$

2.3 Generalization of the Waiting-Time Distributions

In order to generalize to non-Markovian subprocesses, we consider the combination of two stochastic processes whose individual renewal equations are

$$\begin{aligned} q_{k+1}^{(1)}(\boldsymbol{x},t) &= \int_0^t \int_{\mathbb{R}^n} \phi(t-\tau) T(\boldsymbol{x},\boldsymbol{y}) q_k^{(1)}(\boldsymbol{y},\tau) d\boldsymbol{y} d\tau \\ \boldsymbol{q}_{k+1}^{(2)}(t) &= \int_0^t \boldsymbol{K}\boldsymbol{\Lambda}(t-\tau) \boldsymbol{q}_k^{(2)}(\tau) d\tau, \end{aligned}$$

where Λ is a matrix of waiting-time densities for the pure internal state process and $\boldsymbol{K}$ is the matrix of jump probabilities as before. To combine two, possibly non-Markovian, processes into a generalized master equation, we write

$$\boldsymbol{q}_{k+1}(\boldsymbol{x},t) = \int_0^t \int_{\mathbb{R}^n} \boldsymbol{\phi}_c(t-\tau) T(\boldsymbol{x},\boldsymbol{y}) \boldsymbol{q}_k(\boldsymbol{y},\tau) d\boldsymbol{y} d\tau + \int_0^t \boldsymbol{K}\boldsymbol{\Lambda}_c(t-\tau) \boldsymbol{q}_k(\boldsymbol{x},\tau) d\tau, \tag{23}$$

where $\boldsymbol{\phi}_c$ and $\boldsymbol{\Lambda}_c$ are diagonal matrices of waiting-time densities for the composite process that incorporates the probability of no jump in the complementary process, similar to above. Thus

$$\phi_{c_j}(t) = \phi(t)\left(1 - \int_0^t \Lambda_j(\tau)d\tau\right) \quad \text{and} \quad \Lambda_{c,j}(t) = \Lambda_j(t)\left(1 - \int_0^t \phi(\tau)d\tau\right).$$

Next, we sum over $k = 0, \ldots, \infty$ in Eq. (23) to obtain

$$\begin{aligned} \boldsymbol{q}(\boldsymbol{x}, t) = \delta(\boldsymbol{x})\delta(t)S_0 + \int_0^t \int_{\mathbb{R}^n} \boldsymbol{\phi}_c(t-\tau)T(\boldsymbol{x}, \boldsymbol{y})\boldsymbol{q}(\boldsymbol{y}, \tau)d\tau \\ + \int_0^t \boldsymbol{K}\boldsymbol{\Lambda}_c(t-\tau)\boldsymbol{q}(\boldsymbol{x}, \tau)d\tau. \end{aligned}$$

Making use of the relation between $\boldsymbol{p}$ and $\boldsymbol{q}$ in Eq. (8) leads to an integral equation for $\boldsymbol{p}$:

$$\begin{aligned} \boldsymbol{p}(\boldsymbol{x}, t) = \delta(\boldsymbol{x})\hat{\boldsymbol{\Phi}}_c(t)S_0 + \int_0^t \int_{\mathbb{R}^n} \boldsymbol{\phi}_c(t-\tau)T(\boldsymbol{x}, \boldsymbol{y})\boldsymbol{p}(\boldsymbol{y}, \tau)d\tau \\ + \int_0^t \boldsymbol{\kappa}(t-\tau)\boldsymbol{p}(\boldsymbol{x}, \tau)d\tau. \end{aligned}$$

The reaction matrix kernel $\boldsymbol{\kappa}(t)$ introduced in this equation is defined via an inverse Laplace transform:

$$\boldsymbol{\kappa}(t) = \mathscr{L}^{-1}\left[\tilde{\hat{\boldsymbol{\Phi}}}_c(s)\boldsymbol{K}_c\tilde{\boldsymbol{\Lambda}}_c(s)\tilde{\hat{\boldsymbol{\Phi}}}_c^{-1}(s)\right]$$

wherein the diagonal matrix of complementary waiting-time distributions, $\hat{\boldsymbol{\Phi}}_c$ is defined as

$$\hat{\Phi}_{c_i}(t) = 1 - \Phi_{c_i}(t) = 1 - \int_0^t \left(\phi_{c_i}(\tau) + \Lambda_{c_i}(\tau)\right) d\tau.$$

Following the derivation in Eqs. (13)–(15), we obtain the evolution equation for the number density of several non-interacting particles as

$$\begin{aligned} \int_0^t \boldsymbol{\Gamma}(t-\tau)\frac{\partial \boldsymbol{n}}{\partial t}d\tau = -\boldsymbol{n}(\boldsymbol{x}, t) + \int_0^t \int_{\mathbb{R}^n} \boldsymbol{\mu}(t-\tau)T(\boldsymbol{x}, \boldsymbol{y})\boldsymbol{n}(\boldsymbol{y}, \tau)d\boldsymbol{y}d\tau \\ + \int_0^t \boldsymbol{\nu}(t-\tau)\boldsymbol{n}(\boldsymbol{x}, \tau)d\tau \end{aligned} \tag{24}$$

with the kernels, $\boldsymbol{\Gamma}$, $\boldsymbol{\mu}$, and $\boldsymbol{\nu}$ defined in terms of inverse Laplace transformations as follows.

$$\begin{aligned} \boldsymbol{\Gamma}(t) &= \mathscr{L}^{-1}\left[\frac{1}{s}\left[\tilde{\boldsymbol{\phi}}_c(s) + \tilde{\boldsymbol{\Lambda}}_c(s)\right]^{-1}\tilde{\hat{\boldsymbol{\Phi}}}_c(s)\right] \\ \boldsymbol{\mu}(t) &= \mathscr{L}^{-1}\left[\left[\tilde{\boldsymbol{\phi}}_c(s) + \tilde{\boldsymbol{\Lambda}}_c(s)\right]^{-1}\tilde{\boldsymbol{\phi}}_c(s)\right] \\ \boldsymbol{\nu}(t) &= \mathscr{L}^{-1}\left[\left[\tilde{\boldsymbol{\phi}}_c(s) + \tilde{\boldsymbol{\Lambda}}_c(s)\right]^{-1}\tilde{\hat{\boldsymbol{\Phi}}}_c(s)\boldsymbol{K}_c\tilde{\boldsymbol{\Lambda}}_c(s)\tilde{\hat{\boldsymbol{\Phi}}}_c^{-1}(s)\right]. \end{aligned} \tag{25}$$

As shown earlier, when both transport and reaction are governed by Poisson processes, the resulting master equation is a system of ODEs, but what we see here is that non-Poissonian WTDs will generally lead to a more complicated integro-differential equations. We also note that in many cases, $\boldsymbol{\Gamma}$ cannot be defined as a function, but must be understood as a singular distribution (e.g., a Dirac delta function). As an example, the Erlang distribution $\phi(t) = \lambda^2 te^{-\lambda t}$ has a singular memory kernel, $\Gamma(t) = \lambda^{-2}\delta'(t) + 2\lambda^{-1}\delta(t)$ where $\delta'(t)$ is the distributional derivative of the Dirac delta (c.f. Othmer et al. 1988). In particular, if $\phi(0) = 0$, $\Gamma(t)$ will typically be a singular distribution and the evolution equation for $p(x,t)$ will depend on higher-order derivatives of $p(x,t)$ with respect to time.

In what follows, we do not explicitly outline the manipulations needed when combining various stochastic processes. However, in each case in which we introduce a waiting-time distribution, it should be assumed that it is in fact the waiting-time distribution for the combined process whenever needed.

2.4 Lattice Walks

In many problems, the underlying space can be treated as a lattice, obtained for example, by discretizing an underlying continuum space or as a model of cellular tissues. Here we deal with infinite 1D and 2D lattices in which every lattice point has the same number of neighbors. The lattice can be defined by an undirected graph $\mathscr{G}$, all of whose vertices are of the same degree, and we call such graphs regular. The spatial points in an n-dimensional lattice are typically integer multiples of n basis vectors that define the spacing between vertices. The evolution equation for the probability p_i of a walker being at the ith vertex at time t, assuming an exponential WTD, is

$$\frac{dp_i}{dt} = \lambda \sum_{j\in\mathscr{N}(i)} T_{i,j}\left\{p_j(t) - p_i(t)\right\}, \tag{26}$$

where $T_{i,j}$ is the probability of a jump from vertex j to vertex i and $\mathscr{N}(i)$ is the set of all vertices that are connected to i by a single edge. The adjacency matrix $\mathscr{A}$ of graph $\mathscr{G}$ is the doubly infinite matrix whose entries are 0 or 1, with $\mathscr{A}_{ij} = 1$ if vertex i is connected to vertex j by a single edge. Since the graph is undirected, $\mathscr{A}$ is symmetric. If jumps to all incident edges are assumed to be equiprobable, then Eq. (26) can be written

$$\frac{dp_i}{dt} = \frac{\lambda}{d} \sum_{j \in \mathscr{N}(i)} \mathscr{A}_{i.j} \left\{ p_j(t) - p_i(t) \right\}, \tag{27}$$

where d is the degree of any vertex in $\mathscr{G}$.[2] Here we have used the fact that

$$\sum_{j \in \mathscr{N}(i)} \mathscr{A}_{i.j} = d$$

and the fact that $\mathscr{A}$ is symmetric. For example, if $\mathscr{G}$ is a circular ring of N vertices, the evolution of $\boldsymbol{p}(t) = (p_1(t), \cdots, p_N(t))^T$ is the solution of

$$\frac{d\boldsymbol{p}}{dt} = \frac{\lambda}{2} \boldsymbol{\Delta p}, \tag{28}$$

where $\boldsymbol{\Delta}$ is the $N \times N$ graph Laplacian

$$\boldsymbol{\Delta} = \begin{bmatrix} -2 & 1 & 0 & \cdots & 1 \\ 1 & -2 & 1 & \cdots & 0 \\ \vdots & \vdots & \cdots & \vdots & \vdots \\ 1 & 0 & \cdots & 1 & -2 \end{bmatrix}.$$

When there are n_s identical internal states of the walker the state vector is an $n_s N$-component vector and the evolution equation becomes

$$\frac{d\boldsymbol{p}}{dt} = \left\{ \boldsymbol{\Delta} \otimes \frac{\lambda}{2} \boldsymbol{I}_m + \boldsymbol{I}_N \otimes \boldsymbol{K\Lambda}(\mathbf{0}) \right\} \boldsymbol{p}, \tag{29}$$

where I_N and I_{n_s} are identity matrices of size $N \times N$ and $n_s \times n_s$, $\boldsymbol{K}$ and $\boldsymbol{\Lambda}$ have the same meaning as in Sect. 2.1, and $\boldsymbol{A} \otimes \boldsymbol{B}$ is the tensor product of the matrices $\boldsymbol{A}$ and $\boldsymbol{B}$[3].

2.5 Lattices with Multiple Vertex Types and Internal States

In the preceding example, all vertices are equivalent, and the WTD for jumps between them, in what will be called the *primary* undirected graph $\mathscr{G}$, is the same at every vertex. However, many biological systems require more detailed models with more than one type of vertex. For example, consider a horizontal slice through the columnar cells in Fig. 2 and define the primary graph $\mathscr{G}$ by placing a vertex or *junction* in each cell. Secondary structure is introduced by assigning a node to the space between cells,

[2] In a finite regular graph boundary vertices have a different degree, but the boundary conditions determine how the degree changes. See (Othmer and Scriven 1971) for the structurally identical problem of diffusion between coupled cells in a finite regular lattice with various boundary conditions.

[3] The convention used in defining the tensor product is given in "Appendix A" and in Othmer and Scriven (1971).

Fig. 4 Graphical depiction of the setup for a random walk with an SV between junctions with distinct transition rates between the SVs and junctions. Junctions have a waiting-time density $\phi(t) = \lambda e^{-\lambda t}$ while SVs have a waiting-time density $\chi(t) = \mu e^{-\mu t}$. The arrows indicate the transitions, colored according to whether they begin at a junction (blue circle) or SV (red square)

with the result that one obtains a graph in which the edges of $\mathscr{G}$ have vertices between junctions. In general, we call these new vertices *secondary vertices* or SVs, and in effect, introduction of SVs creates a new, extended graph $\mathscr{G}_e$. Later we will show in examples how the primary graph reflects the macrostructure of a system, whereas the extended graph incorporates the microstructure.

To illustrate how the evolution equations for such lattices change, consider the following example.

Example 1: A circular ring with two vertex types

Suppose that $N = 2M$ and that even-numbered vertices are junctions in the primary graph $\mathscr{G}$ while odd-numbered vertices represent the secondary structure. Suppose that all junctions have the same Poisson parameter λ, while the SVs have transition rates labeled μ. Thus there are M pairs of successive vertices, and we label each vertex by V_j^i, $i = 1, 2$ with $j = 1, \ldots, M$. A diagram of the allowable transitions at each point is given in Fig. 4. The evolution equations can be written as

$$\begin{aligned} \frac{dp_j^{(1)}}{dt} &= -\lambda p_j^{(1)} + \frac{\mu}{2} \sum_{k \in \mathscr{N}(j,1)} p_k^{(2)}(t) \\ \frac{dp_j^{(2)}}{dt} &= -\mu p_j^{(2)} + \frac{\lambda}{2} \sum_{k \in \mathscr{N}(j,2)} p_k^{(1)}(t), \end{aligned} \tag{30}$$

where $\mathscr{N}(j, k)$ denotes the neighborhood of vertex k in the jth pair. Define

$$\boldsymbol{p}_i = \begin{pmatrix} p_i^{(1)} \\ p_i^{(2)} \end{pmatrix},$$

and then, (30) can be written

$$\frac{d\boldsymbol{p}_i(t)}{dt} = \boldsymbol{A}\boldsymbol{p}_i + \boldsymbol{B}_{-1}\boldsymbol{p}_{i-1} + \boldsymbol{B}_1\boldsymbol{p}_{i+1}, \tag{31}$$

where

$$\boldsymbol{A} = \begin{bmatrix} -\lambda & \frac{\mu}{2} \\ \frac{\lambda}{2} & -\mu \end{bmatrix} \quad \text{and} \quad \boldsymbol{B}_{-1} = \begin{bmatrix} 0 & \frac{\mu}{2} \\ 0 & 0 \end{bmatrix} \quad \text{and} \quad \boldsymbol{B}_{1} = \begin{bmatrix} 0 & 0 \\ \frac{\lambda}{2} & 0 \end{bmatrix}.$$

Notice that up to a scaling factor, $\boldsymbol{B}_1 = \boldsymbol{B}_{-1}^T$. Equation (31) can be written in block format as

$$\frac{d}{dt}\begin{pmatrix} \boldsymbol{p}_1 \\ \vdots \\ \\ \vdots \\ \boldsymbol{p}_M \end{pmatrix} = \begin{pmatrix} \boldsymbol{A} & \boldsymbol{B}_1 & & & \boldsymbol{B}_{-1} \\ \boldsymbol{B}_{-1} & \boldsymbol{A} & \boldsymbol{B}_1 & & \\ & \boldsymbol{B}_{-1} & \boldsymbol{A} & \boldsymbol{B}_1 & \\ & & \ddots & \ddots & \ddots \\ \boldsymbol{B}_1 & & & \boldsymbol{B}_{-1} & \boldsymbol{A} \end{pmatrix} \begin{pmatrix} \boldsymbol{p}_1 \\ \vdots \\ \\ \vdots \\ \boldsymbol{p}_M \end{pmatrix}. \tag{32}$$

This matrix has a block structure that suggests the discrete Laplacian for the single-state system. In fact, if $\lambda = \mu$ the equation can be written a

$$\frac{d\boldsymbol{P}}{dt} = \frac{\mu}{2}\Delta \boldsymbol{P}$$

where $\boldsymbol{P}$ is the $2M$ vector of states and Δ is the Laplacian for the ring. The structure in (32) will appear in subsequent examples as well, and in this sense, we expect that even a very complex multistate system with many internal states may at a large scale exhibit diffusive behavior. In this simple case, such behavior can further be motivated by considering the case when $\mu = \mu_0\epsilon^{-1}$ and letting ϵ go to zero. In that case, the explicit solution for $p_i^{(2)}(t)$, ignoring initial conditions, can be written as

$$p_i^{(2)}(t) = \lim_{\epsilon \to 0} \int_0^t e^{-\mu_0\epsilon^{-1}(t-\tau)} \frac{\lambda}{2}\left(p_i^{(1)}(t) + p_{i+1}^{(\tau)}(\tau)\right) d\tau = 0.$$

This is merely a reflection of the fact that with a large μ, particles spend negligible amounts of time on the type 2 points. Thus, particles essentially jump directly between type 1 points, and the effect of the intervening points drops out. (A similar mechanism also explains the spikes seen in Fig. 3, and those spike states have slower dynamics than the surrounding states and thus accumulate probability.) In this limit, one can therefore describe the evolution of the systems by

$$\frac{d\boldsymbol{p}^{(1)}}{dt} = 2\lambda\Delta \boldsymbol{p}^{(1)}$$

where $\boldsymbol{p}^{(1)}$ is the vector of all M type 1 points. The factor of 4 when compared with the previous result arises since each jump (really two jumps in rapid succession) now covers twice the distance, and since $D \sim \langle \Delta x^2 \rangle / \Delta t$, doubling the distance per jump quadruples the diffusivity.

Hence, we have found that the dynamics reduce to Laplacian dynamics in both the case that $\lambda = \mu$ and when λ and μ are of greatly different magnitudes. This suggests that perhaps after sufficient time has elapsed to level out local probability gradients, the system behavior may exhibit a diffusive character for arbitrary values of λ and μ. As we show in the subsequent sections, this is indeed true for many systems, and in a rough sense, part of what we discuss is a method of understanding the diffusive behavior of systems as a function of these local jump parameters and later on, the topology of the network. We note, however, that one should not always expect such nice behavior. As demonstrated in the comb example in Sect. 4 (and also many examples in the literature, e.g., Metzler and Klafter 2000; Othmer et al. 1988), non-diffusive behavior can arise too. However, these other limiting behaviors appear naturally in the CTRW framework- no special modifications need be introduced.

The foregoing results can be extended to a system with finitely many types of SVs, each with a different waiting-time distribution. Let $\{\boldsymbol{X}_{\boldsymbol{I}}\}_{\boldsymbol{I} \in \mathbb{Z}^d}$ denote the positions of junctions of a d-dimensional lattice whose connectivity is encoded in the graph $\mathscr{G}$, and suppose that each edge is populated with n_k SVs, which could represent the discretization of the channel joining a pair of tertiary junctions (junctions at which three edges meet) in the hexagonal lattice, that must be traversed in order to move between any given pair of junctions (*cf.* Fig. 5). In biological settings, SVs can be used to represent processes such as diffusion in the space between cells or sites for surface binding and unbinding on a membrane. We assume that each step can be represented as a linear operator, both for simplicity and because the linear case covers a wide variety of applications. Implicit in this assumption is the requirement that particles do not interact with each other. In the context of binding reactions mentioned above, this requires that the substrate which the diffusing particle can bind to is present in great abundance relative to the concentration of the diffusing particles (Lin and Othmer 2017).

In addition to SVs, we also suppose that there are n_s internal states that characterize a walker, as depicted in Fig. 5. These internal states are distinguished from the n_k SVs by the fact that they do not represent changes in position along an edge. Instead, they allow us to consider systems involving first-order reactions of several chemical species or possibly even generalized master-equation-type formulations (Isaacson 2009; Hu and Othmer 2011; Gadgil et al. 2005). We also assume that, there are n_e edges incident to each junction. This results in $n_T = n_s(1 + n_e n_k)$ degrees of freedom associated with each junction.

Recalling the definitions from Sect. 2.1, we introduce the n_T-dimensional vector-valued occupation and arrival probabilities $\boldsymbol{p}(\boldsymbol{X}, t)$ and $\boldsymbol{q}(\boldsymbol{X}, t)$ for particles starting from $\boldsymbol{X} = \boldsymbol{0}$ at $t = 0$. Spatial transitions between adjacent SVs and between SVs and junctions are described by an $n_T \times n_T$-matrix function $\boldsymbol{T}(\boldsymbol{X} - \boldsymbol{X}')$ which is the analog for a multistate lattice, of the spatial jump operator $T(x, y)$ defined in Eq. (2). In the definition of $\boldsymbol{T}(\boldsymbol{X} - \boldsymbol{X}')$, $\boldsymbol{X}$ and $\boldsymbol{X}'$ are junctions on the lattice, and for a given $\boldsymbol{X}$, $\boldsymbol{T}(\boldsymbol{X} - \boldsymbol{X}')$ is typically only nonzero when $\boldsymbol{X}' = \boldsymbol{X}$ or when $\boldsymbol{X}$ and $\boldsymbol{X}'$ are connected by an edge. The waiting-time distributions for each of these jumps form a diagonal matrix, denoted by $\boldsymbol{\phi}(t)$.

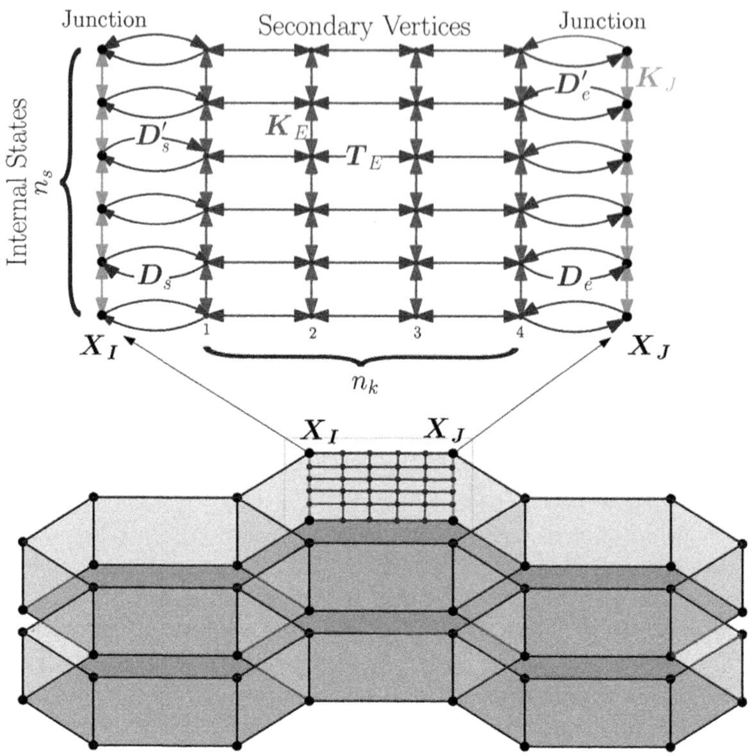

Fig. 5 Transitions allowed in a hexagonal lattice random walk having n_k SVs and n_s internal states. The larger dots represent the internal states at a junction and the smaller dots are the internal states at the n_k SVs. The lattice in this case is a 2D lattice with several internal states, but the edges in 1D or 3D (or higher) lattices behave in the same way with regard to the transitions between spatial states and the n_s internal states. The arrows are color-coded according to the matrix that contains each transition. Note that in practice, the internal states need not correspond with a physical height as they appear in this image

As an example, consider Fig. 4. For this two-state problem, $\boldsymbol{T}$ is nonzero only for $X - X' = \{-1, 0, 1\}$, and for these values,

$$\boldsymbol{T}(0) = \begin{pmatrix} 0 & \frac{1}{2} \\ \frac{1}{2} & 0 \end{pmatrix}, \quad \boldsymbol{T}(-1) = \begin{pmatrix} 0 & \frac{1}{2} \\ 0 & 0 \end{pmatrix}, \quad \boldsymbol{T}(1) = \begin{pmatrix} 0 & 0 \\ \frac{1}{2} & 0 \end{pmatrix}$$

and

$$\boldsymbol{\phi}(t) = \begin{pmatrix} \phi(t) & 0 \\ 0 & \chi(t) \end{pmatrix}.$$

In more general settings, transitions between internal states must also be described, and for this we use matrices $\boldsymbol{K}$ and $\boldsymbol{\Lambda}(t)$. These are generalizations of the $\boldsymbol{K}$ and $\boldsymbol{\Lambda}$ defined above, because they now allow for possibly different reactions to occur at each SV.

Finally, we have the evolution equations:

$$
\begin{aligned}
\boldsymbol{q}\,(\boldsymbol{X},t) &= \delta(t)\boldsymbol{\delta}_{k_0,\ell_0}\delta(\boldsymbol{X}) + \int_0^t \left[\sum_{\boldsymbol{X}'\in\mathcal{N}(\boldsymbol{X})} \boldsymbol{T}(\boldsymbol{X}-\boldsymbol{X}')\boldsymbol{\phi}(t-\tau)\boldsymbol{q}(\boldsymbol{X}',\tau)\right] d\tau \\
&\quad + \int_0^t \boldsymbol{K}\boldsymbol{\Lambda}(t-\tau)\boldsymbol{q}(\boldsymbol{X},\tau)d\tau \\
\boldsymbol{p}\,(\boldsymbol{X},t) &= \hat{\boldsymbol{\Phi}}(t)\boldsymbol{\delta}_{k_0,\ell_0}\delta(\boldsymbol{X}) + \int_0^t \hat{\boldsymbol{\Phi}}(t-\tau)\int_0^\tau \\
&\quad \left[\sum_{\boldsymbol{X}'\in\mathcal{N}(\boldsymbol{X})} \boldsymbol{T}(\boldsymbol{X}-\boldsymbol{X}')\boldsymbol{\phi}(\tau-s)\boldsymbol{q}(\boldsymbol{X}',s)\right] ds\, d\tau \\
&\quad + \int_0^t \hat{\boldsymbol{\Phi}}(t-\tau)\int_0^\tau \boldsymbol{K}\boldsymbol{\Lambda}(\tau-s)\boldsymbol{q}(\boldsymbol{X},s)ds\, d\tau
\end{aligned} \tag{33}
$$

where the vectors $\boldsymbol{\delta}_{k_0,\ell_0}$ specify the initial internal state (ℓ_0) and the initial SV (k_0) for the process, and $\mathcal{N}(\boldsymbol{X})$ is the set of lattice points adjacent to $\boldsymbol{X}$.

The nonzero elements of the diagonal matrix $\hat{\boldsymbol{\Phi}}$ of complementary waiting times (recall Eq. (4)) are defined as

$$
\hat{\Phi}_{k,\ell}(t) = 1 - \int_0^t \psi_{k,\ell}(t) = 1 - \int_0^t \left(\sum_{\boldsymbol{X}'\in\mathcal{N}(\boldsymbol{X})} \mathbf{1}^T\boldsymbol{T}(\boldsymbol{X}')\boldsymbol{\phi}(\tau)\boldsymbol{\delta}_{k,\ell} + \mathbf{1}^T\boldsymbol{K}\boldsymbol{\Lambda}(\tau)\boldsymbol{\delta}_{k,\ell}\right) d\tau, \tag{34}
$$

The factor, $\mathbf{1}$ is a vector of ones of length n_T.

To obtain a differential master equation for this system, a Laplace transformation is applied to Eq. (33) to obtain

$$
\begin{aligned}
&\tilde{\boldsymbol{p}}\,(\boldsymbol{X},s|\,\{k_0,\ell_0\}) \\
&= \tilde{\hat{\boldsymbol{\Phi}}}(s)\boldsymbol{\delta}_{k_0,\ell_0}\delta(\boldsymbol{X}) + \left[\sum_{\boldsymbol{X}'\in\mathcal{N}(\boldsymbol{X})} \tilde{\hat{\boldsymbol{\Phi}}}(s)\boldsymbol{T}(\boldsymbol{X}-\boldsymbol{X}')\tilde{\boldsymbol{\phi}}(s) + \tilde{\hat{\boldsymbol{\Phi}}}(s)\boldsymbol{K}\tilde{\boldsymbol{\Lambda}}(s)\right]\left(\tilde{\hat{\boldsymbol{\Phi}}}(s)\right)^{-1} \\
&\quad \tilde{\boldsymbol{p}}(\boldsymbol{X}',s,\{k_0,\ell_0\}).
\end{aligned} \tag{35}
$$

As long as assumptions about the lack of particle–particle interactions remain valid, an equation for the number density, $\boldsymbol{n}(\boldsymbol{x},t)$ can now be obtained. The manipulations involved are similar to those in Landman et al. (1977), eventually yielding

$$
\begin{aligned}
\tilde{\boldsymbol{\Gamma}}(s)\left(s\tilde{\boldsymbol{n}}(\boldsymbol{X},s) - \boldsymbol{f}(\boldsymbol{X})\right) = -\tilde{\boldsymbol{n}}(\boldsymbol{X},s) &+ \sum_{\boldsymbol{X}'\in\mathcal{N}(\boldsymbol{X})} \tilde{\boldsymbol{\Gamma}}(s)\boldsymbol{T}(\boldsymbol{X}-\boldsymbol{X}')\tilde{\boldsymbol{\phi}}(s)(\tilde{\hat{\boldsymbol{\Phi}}}(s))^{-1}\tilde{\boldsymbol{n}}(\boldsymbol{X}',s) \\
&+ \tilde{\boldsymbol{\Gamma}}(s)\boldsymbol{K}\tilde{\boldsymbol{\Lambda}}(s)\left(\tilde{\hat{\boldsymbol{\Phi}}}(s)\right)^{-1}\tilde{\boldsymbol{n}}(\boldsymbol{X},s)
\end{aligned}
\tag{36}
$$

where $\tilde{\boldsymbol{\Gamma}}(s) = \frac{1}{s}\left(\tilde{\hat{\boldsymbol{\Phi}}}(s)(\tilde{\boldsymbol{\psi}}(s))^{-1}\right)$ is the diagonal memory function matrix. If all of the distributions are Poisson distributed with distinct rates, the above equation simplifies to become

$$
\frac{\partial \boldsymbol{n}}{\partial t} = -(\boldsymbol{\phi}(0) + \boldsymbol{\Lambda}(0))\boldsymbol{n}(\boldsymbol{x},t) + \sum_{\boldsymbol{X}'\in\mathcal{N}(\boldsymbol{X})} \boldsymbol{T}(\boldsymbol{X}-\boldsymbol{X}')\boldsymbol{\phi}(0)\boldsymbol{n}(\boldsymbol{X}',t) + \boldsymbol{K}\boldsymbol{\Lambda}(0)\boldsymbol{n}(\boldsymbol{X},t).
$$

This is the analog of Eq. (22) for multistate lattice random walks.

Since $\boldsymbol{T}$ and $\boldsymbol{\phi}$ fully determine the evolution of $\boldsymbol{p}$, we next focus on their properties through several examples. General forms for the structure of $\boldsymbol{T}$ and $\boldsymbol{K}$ are given in Appendix C.

Example 2: Secondary Vertices

Let us consider an extension of Ex. 1 in Sect. 2.5. That previous system can be thought of as having junctions and SVs with a single SV between each pair of adjacent junctions. Here we extend that example and consider a system with n_k SVs between each junction. The case with $n_k = 3$ is illustrated in Fig. 6. We also now consider the lattice to be of infinite extent rather than over a ring of length $2M$. Although this problem is still relatively straightforward, the goal is to clearly show how different types of transition matrices that arise in more complex problems, namely, $\boldsymbol{T}_{SV}$, $\boldsymbol{D}_s$, $\boldsymbol{D}_e$, $\boldsymbol{D}'_s$, and $\boldsymbol{D}'_e$, defined below, are constructed. As before, let the WTDs for the SVs be $\chi(t) = \mu e^{-\mu t}$ and the WTDs for the junctions be $\phi(t) = \lambda e^{-\lambda t}$. The governing ODEs on each junction and edge are of the form

$$
\begin{aligned}
\begin{pmatrix} \frac{d}{dt}p_i^{(0)} \\ \frac{d}{dt}p_i^{(1)} \\ \vdots \\ \\ \frac{d}{dt}p_i^{(n_k)} \end{pmatrix} = &-\begin{pmatrix} \lambda & & & & \\ & \mu & & & \\ & & \mu & & \\ & & & \ddots & \\ & & & & \mu \end{pmatrix}\begin{pmatrix} p_i^{(0)} \\ p_i^{(1)} \\ p_i^{(2)} \\ \vdots \\ p_i^{(n_k)} \end{pmatrix} + \frac{1}{2}\begin{pmatrix} 0 & \mu & 0 & & \\ \lambda & 0 & \mu & & \\ & \mu & 0 & \mu & \\ & & \ddots & \ddots & \ddots \\ & & & \mu & 0 \end{pmatrix}\begin{pmatrix} p_i^{(0)} \\ p_i^{(1)} \\ p_i^{(2)} \\ \vdots \\ p_i^{(n_k)} \end{pmatrix} \\
&+ \frac{1}{2}\begin{pmatrix} \mu p_{i-1}^{(n_k)} \\ 0 \\ \vdots \\ 0 \\ \lambda p_{i+1}^{(0)} \end{pmatrix}
\end{aligned}
\tag{37}
$$

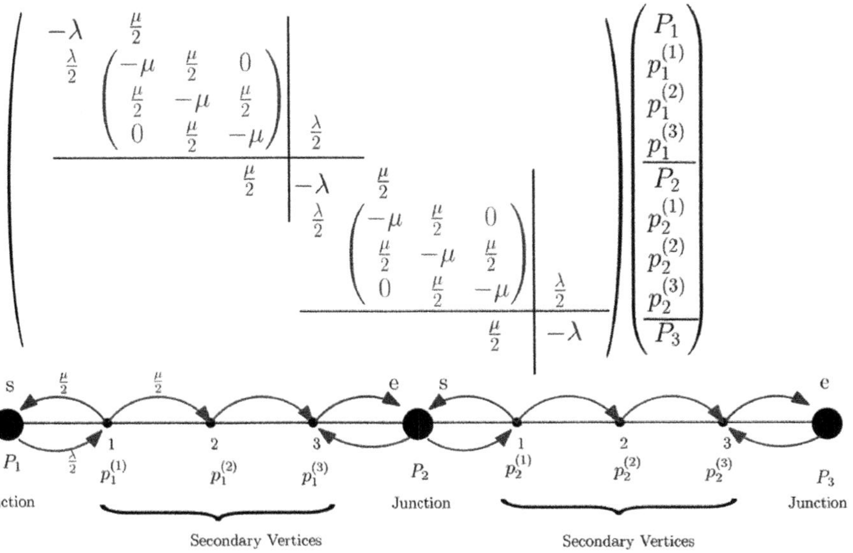

Fig. 6 A depiction of Ex. 2 with $n_k = 3$ over several junctions. Green arrows depict transitions in $\boldsymbol{T}_{SV}$, purple arrows in $\boldsymbol{D}'$ and dark red arrows in $\boldsymbol{D}$. The letters "s" and "e" label the start and end of each edge. The coloring of the entries in the matrix also indicates which types of transitions each entry corresponds to. All blank entries in the matrix are 0

where $p_i^{(0)}(t)$ are the junction probabilities and $p_i^{(k)}(t)$ for $k > 0$ are the probabilities at each SV.

We now rewrite Eq. (37) to highlight certain types of transitions that occur in this system and other problem we will encounter. Let us write the solution vector $\boldsymbol{p}$ as

$$\boldsymbol{p} = \begin{pmatrix} \vdots \\ \boldsymbol{p}_{i-1} \\ \boldsymbol{p}_i \\ \boldsymbol{p}_{i+1} \\ \cdots \end{pmatrix}$$

where each $\boldsymbol{p}_i$ is written in terms of the junction and SV states as

$$\boldsymbol{p}_i = \begin{pmatrix} \boldsymbol{p}_{i,J} \\ \boldsymbol{p}_{i,SV} \end{pmatrix}.$$

Since there is only a single internal state (but many SVs) in this case, there are three types of transitions that must be considered: jumps between adjacent SVs, jumps from an SV to a junction, and jumps from a junction to an SV. We will organize these categories of jumps into transition matrices denoted $\boldsymbol{T}_{SV}$, $\boldsymbol{D}$, and $\boldsymbol{D}'$, respectively. Since transitions from an SV to a junction can occur from either end of an edge ($k = 1$ or $k = n_k$ here), we define two different matrices $\boldsymbol{D}_s$ and

$\boldsymbol{D}_e$ associated with the transitions from the two ends [4]. Since the rate of jumps is μ at each SV, and there is a 1/2 probability of jumping to the right or left, we have $\boldsymbol{D}_{s,e} \in \mathbb{R}^{1\times n_k}$ and (see Fig. 6)

$$\boldsymbol{D}_s = \left(\tfrac{\mu}{2}\ 0 \ldots 0\right), \quad \boldsymbol{D}_e = \left(0 \ldots 0\ \tfrac{\mu}{2}\right).$$

The matrices $\boldsymbol{D}_s$ and $\boldsymbol{D}_e$ can be thought of as linear operators acting on the SVs along an edge to yield the flux of a particle to land on the junctions attached to that edge.

Next, we consider jumps from junctions to SVs. Since each junction is connected to two edges, there are two types of transitions: transitions from junction i to the n_k^{th} SV on edge $i-1$, and transitions from junction i to the first SV on edge i. Thus, we define $\boldsymbol{D}_e'$ to describe jumps to edge $i-1$ and $\boldsymbol{D}_s'$ to describe jumps to edge i, Since the rate of jumping from a junction is λ, and there is a 1/2 probability of jumping right or left, we have $\boldsymbol{D}_{s,e}' \in \mathbb{R}^{n_k\times 1}$ and (see Fig. 6)

$$\boldsymbol{D}_s' = \begin{pmatrix} \frac{\lambda}{2} \\ 0 \\ \vdots \\ 0 \end{pmatrix}, \quad \boldsymbol{D}_e' = \begin{pmatrix} 0 \\ \vdots \\ 0 \\ \frac{\lambda}{2} \end{pmatrix}$$

Similar to $\boldsymbol{D}_{s,e}$, $\boldsymbol{D}_{s,e}'$ may be thought of as linear operators that give the flux of particles from a junction to the SVs on edges attached to it.

Lastly, we must define the matrix of transitions between SVs on the edge. Since transitions between neighboring points are possible in Eq. (37), this yields an $n_k \times n_k$ transition matrix,

$$\boldsymbol{T}_{SV} = \frac{1}{2}\begin{pmatrix} 0 & \mu & 0 & & \\ \mu & 0 & \mu & & \\ & \mu & 0 & \mu & \\ & & \ddots & \ddots & \ddots \\ & & & \mu & 0 \end{pmatrix}$$

which incidentally, is also the transition matrix for a random walk with WTD $\chi(t)$ and absorbing boundaries at $n=0$ and $n=n_k+1$.

The previous matrices describe transitions associated with a single junction and edge, and now, we must incorporate this description into the topology of the primary graph. In particular, we must describe how junction i and edge i are connected to the edges and junctions that they are adjacent to. To formulate this as a matrix problem, define the matrices $\boldsymbol{A}_s$ and $\boldsymbol{A}_e$ as permutations of the

[4] The subscripts s and e are meant to indicate the start ($k=1$) and end ($k=n_k$) of an edge. However, this is merely for labeling purposes and does not mean that the transport is directed.

infinite-dimensional identity matrix over the lattice as follows

$$
\boldsymbol{A}_s = \begin{pmatrix} \ddots & \ddots & & & & \\ & 0 & 1 & & & \\ & & 0 & 1 & & \\ & & & 0 & 1 & \\ & & & & 0 & \ddots \\ & & & & & \ddots \end{pmatrix},
$$

$$
\boldsymbol{A}_e = \boldsymbol{A}_s^T = \begin{pmatrix} \ddots & & & & & \\ \ddots & 0 & & & & \\ & 1 & 0 & & & \\ & & 1 & 0 & & \\ & & & 1 & 0 & \\ & & & & \ddots & \ddots \end{pmatrix}. \tag{38}
$$

From Eq. (37), we see that probability, $p_i^{(n_k)}$, of being located at the n_k^{th} SV of edge i depends directly on $p_{i+1}^{(0)}$. By definition, transitions from the n_k^{th} SV to a junction are described by $\boldsymbol{D}_e$, and $\boldsymbol{A}_e$ describes the connection between edge i and junction $i+1$. Thus, one can see that $\boldsymbol{A}_e \otimes \boldsymbol{D}_e$ is the appropriate matrix to describe these transitions across the entire lattice. Likewise, we see that $\boldsymbol{A}_s \otimes \boldsymbol{D}'_e$ describes jumps from junction i to the n_k^{th} SV on edge $i-1$.
After constructing the transition matrices involving $\boldsymbol{D}_s$ and $\boldsymbol{D}'_s$, and gathering all the terms together, this ultimately leads to a global transition matrix:

$$
\boldsymbol{W} = \left(\begin{array}{cc|cc|cc|cc|c} 0 & \boldsymbol{D}_s & 0 & \boldsymbol{0} & & & & & \\ \boldsymbol{D}'_s & \boldsymbol{T}_{SV} & \boldsymbol{D}'_e & \boldsymbol{0} & & & & & \\ \hline 0 & \boldsymbol{D}_e & 0 & \boldsymbol{D}_s & 0 & \boldsymbol{0} & & & \\ \boldsymbol{0} & \boldsymbol{0} & \boldsymbol{D}'_s & \boldsymbol{T}_{SV} & \boldsymbol{D}'_e & \boldsymbol{0} & & & \\ \hline & & 0 & \boldsymbol{D}_e & 0 & \boldsymbol{D}_s & 0 & \boldsymbol{0} & \\ & & \boldsymbol{0} & \boldsymbol{0} & \boldsymbol{D}'_s & \boldsymbol{T}_{SV} & \boldsymbol{D}'_e & \boldsymbol{0} & \\ \hline & & & & \ddots & & \ddots & & \ddots \end{array}\right) = \boldsymbol{I} \otimes \begin{pmatrix} 0 & \boldsymbol{D}_s \\ \boldsymbol{D}'_s & \boldsymbol{T}_{SV} \end{pmatrix}
$$

$$
+\boldsymbol{A}_s \otimes \begin{pmatrix} 0 & \boldsymbol{0} \\ \boldsymbol{D}'_e & \boldsymbol{0} \end{pmatrix} + \boldsymbol{A}_e \otimes \begin{pmatrix} 0 & \boldsymbol{D}_e \\ \boldsymbol{0} & \boldsymbol{0} \end{pmatrix}
$$

where $\boldsymbol{I}$ is an infinite-dimensional identity matrix on the lattice. We also rewrite the evolution equation in Eq. (37), now over the entire lattice rather than a single junction and edge, as

$$
\frac{d}{dt}\boldsymbol{p} = -\boldsymbol{\Lambda}\boldsymbol{p} + \boldsymbol{W}\boldsymbol{p} \tag{39}
$$

where

$$\boldsymbol{\Lambda} = \boldsymbol{I} \otimes diag(\lambda, \mu, \cdots, \mu).$$

The first term on the right-hand side, $-\boldsymbol{\Lambda p}$, reflects the fact that jumps from junctions occur at a rate λ and from SVs at a rate μ. As in Ex. 1, $-\boldsymbol{\Lambda} + \boldsymbol{W}$ has a kind of block Laplacian type of structure.

Further analysis of this system will be done in Sect. 4.2, but here we describe a few properties that will be clarified later. First, for fixed n_k, if $\mu \gg \lambda$, then we expect that the transitions between the SVs and from SVs to junctions will take a negligible amount of time compared to transitions from junctions to SVs. Given this, the transport at large enough length and timescales will evolve as though it were a random walk with no SVs and a waiting time $\lambda e^{-\lambda t}$ for a particles to jump between two adjacent junctions.

The second property we note is that setting $dp_i^{(k)}/dt = 0$ for all i and k, we can solve for steady-state solutions. If we suppose that the particle is confined to edge i and junction i, the steady-state solution can be found as

$$p_{ss}^{(k)} = \begin{cases} \frac{\mu}{n_k\lambda+\mu} & k = 0 \\ \frac{\lambda}{n_k\lambda+\mu} & k = 1, \ldots, n_k \end{cases}$$

This type of analysis for steady-state solutions has been used elsewhere (Roerdink and Shuler 1985a, b) to obtain the long-time asymptotic mean and variance for certain multistate random walk processes. Accordingly, some of our long-time asymptotics can be found from the analysis presented there. However, the methods developed throughout this paper can also be used to find the full solution, and not just long-time asymptotics.

Although the previous example ultimately just leads to a reformulation of Eq. (37) over the lattice, the categorization of transitions into different classes will prove useful in more complicated problems where the evolution equations are otherwise difficult to write out. The main categories of transitions which may occur are: internal state changes at junctions ($\boldsymbol{K}_J$), junction-to-SV ($\boldsymbol{D}'$), SV-to-junction ($\boldsymbol{D}$), and SV-to-SV spatial and internal jumps ($\boldsymbol{T}_{SV}$ and $\boldsymbol{K}_{SV}$). The transitions at the junctions and SVs described by these matrices are then connected to the topology of the underlying lattice through structural matrices $\boldsymbol{A}_s$ and $\boldsymbol{A}_e$ which are permutations of the infinite-dimensional identity matrix on the lattice. In more general settings, there may be $m = 1, 2, \ldots, n_e$ types of edges each associated with its own structural matrices $\boldsymbol{A}_{s,m}$ and $\boldsymbol{A}_{e,m}$, and $n = 1, 2, \ldots, n_e$ types of vertices each associated with its own $\boldsymbol{K}_J$.

3 Moments of the Density Distribution

In general, equations such as (9) cannot be solved analytically, but often the quantities of interest are the low-order moments of the spatial distribution rather than $p(\boldsymbol{x}, t)$

itself. To illustrate how these can be found, consider a scalar equation in a 1D homogeneous medium. We define the moments of $p(x, t|0)$ as

$$\begin{aligned} m^{(n)}(t) = \langle x^n(t) \rangle &= \int_{-\infty}^{+\infty} x^n p(x, t|0)\, dx \\ &= \int_{-\infty}^{+\infty} \int_0^t \int_{-\infty}^{+\infty} x^n T(x-y)\phi(t-\tau) p(y, \tau|0)\, dy\, d\tau\, dx, \end{aligned} \tag{40}$$

and let

$$m_k = \int_{-\infty}^{+\infty} x^k T(x)\, dx$$

be the kth moment of the jump kernel T about zero. Then after a change of variables, one finds that (40) can be written as

$$m^{(n)}(t) = \int_0^t \sum_{k=0}^{n} \binom{n}{k} m_k \phi(t-\tau) m^{(n)}(t)\, d\tau, \tag{41}$$

and therefore, all the moments of $x(t)$ can be obtained by solving a sequence of linear integral equations of convolution type.

It is often convenient to work with the Laplace transformed equations for the moments since this allows for the use of Laplace transform limit theorems to obtain the asymptotic time dependence of the moments. The Laplace transformed solutions for the first two moments are

$$\begin{aligned} \tilde{m}^{(1)}(s) &= \frac{m_1}{s} \frac{\bar{\phi}(s)}{1-\bar{\phi}(s)} \\ \tilde{m}^{(2)}(s) &= \left(2 m_1 \tilde{m}^{(1)}(s) + \frac{m_2}{s}\right) \frac{\bar{\phi}(s)}{1-\bar{\phi}(s)}. \end{aligned} \tag{42}$$

If the first moment of T vanishes, then these simplify to

$$\begin{aligned} \tilde{m}^{(1)}(s) &= 0 \\ \tilde{m}^{(2)}(s) &= \frac{m_2}{s} \frac{\bar{\phi}(s)}{1-\bar{\phi}(s)}. \end{aligned} \tag{43}$$

In simple cases, the moments can be found directly. For example suppose that $m_1 = 0$ and that $\phi(t) = \lambda e^{-\lambda t}$. Then one finds that

$$m^{(2)}(t) = m_2 \int_0^t \mathscr{L}^{-1}\left(\frac{\lambda}{s}\right) d\tau = m_2 \lambda t, \tag{44}$$

and therefore, the underlying process is asymptotically a diffusion process with diffusion coefficient $D = m_2\lambda/2$. Furthermore, analogous results hold for homogeneous[5] 1D lattice random walks with the integrals in m_k replaced by summation over the lattice.

Anomalous diffusion is said to occur when the large-time limit of the mean square displacement grows either sub- or superlinearly. If the second moment varies as

$$m^{(2)}(t) \sim \gamma t^{\beta}$$

for $\beta \neq 1$ and $t \to \infty$, it is called *subdiffusion* if $\beta < 1$ and *superdiffusion* if $\beta > 1$ (Metzler and Klafter 2000). The former occurs when particles spread slowly, and in particular, if the mean waiting time between jumps is infinite. If $m_1 = 0$, then from (43), if $\tilde{\phi} \sim 1 - \bar{\tau}^{-1} s^{\rho}$ for $\rho \in (0, 1)$, $\bar{\tau} > 0$, and $s \to 0$, then $\langle x^2(t) \rangle \sim m_2 t^{\rho}$ for $t \to \infty$, i.e., movement is asymptotically subdiffusive. An example for which all moments of ϕ are infinite arises from the WTD

$$\phi(t) = \frac{1}{(1+t)^2}.$$

The transform of ϕ is

$$\bar{\phi}(s) = \left(\frac{\pi}{2} - Si(s)\right) \cos s + Ci(s) \sin s$$

where Si and Ci are the sine and cosine integral functions (Ruel 1958). From the asymptotic expansion of the integrals, one finds that

$$m^{(2)}(t) \sim \log t,$$

and thus, the process is also subdiffusive. The superdiffusive case arises when the walk is highly persistent in time, or for walks having a fat-tailed jump distribution. The simplest example of the first case arises when the walker never turns, which leads to a wave equation for which the mean square displacement scales as t^2. An application to bacteria that exhibit long runs is discussed in Matthaus et al. (2009).

3.1 First and Second Moments for Systems

The evolution equations for the first and second moments of a stochastic system consisting of N_c cells coupled by diffusion with first-order reactions in the cells were derived in Gadgil et al. (2005). The deterministic governing equations for this system can be written as

$$\frac{d\mathbf{n}}{dt} = \boldsymbol{\Omega}\mathbf{n},$$

[5] Homogeneous in a lattice as in a continuum, means that $T(x, y)$ and $\phi(t|y)$ do not vary with current position of a particle, as in the continuum case.

where $\mathbf{n}$ is the vector of molecule numbers for all cells, the $n_s N_c \times n_s N_c$ matrix $\boldsymbol{\Omega} \equiv \boldsymbol{\Delta} \otimes \boldsymbol{D} + \boldsymbol{I} \otimes \boldsymbol{K\Lambda}(0)$, and $\boldsymbol{I}$ and $\boldsymbol{\Delta}$ are the identity matrix and graph Laplacian for the underlying lattice, respectively. Evolution equations for the mean and second moments were then derived through a generating function approach.

Generating functions of the form

$$g(z,t) = \sum_{n=0}^{\infty} z^{n_1} z^{n_2} \ldots z^{n_{ns}} p(n_1, n_2, \ldots, n_{n_s}, t)$$

were used in Gadgil et al. (2005) to describe the number of molecules in a particular state. In the present context, we have matrix-valued generating functions for the position and internal state of a single particle, with argument z corresponding to lattice indices rather than particle numbers. For a single-particle, one-dimensional transport problem, we have

$$g_{k,\ell}(z,t) = \sum_{j=-\infty}^{\infty} z^j p_{k,\ell}(j,t), \tag{45}$$

where $j \in \mathbb{Z}$ is a lattice point. The quantities, $p_{k,\ell}(j,t)$, are the probabilities that the particle is located in internal state ℓ at SV k, along edge j of the lattice at time t (e.g., consider the $k = 0, \ldots, n_k$ internal states as a discretization of an edge, as in Fig. 5 or later in Fig. 7). The negative powers are needed as it is assumed that the lattice has infinite extent in both directions. We also note that

$$\lim_{z \to 1^-} g_{k,\ell}(z,t)$$

is the total probability that the particle is located in SV k with internal state ℓ.

Let us assume that jumps only occur between neighboring points. Then for a drift–diffusion process, $\boldsymbol{T}$, which is equal to

$$\boldsymbol{T}(X) = \boldsymbol{D}_0 \delta(X) + \boldsymbol{D}^+ \delta(X - \Delta X) + \boldsymbol{D}^- \delta(X + \Delta X),$$

where $\boldsymbol{D}^+$ and $\boldsymbol{D}^-$ are equal for purely diffusive transport, but differ when there is even a transient drift. As before, the reaction matrix is $\boldsymbol{K}$, or we may write $\boldsymbol{K}\delta(\boldsymbol{X})$ to indicate that no transport occurs during reactions. To construct the generating function, we equate $n\Delta X$ with z^n, where ΔX is the lattice spacing, and write

$$\boldsymbol{\Omega}(z) = \left(\boldsymbol{K\Lambda}(0) + \boldsymbol{D}_0 + \boldsymbol{D}^+ + \boldsymbol{D}^-\right) + \boldsymbol{D}^+ (z-1) + \boldsymbol{D}^- \left(z^{-1} - 1\right).$$

Systems with multiple connections and dimensions require one function like $(z-1)$ or $(z^{-1} - 1)$ for each neighboring junction in the lattice. These functions coordinate the topology of the lattice in the generating function formalism and correspond to $\boldsymbol{A}_{t,m}$, the structural matrices specifying the lattice connectivity used before. Further discussion can be found in the general formulation given in Appendix C.1.

Differentiating the vector $g(z, t)$, whose elements are $g_{k\ell}(z, t)$, with respect to t and z and setting $z \to 1$ yields the spatial moments

$$\frac{d}{dt}\boldsymbol{g} = \lim_{z\to 1} \frac{\partial}{\partial t}\boldsymbol{g}(z,t) = (\boldsymbol{K}\boldsymbol{\Lambda}(0) + \boldsymbol{D}_0 + \boldsymbol{D}^+ + \boldsymbol{D}^-)\boldsymbol{g} \tag{46}$$

$$\frac{d}{dt}\boldsymbol{m}^{(1)} = L \lim_{z\to 1} \frac{\partial^2}{\partial z \partial t}\boldsymbol{g}(z,t) = (\boldsymbol{K}\boldsymbol{\Lambda}(0) + \boldsymbol{D}_0 + \boldsymbol{D}^+ + \boldsymbol{D}^-)\boldsymbol{m} + L(\boldsymbol{D}^+ - \boldsymbol{D}^-)\boldsymbol{g} \tag{47}$$

$$\frac{d}{dt}\boldsymbol{v} = L^2 \lim_{z\to 1} \frac{\partial^3}{\partial z^2 \partial t}\boldsymbol{g}(z,t) = (\boldsymbol{K}\boldsymbol{\Lambda}(0) + \boldsymbol{D}_0 + \boldsymbol{D}^+ + \boldsymbol{D}^-)\boldsymbol{v} + 2L(\boldsymbol{D}^+ - \boldsymbol{D}^-)\boldsymbol{m} + 2L^2\boldsymbol{D}^-\boldsymbol{g} \tag{48}$$

where L is the distance between two junctions, $\boldsymbol{m}$ are the k-th moments, and $\boldsymbol{v}$ is realted to the second moment by

$$\boldsymbol{v}(t) = \boldsymbol{m}^{(2)}(t) - \boldsymbol{m}^{(1)}(t).$$

Note that these spatial moments do not take into account microstructural geometry of edges, specifically the location of SVs relative to the junctions of the lattice. The inclusion of microscale details is discussed in Appendix C.1, although there we use a lattice Fourier transform approach rather than a generating function[6].

We also see that the resulting elements of $\boldsymbol{g}$, $\boldsymbol{m}$, and $\boldsymbol{v}$ are for particles at a specific SV and in a specific internal state. However, experimental measurements are usually not at this level of resolution, and we will discuss this point further in subsequent sections.

Although these results hold for single-particle transport, the multiple particle case described by the general formulation from Gadgil et al. (2005) could in principle be applied. Additionally, for a d-dimensional lattice, the generating function $\boldsymbol{g}$ is of the form

$$\boldsymbol{g}(z_1, \ldots, z_d, t) = \sum_{j_1=-\infty}^{\infty} \cdots \sum_{j_d=-\infty}^{\infty} z_1^{j_1} z_2^{j_2} \ldots z_d^{j_d}\, \boldsymbol{p}(j_1, \ldots, j_d, t)$$

where the elements of $\boldsymbol{g}$ and $\boldsymbol{p}$ again describe SVs and internal states at each junction. The derivations of $\boldsymbol{m}$ and $\boldsymbol{v}$ are similar to the one-dimensional case with one important difference. In multiple dimensions, one gets several distinct vectors, $\boldsymbol{m}_j$ and $\boldsymbol{v}_{jk}$ with $j, k = 1, \ldots d$. This occurs since differentiation is now taken with respect to z_j,

[6] Both are formally equivalent if one replaces z by $e^{i\omega}$ in $\boldsymbol{g}(z, t)$). One advantage of the Fourier transform method is that the lattice spacings can be included as multiplying factors in the Fourier transform, e.g., $e^{i\Delta X\omega}$ whereas the generating function approach is defined only for integer powers of z. What this means is that the moments defined above must be multiplied by appropriate length factors related to the lattice spacing to obtain the correct results, whereas in the Fourier transform approach, these factors are included automatically.

$j = 1, \ldots, d$. Furthermore, $\boldsymbol{v}_{jk}$ is now defined as

$$\boldsymbol{v}_{jk}(t) = \begin{cases} m_{jk}^{(2)}(t) - m_k^{(1)}(t) & j = k \\ m_{jk}^{(2)}(t) & j \neq k. \end{cases}$$

3.2 The Moments in a One-Dimensional System

In the following example, we derive the moment equations for a one-dimensional lattice with SVs.

Example 3

In Ex. 2, we obtained the transition matrix $\boldsymbol{T}$ for that system and the evolution Eq. (39)

$$\frac{d}{dt}\boldsymbol{p} = -\boldsymbol{\Lambda}\boldsymbol{p} + \boldsymbol{W}\boldsymbol{p}. \tag{49}$$

With the definitions of $\boldsymbol{\Lambda}$ and $\boldsymbol{W}$ from Ex. 2 of Sect. 2.5, we can write $\boldsymbol{W}$ as

$$\begin{aligned} \boldsymbol{W}(X) &= (\boldsymbol{K} + \boldsymbol{D}_0)\delta(X) + \boldsymbol{D}^{+}\delta(X - \Delta X) + \boldsymbol{D}^{-}\delta(X + \Delta X) \\ &= \begin{pmatrix} 0 & \boldsymbol{D}_s \\ \boldsymbol{D}_s' & \boldsymbol{T}_E \end{pmatrix}\delta(X) + \begin{pmatrix} 0 & \boldsymbol{D}_e \\ 0 & 0 \end{pmatrix}\delta(X - \Delta X) + \begin{pmatrix} 0 & 0 \\ \boldsymbol{D}_e' & 0 \end{pmatrix}\delta(X + \Delta X). \end{aligned}$$

We now solve for the lower-order moments of this system, and to do so, we use a Fourier–Laplace transform. This yields the 4×4 matrix equation

$$\begin{aligned} &\begin{pmatrix} s\tilde{p}^{(0)}(\omega, s) \\ s\tilde{p}^{(1)}(\omega, s) \\ s\tilde{p}^{(2)}(\omega, s) \\ s\tilde{p}^{(3)}(\omega, s) \end{pmatrix} - \begin{pmatrix} p^{(0)}(0) \\ p^{(1)}(0) \\ p^{(2)}(0) \\ p^{(3)}(0) \end{pmatrix} \\ &= \left[-\boldsymbol{\Lambda} + \begin{pmatrix} 0 & \boldsymbol{D}_s + e^{-i\Delta X\omega}\boldsymbol{D}^e \\ \boldsymbol{D}_s' + e^{i\Delta X\omega}\boldsymbol{D}_e' & \boldsymbol{T}_E \end{pmatrix} \right] \begin{pmatrix} \tilde{p}^{(0)}(\omega, s) \\ \tilde{p}^{(1)}(\omega, s) \\ \tilde{p}^{(2)}(\omega, s) \\ \tilde{p}^{(3)}(\omega, s) \end{pmatrix} \end{aligned}$$

where $\tilde{p}^{(k)}(\omega, s)$ are the Fourier–Laplace transformed probabilities $p^{(k)}(X, t)$. Assuming an initial condition, $p^{(k)}(0) = 1/4$, we now solve for $\tilde{p}^{(k)}(\omega, s)$. The next step is to either invert $\tilde{p}^{(k)}(\omega, s)$ to obtain $p^{(k)}(X, t)$, or if we are interested in the spatial moments, we can make use of certain identities that allow for their direct computation in Fourier space. In particular, applying the operator

$$\left.\left((-i)\frac{\partial}{\partial\omega}\right)^r\right|_{\omega\to 0}$$

to $\tilde{p}^{(k)}$ yields the Laplace transform of the rth spatial moment of $p^{(k)}(x,t)$. For instance, if $r=0$, we obtain the probability of a particle being in SV k on any edge. For $r=1$, we obtain the mean position of particles in SV k, and for $r=2$ the second moment of particles in SV k.

This dependence on the SV state k can be a complication since most experimental results would report a single mean or variance result rather than multiple results for each internal state of some given system. For SVs, we make use of the fact that given a junction at X, the position of SV k is then $X+(k/4)\Delta X$, see Fig. 6. Thus, the overall first moment is

$$m^{(1)}(t)=\sum_{k=0}^{3}\sum_{X\in\mathbb{Z}}\left(X+\frac{k}{4}\Delta X\right)p^{(k)}(X,t)=\sum_{k=1}^{3}\left(m^{(1,k)}(t)+\frac{k}{4}\Delta X m^{(0,k)}(t)\right)$$

where $m^{(r,k)}(t)$ are the rth moments for particles in state k. Note that since there are multiple states, $m^{(0,k)}$, the probability of a particle being in state k is less than unity for each individual SV, but

$$\sum_{k=0}^{3}m^{(0,k)}(t)=1$$

since particles are conserved in this system. Making use of Fourier transform identities, we find that

$$\begin{aligned}\tilde{m}^{(1)}(s)&=\sum_{k=0}^{3}\left((-i)\frac{\partial\tilde{p}^{(k)}(\omega,s)}{\partial\omega}+\frac{k}{4}\Delta X\tilde{p}^{(k)}(0,s)\right)\\&=\frac{3\Delta X}{8s}\end{aligned}\tag{50}$$

with inverse Laplace transform, $m^{(1)}(t)=3\Delta X/8$.

We apply the same technique for the second moment and obtain the following.

$$\begin{aligned}\tilde{m}^{(2)}(s)&=\sum_{k=0}^{3}\sum_{X\in\mathbb{Z}}\left[X^2+2X\frac{k\Delta X}{4}+\left(\frac{k\Delta X}{4}\right)^2\right]p^{(k)}(X,s)\\&=\sum_{k=0}^{3}\left((-i)^2\frac{\partial^2\tilde{p}^{(k)}(\omega,s)}{\partial\omega^2}+2(-i)\frac{k\Delta X}{4}\frac{\partial\tilde{p}^{(k)}(\omega,s)}{\partial\omega}+\left(\frac{k\Delta X}{4}\right)^2\tilde{p}^{(k)}(0,s)\right)\\&=\frac{\Delta X^2\left(14s^3+(39\lambda+31\mu)s^2+4\mu(17\lambda+3\mu)s+8\lambda\mu^2\right)}{32s^2\left(2s^2+2(\lambda+3\mu)s+\mu(3\lambda+\mu)\right)}.\end{aligned}\tag{51}$$

To leading order, this yields

$$m^{(2)}(t) \sim \frac{\Delta X^2}{4}\frac{\lambda\mu}{\lambda+\mu}t = \frac{\Delta X^2}{4}\left(\frac{1}{\mu}+\frac{3}{\lambda}\right)^{-1} \equiv Dt.$$

Thus, to make an analogy to electronic circuits, the individual exit rates for each state combine like resistors in parallel for SVs that are in series. In contrast, for a set of four single-state, 1D lattices aligned in parallel with jump rate λ for the first and μ for the remainder, $m^{(2)}(t) = \frac{1}{4}(\lambda + 3\mu)t$. Thus, SVs in parallel combine like resistors in series. Of course, the analogy becomes harder to follow in cases where there are both internal states and SVs since there are paths in parallel and in series in those cases.

In the above example, the results for all t can also be found explicitly in terms of eigenvalues and eigenfunctions of $\boldsymbol{\Omega}$ (Gadgil et al. 2005). Though the results are somewhat lengthy, the explicit solution, after Fourier and Laplace transformation, for this problem can easily be found by symbolic calculation in Mathematica, for instance. Later, we will discuss a technique that in some cases allows for a simplification of the moment calculation so that it need not be enumerated for every internal state, but can be estimated around each vertex. A general description of the moment computations can be found in Appendix C.1.

4 From "Micro" to "Macro" Moments

As we saw in the preceding section, the macroscale moments, $m^{(r)}$, which are the typical experimental observables, involve sums over the moments for each internal state, $m^{(r,k)}$, and require details of the local geometry. However, this approach becomes cumbersome as more internal states and geometrical features are involved, and suggests that a more direct approach is needed. This is further supported by the observation that, in experiments, it is often difficult to accurately measure concentration distributions over short time and length scales, and thus, in some cases experimental results may only be valid for sufficiently large length and timescales. For instance, tissue-scale diffusivity values would not be expected to hold at very short length scales when a diffusing substance is still confined to a specific cell in the tissue. Thus, a simplified means of obtaining the dynamics over intermediate- and long-term time and length scales is appropriate.

Here we discuss a more direct method of determining these macroscale properties that obviates the need to sum over the SVs. In particular, we reduce the size of the state space of the multistate CTRW by deriving effective waiting-time distributions for jumps between junctions. In this sense, a single jump corresponds to a path beginning at a given junction and terminating at an adjacent junction or returning to the starting point without visiting an adjacent junction. The path is in turn made up of jumps between SVs and internal states on a given edge. As will be shown, the effective waiting-time distributions for these compound jumps are closely related to the solutions of first-passage-time problems.

Example 4: The Alternating 1D Lattice Revisited

To observe how these effective transition rates may be computed, we return to Ex. 1 on the lattice with alternating junctions and SVs (see Fig. 4). Recall that the evolution equations for this random walk are of the form

$$\frac{dp_i^{(1)}}{dt} = -\lambda p_i^{(1)} + \frac{\mu}{2} \sum_{j \in \mathcal{N}(i,1)} p_j^{(2)}(t)$$
$$\frac{dp_i^{(2)}}{dt} = -\mu p_i^{(2)} + \frac{\lambda}{2} \sum_{j \in \mathcal{N}(i,2)} p_j^{(1)}(t),$$

with index $i \in \mathbb{Z}$. As before, we consider the type 1 points to represent the junctions of a graph, and the type 2 points as SVs along edges of this graph. If we assume a particle starts at a type 1 point, then we see from the form of the equations that the occupancy probabilities at a type 2 point depends solely upon the probabilities over time at the two type 1 points it is adjacent to. This is because in order to reach type 2 point i that particle has to first reach one of the neighboring type 1 points at i or $i+1$. Thus, we may determine formulas that give the probability at each 2 point as a function of the probabilities at the adjacent type 1 points. Starting with the evolution equation for $p_i^{(2)}(t)$ which has initial condition, $p_i^{(2)}(0) = 0$ since the particles starts on a type 1 point, we have the following:

$$\frac{dp_i^{(2)}}{dt} = -\mu p_i^{(2)} + \frac{\lambda}{2}(p_{i-1}^{(1)}(t) + p_i^{(1)}(t)) \overset{\mathcal{L}}{\Rightarrow} (s+\mu)\tilde{p}_i^{(2)}(s) = \frac{\lambda}{2}(\tilde{p}_{i-1}^{(1)}(s) + \tilde{p}_i^{(1)}(s)) \tag{52}$$

where the arrow indicates Laplace transformation. For the type 1 points, we have

$$\begin{aligned}\frac{dp_i^{(1)}}{dt} &= -\lambda p_i^{(1)} + \frac{\mu}{2}(p_i^{(2)}(t) + p_{i+1}^{(2)}(t)) \quad \overset{\mathcal{L}}{\Rightarrow} \quad (s+\lambda)\tilde{p}_i^{(1)}(s) \\ &= \delta_{i,0} + \frac{\mu}{2}(\tilde{p}_i^{(2)}(s) + \tilde{p}_{i+1}^{(2)}(s)),\end{aligned} \tag{53}$$

where $\delta_{i,0}$ gives the initial condition of the particle in the type 1 state at $X_0 = 0$. Combining these yields a system of equations for $\tilde{p}_i^{(1)}(s)$ where $\tilde{p}_i^{(2)}(s)$ has been eliminated. After simplifying, we have

$$\tilde{p}_i^{(1)}(s) = \frac{1}{\lambda + s}\delta_{i,0} + \frac{\mu\lambda}{4(s+\mu)(s+\lambda)}\left(\tilde{p}_{i-1}^{(1)} + 2\tilde{p}_i^{(1)} + \tilde{p}_{i+1}^{(1)}\right) \tag{54}$$

or, after inverting the Laplace transform

$$p_i^{(1)}(t) = e^{-\lambda t}\delta_{i,0} + \int_0^t \frac{\lambda\mu}{4(\lambda-\mu)}\left(e^{-\mu(t-\tau)} - e^{-\lambda(t-\tau)}\right)$$
$$\left(p_{i-1}^{(1)}(\tau) + 2p_i^{(1)}(\tau) + p_{i+1}^{(1)}(\tau)\right)d\tau. \tag{55}$$

This is the integral master equation for a random jump process, but while the initial system had two types of points and was Markovian, this new process has a single type of point, and, having a non-Poisson waiting-time distribution, is semi-Markovian. We also see that $p_i^{(1)}$ appears on the right- and left-hand sides of Eq. (55). This obfuscates the meaning of a jump, since jumps that return a particle to its origin could occur in this setting, but the remedy is quite straightforward here – simply rearrange terms in Laplace space before inverting so that p_1^i is isolated on the left. This leads to

$$\tilde{p}_i^{(1)}(s) = \frac{s+\mu}{(s+\lambda)(s+\mu)-\frac{1}{2}\lambda\mu}\delta(X) + \tilde{f}(s)\left(\tilde{p}_{i-1}^{(1)}(s) + \tilde{p}_{i+1}^{(1)}(s)\right) \tag{56}$$

where

$$\tilde{f}(s) = \frac{1}{4}\frac{\mu\lambda}{(s+\lambda)(s+\mu)-\frac{1}{2}\lambda\mu}, \tag{57}$$

or, after inverting the Laplace transform,

$$f(t) = \frac{1}{2}\frac{\lambda\mu e^{-\frac{\lambda+\mu}{2}t}}{\sqrt{\lambda^2+\mu^2}}\sinh\left(\frac{\sqrt{\lambda^2+\mu^2}}{2}t\right). \tag{58}$$

Note that $f(t)$ is not quite a WTD since it integrates to 1/2. In fact, one can think of $f(t)$ as $\frac{1}{2}\phi(t)$ where the factor of 1/2 is from the jump distribution operator (e.g., $T(X_{i\pm1}, X_i) = 1/2$) which accounts for the fact that there is a 50% chance of jumping to the left and 50% to the right at any given step. This in turn leads to an integral master equation of the form

$$p_i^{(1)}(t) = \hat{\Psi}(t)\delta_{i0} + \int_0^t f(t-\tau)\left(p_{i-1}^{(1)}(\tau) + p_{i+1}^{(1)}(\tau)\right)d\tau$$

where

$$\hat{\Psi}(t) = e^{-\frac{\lambda+\mu}{2}t}\left(\cosh\left(\frac{t}{2}\sqrt{\lambda^2+\mu^2}\right) + \frac{\mu-\lambda}{\sqrt{\lambda^2+\mu^2}}\sinh\left(\frac{t}{2}\sqrt{\lambda^2+\mu^2}\right)\right)$$

At first, this last rearrangement may seem like an unnecessary complication, but it enables us to understand the waiting time $f(t)$ in terms of a standard jump process where each jump implies a change of position. Only when self-jumps are removed can the terms in the transition matrix be understood as the waiting times for a jump to occur. As will be discussed in the following section, the modified waiting-time distribution is also equal to the first-passage-time for a particle to travel between type 1 points.

Notice that $\hat{\Psi}(t)$ is *not* the complementary cumulative waiting-time distribution of $f(t)$. This is because even though we have written a master equation for $p^{(1)}(t)$ alone, in reality the system still has two states, and thus, at any given moment, some of the particles will reside in type 1 states, and some in type 2 states. Thus, $\hat{\Psi}$ includes this additional information and differs from $\hat{\Phi}$, as discussed below.

Of course, given $f(t)$, we can compute $\hat{\Phi}$ as

$$\hat{\Phi}(t) = 1 - \int_0^t f(\tau)d\tau = e^{-\frac{\lambda+\mu}{2}t}\left(\cosh\left(\frac{t}{2}\sqrt{\lambda^2+\mu^2}\right)\right.$$
$$\left.+\frac{\lambda+\mu}{\sqrt{\lambda^2+\mu^2}}\sinh\left(\frac{t}{2}\sqrt{\lambda^2+\mu^2}\right)\right).$$

In order to understand the distinction between $\hat{\Psi}$ and $\hat{\Phi}$, we begin by summing over $i \in \mathbb{Z}$ in Eq. (56). This yields an equation in Laplace transform space for the zeroth-order moment, $\tilde{m}^{(0)}(s)$ of $\tilde{p}^{(1)}(s)$. Upon simplifying,

$$\tilde{m}^{(0)}(s) = \frac{\tilde{\hat{\Psi}}(s)}{1-2\tilde{f}(s)}$$

and, after inverse Laplace transformation, $m^{(0)}(t) = 1/(\mu+\lambda)\left(\mu + \lambda e^{-(\mu+\lambda)t/2}\right)$. Thus, the probability of a particle being on a type 1 point varies in time, since, depending on the values of μ and λ, there is a nonzero probability that the particle can be located on a type 2 point at any given time.

On the other hand, if we replace $\tilde{\Psi}$ by $\tilde{\Phi}$, the same steps yield: $m^{(0)}(t) = 1$, or the particle remains on type 1 states with probability 1. At first glance, it appears that we have eliminated (seemingly erroneously) the possibility of a particle to ever land on a type 2 state. However, there is a more subtle interpretation of this change. By going from $\hat{\Psi}$ to $\hat{\Phi}$, we have changed the meaning of the quantity, $p_i^{(1)}(t)$ being computed. Originally, $p_i^{(1)}(t)$ was the probability of being located on the ith type 1 point, but the new probability is that of having reached the ith type 1 point, but not yet having reached an another type 1 point. In other words, with $\hat{\Phi}$, each $p_i(t)$ refers to the combined probability of all paths that occupy the type 1 point at i at time t, and also those that have reached i at a prior time, but not reached the type 1 points at $i \pm 1$.

Depending on the application, it can be useful to use $\hat{\Psi}$ or $\hat{\Phi}$. If we are to reduce this problem solely to a random jump process on type 1 points, then $\hat{\Phi}$ should be used—for instance, if we attempt to compute moments without resorting to details about the type 2 points. On the other hand, if we wish at the end to also compute occupation probabilities for the SVs, then $\hat{\Psi}$ should be used since it includes information about how particles are distributed between type 1 and type 2 points.

Given this explicit solution over the type 1 points, we can also examine how variation of the parameters μ and λ affect the outcome. For instance of $\mu \gg \lambda$, then the process reduces to a single-state process with exit rate λ from each point. For instance, using $\Phi(t)$ and the fact that $\phi(t) = 2f(t)$, we can derive an differential master equation,

$$\frac{1}{\mu\lambda}\frac{d^2 p_i}{dt^2} + \left(\frac{1}{\mu} + \frac{1}{\lambda}\right)\frac{dp_i}{dt} = \frac{1}{2}(p_{i-1}(t) - 2p_i(t) + p_{i+1}(t))$$

This equation appears similar to a discretization of the telegrapher's equation; however, as discussed in Othmer et al. (1988), supposing $\lambda = \mu$, there is no scaling of λ with time that results in the telegrapher's equation. Rather, in this case, diffusive behavior results if we set $\Delta X = X_i - X_{i-1}$ and let $\Delta X \to 0$ while $\lambda \to \infty$ and $\lambda \Delta X^2 = \text{const.}$.

4.1 First-Passage-Time Problems

The first-passage-time (FPT) density of a stochastic process is the probability density function, $f(t, \boldsymbol{x}_1|\boldsymbol{x}_0)$ that characterizes the time it takes the trajectory of a random process, having started at a point $\boldsymbol{x}_0$, to reach a point, $\boldsymbol{x}_1$ for the first time (c.f. Montroll and Weiss 1965). In the special case that $\boldsymbol{x}_0 = \boldsymbol{x}_1$, the FPT density describes the time it takes a particle to return to $\boldsymbol{x}_0$ after having made at least one jump away. We now show how $f(t)$ above can be understood as a FPT density.

In general, FPT densities can be found as the fluxes of particles leaving a domain subject to absorbing boundary conditions (Kampen 1992). Let us consider the first-passage-time from the type 1 point i, to reach the adjacent type 1 points located at X_{i+1} or X_{i-1}. To solve this problem, we apply absorbing boundary conditions at $X_{i\pm1}$, e.g., $p^{(1)}_{i-1}(t) = 0$ and $p^{(1)}_{i+1}(t) = 0$, and write out the evolution equations for the random walk on the interval between these points, which includes an SV on either side of the type 1 point at the origin.

$$\begin{aligned}
\frac{dp_i^{(2)}}{dt} &= -\mu p_i^{(2)} + \frac{\lambda}{2} p_i^{(1)}(t), & p_{i-1}^{(2)}(0) &= 0 \\
\frac{dp_i^{(1)}}{dt} &= -\lambda p_i^{(1)} + \frac{\mu}{2}(p_{i-1}^{(2)}(t) + p_i^{(2)}(t)), & p_i^{(1)}(0) &= 1 \\
\frac{dp_{i+1}^{(2)}}{dt} &= -\mu p_{i+1}^{(2)} + \frac{\lambda}{2} p_i^{(1)}(t), & p_{i+1}^{(2)}(0) &= 0.
\end{aligned} \tag{59}$$

Upon solving Eq. (59) for $p_i^{(1)}$, $p_i^{(2)}$, and $p_{i+1}^{(2)}$, the FPT densities at $X_{i\pm1}$ are then found as the fluxes of particles at the type 1 points at $i \pm 1$. These fluxes are simply the convolution of $\phi(t)$ with $p_i^{(2)}(t)$ and $p_{i+1}^{(2)}(t)$, and therefore,

$$f_{+1}(t) = \frac{\mu}{2}\int_0^t e^{-\mu(t-\tau)} p_{i+1}^{(2)}(\tau)d\tau.$$
$$f_{-1}(t) = \frac{\mu}{2}\int_0^t e^{-\mu(t-\tau)} p_{i}^{(2)}(\tau)d\tau.$$

Due to symmetry, $f_{+1}(t) = f_{-1}(t)$ and we may solve the above equations for the first-passage-time density,

$$f(t) = \frac{1}{2}\frac{\lambda\mu e^{-\frac{\lambda+\mu}{2}t}}{\sqrt{\lambda^2+\mu^2}}\sinh\left(\frac{\sqrt{\lambda^2+\mu^2}}{2}t\right).$$

We see that $f(t)$ is precisely the modified WTD from Eq. (58). The effective WTD for this problem is thus $\phi(t) = 2f(t)$.

Having derived the form of the first-passage-time distribution to travel between adjacent type 1 points, we have reduced the problem to a random walk on a 1D lattice with waiting time $2f(t)$. To find the moments, it is convenient to apply a Laplace transformation to the integral master equation. The result of Eq. (43) can then be applied to yield

$$m^{(2)}(t) = \mathscr{L}^{-1}\left[\frac{L^2}{s}\frac{2f(s)}{1-2f(s)}\right] = \frac{L^2}{2}\frac{\lambda\mu}{\lambda+\mu}t - L^2\frac{\lambda\mu}{(\lambda+\mu)^2}\left(1-e^{-(\lambda+\mu)t}\right).$$

Recall that in previous examples, we had a factor of ΔX^2 multiplying the second moment. The factor of L^2 which appears here reflects the fact that we specify that the lattice spacing $L = \Delta X$, and this appears as the nth power in an nth-order moment.

In Appendices D.1 and 2, we show that the FPT procedure can be generalized to more complicated random walks.

4.2 A 1D CTRW with Many SVs

Let us continue with the preceding example, but now with an arbitrary number, n_k, of SVs, as in Fig. 7. The system of equations describing the probabilities associated with each junction and SV are of the form

$$\begin{pmatrix} \frac{d}{dt}p_i^{(0)} \\ \frac{d}{dt}p_i^{(1)} \\ p_i^{(2)} \\ \vdots \\ \frac{d}{dt}p_i^{(n_k)} \end{pmatrix} = -\begin{pmatrix} \lambda & & & & \\ & \mu & & & \\ & & \mu & & \\ & & & \ddots & \\ & & & & \mu \end{pmatrix}\begin{pmatrix} p_i^{(0)} \\ p_i^{(1)} \\ p_i^{(2)} \\ \vdots \\ p_i^{(n_k)} \end{pmatrix} + \frac{1}{2}\begin{pmatrix} 0 & \mu & 0 & & \\ \lambda & 0 & \mu & & \\ & \mu & 0 & \mu & \\ & & \ddots & \ddots & \ddots \\ & & & \mu & 0 \end{pmatrix}\begin{pmatrix} p_i^{(0)} \\ p_i^{(1)} \\ p_i^{(2)} \\ \vdots \\ p_i^{(n_k)} \end{pmatrix}$$

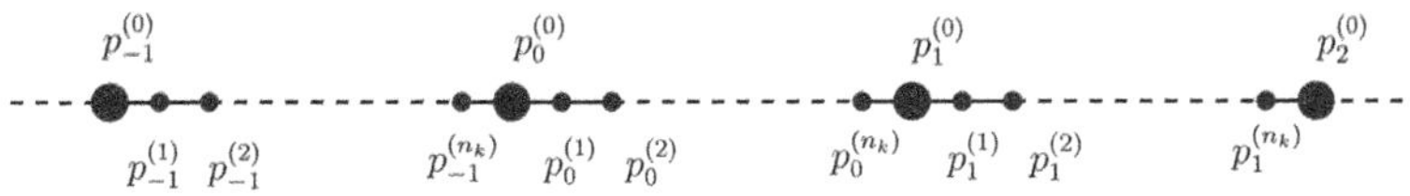

Fig. 7 Depiction of the random walk problem described in Sect. 4.2. Red dots are junctions and blue dots refer to SVs along each edge

$$+\frac{1}{2}\begin{pmatrix} \mu p_{i-1}^{(n_k)} \\ 0 \\ \vdots \\ 0 \\ \lambda p_{i+1}^{(0)} \end{pmatrix} \tag{60}$$

for index, $i \in \mathbb{Z}$.

This problem can serve as a simple model for diffusive transport in a series of cells. The transport of molecules in cell interiors can be modeled as jumps on the n_k SVs, and the junctions represent cell membranes. Since the cell membranes are then represented by a single junction point, this case corresponds to a situation where adjacent cells are closely packed together. Further extensions can be included when the intracellular gap between adjacent cells must be considered, as in Sect. 5.1.

Computing the moments in this case would generally be quite complicated, since it requires averaging over every internal state. Thus, to obtain an approximation for a simplified system, we first compute how long it takes to travel the length of an edge, from junction X_i to X_{i+1}. Once the effective waiting-time distribution for this transport to occur is found, we can then treat this system as a standard one-dimensional CTRW on a lattice.

To solve for the effective WTD at junctions, we compute the first-passage-time density to traverse an edge, i.e. to travel from one side of a cell to the other. Consider a particle starting from membrane i (the left side of cell i, or equivalently the right membrane of cell $i-1$ in this model) and traveling to membrane $i-1$, or $i+1$. Since the process terminates upon arrival at either membrane, we impose absorbing boundary conditions at $X_{i\pm1}$. In terms of the microscale SVs, these absorbing conditions are applied to $p_{i-1}^{(0)}$ and $p_{i+1}^{(0)}$.

We can use a recursion method to find the solution to this problem, as was done in Teimouri and Kolomeisky (2013). We also note that due to the symmetry about X_i, $p_i^{(k)} = p_{i-1}^{(n_k-k)}$ for each k, and without loss of generality, we only have to consider transport from X_i to X_{i+1} with modified boundary condition at X_i. In particular, there is an absorbing boundary condition at X_{i+1}, and a reflecting boundary condition at X_i. The reflecting condition is obtained by simplifying the equation for $p_i^{(0)}$ on the full lattice with the assumption that $p_i^{(1)} = p_{i-1}^{(n_k)}$. That this yields a reflecting boundary can be verified via a method of images argument (Chandrasekhar 1943). The absorbing condition at X_{i+1} and the reflecting condition at X_i can be written as

$$\begin{aligned} \tilde{p}_{i+1}^{(0)} &= 0 \\ \tilde{p}_i^{(0)} &= \frac{1}{s+\lambda} + \frac{\mu}{s+\lambda}\tilde{p}_i^{(1)} \end{aligned} \tag{61}$$

Between X_i and X_{i+1}, Equation (60) can be solved by assuming an *ansatz*, $\tilde{p}_i^{(k)} = a^k$ for $k = 1, 2, 3, \ldots$ and leaving $\tilde{p}_i^0$ unspecified at first. Thus, we start by solving for $\tilde{p}_i^{(k)}$ with

$$\tilde{p}_i^{(1)} = \frac{1}{s+\mu}\left(\frac{\mu}{2}\tilde{p}_i^{(2)} + \frac{\lambda}{2}\tilde{p}_i^{(0)}\right) \qquad k = 1 \tag{62}$$

$$\tilde{p}_{i+1}^{(0)} = 0 \qquad k = 0 \tag{63}$$

After some algebraic simplifications

$$\tilde{p}_i^{(k)} = A_1(s)w_-(s)^k + A_2(s)w_+(s)^k \quad k = 1, 2, 3, \ldots, n_k$$

with

$$w_-(s) = \left(\left(\frac{s}{\mu}+1\right) - \sqrt{\left(\frac{s}{\mu}+1\right)^2 - 1}\right),$$
$$w_+(s) = \left(\left(\frac{s}{\mu}+1\right) + \sqrt{\left(\frac{s}{\mu}+1\right)^2 - 1}\right)$$

and where $A_1(s)$ and $A_2(s)$ are specified by the boundary conditions. Finally, we enforce the boundary condition at $k = 0$ to find $\tilde{p}_i^{(0)}$.

Ultimately, the result only depends on μ, λ, n_k, and s, and if we compute the FPT, $\tilde{f}(s)$, for a particle to arrive at the boundary at X_{i+1}, we obtain

$$\begin{aligned} \tilde{f}(s) &= \frac{\mu}{2} p_i^{n_k}(s) \\ &= \frac{1}{2}\frac{\lambda\mu(w_+ - w_-)w_+^{n_k-1}w_-^{n_k-1}}{w_+^{n_k}\left(2s^2 + 2s(\mu+\lambda) + \mu\lambda - \mu(s+\lambda)w_-\right) - w_-^{n_k}\left(2s^2 + 2s(\mu+\lambda) + \mu\lambda - \mu(s+\lambda)w_+\right)} \end{aligned} \tag{64}$$

Since i was left unspecified, this relation is true for each junction, and we ultimately obtain a system of equations for $\tilde{p}_i^{(0)}$ where the dependence on the SVs has been removed,

$$\tilde{p}_i^{(0)} = \tilde{f}(s)\left(\tilde{p}_{i-1}^{(0)} + \tilde{p}_{i+1}^{(0)}\right).$$

Thus, $\tilde{f}(s)$ is an effective WTD for transitions in a 1D single-state CTRW on $\mathbb{Z}$ with no SVs. It is also useful to note that the mean waiting time to travel between cell, can be found, after much algebra, as

$$\bar{t} = -2 \lim_{s \to 0} \frac{\partial \tilde{f}}{\partial s} = (n_k + 1)\frac{n_k \lambda + \mu}{\lambda \mu}.$$

The factor of 2 appears since if we compute how long it takes to travel from X_i to $X_{i\pm1}$, we must include the rates of particles arriving at X_{i+1} and X_{i-1}, which in this case are both equal to $\tilde{f}(s)$.

The macroscale moments can be found by using the result from Eq. (43) and assuming that $L = \Delta X$ is independent of n_k. The explicit solution can be written out in Laplace space, and the asymptotic result can be found as

$$m^{(2)}(t) = \frac{L^2 \mu \lambda}{(n_k + 1)(n_k \lambda + \mu)} t.$$

Since we regard the SVs as representing a discretization of diffusion within each cell, we expect that the space between each pair of SVs is $\Delta x = L/n_k$. The average time it takes a diffusing particle to travel a distance Δx is $\sqrt{D_m \Delta t}$ where D_m is the microscopic diffusion constant within a cell. Thus, if we set $\mu = D_m/\Delta x^2 = D_m n_k^2/L^2$, we should obtain an approximation of diffusion within each cell. Furthermore, if we let $\lambda = \lambda_\infty n_k$,

$$m^{(2)}(t) = \frac{D_m \lambda_\infty t}{\lambda_\infty + D_m/L^2} \equiv 2 D_M t$$

where D_M may be treated as a macroscale diffusion coefficient. This scaling of λ is necessary because as n_k grows, the number of times a random walker will return to the origin before reaching an adjacent junction increases. If λ does not scale with n_k, this increasing number of returns to the origin leads to arbitrarily long waiting times to reach an adjacent junction, and no macroscale diffusion occurs.

Although we do not consider all of the details here, with suitable scalings of μ and λ as n_k approaches infinity, we can consider the problem not merely as a discretization of diffusion within the cell, but as a continuum problem.

4.3 CTRW with SVs and Internal States

Next we consider an extension of the previous problem in which there are two internal states: a mobile state and an immobile bound state. Let us assume that transitions from the mobile to the bound state occurs at rate k_+ and from bound to mobile at rate k_-. The resulting reaction network is diagrammed in Fig. 8.

From the discussion in Sect. 2.2, we know that Poisson processes can be additively combined to obtain a multistate differential master equation. Thus, we obtain the following system of equations

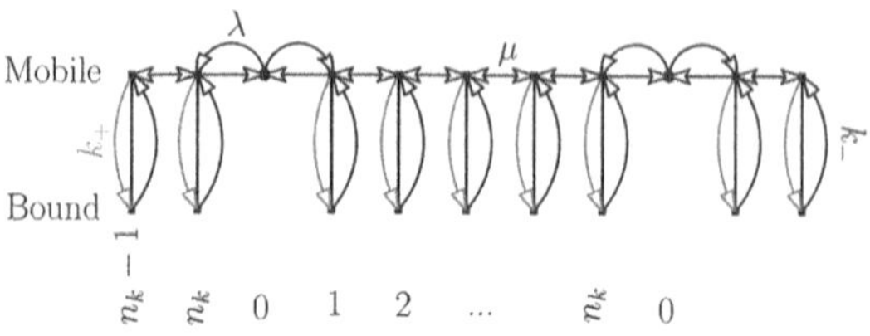

Fig. 8 Diagram of the various transitions allowed in the system discussed in a jump process involving both SVs and internal states. Note that there are no internal state transitions at the junctions since we assume that all particles are in a mobile state there

$$\begin{aligned}
\frac{dp_{i,m}^{(0)}}{dt} &= -\lambda p_{i,m}^{(0)} + \frac{\mu}{2}\left(p_{i-1,m}^{(n_k)} + p_{i,m}^{(1)}\right) && k=0\\
\frac{dp_{i,m}^{(1)}}{dt} &= -(\mu+k_+)p_{i,m}^{(1)} + \frac{1}{2}\left(\lambda p_{i,m}^{(0)} + \mu p_{i,m}^{(2)}\right) + k_- p_{i,b}^{(1)} && k=1\\
\frac{dp_{i,m}^{(k)}}{dt} &= -(\mu+k_+)p_{i,m}^{(k)} + \frac{\mu}{2}\left(p_{i,m}^{(k-1)} + p_{i,m}^{(k+1)}\right) + k_- p_{i,b}^{(k)} && k=2,\ldots,n_k-1 \qquad (65)\\
\frac{dp_{i,m}^{(n_k)}}{dt} &= -(\mu+k_+)p_{i,m}^{(n_k)} + \frac{1}{2}\left(\mu p_{i,m}^{(n_k-1)} + \lambda p_{i+1,m}^{(0)}\right) + k_- p_{i,b}^{(n_k)} && k=n_k\\
\frac{dp_{i,b}^{(k)}}{dt} &= -k_- p_{i,b}^{(k)} + k_+ p_{i,m}^{(k)} && k=1,\ldots,n_k
\end{aligned}$$

where m and b subscripts indicate mobile and bound states. To solve this, we first note that each $p_{i,b}^{(k)}$ can be found in terms of $p_{i,m}^{(k)}$ as

$$p_{i,b}^{(k)} = \int_0^t e^{-k_-(t-\tau)} k_+ p_{i,m}^{(k)}(\tau)d\tau.$$

We can rearrange terms to obtain

$$\frac{dp_{i,m}^{(k)}}{dt} = -(-\mu+k_+)p_{i,m}^{(k)} + \frac{\mu}{2}\left(p_{i,m}^{(k-1)} + p_{i,m}^{(k+1)}\right) + k_-k_+\int_0^t e^{-k_-(t-\tau)} p_{i,m}^{(k)}(\tau)d\tau.$$

In Laplace transform space, we obtain a recursion relation with the same boundary conditions as in the previous section, but with w_+ and w_- now set to

$$w_\pm(s) = \frac{1}{\mu}\left[\left(s+\mu+k_+ - \frac{k_+k_-}{s+k_-}\right) \pm \sqrt{\left(s+\mu+k_+ - \frac{k_+k_-}{s+k_-}\right)^2 - \mu^2}\,\right].$$

We find that to leading order, the FPT in this case is

$$f(s) \sim \frac{1}{2} - \frac{(1+n_k)(k_+n_k\lambda + k_-(n_k\lambda+\mu)}{k_-\lambda\mu}s$$

and the second moment is

$$m^{(2)}(t) \sim \frac{L^2 k_- \lambda \mu}{(n_k + 1)((k_+ + k_-)n_k \lambda + k_- \mu)} t$$

In the limit that $n_k \to \infty$ with $\lambda \sim \lambda_\infty n_k$ and $\mu \sim (D_m/L^2) n_k^2$,

$$m^{(2)}(t) \sim \frac{D_m \lambda_\infty k_-}{(k_+ + k_-)\lambda_\infty + k_- D_m/L^2} t \equiv 2D_M t.$$

We see that if unbinding is much more rapid than binding, i.e. $k_- \gg k_+$, we recover the case with no immobilization.

Another case of interest occurs when $k_- = 0$, i.e. if binding is irreversible, or represents a degradation reaction. In this case, since all particles eventually are degraded, the long-time asymptotic moments vanish. However, if there is a source term, then a steady-state distribution can be found by taking the limit that $s \to 0$ in Laplace space. The solution for the probability of being located in a particle index j is then of the form

$$p(j,s) = \frac{1 - f(s)}{s} \frac{1}{\sqrt{1 - f(s)^2}} \left(\frac{(1 - f(s) - \sqrt{1 - f(s)^2})}{f(s)} \right)^{|j|} h(s)$$

where $h(s)$ is the Laplace transform of a source at $j = 0$. For $h(s) = h_0$, a constant source, the steady-state solution can be found as

$$\lim_{s \to 0} sp(j,s)$$

Of course, there are many variations that can be applied: immobilization and degradation as separate reactions, multiple mobile states with different jump rates, etc.

Example 5: Comb Geometry

Comb models are widely used to study transport phenomenon, especially in percolation theory (Havlin and Ben-Avraham 1987), and in studies of transport in systems where anomalous transport behavior is observed. They also arise in certain polymerization problems and in the study of transport in dendrites (Iomin 2011; Iomin and Méndez 2013; Iomin et al. 2016; Iomin 2019; Berezhkovskii et al. 2014, 2015; Hu and Othmer 2011; Bressloff and Newby 2013).

The basic comb mode shown in Fig. 9 consists of a "backbone" with periodic side branches which effectively act as traps for particles which diffuse along the backbone. If the side branches are infinitely long, the mean waiting time for particles in the side branches to return to the backbone becomes infinite, leading to anomalous transport behavior. If the side branches are of finite length, crossover behaviors occur with an initial anomalous transport phase followed by normal diffusion, albeit with a smaller

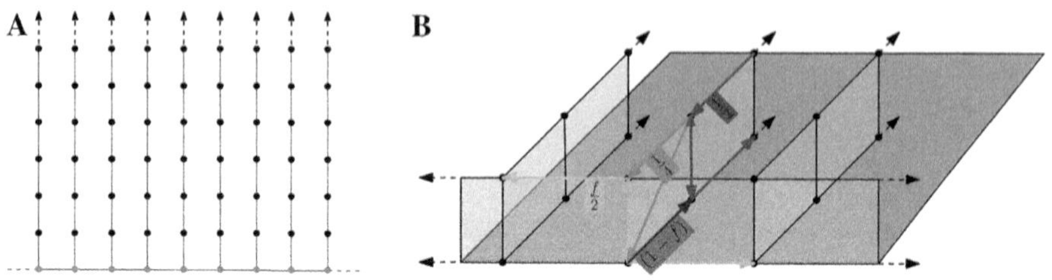

Fig. 9 **A** Basic comb geometry. Green dots represent junctions along the backbone and black dots are secondary spatial states on the teeth. **B** Depiction of the various transitions that can occur from state 1 and state 2 at each junction in the comb model with 2 internal states. The probabilities for each transition to occur are labeled adjacent to the arrows indicating each allowed transition

diffusion coefficient than that which would be observed for a particle traveling on a backbone without traps (Lubashevskii and Zemlyanov 1998).

Rather than simply solve for the density distributions of a random walk along the comb, which is known from many previous studies (Havlin and Ben-Avraham 1987; Weiss and Havlin 1986), we combine the comb geometry with the multistate transport processes discussed earlier. In particular, we will consider a two-state, persistent or biased random walk (Montroll and West 1979). A random walk in 1D will be called persistent if the particles can enter into various internal states that have unequal likelihoods of jumping forward or backwards. In the continuum sense, this leads to telegrapher's equation models in one space dimension (Othmer et al. 1988). In a biological example, this type of transport is believed to play an important role in the putative ability of fingerlike projections, known as cytonemes, to transport morphogens from source cells to target cells (Kornberg and Roy 2014; Kornberg 2014; Roy et al. 2014). Current modeling efforts (Kim and Bressloff 2018; Bressloff and Kim 2018) suggest that this type of transport can be understood in terms of a velocity jump process (Othmer 1983) similar to a telegrapher's equation, but much still remains unknown about this mode of transport and it could be a rich field for modeling.

Here, we combine persistence with a complex geometry and show how our formulation makes this type of otherwise very–complicated problem fairly straightforward to study. For the sake of simplicity, we take a two-state random walk. Junctions are located at integers along the x-axis, and the comb teeth point vertically, emanating from each junction, but only connected horizontally along the x-axis ($y = 0$) as in Fig. 9a.

At each junction, and at SVs along the teeth, we assume that there are two internal states, $\ell = 1$ and $\ell = 2$. At SVs with $y \geq 1$, particles in state 1 only can jump to state 1 at position $y + 1$ or to state 2 while remaining fixed at y. Particles in state 2 can only jump to state 2 at $y - 1$ or transition to state 1. Let us assume that $\phi(t)$ is the WTD for all of the transitions so that we can reduce the number of parameters. This can be generalized if needed. Thus, at each point on a tooth (where $y \geq 1$), we have the following transition matrices:

$$\boldsymbol{T}_{SV}(y, y') = \frac{1}{2}\left[\begin{pmatrix} 1 & 0 \\ 0 & 0 \end{pmatrix}\delta(y - y' - 1) + \begin{pmatrix} 0 & 0 \\ 0 & 1 \end{pmatrix}\delta(y - y' + 1)\right], \qquad \boldsymbol{K}_{SV} = \frac{1}{2}\begin{pmatrix} 0 & 1 \\ 1 & 0 \end{pmatrix}.$$

At $y = 1$, the form of $\boldsymbol{T}(y)$ is different, there is only a $(1 - f)$ probability for a particle traveling along the backbone to enter into a tooth. Thus, for $y = 1$, we write

$$\boldsymbol{T}_{SV}(1, y') = \frac{1}{2}\begin{pmatrix} 0 & 0 \\ 0 & 1 \end{pmatrix} \delta(2 - y')$$

and define

$$\boldsymbol{D}' = \frac{1}{2}\begin{pmatrix} (1-f) & (1-f) \\ 0 & 0 \end{pmatrix}$$

as the transfer matrix for particles jumping from a tooth to the backbone. Similarly, the transfer of particles from the tooth to the backbone is given by

$$\boldsymbol{D} = \begin{pmatrix} 0 & \frac{1}{4} \\ 0 & \frac{1}{4} \end{pmatrix}.$$

It remains to specify $\boldsymbol{K}_J$ and the additional matrix, $\boldsymbol{T}_J$, which describes spatial jumps between adjacent junction points on the backbone. At $y = 0$, let us assume state 1 points can only travel to the right, change state, or travel up to $y = 1$, and a state 2 point can only travel left or change state. In matrix form, we have

$$\boldsymbol{T}_J(x, x') = \frac{1}{2}\left[\begin{pmatrix} f & 0 \\ 0 & 0 \end{pmatrix} \delta(x - x' - 1) + \begin{pmatrix} 0 & 0 \\ 0 & f \end{pmatrix} \delta(x - x' + 1)\right], \qquad \boldsymbol{K}_J = \frac{1}{2}\begin{pmatrix} 0 & f \\ f & 0 \end{pmatrix},$$

where f is the fraction of particles that did not jump to $y = 1$. Putting everything together, we write:

$$\boldsymbol{T} + \boldsymbol{K} = \boldsymbol{I} \otimes \begin{pmatrix} \boldsymbol{K}_J & \boldsymbol{D} \\ \boldsymbol{D}' & \boldsymbol{T}_{SV} + \boldsymbol{K}_{SV} \end{pmatrix} + \boldsymbol{A}_s \otimes \begin{pmatrix} \boldsymbol{T}_J(1, 0) & \boldsymbol{0} \\ \boldsymbol{0} & \boldsymbol{0} \end{pmatrix} + \boldsymbol{A}_e \otimes \begin{pmatrix} \boldsymbol{T}_J(0, 1) & \boldsymbol{0} \\ \boldsymbol{0} & \boldsymbol{0} \end{pmatrix} \tag{66}$$

where $\boldsymbol{A}_s$ and $\boldsymbol{A}_e$ are the infinite-dimensional permutation matrices from Eq. (38). Because of the infinite extent of the teeth, this problem, even after Fourier transformation along the x-direction is still of infinite dimension. Thus, we will make use of the FPT method discussed in Sect. 4.1.

After Laplace transformation of $\phi(t)$, the problem becomes an infinite-dimensional matrix vector problem. To find the relevant FPT, we must invert $\boldsymbol{I} - \tilde{\phi}(s)(\boldsymbol{T}_{SV}(y, y') + \boldsymbol{K}_{SV})$ which, in general is a difficult task when there are infinitely many states involved. However, due to the simple structure of $\boldsymbol{T}_{SV}(y, y') + \boldsymbol{K}_{SV}$, inversion is possible in this case. Let us first consider, instead of a semi-infinite tooth, one that is bidirectionally infinite. Thus, consider $y \in \mathbb{Z}$ rather than just positive-valued. Noting that the transition matrix $\boldsymbol{T}_{SV}(y, y') + \boldsymbol{K}_{SV}$ is translation-invariant along this infinite tooth (i.e. a function only of $y - y'$), then we can take a lattice Fourier transform in the y-direction to obtain that

$$\hat{\boldsymbol{T}}_{SV}(\nu) + \boldsymbol{K}_{SV} = \frac{1}{2}\begin{pmatrix} e^{-i\nu y} & 1 \\ 1 & e^{i\nu y} \end{pmatrix}$$

Accordingly,

$$\left[\boldsymbol{I} - \tilde{\phi}(s)(\hat{\boldsymbol{T}}_{SV}(\nu) + \boldsymbol{K}_{SV})\right]^{-1} = \frac{1}{1-\phi(s)\cos\nu}\begin{pmatrix} 1 - \frac{\tilde{\phi}(s)}{2}e^{i\nu} & \frac{\tilde{\phi}(s)}{2} \\ \frac{\tilde{\phi}(s)}{2} & 1 - \frac{\tilde{\phi}(s)}{2}e^{-i\nu} \end{pmatrix}$$

The inverse lattice Fourier transform is then found as[7]

$$\begin{aligned} \left[\boldsymbol{I} - \tilde{\phi}(s)(\boldsymbol{T}_{SV}(y) + \boldsymbol{K}_{SV})\right]^{-1} &= \frac{1}{2\pi}\int_{-\pi}^{\pi}\left[\boldsymbol{I} - \tilde{\phi}(s)(\hat{\boldsymbol{T}}_{SV}(\nu) + \boldsymbol{K}_{SV})\right]^{-1} e^{-i\nu y} d\nu \\ &= \frac{1}{\sqrt{1-\tilde{\phi}^2}}\zeta^{|y|}\begin{pmatrix} 1 - \frac{\tilde{\phi}}{2}\zeta & \frac{\tilde{\phi}}{2} \\ \frac{\tilde{\phi}}{2} & 1 - \frac{\tilde{\phi}}{2}\zeta \end{pmatrix} \end{aligned} \tag{67}$$

with

$$\zeta = \frac{1-\sqrt{1-\tilde{\phi}^2}}{\tilde{\phi}}.$$

With this solution over an infinite tooth, we can now calculate the first-passage-time distribution for a particle to start at $y = 1$ in state 1 and return to $y = 1$ in state 2. As discussed in Hughes (1996), Montroll and Greenberg (1964), first-passage-time distributions for a discrete time and space random walk can be found by noting that the probability of being located at some point can be calculated as the sum of the probability of arriving at the point for the first time, plus the probability of returning to that point after already having visited it one or more times previously. Recall the definition of $q_n(y, t|y_0)$ as the probability of arriving at y after n steps at time t starting from y_0. We define $f_n(y, t|y_0)$ as the probability of reaching y for the first time after n steps at time t starting from y_0. It has been shown in Montroll and Weiss (1965), Hughes (1996) that the Laplace transforms of these quantities are related by

$$\tilde{q}_n(y, s|y_0) = \delta_{n,0}\delta(y - y_0) + \sum_{m=1}^{n} \tilde{f}_m(y, s|y_0)\tilde{q}_{n-m}(y, s|y).$$

[7] In this sense, the probability $p(y)$ defined at each lattice point can be thought of as the Fourier series for the function $\hat{p}(\nu)$. Thus, the inverse lattice Fourier transform is simply the integral that gives the Fourier coefficients of $p(\nu)$.

One can make use of this result to eventually obtain that the first-passage-time distribution for a CTRW is of the form

$$\tilde{f}(y, s|y_0) = \frac{\tilde{q}(y, s|y_0) - \delta(y - y_0)}{\tilde{q}(y, s|y)}$$

recalling that $q(y, t|y_0)$ is the probability density associated with arrival at y at time t from Eq. (6) in Sect. 2.1. From the derivation in Hughes (1996), we see that this definition extends directly to multistate problems. In the current situation, set

$$\tilde{q}_{ij}(1, s|1) = \boldsymbol{\delta}_{y,1}\delta_i \left[\boldsymbol{I} - \tilde{\phi}(s)(\boldsymbol{T}_{SV} - \boldsymbol{K}_{SV})\right]^{-1} \boldsymbol{\delta}_{y,1}\delta_j$$

where δ_i and δ_j indicate the internal state ($i, j \in \{1, 2\}$) for ending and starting the sojourn on a tooth, and $\boldsymbol{\delta}_{y,1}$ is the Kronecker delta concentrated at $y = 1$. With this, $\tilde{f}_{21}(s)$, which is interpreted as the first-passage-time probability density for reaching state 2 at $y = 1$ after starting at $y = 1$ in state 1, is found as

$$f_{21}(s) = \frac{\tilde{q}_{21}(1, s|1)}{\tilde{q}_{22}(1, s|1)} = \frac{\frac{\tilde{\phi}}{2}}{1 - \frac{\tilde{\phi}}{2}\zeta} = \frac{1}{2}\frac{\tilde{\phi}(s)}{1 + \sqrt{1 - \tilde{\phi}^2(s)}}.$$

The effective transition rate for particles to enter the comb in state 1 and return to the backbone in state 2 is then

$$\begin{aligned}\tilde{\boldsymbol{D}}^{\text{eff}}(s) &= \tilde{\phi}^2(s)\boldsymbol{D}\begin{pmatrix} \tilde{f}_{11}(s) & \tilde{f}_{12}(s) \\ \tilde{f}_{21}(s) & \tilde{f}_{22}(s) \end{pmatrix}\boldsymbol{D}' = \frac{(1 - f)\tilde{\phi}^2(s)\tilde{f}_{21}(s)}{4}\begin{pmatrix} 1 & 1 \\ 1 & 1 \end{pmatrix} \\ &= \frac{1 - f}{8}\frac{\tilde{\phi}^3}{1 + \sqrt{1 - \tilde{\phi}^2}}\begin{pmatrix} 1 & 1 \\ 1 & 1 \end{pmatrix}\end{aligned}$$

where the $\tilde{\phi}^2$ factor comes about due to the waiting times for a particle to jump to the tooth in the first step, and to the backbone in the last step. This formulation yields a problem where we need only consider transport along the backbone. The transition matrix in this case is now

$$\boldsymbol{T}_J(x, x')\tilde{\phi}(s) + \boldsymbol{K}\tilde{\phi}(s) + \tilde{\boldsymbol{D}}^{\text{eff}}(s).$$

After Fourier transformation in the x-direction, we obtain

$$\hat{\boldsymbol{T}}_J(\omega)\tilde{\phi}(s) + \boldsymbol{K}\tilde{\phi}(s) + \tilde{\boldsymbol{D}}^{\text{eff}}(s) = \begin{pmatrix} \frac{f}{2}e^{i\omega} & \frac{1}{2} \\ \frac{1}{2} & \frac{f}{2}e^{-i\omega} \end{pmatrix}\tilde{\phi} + \tilde{\boldsymbol{D}}^{\text{eff}}(s).$$

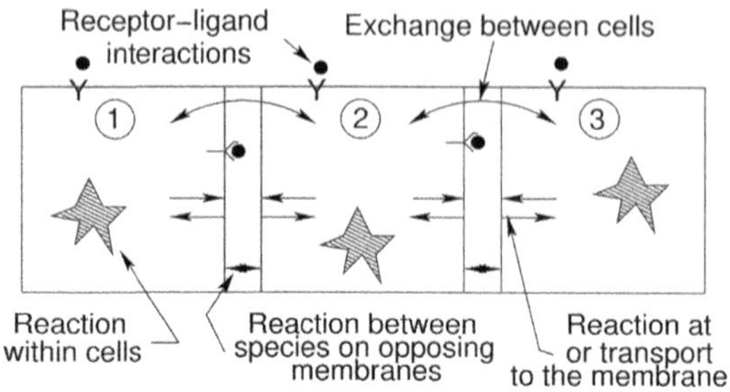

Fig. 10 An elaboration of the transport and kinetics processes involved in the wing disk. In the analysis that follows not all processes are treated, and we identify the membranes on adjacent cells. From Gou et al. (2018)

The moments then follow from Equations (D.2) and (D.3). For $\phi(t) = \rho e^{-\rho t}$, at leading order, we have

$$m^{(1)}(t) \sim \frac{f}{3-f}$$

$$m^{(2)}(t) \sim \sqrt{\frac{2\rho t}{\pi}} \frac{f(3+f)}{(3-f)(1-f)} + \mathscr{O}(1)$$

where the estimates are found using asymptotic results for the Laplace transform. Because the teeth are of infinite extent, the exit time from a tooth may be very long, hindering the transport of the particle along the backbone. This leads to anomalous diffusion and similar anomalous transport results are typical of comb models, as mentioned earlier. Furthermore, recalling the discussion in Sect. 3.1, we see that this is an example of subdiffusion, which is to be expected since the comb teeth act as traps that hinder the particle. Note that the anomalous behavior is dependent on the fact that these traps have an undefined mean waiting time since in Laplace domain space, $\tilde{f}_{21}(s)$ has an unbounded derivative near $s = 0$. When the mean waiting time is finite, the asymptotic long-term behavior is not anomalous, although there can be crossover effects where the system at first appears anomalous but eventually relaxes to normal asymptotic behavior.

5 Transport in a Series of Cells

In this section, we address in more detail the problem that motivated this study, which was to reconcile the vast disparity between experimentally determined parameters for transport in the wing disk. We do this by determining how parameters that describe microscopic processes involved in reaction and transport in the wing disk are reflected in the macroscopic diffusion and decay constants that are estimated from FRAP data. Fig 10 shows the numerous processes that are involved in transport, binding and other reactions in the disk in detail, but as observed earlier, experimentally measured profiles of Dpp measured via FRAP are usually described using a reaction–diffusion model such as Eq. 68, which is based on "free diffusion" and first-order decay (Kicheva et al. 2007; Wartlick et al. 2011).

$$
\begin{aligned}
\frac{\partial c}{\partial t} &= D\frac{\partial^2 c}{\partial x^2} - kc \qquad x \in (0, L_p) \\
-D\frac{\partial c}{\partial x}(x) &= j \qquad x = 0 \\
-D\frac{\partial c}{\partial x}(x) &= 0 \qquad x = L_p
\end{aligned} \tag{68}
$$

Here $x = 0$ ($x = L_p$) is the location of the Dpp source (pouch boundary). For simplicity we consider a line of cells in which there is only axial variation of the components, and we begin with a simple system that only involves diffusion within cells and in the gaps between those cells. A more complicated model might also include membrane binding kinetics, which can be addressed with the method below as well.

5.1 Diffusion with Extracellular Gaps

We begin with a simple problem that involves diffusion only to understand the effect of the gap between cells. Consider the problem with continuum diffusion in a series of cells of width L separated by gaps of width δ, wherein the diffusion coefficients in the cell and in the gap are to be interpreted as those obtained from the earlier analysis of lattice walks. We write the governing equations as

$$
\begin{aligned}
\frac{\partial p_i^{(1)}}{\partial t} &= D_{m,1}\frac{\partial^2 p_i^{(1)}}{\partial x^2}, \qquad X_i < x < \delta + X_i \\
\frac{\partial p_i^{(2)}}{\partial t} &= D_{m,2}\frac{\partial^2 p_i^{(2)}}{\partial x^2}, \qquad X_i + \delta < x < X_i + L \\
p_i^{(1)}(X_i + \delta, t) &= p_i^{(2)}(X_i + \delta, t) \\
D_{m,1}\left.\frac{\partial p_i^{(1)}}{\partial x}\right|_{x=X_i+\delta} &= D_{m,2}\left.\frac{\partial p_i^{(2)}}{\partial x}\right|_{x=X_i+\delta}
\end{aligned} \tag{69}
$$

where $p_i^{(1)}(x, t)$ is the probability density for a particle to be located in the ith gap, and $p_i^{(2)}(x, t)$ the probability density for a particle to be located in the ith cell. Consider the left-hand cell membrane of each cell (e.g., $X_i = iL$) the junctions in this system and the right-hand cell membrane as a secondary vertex between each pair of junctions.

Then the WTDs for particles to jump between the two types of vertices are the respective FPTs for a particle starting at a membrane to reach a membrane to its left or right. The two FPTs (to travel across a gap, and to travel across a cell) are found as the fluxes to the left and right of the following system:

$$
\begin{aligned}
\frac{\partial G^{(1)}}{\partial t} &= D_{m,1}\frac{\partial^2 G^{(1)}}{\partial x^2} && 0 < x < \delta \\
\frac{\partial G^{(2)}}{\partial t} &= D_{m,2}\frac{\partial^2 \partial G^{(2)}}{\partial x^2} && \delta < x < L \\
G^{(1)}(\delta, t) &= G^{(2)}(\delta, t) \\
D_{m,1}\frac{\partial G^{(1)}}{\partial x}\bigg|_{x=\delta} &= \delta(t) + D_{m,2}\frac{\partial G^{(2)}}{\partial x}\bigg|_{x=\delta}
\end{aligned}
\tag{70}
$$

which can be solved after Laplace transformation. The resulting fluxes at $x = 0$ and $x = L$ are then

$$
\begin{aligned}
\tilde{f}_0(s) &= \frac{\sqrt{D_{m,1}}\operatorname{csch}\left(\sqrt{\frac{s}{D_{m,1}}}\delta\right)}{\sqrt{D_{m,1}s}\coth\left(\sqrt{\frac{s}{D_{m,1}}}\delta\right) + \sqrt{D_{m,2}s}\coth\left(\sqrt{\frac{s}{D_{m,2}}}(L-\delta)\right)} \\
\tilde{f}_L(s) &= \frac{\sqrt{D_{m,2}}\operatorname{csch}\left(\sqrt{\frac{s}{D_{m,2}}}(L-\delta)\right)}{\sqrt{D_{m,1}s}\coth\left(\sqrt{\frac{s}{D_{m,1}}}\delta\right) + \sqrt{D_{m,2}s}\coth\left(\sqrt{\frac{s}{D_{m,2}}}(L-\delta)\right)}
\end{aligned}
\tag{71}
$$

Making use of this result, we can simplify Eq. (69) to obtain, after Laplace transformation, that

$$
\begin{aligned}
\tilde{P}_i^{(L)} &= \tilde{f}_0(s)\tilde{P}_{i-1}^{(R)} + \tilde{f}_L(s)\tilde{P}_i^{(R)} \\
\tilde{P}_i^{(R)} &= \tilde{f}_0(s)\tilde{P}_i^{(L)} + \tilde{f}_L(s)\tilde{P}_{i+1}^{(L)}
\end{aligned}
\tag{72}
$$

where $\tilde{P}_i^{(L,R)}(s)$ represent the probability densities at the left and right cell membrane of each cell.[8]

Already, we have reduced the problem to essentially a two-state CTRW, but before obtaining the diffusivity, we further collapse the system, by solving for the FPT to go between $P_i^{(L)}$ and $P_{i\pm1}^{(L)}$. This yields

$$
P_i^{(L)} = \frac{\tilde{f}_0(s)\tilde{f}_L(s)}{1 - \tilde{f}_0^2(s) - \tilde{f}_L^2(s)}\left(P_{i-1}^{(L)} + P_{i+1}^{(L)}\right)
$$

We have at this point transformed what was initially a continuum problem into a single-state spatially discrete CTRW for transport between cell membranes. The waiting-time distribution is the first-passage-time for a particle undergoing heterogeneous diffusion to travel a distance L after starting at $x = 0$. As such, Eq. (43) can be used to show

[8] These two quantities technically represent the probabilities associated with particles having reached membrane i (right or left) and not having yet reached a subsequent cell membrane.

that the asymptotic second moment at large t is

$$m^{(2)}(t) = \frac{2t}{\frac{1-\eta}{D_{m,2}} + \frac{\eta}{D_{m,1}}} \equiv 2D_M t$$

where $\eta = \delta/L$. This is equivalent to equation (42) in Othmer (1983) with $\lambda = 1$ (in fact the more general case with $\lambda \neq 1$ can analyzed with FPTs as well).

5.2 Macroscale Equations

With the result for pure diffusion at hand, we consider a line of cells as shown in Fig. 10, with internal reactions and degradation, and derive a macroscale equation that describes the overall process at large enough time and space scales. Within each cell there is diffusion, binding to and release from an immobile site, and degradation of the immobilized particle. The probability densities of states in cell i evolve according to the following equations,

$$\frac{\partial p_i^{(2,1)}}{\partial t} = D_{m,1}\frac{\partial^2 p_i^{(2,1)}}{\partial x^2} + k_- p_i^{(2,2)} - k_+ p_i^{(2,1)} \tag{73}$$

$$\frac{dp_i^{(2,2)}}{dt} = -k_- p_i^{(2,2)} + k_+ p_i^{(2,1)} - k_d p_i^{(2,2)}. \tag{74}$$

$$\frac{\partial p_i^{(1)}}{\partial t} = D_{m,2}\frac{\partial^2 p_i^{(1)}}{\partial x^2} \tag{75}$$

where $p^{(2,1)}$ and $p^{(2,2)}$ represent mobile and bound states in the intracellular domains, and $p^{(1)}$ represents the (mobile) extracellular state. At the boundaries between adjacent cells, the boundary conditions on $p_i^{(1)}$ and $p_i^{(2,1)}$ are the same as those discussed in the previous section.

Eq. (74) can be eliminated by solving for $p_i^{(2,2)}$ in terms of $p_i^{(2,1)}$, and as a result, Eq. (73) becomes

$$\frac{\partial p_i^{(2,1)}}{\partial t} = D_m \frac{\partial^2 p_i^{(2,1)}}{\partial x^2} + k_- k_+ \int_0^t e^{-(k_- + k_d)(t-\tau)} p_i^{(2,1)}(\tau)d\tau - k_+ p_i^{(2,1)}.$$

To further simplify this, we consider the FPT for a particle starting in state 1 at a membrane to reach an adjacent membrane in state 1. After applying a Laplace transformation, the FPT can be found as in the previous example, but with s replaced by

$$\nu(s) = s - k_+ \left(\frac{k_-}{s + k_- + k_d} - 1 \right),$$

in the Green's function[9]. Thus, the FPTs for a particle starting at a membrane to reach an adjacent membrane (to the left or right) are

$$\tilde{f}_0(s) = \frac{\sqrt{D_{m,1}s}\,\mathrm{csch}\left(\sqrt{\frac{s}{D_{m,1}}}\delta\right)}{\sqrt{D_{m,1}s}\coth\left(\sqrt{\frac{s}{D_{m,1}}}\delta\right) + \sqrt{D_{m,2}\nu(s)}\coth\left(\sqrt{\frac{\nu(s)}{D_{m,2}}}(L-\delta)\right)}$$
$$\tilde{f}_L(s) = \frac{\sqrt{D_{m,2}\nu(s)}\,\mathrm{csch}\left(\sqrt{\frac{\nu(s)}{D_{m,2}}}(L-\delta)\right)}{\sqrt{D_{m,1}s}\coth\left(\sqrt{\frac{s}{D_{m,1}}}\delta\right) + \sqrt{D_{m,2}\nu(s)}\coth\left(\sqrt{\frac{\nu(s)}{D_{m,2}}}(L-\delta)\right)} \tag{76}$$

and as in the previous section, we obtain an equation for the membrane concentrations, $P_i^{(L)}(s)$,

$$\tilde{P}_i^{(L)}(s) = \frac{\tilde{f}_0(s)\tilde{f}_L(s)}{1-\tilde{f}_0^2(s)-\tilde{f}_L^2(s)}\left(\tilde{P}_{i-1}^{(L)}(s)+\tilde{P}_{i+1}^{(L)}(s)\right)+\hat{\Phi}(s)\delta_{i0} \tag{77}$$

with WTD

$$\tilde{F}(s) = \frac{2\tilde{f}_0(s)\tilde{f}_L(s)}{1-\tilde{f}_0^2(s)-\tilde{f}_L^2(s)}.$$

We leave the complementary waiting-time distribution, $\hat{\Phi}(t)$, which gives the probability that no jumps have occurred by time t, undefined for a moment.

We know that $\tilde{F}(s)$ is the WTD for a process with both diffusion and degradation, and in order to see how the diffusion and degradation terms appear on the macroscale, we write,

$$F(t) = \alpha g(t). \tag{78}$$

Here α is the overall probability that a particle leaving membrane i eventually reaches membrane $i+1$ or $i-1$, and $g(t)$ is the unitary WTD (i.e. it integrates to 1) for this to occur. Since degradation is involved, $\alpha < 1$, but since $g(t)$ is a unitary WTD, we can define $\hat{\Phi}$ in terms of $\tilde{g}(s)$ as

$$\hat{\Phi}(s) = \frac{1-\tilde{g}(s)}{s}.$$

The reason we use $\tilde{g}(s)$ rather than $\tilde{F}(s)$ here is that a particle must eventually jump or be degraded. Defining $\hat{\Phi}$ in terms of $F(t)$, which is non-unitary, would imply that there is a $(1-\alpha)$ chance for a particle to simply never jump, which is inconsistent with Eqs. (73) and (74). The use of a unitary WTD, $g(t)$, resolves this inconsistency. On the other hand, the presence of decay implies that the use of $g(t)$ involves a slight

[9] If particle immobilization is irreversible the following analysis simplifies somewhat, for Eq. (73) is then independent of $p_i^{(2,2)}$ and k_+ takes the place of k_d in the macroscopic parameters.

approximation since there is a nonzero probability of degradation to occur before reaching a junction. However, since we are ultimately interested in asymptotic limits that are approached only after many jumps occur, the use of an approximate WTD for the first step of the transport process has a negligible effect.

With $\hat{\Phi}$ specified, we rearrange Eq. (77), following the derivation in Sect. 2.1, to obtain the Laplace transform of the differential form of the master equation,

$$\frac{1-\tilde{g}(s)}{s\tilde{g}(s)}\left(s\tilde{p}_i - \delta_{i0}\right) = -(1-\alpha)\tilde{p}_i + \frac{\alpha}{2}\left(\tilde{p}_{i-1} - 2\tilde{p}_i + \tilde{p}_{i+1}\right). \tag{79}$$

If $g(t)$ were a Poisson distribution of rate θ, this could be inverted to obtain

$$\frac{dp_i}{dt} = -\theta(1-\alpha)p_i + \frac{\alpha\theta}{2}\left(p_{i-1} - 2p_i + p_{i+1}\right),$$

and this leads to an interpretation of $\alpha\theta/2$ as a discrete diffusivity and $\theta(1-\alpha)$ as a degradation rate.

However, $g(t)$ as defined via $F(t)$ is a more complex distribution whose explicit time-dependent form cannot be obtained analytically. Nonetheless, at leading order we may still approximate $g(t)$ by a Poisson distribution. To do so, we compute the Taylor expansion of $\tilde{g}(s)$ about $s=0$ up to first order as

$$\tilde{g}(s) \approx \tilde{g}(0) + \tilde{g}'(0)s = 1 - g'(0)s = \left(\int_0^\infty g(t)dt\right) - \left(\int_0^\infty tg(t)dt\right)s.$$

We then note that by equating this expansion with the first-order Taylor expansion of the Laplace transform of a Poisson distribution, we may approximate $g(t)$ by a Poisson distribution with rate $\theta = -1/\tilde{g}'(0)$. By the same Taylor expansion approach, we also see that $\alpha = \tilde{F}(0)$. With $\tilde{F}(s)$ as defined above, we see that

$$\begin{aligned}\alpha &= \frac{2D_{m,2}}{2D_{m,2}\cosh\left(\sqrt{\frac{\nu(0)}{D_{m,1}}}(L-\delta)\right) + \sqrt{\nu(0)D_{m,1}}\,\delta\sinh\left(\sqrt{\frac{\nu(0)}{D_{m,1}}}(L-\delta)\right)},\\ \theta &= \frac{2}{\left(\frac{(1-\eta)}{D_{m,2}} + \frac{\eta}{\eta D_{m,1}}\right)\left(\eta + (1-\eta)\nu'(0)\right)L^2} + \mathcal{O}(1)\end{aligned} \tag{80}$$

where

$$\nu(0) = k_+k_d/(k_d+k_-), \quad \nu'(0) = 1 + \frac{k_+k_m}{(k_d+k_-)^2}.$$

On a tissue-level scale, we are concerned with the evolution of p_i at length scales $L_\infty \gg L$: thus, we set $L = \epsilon L_\infty$. In the limit $\epsilon \to 0$, this leads to

$$(1-\alpha)\theta = \frac{(1-\eta)\nu(0)}{\eta + (1-\eta)\nu'(0)} = \frac{(1-\eta)k_dk_+(k_d+k_-)}{(k_d+k_-)^2 + (1-\eta)k_-k_+}, \tag{81}$$

$$\alpha\theta = \frac{2}{\left(\frac{(1-\eta)}{D_{m,2}} + \frac{\eta}{\eta D_{m,1}}\right)(\eta + (1-\eta)\nu'(0))\,\epsilon^2 L_\infty^2}$$
$$= \frac{2}{\left(\frac{(1-\eta)}{D_{m,2}} + \frac{\eta}{D_{m,1}}\right)\left(\eta + (1-\eta)\left(1 + \frac{k_- k_+}{(k_d+k_-)^2}\right)\right)\epsilon^2 L_\infty^2}. \tag{82}$$

Furthermore, in this limit $p_i(t)$ can be treated as a function defined with respect to a continuous variable x (i.e. $p(x,t)$) since $\Delta X = X_i - X_{i-1} = \epsilon^2 L_\infty$ becomes very small compared to L_∞, the length scale of interest over the tissue. Noting that $1/(\epsilon^2 L_\infty^2)$ times the discrete Laplacian approaches the continuum Laplacian, we can write

$$\begin{aligned}\frac{\partial p(x,t)}{\partial t} &= -\frac{(1-\eta)\nu(0)}{\eta + (1-\eta)\nu'(0)} p(x,t) + \frac{1}{\left(\frac{(1-\eta)}{D_{m,2}} + \frac{\eta}{D_{m,1}}\right)(\eta + (1-\eta)\nu'(0))} \frac{\partial^2 p(x,t)}{\partial x^2} \\ &= -\left[\frac{(1-\eta)k_d k_+ (k_d + k_-)}{(k_d + k_-)^2 + (1-\eta)k_- k_+}\right] p(x,t) \\ &\quad + \left[\frac{1}{\left(\frac{(1-\eta)}{D_{m,2}} + \frac{\eta}{D_{m,1}}\right)\left(\eta + (1-\eta)\left(1 + \frac{k_- k_+}{(k_d+k_-)^2}\right)\right)}\right] \frac{\partial^2 p(x,t)}{\partial x^2} \\ &= -K_M p(x,t) + D_M \frac{\partial^2}{\partial x^2} p(x,t).\end{aligned} \tag{83}$$

Thus the macroscale diffusion and degradation coefficients

$$\begin{aligned}K_M &\equiv \frac{(1-\eta)k_d k_+ (k_d + k_-)}{(k_d + k_-)^2 + (1-\eta)k_- k_+}, \quad \text{and} \\ D_M &\equiv \frac{1}{\left(\frac{(1-\eta)}{D_{m,2}} + \frac{\eta}{D_{m,1}}\right)\left(\eta + (1-\eta)\left(1 + \frac{k_- k_+}{(k_d+k_-)^2}\right)\right)}\end{aligned} \tag{84}$$

are complex functions of the microscale parameters.

Furthermore, since there are more microscopic parameters than macroscale coefficients, there will be many combinations of the microscopic parameters that yield the same macroscopic parameters. Thus, if one attempts to fit data on the macroscale to a "standard" diffusion–degradation equation, the resulting diffusion and degradation coefficients cannot be interpreted as having simple meaning in terms of microscale processes. This presents a real challenge to those in the field, since it certainly seems possible that many microscale models could be adapted to macroscale diffusion–reaction equations as we have done here, but in each case, the macroscale diffusion and degradation coefficients are found as different functions of the microscale parameters.

In certain limits, the multiplicity is reduced somewhat. For example, if k_- and k_+ both tend to infinity at a fixed finite ratio $k_+/k_- = \beta$ (e.g., rapid equilibration of diffusing and immobile species), the foregoing reduces to

$$K_M = \frac{k_d\beta(1-\eta)}{1+(1-\eta)\beta} \qquad D_M = \frac{1}{(1+(1-\eta)\beta)\left[\frac{\eta}{D_{m,1}} + \frac{1-\eta}{D_{m,2}}\right]} \tag{85}$$

and there is one less microscopic parameter. In any case, if the microscale parameters can be perturbed, one can observe the macroscale behavior, and given the functional dependence of K_M and D_M on microscale parameters, this may prove to be a useful way of ascertaining whether a given model is plausible.

As an application of this analysis, we consider the model for Dpp transport in the Drosophila wing disk discussed in Kicheva et al. (2007). There D_M and K_M are estimated from FRAP data to be

$$D_M = 0.10 \pm 0.05\mu m/s, \qquad K_M = 2.52 \times 10^{-4} \pm 1.29^{-4} s^{-1}.$$

The authors also note that there is a slow "irreversible" binding process, and an immobile fraction of 62.8% of all Dpp molecules. Since the microscale model has 6 unknown parameters—$D_{m,1}$, $D_{m,2}$, k_d, k_+, k_-, and η—but only three parameters—D_M, K_M, and the immobile fraction—were measured experimentally, there can be many combinations of the microscale parameters that yield the observed macroscale parameter values. In addition, a more complete description of the processes in Fig. 10 would introduce other microscopic processes and the attendant parameters, and make the problem of connecting macroscopic and microscopic parameters even more difficult. While parameter estimates from other contexts and limiting cases discussed earlier may be useful, further experiments are needed to connect the micro- and macroprocesses.

Finally, we see that by deriving the macroscale equation from a starting point that involves microscope-level details, assumptions that are needed for the validity of the macroscopic equation can be determined. For instance, in order to justify the limit $\epsilon \to 0$, it is necessary that sufficient time has elapsed so that the waiting time for a single jump between adjacent cells is small. Otherwise, the discrete nature of the system will dominate and the continuum description will not be appropriate.

6 The Hexagonal Lattice

In this section, we return to a discussion of spatially discrete CTRWs. The goal here is to formulate solutions in hexagonal lattices, which are topologically more complex. However, note that in some cases one can use the FPTs of continuum processes as WTDs for a discrete CTRW, as was done in the preceding section.

Hexagonal lattices are of particular importance since cells of many tissues tend to pack into a arrays that can be approximated by hexagonal grids. There are in fact two types of arrays that can be described in this context: those where transport occurs directly between neighboring, hexagonally shaped cells, and those in which transport occurs around the borders of such cells. The two topologies thus defined for the primary

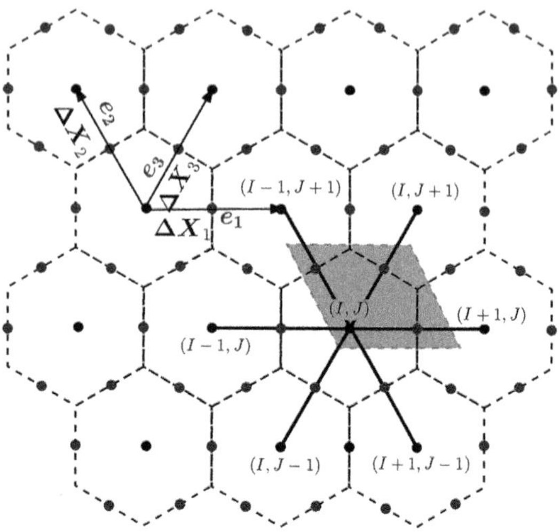

Fig. 11 Interior hexagon lattice that connects adjacent hexagon centers to one another. Each center is connected by an edge to six neighbors on adjacent edges. The vectors in the upper left show the displacements between adjacent junctions and the three types of edges are labeled as e_i. The red shaded region contains all junctions and SVs with the same index (I, J)

graphs lead to what we call the interior and exterior hexagonal lattice problems. They are related in the sense that the exterior lattice is the Voronoi diagram of the interior hexagonal lattice, and thus, the interior lattice is the Delaunay tessellation of the exterior lattice (Aurenhammer et al. 2013).

The interior lattice can be described as in Othmer and Scriven (1971) (see Fig. 11). Each junction is connected by a single edge to six adjacent junctions. The edges can be classified into three types distinguished by their orientation relative to the x-axis. In order to describe this connectivity, some labeling convention must be introduced that maps each lattice point to the edges it is linked to and each edge with the lattice points at which it terminates. One such convention is to assume that for each junction (labeled by $\boldsymbol{X}_{\boldsymbol{I}}$ with $\boldsymbol{I} = (I, J) \in \mathbb{Z}^2$), there are three edges (one of each type) with index $\boldsymbol{I}$ as well that connect to junction $\boldsymbol{I}$. As shown in Fig. 11, one can consistently label all lattice points and edges in this way. It turns out then, that junction $\boldsymbol{I}$ is connected to the three edges of label $\boldsymbol{I}$ and the type I (horizontal) edge of label $(I-1, J)$, the type II (vertical) edge of label $(I, J-1)$ and the type III (diagonal) edge of $(I-1, J-1)$. To make this consistent, it then must be the case that each edge of type I of label (I, J) is attached to lattice points (I, J) and $(I+1, J)$, and each edge of type II with label (I, J) is attached to junctions (I, J) and $(I, J+1)$. Finally, each type III edge of label (I, J) is attached to junctions (I, J) and $(I+1, J+1)$.

With this description of the geometry, let us consider an example that makes use of the interior lattice. Suppose that we have a hexagonal array of cells and some substance that can diffuse through the interior and can be transported through the boundaries of adjacent cells. Let us also assume that this substance can transiently enter an immobile state when on the cell boundaries. In a simple model, we can describe the key steps in the transport process in terms of a few waiting-time distributions. Let $\psi(t)$ be the WTD for the substance to diffuse through the cell and reach a boundary. Since this process involves diffusion within the cell, $\psi(t)$ must be considered here as a spatial average of waiting times for particles starting at different positions within the cell. Let

$\phi(t)$ be the WTD for a particle to be transported across the boundary to an adjacent cell, and $\chi(t)$ be the WTD for transitions to and from the immobile state which we will suppose only occurs on the boundary. This model can be thought of as having $n_k = 1$ (one boundary point between each adjacent pair of junctions) and $n_s = 2$ (a mobile and an immobile state). For simplicity, we will assume that all three WTDs (ψ, ϕ, and χ) are Poisson distributions with parameters λ, μ, and ξ resp. The objective is now to pose this problem in the framework of the earlier sections and solve for the probability distributions.

The matrix, $\boldsymbol{K\Lambda}$, from Equation (33) describes all internal state changes. For a single junction and set of edges, it can be written as

$$\boldsymbol{K\Lambda} = \begin{pmatrix} \boldsymbol{K}_J\boldsymbol{\Lambda}_J & \mathbf{0} & \mathbf{0} & \mathbf{0} \\ \mathbf{0} & \boldsymbol{K}_{SV}\boldsymbol{\Lambda}_{SV} & \mathbf{0} & \mathbf{0} \\ \mathbf{0} & \mathbf{0} & \boldsymbol{K}_{SV}\boldsymbol{\Lambda}_{SV} & \mathbf{0} \\ \mathbf{0} & \mathbf{0} & \mathbf{0} & \boldsymbol{K}_{SV}\boldsymbol{\Lambda}_{SV} \end{pmatrix}$$

where $\boldsymbol{K}_J\boldsymbol{\Lambda}_J$ describes internal state transitions at junctions, and $\boldsymbol{K}_{SV}\boldsymbol{\Lambda}_{SV}$ describes internal state transitions at SVs on each edge. Since the junctions are placed in the cell centers, transitions between mobile and immobile states inside the cell are described by $\boldsymbol{K}_J$. However, we have assumed that immobilization only occurs on the boundary, and therefore $\boldsymbol{K}_J$ is simply a 2×2 matrix of zeros. Transitions between internal states at SVs on each edge are described by $\boldsymbol{K}_{SV}$. There are two types of transitions that occur at an SV: 1) mobile to immobile and 2) immobile to mobile. Defining the probability of each state as $\boldsymbol{p} = (p_m, p_i)$ with m = mobile and i = immobile, $\boldsymbol{K}_{SV}$ can be written in matrix form as

$$\boldsymbol{K}_{SV}\boldsymbol{\Lambda}_{SV} = \begin{pmatrix} 0 & 1 \\ (1-f) & 0 \end{pmatrix} \begin{pmatrix} \phi & 0 \\ 0 & \chi \end{pmatrix} = \begin{pmatrix} 0 & \chi \\ (1-f)\phi & 0 \end{pmatrix}.$$

where $(1 - f)$ is the probability of transitioning to an immobile state.

The next step is to define the transition matrices $\boldsymbol{D}$, $\boldsymbol{D}'$, $\boldsymbol{T}_{SV}$ and describing spatial jumps between junctions and SVs and between adjacent SVs. Since there is only a single SV on each edge, all spatial jumps involve transitions between edges and junctions, thus $\boldsymbol{T}_{SV} = \mathbf{0}$. Furthermore, since the immobile state cannot change position, the rows of $\boldsymbol{D}$ and $\boldsymbol{D}'$ corresponding to the immobile state will be all zeros. Transitions from junctions to SVs are governed by ψ and are written in matrix form in terms of $\boldsymbol{D}'$. With $\boldsymbol{p} = (p_m, p_i)$ as above, we can write

$$\boldsymbol{D}' = \begin{pmatrix} \frac{1}{6}\psi & 0 \\ 0 & 0 \end{pmatrix}.$$

The factor of $1/6$ comes from the fact that each cell center is linked to 6 SVs, and each of them is equally likely to receive a particle from the cell center. In this case, since $n_k = 1$, the start and end SV on each edge coincide, and we do not need to distinguish between $\boldsymbol{D}'_s$ and $\boldsymbol{D}'_e$, or $\boldsymbol{D}_s$ and $\boldsymbol{D}_e$ as we did in Ex. 2. Since transitions from the cell membrane to the cell center are governed by WTD ϕ, $\boldsymbol{D}$ takes on a similar form with

ϕ in place of ψ. In particular,

$$D = \begin{pmatrix} \frac{f}{2}\phi & 0 \\ 0 & 0 \end{pmatrix}.$$

The factor of $f/2$ accounts for the probability of particles *not* being immobilized, and that a mobile particle has a 50% chance of going to either of the junctions that the edge is connected to. This concludes the enumeration of all the transitions relevant on the edges and in the cell centers. What remains is to connect the current transition matrices with the topological structure of the hexagonal lattice, which is done via $\boldsymbol{A}_m$ and $\boldsymbol{A}'_m$.

Since $\boldsymbol{K}_J$ and $\boldsymbol{K}_{SV}$ do not involve changes of position, the relevant matrices for internal state changes over the entire lattice are formed by taking Kronecker products of these matrices with the lattice identity matrix, e.g., $\boldsymbol{I} \otimes \boldsymbol{K}_J$. The transitions in $\boldsymbol{D}$ and $\boldsymbol{D}'$ involve changes in position of the particles, and thus require non-trivial structural matrices. Generally, $\boldsymbol{A}_m$ and $\boldsymbol{A}'_m$ can be written as permutations of the identity matrix over the junctions. However, they can also be thought of as linear operators on the lattice points. In particular, for $m = 1, 2, 3$, define

$$\boldsymbol{A}_m(\boldsymbol{X}, \boldsymbol{X}') = \delta(\boldsymbol{X} - \boldsymbol{X}' - \boldsymbol{\Delta X}_m),$$

where $\delta(\cdot)$ is a Kronecker delta over the junctions, e.g., $\delta(\boldsymbol{X}) = 0$ for $\boldsymbol{X} \neq \boldsymbol{0}$ and 1 for $\boldsymbol{X} = \boldsymbol{0}$ where $\boldsymbol{X}$ is a lattice point in the hexagonal lattice. Similarly,

$$\boldsymbol{A}'_m(\boldsymbol{X}, \boldsymbol{X}') = \delta(\boldsymbol{X} - \boldsymbol{X}' + \boldsymbol{\Delta X}_m).$$

The displacements $\boldsymbol{\Delta X}_m$ are those depicted in Fig. 11. From geometric considerations, these are found as

$$\boldsymbol{\Delta X}_1 = (1, 0), \quad \boldsymbol{\Delta X}_2 = \left(\frac{1}{2}, \frac{\sqrt{3}}{2}\right), \quad \boldsymbol{\Delta X}_3 = \left(-\frac{1}{2}, \frac{\sqrt{3}}{2}\right). \tag{86}$$

Although $\boldsymbol{A}_m$ can also be written as matrices, in complicated problems it seems that the operator approach may be more intuitive since it reflects the problem geometry rather than a particular ordering of the degrees of freedom.

Putting all of the pieces together, we obtain the overall transition matrix $\boldsymbol{T}$. Let $\boldsymbol{q} = (\boldsymbol{q}_J, \boldsymbol{q}_{SV_1}, \boldsymbol{q}_{SV_2}, \boldsymbol{q}_{SV_3})$ be the probabilities for a particle starting in cell $(I_0, J_0) = (0, 0)$ to arrive in some other cell (I, J) at time t in the cell interior ($\boldsymbol{q}_J$) or at the portion of the cell membrane attributed to an edge of type i, ($\boldsymbol{q}_{SV_i}$). After Laplace transformation, we write Eq. (33) for this system:

$$\begin{pmatrix} \tilde{q}_J \\ \tilde{q}_{SV_1} \\ \tilde{q}_{SV_2} \\ \tilde{q}_{SV_3} \end{pmatrix} = \begin{pmatrix} \delta(X) \\ 0 \\ 0 \\ 0 \end{pmatrix} + I \otimes \left[\begin{pmatrix} K_J \Lambda_J & D & D & D \\ D' & K_{SV}\Lambda_{SV} & 0 & 0 \\ D' & 0 & K_{SV}\Lambda_{SV} & 0 \\ D' & 0 & 0 & K_{SV}\Lambda_{SV} \end{pmatrix} + \sum_{m=1}^{3} \left(A_m \otimes D_m + A'_m D'_m \right) \right] \begin{pmatrix} \tilde{q}_J \\ \tilde{q}_{SV_1} \\ \tilde{q}_{SV_2} \\ \tilde{q}_{SV_3} \end{pmatrix} \tag{87}$$

where

$$D_m = \begin{pmatrix} \cdot & D\delta_{m,1} & D\delta_{m,2} & D\delta_{m,3} \\ \cdot & \cdot & \cdot & \cdot \\ \cdot & \cdot & \cdot & \cdot \\ \cdot & \cdot & \cdot & \cdot \end{pmatrix}, \quad D'_m = \begin{pmatrix} \cdot & \cdots \\ D'\delta_{m,1} & \cdots \\ D'\delta_{m,2} & \cdots \\ D'\delta_{m,3} & \cdots \end{pmatrix},$$

with δ_{mn} the Kronecker delta concentrated at $m = n$.

To solve this infinite-dimensional matrix equation, we make use of a lattice Fourier transform to block-diagonalize the problem. For any function $f(X)$ defined over the lattice, the lattice Fourier transform is defined as

$$\hat{f}(\omega) = \sum_{X \in \mathcal{G}} e^{iX\cdot\omega} f(X). \tag{88}$$

This definition holds whether $f(X)$ is scalar-, vector-, or matrix-valued. Since the space between junctions on a lattice is fixed, the maximum wave number that can be represented on the junctions is $\omega_{max} = (2\pi)/|\Delta X|$ where ΔX is the edge length. Additionally, on an infinite lattice, ω is defined over a continuum with each element of ω in the range $-\pi/\Delta X_i < \omega_i < \pi/\Delta X_i$. This differs from Fourier analysis of discretizations which are being refined, where ΔX is not fixed. In those cases, the maximum wave number which can be represented by a lattice function increases as the discretization is refined.

Applying the lattice Fourier transform to the evolution to $\tilde{q}$, we obtain an 8×8 system:

$$\begin{pmatrix} \hat{q}_J \\ \hat{q}_{SV_1} \\ \hat{q}_{SV_2} \\ \hat{q}_{SV_3} \end{pmatrix} = \begin{pmatrix} 1 \\ 0 \\ 0 \\ 0 \end{pmatrix} + I \otimes \begin{pmatrix} K_J & (1+e^{-i\Delta X_1\cdot\omega})D & (1+e^{-i\Delta X_2\cdot\omega})D & (1+e^{-i\Delta X_3\cdot\omega})D \\ (1+e^{i\Delta X_1\cdot\omega})D' & K_{SV} & 0 & 0 \\ (1+e^{i\Delta X_2\cdot\omega})D' & 0 & K_{SV} & 0 \\ (1+e^{i\Delta X_3\cdot\omega})D' & 0 & 0 & K_{SV} \end{pmatrix}$$

$$\begin{pmatrix} \hat{q}_J \\ \hat{q}_{SV_1} \\ \hat{q}_{SV_2} \\ \hat{q}_{SV_3} \end{pmatrix}.$$

This matrix equation can be solved directly, but can be slightly simplified. First, $\boldsymbol{K}_J = \boldsymbol{0}$, and second, the immobile state associated with $\hat{\boldsymbol{q}}_J$ has no transitions to or from any other states, and thus can be removed to yield a 7×7 system.

The solution here yields $\hat{\boldsymbol{q}}$ from Eq. (33). However, the most important quantity is typically $\boldsymbol{p}(\boldsymbol{X}, t)$, the probability of being located at $\boldsymbol{X}$ at time t. Inversion of the Fourier and Laplace transforms is often only feasible numerically, but the solution for $\hat{\boldsymbol{p}}(\boldsymbol{\omega}, s)$ is possible once we have found $\hat{\boldsymbol{\Phi}}$ defined in Eq. (34). After Fourier and Laplace transform, one can show that the nonzero entries are

$$\hat{\boldsymbol{\Phi}}_{k,\ell}(s) = \frac{1}{s}\left(1 - \boldsymbol{1}^T \boldsymbol{K} \boldsymbol{\delta}_{k,\ell} - \boldsymbol{1}^T \hat{\boldsymbol{T}}(\boldsymbol{\omega} = \boldsymbol{0}, s)\boldsymbol{\delta}_{k,\ell}\right). \tag{89}$$

In this example, after simplifying (89), we obtain diagonal entries:

$$\hat{\boldsymbol{\Phi}} = \operatorname{diag}\left(\tfrac{1}{\lambda+s}\ \tfrac{1}{\mu+s}\ \tfrac{1}{\xi+s}\ \tfrac{1}{\mu+s}\ \tfrac{1}{\xi+s}\ \tfrac{1}{\mu+s}\ \tfrac{1}{\xi+s}\right)$$

Finally, we may solve (33) for $\boldsymbol{p}$. The exact solutions may be found using Mathematica's symbolic solver; however, it is often useful to obtain simpler summary statistics that describe the distribution. Here we give one sample that is straightforward to compute. The proportion of particles in the cell centers can be found by summing $\boldsymbol{p}_c$ over the junctions. In Fourier space,

$$\sum_{\boldsymbol{X}} p_c(\boldsymbol{X}, t) = \lim_{\boldsymbol{\omega} \to \boldsymbol{0}} \hat{p}_c(\boldsymbol{\omega}, t).$$

Applying this result to our solution in the Laplace domain,

$$\lim_{\boldsymbol{\omega} \to \boldsymbol{0}} \hat{p}_c(\boldsymbol{\omega}, s) = \frac{1}{s} \frac{2s(\xi + s) + \mu(\xi + 2s)}{2s(\xi + s) + \mu(\xi + 2s) + \lambda(\mu + 2(\xi + s))}$$

The steady-state limit may be found using the Laplace transform final value theorem (Bracewell et al. 1986) as

$$\lim_{s \to 0} sp(\boldsymbol{0}, s) = \frac{\mu\xi}{\mu\xi + \lambda\mu + 2\lambda\xi}.$$

We can observe how this compares with the steady-state probabilities, $p_{c,ss}$, reported for a segment in Ex. 2. We see here that the analogous result is obtained when $\xi \to \infty$,

$$p_{c,ss} = \frac{\mu}{\mu + 2\lambda},$$

or, in other words, when mobilization (unbinding) occurs rapidly compared to the other transitions. However, because of the hexagonal topology, in that case even though there is only a single SV per edge, the denominator has 2λ rather than λ as it would for the 1D case. This same result is observed for a square lattice with a single SV and the same immobilization/mobilization transition at that SV.

A key point here is that by describing a complicated transport process in terms of fundamental steps, we found results that reflect how macroscale observables, here the steady-state solution, depend on microstructural mechanisms. Next, we turn to computation of the mean and variance of the hexagonal lattice problem.

6.1 The Moments for Transport in the Hexagonal Lattice

For the interior hexagonal lattice in Fig. 11, let us assume that a particle starts inside a cell (that is, on a junction). Then we may use Eq. (C.7) to find the first moment of the transport process, which due to the symmetry of this problem should be zero. This fact is confirmed upon computation.

The second moment, found by using Eq. (C.9), is nonzero. In order to compute this result, we must specify how each internal state and each SV are positioned relative to the junction that shares the same index, (I, J). To determine this, we use $\boldsymbol{\Delta X}_i$ from above, and note that the SVs are located precisely halfway along each edge. This gives us sufficient information to compute $\boldsymbol{x}$ for each SV and internal state. All that remains is to apply Eq. (C.9) and substitute each term with one specific to the hexagonal lattice system at hand. The large-t asymptotic result for the second moment is

$$m^{(2)}(t) = \frac{\lambda\mu f\xi t}{4\left(f\mu\xi + \lambda((1-f)\mu + \xi)\right)}$$

for both x-direction transport and y-direction transport. The moments for particles to be in the mobile or immobile state can easily be found by summing over the degrees of freedom associated with the mobile or immobile states.

Recall that λ, μ, and ξ are the rate constants associated with diffusion, boundary binding, and immobilization. Thus, we see that if mobilization/immobilization happens rapidly compared to diffusion and binding, e.g., that $\xi \to \infty$, then

$$m^{(2)}(t) = \frac{\lambda\mu f t}{4\left(f\mu + \lambda\right)}.$$

Similarly, if diffusion is very rapid compared to binding and immobilization, then $\lambda \to \infty$ and

$$m^{(2)}(t) = \frac{\mu\xi f t}{4((1-f)\mu + \xi)}.$$

If binding is very rapid compared to immobilization and diffusion, $\mu \to \infty$ and

$$m^{(2)}(t) = \frac{\lambda\xi f t}{4(f\xi + (1-f)\lambda)}.$$

Finally, if binding is rapid compared with diffusion and very little immobilization occurs, e.g., $f = 1$, we obtain

$$m^{(2)}(t) = \frac{\lambda t}{4}$$

independent of ξ. This is the result for pure diffusion in this lattice structure.

For the exterior hexagonal lattice, the local displacements are found by similar calculations since $\boldsymbol{\Delta X}_i$ are equivalent to those for the interior lattice. The details are described in "Appendix B". In that case, consider an initial condition with a particle at junction $\boldsymbol{X} = \boldsymbol{0}$. This condition leads to a first moment equal to zero, and the asymptotic second moment is found using Eq. (C.9):

$$m^{(2)}(t) = \frac{\lambda \mu f \xi t}{12\left(f \mu \xi + \lambda((1-f)\mu + \xi)\right)}.$$

Remark 1 Note that the scale factors $1/4$ and $1/12$ corresponds to the case that $\boldsymbol{\Delta X}_i$ are the same length for the interior and exterior lattices (see Fig. 11). However, in the exterior hexagonal lattice with $n_k = 1$, the length of each jump is $1/\sqrt{3}$ times the length of each jump in the interior lattice. In a diffusive process, the average time $\bar{t}$ required to travel a distance $\bar{x}$ scales quadratically, e.g., $D\bar{t} \sim \bar{x}^2$. If one rescales λ, μ, and ξ in the exterior lattice by a factor of 3 to account for shorter distances, one obtains precisely the result on the interior lattice. However, the rescaling we have supposed is only sensible for μ and λ. The coefficient, ξ was not hypothesized to depend on diffusion, but instead is a kinetic immobilization rate constant. Thus, although there may be some analogies between transport on interior and exterior hexagonal lattices, there are situations where the analogy breaks down.

Remark 2 In general, the second moment and the variance can differ since the variance involves both the first and second moment as

$$\sigma^2(t) = m^{(2)}(t) - \left(m^{(1)}(t)\right)^2.$$

This distinction is especially important in biased CTRWs (Shlesinger 1974; Othmer et al. 1988); however, when the transport is unbiased, then $m^{(1)}(t) = 0$ and the variance and second moment are equal.

Remark 3 With the use of first-passage-times, the moment calculation can be simplified somewhat. Rather than needing to handle details about the positions of SVs relative to each other, FPTs can be used to reduce this system to a single-state CTRW (for the junctions). Of course, collapsing all the information about the local geometry leads to a slight approximation; however, if the timescales of interest are larger than λ, μ, and ξ, the approximation will be accurate.

7 Conclusion

Transport processes in complex media such as biological tissues often involve several sub-processes that play a crucial role in setting the experimental data, and without incorporating the microscopic processes, macroscopic models may shed little light on the meaning of the data. As evidenced by the example involving transport in the wing disk, the macroscopic coefficients have detailed information embedded in them, and interpretation of the experimental data must be based on the microscopic processes. However, the complexity of the geometry in tissues often makes it difficult to do this using continuum reaction–diffusion equations. For example, developing an analytically tractable model of transport, binding, and degradation of Dpp in the cells and extracellular space of the wing disk is nigh impossible, but as we have shown herein, approaching this problem from a graph-based model of the geometry, coupled to a discrete description of spatial and internal states of a walker, enables one to obtain the coefficients of a macroscopic reaction–transport model based on the underlying microscopic processes. Our analysis is based on lattices in which all junctions have the same degree, all edge lengths are equal, and the lattice is infinite, which allows the use of Fourier transforms to analyze the resulting equations, but the incorporation of secondary vertices along edges and internal states of the walkers provides the means for describing several levels of microstructure in the models. In this description, the extracellular space in the wing disk is approximated by the set of edges in a graph, which simplifies the description, and while this does not capture all the details of the underlying structure, it is a significant step toward "lifting" microscopic information to a macroscopic description.

One potential complication is that microscale models are often difficult to define, and are not necessarily "better" or "worse" than an ad hoc macroscale model. However, when there is uncertainty about the macroscale model, understanding how microscale details eventually coalesce into readily observable macroscale behavior may help to separate promising theories from specious ones. In the wing disk, there are many experimental tools available to perturb various cellular-level biochemical and mechanical cues that are involved in normal growth and development. By perturbing these microscale parameters through genetic modifications or other means, and observing the resulting Dpp distribution, one can determine how K_M and D_M should depend on those parameters. These results can then be compared with the K_M and D_M determined via the methods presented here to assess whether the model is sensible. For instance, it is not immediately obvious that D_M in Eq. (84) should depend on the degradation rate, k_d, but retrospectively this is easily explained: rapid degradation will rapidly remove immobile particles leading to a larger proportion of mobile particles. This increases the macroscale diffusivity and the macroscale degradation coefficient.

Several previous studies (Goldhirsch and Gefen 1986, 1987) have presented ideas related to the development here with regard to compressing degrees of freedom in a random walk and computing moments. However, in those investigations the internal states consisted only of discrete spatial positions within various blocks making up a larger network. Neither the continuous-time aspects, nor the multistate aspects were studied. Moments of multistate random walks were also in Landman and Shlesinger (1979b), Landman and Shlesinger (1979a), Roerdink and Shuler (1985b), Roerdink and Shuler (1985a), Scher and Wu (1981), and Gadgil et al. (2005). Here we have extended the approaches in these studies and have shown how to systematically work with multistate systems that would be quite cumbersome to deal with in the frameworks developed previously. We have also shown how the theory developed in Othmer and Scriven (1971) can be applied to random processes that are not locally homogeneous.

The connection between micro- and macroscale properties is also extensively studied in homogenization theory (Pavliotis and Stuart 2008) and renewal theory (Cox 1967). In fact many of the problems we analyzed can be posed as systems of renewal equations, and some recent results have been obtained regarding the derivation of macroscale transport coefficients (Ciocanel et al. 2020). One difference between our method and classically homogenization theory is in the type of approximation used to extract the macroscale dynamics. In classical homogenization theory, one often *begins* by assuming an *ansatz* of the form

$$p_\epsilon(t, x, x/\epsilon, \dots) = \sum_{i=0}^{\infty} \epsilon^i p_i(t, x, x/\epsilon, \dots)$$

and then obtaining limiting equations as $\epsilon \to 0$. In a number of cases, this process has been rigorously verified. On the other hand, it is not true in all cases that the *ansatz* is valid, or frequently, the presence of boundary layers means that the expansion is only valid up to low order. In contrast, with our method, the asymptotic limit only need be considered in the last step of our method after deriving first-passage-time distributions which preserve the exact internal dynamics.

Finally, in this study we primarily considered examples in which the local state space of the random walk was discrete in nature, even when infinitely many states existed, such as in the comb problem. In a sequel, to this paper we will develop a detailed continuum computational model of the wing disk, which will enable us to compare predictions made herein with a microscopic continuum description. In effect, we will be able to model the time evolution of the Dpp distribution in the disk and determine in what parameter regimes the macroscopic parameters determined from the graph-based model coincide with those of the continuum model. It will also enable us to study variations in cell size and other factors that may affect morphogen distributions, and to determine when such problems can be reduced to lattice-based random walks by solving for certain continuum first-passage-time distributions.

Acknowledgements Supported in part by NIH Grants # GM29123 and 54-CA-210190 and NSF Grants DMS # 178743 and 185357.

Appendix A: Definition of the Kronecker Product

We define the tensor product of vectors as

$$\mathbf{x} \otimes \mathbf{y} \equiv (x_1\mathbf{y}, \dots, x_N\mathbf{y})^T = (x_1 y_1, \dots, x_1 y_n \dots, x_N y_1, \dots, x_N y_n)^T,$$

and if **R** is an $N \times N$ matrix and **T** an $n \times n$ one, their tensor product is the $Nn \times Nn$ matrix

$$\mathbf{R} \otimes \mathbf{T} = \begin{bmatrix} R_{11}T & \dots & R_{1N}T \\ \cdot & & \cdot \\ \cdot & & \cdot \\ \cdot & & \cdot \\ R_{N1}T & \dots & R_{NN}T \end{bmatrix}.$$

As we have seen throughout, multistate random walks can be conveniently written as matrix–vector problems involving Kronecker products. An important property of matrices and vectors formed as Kronecker products is that one can compute Fourier transforms on the first and second terms of the Kronecker product separately. For instance if $\boldsymbol{F}$ is a discrete Fourier transform matrix with Hermitian adjoint $\boldsymbol{F}'$,

$$(\boldsymbol{F} \otimes \boldsymbol{I})(\boldsymbol{A} \otimes \boldsymbol{B})\left(\boldsymbol{F}' \otimes \boldsymbol{I}\right) = \boldsymbol{F}\boldsymbol{A}\boldsymbol{F}' \otimes \boldsymbol{B} = \tilde{\boldsymbol{A}} \otimes \boldsymbol{B}.$$
$$(\boldsymbol{I} \otimes \boldsymbol{F})(\boldsymbol{A} \otimes \boldsymbol{B})\left(\boldsymbol{I} \otimes \boldsymbol{F}'\right) = \boldsymbol{A} \otimes \boldsymbol{F}\boldsymbol{B}\boldsymbol{F}' = \boldsymbol{A} \otimes \tilde{\boldsymbol{B}}.$$

Kronecker products provide an easy way to describe certain multidimensional problems in terms of simpler one-dimensional problems (Othmer and Scriven 1971). Each additional dimension is, roughly speaking, included by appending an additional Kronecker product to the previous transition matrix. This also applies in cases where the dimension of the state space is increased by adding internal state transitions to a spatial jump process.

Appendix B: The Exterior Hexagonal Lattice

In an exterior hexagonal lattice, there are three orientations of edges that occur, and two types of junctions: those centered at upwards facing trijunctions (type I), and those at downwards facing trijunctions (type II), see Fig. 12. Let us assume here that $\boldsymbol{T}_{SV}$ and $\boldsymbol{K}_{SV}$ are the same for each edge, and that $\boldsymbol{K}_J$ is the same for all junctions. Likewise we assume that $\boldsymbol{D}_s$, $\boldsymbol{D}_e$, $\boldsymbol{D}'_e$, and $\boldsymbol{D}'_s$ do not vary depending on the edge or junction being considered.

With two types of junctions that alternate, an easy way of assigning which edges start or end at which junctions is to require that all edges start at a type I junction and end at a type II junction (see the edges labeled "e" and "s" in Fig. 12). It is also helpful to specify some type of labeling system on the junctions and edges. To do so, we consider the combination of three edges around a type I vertex, and the type II junction attached at the end of the vertical edges to be labeled with the same label, (I, J) (see Fig. 12). The lattice position associated with this structure is given as $\boldsymbol{X}$, the position of the type I vertex. Of course, the positions of each SV on an edge can be found as some displacement, $\boldsymbol{x}$ from $\boldsymbol{X}$.

With this description, type I edges are vertical edges that pair type I and type II points with the same lattice index, e.g., $\boldsymbol{\Delta I}_1 = \boldsymbol{0}$. Type II edges are diagonal edges that pair type I lattice points and type II lattice points that differ with $\boldsymbol{\Delta I}_2 = (0, -1)$ from

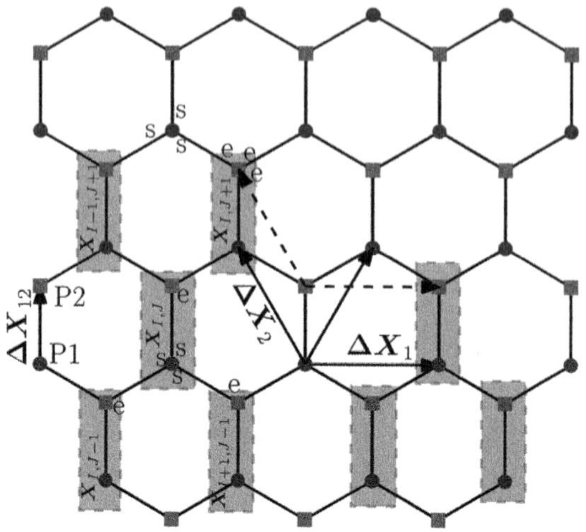

Fig. 12 Geometric quantities associated with the hexagonal lattice are depicted. Blue points are type I and Red points are type II. The quantity $\boldsymbol{\Delta X}_{12}$ is the displacement vector between a type I and type II point, and $\boldsymbol{\Delta X}_1$, $\boldsymbol{\Delta X}_2$, and $\boldsymbol{\Delta X}_3$ represent lattice displacement vectors between adjacent cells on the lattice. The red boxes indicate pairs of type I and type II points with the same lattice index, (I, J). The labels "s" and "e" indicate the starts and ends of edges

start to end. Finally, type III edges pair type I and type II points with $\boldsymbol{\Delta I}_3 = (1, -1)$ from start to end.

Since type I points are attached to the start of each edge and type II to the end of each edge, if we structure the probability density vector as $\boldsymbol{p} = (\boldsymbol{p}_I, \boldsymbol{p}_{II}, \boldsymbol{p}_{e_1}, \boldsymbol{p}_{e_2}, \boldsymbol{p}_{e_3})^T$, we may write the overall transition matrix $\boldsymbol{T} + \boldsymbol{K}$ as

$$
\begin{aligned}
\boldsymbol{T} + \boldsymbol{K} = \boldsymbol{I} \otimes & \left(\begin{array}{cc|ccc} \boldsymbol{K}_J & \boldsymbol{0} & \boldsymbol{D}_s & \boldsymbol{D}_s & \boldsymbol{D}_s \\ \boldsymbol{0} & \boldsymbol{K}_J & \boldsymbol{D}_e & \boldsymbol{0} & \boldsymbol{0} \\ \hline \boldsymbol{D}_s' & \boldsymbol{D}_e' & \boldsymbol{T}_{SV}+\boldsymbol{K}_{SV} & \boldsymbol{0} & \boldsymbol{0} \\ \boldsymbol{D}_s' & \boldsymbol{0} & \boldsymbol{0} & \boldsymbol{T}_{SV}+\boldsymbol{K}_{SV} & \boldsymbol{0} \\ \boldsymbol{D}_s' & \boldsymbol{0} & \boldsymbol{0} & \boldsymbol{0} & \boldsymbol{T}_{SV}+\boldsymbol{K}_{SV} \end{array}\right) \\
& + \boldsymbol{A}_{e,1} \otimes \left(\begin{array}{c|c} \boldsymbol{0} & \begin{matrix} \boldsymbol{0} & \boldsymbol{0} & \boldsymbol{0} \\ \boldsymbol{0} & \boldsymbol{D}_e & \boldsymbol{0} \end{matrix} \\ \hline \boldsymbol{0} & \boldsymbol{0} \end{array}\right) + \boldsymbol{A}_{e,1}^T \otimes \left(\begin{array}{c|c} \boldsymbol{0} & \boldsymbol{0} \\ \hline \begin{matrix} \boldsymbol{0} & \boldsymbol{0} \\ \boldsymbol{0} & \boldsymbol{D}_e' \\ \boldsymbol{0} & \boldsymbol{0} \end{matrix} & \boldsymbol{0} \end{array}\right) \\
& + \boldsymbol{A}_{e,2} \otimes \left(\begin{array}{c|c} \boldsymbol{0} & \begin{matrix} \boldsymbol{0} & \boldsymbol{0} & \boldsymbol{0} \\ \boldsymbol{0} & \boldsymbol{0} & \boldsymbol{D}_e \end{matrix} \\ \hline \boldsymbol{0} & \boldsymbol{0} \end{array}\right) + \boldsymbol{A}_{e,2}^T \otimes \left(\begin{array}{c|c} \boldsymbol{0} & \boldsymbol{0} \\ \hline \begin{matrix} \boldsymbol{0} & \boldsymbol{0} \\ \boldsymbol{0} & \boldsymbol{0} \\ \boldsymbol{0} & \boldsymbol{D}_e' \end{matrix} & \boldsymbol{0} \end{array}\right)
\end{aligned}
\tag{B.1}
$$

We apply a lattice Fourier transformation to $\boldsymbol{X}$ to obtain

$$
\hat{\boldsymbol{T}}(\boldsymbol{\omega}) + \boldsymbol{K} = \boldsymbol{I} \otimes \left(\begin{array}{cc|ccc} \boldsymbol{K}_J & \boldsymbol{0} & \boldsymbol{D}_s & \boldsymbol{D}_s & \boldsymbol{D}_s \\ \boldsymbol{0} & \boldsymbol{K}_J & \boldsymbol{D}_e & e^{i\boldsymbol{\Delta X}_2\cdot\boldsymbol{\omega}}\boldsymbol{D}_e & e^{i\boldsymbol{\Delta X}_3\cdot\boldsymbol{\omega}}\boldsymbol{D}_e \\ \hline \boldsymbol{D}_s' & \boldsymbol{D}_e' & \boldsymbol{T}_{SV} + \boldsymbol{K}_{SV} & \boldsymbol{0} & \boldsymbol{0} \\ \boldsymbol{D}_s' & e^{-i\boldsymbol{\Delta X}_2\cdot\boldsymbol{\omega}} & \boldsymbol{0} & \boldsymbol{T}_{SV} + \boldsymbol{K}_{SV} & \boldsymbol{0} \\ \boldsymbol{D}_s' & e^{-i\boldsymbol{\Delta X}_3\cdot\boldsymbol{\omega}}\boldsymbol{D}_e' & \boldsymbol{0} & \boldsymbol{0} & \boldsymbol{T}_{SV} + \boldsymbol{K}_{SV} \end{array}\right)
\tag{B.2}
$$

where $\boldsymbol{\Delta X}_m$ are as defined in Eq. (86). Once the local transitions in $\boldsymbol{K}_J$, $\boldsymbol{K}_{SV}$, $\boldsymbol{T}_{SV}$, $\boldsymbol{D}_m$, and $\boldsymbol{D}_m'$ are specified, substituting this matrix into Eq. (33) and solving for $\boldsymbol{q}$ and $\boldsymbol{p}$ will yield a solution for the spatial distribution of a particle over the hexagonal lattice with possible internal states and SVs between junction points. If we make use of the same definitions for $\boldsymbol{D}$, $\boldsymbol{D}'$, $\boldsymbol{T}_{SV}$, $\boldsymbol{K}_{SV}$ and $\boldsymbol{K}_J$ from the interior lattice example in Sect. 6, with $n_k = 1$ and $n_s = 2$, we obtain a 8×8 matrix (after removing the rows and columns associated with immobile states at the junctions as in Sect. 6) that can be inverted to solve for $\boldsymbol{p}$ in Fourier–Laplace transform space. For instance, with $1 - f$ the proportion of particles that are immobilized at the boundary, the steady-state concentration at the junctions can be found as

$$\lim_{\boldsymbol{\omega}\to\boldsymbol{0}} \lim_{s\to 0} s\left(p_I(\boldsymbol{\omega}, s) + p_{II}(\boldsymbol{\omega}, s)\right) = \frac{f\mu\xi}{f\mu\xi + d((1-f)\mu + \xi)}.$$

In closing this section, we note that random walks on exterior hexagonal lattices have been studied several times previously (Montroll 1969; Hughes 1996; Henyey and Seshadri 1982); however, in those cases, no internal states were considered. Thus, our formulation here provides a straightforward way to extend these previous results to more complicated transport processes. We also note that the lattice Green's function for the exterior hexagonal lattice with no internal states or SVs is known (Henyey and Seshadri 1982; Hughes 1996). In Fourier transform space, the Green's function for a hexagonal lattice where each edge is taken in one step, and the transition probabilities at each intersection are all 1/3, is of the form

$$\boldsymbol{p}^{(H)}(\boldsymbol{\omega}, z) = \frac{\begin{pmatrix} 1 & \frac{z}{3}\left(1 + e^{i\boldsymbol{\Delta X}_2\cdot\boldsymbol{\omega}} + e^{i\boldsymbol{\Delta X}_3\cdot\boldsymbol{\omega}}\right) \\ \frac{z}{3}\left(1 + e^{-i\boldsymbol{\Delta X}_2\cdot\boldsymbol{\omega}} + e^{-i\boldsymbol{\Delta X}_3\cdot\boldsymbol{\omega}}\right) & 1 \end{pmatrix}}{1 - \frac{z^2}{9}\left(3 + 2\cos\omega_x + 4\cos\left(\frac{\omega_x}{2}\right)\cos\left(\frac{\sqrt{3}\omega_y}{2}\right)\right)} \quad \text{(B.3)}$$

Of course, when we set $n_k = 1$, $n_s = 1$, and replace $\psi(t)$ with z in Eq. (B.2), we would obtain this result. The ij^{th} element of $\boldsymbol{p}^H$ gives the occupation probability for a random walk that started on a type j vertex and is currently on a type i vertex.

Our result here appears to differ slightly from the previous results for this Green's function. This because we have directly included the displacement, $\boldsymbol{\Delta X}_i$ into the computation, rather than just using changes in index $\boldsymbol{\Delta I}_i$, which yield precisely the result previously reported. In other words, Eq. (B.3) gives the probability of being located at position $\boldsymbol{X}_{\boldsymbol{I}}$, whereas previous results give the probability of being located at index $\boldsymbol{I}$.

B.1 The Exterior Hexagonal Lattice with Arbitrary n_k

Here, we consider the hexagonal lattice with n_k SVs and $n_s = 1$. The FPT method from Sect. 4.1 becomes essential here as with arbitrary n_k, directly solving the resulting matrices is non-trivial.

From Appendix B, we found that the form of $\hat{\boldsymbol{T}}(\omega)+\hat{\boldsymbol{K}}(\omega)$ for the exterior hexagonal lattice. In this example, $n_s = 1$, so $\boldsymbol{K}_{SV} = \boldsymbol{K}_J = \boldsymbol{0}$. We also have that

$$\begin{aligned} \boldsymbol{D}_s &= \frac{\psi}{2}\begin{pmatrix}1 & 0 & \ldots & 0\end{pmatrix} \\ \boldsymbol{D}_e &= \frac{\psi}{2}\begin{pmatrix}0 & \ldots & 0 & 1\end{pmatrix} \end{aligned} \tag{B.4}$$

and

$$\boldsymbol{D}'_s = \frac{\psi}{3}\begin{pmatrix}1\\0\\\vdots\\0\end{pmatrix}, \quad \boldsymbol{D}'_e = \frac{\psi}{3}\begin{pmatrix}0\\\vdots\\0\\1\end{pmatrix}.$$

Lastly, $\boldsymbol{T}_{SV}$ is the matrix for a 1D random walk on a segment of length n_k with absorbing boundaries. This can be written as

$$\boldsymbol{T}_{SV} = \frac{\psi}{2}\begin{pmatrix} 0 & 1 & 0 & \ldots & & & & \\ 1 & 0 & 1 & 0 & \ldots & & & \\ 0 & 1 & 0 & 1 & 0 & \ldots & & \\ \vdots & & \ddots & \ddots & \ddots & 0 & \ldots & \\ 0 & & \ldots & 0 & 1 & 0 & 1 & \\ 0 & \ldots & & & 0 & 1 & 0. & \end{pmatrix}$$

Since there is one internal state, $\boldsymbol{K}_{SV} = \boldsymbol{0}$. Since each edge has the same $\boldsymbol{T}_{SV}$, $\boldsymbol{D}$ and $\boldsymbol{D}'$, we may use the formulation in Appendix D.2 to obtain the effective transition rates, $\boldsymbol{T}^{\text{eff}}_{2,r}$.

To do so, start by noting that since $n_s = 1$, $\boldsymbol{T}^{\text{eff}}_{1,r}$ and $\boldsymbol{K}_r$ are both equal to 0 and $\boldsymbol{T}^{\text{eff}}_{2,r}$ is a scalar-valued function. Since the type I and type II junctions are equivalent in terms of the transitions that occur around them, $\boldsymbol{T}^{\text{eff}}_{2,r}$ is found independent of whether the particle is at a type I or type II junction.

This leaves us with the following matrix equation:

$$\left(\begin{pmatrix} 1 & 0 & 0 & \ldots \\ 0 & 1 & 0 & \ldots \\ \vdots & & \ddots & \\ 0 & \ldots & & 1 \end{pmatrix} - \begin{pmatrix} 0 & \left(\frac{\psi}{2}\ 0 \ldots\right) \\ \begin{pmatrix}\psi\\0\\\vdots\end{pmatrix} & \boldsymbol{T}_{SV} \end{pmatrix}\right)\begin{pmatrix}\tilde{f}_1\\\tilde{f}_2\\\vdots\\\tilde{f}_{n_k}\end{pmatrix} = \begin{pmatrix}1\\0\\\vdots\\0\end{pmatrix}$$

for $\tilde{f}(s)$. Solving for $\tilde{f}_1$ via a Schur complement yields,

$$\tilde{f}_1 = 1 - 3\boldsymbol{D}_s(\boldsymbol{I} - \boldsymbol{T}_{SV})^{-1}\boldsymbol{D}_s,$$

and the remaining elements of $\tilde{f}$ are found as

$$\tilde{f} = \frac{(\boldsymbol{I} - \boldsymbol{T}_{SV})^{-1}\boldsymbol{D}_s'}{\tilde{f}_1}.$$

Finally,

$$\boldsymbol{T}_{2,r}^{\text{eff}} = \boldsymbol{D}_e\tilde{f} = \frac{\boldsymbol{D}_e(\boldsymbol{I} - \boldsymbol{T}_{SV})^{-1}\boldsymbol{D}_s'}{1 - 3\boldsymbol{D}_s'(\boldsymbol{I} - \boldsymbol{T}_{SV})^{-1}\boldsymbol{D}_s} = \frac{\psi}{2}\tilde{f}_{n_k}.$$

The next step is of course to explicitly solve for this FPT. Since each edge of the lattice is simply a 1D segment with absorbing boundaries, the discussion in Appendix E provides us with an explicit form of the solution to $(\boldsymbol{I} - \boldsymbol{T}_{SV})^{-1}$. After some algebraic simplifications, we find that

$$\boldsymbol{T}_{2,r}^{\text{eff}} = \frac{2}{3}\frac{1}{\left(\frac{1-\sqrt{1-\psi^2}}{\psi}\right)^{n_k+1} + \left(\frac{1-\sqrt{1-\psi^2}}{\psi}\right)^{-n_k-1}}$$

Since $\boldsymbol{K} = 0$ and $\boldsymbol{T}_{1,r}^{\text{eff}} = 0$, this reduces the problem to a single-state CTRW on a hexagonal lattice. Thus, we can make use of formula in Eq. (B.3) with z replaced by $\psi^{\text{eff}} = 3\boldsymbol{T}_{2,r}^{\text{eff}}$. The factor of 3 comes from the fact that $\boldsymbol{T}_{2,r}^{\text{eff}}$ is in fact the effective waiting time distribution multiplied by the probability that the particle actually travels down a particular edge. This probability is 1/3 since each junction connects three edges in the exterior hexagonal lattice, and each edge has the same probability of being traveled on.

With this result, it is possible to directly compute the moments for the hexagonal lattice with n_K SVs. However, assuming each segment has a fixed total length of 1 independent of n_k, the exact formulas become

$$\tilde{m}^{(1)}(s) = 0$$

$$\tilde{m}^{(2)}(s) = \frac{1-\psi^{\text{eff}}}{s}\frac{\psi^{\text{eff}}}{2(1-\psi)^2} = \frac{1}{s}\frac{1-\psi^{\text{eff}}}{(a^{n_k+1} + a^{-n_k-1}) + 4\frac{1}{a^{n_k+1}+a^{-n_k-1}} - 4}$$

with $a = (1 - \sqrt{1-\psi^2})\psi^{-1}$. For a Poisson distributed waiting time $\psi = \lambda e^{-\lambda t}$, we obtain the following results for the first several $n_k = 1, 2, \ldots$

$$m^{(2)}(t) = \frac{1}{2}\lambda t \qquad n_k = 0$$

$$m^{(2)}(t) = \frac{1}{2}\frac{\lambda t}{4} - \frac{1}{16}\left(1 - e^{-2\lambda t}\right) \qquad n_k = 1$$

$$m^{(2)}(t) = \frac{1}{2}\frac{\lambda t}{9} - \frac{2}{27} + \frac{e^{-3\lambda t}}{54}(4 + 3\lambda t) \qquad n_k = 2$$

$$\vdots$$

$$m^{(2)}(t) = \frac{1}{2}\frac{\lambda t}{n_k^2} + \ldots$$

In these results, λ and n_k are independent. However, if this hexagonal lattice is instead interpreted as a discretization of a continuum problem with 1D diffusion along segments of the lattice, then $\lambda \sim n_k^2 D$ where D is the diffusivity along the segment. In that case, the leading-order term for any n_k is

$$m^{(2)}(t) = \frac{1}{2}Dt$$

Although we did not specify the direction of diffusion here, it turns out that for the hexagonal lattice, whether diffusion is considered along the x or y direction, the results are equivalent.

Note that the long-term asymptotics may be obtained easily, at least for a single internal state, from the theory developed in Roerdink and Shuler (1985a). However, the full time dependence is not computable under that theory.

Appendix C: General Form of T and K

Recall from Eqs. (33)–(36) that the solution $\boldsymbol{p}(\boldsymbol{X}, t|\{k_0, \ell_0\})$, to a transport problem depends on the initial concentration distribution, and on the form of the transition matrix. Equation (33) may be written as,

$$\boldsymbol{q}\,(\boldsymbol{X}, t) = \delta(t)\boldsymbol{\delta}_{k_0,\ell_0}\delta(\boldsymbol{X}) + \int_0^t \left[\sum_{\boldsymbol{X}'\in\mathcal{N}(\boldsymbol{X})} \boldsymbol{T}(\boldsymbol{X}-\boldsymbol{X}')\boldsymbol{\phi}(t-\tau)\boldsymbol{q}(\boldsymbol{X}', \tau)\right] d\tau$$

$$+ \int_0^t \boldsymbol{K\Lambda}(t-\tau)\boldsymbol{q}(\boldsymbol{X}, \tau, \{k_0, \ell_0\}\, d\tau$$

$$\boldsymbol{p}\,(\boldsymbol{X}, t) = \hat{\boldsymbol{\Phi}}(t)\boldsymbol{\delta}_{k_0,\ell_0}\delta(\boldsymbol{X}) + \int_0^t \hat{\boldsymbol{\Phi}}(t-\tau)\int_0^\tau \Bigg[\sum_{\boldsymbol{X}'\in\mathcal{N}(\boldsymbol{X})} \boldsymbol{T}(\boldsymbol{X}-\boldsymbol{X}')\boldsymbol{\phi}(\tau$$

$$-s)\boldsymbol{q}(\boldsymbol{X}', s,)\Big]\, ds\, d\tau + \int_0^t \hat{\boldsymbol{\Phi}}(t-\tau)\int_0^\tau \boldsymbol{K\Lambda}(\tau-s)\boldsymbol{q}(\boldsymbol{X}, s) ds\, d\tau$$

with

$$\hat{\Phi}_{k,\ell}(t) = 1 - \int_0^t \psi_{k,\ell}(t) = 1 - \int_0^t \left(\sum_{\boldsymbol{X}'\in\mathcal{N}(\boldsymbol{X})} \boldsymbol{1}^T\boldsymbol{T}(\boldsymbol{X}')\boldsymbol{\phi}(\tau)\boldsymbol{\delta}_{k,\ell} + \boldsymbol{1}^T\boldsymbol{K\Lambda}(\tau)\boldsymbol{\delta}_{k,\ell}\right) d\tau.$$

We now present a general structure for matrices $\boldsymbol{T}$ and $\boldsymbol{K}$ that can include the examples studied, along with other examples that may arise in a variety of applications.

Recall that the number of SVs is n_k, the number of types of edges, n_e, and the number of internal states, n_s. The global matrices $\boldsymbol{T\phi}$ and $\boldsymbol{K\Lambda}$ can be written in block-structured formats, and will be understood here as functions of the Laplace transform variable s. We also will abuse notation here and write $\boldsymbol{T}$ and $\boldsymbol{K}$ rather than $\boldsymbol{T\phi}$ and $\boldsymbol{K\Lambda}$. Recall also that $\boldsymbol{K}$ involves only transitions between internal states, and no spatial movement whereas $\boldsymbol{T}$ involves spatial movement without any change in the internal state. Let $\boldsymbol{K}_J \in \mathbb{R}^{n_s \times n_s}$, $\boldsymbol{T}_{SV}$, $\boldsymbol{K}_{SV} \in \mathbb{R}^{(n_k n_s)\times(n_k n_s)}$, $\boldsymbol{D}'_s$, $\boldsymbol{D}'_e \in \mathbb{R}^{(n_k n_s)\times n_s}$, and $\boldsymbol{D}_s$, $\boldsymbol{D}_e \in \mathbb{R}^{n_s \times (n_k n_s)}$ be various blocks of $\boldsymbol{T}$ and $\boldsymbol{K}$. In particular,

- $\boldsymbol{K}_J$ characterizes transitions between the internal states at each junction
- $\boldsymbol{K}_{SV}$ characterizes transitions between internal states associated with the SVs along each edge
- $\boldsymbol{T}_{SV}$ characterizes jumps between SVs on an edge
- $\boldsymbol{D}_s$ and $\boldsymbol{D}_e$ characterize transitions from the SVs to a vertex
- $\boldsymbol{D}'_s$ and $\boldsymbol{D}'_e$ characterize transitions from a junction to an SV

The transitions represented by these matrices are diagrammed in Fig. 5. The indices s and e on $\boldsymbol{D}$ and $\boldsymbol{D}'$ are used to distinguish whether the transfers occur at the "start" or "end" of each edge (see Fig. 5). This is necessary since when the SVs are labeled from $k = 1$ to $k = n_k$, different (but closely related) matrices are needed to describe transfers from $k = 1$ (start) and $k = n_k$ (end). Recall Ex. 2 in Sect. 2.5 where $\boldsymbol{D}_s$ and $\boldsymbol{D}_e$ were explicitly constructed. This labeling does not imply that an edge is directed or undirected, but merely serves as a way of specifying the exact positions of SVs along the edge.

Next, define $\boldsymbol{I}$ as the infinite-dimensional identity matrix over the lattice positions, $\boldsymbol{X}$, and let $\boldsymbol{A}_{s,m}$ and $\boldsymbol{A}_{e,m}$ be infinite-dimensional adjacency matrices that specify which edges start and end at a given vertex. Similarly, let $\boldsymbol{A}'_{s,m}$ and $\boldsymbol{A}'_{e,m}$ specify the vertices that each edge is attached to. Often, $\boldsymbol{A}'_{e,m}$ is the transpose of $\boldsymbol{A}_{s,m}$, and likewise for $\boldsymbol{A}'_{s,m}$ and $\boldsymbol{A}_{e,m}$, so long as the graph is undirected. The index $m = 1, \ldots, n_e$ is used to distinguish between edges of different orientations (e.g., vertical or horizontal), and n_e is the number of different types of edges. For instance, in a square lattice, horizontal edges connect junctions that differ in their x-coordinate, but the vertical edges connect junctions that differ in their y-coordinate. Even if the same types of transition occur regardless of the edge orientation, distinguishing the orientation of each edge is crucial for obtaining spatial moments of the distribution. Furthermore, this formulation lends itself to extensions where not all edges have the same internal transition matrices.

The overall transition matrix is of the form

$$\boldsymbol{T} = \boldsymbol{I} \otimes \begin{pmatrix} \boldsymbol{0} & & & \\ & \boldsymbol{T}_{SV} & & \\ & & \ddots & \\ & & & \boldsymbol{T}_{SV} \end{pmatrix} + \sum_{k=1}^{n_e} \left(\boldsymbol{A}_{s,k} \otimes \bar{\boldsymbol{D}}_{s,k} + \boldsymbol{A}_{e,k} \otimes \bar{\boldsymbol{D}}_{e,k} \right)$$

$$+ \sum_{k=1}^{n_e} \left(\boldsymbol{A}^T_{s,k} \otimes \bar{\boldsymbol{D}}'_{s,k} + \boldsymbol{A}^T_{e,k} \otimes \bar{\boldsymbol{D}}'_{e,k} \right) \quad \text{(C.1)}$$

$$\boldsymbol{K} = \boldsymbol{I} \otimes \begin{pmatrix} \boldsymbol{K}_J & & & \\ & \boldsymbol{K}_{SV} & & \\ & & \ddots & \\ & & & \boldsymbol{K}_{SV} \end{pmatrix} \quad \text{(C.2)}$$

The blocks $\bar{\boldsymbol{D}}_{t,k}$ and $\bar{\boldsymbol{D}}'_{t,k}$ for $k = 1, \ldots, n_e$ and $t = \{s, e\}$ are themselves block matrices of the form

$$\bar{\boldsymbol{D}}_{t,1} = \left(\begin{array}{c|cccc} \mathbf{0} & \boldsymbol{D}_{t,1} & \mathbf{0} & \ldots & \\ \hline \mathbf{0} & \mathbf{0} & & & \\ \mathbf{0} & & \mathbf{0} & & \\ \vdots & & & \ddots & \\ \mathbf{0} & & & & \mathbf{0} \end{array}\right), \quad \bar{\boldsymbol{D}}_{t,2} = \left(\begin{array}{c|ccccc} \mathbf{0} & \mathbf{0} & \boldsymbol{D}_{t,2} & \mathbf{0} & \ldots & \\ \hline \mathbf{0} & \mathbf{0} & & & & \\ \mathbf{0} & & \mathbf{0} & & & \\ \vdots & & & \ddots & & \\ \mathbf{0} & & & & & \mathbf{0} \end{array}\right), \quad \text{etc.}$$

$$\bar{\boldsymbol{D}}'_{t,1} = \left(\begin{array}{c|cccc} \mathbf{0} & \mathbf{0} & \mathbf{0} & \ldots & \mathbf{0} \\ \hline \boldsymbol{D}'_{t,1} & \mathbf{0} & & & \\ \mathbf{0} & & \mathbf{0} & & \\ \vdots & & & \ddots & \\ \mathbf{0} & & & & \mathbf{0} \end{array}\right), \quad \bar{\boldsymbol{D}}'_{t,2} = \left(\begin{array}{c|cccc} \mathbf{0} & \mathbf{0} & \mathbf{0} & \ldots & \mathbf{0} \\ \hline \mathbf{0} & \mathbf{0} & & & \\ \boldsymbol{D}'_{t,2} & & \mathbf{0} & & \\ \mathbf{0} & & & \ddots & \\ \vdots & & & & \mathbf{0} \end{array}\right), \quad \text{etc.}$$

Note that for each type of edge in $k = 1, \ldots, n_e$, it is often possible to label the lattice so that edge $\boldsymbol{J}$ of that type connects to junction $\boldsymbol{J}$ at one terminus. With this in mind, we assume that for each edge, $\boldsymbol{J}$ of type k, the s-end of the edge border junction $\boldsymbol{J}$. Thus, $\boldsymbol{A}_{s,k} = \boldsymbol{A}'_{s,k} = \boldsymbol{I}$, where $\boldsymbol{I}$ is the identity operator on the lattice. If we further assume that the matrices $\boldsymbol{D}_k$ are all equal, then $\boldsymbol{T}$ simplifies nicely,

$$\boldsymbol{T} = \boldsymbol{I} \otimes \begin{pmatrix} \mathbf{0} & \boldsymbol{D}_s & \ldots & \boldsymbol{D}_s \\ \boldsymbol{D}'_s & \boldsymbol{T}_{SV} & & \\ \vdots & & \ddots & \\ \boldsymbol{D}'_s & & & \boldsymbol{T}_{SV} \end{pmatrix} + \sum_{k=1}^{n_e} \left(\boldsymbol{A}_{e,k} \otimes \boldsymbol{D}_{e,k} + \boldsymbol{A}^T_{e,k} \otimes \boldsymbol{D}'_{e,k} \right) \tag{C.3}$$

which resembles the Laplacian-type structure observed previously.

Although $\boldsymbol{I}$, $\boldsymbol{A}_{t,m}$, and $\boldsymbol{A}'_{t,m}$ with $t = \{s, e\}$ were defined as infinite-dimensional matrices, they can also be understood as linear operators over the infinite lattice, e.g., $\boldsymbol{A}_{t,m} = A_{t,m}(\boldsymbol{X}, \boldsymbol{Y})$ where $\boldsymbol{X}$ and $\boldsymbol{Y}$ are two junctions. In all the cases, we have considered, for each fixed $\boldsymbol{Y}$, the lattice operators are only nonzero over a few (or one) values of $\boldsymbol{X}$. The interpretation of $\boldsymbol{A}_{t,m}$ as lattice operators also gives meaning to the notation $\boldsymbol{T}(\boldsymbol{X} - \boldsymbol{X}', s)$ used in Eqs. (33)–(36).

Remark

In some cases there can be more than one type of junction. An example is the exterior hexagonal lattice, which has two types of junctions. When this occurs, additional blocks of rows and columns will augment Eqs. (C.1), (C.2), and the definitions of $\boldsymbol{D}$ and $\boldsymbol{D}'$ to describe internal transitions at each type of junction, transitions from each type of junction to each type of edge, and transitions from each type of edge to each type of junction, c.f. Eq. (B.1).

We see that the dimension of the state-space associated with each junction, $\boldsymbol{X}$ is $n_T = n_s(n_v + n_e n_k)$ where n_v is the number of junction types, and n_e the number of edge types.

C.1 Computation of Spatial Moments in the General Case

We now turn our attention toward computation of the moments for the general evolution equations in Eq. (33). Many properties of CTRWs are most easily analyzed in Laplace transform space. With lattice random walks, it is also often convenient to apply a lattice Fourier transform over $\boldsymbol{X}$, the lattice variable.

To compute the spatial moments (as functions of s, the Laplace transform variable), we start by computing the lattice Fourier transform of Eqs. (C.3) and (C.2) over the lattice spatial variable, $\boldsymbol{X}$, obtaining

$$\hat{\boldsymbol{T}} + \hat{\boldsymbol{K}} = (\mathscr{F} \otimes \boldsymbol{I})\,(\boldsymbol{T} + \boldsymbol{K})\left(\mathscr{F}^{-1} \otimes \boldsymbol{I}\right) \tag{C.4}$$

$$= \boldsymbol{I} \otimes \left[\begin{pmatrix} \boldsymbol{K}_J & \boldsymbol{D}_e + \boldsymbol{D}_s & \dots & \boldsymbol{D}_e + \boldsymbol{D}_s \\ \boldsymbol{D}'_e + \boldsymbol{D}'_s & \boldsymbol{T}_{SV} + \boldsymbol{K}_{SV} & \boldsymbol{0} & \dots \\ \vdots & \boldsymbol{0} & \ddots & \boldsymbol{0} \\ \boldsymbol{D}'_e + \boldsymbol{D}'_s & \vdots & \boldsymbol{0} & \boldsymbol{T}_{SV} + \boldsymbol{K}_{SV} \end{pmatrix}\right.$$

$$\left. + \begin{pmatrix} \boldsymbol{0} & \alpha_{e,1}(\boldsymbol{\omega})\boldsymbol{D}_e & \dots & \alpha_{e,m}(\boldsymbol{\omega})\boldsymbol{D}_e \\ \beta_{e,1}(\boldsymbol{\omega})\boldsymbol{D}'_e & \boldsymbol{0} & \dots & \\ \vdots & \vdots & & \\ \beta_{e,m}(\boldsymbol{\omega})\boldsymbol{D}'_e & \boldsymbol{0} & \dots & \end{pmatrix}\right] \tag{C.5}$$

$$= \boldsymbol{I} \otimes \left[\boldsymbol{K}' + \boldsymbol{T}'(\boldsymbol{\omega})\right] \tag{C.6}$$

where

$$\alpha_{e,m}(\boldsymbol{\omega}) = (e^{i\Delta \boldsymbol{X}_m \cdot \boldsymbol{\omega}} - 1), \qquad \beta_{e,m}(\boldsymbol{\omega}) = (e^{-i\Delta \boldsymbol{X}_m \cdot \boldsymbol{\omega}} - 1)$$

and $\mathscr{F}$ is the Fourier transform operator on the lattice. The vectors $\Delta \boldsymbol{X}_m$ specify the displacements between adjacent junctions along a type m edge. The functions $\alpha_e(\boldsymbol{\omega})$ and $\beta_e(\boldsymbol{\omega})$ are analogous to the functions $(z-1)$ and $(z^{-1}-1)$ discussed in Sect. 3.1.

Since we use a lattice Fourier transform as opposed to a generating function formalism here, we have replaced z by $e^{i\omega}$. With the lattice Fourier transform, it seems somewhat easier to handle cases where $\Delta \boldsymbol{X}_{t,m}$ are not all of the same length for every $t = \{s, e\}$ or $m = 1, \dots, n_s$, although these minutiae could likely be included in a generating function approach with only a few modifications. Lastly, the matrix $\boldsymbol{K}'$ corresponds with $\boldsymbol{K}_0 + \boldsymbol{D}^+ + \boldsymbol{D}^-$, and $\boldsymbol{T}'(\boldsymbol{\omega})$ corresponds with $(z-1)\boldsymbol{D}^+ + (1/z-1)\boldsymbol{D}^-$ from Sect. 3.2.

Due to the translation invariance of the lattice points, we see that the Fourier transform diagonalizes the adjacency matrices, $\boldsymbol{A}_{e,m}$ and $\boldsymbol{A}'_{e,m}$. This reduces an infinite-dimensional matrix problem to an $n_T \times n_T$ problem parameterized by a variable, $\boldsymbol{\omega}$. The spatial moments over the junctions are found by considering derivatives of $(\boldsymbol{I} - \hat{\boldsymbol{K}} - \hat{\boldsymbol{T}})^{-1}$ with respect to $\boldsymbol{\omega}$ in the limit that $\boldsymbol{\omega} \to \boldsymbol{0}$. For instance, on an infinite one-dimensional lattice, the derivative of the lattice Fourier transform of a function $f(x_k)$ is

$$\lim_{\omega\to 0} \frac{\partial^j \hat{f}(\omega)}{\partial \omega^j} = \sum_{k=-\infty}^{\infty} (-ix)^j f(x_k) = (-i)^j m^{(j)} \qquad j = 0, 1, \ldots$$

Given $\hat{\boldsymbol{T}}$ and $\hat{\boldsymbol{K}}$, in the limit $\boldsymbol{\omega} \to \boldsymbol{0}$, we have

$$\frac{\partial}{\partial \omega_j} \left[\boldsymbol{I} - \boldsymbol{K}' - \boldsymbol{T}'(\boldsymbol{\omega})\right]^{-1} = \left[\boldsymbol{I} - \boldsymbol{K}'\right]^{-1} \left(\frac{\partial \boldsymbol{T}'}{\partial \omega_j}\right) \left[\boldsymbol{I} - \boldsymbol{K}'\right]^{-1}$$

$$\frac{\partial^2}{\partial \omega_j \partial \omega_k} \left[\boldsymbol{I} - \boldsymbol{K}' - \boldsymbol{T}'(\boldsymbol{\omega})\right]^{-1} = 2\left[\boldsymbol{I} - \boldsymbol{K}'\right]^{-1} \left(\frac{\partial \boldsymbol{T}'}{\partial \omega_j}\right) \left[\boldsymbol{I} - \boldsymbol{K}'\right]^{-1} \left(\frac{\partial \boldsymbol{T}'}{\partial \omega_k}\right) \left[\boldsymbol{I} - \boldsymbol{K}'\right]^{-1} + \left[\boldsymbol{I} - \boldsymbol{K}'\right]^{-1} \left(\frac{\partial^2 \boldsymbol{T}'}{\partial \omega_j \partial \omega_k}\right) \left[\boldsymbol{I} - \boldsymbol{K}'\right]^{-1}$$

where derivatives, $\frac{\partial}{\partial \omega_j}$, are taken with respect to $\boldsymbol{\omega}$ and are understood to be evaluated at $\boldsymbol{\omega} = \boldsymbol{0}$. Since $\boldsymbol{T}'(\boldsymbol{\omega}) = \boldsymbol{0}$ as $\boldsymbol{\omega} \to \boldsymbol{0}$, the only matrix inverse required is $\left[\boldsymbol{I} - \boldsymbol{K}'\right]^{-1}$. Averaging over the internal states within a cell, we can compute the average spatial moments of a process with transition matrix, $\hat{\boldsymbol{T}}$ as (c.f. Landman and Shlesinger 1979a)

$$\boldsymbol{m}_j^{(1)} = \left\langle (-i)\hat{\boldsymbol{\Phi}} \left[\boldsymbol{I} - \boldsymbol{K}'\right]^{-1} \left(\frac{\partial \boldsymbol{T}'}{\partial \omega_j}\right) \left[\boldsymbol{I} - \boldsymbol{K}'\right]^{-1} + x_j \hat{\boldsymbol{\Phi}} \left[\boldsymbol{I} - \boldsymbol{K}'\right]^{-1} \right\rangle \tag{C.7}$$

$$\boldsymbol{m}_{jk}^{(2)} = \left\langle -\hat{\Phi} \left[\boldsymbol{I} - \boldsymbol{K}'\right]^{-1} \left(2\left(\frac{\partial \boldsymbol{T}'}{\partial \omega_j}\right) \left[\boldsymbol{I} - \boldsymbol{K}'\right]^{-1} \left(\frac{\partial \boldsymbol{T}'}{\partial \omega_k}\right) + \left(\frac{\partial^2 \boldsymbol{T}'}{\partial \omega_j \partial \omega_k}\right)\right) \left[\boldsymbol{I} - \boldsymbol{K}'\right]^{-1} \right. \tag{C.8}$$

$$\left. + 2(-i) x_j \hat{\boldsymbol{\Phi}} \left[\boldsymbol{I} - \boldsymbol{K}'\right]^{-1} \left(\frac{\partial \boldsymbol{T}'}{\partial \omega_k}\right) \left[\boldsymbol{I} - \boldsymbol{K}'\right]^{-1} + x_j x_k \hat{\boldsymbol{\Phi}} \left[\boldsymbol{I} - \boldsymbol{K}'\right]^{-1} \right\rangle, \tag{C.9}$$

where $\boldsymbol{x} = (x_1, x_2, \ldots, x_d)$ describes the position of SVs relative to a junction and $\langle \ldots \rangle$ is a suitable spatial-summation operator acting over the SVs. In this context, elements of $\hat{\Phi}$ from Eq. (34) are written as

$$\hat{\Phi}_i = \frac{1}{s}\left(1 - \sum_j \boldsymbol{K}'_{ij}(s)\right). \tag{C.10}$$

Many of the same points carry through if the transport process is posed as a generalized differential master equation rather than as an integral master equation. In the differential master equation setting, after Fourier and Laplace transformation, we write

$$\left(s\boldsymbol{\Gamma} + \boldsymbol{I} - \boldsymbol{H} - \boldsymbol{W}(\boldsymbol{\omega})\right) \boldsymbol{P} = \boldsymbol{\Gamma}\boldsymbol{\delta}$$

with (recall we have set $\boldsymbol{T\phi} \mapsto \boldsymbol{T}$ and $\boldsymbol{K\Lambda} \mapsto \boldsymbol{K}$ in this section)

$$\boldsymbol{W}(\omega) = (\mathscr{F} \otimes \boldsymbol{I})\,\boldsymbol{\Gamma T}\left(\hat{\boldsymbol{\Phi}}\right)^{-1}(\mathscr{F} \otimes \boldsymbol{I}), \quad \boldsymbol{H} = \boldsymbol{\Gamma K}\left(\hat{\boldsymbol{\Phi}}\right)^{-1}$$

Similar to the results above, we define the matrix $\boldsymbol{S} = (s\boldsymbol{\Gamma} + \boldsymbol{I} - \boldsymbol{H} - \boldsymbol{W}(\boldsymbol{0}))^{-1}$, and then, we may compute the moments using the matrix derivative formulas above as

$$\begin{aligned}
\boldsymbol{m}_j^{(1)} &= (-i)\boldsymbol{S}^{-1}\left(\frac{\partial \boldsymbol{W}}{\partial \omega_j}\right)\boldsymbol{S}^{-1}\boldsymbol{\Gamma}\boldsymbol{\delta} + \boldsymbol{x}\boldsymbol{S}\boldsymbol{\Gamma}\boldsymbol{\delta} \\
\boldsymbol{m}_{jk}^{(2)} &= \Bigg[-\left(2\boldsymbol{S}\left(\frac{\partial \boldsymbol{W}}{\partial \omega_j}\right)\boldsymbol{S}\left(\frac{\partial \boldsymbol{W}}{\partial \omega_k}\right)\boldsymbol{S} + \boldsymbol{S}\left(\frac{\partial^2 \boldsymbol{W}}{\partial \omega_j \partial \omega_k}\right)\boldsymbol{S}\right) - 2ix_j\boldsymbol{S}\left(\frac{\partial \boldsymbol{W}}{\partial \omega_k}\right)\boldsymbol{S} \\
&\quad + x_j x_k \boldsymbol{S}\Bigg]\boldsymbol{\Gamma}\boldsymbol{\delta}.
\end{aligned}$$

Appendix D: General First-Passage-Time Problems and Secondary Vertex Reductions

In Sect. 4.1, we discussed how the SVs can be eliminated from some problems to form a CTRW involving only the junctions. Here, we discuss how this is done in the general case using the description in See Appendix C and C.1.

D.1 Reduction of SVs in the General Case

Recall the general integral master equation formulation in Eq. (36) with the $n_T \times n_T$ matrices $\boldsymbol{T}$ and $\boldsymbol{K}$ defined in Eqs. (C.1) and (C.2). What we show in this section is that the effective transition rates derived in the previous examples applies more generally as well. In particular, by obtaining effective transition rates, the evolution of a system that has arbitrarily many SVs can be described by a reduced system that only involves internal states at junctions. We first give the resulting evolution equations and moment formulas, and then proceed to describe how to derive the elements of the effective transition matrices that appear in those equations.

In Laplace transform space, we write the n_s-dimensional integral master equation for the reduced system as

$$\begin{aligned}
\boldsymbol{q}_J(\boldsymbol{X}, s) &= \delta(\boldsymbol{X})\delta(t)\delta_k + \sum_{\boldsymbol{X}'} \boldsymbol{A}_2(\boldsymbol{X} - \boldsymbol{X}') \otimes \boldsymbol{T}_2^{\text{eff}} \boldsymbol{q}_J(\boldsymbol{X}', s) \\
&\quad + \left(\boldsymbol{I} \otimes \boldsymbol{K}_J + \boldsymbol{A}_1 \otimes \boldsymbol{T}_1^{\text{eff}}\right)\boldsymbol{q}_J(\boldsymbol{X}, s) \\
\boldsymbol{p}_J(\boldsymbol{X}, s) &= \hat{\boldsymbol{\Phi}}_0(s)\boldsymbol{q}_J(\boldsymbol{X}, s)
\end{aligned} \tag{D.1}$$

where each nonzero element of the diagonal matrix $\hat{\boldsymbol{\Phi}}_0$ is

$$\hat{\Phi}_{0,k} = \frac{1}{s}\left(1 - \mathbf{1}^T\left(\boldsymbol{I} \otimes \boldsymbol{K}_J + \boldsymbol{A}_1 \otimes \boldsymbol{T}_1^{\text{eff}}\right)\delta_k - \sum_{X} \mathbf{1}^T \boldsymbol{A}_2(\boldsymbol{X} - \boldsymbol{X}') \otimes \boldsymbol{T}_2^{\text{eff}}\delta_k\right).$$

The $n_s \times n_s$ matrices $\boldsymbol{T}_1^{\text{eff}}$ and $\boldsymbol{T}_2^{\text{eff}}$ describe effective transition rates for transitions within a junction, and for travel between adjacent junctions. They will be defined precisely in Appendix D.2. The matrix, $\boldsymbol{K}_J$ is the same as that in Eq. (C.2), and it also describes internal transitions at a junction. The infinite-dimensional adjacency matrices $\boldsymbol{A}_1$ and $\boldsymbol{A}_2$ are defined as[10]

$$\boldsymbol{A}_1 = \left(\sum_{m=1}^{n_e} \boldsymbol{A}_{e,m}\boldsymbol{A}'_{e,m} + \boldsymbol{A}_{s,m}\boldsymbol{A}'_{s,m}\right) = 2n_e\boldsymbol{I}$$

$$\boldsymbol{A}_2 = \left(\sum_{m=1}^{n_e} \boldsymbol{A}_{s,m}\boldsymbol{A}'_{e,m} + \boldsymbol{A}_{e,m}\boldsymbol{A}'_{s,m}\right),$$

where n_e is the number of different types of edges in the lattice. The adjacency matrices $\boldsymbol{A}_{e,m}$ and $\boldsymbol{A}_{s,m}$ are those that are introduced in Appendix C, and m ranges over the various types of edges in the system. As described in Appendix C, these matrices can also be understood as operators over the lattice. This leads to the notation in Eq. (D.1) where $\boldsymbol{A}_2(\boldsymbol{X} - \boldsymbol{X}')$ is written as a function of the lattice points.

From Eq. (D.1), the integral master equation for $\boldsymbol{p}_J$ can be derived as

$$\begin{aligned}\boldsymbol{p}_J(\boldsymbol{X}, s) = \hat{\boldsymbol{\Phi}}_0\delta(\boldsymbol{X})\delta_k + \sum_{X'} \hat{\boldsymbol{\Phi}}_0\left(\boldsymbol{A}_2(\boldsymbol{X} - \boldsymbol{X}') \otimes \boldsymbol{T}_2^{\text{eff}}\right)\hat{\boldsymbol{\Phi}}_0^{-1}\boldsymbol{p}_J(\boldsymbol{X}', s)\\ + \hat{\boldsymbol{\Phi}}_0\left(\boldsymbol{I} \otimes \boldsymbol{K}_J + \boldsymbol{A}_1 \otimes \boldsymbol{T}_1^{\text{eff}}\right)\hat{\boldsymbol{\Phi}}_0^{-1}\boldsymbol{p}_J(\boldsymbol{X}, s).\end{aligned}$$

Normally, each transition in a random jump process implies a change of internal state or position; however, with effective transitions, we must be able to account for paths where a particle, starting from a junction in internal state ℓ, makes a number of jumps along an edge and then returns to that same junction in state ℓ. The WTDs for such events to occur are given by the diagonal elements of $\boldsymbol{T}_1^{\text{eff}}$, and they involve no change in position or internal state. We will later discuss how to reformulate the effective transition rates so that these self-jumps need not be considered as "transitions" in $\boldsymbol{T}_1^{\text{eff}}$.

It also is the case that unlike $\boldsymbol{T}_J$ and $\boldsymbol{T}_{SV}$ which did not invoke changes in internal states, $\boldsymbol{T}_2^{\text{eff}}$ can include transitions that simultaneously change the position and internal state of a particle. This is because each element of $\boldsymbol{T}_2^{\text{eff}}$ is the aggregate of a number of different steps on an edge. Even though each individual transition on the edge involves only changes in state ($\boldsymbol{K}_{SV}$) or position ($\boldsymbol{T}_{SV}$), the aggregate of many transitions required for a particle to travel between junctions can change both.

[10] The reduction of $\boldsymbol{A}_1$ to $2n_e\boldsymbol{I}$ is valid only under the restrictions discussed below.

Each element of $\boldsymbol{T}_1^{\mathrm{eff}}$ and $\boldsymbol{T}_2^{\mathrm{eff}}$ is found by solving a first-passage-time problem to obtain the distribution of times for a particle to arrive in internal state ℓ at a junction after having started in state ℓ_0 at that junction or an adjacent one without having visited any other junction states in the intervening time. We will discuss the solution to these problems in the next section, but first turn to the computation of the spatial moments for the reduced system.

Note that aside from the exterior hexagonal lattice and a remark in Appendix C, we have only considered lattices with a single type of junction [11] (e.g., the exterior hexagonal lattice does *not* fall into this category since it has type I and type II junctions). This restriction is mostly done for ease of notation and clarity, as there do not appear to be theoretical difficulties with considering multiple types of junctions. In this section, we will make two additional restrictions, also to ensure clarity of the arguments that follow.

Consider a lattice with a single type of junction and n_e types of edges, each type distinguished from the others only by its orientation relative to the x-axis. The first additional restriction is that the transport along each edge is undirected. The second restriction is that we now suppose that the lattice is constructed so that at each junction, one edge of each type starts and one edge of each type ends at that junction. Note that this restriction is also used in Appendix C. Aside from the exterior hexagonal lattice and persistent comb random walk, the examples discussed in this paper have these properties. As an example, in a square lattice, at each junction, one vertical edge and one horizontal edge start (have their s-terminus) at that junction, and one vertical and one horizontal edge end (have their e-terminus) at that junction. Thus, there are $n_e = 2$ types of edges (vertical and horizontal), and each junction is connected to $4 = 2n_e$ edges in the square lattice.

Recall that the lattice Fourier transform of a function is

$$\hat{f}(\boldsymbol{\omega}) = \sum_{\boldsymbol{X}\in\mathscr{G}} f(\boldsymbol{X})e^{i\boldsymbol{X}\cdot\boldsymbol{\omega}}$$

with $\boldsymbol{\omega}$ defined over a d-dimensional cube, $\left[-\frac{\pi}{|\Delta X|}, \frac{\pi}{|\Delta X|}\right]^d$. Here, d is the dimension of the space and $\boldsymbol{\Delta X}$ the lattice spacing.

Applying the lattice Fourier transform to the matrix,

$$\left(\boldsymbol{I}\otimes\boldsymbol{K}_J + \boldsymbol{A}_1\otimes\boldsymbol{T}_1^{\mathrm{eff}}\right) + \boldsymbol{A}_2(\boldsymbol{X})\otimes\boldsymbol{T}_2^{\mathrm{eff}},$$

from Eq. (D.1), and keeping in mind the restrictions above, we obtain

$$\boldsymbol{I}\otimes\left[\boldsymbol{K}_J + 2n_e(\boldsymbol{T}_1^{\mathrm{eff}} + \boldsymbol{T}_2^{\mathrm{eff}})\right] + \boldsymbol{I}\otimes\left[\sum_{j=1}^{2n_e}(e^{i\Delta\boldsymbol{X}_j\cdot\boldsymbol{\omega}} - 1)\boldsymbol{T}_2^{\mathrm{eff}}\right]$$

[11] Each junction can still have arbitrarily many internal states. A *single type* of junction refers to all junctions in a lattice having the same connectivity, via a single edge, to other junctions in that lattice. For example, in a square lattice, each junction is connected to its neighbors to the north, south, east, and west. This differs from the exterior hexagonal lattice where type I junctions are connected to their neighbors to the north, southeast, and southwest; and type II junctions to neighbors to their south, northeast, and northwest.

$$\equiv \boldsymbol{I} \otimes \boldsymbol{K}_0 + \boldsymbol{I} \otimes \boldsymbol{T}_0^{\text{eff}}(\boldsymbol{\omega}).$$

To find the moments, we differentiate in Fourier space as in Appendix C.1. Differentiating $\tilde{\boldsymbol{p}}_J(\boldsymbol{\omega}, s)$, we obtain

$$\tilde{\boldsymbol{m}}_j^{(1)} = (-i)\hat{\Phi}_0 \left[\boldsymbol{I} - \boldsymbol{K}_0\right]^{-1} \left(\frac{\partial \boldsymbol{T}_0^{\text{eff}}}{\partial \omega_j}\right) \left[\boldsymbol{I} - \boldsymbol{K}_0\right]^{-1} \tag{D.2}$$

$$\tilde{\boldsymbol{m}}_{jk}^{(2)} = -\hat{\Phi}_0 \left[\boldsymbol{I} - \boldsymbol{K}_0\right]^{-1} \left(2\left(\frac{\partial \boldsymbol{T}_0^{\text{eff}}}{\partial \omega_j}\right) \left[\boldsymbol{I} - \boldsymbol{K}_0\right]^{-1} \left(\frac{\partial \boldsymbol{T}_0^{\text{eff}}}{\partial \omega_k}\right) + \left(\frac{\partial^2 \boldsymbol{T}_0^{\text{eff}}}{\partial \omega_j \partial \omega_k}\right)\right) \left[\boldsymbol{I} - \boldsymbol{K}_0\right]^{-1} \tag{D.3}$$

with $j, k = 1, 2, \ldots, d$. Notice that there is no summation over the n_k SVs since details regarding the SVs have been condensed into the elements of $\boldsymbol{T}_2^{\text{eff}}$ which describes transfers between adjacent junctions. Likewise we do not need any information about how the internal states are situated relative to a junction. We can also write $\hat{\boldsymbol{\Phi}}_0$ from above as

$$\left(\hat{\Phi}_0(s)\right)_i = \frac{1}{s}\left(1 - \sum_j \boldsymbol{K}_{0,ij}(s)\right).$$

Unlike the SVs, the internal states have not been summed over, so the moments are still written as vectors over these states.

Remark 1 It is important to note that since no summation is done over the SVs, the value of $\boldsymbol{p}_J(\boldsymbol{X}, t)$ at junction $\boldsymbol{X}$ is now the aggregate probability density of all particles that have reached that junction, but not yet arrived at any other junctions. In other words, a particle located on an edge after exiting junction $\boldsymbol{X}$, but not having yet reached an adjacent junction, would contribute to $\boldsymbol{p}_J(\boldsymbol{X}, t)$. This is a reflection of the fact that in the reduced process, the fine details of the exact location of a particle along an edge have been suppressed. Nonetheless, if one uses the elements of $\hat{\boldsymbol{\Phi}}$ associated with junctions from the full system (e.g., Eq. (34)) rather than $\hat{\boldsymbol{\Phi}}_0$ for the reduced system, the probabilities *at* junction $\boldsymbol{X}$ are obtained (c.f. the difference between $\hat{\boldsymbol{\Phi}}$ and $\hat{\boldsymbol{\Psi}}$ in Ex. 4).

Remark 2 When the FPT procedure is applied to a Markov process, this result is typically a non-Markovian process with a reduced number of degrees of freedom (Kampen 1992). The non-Markovian character arises since even if all the transitions in the original process have Poisson distributed waiting times, the aggregate waiting-time distribution for several sequential Poisson processes to occur is not a Poisson distribution. As noted above, this leads to an overall non-Markovian process since the only continuous waiting-time distributions which leads to a Markov process are Poisson distributions.

D.2 General First-Passage-Time Problems

It now remains to describe the general first-passage-time (FPT) problem which must be solved to obtain $\boldsymbol{T}_1^{\text{eff}}$ and $\boldsymbol{T}_2^{\text{eff}}$ in Eq. (D.1). The elements of $\boldsymbol{T}_{1,2}^{\text{eff}}$ are closely related to FPTs for a random walk to reach state ℓ at a junction given that it started in state ℓ_0 at one of the junctions bordering that edge and did not leave the edge yet.

Recall that we assume that all edges have the same set internal transitions and spatial jump transitions. Also recall that the transition matrices for internal state changes and spatial jumps on an edge are $\boldsymbol{K}_{SV}$ and $\boldsymbol{T}_{SV}$, respectively. Since a particle starting out on an edge can eventually jump to a vertex at the "s" or "e" end of that edge, the matrix $\boldsymbol{T}_{SV} + \boldsymbol{K}_{SV}$ cannot conserve particle number. Thus, the matrix $\boldsymbol{T}_{SV} + \boldsymbol{K}_{SV}$ is the transition matrix for a random walk confined to an edge and subject to absorbing boundary conditions. Since particles are absorbed at the boundaries, the rate at which particles are absorbed can be defined. In the full matrix system, these absorption rates correspond to the rate at which particles arrive at the vertices at either end of an edge.

In order to define the arrival rates, we consider the nonzero elements of $\boldsymbol{D}_e$ and $\boldsymbol{D}_s$ which describe transitions from an edge to a vertex. As depicted in Fig. 5, there are typically at most n_s distinct transitions from each end of an edge, one from each internal state at $k = 1$ and $k = n_k$. Thus, there are typically at most n_s nonzero entries in $\boldsymbol{D}_s$ and $\boldsymbol{D}_e$. Considering $\boldsymbol{D}_s$ first, each nonzero element of $\boldsymbol{D}_s$ can be written as $d_\ell \psi_{d,\ell}(t)$. Here, d_ℓ is the probability for a particle at the first SV ($k = 1$) in state ℓ to jump to state ℓ at the adjacent junction, and $\psi_{d,\ell}(t)$ is the WTD for the jump when it does occur. The same description holds for nonzero elements of $\boldsymbol{D}_e$, except in that case the particle is jumping from the n_kth SV, $k = n_k$ rather than $k = 1$.

Then, the Laplace transform of the FPT density for a particle to reach the junction attached to the "s"-end of an edge in state ℓ after starting in state (k_0, ℓ_0) on the edge is of the form

$$\tilde{f}_\ell(s|\{k_0, \ell_0\}) = d_\ell \psi_{d,\ell}(s) \boldsymbol{\delta}_{1,\ell}^T (\boldsymbol{I} - \boldsymbol{T}_{SV} - \boldsymbol{K}_{SV})^{-1} \boldsymbol{\delta}_{k_0,\ell_0}$$

where $\boldsymbol{\delta}_{k_0,\ell_0}$ and $\boldsymbol{\delta}_{1,\ell}$ are Kronecker deltas over the state-space of the edge. But $\tilde{f}_\ell(s|\{k_0, \ell_0\})$ is just the ℓth component of a vector $\tilde{\boldsymbol{f}}(s|\{k_0, \ell_0\})$ over the internal states at a junction. The FPT densities to reach each of the n_s internal states at a junction defined easily by decomposing $\boldsymbol{D}_s$ as a sum of rank one matrices,

$$\boldsymbol{D}_s = \sum_{\ell=1}^{n_s} d_\ell \psi_{d,\ell}(s) \boldsymbol{\delta}_\ell \boldsymbol{\delta}_{1,\ell}^T$$

and writing

$$\tilde{\boldsymbol{f}}(s|\{k_0, \ell_0\}) = \boldsymbol{D}_s (\boldsymbol{I} - \boldsymbol{T}_{SV} - \boldsymbol{K}_{SV})^{-1} \boldsymbol{\delta}_{k_0,\ell_0}.$$

This describes the FPT densities to reach the vertex adjacent to the "s"-end of the edge for a particle starting at some SV on edge, and the FPT densities at the "e"-end can be found similarly. However, the effective transition rates we seek describe how long it takes a particle to jump between vertices, not from an SV to a vertex.

Let us consider the n_k^{th} SV on an edge. Each internal state associated with this SV is directly connected to the junction at the "e"-end of the edge, and jumps from that junction to the edge are governed by $\boldsymbol{D}_e'$. Like $\boldsymbol{D}_e$, $\boldsymbol{D}_e'$ typically has at most n_s nonzero entries and we can write $\boldsymbol{D}_e'$ as

$$\boldsymbol{D}_e' = \sum_{\ell=1}^{n_s} d_\ell' \psi_{d,\ell}'(s) \boldsymbol{\delta}_{n_k,\ell} \boldsymbol{\delta}_\ell^T$$

where d_ℓ' is the probability of a particle in state ℓ at the junction jumping to the edge, and $\psi_{d,\ell}'$ is the WTD when the jump occurs.

With this, the elements of the vector,

$$\tilde{\boldsymbol{f}}(s|\ell_0\}) = \boldsymbol{D}_s \left(\boldsymbol{I} - \boldsymbol{T}_{SV} - \boldsymbol{K}_{SV}\right)^{-1} \boldsymbol{\delta}_{n_k,\ell_0} \psi_{d,\ell_0}'$$

are the FPTs for a particle jumping from state ℓ_0 at the junction at the "e"-end of an edge to reach state ℓ at the junction at the "s"-end of the same edge. Elements of $\boldsymbol{T}_2^{\text{eff}}$ are then found by multiplying these FPTs by d_{ℓ_0}' which gives the probability of the jump occurring. Collecting all of these FPTs into a matrix, we obtain,

$$\boldsymbol{T}_2^{\text{eff}} = \boldsymbol{D}_s \left(\boldsymbol{I} - \boldsymbol{T}_{SV} - \boldsymbol{K}\right)^{-1} \boldsymbol{D}_e' \tag{D.4}$$

By similar arguments, we find the elements of $\boldsymbol{T}_1^{\text{eff}}$ by replacing $\boldsymbol{D}_s$ with $\boldsymbol{D}_e$,

$$\boldsymbol{T}_1^{\text{eff}} = \boldsymbol{D}_e \left(\boldsymbol{I} - \boldsymbol{T}_{SV} - \boldsymbol{K}\right)^{-1} \boldsymbol{D}_e'. \tag{D.5}$$

Notice that we could have replaced $\boldsymbol{D}_s$ and $\boldsymbol{D}_e'$ by $\boldsymbol{D}_e$ and $\boldsymbol{D}_s'$ in the computation of $\boldsymbol{T}_2^{\text{eff}}$. Either combination is valid since we have assumed that transport along each edge is undirected. Note that many useful examples involve just two types of effective transitions (return to origin, and jump to other states) in which case the matrices above will be rank two.

The presence of condensed paths over the SVs that return a particle to its starting state prior to jumping is a peculiarity that arises here. However, as we will now show, these self-jumps can be eliminated by solving an extended first-passage-time problem which results in a rescaling of the values of $\boldsymbol{T}_2^{\text{eff}}$ and the off-diagonal elements of $\boldsymbol{T}_1^{\text{eff}}$ to account for the contingency of multiple returns to the starting point. This was done implicitly in Ex. 4.2 when we solved for the FPT to reach $X_{i\pm1}$ from X_i in the a 1D lattice. In that case, no self-jumps were possible since the effective WTD was defined such that it described sojourns on edges that started (at X_i) and ended (at $X_{i\pm1}$) at different points in space. This same principle holds in general.

It turns out the form of the rescaled transition rates is relatively simple. We first give the result and then discuss its derivation. To obtain the rescaled transition rates,

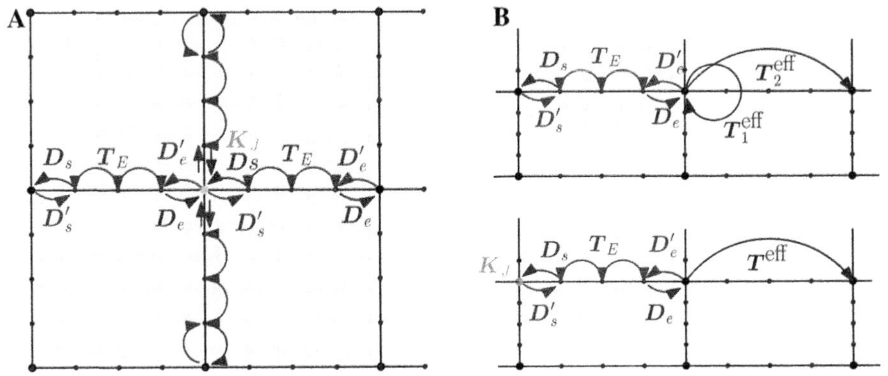

Fig. 13 Random walk first-passage-time problems with $n_\ell = 1$. **A** The most general type of FPT problem involving the full set of states in between any two adjacent lattice states. **B** The two simplifications. In the first simplification (top), only the internal states within an edge are considered, the FPT matrix here is $\boldsymbol{D}\,(\boldsymbol{I} - \boldsymbol{T}_{SV} - \boldsymbol{K}_{SV})^{-1}\,\boldsymbol{D}'$. Self-returns and jumps to adjacent junctions are highlighted. In the second case, the junction is considered as well. This rescales the problem to eliminate self-jumps. It can be understood as equivalent to the general FPT problem when all edges exhibit the same set of transitions including those to and from a lattice point. In (a) and (b), the highlighted boxes are used to indicate the states involved in the FPT computation

we additively decompose $\boldsymbol{T}_1^{\mathrm{eff}}$ into diagonal and off-diagonal components,

$$\boldsymbol{T}_1^{\mathrm{eff}} = \boldsymbol{T}_{1,d}^{\mathrm{eff}} + \boldsymbol{T}_{1,o}^{\mathrm{eff}}.$$

The subscripts d and o indicate the diagonal and off-diagonal components of $\boldsymbol{T}_1^{\mathrm{eff}}$. The self-jumps are characterized by $\boldsymbol{T}_{1,d}^{\mathrm{eff}}$, and jumps that change the state or position of a particle are given by $\boldsymbol{K}$, $\boldsymbol{T}_{1,o}^{\mathrm{eff}}$, and $\boldsymbol{T}_2^{\mathrm{eff}}$. The rescaled internal transitions and first-passage-times are then of the form

$$\boldsymbol{K}_r = \boldsymbol{K}\left(\boldsymbol{I} - 2n_e\boldsymbol{T}_{1,d}^{\mathrm{eff}}\right)^{-1}, \quad \boldsymbol{T}_{1,r}^{\mathrm{eff}} = 2n_e\boldsymbol{T}_{1,o}^{\mathrm{eff}}\left(\boldsymbol{I} - 2n_e\boldsymbol{T}_{1,d}^{\mathrm{eff}}\right)^{-1},$$
$$\boldsymbol{T}_{2,r}^{\mathrm{eff}} = \boldsymbol{T}_2^{\mathrm{eff}}\left(\boldsymbol{I} - 2n_e\boldsymbol{T}_{1,d}^{\mathrm{eff}}\right)^{-1} \tag{D.6}$$

As noted before, this rescaling is equivalent to the solution of an extended first-passage-time problem that does not terminate when a particle returns to its starting state. Consider a particle starting at state ℓ at vertex $\boldsymbol{X}$. To exclude self-jumps, consider the transition matrix for a particle in a system that includes state ℓ at vertex $\boldsymbol{X}$ and all of the SVs and internal states on the edges that are attached to $\boldsymbol{X}$ (see Fig. 13).

The transition matrices for each ℓ are of the form

$$\boldsymbol{T}_{r,\ell}+\boldsymbol{K}_{r,\ell} = \begin{pmatrix} 0 & \boldsymbol{\delta}_\ell^T\boldsymbol{D}_{t_1} & \dots & \boldsymbol{\delta}_\ell^T\boldsymbol{D}_{t_{2n_e}} \\ \boldsymbol{D}'_{t_1}\boldsymbol{\delta}_\ell & \boldsymbol{T}_{SV}+\boldsymbol{K}_{SV} & \boldsymbol{0}\dots & \\ \vdots & \vdots & \ddots & \vdots \\ \boldsymbol{D}'_{t_{2n_e}}\boldsymbol{\delta}_\ell & \boldsymbol{0} & \dots & \boldsymbol{T}_{SV}+\boldsymbol{K}_{SV} \end{pmatrix}, \quad \ell = 1,\dots,N_\ell \tag{D.7}$$

where $t_k = \{s, e\}$ specifies which end of edge k connects to vertex, $\boldsymbol{X}$, see Fig. 13a. We see that a particle starting on $\boldsymbol{X}$ in state ℓ can exit the system by 1) reaching $\boldsymbol{X}$ in any internal state except for ℓ, or 2) by reaching a junction adjacent to $\boldsymbol{X}$ in any internal state including ℓ. Since the particle cannot exit this system from the same position and state at which it started, no effective self-jumps occur.

The resulting matrix for the FPT problem is much larger in this case than the matrix $\boldsymbol{T}_e + \boldsymbol{K}_e$ which was needed to find $\boldsymbol{T}^{\text{eff}}_{1,2}$. This is disadvantageous since solutions may be more difficult to compute. However, a significant simplification is possible due to assumptions we have made. In the case, we have been considering, all edges have the same set of transitions and WTDs. Given this, we find that since the particle starts at the junction, the probabilities, $\tilde{p}_{k,\ell}(s)$ along each edge are equal. This allows us to write

$$\boldsymbol{T}_{r_0,\ell} + \boldsymbol{K}_{r_0,\ell} = \begin{pmatrix} 0 & (\boldsymbol{D}_{t_1})_\ell \\ 2n_e(\boldsymbol{D}'_{t_1})^T_\ell & \boldsymbol{T}_{SV} + \boldsymbol{K}_{SV} \end{pmatrix}.$$

Given this matrix, FPTs to reach an adjacent vertex or return to $\boldsymbol{X}$ in a state other than ℓ may be found using the techniques discussed above. After simplification, one can show that Eq. D.7 are the resulting solutions.

Likewise, one can substitute $\boldsymbol{T}^{\text{eff}}_{1,r}$ and $\boldsymbol{T}^{\text{eff}}_{2,r}$ into Eqs. (D.2) and (D.3) for the spatial moments. However, since both approaches end up computing approximate moments for the same initial system which includes SVs and junctions, they must both yield very similar results. The distinction is that in the latter case, the definition of a single jump has been redefined to only include jumps where the position or state of the particle changes. This distinction is important in regard to interpreting the elements of $\boldsymbol{T}^{\text{eff}}$. In the former case, they cannot be considered as jump probabilities multiplying waiting-time distributions in the standard sense, but after rescaling, they can.

Appendix E: Random Walks on Segments with Various Boundary Conditions

Consider a random walk on $\mathbb{Z}$ where at each $k \in \mathbb{Z}$, there is a 1/2 probability of jumping to the left or right. If the random walk starts at $\ell \in \mathbb{Z}$, the probability generating function for the probability of finding the walker at m is (Hughes1996)

$$u^{(F)}(z, m|\ell) = \sum_{n=0}^{\infty} z^n p_n(m|\ell) = (1 - z^2)^{-1/2} x^{|m-\ell|} \tag{E.1}$$

with

$$x \equiv z^{-1}\left(1 - \sqrt{1 - z^2}\right)$$

and $p_n(m|\ell)$ being the probability of reaching m after starting at ℓ on the nth step of the walk. For a periodic segment with $m, \ell = 0, 1, \ldots, N$, the solution is given by using the method of images (Montroll and Greenberg 1964) as

$$u^{(P)}(z,m|\ell) = \sum_{k=-\infty}^{\infty} u^{(F)}(z,m+Nk|\ell) = \left(\frac{x^{m-\ell}+x^{N-(m-\ell)}}{1-x^N}\right)(1-z^2)^{-1/2}.$$

It is also possible to find solutions with other types of boundary conditions (Hughes 1996; Weiss and Rubin 2007). For instance, the image points for a random walk on a segment with two absorbing boundaries are of the form $\{2kN \pm \ell\}$, and the method of images solution yields

$$u^{(AA)}(z,m|\ell) = \frac{x^{|m-\ell|} - x^{|m+\ell|} + x^{2N}(x^{-|m-\ell|} - x^{-|m+\ell|})}{\sqrt{1-z^2}\left(1-x^{2N}\right)}. \tag{E.2}$$

Moments of $u^{(F)}(z,\cdot|\cdot)$ can be found by summing over m and using identities related to geometric series:

$$\sum_{n=0}^{N} r^n = \frac{1-r^{N+1}}{1-r}$$

$$\sum_{n=0}^{N} nr^n = r\frac{d}{dr}\left[\frac{1-r^{N+1}}{1-r}\right] = r\frac{1-(N+1)r^N+Nr^{N+1}}{(1-r)^2}$$

$$\sum_{n=0}^{N} n(n-1)r^n = r^2\frac{d^2}{dr^2}\left[\frac{1-r^{N+1}}{1-r}\right]$$

$$= r^2\frac{2-N(1+N)r^{N-1}+2(N^2-1)r^N-N(N-1)r^{1+N}}{(1-r)^3}$$

$$\sum_{n=0}^{N} n(n-1)\dots(n-k)r^n = r^k\frac{d^k}{dr^k}\left[\frac{1-r^{N+1}}{1-r}\right]$$

For finite N, these formulas are valid (if one uses L'Hopitals rule at $r = 1$) for all r, and in the case that $N \to \infty$, the results are valid if $0 \le r < 1$. In particular, the variance of $u^{(F)}$ over the integers is

$$\sigma^2(z) = \frac{z}{(1-z)^2}. \tag{E.3}$$

In the CTRW setting, one can find the arrival probabilities, $q(x,t|y)$ by substituting the Laplace transform of $\psi(t)$ for z in the formulas in this section. Likewise, upon multiplying by $\hat{\Phi}(s)$, the probability, $p(x,t|y)$ and time-dependent moments can be found. For instance, for an unbiased 1D CTRW with nearest neighbor jumps,

$$\sigma(s) = L^2\frac{1-\psi}{s}\frac{\psi}{(1-\psi)^2} = \frac{L^2}{s}\frac{\psi}{1-\psi}$$

as was found in Eq. (43) by different techniques.

The first-passage-time probability generating function may also be found by making use of the definition in Eq. (E.1) for $u^{(F)}$. As derived in Montroll and Weiss (1965),

$$f^{(F)}(z, m|\ell) = \frac{u^{(F)}(z, m|\ell) - \delta_{m,\ell}}{u^{(F)}(z, 0|0)}.$$

Furthermore, this result is easily extended to much more general cases such as multi-state random walks, and walks on complicated structures (Hughes 1996).

References

Akiyama T, Kamimura K, Firkus C, Takeo S, Shimmi O, Nakato H (2008) Dally regulates Dpp morphogen gradient formation by stabilizing Dpp on the cell surface. Dev Biol 313(1):408–419

Iomin A (2011) Subdiffusion on a fractal comb. Phys Rev E 83(5):052106

Iomin A (2019) Richardson diffusion in neurons. Phys Rev E 100(1):010104

Iomin A, Zaburdaev V, Pfohl T (2016) Reaction front propagation of actin polymerization in a comb-reaction system. Chaos Solitons Fract 92:115–122

Kicheva A, Pantazis P, Bollenbach T, Kalaidzidis Y, Bittig T, Jülicher F, Gonzalez-Gaitan M (2007) Kinetics of morphogen gradient formation. Science 315(5811):521–525

Berezhkovskii AM, Dagdug L, Bezrukov SM (2014) From normal to anomalous diffusion in comb-like structures in three dimensions. J Chem Phys 141(5):054907

Berezhkovskii AM, Dagdug L, Bezrukov SM (2015) Biased diffusion in three-dimensional comb-like structures. J Chem Phys 142(13):134101

Bracewell RN (1986) The Fourier transform and its applications, vol 31999. McGraw-Hill, New York

Bressloff PC, Kim H (2018) Bidirectional transport model of morphogen gradient formation via cytonemes. Phys Biol 15(2):

Bressloff PC, Newby JM (2013) Stochastic models of intracellular transport. Rev Mod Phys 85(1):135

Gadgil C, Lee CH, Othmer HG (2005) A stochastic analysis of first-order reaction networks. Bull Math Biol 67:901–946

Cox DR (1967) Renewal theory, vol 1. Methuen, London

Entchev EV, Schwabedissen A, Gonzalez-Gaitan M (2000) Gradient formation of the TGF-beta homolog Dpp. Cell 103(6):981–91

Feller W (1968) An introduction to probability theory. Wiley, New York

Mainardi F, Raberto M, Gorenflo R, Scalas E (2000) Fractional calculus and continuous-time finance II: the waiting-time distribution. Phys A 287(3–4):468–481

Aurenhammer F, Klein R, Lee DT (2013) Voronoi diagrams and Delaunay triangulations. World Scientific Publishing Company, Singapore

Gibson WT, Gibson MC (2009) Cell topology, geometry, and morphogenesis in proliferating epithelia. Curr Topics Dev Biol 89:87–114

Gibson MC, Lehman DA, Schubiger G (2002) Lumenal transmission of Decapentaplegic in Drosophila imaginal discs. Dev Cell 3(3):451–60

Goldhirsch I, Gefen Y (1987) Biased random walk on networks. Phys Rev A 35(3):1317

Haerry TE, Khalsa O, O'Connor MB, Wharton KA (1998) Synergistic signaling by two BMP ligands through the SAX and TKV receptors controls wing growth and patterning in Drosophila. Development 125(20):3977–3987

Hamid T, Kolomeisky AB (2013) All-time dynamics of continuous-time random walks on complex networks. J Chem Phys 138(8):084110

Harris TJC, Tepass U (2010) Adherens junctions: from molecules to morphogenesis. Nat Revs Mol Cell Biol 11(7):502–514

Henyey FS, Seshadri V (1982) On the number of distinct sites visited in 2 d lattices. J Chem Phys 76(11):5530–5534

Hu J, Othmer HG (2011) A theoretical analysis of filament length fluctuations in actin and other polymers. J Math Biol 63(6):1001–1049

Hughes BD (1996) Random walks and random environments. Clarendon Press, Oxford
Iomin A, Méndez V (2013) Reaction-subdiffusion front propagation in a comblike model of spiny dendrites. Phys Rev E 88(1):
Goldhirsch I, Gefen Y (1986) Analytic method for calculating properties of random walks on networks. Phys Rev A 33(4):2583
Isaacson SA (2009) The reaction-diffusion master equation as an asymptotic approximation of diffusion to a small target. SIAM J Appl Math 70(1):77–111
Gou J, Lin L, Othmer HG(2018) A model for the Hippo pathway in the Drosophila wing disc. Biophys J 115(4):737–747 PMID: 30041810
Gou J, Stotsky JA, Othmer HG (2020) Growth control in the Drosophila wing disk. Wiley Interdisciplinary Reviews, New York, Systems biology and medicine, p e1478
Kang HW, Zheng L, Othmer HG (2012) A new method for choosing the computational cell in stochastic reaction-diffusion systems. J Math Biol 2012:1017–1099
Karlin S, Taylor HM (1975) A first course in stochastic processes, 2nd edn. Academic Press, New York
Kim H, Bressloff PC (2018) Direct vs. synaptic coupling in a mathematical model of cytoneme-based morphogen gradient formation. SIAM J Appl Math 78(5):2323–2347
Kornberg TB (2014) Cytonemes and the dispersion of morphogens. Wiley Interdiscip Rev Dev Biol 3(6):445–463
Kornberg TB, Roy S (2014) Cytonemes as specialized signaling filopodia. Development 141(4):729–736
Choi KW (2018) Upstream paths for Hippo signaling in Drosophila organ development. BMB Reports 51(3):134
Lin L, Othmer HG (2017) Improving parameter inference from frap data: an analysis motivated by pattern formation in the Drosophila wing disc. B Math Biol 79(3):448–497
Lubashevskii IA, Zemlyanov AA (1998) Continuum description of anomalous diffusion on a comb structure. J Exp Theor Phys 87(4):700–713
Ciocanel MV, Fricks J, Kramer PR, McKinley SA (2020) Renewal reward perspective on linear switching diffusion systems in models of intracellular transport. Bull Math Biol 82(10):1–36
Matthaus F, Jagodic M, Dobnikar J (2009) E. coli superdiffusion and chemotaxis-search strategy, precision, and motility. Biophys J 97(4):946–957
Metzler R, Klafter J (2000) The random walk's guide to anomalous diffusion: a fractional dynamics approach. Phys Reports 339(1):1–77
Montroll EW, Weiss GH (1965) Random walks on lattices II. J Math Phys 6(2):167–181
Montroll EW, West BJ (1979) On an enriched collection of stochastic processes. Fluct Phen 66:61
Montroll EW (1969) Random walks on lattices. III. calculation of first-passage times with application to exciton trapping on photosynthetic units. J Math Phys 10(4):753–765
Montroll EW, Greenberg JM (1964) Proceedings of the symposium on applied mathematics. Am Math Soc Providence 16:193
Mundt MG (2013) Characterization of a unique basolateral targeting domain in the Drosophila TGF-β type II receptor punt. Master's thesis, University of Minnesota
Othmer HG, Painter K, Umulis D, Xue C (2009) The intersection of theory and application in biological pattern formation. Math Mod Nat Phenom 4(4):3–82
Othmer HG (1983) A continuum model for coupled cells. J Math Biol 17:351–369
Othmer HG, Scriven LE (1971) Instability and dynamic pattern in cellular networks. J Theor Biol 32:507–537
Othmer HG, Dunbar SR, Alt W (1988) Models of dispersal in biological systems. J Math Biol 26(3):263–298
Pavliotis G, Stuart A (2008) Multiscale methods: averaging and homogenization. Springer, Berlin
Roerdink JBTM, Shuler KE (1985) Asymptotic properties of multistate random walks. I. theory. J Stat Phys 40(1):205–240
Roerdink JBTM, Shuler KE (1985) Asymptotic properties of multistate random walks. II. applications to inhomogeneous periodic and random lattices. J Stat Phys 41(3):581–606
Churchill RV (1958) Operational Mathematics. McGraw-Hill
Scher H, Wu CH (1981) Random walk theory of a trap-controlled hopping transport process. Proc Natl Acad Sci 78(1):22–26
Zhou S, Lo WC, Suhalim JL, Digman MA, Enrico G, Qing N, Lander AD (2012) Free extracellular diffusion creates the Dpp morphogen gradient of the Drosophila wing disc. Curr Biol 22(8):668–675
Shlesinger MF (1974) Asymptotic solutions of continuous-time random walks. J Stat Phys 10(5):421–434
Havlin S, Ben-Avraham D (1987) Diffusion in disordered media. Adv Phys 36(6):695–798

Roy S, Huang H, Liu S, Kornberg TB (2014) Cytoneme-mediated contact-dependent transport of the Drosophila decapentaplegic signaling protein. Science 343(6173):1244624
Harmansa S, Alborelli I, Dimitri B, Caussinus E, Affolter M (2017) A nanobody-based toolset to investigate the role of protein localization and dispersal in Drosophila. Elife 6:
Subrahmanyan C (1943) Stochastic problems in physics and astronomy. Rev Mod Phys 15(1):1
Landman U, Shlesinger MF (1977) Cluster motion on surfaces: a stochastic model. Phys Rev B 16(8):3389
Landman U, Shlesinger MF (1979) Stochastic theory of multistate diffusion in perfect and defective systems. I. mathematical formalism. Phys Rev B 19(12):6207
Landman U, Shlesinger MF (1979) Stochastic theory of multistate diffusion in perfect and defective systems. II. case studies. Phys Rev B 19(12):6220
Landman U, Montroll EW, Shlesinger MF (1977) Random walks and generalized master equations with internal degrees of freedom. Proc Natl Acad Sci 74(2):430–433
Van Kampen NG (1992) Stochastic processes in physics and chemistry, vol 1. Elsevier, London
Wartlick O, Mumcu P, Jülicher F, Gonzalez-Gaitan M (2011) Understanding morphogenetic growth control - lessons from flies. Nat Rev Mol Cell Biol 12(9):594–604
Weiss GH, Havlin S (1986) Some properties of a random walk on a comb structure. Phys A 134(2):474–482
Weiss GH, Rubin RJ (2007) Random walks: theory and selected applications. Wiley-Blackwell, London, pp 363–505
Widmann TJ, Dahmann C (2009) Wingless signaling and the control of cell shape in Drosophila wing imaginal discs. Dev Biol 334(1):161–173
Yamazaki Y, Palmer L, Alexandre C, Kakugawa S, Beckett K, Gaugue I, Palmer RH, Vincent JP (2016) Godzilla-dependent transcytosis promotes Wingless signalling in Drosophila wing imaginal discs. Nat Cell Biol 18(4):451–457

Bulletin of Mathematical Biology (2021) 83:80
https://doi.org/10.1007/s11538-021-00912-5

Society for Mathematical Biology

ORIGINAL ARTICLE

Mechanical Models of Pattern and Form in Biological Tissues: The Role of Stress–Strain Constitutive Equations

Chiara Villa[1] · Mark A. J. Chaplain[1] · Alf Gerisch[2] · Tommaso Lorenzi[3]

Received: 23 September 2020 / Accepted: 11 May 2021 / Published online: 26 May 2021

Abstract

Mechanical and mechanochemical models of pattern formation in biological tissues have been used to study a variety of biomedical systems, particularly in developmental biology, and describe the physical interactions between cells and their local surroundings. These models in their original form consist of a balance equation for the cell density, a balance equation for the density of the extracellular matrix (ECM), and a force-balance equation describing the mechanical equilibrium of the cell-ECM system. Under the assumption that the cell-ECM system can be regarded as an isotropic linear viscoelastic material, the force-balance equation is often defined using the Kelvin–Voigt model of linear viscoelasticity to represent the stress–strain relation of the ECM. However, due to the multifaceted bio-physical nature of the ECM constituents, there are rheological aspects that cannot be effectively captured by this model and, therefore, depending on the pattern formation process and the type of biological tissue considered, other constitutive models of linear viscoelasticity may be better suited. In this paper, we systematically assess the pattern formation potential of different stress–strain constitutive equations for the ECM within a mechanical model of pattern formation in biological tissues. The results obtained through linear stability analysis and the dispersion relations derived therefrom support the idea that fluid-like con-

✉ Mark A. J. Chaplain
majc@st-andrews.ac.uk

Chiara Villa
cv23@st-andrews.ac.uk

Alf Gerisch
gerisch@mathematik.tu-darmstadt.de

Tommaso Lorenzi
tommaso.lorenzi@polito.it

1 School of Mathematics and Statistics, University of St Andrews, St Andrews 16 9SS, UK

2 Fachbereich Mathematik, Technische Universität Darmstadt, Dolivostr. 15, 64293 Darmstadt, Germany

3 Department of Mathematical Sciences "G. L. Lagrange", Dipartimento di Eccellenza 2018-2022, Politecnico di Torino, 10129 Torino, Italy

stitutive models, such as the Maxwell model and the Jeffrey model, have a pattern formation potential much higher than solid-like models, such as the Kelvin–Voigt model and the standard linear solid model. This is confirmed by the results of numerical simulations, which demonstrate that, all else being equal, spatial patterns emerge in the case where the Maxwell model is used to represent the stress–strain relation of the ECM, while no patterns are observed when the Kelvin–Voigt model is employed. Our findings suggest that further empirical work is required to acquire detailed quantitative information on the mechanical properties of components of the ECM in different biological tissues in order to furnish mechanical and mechanochemical models of pattern formation with stress–strain constitutive equations for the ECM that provide a more faithful representation of the underlying tissue rheology.

Keywords Pattern formation · Mechanical models · Murray–Oster theory · Biological tissues · Stress–strain constitutive equations · Linear viscoelasticity

1 Introduction

Pattern formation resulting from spatial organisation of cells is at the basis of a broad spectrum of physiological and pathological processes in living tissues (Jernvall et al. 2003). While the first formal exploration of pattern and form from a mathematical (strictly speaking, geometrical) perspective goes back over a century to D'Arcy Thompson's "*On Growth and Form*" (Thompson 1917), the modern development of mathematical models for this biological phenomenon started halfway through the twentieth century to elucidate the mechanisms that underly morphogenesis and embryogenesis (Maini 2005). Since then, a number of mathematical models for the formation of cellular patterns have been developed (Urdy 2012). Amongst these, particular attention has been given to reaction–diffusion models and mechanochemical models of pattern formation (Murray 2001).

Reaction–diffusion models of pattern formation, first proposed by Turing in his seminal 1952 paper (Turing 1952) and then further developed by Gierer and Meinhardt (Gierer and Meinhardt 1972; Meinhardt 1982), apply to scenarios in which the heterogeneous spatial distribution of some chemicals (i.e. morphogens) acts as a template (i.e. a pre-pattern) according to which cells organise and arrange themselves in different sorts of spatial patterns. These models are formulated as coupled systems of reaction–diffusion equations for the space-time dynamics of the concentrations of two morphogens, with different reaction kinetics depending on the biological problem at stake. Such systems exhibit diffusion-driven instability whereby homogenous steady states are driven unstable by diffusion, resulting in the formation of pre-patterns, provided that the diffusion rate of one of the morphogens is sufficiently higher than the other (Maini et al. 1997; Maini and Woolley 2019; Maini et al. 2012; Murray 1981).

On the other hand, mechanochemical models of pattern formation, first proposed by Murray, Oster and coauthors in the 1980s (Murray and Oster 1984a, b; Murray et al. 1983; Oster et al. 1983), describe spatial organisation of cells driven by the mechanochemical interaction between cells and the extracellular matrix (ECM)—i.e. the substratum composed of collagen fibres and various macromolecules, partly produced by the cells themselves, in which cells are embedded (Harris 1984; Harris et al. 1981). These models in their original form consist of systems of partial differential equations (PDEs) comprising a balance equation for the cell density, a balance equation for the ECM density, and a force-balance equation describing the mechanical equilibrium of the cell-ECM system (Murray and Maini 1989; Murray et al. 1988). When chemical processes are neglected, these models reduce to mechanical models of pattern formation (Byrne and Chaplain 1996; Murray and Maini 1989; Murray et al. 1988).

While reaction–diffusion models well explain the emergence and characteristics of patterns arising during chemical reactions (Castets et al. 1990; Maini et al. 1997; Maini and Woolley 2019), as well as pigmentation patterns found on shells (Meinhardt 2009) or animal coatings (Kondo and Asai 1995; Murray 2001), various observations seem to suggest they may not always be the most suited models to study morphogenic pattern formation (Bard and Lauder 1974; Brinkmann et al. 2018; Maini and Woolley 2019). For instance, experiments up to this day seem to fail in the identification of appropriate morphogens and overall molecular interactions predicted by Turing models in order for *de novo* patterns to emerge may be too complex. In addition, unrealistic parameter values would be required in order to reproduce experimentally observable patterns and the models appear to be too sensitive to parameter changes, hence lacking the robustness required to capture precise patterns. These considerations indicate that other mechanisms, driven for instance by significant mechanical forces, should be considered since solely chemical interactions may not suffice in explaining the emergence of patterns during morphogenesis. Hence, mechanochemical models may be better suited. Interestingly, this need to change modelling framework sometimes arises within the same biological application as time progresses. For instance, supracellular organisation in the early stages of embryonic development closely follows morphogenic chemical patterns, but further tissue-level organisation requires additional cooperation of osmotic pressures and mechanical forces (Petrolli et al. 2019). Similarly, pattern formation during vasculogenesis is generally divided into an early stage highly driven by cell migration following chemical cues, and a later one dominated by mechanical interactions between the cells and the ECM (Ambrosi et al. 2005; Scianna et al. 2013; Tosin et al. 2006). Finally, purely mechanical models are a useful tool for studying the isolated role of mechanical forces and can capture observed phenomena without the inclusion of chemical cues (Petrolli et al. 2019; Serra-Picamal et al. 2012; Tlili et al. 2018).

Over the years, mechanochemical and mechanical models of pattern formation in biological tissues have been used to study a variety of biomedical problems, including morphogenesis and embryogenesis (Brinkmann et al. 2018; Cruywagen and Murray 1992; Maini and Murray 1988; Murray and Maini 1986; Murray et al. 1988; Murray and Oster 1984a, b; Murray et al. 1983; Oster et al. 1983; Perelson et al. 1986), angiogenesis and vasculogenesis (Manoussaki 2003; Scianna et al. 2013; Tranqui and Tracqui 2000), cytoskeleton reorganisation (Alonso et al. 2017; Lewis and Murray 1991), wound healing and contraction (Javierre et al. 2009; Maini et al. 2002; Olsen et al. 1995; Tranquillo and Murray 1992), and stretch marks (Gilmore et al. 2012). These models have also been used to estimate the values of cell mechanical parameters, with a particular focus on cell traction forces (Barocas et al. 1995; Barocas and

Tranquillo 1994; Bentil and Murray 1991; Ferrenq et al. 1997; Moon and Tranquillo 1993; Perelson et al. 1986). The roles that different biological processes play in the formation of cellular patterns can be disentangled via linear stability analysis (LSA) of the homogenous steady states of the model equations—i.e. investigating what parameters of the model, and thus what biological processes, can drive homogenous steady states unstable and promote the emergence of cell spatial organisation. Further insight into certain aspects of pattern formation in biological tissues can also be provided by nonlinear stability analysis of the homogenous steady states (Cruywagen and Murray 1992; Lewis and Murray 1991; Maini and Murray 1988).

These models usually rely on the assumption that the cell-ECM system can be regarded as an isotropic linear viscoelastic material. This is clearly a simplification due to the nonlinear viscoelasticity and anisotropy of soft tissues (Bischoff et al. 2004; Huang et al. 2005; Liu and Bilston 2000; Nasseri et al. 2002; Snedeker et al. 2005; Valtorta and Mazza 2005; Verdier 2003), a simplification that various rheological tests conducted on biological tissues have nonetheless shown to be justified in the regime of small strains (Bilston et al. 1997; Liu and Bilston 2000; Nasseri et al. 2002; Valtorta and Mazza 2005), which is the one usually of interest in the applications of such models. Under this assumption, the force-balance equation for the cell-ECM system is often defined using the Kelvin–Voigt model of linear viscoelasticity to represent the stress–strain relation of the ECM (Byrne and Chaplain 1996; Murray et al. 1988; Oster et al. 1983). However, due to the multifaceted bio-physical nature of the ECM constituents, there are rheological aspects that cannot be effectively captured by the Kelvin–Voigt model and, therefore, depending on the pattern formation process and the type of biological tissue considered, other constitutive models of linear viscoelasticity may be better suited (Barocas and Tranquillo 1994). In this regard, Byrne and Chaplain (1996) demonstrated that, *ceteris paribus*, using the Maxwell model of linear viscoelasticity to describe the stress–strain relation of the ECM in place of the Kelvin–Voigt model can lead to different dispersion relations with a higher pattern formation potential. This suggests that a more thorough investigation of the capability of different stress–strain constitutive equations of producing spatial patterns is required.

With this aim, here we complement and further develop the results presented in Byrne and Chaplain (1996) by systematically assessing the pattern formation potential of different stress–strain constitutive equations for the ECM within a mechanical model of pattern formation in biological tissues (Byrne and Chaplain 1996; Murray et al. 1988; Oster et al. 1983). Compared to the work of Byrne and Chaplain (1996), here we consider a wider range of constitutive models, we allow cell traction forces to be reduced by cell–cell contact inhibition, and undertake numerical simulations of the model equations showing the formation of cellular patterns both in one and in two spatial dimensions. A related study has been conducted by Alonso et al. (2017), who considered a mathematical model of pattern formation in the cell cytoplasm.

The paper is structured as follows. In Sect. 2, we recall the essentials of viscoelastic materials and provide a brief summary of the one-dimensional stress–strain constitutive equations that we examine. In Sect. 3, we describe the one-dimensional mechanical model of pattern formation in biological tissues that is used in our study, which follows closely the one considered in Byrne and Chaplain (1996); Murray et al. (1988); Oster et al. (1983). In Sect. 4, we carry out a linear stability analysis (LSA) of a biologically relevant homogeneous steady state of the model equations, derive dispersion relations when different stress–strain constitutive equations for the ECM are used and investigate how the model parameters affect the dispersion relations obtained. In Sect. 5, we verify key results of LSA via numerical simulations of the model equations. In Sect. 6, we complement these findings with the results of numerical simulations of a two-dimensional version of the mechanical model of pattern formation considered in the previous sections. Section 7 concludes the paper and provides an overview of possible research perspectives.

2 Essentials of Viscoelastic Materials and Stress–Strain Constitutive Equations

In this section, we first recall the main properties of viscoelastic materials (see Sect. 2.1). Then, we briefly present the one-dimensional stress–strain constitutive equations that are considered in our study and summarise the main rheological properties of linear viscoelastic materials that they capture (see Sect. 2.2). Most of the contents of this section can be found in standard textbooks, such as Findley et al. (1976) [chapters 1 and 5] and Mase (1970), and are reported here for the sake of completeness. Specific considerations of and applications to living tissues can be found in Fung (1993).

2.1 Essentials of Viscoelastic Materials

As the name suggests, viscoelastic materials exhibit both viscous and elastic characteristics, and the interplay between them may result in a wide range of rheological properties that can be examined through creep and stress relaxation tests. During a creep test, a constant stress is first applied to a specimen of material and then removed, and the time dynamic of the correspondent strain is tracked. During a stress relaxation test, a constant strain is imposed on a specimen of material and the evolution in time of the induced stress is observed (Findley et al. 1976).

Here, we list the main properties of viscoelastic materials that may be observed during the first phase of a creep test (see properties 1a–1c), during the recovery phase, that is, when the constant stress is removed from the specimen (see properties 2a–2c), and during a stress relaxation test (see property 3).

1a *Instantaneous elasticity.* As soon as a stress is applied, an instantaneous corresponding strain is observed.
1b *Delayed elasticity.* While the instantaneous elastic response to a stress is a purely elastic behaviour, due to the viscous nature of the material a delayed elastic response may also be observed. In this case, under constant stress the strain slowly and continuously increases at decreasing rate.
1c *Viscous flow.* In some viscoelastic materials, under a constant stress, the strain continues to grow within the viscoelastic regime (i.e. before plastic deformation). In particular, viscous flow occurs when the strain increases linearly with time and stops growing at removal of the stress only.

2a *Instantaneous recovery.* When the stress is removed, an instantaneous recovery (i.e. an instantaneous strain decrease) is observed because of the elastic nature of the material.
2b *Delayed recovery.* Upon removal of the stress, a delayed recovery (i.e. a continuous decrease in the strain at decreasing rate) occurs.
2c *Permanent set.* While elastic strain is reversible, in viscoelastic materials a nonzero strain, known as "permanent set" or "residual strain", may persist even when the stress is removed.
3 *Stress relaxation.* Under constant strain, gradual relaxation of the induced stress occurs. In some cases, this may even culminate in total stress relaxation (i.e. the stress decays to zero).

The subset of these properties exhibited by a viscoelastic material will depend on—and hence define—the type of material being tested. Moreover, during each phase of the creep test, more than one of the above properties may be observed. For instance, a Maxwell material under constant stress will exhibit instantaneous elasticity followed by viscous flow.

2.2 One-Dimensional Stress–Strain Constitutive Equations Examined in Our Study

In this section, we briefly describe the different constitutive equations that are used in our study to represent the stress–strain relation of the ECM. In general, these equations can be used to predict how a viscoelastic material will react to different loading conditions, in one spatial dimension, and rely on the assumption that viscous and elastic characteristics of the material can be modelled, respectively, via linear combinations of dashpots and springs, as illustrated in Fig. 1. Different stress–strain constitutive equations correspond to different arrangements of these elements and capture different subsets of the rheological properties summarised in the previous section (see Table 2). In the remainder of this section, we will denote the stress and the strain at position x and time t by $\sigma(t, x)$ and $\varepsilon(t, x)$, respectively.
Linear elastic model. When viscous characteristics are neglected, a linear viscoelastic material can be modelled as a purely elastic spring with elastic modulus (*i.e.* Young's modulus) $E > 0$, as illustrated in Fig. 1a. In this case, the stress–strain constitutive equation is given by Hooke's spring law for continuous media, that is,

$$\sigma = E\varepsilon \,. \tag{1}$$

Linear viscous model. When elastic characteristics are neglected, a linear viscoelastic material can be modelled as a purely viscous damper of viscosity $\eta > 0$, as illustrated in Fig. 1b. In this case, the stress–strain constitutive equation is given by Newton's law of viscosity, that is,

$$\sigma = \eta \, \partial_t \varepsilon \,. \tag{2}$$

Kelvin–Voigt model. The Kelvin–Voigt model, also known as the Voigt model, relies on the assumption that viscous and elastic characteristics of a linear viscoelastic material

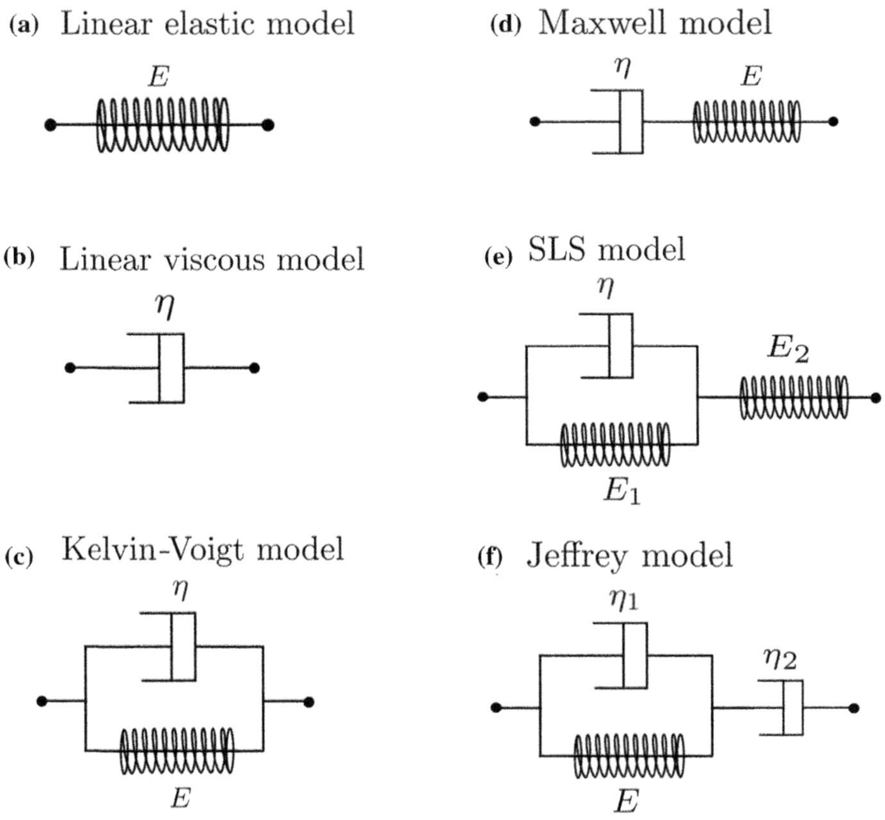

Fig. 1 Combinations of elastic springs and viscous dampers, together with the associated elastic (E, E_1, E_2) and viscous moduli (η, η_1, η_2), for the models of linear viscoelasticity considered in this work: the linear elastic model (**a**), the linear viscous model (**b**), the Kelvin–Voigt model (**c**), the Maxwell model (**d**), the SLS model (**e**), and the Jeffrey model (**f**)

can simultaneously be captured by considering a purely elastic spring with elastic modulus E and a purely viscous damper of viscosity η in parallel, as illustrated in Fig. 1c. The corresponding stress–strain constitutive equation is

$$\sigma = E\varepsilon + \eta\,\partial_t\varepsilon\,. \tag{3}$$

Maxwell model. The Maxwell model relies on the assumption that viscous and elastic characteristics of a linear viscoelastic material can be captured by considering a purely elastic spring with elastic modulus E and a purely viscous damper of viscosity η in series, as illustrated in Fig. 1d. The corresponding stress–strain constitutive equation is

$$\frac{1}{E}\,\partial_t\sigma + \frac{\sigma}{\eta} = \partial_t\varepsilon\,. \tag{4}$$

Standard linear solid (SLS) model. The SLS model, also known as the Kelvin model, relies on the assumption that viscous and elastic characteristics of a linear viscoelastic material can be captured by considering a Kelvin arm of elastic modulus E_1 and viscosity η in series with a purely elastic spring of elastic modulus E_2, as illustrated in Fig. 1e. The corresponding stress–strain constitutive equation is Mase (1970)

$$\frac{1}{E_2}\,\partial_t\sigma + \frac{1}{\eta}\left(1 + \frac{E_1}{E_2}\right)\sigma = \partial_t\varepsilon + \frac{E_1}{\eta}\varepsilon\,. \tag{5}$$

Jeffrey model. The Jeffrey model, also known as the Oldroyd-B or 3-parameter viscous model, relies on the assumption that viscous and elastic characteristics of a linear viscoelastic material can be captured by considering a Kelvin arm of elastic modulus E and viscosity η_1 in series with a purely viscous damper of viscosity η_2, as illustrated in Fig. 1f. The corresponding stress–strain constitutive equation is

$$\left(1 + \frac{\eta_1}{\eta_2}\right)\partial_t\sigma + \frac{E}{\eta_2}\sigma = \eta_1\partial^2_{tt}\varepsilon + E\,\partial_t\varepsilon\,. \tag{6}$$

Generic 4-parameter model. The following stress–strain constitutive equation encompasses all constitutive models of linear viscoelasticity whereby a combination of purely elastic springs and purely viscous dampers, up to a total of four elements, and therefore 4 parameters, is considered

$$a_2\partial^2_{tt}\sigma + a_1\partial_t\sigma + a_0\sigma = b_2\partial^2_{tt}\varepsilon + b_1\partial_t\varepsilon + b_0\varepsilon\,. \tag{7}$$

Here, the non-negative, real parameters $a_0, a_1, a_2, b_0, b_1, b_2$ depend on the elastic moduli and the viscosities of the underlying combinations of springs and dampers. When these parameters are defined as in Table 1, the generic 4-parameter constitutive model (7) reduces to the specific stress–strain constitutive equations (1)–(6). For convenience of notation, we define the differential operators

$$\mathcal{L}_a := a_2\partial^2_{tt} + a_1\partial_t + a_0 \quad \text{and} \quad \mathcal{L}_b := b_2\partial^2_{tt} + b_1\partial_t + b_0 \tag{8}$$

so that the stress–strain constitutive equation (7) can be rewritten in the following compact form

$$\mathcal{L}_a[\,\sigma\,] = \mathcal{L}_b[\,\varepsilon\,]\,. \tag{9}$$

A summary of the rheological properties of linear viscoelastic materials listed in Sect. 2.1 that are captured by the one-dimensional stress–strain constitutive Eqs. (1)–(6) is provided in Table 2. These properties can be examined through mathematical procedures that mimic creep and stress relaxation tests (Findley et al. 1976). Notice that, for all these constitutive models, instantaneous elasticity correlates with instantaneous recovery, delayed elasticity correlates with delayed recovery, and viscous flow correlates with permanent set. Materials are said to be more solid-like when their elastic response dominates their viscous response, and more fluid-like in the opposite

Table 1 Relations between the generic 4-parameter model (7) and the stress–strain constitutive Eqs. (1)–(6)

Generic 4-parameters model	a_2	a_1	a_0	b_2	b_1	b_0
Linear elastic model	0	0	1	0	0	E
Linear viscous model	0	0	1	0	η	0
Kelvin-Voigt model	0	0	1	0	η	E
Maxwell model	0	$\frac{1}{E}$	$\frac{1}{\eta}$	0	1	0
SLS model	0	$\frac{1}{E_2}$	$\frac{1}{\eta}\left(1+\frac{E_1}{E_2}\right)$	0	1	$\frac{E_1}{\eta}$
Jeffrey model	0	$1+\frac{\eta_1}{\eta_2}$	$\frac{E}{\eta_2}$	η_1	E	0

case (Nargess and Yanlan 2021). For this reason, models of linear viscoelasticity that capture viscous flow and, as a consequence, permanent set—such as the Maxwell model and the Jeffrey model—are classified as "fluid-like models", while those which do not— such as the Kelvin–Voigt model and the SLS model—are classified as "solid-like models". In the remainder of the paper, we are going to include the linear viscous model in the fluid-like class and the linear elastic model in the solid-like class, as they capture—or do not capture—the relevant properties, even if they are not models of viscoelasticity *per se*.

3 A One-Dimensional Mechanical Model of Pattern Formation

We consider a one-dimensional region of tissue and represent the normalised densities of cells and ECM at time $t \in [0, T]$ and position $x \in [\ell, L]$ by means of the non-negative functions $n(t, x)$ and $\rho(t, x)$, respectively. We let $u(t, x)$ model the displacement of a material point of the cell-ECM system originally at position x, which is induced by mechanical interactions between cells and the ECM—i.e. cells pull on the ECM in which they are embedded, thus inducing ECM compression and densification which in turn cause a passive form of cell repositioning (Van Helvert et al. 2018).

3.1 Dynamics of the Cells

Following Murray et al. (1988); Oster et al. (1983), we consider a scenario where cells change their position according to a combination of: (i) undirected, random movement, which we describe through Fick's first law of diffusion with diffusivity (i.e. cell motility) $D > 0$; (ii) haptotaxis (i.e. cell movement up the density gradient of the ECM) with haptotactic sensitivity $\alpha > 0$; (iii) passive repositioning caused by mechanical interactions between cells and the ECM, which is modelled as an advection with velocity field $\partial_t u$. Moreover, we model variation of the normalised cell density caused by cell proliferation and death via logistic growth with intrinsic growth rate $r > 0$ and unitary local carrying capacity. Under these assumptions, we describe cell

Table 2 Properties of linear viscoelastic materials captured by the stress–strain constitutive Eqs. (1)–(6)

	Instantaneous elasticity	Delayed elasticity	Viscous flow	Instantaneous recovery	Delayed recovery	Permanent set	Stress relaxation
Linear elastic model	✓			✓			
Linear viscous model			✓			✓	N. A.
Kelvin-Voigt model		✓			✓		
Maxwell model	✓		✓	✓		✓	✓
SLS model	✓	✓		✓	✓		✓
Jeffrey model		✓	✓		✓	✓	✓

dynamics through the following balance equation for $n(t, x)$

$$\partial_t n = \partial_x [D \partial_x n - n (\alpha \partial_x \rho + \partial_t u)] + r n(1 - n) \tag{10}$$

subject to suitable initial and boundary conditions.

3.2 Dynamics of the ECM

As was done for the cell dynamics, in a similar manner we model compression and densification of the ECM induced by cell-ECM interactions as an advection with velocity field $\partial_t u$. Furthermore, as in Murray et al. (1988); Oster et al. (1983), we neglect secretion of ECM components by the cells since this process occurs on a slower time scale compared to mechanical interactions between cells and the ECM. Under these assumptions, we describe the cell dynamics through the following transport equation for $\rho(t, x)$

$$\partial_t \rho = -\partial_x (\rho \partial_t u) \tag{11}$$

subject to suitable initial and boundary conditions.

3.3 Force-Balance Equation for the Cell-ECM System

Following Murray et al. (1988); Oster et al. (1983), we represent the cell-ECM system as a linear viscoelastic material with low Reynolds number (i.e. inertial terms are negligible compared to viscous terms) and we assume the cell-ECM system to be in mechanical equilibrium (i.e. traction forces generated by the cells are in mechanical equilibrium with viscoelastic restoring forces developed in the ECM and any other external forces). Under these assumptions, the force-balance equation for the cell-ECM system is of the form

$$\partial_x (\sigma_c + \sigma_m) + \rho F = 0, \tag{12}$$

where $\sigma_m(t, x)$ is the contribution to the stress of the cell-ECM system coming from the ECM, $\sigma_c(t, x)$ is the contribution to the stress of the cell-ECM system coming from the cells, and $F(t, x)$ is the external force per unit matrix density, which comes from the surrounding tissue that constitutes the underlying substratum to which the ECM is attached.

The stress σ_c is related to cellular traction forces acting on the ECM and is defined as

$$\sigma_c := \tau f(n) n \left(\rho + \beta \partial^2_{xx} \rho \right) \quad \text{with} \quad f(n) := \frac{1}{1 + \lambda n^2}. \tag{13}$$

Definition (13) relies on the assumption that the stress generated by cell traction on the ECM is proportional to the cell density n and—in the short range—the ECM density ρ, while the term $\beta \partial^2_{xx} \rho$ accounts for long-range cell traction effects, with β being the long-range traction proportionality constant. The factor of proportionality is given by a positive parameter, τ, which measures the average traction force generated by a cell, multiplied by a non-negative and monotonically decreasing function of the cell density, $f(n)$, which models the fact that the average traction force generated by a cell is reduced by cell-cell contact inhibition (Murray 2001). The parameter $\lambda \geq 0$ measures the level of cell traction force inhibition and assuming $\lambda = 0$ corresponds to neglecting the reduction in the cell traction forces caused by cellular crowding.

The stress σ_m is given by the stress–strain constitutive equation that is used for the ECM, which we choose to be the general constitutive model (9) with the strain $\varepsilon(t, x)$ being given by the gradient of the displacement $u(t, x)$, that is, $\varepsilon = \partial_x u$. Therefore, we define the stress–strain relation of the ECM via the following equation

$$\mathcal{L}_a[\, \sigma_m \,] = \mathcal{L}_b[\, \partial_x u \,] \,, \tag{14}$$

where the differential operators $\mathcal{L}_a$ and $\mathcal{L}_b$ are defined according to (8).

Assuming the surrounding tissue to which the ECM is attached to be a linear elastic material (Murray 2001), the external body force F can be modelled as a restoring force proportional to the cell-ECM displacement, that is,

$$F := -s\, u \,. \tag{15}$$

Here, the parameter $s > 0$ represents the elastic modulus of the surrounding tissue.

In order to obtain a closed equation for the displacement $u(t, x)$, we apply the differential operator $\mathcal{L}_a[\,\cdot\,]$ to the force-balance Eq. (12) and then substitute (13)–(15) into the resulting equation. In so doing, we find

$$\begin{aligned}
&\mathcal{L}_a\,[\,\partial_x\,(\sigma_m + \sigma_c)\,] = -\mathcal{L}_a\,[\,\rho\, F\,] \\
&\Leftrightarrow \mathcal{L}_a\,[\,\partial_x\,\sigma_m\,] + \mathcal{L}_a\,[\,\partial_x\,\sigma_c\,] = \mathcal{L}_a\,[\,s\rho u\,] \\
&\Leftrightarrow \partial_x\,\mathcal{L}_a\,[\,\sigma_m\,] = \mathcal{L}_a\,[\,s\rho u\,] - \mathcal{L}_a\,[\,\partial_x\,\sigma_c\,] \\
&\Leftrightarrow \partial_x\,\mathcal{L}_b\,[\,\partial_x u\,] = \mathcal{L}_a\,[\,s\rho u - \partial_x\sigma_c\,] \\
&\Leftrightarrow \mathcal{L}_b\,[\,\partial^2_{xx} u\,] = \mathcal{L}_a\,[\,s\rho u - \partial_x\sigma_c\,]\,,
\end{aligned}$$

that is,

$$\mathcal{L}_b\,[\,\partial^2_{xx} u\,] = \mathcal{L}_a\left[\, s\rho u - \partial_x\left(\frac{\tau n}{1+\lambda n^2}(\rho + \beta\partial^2_{xx}\rho)\right)\right]. \tag{16}$$

Finally, to close the system, Eq. (16) needs to be supplied with suitable initial and boundary conditions.

3.4 Boundary Conditions

We close our mechanical model of pattern formation defined by the system of PDEs (10), (11) and (16) with the following boundary conditions

$$\begin{cases} n(t,\ell) = n(t,L)\,, & \partial_x n(t,\ell) = \partial_x n(t,L)\,, \\ \rho(t,\ell) = \rho(t,L)\,, & \partial^2_{xx}\rho(t,\ell) = \partial^2_{xx}\rho(t,L)\,, \\ u(t,\ell) = u(t,L)\,, & \partial_x u(t,\ell) = \partial_x u(t,L), \end{cases} \qquad \text{for all } t \in [0,T]\,. \tag{17}$$

Here, the conditions on the derivatives of n, ρ and u ensure that the fluxes in Eqs. (10) and (11), and the overall stress $(\sigma_m + \sigma_c)$ in Eq. (16), are periodic on the boundary, *i.e.* they ensure that

$$\begin{cases} [D\,\partial_x n - n\,(\alpha\,\partial_x\rho + \partial_t u)]_{x=\ell} = [D\,\partial_x n - n\,(\alpha\,\partial_x\rho + \partial_t u)]_{x=L}\,, \\ [n\,\partial_t u]_{x=\ell} = [n\,\partial_t u]_{x=L}\,, \\ \left[\tau\dfrac{n}{(1+\lambda^2)}\,(\rho + \beta\,\partial^2_{xx}\rho) + \sigma_m\right]_{x=\ell} = \left[\tau\dfrac{n}{(1+\lambda^2)}\,(\rho + \beta\,\partial^2_{xx}\rho) + \sigma_m\right]_{x=L}\,, \end{cases} \qquad \text{for all } t \in [0,T]\,,$$

with σ_m given as a function of $\partial_x u$ in Eq. (14), according to the selected constitutive model. The periodic boundary conditions (17) reproduce a biological scenario in which the spatial region considered is part of a larger area of tissue whereby similar dynamics of the cells and the ECM occur.

4 Linear Stability Analysis and Dispersion Relations

In this section, we carry out LSA of a biologically relevant homogeneous steady state of the system of PDEs (10), (11) and (16) (see Sect. 4.1) and we compare the dispersion relations obtained when the constitutive models (1)–(6) are alternatively used to represent the contribution to the overall stress coming from the ECM, in order to explore the pattern formation potential of these stress–strain constitutive equations (see Sect. 4.2).

4.1 Linear Stability Analysis

Biologically relevant homogeneous steady state. All non-trivial homogeneous steady states $(\bar{n}, \bar{\rho}, \bar{u})^{\mathsf{T}}$ of the system of PDEs (10), (11) and (16) subject to boundary conditions (17) have components $\bar{n} \equiv 1$ and $\bar{u} \equiv 0$, and we consider the arbitrary non-trivial steady state $\bar{\rho} \equiv \rho_0 > 0$ amongst the infinite number of possible homogeneous steady states of the transport Eq. (11) for the normalised ECM density ρ. Hence, we focus our attention on the biologically relevant homogeneous steady state $\bar{\mathbf{v}} = (1, \rho_0, 0)^{\mathsf{T}}$.
Linear stability analysis to spatially homogeneous perturbations. In order to undertake linear stability analysis of the steady state $\bar{\mathbf{v}} = (1, \rho_0, 0)^{\mathsf{T}}$ to spatially homogeneous perturbations, we make the ansatz $\mathbf{v}(t,x) \equiv \bar{\mathbf{v}} + \tilde{\mathbf{v}}(t)$, where the vector $\tilde{\mathbf{v}}(t) = (\tilde{n}(t), \tilde{\rho}(t), \tilde{u}(t))^{\mathsf{T}}$ models small spatially homogeneous perturbations and linearise the system of PDEs (10), (11) and (16) about the steady state $\bar{\mathbf{v}}$. Assuming

$\tilde{n}(t)$, $\tilde{\rho}(t)$ and $\tilde{u}(t)$ to be proportional to $\exp(\psi t)$, with $\psi \neq 0$, one can easily verify that ψ satisfies the algebraic equation $\psi(\psi + r)(\psi^2 a_2 + \psi a_1 + a_0) = 0$. Since r is positive and the parameters a_0, a_1 and a_2 are all non-negative, the solution ψ of such an algebraic equation is necessarily negative and, therefore, the small perturbations $\tilde{n}(t)$, $\tilde{\rho}(t)$ and $\tilde{u}(t)$ will decay to zero as $t \to \infty$. This implies that the steady state $\bar{\mathbf{v}}$ will be stable to spatially homogeneous perturbations for any choice of the parameter a_0, a_1, a_2, b_0, b_1 and b_2 in the stress–strain constitutive equation (14) (i.e. for all constitutive models (1)–(6)).

Linear stability analysis to spatially inhomogeneous perturbations. In order to undertake linear stability analysis of the steady state $\bar{\mathbf{v}} = (1, \rho_0, 0)^\intercal$ to spatially inhomogeneous perturbations, we make the ansatz $\mathbf{v}(t,x) = \bar{\mathbf{v}} + \tilde{\mathbf{v}}(t,x)$, where the vector $\tilde{\mathbf{v}}(t,x) = (\tilde{n}(t,x), \tilde{\rho}(t,x), \tilde{u}(t,x))^\intercal$ models small spatially inhomogeneous perturbations and linearise the system of PDEs (10), (11) and (16) about the steady state $\bar{\mathbf{v}}$. Assuming $\tilde{n}(t,x)$, $\tilde{\rho}(t,x)$ and $\tilde{u}(t,x)$ to be proportional to $\exp(\psi t + ikx)$, with $\psi \neq 0$ and $k \neq 0$, we find that ψ satisfies the following equation

$$\psi\left[c_3(k^2)\psi^3 + c_2(k^2)\psi^2 + c_1(k^2)\psi + c_0(k^2)\right] = 0\,, \tag{18}$$

with

$$c_3(k^2) := a_2\tau\lambda_1\beta\, k^4 + \left[b_2 - a_2\tau(\lambda_1 + \lambda_2\rho_0)\right]k^2 + a_2 s\rho_0 \tag{19}$$

$$c_2(k^2) := a_2\tau\lambda_1 D\beta\, k^6 + \left[b_2 D - a_2\tau(\lambda_2\rho_0\alpha + D\lambda_1 - r\lambda_1\beta) + a_1\tau\lambda_1\beta\right]k^4 + \left[b_2 r + b_1 + a_2(Ds\rho_0 - r\tau\lambda_1) - a_1\tau(\lambda_1 + \lambda_2\rho_0)\right]k^2 + (a_1 + a_2 r)s\rho_0 \tag{20}$$

$$c_1(k^2) := a_1\tau\lambda_1 D\beta\, k^6 + \left[b_1 D - a_1\tau(\lambda_2\rho_0\alpha + D\lambda_1 - r\lambda_1\beta) + a_0\tau\lambda_1\beta\right]k^4 + \left[b_1 r + b_0 + a_1(Ds\rho_0 - r\tau\lambda_1) - a_0\tau(\lambda_1 + \lambda_2\rho_0)\right]k^2 + (a_0 + a_1 r)s\rho_0 \tag{21}$$

and

$$\begin{aligned} c_0(k^2) := {} & a_0\tau\lambda_1 D\beta\, k^6 + \left[b_0 D - a_0\tau(\lambda_2\rho_0\alpha + D\lambda_1 - r\lambda_1\beta)\right]k^4 \\ & + \left[b_0 r + a_0(Ds\rho_0 - r\tau\lambda_1)\right]k^2 + a_0 r s\rho_0 \end{aligned} \tag{22}$$

where

$$\lambda_1 := \frac{1}{1+\lambda} \quad \text{and} \quad \lambda_2 := \frac{(1-\lambda)}{(1+\lambda)^2}\,.$$

Equation (18) has multiple solutions $(\psi(k^2))$ for each k^2, and we denote by $\mathrm{Re}(\cdot)$ the maximum real part of all these solutions. For cell patterns to emerge, we need the non-trivial homogeneous steady state $\bar{\mathbf{v}}$ to be unstable to spatially inhomogeneous perturbations, that is, we need $\mathrm{Re}(\psi(k^2)) > 0$ for some $k^2 > 0$. Notice that a necessary condition for this to happen is that at least one amongst $c_0(k^2)$, $c_1(k^2)$, $c_2(k^2)$ and $c_3(k^2)$ is negative for some $k^2 > 0$. Hence, the fact that if $\tau = 0$, then $c_0(k^2)$, $c_1(k^2)$, $c_2(k^2)$ and $c_3(k^2)$ are all non-negative for any value of k^2 allows us to conclude that having $\tau > 0$ is a necessary condition for pattern formation to occur. This was expected based on the results presented in Murray (2001) and references therein.

In the case, where the model parameters are such that $c_2(k^2) = 0$ and $c_3(k^2) = 0$, solving Eq. (18) for ψ gives the following dispersion relation

$$\psi(k^2) = -\frac{c_0(k^2)}{c_1(k^2)} \tag{23}$$

and for the condition $\text{Re}(\psi(k^2)) > 0$ to be met it suffices that, for some $k^2 > 0$,

$$c_0(k^2) > 0 \quad \text{and} \quad c_1(k^2) < 0 \quad \text{or} \quad c_0(k^2) < 0 \quad \text{and} \quad c_1(k^2) > 0\,.$$

On the other hand, when the model parameters are such that only $c_3(k^2) = 0$, from Eq. (18) we obtain the following dispersion relation

$$\psi(k^2) = \frac{-c_1(k^2) \pm \sqrt{\left(c_1(k^2)\right)^2 - 4c_2(k^2)c_0(k^2)}}{2c_2(k^2)}\,, \tag{24}$$

and for the condition $\text{Re}(\psi(k^2)) > 0$ to be satisfied it is sufficient that one of the following four sets of conditions holds

$$c_2(k^2) > 0 \quad \text{and} \quad c_0(k^2) < 0 \quad \text{or} \quad c_2(k^2) > 0\,, \quad c_1(k^2) < 0 \quad \text{and} \quad c_0(k^2) > 0$$

or

$$c_2(k^2) < 0 \quad \text{and} \quad c_0(k^2) > 0 \quad \text{or} \quad c_2(k^2) < 0\,, \quad c_1(k^2) > 0 \quad \text{and} \quad c_0(k^2) < 0\,.$$

Finally, in the general case where the model parameters are such that $c_3(k^2) \neq 0$ as well, from Eq. (18) we obtain the following dispersion relation

$$\psi(k^2) = \left\{q + \left[q^2 + \left(m - p^2\right)^3\right]^{1/2}\right\}^{1/3} + \left\{q - \left[q^2 + \left(m - p^2\right)^3\right]^{1/2}\right\}^{1/3} + p\,, \tag{25}$$

where $p \equiv p(k^2)$, $q \equiv q(k^2)$ and $m \equiv m(k^2)$ are defined as

$$p := -\frac{c_2}{3c_3}\,, \quad q := p^3 + \frac{c_2c_1 - 3c_3c_0}{6c_3^2}\,, \quad m := \frac{c_1}{3c_3}\,.$$

In this case, identifying sufficient conditions to ensure that the real part of $\psi(k^2)$ is positive for some $k^2 > 0$ requires lengthy algebraic calculations. We refer the interested reader to Gilmore et al. (2012), where the Routh–Hurwitz stability criterion was used to analyse this general case and obtain more explicit conditions on the model parameters under which pattern formation occurs.

4.2 Dispersion Relations

Substituting the definitions of a_0, a_1, a_2, b_0, b_1 and b_2 corresponding to the stress–strain constitutive equations (1)–(6), which are reported in Table 1, into definitions (19)–(22) for $c_0(k^2)$, $c_1(k^2)$, $c_2(k^2)$ and $c_3(k^2)$, and then using the dispersion relation given by formula (23), (24) or (25) depending on the values of $c_2(k^2)$ and $c_3(k^2)$ so obtained, we derive the dispersion relation for each of the constitutive models (1)–(6). In particular, we are interested in whether the real part of each dispersion relation is positive, so whenever multiple roots are calculated—for instance using (24)—the largest root is considered. In addition, dispersion relations throughout this section are plotted against the quantity k/π, which directly correlates with perturbation modes and can therefore better highlight mode selection during the sensitivity analysis.
Base-case dispersion relations. Figure 2 displays the dispersion relations obtained for the stress–strain constitutive equations (1)–(6) under the following base-case parameter values

$$E = 1\,, \quad E_1 = E_2 = \frac{1}{2}E = 0.5\,, \quad \eta = 1\,, \quad \eta_1 = \eta_2 = \frac{1}{2}\eta = 0.5\,, \quad D = 0.01\,, \tag{26}$$

$$\rho_0 = 1\,, \quad \alpha = 0.05\,, \quad r = 1\,, \quad s = 10\,, \quad \lambda = 0.5\,, \quad \tau = 0.2 \quad \beta = 0.005\,. \tag{27}$$

The parameter values given by (26) and (27) are chosen for illustrative purposes, in order to highlight the different qualitative behaviour of the dispersion relations obtained using different models and are comparable with nondimensional parameter values that can be found in the extant literature (see Appendix A for further details). A comparison between the plots in Fig. 2 reveals that fluid-like models, that is, the linear viscous model (2), the Maxwell model (4), and the Jeffrey model (6) (*cf.* Table 2), have a higher pattern formation potential than solid-like models, since under the same parameter set they exhibit a range—or, more precisely, they exhibit the same range—of unstable modes (*i.e.* $\mathrm{Re}(\psi(k^2)) > 0$ for a range of values of k/π), while the others have no unstable modes.

We now undertake a sensitivity analysis with respect to the different model parameters and discuss key changes that occur in the base-case dispersion relations displayed in Fig. 2.
ECM elasticity. The plots in Fig. 3 illustrate how the base-case dispersion relations displayed in Fig. 2 change when different values of the parameter E, and therefore also E_1 and E_2 (*i.e.* the parameters modelling ECM elasticity), are considered. These plots show that lower values of these parameters correlate with overall larger values of $\mathrm{Re}(\psi(k^2))$ for all constitutive models, except for the linear viscous one, which corresponds to speeding up the formation of spatial patterns, when these may form. In addition, sufficiently small values of the parameters E, E_1, and E_2 allow the linear elastic model (1), the Kelvin–Voigt model (3), and the SLS model (5) to exhibit unstable modes. However, further lowering the values of these parameters appears to lead to singular dispersion relations (*cf.* the plots for the linear elastic model (1), the Maxwell model (4), and the SLS model (5) in Fig. 3), which suggests that linear stability theory may fail in the regime of low ECM elasticity.

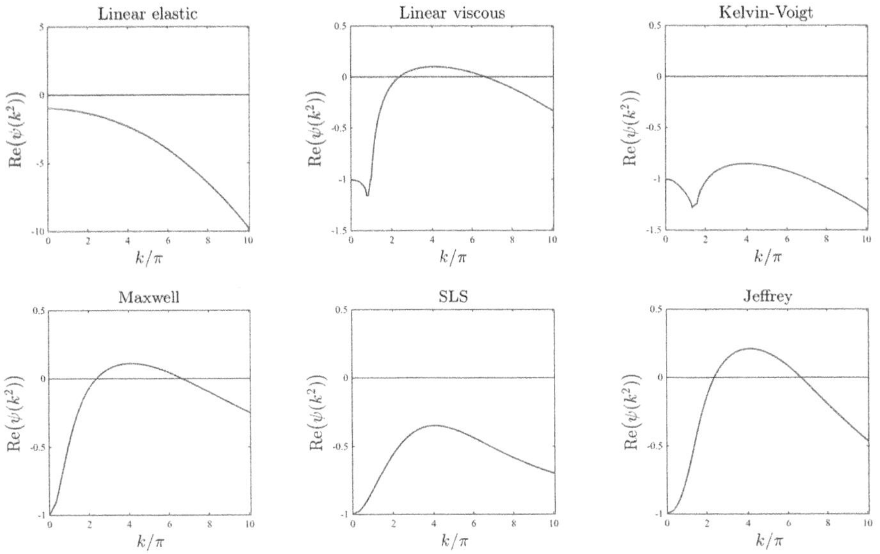

Fig. 2 Base-case dispersion relations. Dispersion relations corresponding to the stress–strain constitutive equations (1)–(6) for the base-case set of parameter values given by (26) and (27)

ECM viscosity. The plots in Fig. 4 illustrate how the base-case dispersion relations displayed in Fig. 2 change when different values of the parameter η, and therefore also η_1 and η_2 (i.e. the parameters modelling ECM viscosity), are considered. These plots show that larger values of these parameters leave the range of modes for which $\mathrm{Re}(\psi(k^2)) > 0$ unchanged but reduce the values of $\mathrm{Re}(\psi(k^2))$. This supports the idea that a higher ECM viscosity may not change the pattern formation potential of the different constitutive models but may slow down the corresponding pattern formation processes.

Cell motility. The plots in Fig. 5 illustrate how the base-case dispersion relations displayed in Fig. 2 change when different values of the parameter D (i.e. the parameter modelling cell motility) are considered. These plots show that larger values of this parameter may significantly shrink the range of modes for which $\mathrm{Re}(\psi(k^2)) > 0$. In particular, with the exception of the linear elastic model, all constitutive models exhibit: infinitely many unstable modes when $D \to 0$; a finite number of unstable modes for intermediate values of D; no unstable modes for sufficiently large values of D. This is to be expected due to the stabilising effect of undirected, random cell movement and indicates that higher cell motility may correspond to lower pattern formation potential.

Intrinsic growth rate of the cell density and elasticity of the surrounding tissue. The plots in Figs. 6 and 7 illustrate how the base-case dispersion relations displayed in Fig. 2 change when different values of the parameter r (*i.e.* the intrinsic growth rate of the cell density) and the parameter s (*i.e.* the elasticity of the surrounding tissue) are, respectively, considered. These plots show that considering larger values of these parameters reduces the values of $\mathrm{Re}(\psi(k^2))$ for all constitutive models, and in particular it shrinks the range of unstable modes for the linear viscous model (2), the Maxwell model (4), and the Jeffrey model (6), which can become stable for values of r or s sufficiently large. This supports the idea that higher growth rates of the cell density (i.e. faster cell proliferation and death), and higher substrate elasticity (*i.e.* stronger external tethering force) may slow down pattern formation processes and overall reduce the pattern formation potential for all constitutive models. Moreover, the plots in Fig. 7 indicate that higher values of s may in particular reduce the pattern formation potential of the different constitutive models by making it more likely that $\text{Re}(\psi(k^2)) < 0$ for smaller values of k/π (i.e. low-frequency perturbation modes will be more likely to vanish).

Level of contact inhibition of the cell traction forces and long-range cell traction forces. The plots in Figs. 8 and 9 illustrate how the base-case dispersion relations displayed in Fig. 2 change when different values of the parameter λ (i.e. the level of cell-cell contact inhibition of the cell traction forces) and the parameter β (i.e. the long-range cell traction forces) are, respectively, considered. Considerations similar to those previously made about the dispersion relations obtained for increasing values of the parameters r and s apply to the case where increasing values of the parameter λ and the parameter β are considered. In addition to these considerations, the plots in Figs. 8 and 9 indicate that for small enough values of λ or β the SLS model (5) can exhibit unstable modes, which further suggests that weaker contact inhibition of cell traction forces and lower long-range cell traction forces foster pattern formation. Moreover, the plots in Fig. 9 indicate that in the asymptotic regime $\beta \to 0$ we may observe infinitely many unstable modes (*i.e.* $\text{Re}(\psi(k^2)) > 0$ for arbitrarily large wavenumbers), exiting the regime of physically meaningful pattern forming instabilities (Moreo et al. 2010; Perelson et al. 1986).

Cell haptotactic sensitivity and cell traction forces. The plots in Figs. 10 and 11 illustrate how the base-case dispersion relations displayed in Fig. 2 change when different values of the parameter α (*i.e.* the cell haptotactic sensitivity) and the parameter τ (*i.e.* the cell traction force) are, respectively, considered. As expected (Murray 2001), larger values of these parameters overall increase the value of $\text{Re}(\psi(k^2))$ and broaden the range of values of modes for which $\text{Re}(\psi(k^2)) > 0$, so that for large enough values of these parameters the linear viscous model (2), the Kelvin–Voigt model (3) and the SLS model (5) can exhibit unstable modes. However, sufficiently large values of τ appear to lead to singular dispersion relations (*cf.* the plots for the linear elastic model (1), the Maxwell model (4), and the SLS model (5) in Fig. 11), which suggests that linear stability theory may fail in the regime of high cell traction for certain constitutive models, as previously observed in (Byrne and Chaplain 1996).

Initial ECM density. The plots in Fig. 12 illustrate how the base-case dispersion relations displayed in Fig. 2 change when different values of the parameter ρ_0 (*i.e.* the initial ECM density) are considered. Considerations similar to those previously made about the dispersion relations obtained for increasing values of the parameter α apply to the case where increasing values of the parameter ρ_0 are considered. In addition to these considerations, the plots in Fig. 12 indicate that smaller values of the parameter ρ_0, specifically $\rho_0 < 1$, correlate with a shift in mode selection towards lower modes (*cf.* the plots for the linear viscous model (2), the Maxwell model (4) and the Jeffrey model (6) in Fig. 12).

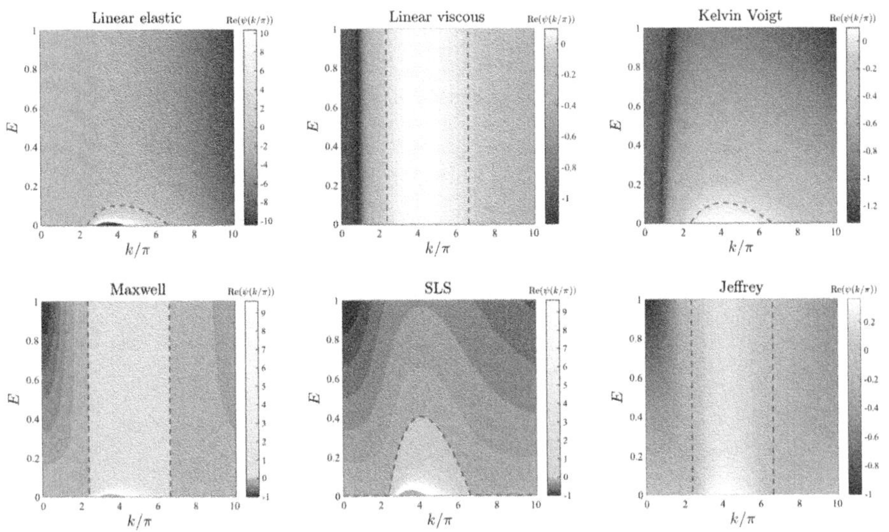

Fig. 3 Effects of varying the ECM elasticity. Dispersion relations corresponding to the stress–strain constitutive equations (1)–(6) for increasing values of the ECM elasticity, that is for $E \in [0, 1]$. The values of the other parameters are given by (26) and (27). White regions in the plots related to the linear elastic model, the Maxwell model and the SLS model correspond to $\mathrm{Re}(\psi(k^2)) > 10$ (i.e. a vertical asymptote is present in the dispersion relation). Red dashed lines mark contour lines where $\mathrm{Re}(\psi(k^2)) = 0$

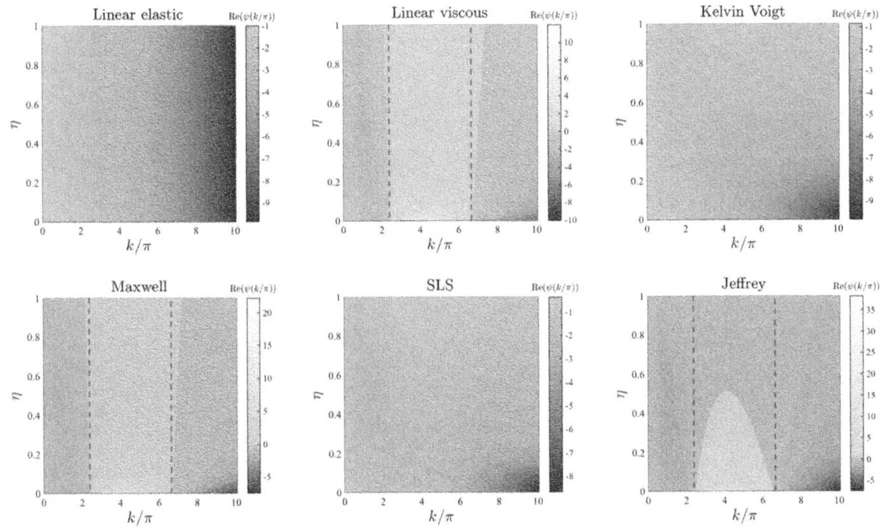

Fig. 4 Effects of varying the ECM viscosity. Dispersion relations corresponding to the stress–strain constitutive equations (1)–(6) for increasing values of the ECM viscosity, that is for $\eta \in [0, 1]$. The values of the other parameters are given by (26) and (27). Red dashed lines mark contour lines where $\mathrm{Re}(\psi(k^2)) = 0$

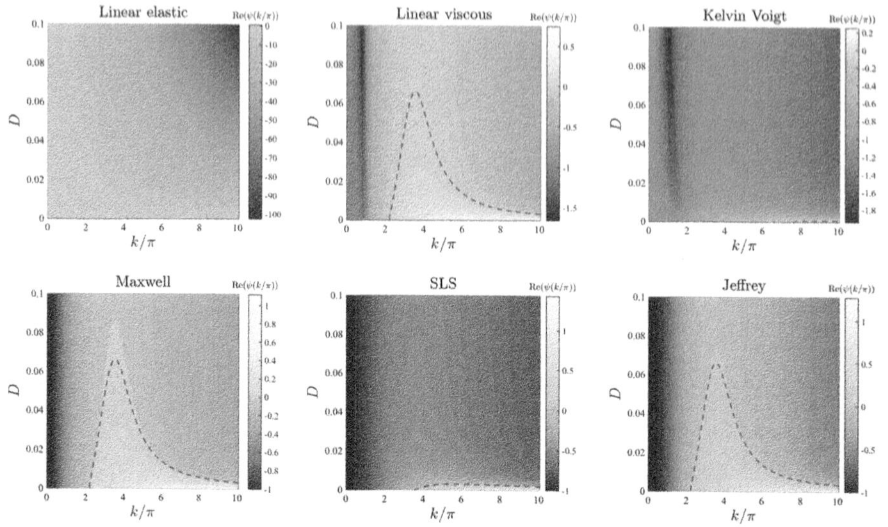

Fig. 5 Effects of varying the cell motility. Dispersion relations corresponding to the stress–strain constitutive equations (1)-(6) for increasing values of the cell motility, that is for $D \in [0, 0.1]$. The values of the other parameters are given by (26) and (27). Red dashed lines mark contour lines where $\mathrm{Re}(\psi(k^2)) = 0$

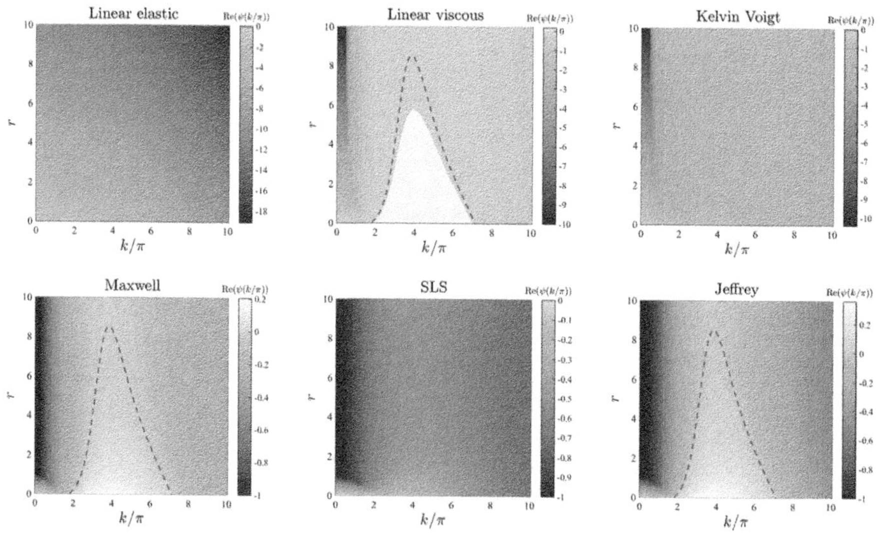

Fig. 6 Effects of varying the intrinsic growth rate of the cell density. Dispersion relations corresponding to the stress–strain constitutive equations (1)–(6) for increasing values of the intrinsic growth rate of the cell density, that is for $r \in [0, 10]$. The values of the other parameters are given by (26) and (27). Red dashed lines mark contour lines where $\mathrm{Re}(\psi(k^2)) = 0$

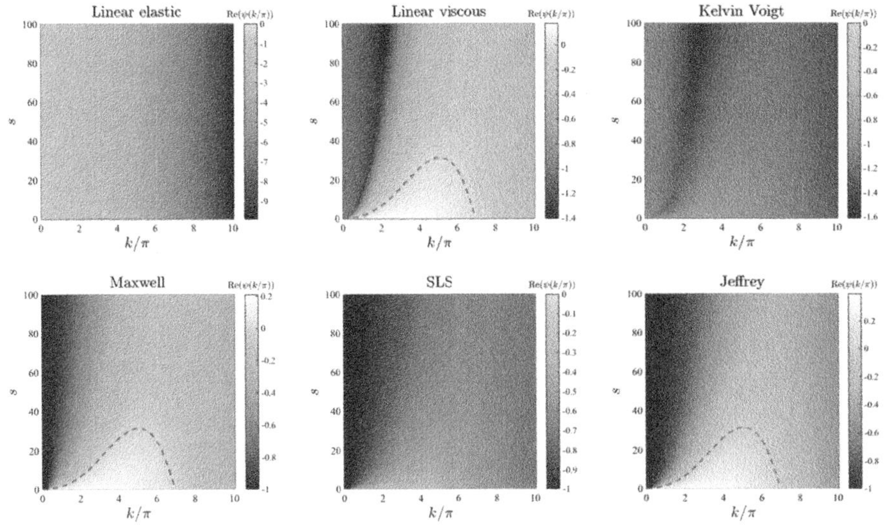

Fig. 7 Effects of varying the elasticity of the surrounding tissue. Dispersion relations corresponding to the stress–strain constitutive equations (1)–(6) for increasing values of the elasticity of the surrounding tissue, that is for $s \in [0, 100]$. The values of the other parameters are given by (26) and (27). Red dashed lines mark contour lines where $\mathrm{Re}(\psi(k^2)) = 0$

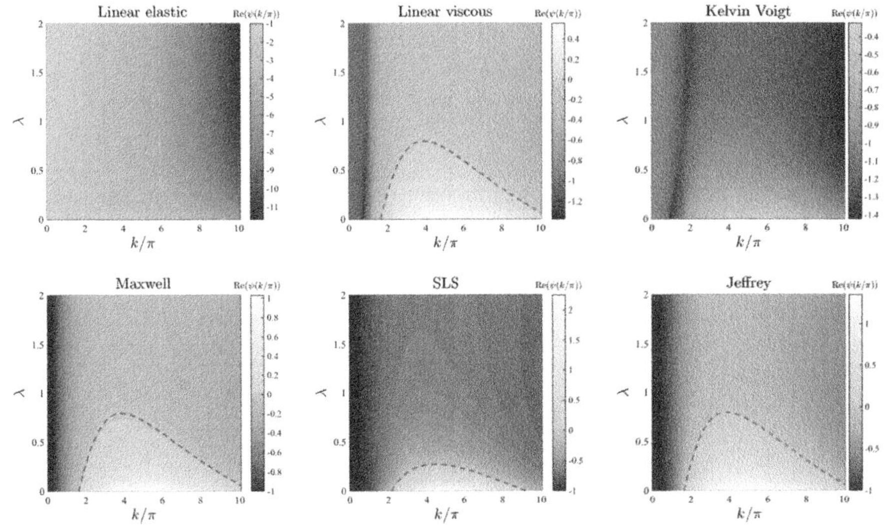

Fig. 8 Effects of varying the level of cell–cell contact inhibition of the cell traction forces. Dispersion relations corresponding to the stress–strain constitutive equations (1)–(6) for increasing levels of cell–cell contact inhibition of the cell traction forces, that is for $\lambda \in [0, 2]$. The values of the other parameters are given by (26) and (27). Red dashed lines mark contour lines where $\mathrm{Re}(\psi(k^2)) = 0$

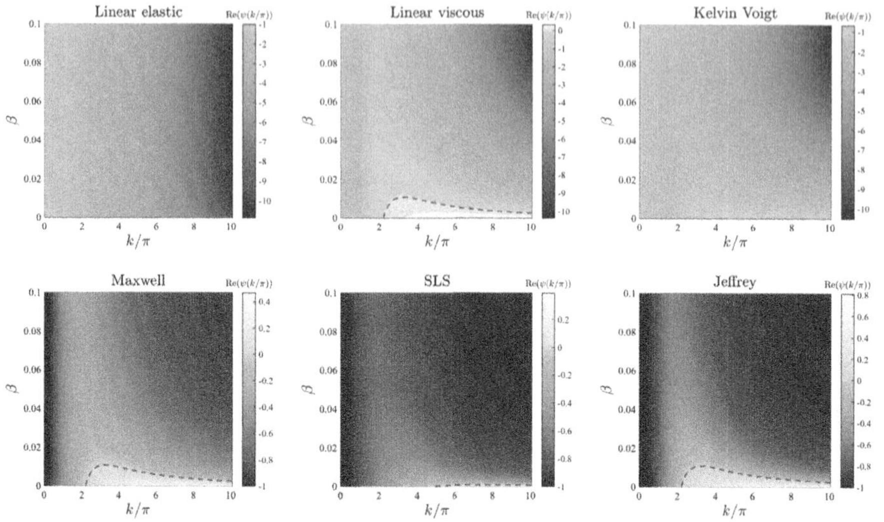

Fig. 9 Effects of varying the long-range cell traction forces. Dispersion relations corresponding to the stress–strain constitutive equations (1)–(6) for increasing long-range cell traction forces, that is for $\beta \in [0, 0.1]$. The values of the other parameters are given by (26) and (27). Red dashed lines mark contour lines where $\mathrm{Re}(\psi(k^2)) = 0$

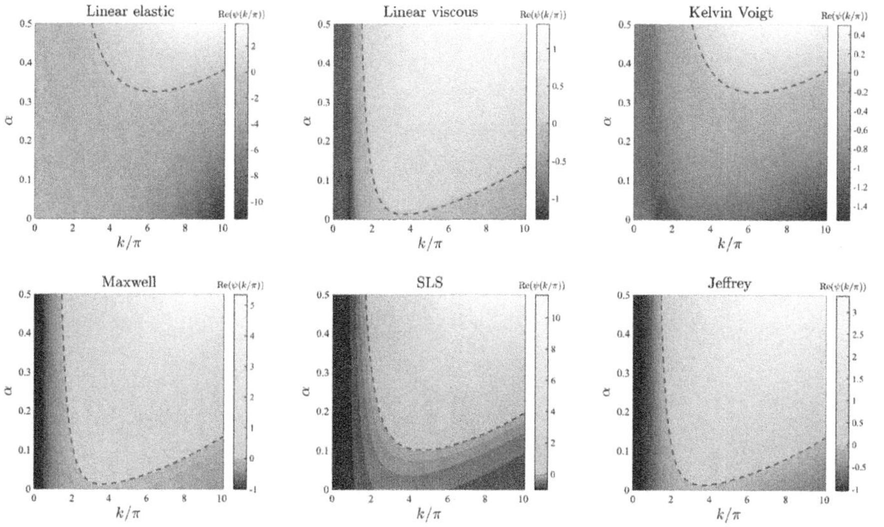

Fig. 10 Effects of varying the cell haptotactic sensitivity. Dispersion relations corresponding to the stress–strain constitutive equations (1)–(6) for increasing values of the cell haptotactic sensitivity, that is for $\alpha \in [0, 0.5]$. The values of the other parameters are given by (26) and (27). Red dashed lines mark contour lines where $\mathrm{Re}(\psi(k^2)) = 0$

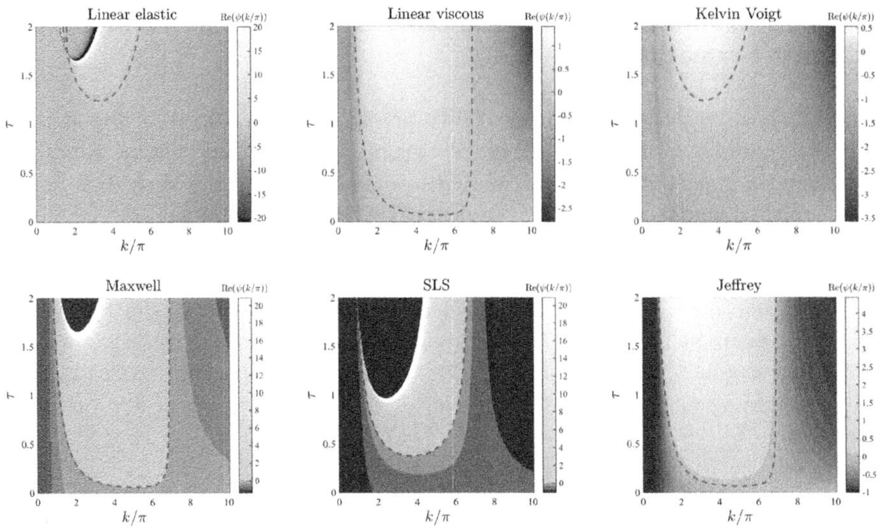

Fig. 11 Effects of varying the cell traction forces. Dispersion relations corresponding to the stress–strain constitutive equations (1)–(6) for increasing cell traction forces, that is for $\tau \in [0, 2]$. The values of the other parameters are given by (26) and (27). White and black regions in the plots related to the linear elastic model, the Maxwell model and the SLS model correspond, respectively, to $\mathrm{Re}(\psi(k^2)) > 20$ and $\mathrm{Re}(\psi(k^2)) < -20$ (*i.e.* a vertical asymptote is present in the dispersion relation). Red dashed lines mark contour lines where $\mathrm{Re}(\psi(k^2)) = 0$

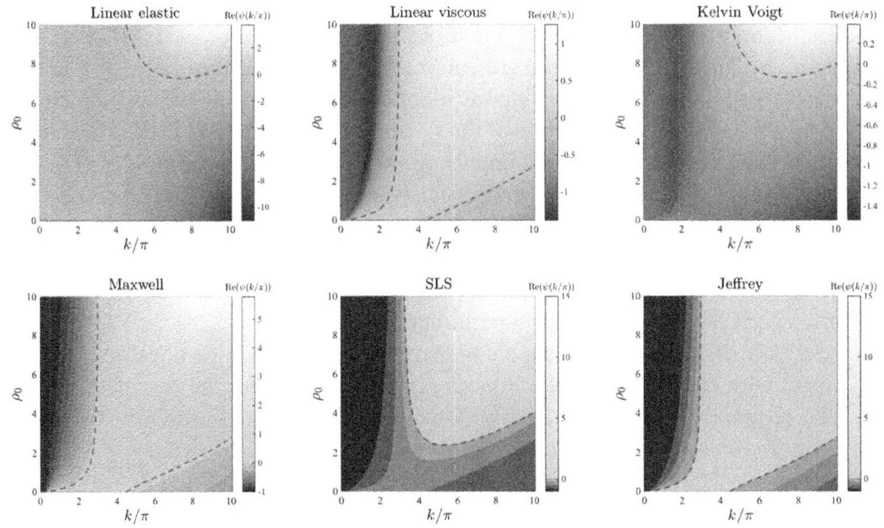

Fig. 12 Effects of varying the initial ECM density. Dispersion relations corresponding to the stress–strain constitutive equations (1)–(6) for increasing values of the initial ECM density, that is for $\rho_0 \in [0, 10]$. The values of the other parameters are given by (26) and (27). Red dashed lines mark contour lines where $\mathrm{Re}(\psi(k^2)) = 0$

5 Numerical Simulations of a One-Dimensional Mechanical Model of Pattern Formation

In this section, we verify key results of LSA presented in Sect. 4 by solving numerically the system of PDEs (10), (11) and (16) subject to boundary conditions (17). In particular, we report on numerical solutions obtained in the case where Eq. (16) is complemented with the Kelvin–Voigt model (3) or the Maxwell model (4). A detailed description of the numerical schemes employed is provided in the Supplementary Material (see file 'SuppInfo').

Set-up of numerical simulations. We carry out numerical simulations using the parameter values given by (26) and (27). We choose the endpoints of the spatial domain to be $\ell = 0$ and $L = 1$, and the final time T is chosen sufficiently large so that distinct spatial patterns can be observed at the end of simulations. We consider the initial conditions

$$n(0, x) = 1 + 0.01\,\epsilon(x)\,, \quad \rho(0, x) \equiv \rho_0\,, \quad u(0, x) \equiv 0\,, \tag{28}$$

where $\epsilon(x)$ is a normally distributed random variable with mean 0 and variance 1 for every $x \in [0, 1]$. Initial conditions (28) model a scenario where random small perturbations are superimposed to the cell density corresponding to the homogeneous steady state of components $\overline{n} = 1, \overline{\rho} = \rho_0$ and $\overline{u} = 0$. This is the steady state considered in the LSA undertaken in Sect. 4.1. Consistent initial conditions for $\partial_t n(0, x)$, $\partial_t \rho(0, x)$ and $\partial_t u(0, x)$ are computed numerically —details provided in the Supplementary Material (see file 'SuppInfo'). Numerical computations are performed in MATLAB.

Main results. The results obtained are summarised by the plots in Fig. 13, together with the corresponding videos provided as supplementary material. The supplementary video 'MovS1' displays the solution of the system of PDEs (10), (11) and (16) subject to the boundary conditions (17) and initial conditions (28) for the Kelvin–Voigt model and the Maxwell model from $t = 0$ until a steady state, displayed in Fig. 13, is reached. The supplementary videos 'MovS2', 'MovS3', and 'MovS4' display the solution of the same system of PDEs for the Maxwell model under alternative initial perturbations in the cell density, *i.e.* randomly distributed ('MovS2'), periodic ('MovS3') or randomly perturbed periodic ('MovS4') initial perturbations.

The results in Fig. 13 and the supplementary video 'MovS1' demonstrate that, in agreement with the dispersion relations displayed in Fig. 2, for the parameter values given by (26) and (27), small randomly distributed perturbations present in the initial cell density:

- vanish in the case of the Kelvin–Voigt model, thus leading the cell density to relax to the homogeneous steady state $\overline{n} = 1$ and attain numerical equilibrium at $t = 100$ while leaving the ECM density unchanged;
- grow in the case of the Maxwell model, resulting in the formation of spatial patterns both in the cell density n and in the ECM density ρ, which attain numerical equilibrium at $t = 500$.

Notice that the formation of spatial patterns correlates with the growth of the cell-ECM displacement u. In fact, the displacement remains close to zero (*i.e.* $\sim O(10^{-11})$)

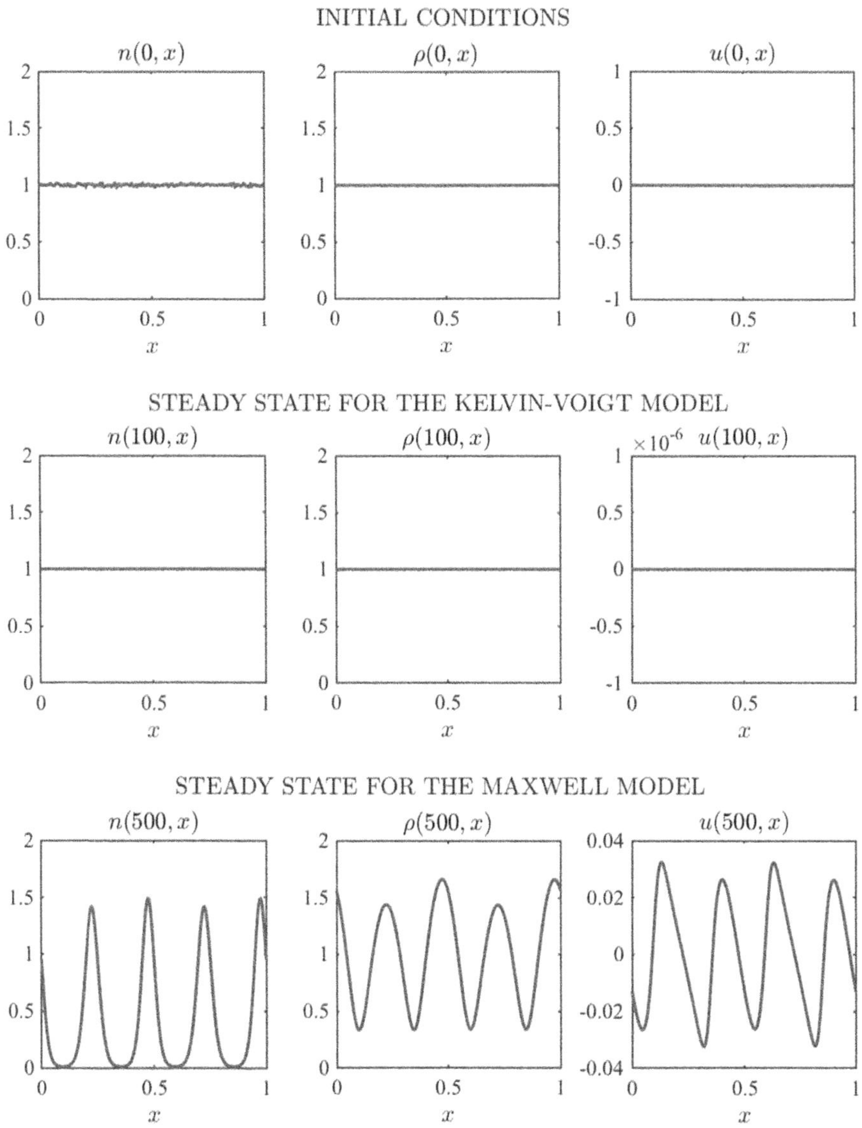

Fig. 13 Simulation results for the Kelvin–Voigt model (3) and the Maxwell model (4) under initial conditions (28). Cell density $n(t, x)$ (left), ECM density $\rho(t, x)$ (centre) and cell-ECM displacement $u(t, x)$ (right) at $t = 0$ (first row) and at steady state obtained solving numerically the system of PDEs (10), (11) and (16) complemented with the Kelvin–Voigt model (3) (second row) and with the Maxwell model (4) (third row), respectively, subject to boundary conditions (17) and initial conditions (28), for the parameter values given by (26) and (27)

for the Kelvin–Voigt model, whereas it grows with time for the Maxwell model. In addition, the steady state obtained for the Maxwell model in Fig. 13, together with those obtained when considering alternative initial perturbations (see supplementary videos 'MovS2', 'MovS3', and 'MovS4'), demonstrate that, in agreement with the dispersion relation displayed in Fig. 2 for the Maxwell model, for the parameter values given by (26) and (27), under small perturbations in the cell density, be they randomly distributed (*cf.* supplementary video 'MovS2'), randomly perturbed periodic (*cf.* supplementary video 'MovS3') or periodic (*cf.* supplementary video 'MovS4'), the fourth mode is the fastest growing one within the range of unstable modes (*cf.* $\mathrm{Re}(\psi(k^2)) > 0$ for k/π between 2 and 6, with $\max\left(\mathrm{Re}(\psi(k^2))\right) \approx 4$ in Fig. 2 for the Maxwell model). In addition, the cellular pattern observed at steady state exhibits 4 large and equally spaced peaks independently of the initial perturbation (*cf.* supplementary videos 'MovS1', 'MovS2', 'MovS3', and 'MovS4'). Moreover, all the obtained cellular patterns at steady state exhibit the same structure—up to a horizontal shift—consisting of four large peaks, independently of the initial conditions that is used (*cf.* left panel in the bottom row of Fig. 13 and supplementary videos 'MovS2', 'MovS3', and 'MovS4'). This indicates robustness and consistency in the nature of the saturated nonlinear steady state under specific viscoelasticity assumptions and parameter choices.

6 Numerical Simulations of a Two-Dimensional Mechanical Model of Pattern Formation

In this section, we complement the results presented in the previous sections with the results of numerical simulations of a two-dimensional mechanical model of pattern formation in biological tissues. In particular, we report on numerical solutions obtained in the case where the two-dimensional analogue of the system of PDEs (10), (11) and (16) is complemented with a two-dimensional version of the one-dimensional Kelvin–Voigt model (3) or a two-dimensional version of the one-dimensional Maxwell model (4).

A two-dimensional mechanical model of pattern formation. The mechanical model of pattern formation defined by the system of PDEs (10), (11) and (12) posed on a two-dimensional spatial domain represented by a bounded set $\Omega \subset \mathbb{R}^2$ with smooth boundary $\partial\Omega$ reads as

$$\begin{cases} \partial_t n = \operatorname{div}\left[D\,\nabla n \; - n\;(\alpha\,\nabla\rho + \partial_t \boldsymbol{u})\right] + r\,n(1-n)\,, \\ \partial_t \rho = -\operatorname{div}(\rho\,\partial_t \boldsymbol{u})\,, \\ \operatorname{div}(\boldsymbol{\sigma}_m + \boldsymbol{\sigma}_c) + \rho \boldsymbol{F} = 0\,, \end{cases} \tag{29}$$

with $t \in (0, T]$, $\boldsymbol{x} = (x_1, x_2)^\intercal \in \Omega$ and $\boldsymbol{u} = (u_1, u_2)^\intercal$. We close the system of PDEs (29) imposing the two-dimensional version of the periodic boundary conditions (17) on $\partial\Omega$. Furthermore, we use the following two-dimensional analogues of

Table 3 Relations between the parameters in the generic two-dimensional stress–strain constitutive equation (31) and those in the two-dimensional constitutive equations for the Kelvin–Voigt model and the Maxwell model.

Generic two-dimensional model	a_1	a_0	b_1	b_0	c_1	c_0
Kelvin-Voigt model	0	$\frac{1}{\eta}$	1	$\frac{E'}{\eta}$	ν'	$\frac{E'\nu'}{\eta}$
Maxwell model	$\frac{1}{E'}$	$\frac{1}{\eta}$	1	0	ν'	0

definitions (13) and (15)

$$\boldsymbol{\sigma}_c := \frac{\tau n}{1+\lambda n^2}\Big(\rho + \beta \Delta \rho\Big)\boldsymbol{I} \quad \text{and} \quad \boldsymbol{F} := -s\,\boldsymbol{u}\,, \tag{30}$$

where $\boldsymbol{I}$ is the identity tensor. Moreover, in analogy with the one-dimensional case, we define the stress tensor $\boldsymbol{\sigma}_m$ via the two-dimensional constitutive model that is used to represent the stress–strain relation of the ECM. In particular, we consider the following generic two-dimensional constitutive equation

$$a_1 \partial_t \boldsymbol{\sigma} + a_0 \boldsymbol{\sigma} = b_1 \partial_t \boldsymbol{\varepsilon} + b_0 \boldsymbol{\varepsilon} + c_1 \partial_t \theta \boldsymbol{I} + c_0 \theta \boldsymbol{I}\,. \tag{31}$$

This constitutive equation, together with the associated parameter choices reported in Table 3, summarises the two-dimensional version of the one-dimensional Kelvin–Voigt model (3) and the two-dimensional version of the one-dimensional Maxwell model (4) that are considered, which are derived in Appendix B.

Here, the strain $\boldsymbol{\varepsilon}(t, \boldsymbol{x})$ and the dilation $\theta(t, \boldsymbol{x})$ are defined in terms of the displacement $\boldsymbol{u}(t, \boldsymbol{x})$ as

$$\boldsymbol{\varepsilon} = \frac{1}{2}\big(\nabla \boldsymbol{u} + \nabla \boldsymbol{u}^{\intercal}\big) \quad \text{and} \quad \theta = \nabla \cdot \boldsymbol{u}\,. \tag{32}$$

Notice that both $\boldsymbol{\varepsilon}$ and θ reduce to $\varepsilon = \partial_x u$ in the one-dimensional case. Amongst the parameters in the stress–strain constitutive equation (31) reported in Table 3 for the two-dimensional Kelvin–Voigt and Maxwell models, η is the shear viscosity,

$$E' := \frac{E}{1+\nu} \quad \text{and} \quad \nu' := \frac{\nu}{1-2\nu}\,, \tag{33}$$

where ν is Poisson's ratio and E is Young's modulus. As clarified in Appendix B, the two-dimensional Maxwell model in the form (31) holds under the simplifying assumption that the quotient between the bulk viscosity and the shear viscosity of the ECM is equal to ν'.

Set-up of numerical simulations. We solve numerically the system of PDEs (29) subject to the two-dimensional version of the periodic boundary conditions (17) and complemented with (30)-(33). Numerical simulations are carried out using the following parameter values

$$E = 1\,,\quad \eta = 1\,,\quad D = 0.01\,,\quad \nu = 0.25\,, \tag{34}$$

$$\alpha = 0.05\,,\quad r = 1\,,\quad s = 10\,,\quad \lambda = 0.5\,,\quad \tau = 0.2\quad \beta = 0.005\,, \tag{35}$$

which are chosen for illustrative purposes and are comparable with nondimensional parameter values that can be found in the extant literature (see Appendix A for further details). We choose $\Omega = [0, 1] \times [0, 1]$ and the final time T is chosen sufficiently large so that distinct spatial patterns can be observed at the end of simulations. We consider first the following two-dimensional analogue of initial conditions (28)

$$n(0, x_1, x_2) = 1 + 0.01\,\epsilon(x_1, x_2)\,,\quad \rho(0, x_1, x_2) \equiv 1\,,\quad \boldsymbol{u}(0, x_1, x_2) \equiv \boldsymbol{0}\,, \tag{36}$$

where $\epsilon(x_1, x_2)$ is a normally distributed random variable with mean 0 and variance 1 for each $(x_1, x_2) \in [0, 1] \times [0, 1]$. Consistent initial conditions for $\partial_t n(0, x_1, x_2)$, $\partial_t \rho(0, x_1, x_2)$ and $\partial_t \boldsymbol{u}(0, x_1, x_2)$ are computed numerically, as similarly done in the one-dimensional case, and numerical computations are performed in MATLAB with a numerical scheme analogous to that employed in the one-dimensional case— details provided in the Supplementary Material (see file 'SuppInfo').

Main results. The results obtained are summarised by the plots in Figures 14 and 15, together with the corresponding videos provided as supplementary material. Solutions of the system of PDEs (29), together with (30)-(33), subject to initial conditions (36) and periodic boundary conditions, for the parameter values given by (34) and (35), are calculated both for the Kelvin–Voigt model (see supplementary video 'MovS5') and the Maxwell model (see supplementary video 'MovS6') according to the parameter changes summarised in Table 3. The randomly generated initial perturbation in the cell density, together with the cell density at $t = 200$ both for the Kelvin–Voigt and the Maxwell model are displayed in Fig. 14, while the solution to the Maxwell model is plotted at a later time in Fig. 15. Overall, these results demonstrate that, in the scenarios considered here, which are analogous to those considered for the corresponding one-dimensional models, small randomly distributed perturbations present in the initial cell density (*cf.* first panel in Fig. 14):

- vanish in the case of the Kelvin–Voigt model, thus leading the cell density to relax to the homogeneous steady state $\overline{n} = 1$ and attain numerical equilibrium at $t = 260$ (*cf.* second panel of Fig. 14) while leaving the ECM density unchanged (see supplementary video 'MovS5');
- grow in the case of the Maxwell model, leading to the formation of spatio-temporal patterns both in the cell density n and in the ECM density ρ (*cf.* third panel of Figure 14, Fig. 15 and supplementary video 'MovS6'), capturing spatio-temporal dynamic heterogeneity arising in the system.

Similarly to the one-dimensional case, the formation of spatial patterns correlates with the growth of the cell-ECM displacement $\boldsymbol{u}$. In fact, the displacement remains close to zero (*i.e.* $\sim O(10^{-11})$) for the Kelvin–Voigt model (see supplementary video 'MovS5'), whereas it grows with time for the Maxwell model (see Figure 15 and supplementary video 'MovS6').

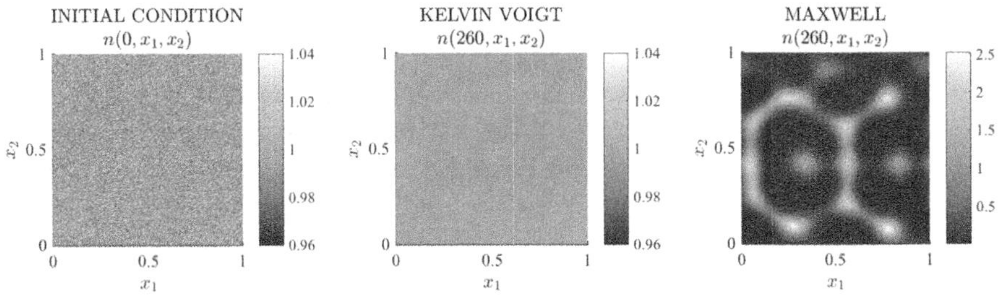

Fig. 14 Simulation results for the two-dimensional Kelvin–Voigt and Maxwell models (31) under initial conditions (36). Cell density $n(t, x_1, x_2)$ at $t = 0$ (left panel) and at $t = 260$ for the Kevin–Voigt model (central panel) and the Maxwell model (right panel) obtained solving numerically the system of PDEs (29) subject to the two-dimensional version of the periodic boundary conditions (17) and initial conditions (36), complemented with (30)–(33), for the parameter values given by (34) and (35)

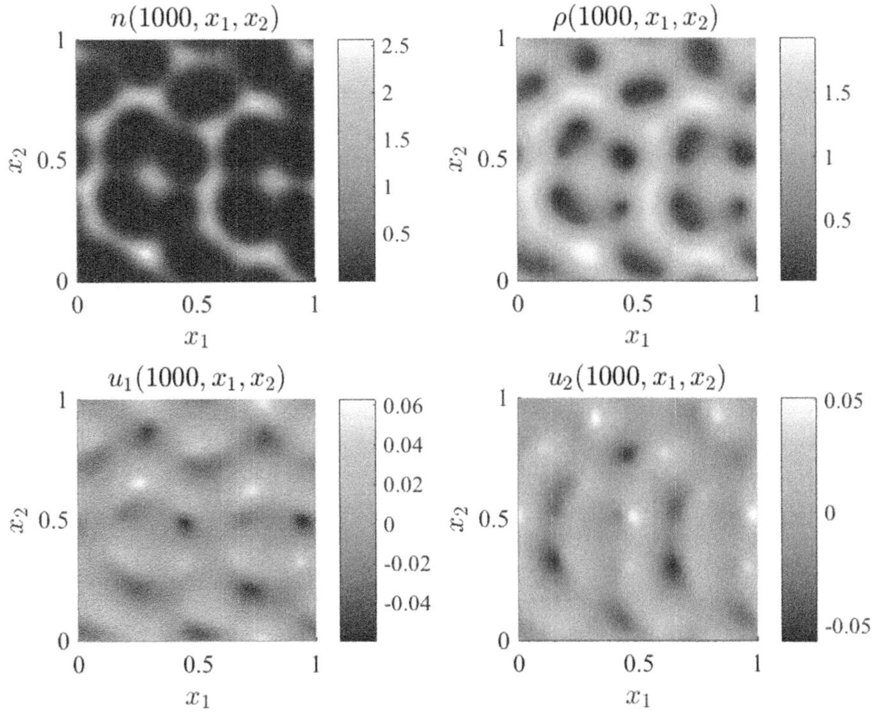

Fig. 15 Simulation results for the two-dimensional Maxwell model (31) under initial conditions (36). Cell density $n(t, x_1, x_2)$ (top row, left panel), ECM density $\rho(t, x_1, x_2)$ (top row, right panel), first and second components of the cell-ECM displacement $\boldsymbol{u}(t, x_1, x_2)$ (bottom row, left panel and right panel, respectively) at $t = 1000$ for the Maxwell model obtained solving numerically the system of PDEs (29) subject to the two-dimensional version of the periodic boundary conditions (17) and initial conditions (36), complemented with (30)–(33), for the parameter values given by (34) and (35). The random initial perturbation of the cell density is displayed in the left panel of Fig. 14

7 Conclusions and Research Perspectives

Conclusions. We have investigated the pattern formation potential of different stress–strain constitutive equations for the ECM within a one-dimensional mechanical model of pattern formation in biological tissues formulated as the system of implicit PDEs (10), (11), and (16).

The results of linear stability analysis undertaken in Sect. 4 and the dispersion relations derived therefrom support the idea that fluid-like stress–strain constitutive equations (*i.e.* the linear viscous model (2), the Maxwell model (4) and the Jeffrey model (6)) have a pattern formation potential much higher than solid-like constitutive equations (*i.e.* the linear elastic model (1), the Kelvin–Voigt model (3) and the SLS model (5)). This is confirmed by the results of numerical simulations presented in Sect. 5, which demonstrate that, all else being equal, spatial patterns emerge in the case where the Maxwell model (4) is used to represent the stress–strain relation of the ECM, while no patterns are observed when the Kelvin–Voigt model (3) is employed. In addition, the structure of the spatial patterns presented in Sect. 5 for the Maxwell model (4) is consistent with the fastest growing mode predicted by linear stability analysis. In Sect. 6, as an illustrative example, we have also reported on the results of numerical simulations of a two-dimensional version of the model, which is given by the system of PDEs (29) complemented with the two-dimensional Kelvin–Voigt and Maxwell models (31). These results demonstrate that key features of spatial pattern formation observed in one spatial dimension carry through when two spatial dimensions are considered, thus conferring additional robustness to the conclusions of our work.

Our findings corroborate the conclusions of Byrne and Chaplain (1996), suggesting that prior studies on mechanochemical models of pattern formation relying on the Kelvin–Voigt model of viscoelasticity may have underestimated the pattern formation potential of biological tissues and advocating the need for further empirical work to acquire detailed quantitative information on the mechanical properties of single components of the ECM in different biological tissues, in order to furnish such models with stress–strain constitutive equations for the ECM that provide a more faithful representation of tissue rheology, *cf.* Fung (1993).

Research perspectives. The dispersion relations given in Sect. 4 indicate that there may be parameter regimes whereby solid-like constitutive models of linear viscoelasticity give rise to dispersion relations which exhibit a range of unstable modes, while the dispersion relations obtained using fluid-like constitutive models exhibit singularities, exiting the regime of validity of linear stability analysis. In this regard, it would be interesting to consider extended versions of the mechanical model of pattern formation defined by the system of PDEs (10), (11) and (16), in order to re-enter the regime of validity of linear stability analysis for the same parameter regimes and verify that in such regimes all constitutive models can produce patterns. For instance, it is known that including long-range effects, such as long-range diffusion or long-range haptotaxis, can promote the formation of stable spatial patterns (Moreo et al. 2010; Oster et al. 1983), which could be explored through nonlinear stability analysis, as previously done for the case in which the stress–strain relation of the ECM is represented by the Kelvin–Voigt model (Cruywagen and Murray 1992; Lewis and Murray 1991; Maini and Murray 1988). In particular, weakly nonlinear analysis could provide information on the existence and stability of saturated nonlinear steady states, supercritical bifurcations or subcritical bifurcations, which may exist even when the homogeneous steady states are stable to small perturbations according to linear stability analysis (Cross and Greenside 2009). Nonlinear analysis would further enable exploring the existence of possible differences in the spatial patterns obtained when different stress–strain constitutive equations for the ECM are used—such as amplitude of patterns, perturbation mode selection, and geometric structure in two spatial dimensions. In particular, the base-case dispersion relations given in Sect. 4 for different fluid-like models of viscoelasticity displayed the same range of unstable modes. This suggests that the investigation of similarities and differences in mode selection between the various models of viscoelasticity could yield interesting results. It would also be interesting to construct numerical solutions for the mechanical model defined by the system of PDEs (10), (11) and (16) complemented with the Jeffrey model (6). For this to be done, suitable extensions of the numerical schemes presented in the Supplementary Material (see file 'SuppInfo') need to be developed.

It would also be relevant to systematically assess the pattern formation potential of different constitutive models of viscoelasticity in two spatial dimensions. This would require to relax the simplifying assumption (A.4) on the shear and bulk viscosities of the ECM, which we have used to derive the two-dimensional Maxwell model in the form of (31), and, more in general, to find analytically and computationally tractable stress–strain-dilation relations, which still remains an open problem (Birman et al. 2002; Haghighi-Yazdi and Lee-Sullivan 2011). In order to solve this problem, new methods of derivation and parameterisation for constitutive models of viscoelasticity might need to be developed (Valtorta and Mazza 2005).

As previously mentioned, the values of the model parameters used in this paper have been chosen for illustrative purposes only. Hence, it would be useful to re-compute the dispersion relations and the numerical solutions presented here for a calibrated version of the model based on real biological data. On a related note, there exists a variety of interesting applications that could be explored by varying parameter values in the generic constitutive equation (16) both in space and time. For instance, cell monolayers appear to exhibit solid-like behaviours on small time scales, whereas they exhibit fluid-like behaviours on longer time scales (Tlili et al. 2018), and spatio-temporal changes in basement membrane components are known to affect structural properties of tissues during development or ageing, as well as in a number of genetic and autoimmune diseases (Khalilgharibi and Mao 2021). Amongst these, remarkable examples are Alport's syndrome, characterised by changes in collagen IV network due to genetic mutations associated with the disease, diabetes mellitus, whereby high levels of glucose induce significant basement membrane turnover, and cancer. In particular, cancer-associated fibrosis is a disease characterised by an excessive production of collagen, elastin, and proteoglycans, which directly affects the structure of the ECM resulting in alterations of viscoelastic tissue properties (Ebihara et al. 2000). Such alterations in the ECM may facilitate tumour invasion and angiogenesis. Considering a calibrated mechanical model of pattern formation in biological tissues, whereby the values of the parameters in the stress–strain constitutive equation for the ECM change during fibrosis progression, may shed new light on the existing connections between structural changes in the ECM components and higher levels of malignancy in cancer (Chandler et al. 2019; Park et al. 2001).

Supplementary Information The online version contains supplementary material available at https://doi.org/10.1007/s11538-021-00912-5.

Acknowledgements The authors wish to thank the two anonymous Reviewers for their thoughtful and constructive feedback on the manuscript. MAJC gratefully acknowledges the support of EPSRC Grant No. EP/S030875/1 (EPSRC SofTMech^MP Centre-to-Centre Award).

Declarations

Conflicts of interest The authors declare that they have no conflict of interest.

A Choice of the Parameter Values for the Baseline Parameter Sets (26)–(27) and (34)–(35)

In order not to limit the conclusions of our work by selecting a specific biological scenario, we identified possible ranges of values for each parameter of our model on the basis of the existing literature on mechanochemical models of pattern formation and then define our baseline parameter set by selecting values in the middle of such ranges. In the sensitivity analysis presented in Section 4.2, we then consider the effect of varying the parameter values within an appropriate range. We first consider the parameters appearing in equations (10), (11) and (16), as well as in the initial conditions (28) and then consider additional parameters appearing in the two-dimensional system (29)–(33), and the associated initial conditions (36).
Parameters in the balance equation (10) Nondimensional parameter values for the cell motility coefficient D in the literature appear as low as $D = 10^{-8}$ (Gilmore et al. 2012), and as high as $D = 10$ (Murray and Oster 1984b), but are generally taken in the range $[10^{-5}, 1]$ (Bentil and Murray 1991; Byrne and Preziosi 2003; Cruywagen and Murray 1992; Ferrenq et al. 1997; Maini et al. 2002; Murray et al. 1988; Namy et al. 2004; Olsen et al. 1995; Perelson et al. 1986). Hence, we take $D = 0.01$ for our baseline parameter set. The nondimensional haptotactic sensitivity of cells α takes values in the range $[10^{-5}, 5]$ (Bentil and Murray 1991; Cruywagen and Murray 1992; Gilmore et al. 2012; Murray et al. 1988; Namy et al. 2004; Olsen et al. 1995; Perelson et al. 1986), and we take $\alpha = 0.05$ for our baseline parameter set. While most authors ignore cell proliferation dynamics, *i.e.* consider $r = 0$ (Ambrosi et al. 2005; Byrne and Preziosi 2003; Gilmore et al. 2012; Murray et al. 1988; Perelson et al. 1986), when present, the rate of cell proliferation takes nondimensional value in the range [0.02, 5]

(Cruywagen and Murray 1992; Olsen et al. 1995; Perelson et al. 1986). Hence, we choose $r = 1$ for our baseline parameter set.

Parameters in the balance equation (11) While no parameters appear in the balance equation (11), the value of the parameter ρ_0 introduced in Sect. 4 as the spatially homogenous steady state $\bar{\rho} = \rho_0$, and successively specified to be the initial ECM density in (28) for our numerical simulations, stems from neglected terms in Eq. (11). With the exception of Cruywagen and Murray (1992) and Maini et al. (2002) who, respectively, have $\rho_0 = 100.2$ and $\rho_0 = 0.1$, this parameter is usually taken to be $\rho_0 = 1$ in mechanochemical models ignoring additional ECM dynamics (Bentil and Murray 1991; Cruywagen and Murray 1992; Harris et al. 1981; Manoussaki 2003; Moreo et al. 2010; Murray and Oster 1984a, b; Olsen et al. 1995; Oster et al. 1983; Perelson et al. 1986).This is generally justified by assuming the steady state ρ_0 of equation (11) that is introduced by the additional term, say $S(n, \rho)$, is itself used to nondimensionalise ρ, before assuming the dynamics modelled by $S(n, \rho)$ to occur on a much slower timescale than convection driven by the cell-ECM displacement, thus neglecting this term (Murray 2001, p.328), resulting in the nondimensional parameter $\hat{\rho}_0 = 1$. Hence, we take $\rho_0 = 1$.

Parameters in the force balance Eq. (16) The elastic modulus, or Young modulus, E is usually itself used to nondimensionalise the other parameters in the dimensional correspondent of equation (16) and, therefore, does not appear in the nondimensional system (Bentil and Murray 1991; Gilmore et al. 2012; Murray and Oster 1984b, b; Murray et al. 1988; Olsen et al. 1995; Perelson et al. 1986). This corresponds to the nondimensional value $E = 1$, which is what we take for our baseline parameter set. The viscosity coefficient η has been taken with nondimensional values in low orders of magnitude, such as $\eta \sim 10^{-3}$–10^{-1} (Bentil and Murray 1991; Cruywagen and Murray 1992; Gilmore et al. 2012; Perelson et al. 1986), as well as in high orders of magnitude, such as $\eta \sim 10^2 - 10^3$ (Gilmore et al. 2012; Olsen et al. 1995). It is, however, generally taken to be $\eta = 1$ (Bentil and Murray 1991; Byrne and Chaplain 1996; Cruywagen and Murray 1992; Murray and Oster 1984b; Murray et al. 1988; Perelson et al. 1986), which is what we choose for our baseline parameter set. When the constitutive model includes two elastic moduli, *i.e.* for the SLS model (5), or two viscosity coefficients, *i.e.* for the Jeffrey model (6), we take $E_1 = E_2 = E/2 = 0.5$ and $\eta_1 = \eta_2 = \eta/2 = 0.5$ as done by Alonso et al. (2017). The cell traction parameter τ takes nondimensional values spanning many orders of magnitude: it can be found as low as $\tau = 10^{-5}$ (Ferrenq et al. 1997) and as high as $\tau = 10$ (Bentil and Murray 1991; Cruywagen and Murray 1992; Perelson et al. 1986), but it is generally taken to be of order $\tau \sim 1$ (Bentil and Murray 1991; Byrne and Chaplain 1996; Gilmore et al. 2012; Murray et al. 1988; Perelson et al. 1986) and many works consider $\tau \sim 10^{-2} - 10^{-1}$ (Byrne and Chaplain 1996; Ferrenq et al. 1997; Murray and Oster 1984b; Olsen et al. 1995). Hence, for our baseline parameter set we choose $\tau = 0.2$. The cell–cell contact inhibition parameter λ generally takes nondimensional values in the range $[10^{-2}, 1]$ (Bentil and Murray 1991; Byrne and Chaplain 1996; Murray et al. 1988; Perelson et al. 1986), so we choose $\lambda = 0.5$ for our baseline parameter set. The long-range cell traction parameter β, when present, takes nondimensional values in the range $[10^{-3}, 10^{-2}]$ (Bentil and Murray 1991; Cruywagen and Murray 1992; Gilmore et al. 2012; Moreo et al. 2010; Murray et al. 1988; Perelson et al. 1986) so we choose $\beta = 0.005$ for our baseline parameter set. The elasticity of the external elastic substratum s, which is sometimes ignored or substituted with a viscous drag, has been taken to have nondimensional values as low as $s \in [10^{-1}, 1]$ (Byrne and Chaplain 1996; Murray and Oster 1984b; Olsen et al. 1995) but is generally chosen in the range $[10, 400]$ (Bentil and Murray 1991; Gilmore et al. 2012; Murray et al. 1988; Perelson et al. 1986). Hence, we take $s = 10$ for our baseline parameter set.

Parameters in the 2D system (29)–(33) For the parameters in the 2D system (29)–(33) and initial condition (36) that also appear in the equations (10), (11), (16) and initial conditions (28), we make use of the same nondimensional values selected in the one-dimensional case (see previous paragraphs). The Poisson ratio ν, which can only take values in the range $[0.1, 9.45]$, has been estimated to be in the range $[0.2, 0.3]$ for the biological tissue considered in mechanochemical models in the current literature (Ambrosi et al. 2005; Cruywagen and Murray 1992; Manoussaki 2003; Moreo et al. 2010). Hence, we choose $\nu = 0.25$ for our baseline parameter set. This results in $E' = E/(1+\nu) = 0.8$ and $\nu' = \nu/(1-2\nu) = 0.5$ according to definitions (33). In addition, under the simplifying assumption (A.4) introduced in Appendix B, the bulk viscosity takes the value $\mu = \nu'\eta = 0.5\eta = 0.5$, which is in agreement with the fact that the bulk and shear viscosities are usually assumed to take values of a similar order of magnitude in the extant literature (Ambrosi et al. 2005; Manoussaki 2003; Moreo et al. 2010; Murray 2003).

B Derivation of the Two-Dimensional Kelvin–Voigt and Maxwell Models (31)

Landau & Lifshitz derived from first principles the stress–strain relations that give the two-dimensional versions of the linear elastic model (1) and of the linear viscous model (2) in isotropic materials (Landau and Lifshitz 1970), which read, respectively, as

$$\boldsymbol{\sigma}_e = \frac{E}{1+\nu}\left(\boldsymbol{\varepsilon}_e + \frac{\nu}{1-2\nu}\theta_e \boldsymbol{I}\right) \quad \text{and} \quad \boldsymbol{\sigma}_v = \eta\, \partial_t \boldsymbol{\varepsilon}_v + \mu\, \partial_t \theta_v \boldsymbol{I}\,. \tag{A.1}$$

Here, E is Young's modulus, ν is Poisson's ratio, $\boldsymbol{I}$ is the identity tensor, η is the shear viscosity and μ is the bulk viscosity. Moreover, $\boldsymbol{\varepsilon}_e$ and θ_e are the strain and dilation under a purely elastic deformation $\boldsymbol{u}_e$ while $\boldsymbol{\varepsilon}_v$ and θ_v are the strain and dilation under a purely viscous deformation $\boldsymbol{u}_v$, which are all defined via (32).

In the case of a linearly viscoelastic material satisfying Kelvin–Voigt model, the two-dimensional analogue of (3) is simply given by

$$\boldsymbol{\sigma} = \boldsymbol{\sigma}_e + \boldsymbol{\sigma}_v = E'\boldsymbol{\varepsilon} + E'\nu'\theta \boldsymbol{I} + \eta\, \partial_t \boldsymbol{\varepsilon} + \mu\, \partial_t \theta \boldsymbol{I}\,. \tag{A.2}$$

Here, E' and ν' are defined via (33) and there is no distinction between the strain or dilation associated with each component (*i.e.* $\boldsymbol{\varepsilon} = \boldsymbol{\varepsilon}_e = \boldsymbol{\varepsilon}_v$ and $\theta = \theta_e = \theta_v$), as the viscous and elastic components are connected in parallel. This is the stress–strain constitutive equation that is typically used to describe the contribution to the stress of the cell-ECM system coming from the ECM in two-dimensional mechanochemical models of pattern formation (Cruywagen and Murray 1992; Ferrenq et al. 1997; Javierre et al. 2009; Maini and Murray 1988; Manoussaki 2003; Murray 2001; Murray et al. 1988; Murray and Oster 1984a, b; Murray et al. 1983; Olsen et al. 1995; Oster et al. 1983; Perelson et al. 1986).

On the other hand, deriving the two-dimensional analogues of Maxwell model (4), of the SLS model (5) and of the Jeffrey model (6) is more complicated due to the presence of elements connected in series. In the case of Maxwell model, using the fact that the overall strain and dilation will be distributed over the different components (*i.e.* $\boldsymbol{\varepsilon} = \boldsymbol{\varepsilon}_e + \boldsymbol{\varepsilon}_v$ and $\theta = \theta_e + \theta_v$) along with the fact that the stress on each component will be the same as the overall stress (*i.e.* $\boldsymbol{\sigma} = \boldsymbol{\sigma}_e = \boldsymbol{\sigma}_v$), one finds

$$\frac{1}{\eta}\boldsymbol{\sigma} + \frac{1}{E'}\partial_t \boldsymbol{\sigma} = \partial_t \boldsymbol{\varepsilon} + \nu' \partial_t \theta \boldsymbol{I} + \left(\frac{\mu}{\eta} - \nu'\right) \partial_t \theta_v \mathbf{I}, \tag{A.3}$$

with E' and ν' being defined via (33). Under the simplifying assumption that

$$\frac{\mu}{\eta} = \nu' \tag{A.4}$$

the stress–strain constitutive equation (A.3) can be rewritten in the form given by the generic two-dimensional constitutive equation (31) under the parameter choices reported in Table 3. Dividing (A.2) by η, under the simplifying assumption (A.4), the stress–strain constitutive equation for the Kelvin–Voigt model (A.2) can be rewritten as

$$\frac{1}{\eta}\boldsymbol{\sigma} = \frac{E'}{\eta}\boldsymbol{\varepsilon} + \frac{E'\nu'}{\eta}\theta \boldsymbol{I} + \partial_t \boldsymbol{\varepsilon} + \nu' \, \partial_t \theta \boldsymbol{I} \,,$$

which is in the form given by the generic two-dimensional constitutive equation (31) under the parameter choices reported in Table 3.

References

Alonso S, Radszuweit M, Engel H, Bär M (2017) Mechanochemical pattern formation in simple models of active viscoelastic fluids and solids. J Phys D Appl Phys 50(43):434004

Ambrosi D, Bussolino F, Preziosi L (2005) A review of vasculogenesis models. J Theor Med 6(1):1–19

Bard J, Lauder I (1974) How well does turing's theory of morphogenesis work? J Theor Biol 45(2):501–531

Barocas VH, Moon AG, Tranquillo RT (1995) The fibroblast-populated collagen microsphere assay of cell traction force–part 2: measurement of the cell traction parameter. J Biomech Eng 117(2):161–170

Barocas, V. H., Tranquillo, R. T., 1994. Biphasic theory and in vitro assays of cell-fibril mechanical interactions in tissue-equivalent gels. In: Cell Mechanics and Cellular Engineering. Springer, pp 185–209

Bentil DE, Murray JD (1991) Pattern selection in biological pattern formation mechanisms. Appl Math Lett 4(3):1–5

Bilston LE, Liu Z, Phan-Thien N (1997) Linear viscoelastic properties of bovine brain tissue in shear. Biorheology 34(6):377–385

Birman VB, Binienda WK, Townsend G (2002) 2d maxwell model. J Macromol Sci Part B 41(2):341–356

Bischoff JE, Arruda EM, Grosh K (2004) A rheological network model for the continuum anisotropic and viscoelastic behavior of soft tissue. Biomech Model Mechanobiol 3(1):56–65

Brinkmann F, Mercker M, Richter T, Marciniak-Czochra A (2018) Post-turing tissue pattern formation: advent of mechanochemistry. PLoS Comput Biol 14(7):e1006259
Byrne H, Preziosi L (2003) Modelling solid tumour growth using the theory of mixtures. Math Med Biol J IMA 20(4):341–366
Byrne HM, Chaplain MAJ (1996) The importance of constitutive equations in mechanochemical models of pattern formation. Appl Math Lett 9(6):85–90
Castets V, Dulos E, Boissonade J, De Kepper P (1990) Experimental evidence of a sustained standing turing-type nonequilibrium chemical pattern. Phys Rev Lett 64(24):2953
Chandler C, Liu T, Buckanovich R, Coffman L (2019) The double edge sword of fibrosis in cancer. Trans Res 209:55–67
Cross M, Greenside H (2009) Pattern formation and dynamics in nonequilibrium systems. Cambridge University Press, Cambridge
Cruywagen GC, Murray JD (1992) On a tissue interaction model for skin pattern formation. J Nonlinear Sci 2(2):217–240
Ebihara T, Venkatesan N, Tanaka R, Ludwig MS (2000) Changes in extracellular matrix and tissue viscoelasticity in bleomycin-induced lung fibrosis: Temporal aspects. Am J Respirat Crit Care Med 162(4):1569–1576
Ferrenq I, Tranqui L, Vailhe B, Gumery PY, Tracqui P (1997) Modelling biological gel contraction by cells: mechanocellular formulation and cell traction force quantification. Acta Biotheoretica 45(3–4):267–293
Findley WN, Lai JS, Onaran K (1976) Creep and relaxation of nonlinear viscoelastic materials—with an introduction to linear viscoelasticity. Dover Publications
Fung YC (1993) Biomechanics mechanical properties of living tissues, 2nd edn. Springer, New York
Gierer A, Meinhardt H (1972) A theory of biological pattern formation. Kybernetik 12:30–39
Gilmore SJ, Vaughan BL Jr, Madzvamuse A, Maini PK (2012) A mechanochemical model of striae distensae. Math Biosci 240(2):141–147
Haghighi-Yazdi M, Lee-Sullivan P (2011) Modeling linear viscoelasticity in glassy polymers using standard rheological models
Harris AK (1984) Tissue culture cells on deformable substrata: biomechanical implications. J Biomech Eng 106(1):19–24
Harris AK, Stopak D, Wild P (1981) Fibroblast traction as a mechanism for collagen morphogenesis. Nature 290(5803):249–251
Huang Y-P, Zheng Y-P, Leung S-F (2005) Quasi-linear viscoelastic properties of fibrotic neck tissues obtained from ultrasound indentation tests in vivo. Clin Biomech 20(2):145–154
Javierre E, Moreo P, Doblaré M, García-Aznar JM (2009) Numerical modeling of a mechano-chemical theory for wound contraction analysis. Int J Solids Struct 46(20):3597–3606
Jernvall J, Newman SA et al (2003) Mechanisms of pattern formation in development and evolution. Development 130(10):2027–2037
Khalilgharibi N, Mao Y (2021) To form and function: on the role of basement membrane mechanics in tissue development, homeostasis and disease. Open Biol 11(2):200360
Kondo S, Asai R (1995) A reaction-diffusion wave on the skin of the marine angelfish pomacanthus. Nature 376(6543):765–768
Landau LD, Lifshitz EM (1970) Theory of elasticity. Pergamon Press, Cambridge
Lewis MA, Murray JD (1991) Analysis of stable two-dimensional patterns in contractile cytogel. J Nonlinear Sci 1(3):289–311
Liu Z, Bilston L (2000) On the viscoelastic character of liver tissue: experiments and modelling of the linear behaviour. Biorheology 37(3):191–201
Maini P, Painter K, Chau HP (1997) Spatial pattern formation in chemical and biological systems. J Chem Soc Faraday Trans 93(20):3601–3610
Maini PK (2005) Morphogenesis, biological. In: Scott A (ed) Morphogenesis. Routledge, Biological, Encyclopedia of Nonlinear Science, pp 587–589
Maini PK, Murray JD (1988) A nonlinear analysis of a mechanical model for biological pattern formation. SIAM J Appl Math 48(5):1064–1072
Maini PK, Olsen L, Sherratt JA (2002) Mathematical models for cell-matrix interactions during dermal wound healing. Int J Bifurc Chaos 12(09):2021–2029
Maini PK, Woolley TE (2019) The Turing model for biological pattern formation. In: The dynamics of biological systems. Springer, pp 189–204

Maini PK, Woolley TE, Baker RE, Gaffney EA, Lee SS (2012) Turing's model for biological pattern formation and the robustness problem. Interface Focus 2(4):487–496

Manoussaki D (2003) A mechanochemical model of angiogenesis and vasculogenesis. ESAIM Math Modell Numer Anal 37(4):581–599

Mase GE (1970) Continuum mechanics. McGraw-Hill, New York

Meinhardt H (1982) Models of biological pattern formation. Academic Press, London

Meinhardt H (2009) The algorithmic beauty of sea shells. Springer, Berlin

Moon AG, Tranquillo RT (1993) Fibroblast-populated collagen microsphere assay of cell traction force: Part 1. continuum model. AIChE J 39(1):163–177

Moreo P, Gaffney EA, Garcia-Aznar JM, Doblaré M (2010) On the modelling of biological patterns with mechanochemical models: insights from analysis and computation. Bull Math Biol 72(2):400–431

Murray JD (1981) A pre-pattern formation mechanism for animal coat markings. J Theor Biol 88:161–199

Murray JD (2001) Mathematical biology. II Spatial models and biomedical applications {Interdisciplinary Applied Mathematics V. 18}. Springer, New York Incorporated New York

Murray JD (2003) On the mechanochemical theory of biological pattern formation with application to vasculogenesis. Comptes Rendus Biologies 326(2):239–252

Murray JD, Maini PK (1986) A new approach to the generation of pattern and form in embryology. Scientific Programme, Oxford

Murray JD, Maini PK (1989) Pattern formation mechanisms–a comparison of reaction-diffusion and mechanochemical models. In: Cell to Cell Signalling. Elsevier, pp 159–170

Murray JD, Maini PK, Tranquillo RT (1988) Mechanochemical models for generating biological pattern and form in development. Phys Rep 171(2):59–84

Murray JD, Oster GF (1984a) Cell traction models for generating pattern and form in morphogenesis. J Math Biol 19(3):265–279

Murray JD, Oster GF (1984b) Generation of biological pattern and form. Math Med Biol J IMA 1(1):51–75

Murray JD, Oster GF, Harris AK (1983) A mechanical model for mesenchymal morphogenesis. J Math Biol 17(1):125–129

Namy P, Ohayon J, Tracqui P (2004) Critical conditions for pattern formation and in vitro tubulogenesis driven by cellular traction fields. J Theor Biol 227(1):103–120

Nargess K, Yanlan M (2021) To form and function: on the role of basement membrane mechanics in tissue development, homeostasis and disease. Open Biol 11(2):200360

Nasseri S, Bilston LE, Phan-Thien N (2002) Viscoelastic properties of pig kidney in shear, experimental results and modelling. Rheologica Acta 41(1–2):180–192

Olsen L, Sherratt JA, Maini PK (1995) A mechanochemical model for adult dermal wound contraction and the permanence of the contracted tissue displacement profile. J Theor Biol 177(2):113–128

Oster GF, Murray JD, Harris AK (1983) Mechanical aspects of mesenchymal morphogenesis. Development 78(1):83–125

Park J, Kim DS, Shim TS, Lim CM, Koh Y, Lee SD, Kim WS, Kim WD, Lee JS, Song KS (2001) Lung cancer in patients with idiopathic pulmonary fibrosis. Eur Respirat J 17(6):1216–1219

Perelson AS, Maini PK, Murray JD, Hyman JM, Oster GF (1986) Nonlinear pattern selection in a mechanical model for morphogenesis. J Math Biol 24(5):525–541

Petrolli V, Le Goff M, Tadrous M, Martens K, Allier C, Mandula O, Hervé L, Henkes S, Sknepnek R, Boudou T et al (2019) Confinement-induced transition between wavelike collective cell migration modes. Phys Rev Lett 122(16):168101

Scianna M, Bell CG, Preziosi L (2013) A review of mathematical models for the formation of vascular networks. J Theor Biol 333:174–209

Serra-Picamal X, Conte V, Vincent R, Anon E, Tambe DT, Bazellieres E, Butler JP, Fredberg JJ, Trepat X (2012) Mechanical waves during tissue expansion. Nat Phys 8(8):628–634

Snedeker JG, Niederer P, Schmidlin F, Farshad M, Demetropoulos C, Lee J, Yang K (2005) Strain-rate dependent material properties of the porcine and human kidney capsule. J Biomech 38(5):1011–1021

Thompson DW (1917) On growth and form. Cambridge University Press, Cambridge

Tlili S, Gauquelin E, Li B, Cardoso O, Ladoux B, Delanoë-Ayari H, Graner F (2018) Collective cell migration without proliferation: density determines cell velocity and wave velocity. R Soc Open Sci 5(5):172421

Tosin A, Ambrosi D, Preziosi L (2006) Mechanics and chemotaxis in the morphogenesis of vascular networks. Bull Math Biol 68(7):1819–1836

Tranqui L, Tracqui P (2000) Mechanical signalling and angiogenesis. the integration of cell–extracellular matrix couplings. Comptes Rendus de l'Académie des Sciences-Series III-Sciences de la Vie 323 (1), 31–47
Tranquillo RT, Murray JD (1992) Continuum model of fibroblast-driven wound contraction: inflammation-mediation. J Theor Biol 158(2):135–172
Turing AM (1952) The chemical basis of morphogenesis. Philos Trans R Soc London B 237:37–72
Urdy S (2012) On the evolution of morphogenetic models: mechano-chemical interactions and an integrated view of cell differentiation, growth, pattern formation and morphogenesis. Biol Rev 87(4):786–803
Valtorta D, Mazza E (2005) Dynamic measurement of soft tissue viscoelastic properties with a torsional resonator device. Med Image Anal 9(5):481–490
Van Helvert S, Storm C, Friedl P (2018) Mechanoreciprocity in cell migration. Nat Cell Biol 20(1):8–20
Verdier C (2003) Rheological properties of living materials from cells to tissues. Comput Math Methods Med 5(2):67–91

Bulletin of Mathematical Biology (2021) 83:73
https://doi.org/10.1007/s11538-021-00877-5

Society for Mathematical Biology

SPECIAL ISSUE: CELEBRATING J. D. MURRAY

Generalized Stoichiometry and Biogeochemistry for Astrobiological Applications

Christopher P. Kempes[1] · Michael J. Follows[2] · Hillary Smith[3] · Heather Graham[4,6] · Christopher H. House[3] · Simon A. Levin[5]

Received: 1 October 2020 / Accepted: 25 February 2021 / Published online: 18 May 2021

Tribute to J.D. Murray

James Murray is a true giant in applied mathematics, and especially in mathematical biology. His career has spanned an enormous range of topics, and his monumental "Mathematical Biology" is one of the foundational works in the field. His brilliant career has been influenced heavily by his research, pedagogy, mentorship, and collegiality. We are delighted at the chance to dedicate this paper to Jim.

Abstract

A central need in the field of astrobiology is generalized perspectives on life that make it possible to differentiate abiotic and biotic chemical systems McKay (2008). A key component of many past and future astrobiological measurements is the elemental ratio of various samples. Classic work on Earth's oceans has shown that life displays a striking regularity in the ratio of elements as originally characterized by Redfield (Redfield 1958; Geider and La Roche 2002; Eighty years of Redfield 2014). The body of work since the original observations has connected this ratio with basic ecological dynamics and cell physiology, while also documenting the range of elemental ratios found in a variety of environments. Several key questions remain in considering how to best apply this knowledge to astrobiological contexts: How can the observed variation of the elemental ratios be more formally systematized using basic biological physiology and ecological or environmental dynamics? How can these elemental ratios be generalized beyond the life that we have observed on our own planet? Here, we expand recently developed generalized physiological models (Kempes et al. 2012,

✉ Christopher P. Kempes
ckempes@santafe.edu

1 The Santa Fe Institute, Santa Fe, NM, USA

2 Department of Earth, Atmospheric, and Planetary Sciences, Massachusetts Institute of Technology, Cambridge, MA, USA

3 Department of Geosciences, Pennsylvania State University, University Park, PA, USA

4 NASA Goddard Spaceflight Center, Greenbelt, MD, USA

5 Department of Ecology and Evolutionary Biology, Princeton University, Princeton, NJ, USA

6 Catholic University of America, Washington, DC, USA

2016, 2017, 2019) to create a simple framework for predicting the variation of elemental ratios found in various environments. We then discuss further generalizing the physiology for astrobiological applications. Much of our theoretical treatment is designed for *in situ* measurements applicable to future planetary missions. We imagine scenarios where three measurements can be made—particle/cell sizes, particle/cell stoichiometry, and fluid or environmental stoichiometry—and develop our theory in connection with these often deployed measurements.

Introduction

Since the recognition that life on Earth is characterized by a striking regularity in the ratio of elements, as originally characterized by Redfield (Redfield 1958; Geider and La Roche 2002; Eighty years of Redfield 2014), stoichiometric ratios have been a primary target of astrobiological measurements and theories (Elser 2003; Young et al. 2014). From an astrobiological perspective the natural questions that emerge are how much variation exists in these ratios across the range of environments and biological diversity on Earth, how different ratios could have been in time, how different they could be for non-Terran life, and how they depend on planetary composition (Elser 2003; Young et al. 2014; Anbar 2008; Chopra and Lineweaver 2008; Lineweaver and Chopra 2012; Neveu et al. 2016; Wang et al. 2018; Geider and La Roche 2002). On Earth the Redfield ratio is known to vary significantly due to both environmental and physiological effects that have been considered in ecological and biogeochemical theories (e.g. Geider and La Roche 2002; Klausmeier et al. 2004a, b, 2008; Loladze and Elser 2011; Neveu et al. 2016; Sterner et al. 2008; Vrede et al. 2004; Elser et al. 2000; Kerkhoff et al. 2005; Elser et al. 2010; Liefer et al. 2019; Finkel et al. 2016a, b). For life with a different evolutionary history we need new approaches that are able to generalize organism physiology and define when the stoichiometric ratios associated with life are distinct and distinguishable from the environment.

Our general approach here is to first focus on the macromolecules and physiology shared by all of life on Earth. For the macromolecules we are interested in components like proteins, nucleic acids, and cell membranes. For the shared physiology we consider processes such as growth rates, nutrient uptake, and nutrient storage, some of which are derivable from the macromolecular composition of cells. In thinking about the applicability of these two perspectives to life anywhere in the universe it is important to note that the specific set of macromolecules might vary significantly while the general physiological processes might be more conserved. However, our treatment of the macromolecules is easily generalized if one makes two assumptions: 1) that life elsewhere shares a set of macromolecules, even if that set is very different from Terran life, and 2) that those macromolecules fall along systematic scaling relationships. Throughout this paper we operate within these two assumptions and first address the observation and implications of (2), before moving on to a general treatment of physiological scaling which abstracts the underlying details of (1). Throughout we go back and forth between the patterns observed across single organisms of different size and the aggregate results for entire ecosystems composed of diverse organisms, which we characterize by a distribution of cell sizes.

We first consider how to systematize stoichiometry across the diversity of microbial life using scaling laws based on cell size. We then combine these with abundance distributions to obtain a simple perspective on the bulk stoichiometry expected for a population of various cell sizes, and we demonstrate the impact that size distributions can have on these bulk stoichiometries. We then turn to a simple chemostat model of biogeochemistry where nutrients flow into an environment and interact with cellular physiology. Here, we consider the differences in cellular and fluid stoichiometry in an ecosystem composed first of a single cell size, and then of many cell sizes. This approach relies on the scaling of bulk physiological characteristics, such as nutrient quotas, with cell size, and we end by generalizing the exponents of these scaling relationships and showing the consequences this has on differences between the particulate and fluid stoichiometry. Throughout we discuss the general signatures of life that exist at either the cell and ecosystem level.

Deriving Elemental Ratios Across Cell Size

Our interest here is in generalizing organism physiology and connecting it to stoichiometric ratio measurements that could be performed as part of astrobiological explorations of other planets using current or near-future instrumentation. Stoichiometry could be used as a relatively simple biosignature and, when considered within the context of the stoichiometry of the environment surrounding the particle/cell, could serve as a universal or agnostic biosignature. Agnostic biosignatures aim to identify patterns of living systems that may not necessarily share the same biochemical machinery as life on Earth. The need for reliable agnostic biosignatures increases as we examine planets deeper in the Solar System where common heritage with life on Earth is less likely.

Recently a variety of biological regularities have been discovered for life on Earth that show that organism physiology can be characterized by systematic trends across diverse organisms (e.g. Andersen et al. 2016; Brown et al. 2004; West and Brown 2005; Savage et al. 2004). These trends are often power-law relationships between organism size and a variety of physiological and metabolic features, and are derivable from a small set of physical and biological constraints (Kempes et al. 2019). Both physiological features and bulk organism stoichiometries have been previously shown to follow allometric scaling relationships for diverse organisms ranging from bacteria to multicellular plants (Elser et al. 1996, 2000; Vrede et al. 2004; Kerkhoff et al. 2005; Elser et al. 2010; DeLong et al. 2010; Kempes et al. 2012; Edwards et al. 2012; Kempes et al. 2016, 2017; Finkel et al. 2004, 2016a, b; Finkel 2001; Beardall et al. 2009; Tang 1995; West and Brown 2005), and so the allometric perspective taken here on stoichiometry could be applied to many levels of biological organization including entire ecosystems (e.g. Elser et al. 2000). Intuitively, these relationships can be viewed as the optimization of physiological function under fixed constraints through evolutionary processes (Kempes et al. 2019). As such, in many contexts these scaling relationships may represent universal relationships connected to fundamental physical laws such as diffusive constraints. However, in many cases the cross-species scaling may reflect emergent and interconnected constraints of the physiology itself or of evolutionary history and contingency, in which case we might expect these scaling relationships to vary across life on diverse worlds. For example, changes in the network architecture of the metabolism with size (Kim et al. 2019) could be governed by the likelihood of cross-reactivity between molecules, which could depend on what types of molecules are being employed. In general, the possibility of contingent and emergent constraints is an important consideration for astrobiology.

In microbial life, classic and recent work has systematized macromolecular abundances in terms of key properties of organisms such as overall growth rate or cell size (e.g. Shuler et al. 1979; Vrede et al. 2004; Loladze and Elser 2011; DeLong et al. 2010; Kempes et al. 2012; Edwards et al. 2012; Kempes et al. 2016, 2017; Savage et al. 2004; Tang 1995; West and Brown 2005). For example the connection between cellular growth rate and RNA and protein abundances has long been documented with various models proposing mechanisms for predicting these trends (Shuler et al. 1979; Klausmeier et al. 2004b; Vrede et al. 2004; Loladze and Elser 2011; Kempes et al. 2016). Here, we rely on work that has systematized various physiological processes and interconnections among macromolecular abundance in terms of cell size (e.g. DeLong et al. 2010; Kempes et al. 2012, 2016; Finkel et al. b; Tang 1995; West and Brown 2005), where often the interconnection between features can be systematically derived. For example, models have derived the dependence of growth rate on cell size from the cross-species scaling of metabolic rate with cell size (Kempes et al. 2012), and in turn, the ribosomal requirements given this growth rate and the scaling of protein abundance (Kempes et al. 2016). Not all of these scaling laws are understood from first principles, but they do provide a way to systematically determine macromolecular abundances from organism size. For example, bacteria follow a systematic set of scaling relationships where protein concentrations are decreasing with increasing cell size and RNA components are increasing in concentration (Kempes et al. 2016).

From the broad set of macromolecular scaling relationships it is possible to derive the elemental ratio of a cell of a given size simply by considering the abundance and elemental composition of each component. The elemental ratio of the entire ecosystem is then found by considering the size distribution of organisms.

We calculate the total elemental abundances for a cell by knowing the elemental composition of a component, c_i (e.g. N/protein), and the total quantity of that component, $n_i\left(V_c\right)$, in a cell of a given size V_c. The total abundance of one element, E $(mol/cell)$, is equal to the sum across all cellular components

$$E\left(V_c\right)=\sum_i c_i n_i\left(V_c\right), \tag{1}$$

where the components are major categories of macromolecules such as proteins, ribosomes, and mRNA. Each of these components has a known scaling with cell size given in Box 1. As an example, the total nitrogen content in bacteria is given by

$$\begin{aligned}E\left(V_c\right)= & c_{N,p} n_{protein}+c_{N,DNA} n_{DNA}+c_{N,mRNA} n_{mRNA} \\ & +c_{N,tRNA} n_{tRNA}+c_{N,ribo} n_{ribosomes}+c_{N,l} n_l+c_{N,e} n_e\end{aligned} \tag{2}$$

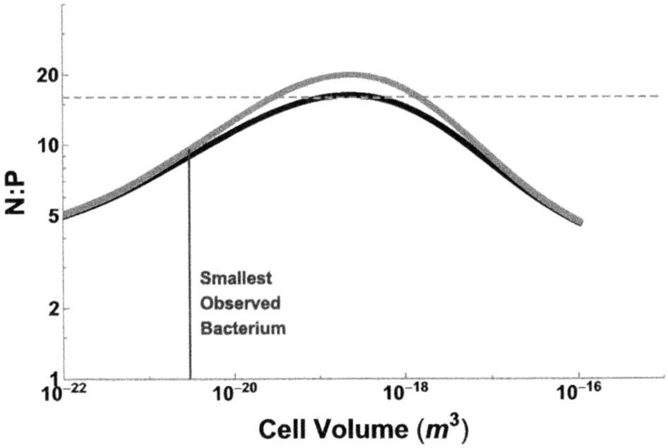

Fig. 1 Elemental ratios as a function of bacterial cell size showing a non-constant stoichiometry that often differs from the Redfield ratio (e.g. N:P of 16:1 indicated by the dashed line) for many cell sizes. The black curve is for gram-negative bacteria and gray is for gram-positive bacteria

where $c_{N,p}$ is the average number of N in protein, $c_{N,DNA}$, $c_{N,mRNA}$, $c_{N,tRNA}$, and $c_{N,ribo}$ are the average N in various types of DNA and RNA, $c_{N,l}$ is the N in lipids, and $c_{N,e}$ is the N in energy storage molecules such as ATP and carbohydrates. The counts of the macromolecules are given by $n_{protein}$, $n_{ribosomes}$, n_{DNA},n_{tRNA}, n_{mRNA}, n_l, and n_e which represent the numbers of proteins, ribosomes, DNA, tRNA, mRNA, lipids, and energy storage molecules in the cell, all of which depend on cell size (Box 1). For our analysis here we focus on N:P as an illustrative case, and thus typically ignore carbohydrates and lipids, which are minor cellular sources of these elements.

Using typical values for the elemental composition of each component Geider and La Roche (2002), Fig. 1 gives the ratio of elements with overall cell size. This result shows that the elemental ratios agree with Redfield for some cell sizes but deviate significantly for most bacterial cell sizes. Both small and large bacterial cells have a decreased ratio of N to P compared with the Redfield ratio. It should be noted that the Redfield ratio is known to vary widely, and do so in ways that are ecologically meaningful from a resource competition perspective (e.g. Geider and La Roche 2002; Klausmeier et al. 2004b, a, 2008). We discuss these points in greater detail below.

These observations also show that one possible agnostic biosignature is non-constant elemental ratios across particle sizes. This is the result of evolution optimizing organism physiology at different scales (Kempes et al. 2019) which will lead to different ratios of macromolecules and thus different elemental ratios at each particle size. This is true even when the set of macromolecules is largely conserved across many species but the relative ratio of these macromolecules changes due to scaling laws with cell size, as is the case with ribosomes, DNA, and proteins. The strong and consistent trend of elemental ratios with cell size should be distinctly different from the patterns of abiotic particles.

Within this overall framework it is important to consider the assumptions made by a particular model as differences in these assumptions will give rise to a variety of scaling relationships for macromolecular abundances with cell size or growth rate. For example, considering models of RNA and protein abundance, the set of past models often focuses on the tradeoffs and interconnected requirements for ribosomes and all other functional proteins (Shuler et al. 1979; Klausmeier et al. 2004b; Loladze and Elser 2011; Kempes et al. 2016). Klausmeier et al. consider the tradeoffs associated with the investment in resource acquisition or biosynthesis out of a fixed abundance of proteins in a model that couples physiology to the environment (Klausmeier et al. 2004b). This model shows that there is a different optimum for the number of ribosomes under exponential growth compared with a population that is at competitive equilibrium. Loladze and Elser consider exponential growth and define a reciprocal feedback between ribosomes and proteins, where RNA drives the rate of protein synthesis, and protein abundance drives the rate of rRNA production through RNA polymerase (Loladze and Elser 2011). This reciprocal dynamic leads to the prediction of a single homeostatic ratio of protein:rRNA, which can be calculated from biochemical parameters and where the prediction agrees with the data for several species. Kempes et al. focus on the requirement that the ribosomes replicate all proteins (including ribosomal proteins) in the time that the cell divides and takes the cellular growth rate and protein abundances (both of which systematically scale with cell size) as inputs to predict the ribosome requirement (Kempes et al. 2016). This result differs from Loladze and Elser in that it allows for a non-constant protein:rRNA ratio that depends on the distinct scaling of growth rate and protein abundance, where it is important to note that the total quantity of RNA polymerase and total quantity of all proteins could each have a distinct scaling with cell size.

For future work aimed at building general models of cell physiologies for astrobiology it is important to consider both how differences in assumptions and model complexity – which could range from the simple coupled dynamics of protein and RNA production, to whole-cell models which consider much more complicated interconnections amongst transport, metabolic, and synthesis processes (e.g. Shuler et al. 1979; Karr et al. 2012) – will lead to different predictions. For our purposes here it is sufficient to rely on models, or empirical descriptions, that match the known interspecific scaling in macromolecular abundance.

It should also be noted from a practical perspective that sampling issues may still exist. For example, it can be hard to separate biotic from abiotic particles in the Earth's oceans using known devices (Andersson and Rudehäll 1993). However, the stoichiometry of these particles once sorted are expected to radically differ, which should be systematically verified. Addressing these issues is an important topic of future work. In addition, it is important to note that these results are based on the macromolecular abundances of cells growing at maximum rate under optimal nutrient conditions, and cells are known to respond to environmental conditions by shifting macromolecular ratios and elemental abundances (Elrifi and Turpin 1985; Healey 1985; Rhee 1978). We address these processes of acclimation in our coupled biogeochemical model.

Box 1 Equations governing macromolecular content in cells

Many features of the cell have been previously shown to scale with overall cell size Kempes et al. (2016). The scaling relationships for counts of the main macromolecular components follow

$$n_{protein} = p_0 V_c^{\beta_p} \tag{3}$$

$$n_{DNA} = d_0 V_c^{\beta_d} \tag{4}$$

$$n_{ribo} = \frac{\bar{l}_p n_{protein} \left(\frac{\phi}{\mu} + 1\right)}{\frac{\bar{r}_r}{\mu} - \bar{l}_r \left(\frac{\eta}{\mu} + 1\right)} \tag{5}$$

$$n_{tRNA} = t_0 n_{ribo}^{\beta_t} \tag{6}$$

$$n_{mRNA} = m_0 n_{ribo}^{\beta_m} \tag{7}$$

where l_r is the average length of a ribosome in base pairs, r_r (bp s^{-1}) is the maximum base pair processing rate of the ribosome which is assumed to be constant across both taxa and cell size, η (s^{-1}) and ϕ (s^{-1}) are specific degradation rates for ribosomes and proteins respectively, and the μ is the growth rate of the cell. Some of these relationships are phenomenological, such as the scaling of protein content, while others can be derived from simple models. For example, the number of ribosomes is found using the coupled dynamics of protein and ribosome replication:

$$\frac{dn_{ribo}}{dt} = \gamma \frac{r_r}{l_r} n_{ribo} - \eta n_{ribo} \tag{8}$$

$$\frac{dn_{protein}}{dt} = (1 - \gamma) \frac{r_r}{l_p} n_{ribo} - \phi n_{protein} \tag{9}$$

where γ is the fraction of ribosomes making ribosomal proteins. These equations can be solved analytically, where γ can be found by enforcing that both the ribosomal and protein pools double at the same time, and Equation 4 is given by the lifetime average of this solution. In addition to the average protein content of the cell, the ribosomal model also requires that we know the growth rate of an organism, which has also been shown to change with cell size Kempes et al. (2012) based on the following simple model of energetic partitioning of total metabolism of a given cell size, $B_0 V_c^{\beta_B}$, into growth and repair:

$$B_0 V_c^{\beta_B} = E_V \frac{dV_c}{dt} + B_V V_c \tag{10}$$

where B_V (W m^{-3}) is unit maintenance metabolism, E_m (J m^{-3}) is the unit cost of biosynthesis, $\beta_B \approx 1.7$ is the scaling exponent of metabolic rate for bacteria, B_0 is a metabolic normalization constant with units W (m^3)$^{-\beta_B}$. This equation can be solved for $V_c(t)$, the temporal growth trajectory of a cell, from which a time to reproduce can be found, which in turn gives the population growth rate as

$$\mu = \frac{(B_V/E_V)(1-\beta_B)\ln[\epsilon]}{\ln\left[\frac{1-(B_V/B_0)V_c^{1-\beta_B}}{1-\epsilon^{1-\beta_B}(B_V/B_0)V_c^{1-\beta_B}}\right]}. \tag{11}$$

This growth rate is found solving for the time to divide, t_d in the equation $V_c(t_d) = \epsilon V_0$, where $\epsilon \approx 2$ is the ratio of the cell size at division compared to its initial size, V_0, and where $\mu = \ln(\mu)/t_d$. Finally, for the energy storage component of the macromolecular pool we should focus on ATP and ignore carbohydrates since we are concerned primarily with N:P ratios in this paper. The previous work cited above does provide scaling relationships or models for ATP, but Figure 2 gives the dependence of total ATP on cell volume from data for marine bacteria Hamilton and Holm-Hansen (1967) which is fit well by

$$n_a = a_0 V_c^{\beta_a}. \tag{12}$$

It should be noted that some cells are known to have inorganic stores of phosphate and nitrate (Rhee 1973; Galbraith and Martiny 2015), and our treatment here does not account for such storage which is not characterized systematically for diverse bacteria.

Deriving Elemental Ratios in Environments From Size Distributions

Our derivations and calculations above focus on measurements of single cells along with their size, however the most common measurement of stoichiometry, including the original Redfield measurements (Redfield 1958), is of the bulk properties of all filtered particles. Thus, it is useful to translate the above cell-level N:P ratios to whole-environment values. Here, we will consider the value found from aggregating all particulate matter, later we will address both the aggregate particulate and surrounding fluid. It is important to note that these considerations of particulate stoichiometry only account for living cells and that a complete model would need to add the contribution from abiotic particles and detritus.

Given the strong connection between cell size and elemental ratios we can determine the aggregate elemental ratio within a microbial ecosystem by simply knowing the cell-size distribution. The total concentration of one element in an environment is

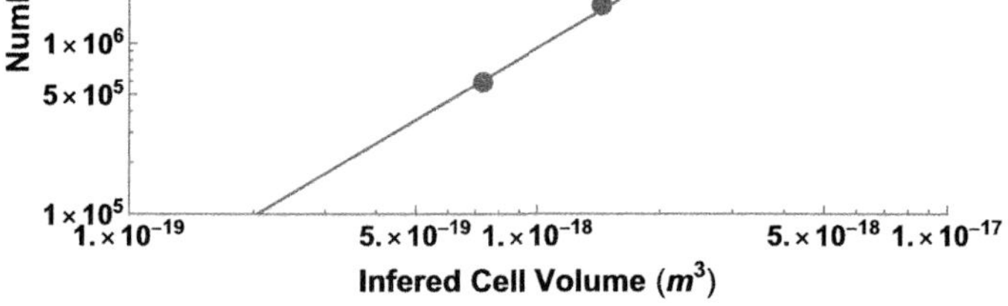

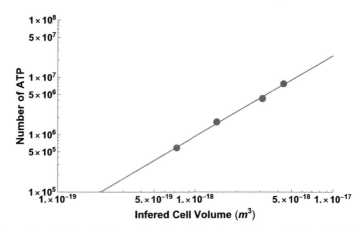

Fig. 2 The number of ATP molecules as a function of cell volume in bacteria. The original data are from Hamilton and Holm-Hansen (1967) where the original measurement of carbon content of a cell has been converted to cell volume using the relationship in Løvdal et al. (2008), and ATP mass per cell has been converted to counts per cell. The data follow $n_a = a_0 V_c^{\beta_a}$ with $\beta_a = 1.41 \pm 0.22$

given by

$$E_{tot} = \int_{V_{min}}^{V_{max}} E\left(V_c\right) \mathcal{N}\left(V_c\right) dV_c \tag{13}$$

where $\mathcal{N}\left(V_c\right)$ $(cells/m^3$ per increment of cell size) is the concentration of individuals of size V_c in the environment (note that this equation holds for concentrations or frequencies of individuals), and V_{min} and V_{max} give the smallest and largest sizes, respectively.

To compare the elemental ratios we must first specify the frequency of individuals of different size. The distribution of individual sizes, often referred to as the size-spectrum, has been previously investigated in detail (e.g. Sheldon and Parsons 1967; Cavender-Bares et al. 2001; Cuesta et al. 2018; Ward et al. 2012; Taniguchi et al. 2014; Irwin et al. 2006), and is observed to follow a variety of functional forms. One commonly observed relationship is a negative power law between cell size and abundance in an environment of the form $\mathcal{N}\left(V_c\right) = CV_c^{-\alpha}$, where (Cavender-Bares et al. 2001) showed that exponents typically vary between $\alpha = -0.95$ and $\alpha = -1.35$ using logarithmic binning (see Fig. 3b for an example abundance relationship).

Using a logarithmically-binned discrete version of Eq. 13 with the elemental relationships $E\left(V_c\right)$ from the previous section, and taking $\mathcal{N}\left(V_c\right) = CV_c^{-\alpha}$ we can explore the range of elemental ratios as a function of α, where the value of α adjusts which cell sizes are being more heavily weighted in the integral. More specifically, $\alpha = 0$ weights all cell sizes equally, more negative exponents increasingly weight smaller cells, and more positive exponents increasingly weight larger cells. In Fig. 3a we have plotted the range of elemental ratios as a function of α, where we find that only certain size distributions would produce values close to the typical Redfield ratio at the scale of an entire environment. Specifically, for $\alpha < 0$, we find values that vary between 5:1 and 15:1 in the N:P ratio. The values most closely match the Redfield ratio of 16:1 for $\alpha = -1.35$ which differs slightly from the best fit exponent of $\alpha = -1.07 \pm 0.05$ (Cavender-Bares et al. 2001) (Fig. 3b). However, it should be noted that characterizing the distributions of cell sizes as a power law is a simplification of more complicated distributions which often have a maximum abundance at an intermediate size (Sheldon and Parsons 1967; Cavender-Bares et al. 2001). The maximum abundance can be seen at the far left of Fig. 3b where the peaked function is well approximated by a piecewise power law with a positive exponent on the left and negative exponent on the right. If we use the exact empirical function for $\mathcal{N}\left(V_c\right)$ over the range of bacterial sizes we calculate an N:P of 12.84 for gram-negative bacteria and 14.65 for gram-positive bacteria, which closely match the Redfield ratio.

These results show that our procedure generates *a priori* expectations of whole-environment stoichiometries from particle-size distributions and known organism physiology, and could be generalized to any distribution of cell sizes and any systematic physiology (e.g. presently unknown living systems that use a set of P and N-containing biomolecules different than Earth's proteins and nucleic acids). Even without generalizing physiology the variation in size distribution leads to a variety of total biomass N:P ratios. No single ratio can be relied on as a distinct biosignature.

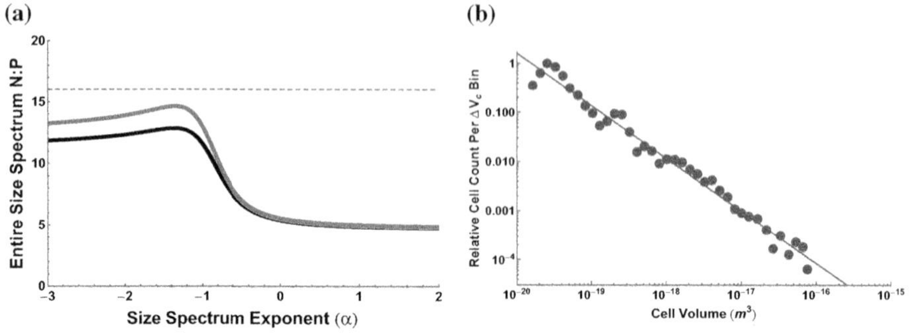

Fig. 3 **a** Elemental ratios for an entire ecosystem given a cell size distribution characterized by $\mathcal{F} \propto V_C^{-\alpha}$ where $\mathcal{F}$ is frequency and V_C is cell size. The dashed lined is the standard Redfield N:P ratio of 16:1. The black curve is for gram-negative bacteria and gray is for gram-positive bacteria. For reference, **b** gives a measured size spectrum from Cavender-Bares et al. (2001) with a fitted exponent of $\alpha = -1.07 \pm 0.05$. Here the size bin is defined by $i \leq \log_{10} V_C < i + \Delta$ with i taken in steps of $\Delta = 0.10$. The comprehensive data from Cavender-Bares et al. (2001) show exponents that vary between $\alpha = -0.95$ and -1.35. In **a** we used a discrete logarithmic summation to obtain total N and P concentrations.

Generalized Physiological and Ecological Models of Biogeochemistry

Thus far we have seen that strong trends in N:P with particle size could be an indicator of life, but that the total stoichiometric ratio of all biomass (filtered particles) does not have a single reliable value as a biosignature because this depends on the distribution of cell sizes. This approach also considers the entire set of particulate matter in isolation without considerations of environmental conditions. We need measurements that assess the "livingness" of a particular sample in the context of its environment, and one possibility is to simultaneously measure both the particulate and environmental (fluid) stoichiometries. It is also important to consider that macromolecular and elemental abundances in cells change as cells acclimate to environmental constraints, where there are known physiological optima based on environmental conditions (Burmaster 1979; Legović and Cruzado 1997; Klausmeier et al. 2004a, b, 2007, 2008), and which is the topic of the following chemostat models for microbial life living in aquatic environments.

A variety of efforts have shown how steady-state elemental ratios can be derived from physiological models coupled to flow rates in an environment (Legović and Cruzado 1997; Klausmeier et al. 2004a, b, 2007, 2008). These chemostat models contain the simplest components of a biogeochemical model: the influx of inorganic nutrients, consumption and transformation of nutrients into cellular materials as cells grow, and the loss of both biomass and inorganic nutrients from the system. Such models are typically written as

$$\frac{dR_i}{dt} = a\left(R_{i,0} - R_i\right) - f_i\left(R_i\right)\mathcal{N} \tag{14}$$

$$\frac{dQ_i}{dt} = f_i\left(R_i\right) - \mu(\vec{Q})Q_i \tag{15}$$

$$\frac{d\mathcal{N}}{dt} = \mu(\vec{Q})\mathcal{N} - m\mathcal{N} \tag{16}$$

where $\mu(\vec{Q})$ is the growth rate as a function of all of the existing elemental quotas (cellular quantities), and is typically given by

$$\mu(\vec{Q}) = \mu_\infty \min\left(1 - \frac{Q_{1,min}}{Q_1}, 1 - \frac{Q_{2,min}}{Q_2}, ..., 1 - \frac{Q_{q,min}}{Q_q}\right) \quad (17)$$

where q is the total number of limiting elements (Legović and Cruzado 1997; Klausmeier et al. 2004a, b, 2007, 2008). The function $f_i\ (R_i)$ is the uptake rate for a given nutrient. The terms Q_i, μ_∞, and $f_i\ (R_i)$ are all known to systematically change with cell size (see Box 1), where commonly the uptake function is given by

$$f_i = U_{max} \frac{R_i}{K_i + R_i} \quad (18)$$

given the half-saturation constant K_i and the maximum uptake rate U_{max} (Burmaster 1979). In this model a is the flow rate of the system, which affects both the inflow of nutrients from outside the system where $R_{i,0}$ is the concentration outside the system, and the loss of the nutrients from the system. Similarly, m is the mortality rate of the cells and is often taken to be equal to the flow rate a (Klausmeier et al. 2004a, b, 2007, 2008). In this system one nutrient is typically limiting because of the minimum taken in Eq. 17, and thus the equilibria of the system are typically dictated by the exhaustion and limitation of one nutrient. Previous work has shown that growth can be maximized in this framework by considering the allocation of resources to different cellular machinery, and that this leads to two optimum physiologies, one where maximum growth rate is optimized, and another where all of the resource equilibrium values are simultaneously minimized leading to resource colimitation and neutral competitiveness with all other species (Klausmeier et al. 2004a, b).

In this model the steady-state biomass, $\mathcal{N}^*$, limiting resource, R^*, and quota of the limiting resource, Q^*, are given by

$$\mathcal{N}^* = \frac{a\,(R_{in} - R^*)\,(\mu_\infty - m)}{Q_{min}\mu_\infty m} \quad (19)$$

$$R^* = \frac{Q_{min} m \mu_\infty K}{U_{max}\,(\mu_\infty - m) - Q_{min}\mu_\infty m} \quad (20)$$

$$Q^* = Q_{min}\frac{\mu_\infty}{\mu_\infty - m}, \quad (21)$$

(Legović and Cruzado 1997; Klausmeier et al. 2004a, b, 2007, 2008) where, for extant life, the physiological features are known to depend on size according to

$$U_{max} = U_0 V_c^{\zeta} \quad (22)$$

$$K = K_0 V_c^{\beta} \quad (23)$$

$$Q_{min} = Q_0 V_c^{\gamma} \quad (24)$$

$$\mu_\infty = \mu_0 V_c^{\eta}. \quad (25)$$

where the empirical values for the exponents and normalization constants are provided in Box 2. Given these general physiological scaling relationships the steady states are

$$\mathcal{N}^*(V_c) = \frac{a\left(R_{in} - R^*\right)\left(\mu_0 V^\eta - m\right)}{m Q_0 \mu_0 V^{\gamma+\eta}} \tag{26}$$

$$R^*(V_c) = \frac{m Q_0 \mu_0 K_0 V^{\gamma+\eta+\beta}}{U_0 V^\zeta \left(\mu_0 V^\eta - m\right) - m Q_0 \mu_0 V^{\gamma+\eta}} \tag{27}$$

$$Q^*(V_c) = \frac{Q_0 \mu_0 V^{\gamma+\eta}}{\mu_0 V^\eta - m} \tag{28}$$

where it is important to note that these equations provide results for a single cell size considered in isolation. Below we first consider how these functions change due to cell size using known physiological scaling and then general exponents, and then we derive an ecosystem-level perspective from these results and discuss potential biosignatures under a range of exponent values.

Box 2 Standard scaling relationships for physiological features

A wide variety of organism features are known to depend on overall size for various taxa (e.g. Andersen et al. 2016; Brown et al. 2004; West and Brown 2005; Savage et al. 2004), including the key features for biogeochemical considerations (Edwards et al. 2012; Litchman et al. 2007; Verdy et al. 2009). The physiological features of the coupled model are given by

		Nitrogen	Phosphorous
$U_{max} = U_t V_c^{\zeta_t}$	$\zeta_t = 0.67$	$U_t = 1.04 \times 10^4$	$U_t = 3.77 \times 10^2$
$K = K_t V_c^{\beta_t}$	$\beta_t = 0.27$	$K_t = 1.23 \times 10^4$	$K_t = 4.40 \times 10^2$
$Q_{min} = Q_t V_c^{\gamma_t}$	$\gamma_t = 0.77$	$Q_t = 9.85 \times 10^4$	$Q_t = 3.56 \times 10^3$
$\mu_\infty = \mu_t V_c^{\eta_t}$	$\eta_t = 0.65$	$\mu_t = 4.02 \times 10^{12}$	

where the t subscript indicates that these are the known values for extant Terran life. It should be noted that these values are for eukaryotic organisms compared with the earlier physiological models for bacteria.

Single-species Biogeochemistry for Extant Life

The above model provides a simple but general biogeochemical system where cellular physiology is coupled to an environment, and can be deployed to address ecosystems of various ecological complexity. First we consider the case where an environment is dominated by a single species, which would correspond to the measurement of a consistent particle size in our framework. Taking the known physiological scaling relationships for extant life (Box 2) we find that the size of the organism has a strong effect on the stoichiometric ratios of both the particles and fluid. Figure 4 gives the steady state N:P of cells as a function of steady state environmental N:P and cell size. The variation in the steady-state environmental and cellular N:P was achieved by varying the inflow concentrations $R_{i,0}$.

We find that the largest cells will show the greatest deviation from the environmental concentration for most environmental ratios. Differences between the fluid and particle stoichiometry may define a biosignature, and these will be most noticeable

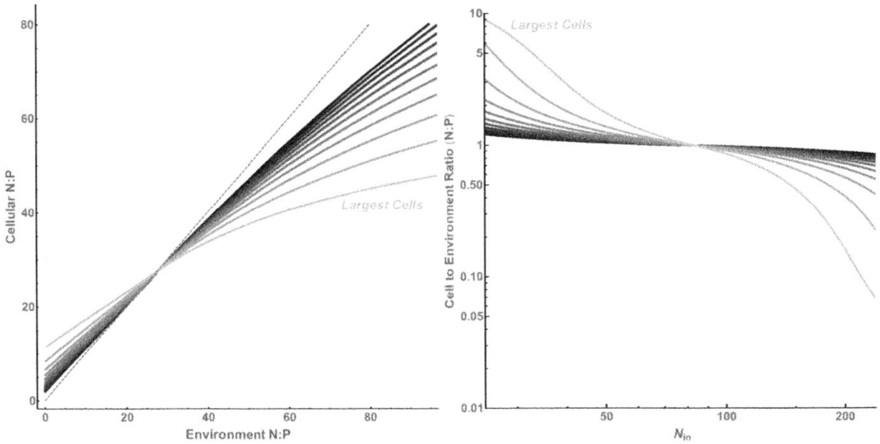

Fig. 4 **a** Elemental ratios within cells as a function of the environmental ratio and cell size (dark blue is the smallest and light yellow is the largest cells), where the ecosystem is composed of only a single cell size. The dashed line is the one-to-one line. **b** The differences between cell stoichiometric ratios and the environment as a function of the nitrogen inflow, N_{in}, which is also the parameter being varied in **a**

for environments dominated by the largest cells. It should be noted that these results depend on the specific scaling relationships of the physiological features given in Box 2, which could greatly vary for life beyond Earth and are even known to vary across taxa for extant Terran life (DeLong et al. 2010; Kempes et al. 2012).

Generalized ecosystem biogeochemistry

The above coupling of cells to an environment considers a biogeochemical dynamic with only a single species that is characterized by a given cell size. This is the most rudimentary possibility for an ecosystem and is generally unlikely, but considering the full range of possibilities for life in the universe, could be of relevance to particular astrobiological contexts such as environments with low energy flux and characterized by a single resource limitation. However, we would like to expand this perspective to more complicated ecosystems with greater diversity as represented by a variety of cell sizes.

Classic resource competition theory in equilibrium (e.g. Tilman 1982; Levin 1970; Hutchinson 1953, 1957; Volterra 1927, 1931) indicates that for multiple species, in our case multiple cell sizes, to coexist on a single limiting resource they must all share the same R^* value. This is not naturally the case given the physiological scaling relationships outlined above, or the unlikelihood that many species will have identical physiological parameter values. In general, at most x number of species can coexist in equilibrium if there are x independent limiting factors (Levin 1970), and in our framework we can adjust the mortality rate, m, to abstractly represent the combination of many factors and to obtain coexistence. This adjustment could be the consequence of a variety of other factors such as variable predation, sinking rates, phage susceptibility, or intrinsic death. For our purposes this approach allows us to obtain a spectrum of cell sizes in connection with our earlier focus.

To enforce coexistance we take $R^*(V_c) = R_c$, where R_c is a constant, in which case the required mortality rate is given by

$$m = \frac{R_c U_{max} \mu_\infty}{R_c U_{max} + Q_{min} \mu_\infty (K + R_c)} \tag{29}$$

$$= \frac{R_c U_0 \mu_0 v^{\zeta+\eta}}{R_c U_0 v^\zeta + Q_0 \mu_0 v^{\gamma+\eta} \left(R_c + K_0 v^\beta\right)}. \tag{30}$$

This function for m should be seen as the consequence of the complicated evolutionary dynamics of many species living in a coupled ecosystem where prey and predator traits have evolved over time and new effective niches have emerged. It should also be noted m is now size dependent compared with being set to constant value which was the case for the earlier results.

Our mortality relationship can be incorporated into $\mathcal{N}^*$ to give the scaling of biomass concentration for each cell size:

$$\mathcal{N}^*(V_c) = \frac{a(R_{in} - R_c) V_c^{-\zeta} \left(R_c + K_0 v^\beta\right)}{R_c U_0}. \tag{31}$$

This result has two important limits, where either the half-saturation constant is much smaller than the equilibrium value of nutrient in the environment, $K_0 v^\beta \ll R_c$, or is much bigger than this environmental concentration, which leads to

$$\mathcal{N}^*(v) = \begin{cases} \propto V_c^{-\zeta} & K_0 V_c^\beta \ll R_c \\ \propto V_c^{\beta-\zeta} & K_0 V_c^\beta \gg R_c \end{cases} \tag{32}$$

These two relationships provide nice bounds on the scaling of $\mathcal{N}$ given the underlying physiological dependencies.

Similarly, the quota is given by

$$Q^* = Q_0 V_c^\gamma + \frac{R_c v^{\zeta-\eta} U_0}{\mu_0 \left(R_c + K_0 v^\beta\right)}. \tag{33}$$

which implies that the ratio of particle to fluid elemental abundance for the limiting nutrient is the following function of cell size

$$\frac{\mathcal{N}^* Q^*}{R^*} = \begin{cases} \dfrac{a(R_{in}-R_c)\left(Q_0 V_c^{\gamma-\zeta} + \frac{U_0 V_c^{-\eta}}{\mu_0}\right)}{R_c U_0} & K_0 V_c^\beta \ll R_c \\ \dfrac{a(R_{in}-R_c)\left(K_0 Q_0 \mu_0 V_c^{\beta+\gamma-\zeta} + R_c U_0 V_c^{-\eta}\right)}{R_c^2 U_0 \mu_0} & K_0 V_c^\beta \gg R_c \end{cases} \tag{34}$$

This relationship is similar to the types of results shown in Fig. 4, but gives the ratio between cell and environment concentrations for a single element of interest (rather than as comparisons of ratios of elements), and importantly, does so under the constraints of coexistence. This result leads to particular biosignature possibilities when measuring only a single element, and does so for the more realistic ecosystem conditions of coexistence. If we measure the particle size distribution in an environment, then this is enough to specify the value of $\alpha = -\zeta$ or $\alpha = \beta - \zeta$ from Eq. 34, leaving us with γ and η to determine the element ratio scaling between cells and the environment as a function of particle size.

From a biosignatures perspective, the most ambiguous measurement would be particles that perfectly mirror the environmental stoichiometry where $N^* Q^* / R^*$ equals a constant for all particle sizes. In the first limit, $K_0 V_c^{\beta} \ll R_c$, this would require $\zeta = \gamma = -\alpha$ and $\eta = 0$. This result would imply that the quota and uptake rates would need to scale with the same exponent and as the negative value of the size exponent, both of which are consistent with the observations of Box 2 and Fig. 3b for extant life. However, this result also requires that there would be no change in growth rate with cell size, which is very unlikely from a variety of biophysical arguments.

In the second limit, $K_0 V_c^{\beta} \gg R_c$, a constant value of $N^* Q^* / R^*$ requires that $\zeta - \beta = \gamma = -\alpha$ and $\eta = 0$. Again the absence of changes in growth rate connected with $\eta = 0$ is unlikely. In addition, under this scenario the difference in the uptake and half-saturation scaling, represented by $\zeta - \beta$, must equal the scaling of the quota and take the opposite value as the size-spectrum scaling, which is a combination that is not consistent with extant life and is a very special case in general. Thus, under both limits $N^* Q^* / R^*$ is unlikely to have a constant value as a function of cell size, and an observed scaling in this ratio forms a likely biosignature.

This potential biosignature still requires one to measure the cell-size spectrum in detail, which may be challenging in certain settings or with certain devices. However, these relationships can be easily translated into an aggregate ecosystem-level measurement by averaging over all coexisting cells, where the average is given by

$$\left\langle \frac{\mathcal{N}^* Q^*}{R^*} \right\rangle = \frac{1}{V_{max} - V_{min}} \int_{V_{min}}^{V_{max}} \frac{\mathcal{N}^* (V) \, Q^* (V)}{R^*} dV \tag{35}$$

$$= \left. \frac{a \left(R_{in} - R_c\right) V_c \left(\frac{R_c U_0 V_c^{-\eta}}{1-\eta} + \frac{R_c Q_0 \mu_0 V_c^{\gamma-\zeta}}{1+\gamma-\zeta} + \frac{Q_0 K_0 \mu_0 V_c^{\beta+\gamma-\zeta}}{1+\beta+\gamma-\zeta} \right)}{\left(V_{max} - V_{min}\right) R_c^2 U_0 \mu_0} \right|_{V_c=V_{min}}^{V_c=V_{max}} \tag{36}$$

which, considering the two approximations for $\mathcal{N}$, becomes

$$\left\langle \frac{\mathcal{N}^* Q^*}{R^*} \right\rangle = \begin{cases} \left. \dfrac{a(R_{in}-R_c)V_c\left(\frac{U_0 V^{-\eta}}{\mu_0(1-\eta)} + \frac{Q_0 V_c^{\gamma-\zeta}}{1+\gamma-\zeta}\right)}{(V_{max}-V_{min})R_c U_0} \right|_{V_c=V_{min}}^{V_c=V_{max}} & K_0 V_c^{\beta} \ll R_c \\ \left. \dfrac{a(R_{in}-R_c)V_c\left(\frac{R_c U_0 V_c^{-\eta}}{1-\eta} + \frac{Q_0 K_0 \mu_0 V_c^{\beta+\gamma-\zeta}}{1+\beta+\gamma-\zeta}\right)}{(V_{max}-V_{min})R_c^2 U_0 \mu_0} \right|_{V_c=V_{min}}^{V_c=V_{max}} & K_0 V_c^{\beta} \gg R_c \end{cases} \tag{37}$$

To fully specify this community level ratio for generalized life we need to constrain the normalizations constants, Q_0, k_0, U_0, and μ_0 given any choice of the exponents. A reasonable way to determine the values of these constants is to match the generalized rates to the observed Terran rates from Box 2 at a particular reference size, V_r, which leads to

$$U_0 = U_t V_r^{\zeta_t - \zeta} \tag{38}$$

$$K_0 = K_t V_r^{\beta_t - \beta} \tag{39}$$

$$Q_0 = Q_t V_r^{\gamma_t - \gamma} \tag{40}$$

$$\mu_0 = \mu_t V_r^{\eta_t - \eta}. \tag{41}$$

After calibrating the constants to an intermediate cell size of $V_r = 10^{-18}$ (m^3), Fig. 5 gives the community level $\langle \mathcal{N}^* Q^* / R^* \rangle$ as a function of the scaling exponents. When $K_0 v^{\beta} \ll R_c$ the size exponent α specifies $-\zeta$, and when $K_0 v^{\beta} \gg R_c$ then α specifies $\beta - \zeta$. In both approximations we plot $\langle \mathcal{N}^* Q^* / R^* \rangle$ as a function of η and γ for a range of α values.

We find that typically $\langle \mathcal{N}^* Q^* / R^* \rangle$ differs from 1 for a wide range of α, η and γ values. This is true under both limits. For fixed values of α the $\langle \mathcal{N}^* Q^* / R^* \rangle = 1$ line is a closed curve as a function of η and γ (Fig. 5). This curve defines the regime within which it is possible to find $\langle \mathcal{N}^* Q^* / R^* \rangle = 1$ for any value of either η and γ, and this region covers a wide range of exponent values. However, the known values of η and γ for extant life occur fairly far from this curve and would show elemental concentrations that are distinguishable from the environment. It is likely that the full range of α, η and γ combinations explored here are precluded for biophysical reasons, but this requires more detailed work in the future. Finally, it is important to note that most α, η and γ combinations would yield cell-to-environment ratios that significantly differ from 1, and that the gradients are very steep around the $\langle \mathcal{N}^* Q^* / R^* \rangle$=1 line. Thus, it is a fairly safe assumption that the elemental abundances of cells should differ from the environment as this would be the expectation for physiological scaling chosen at random.

Box 3 Summary of potential biosignatures

1. Systematic shifts in stochiometry with particle size
2. Particle sizes that follow a power law distribution for abundance
3. Systematic shifts in the ratio of particle to fluid elemental abundance as a function of particle size

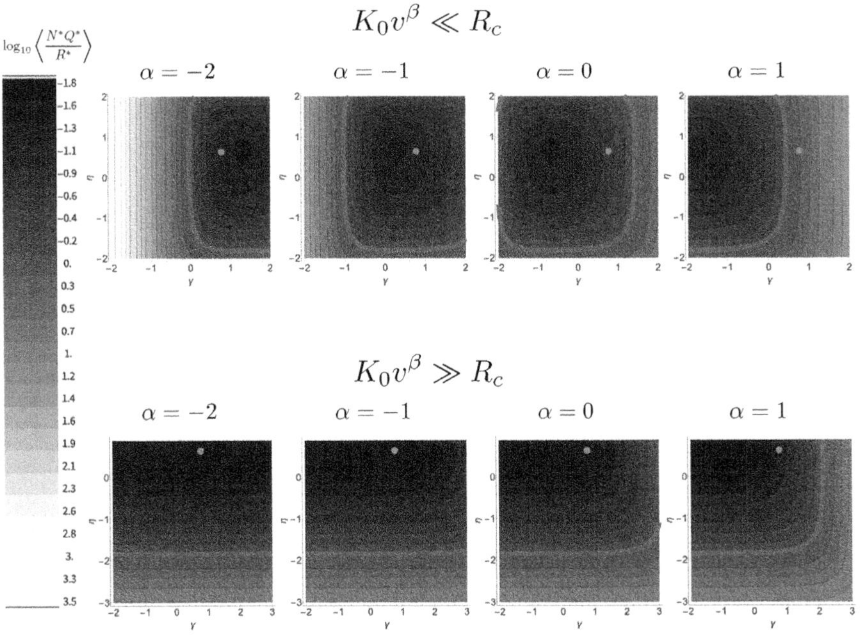

Fig. 5 The ratio of the cellular to environmental nitrogen, $\langle \mathcal{N}^* Q^* / R^* \rangle$, as function of the size spectrum exponent α, the minimum quota (cellular requirement) scaling exponent, γ, and the growth rate scaling exponent, η. We have shown the results for the two approximations $K_0 v^\beta \ll R_c$ and $K_0 v^\beta \gg R_c$. In each plot the green line represents $\langle \mathcal{N}^* Q^* / R^* \rangle = 1$, or $\log_{10} \langle \mathcal{N}^* Q^* / R^* \rangle = 0$. The orange point represents the known γ and η exponent values for extant life from Box 2

Discussion

The general framework provided here should make it possible to assess biosignatures for a wide diversity of potential life (see Box 3 for a summary). We focused on bacterial life as an example of what we would expect in ecosystems dominated by the simplest life. However, all of our results could be tuned to other classes of organisms with the appropriate changes in scaling relationships for macromolecular content and abundance distributions. Our results do this generally for any life that is governed by a set of physiological scaling relationships, where, for example, the nutrient quotas are abstracting the underlying changes in macromolecules and could represent a diverse set of alternate physiologies and sets of macromolecules for alternate evolutionary histories or origins of life. Even for extant life on Earth the typical stoichiometry varies significantly within and across taxa (Geider and La Roche 2002; Klausmeier et al. 2004b, a, 2008; Loladze and Elser 2011; Neveu et al. 2016; Sterner et al. 2008; Vrede et al. 2004; Elser et al. 2000; Liefer et al. 2019; Finkel et al. 2016a, b), for example, plant leaves have an N:P of 30 rather than 16 (Elser et al. 2010; Kerkhoff et al. 2005). However, the main assumption in our generalized physiological model is that life will fall along allometric scaling relationships, which occurs for multiple taxa on Earth and has good justification from various arguments connected with universal physical constraints.

It should also be noted that many of the physiological scaling relationships have strong physical principles motivating the exponents and the wide variation taken in the generalized equations may not be realizable by life anywhere in the universe. Thus observed biosignatures may be much more similar to our analyses in Figs. 1, 3, and 4 than the possibilities encapsulated in our generalized physiological model as capture in Figure 5.

In addition, our efforts here have often focused on the assumption of one limiting nutrient. However, this scenario of a single resource typically does not lead to coexistence (e.g. Tilman 1982; Levin 1970; Hutchinson 1953, 1957; Volterra 1927, 1931). The problem of coexistence can be solved by many additional considerations such as environmental stochasticity, the addition of spatial dynamics, or species adaptation (e.g. Hutchinson 1953; Levins and Culver 1971; Klausmeier and Tilman 2002; Kremer and Klausmeier 2013), all of which could be important for future modeling efforts or for measurements of the spatial variation in stoichometry. However, our solution for mortality allows for coexistence in a single environment and our model is compatible with measurements made at a single or coarse-grained location which may be typical of many future astrobiological measurements. Our general physiological perspective should be combined with more advanced biogeochemical models that consider many nutrients, including trace elements, and more complex ecological and evolutionary dynamics — many of which can be connected systematically with size (Andersen et al. 2016; Kempes et al. 2019) — to fully explore the range of particle size distributions, and the particle to fluid stoichiometric differences that can be reasonably expected to represent biosignatures. Finally, since our approach only considers the living component of particulate matter future models should incorporate the stoichiometric contributions of abiotic particles and detritus along with more complex geochemistry and ask how much this addition can shift the general biosignatures presented here.

Acknowledgements The authors thank Natalie Grefenstette for useful discussions and comments. This work was supported by a grant from the Simons Foundation (#395890, Simon Levin), and grants from the National Aeronautics Space Administration (80NSSC18K1140, Christopher Kempes, Hillary Smith, Heather Graham, and Christopher House), and the National Science Foundation (OCE-1848576, Simon Levin).

Appendix: Macromolecular Constants for Stoichiometry

One of our overall goals is to assess the size-dependence of elemental ratios. To do this we relied on average elemental abundances of various macromolecules coupled with the number scaling of those macromolecules with cell size. Here, we provide more details on the conversion of macromolecular quantities and some novel quantification of the scaling of macromolecular abundance. For all of the elemental abundances of N and P in particular macromolecules we rely on Geider and La Roche (2002) unless otherwise indicated. Most of the scaling relationships follow from Box 1 converted into molecule numbers. It should be noted that several additional features could shift the following calculations, such as gram-positive vs. gram-negative architectures, or the observation of scaling relationships between the total membrane-bound proteins and cell size. We have included a gram-negative and gram-positive comparison in Fig. 1 to provide a sense of the range of values that these variations can cause.

Amino Acids We used the scaling of total protein mass for bacteria (Kempes et al. 2016) combined with the average mass of an individual amino acid, 1.79×10^{-22} g (Bremer et al. 1996), to obtain the total number of amino acids in the cell.

Nucleotides We used the total ribosome number count from Kempes et al. (2016) combined with the number of RNA nucleotides in a ribosome, 4566 (Bremer et al. 1996), to obtain the total number of RNA molecules contributed from ribosomes. For the mRNA we used the number of mRNA per ribosome, 1.08 (mRNA/ribosome) (Kempes et al. 2016), the average length an mRNA, 975 nucleotides (see Kempes et al. 2016 for a review), and the number of ribosomes to estimate the total abundance. Similarly, for the tRNA we used the tRNA per ribosome of 9.3 (Bremer et al. 1996), and the average length of a tRNA, 80 nucleotides (Bremer et al. 1996). For DNA we used the base pair count from Kempes et al. (2016).

Phospholipids and Peptidoglycan For the phospholipids we consider that $p = 0.30$ of outer membrane is composed of proteins (Szalontai et al. 2000). For the remaining surface area we consider that a single phospholipid occupies $s = 55 \times 10^{-20}$ m^2 (Nichols and Deamer 1980) of surface area, such that for a gram-negative bacterium the total number of phospholipids is given by $8\pi(1-p)\left(r_c^2 + (r_c - \delta)^2\right)/s$, where r_c is the radius of the cell and $\delta = 3.43 \times 10^{-8}$ m is the distance between the plasma and outer membranes (Meroueh et al. 2006). For a gram positive bacterium the total number of phospholipids can be approximated by $4\pi(1-p)\left(r_c - t_p\right)^2/s$, where $t_p = 6.4 \times 10^{-9}$ m is the thickness of the peptidoglycan layer (Meroueh et al. 2006).

For the peptidoglycan layer we take the basic monomer of NAG-NAM and associated peptides from Meroueh et al. (2006), which contains 8 N and no P and is arranged into a unit volume defined by 6 strands each of which is 8 monomers long such that the unit volume contains 384 N. This unit volume is an approximate cylinder with a diameter of $\approx 7 \times 10^{-9}$ m and a height of $\approx 1 \times 10^{-8}$ m (Meroueh et al. 2006). The total volume of peptidoglycan in the cell is then given by $4/3\pi\left((r_c - t_0 - t_{peri})^3 - (r_c - t_o - t_{peri} - t_p)^3\right)$ for a gram-negative bacterium, where $t_o = 6.9 \times 10^{-9}$ m is the thickness of the outer membrane and $t_{peri} = 2.1 \times 10^{-8}$ m is the thickness of the periplasm (Meroueh et al. 2006). For a gram-positive bacterium the total peptidoglycan volume is approximately $4/3\pi\left(r_c^3 - (r_c - t_p)^3\right)$. Both volumes allow us to count the number of unit volumes and associated elemental abundances.

ATP For the ATP we use the scaling from Fig. 2.

References

Anbar AD (2008) Elements and Evolution. Science 322(5907):1481–1483

Andersen KH, Berge T, Gonçalves RJ, Hartvig M, Heuschele J, Hylander S, Jacobsen NS, Lindemann C, Martens EA, Neuheimer AB et al (2016) Characteristic sizes of life in the oceans, from bacteria to whales. Ann Rev Mar Sci 8:217–241

Andersson A, Rudehäll Å. (1993) Proportion of plankton biomass in particulate organic carbon in the northern Baltic Sea. Mar Ecol Prog Series 95(1/2):133–139

Beardall J, Allen D, Bragg J, Finkel ZV, Flynn KJ, Quigg A, Rees TAV, Richardson A, Raven JA (2009) Allometry and stoichiometry of unicellular, colonial and multicellular phytoplankton. New Phytol 181(2):295–309

Bremer H, Dennis PP, Neidhardt F Eds (1996) Modulation of chemical composition and other parameters of the cell by growth rate. In: Escherichia coli and Salmonella typhimurium. Cellular and molecular biology: Chapter 96. Second Edition. American society for microbiology

Brown J, Gillooly J, Allen A, Savage V, West G (2004) Toward a metabolic theory of ecology. Ecology 85(7):1771–1789

Burmaster DE (1979) The continuous culture of phytoplankton: mathematical equivalence among three steady-state models. Am Nat 113(1):123–134

Cavender-Bares KK, Rinaldo A, Chisholm SW (2001) Microbial size spectra from natural and nutrient enriched ecosystems. Limnol Oceanogr 46(4):778–789

Chopra A, Lineweaver CH (2008) The major elemental abundance differences between life, the oceans and the sun. In: Proceedings of the 8th Australian space science conference, Canberra, pp. 49–55

Cuesta JA, Delius GW, Law R (2018) Sheldon spectrum and the plankton paradox: two sides of the same coin: a trait-based plankton size-spectrum model. J Math Biol 76(1):67–96

DeLong J, Okie J, Moses M, Sibly R, Brown J (2010) Shifts in metabolic scaling, production, and efficiency across major evolutionary transitions of life. Proceed Nat Acad Sci 107(29):12941–12945

Edwards KF, Thomas MK, Klausmeier CA, Litchman E (2012) Allometric scaling and taxonomic variation in nutrient utilization traits and maximum growth rate of phytoplankton. Limnol Oceanogr 57(2):554–566

Nature Geoscience Editorial (2014) Eighty years of Redfield.Nat. Geosci 7: 849

Elrifi IR, Turpin DH (1985) Steady–state luxury consumption and the concept of optimum nutrient ratios: a study with phosphate and nitrate limited *Selenastrum minutum* (chlorophyta). J Phycol 21(4):592–602

Elser JJ (2003) Biological stoichiometry: a theoretical framework connecting ecosystem ecology, evolution, and biochemistry for application in astrobiology. Int J Astrobiol 2(3):185–193

Elser JJ, Dobberfuhl DR, MacKay NA, Schampel JH (1996) Organism size, life history, and N: P stoichiometry: toward a unified view of cellular and ecosystem processes. BioScience 46(9):674–684

Elser J, Sterner RW, Gorokhova E, Fagan W, Markow T, Cotner JB, Harrison J, Hobbie SE, Odell G, Weider L (2000) Biological stoichiometry from genes to ecosystems. Ecol Lett 3(6):540–550

Elser JJ, Fagan WF, Kerkhoff AJ, Swenson NG, Enquist BJ (2010) Biological stoichiometry of plant production: metabolism, scaling and ecological response to global change. New Phytol 186(3):593–608

Finkel ZV (2001) Light absorption and size scaling of light-limited metabolism in marine diatoms. Limnol Oceanogr 46(1):86–94

Finkel ZV, Irwin AJ, Schofield O (2004) Resource limitation alters the 3/4 size scaling of metabolic rates in phytoplankton. Marine Ecol Progress Series 273:269–279

Finkel ZV, Follows MJ, Liefer JD, Brown CM, Benner I, Irwin AJ (2016) Phylogenetic diversity in the macromolecular composition of microalgae. PLoS One 11(5):e0155977

Finkel Z, Follows M, Irwin A (2016) Size-scaling of macromolecules and chemical energy content in the eukaryotic microalgae. J Plankton Res 38(5):1151–1162

Galbraith ED, Martiny AC (2015) A simple nutrient-dependence mechanism for predicting the stoichiometry of marine ecosystems. Proceed Nat Acad Sci 112(27):8199–8204

Geider RJ, La Roche J (2002) Redfield revisited: variability of C:N: P in marine microalgae and its biochemical basis. Eur J Phycol 37(1):1–17

Hamilton RD, Holm-Hansen O (1967) Adenosine triphosphate content of marine bacteria. Limnol Oceanogr 12(2):319–324

Healey FP (1985) Interacting effects of light and nutrient limitation on the growth rate of *Synechococcus linearis* (cyanophyceae). J Phycol 21(1):134–146

Hutchinson GE (1957) Concluding remarks: Cold Spring Harbor Symposium. Quant Biol 22:415–427
Hutchinson GE (1953) The concept of pattern in ecology. Proceed Acad Nat Sci Philadelphia 105:1–12
Irwin AJ, Finkel ZV, Schofield OM, Falkowski PG (2006) Scaling-up from nutrient physiology to the size-structure of phytoplankton communities. J Plankton Res 28(5):459–471
Karr JR, Sanghvi JC, Macklin DN, Gutschow MV, Jacobs JM, Bolival B Jr, Assad-Garcia N, Glass JI, Covert MW (2012) A whole-cell computational model predicts phenotype from genotype. Cell 150(2):389–401
Kempes CP, Dutkiewicz S, Follows MJ (2012) Growth, metabolic partitioning, and the size of microorganisms. Proceed Nat Acad Sci 109(2):495–500
Kempes CP, Wang L, Amend JP, Doyle J, Hoehler T (2016) Evolutionary tradeoffs in cellular composition across diverse bacteria. ISME J 10(9):2145–2157
Kempes CP, van Bodegom PM, Wolpert D, Libby E, Amend J, Hoehler T (2017) Drivers of bacterial maintenance and minimal energy requirements. Front Microbiol 8:31
Kempes CP, West GB, Koehl M (2019) The scales that limit: the physical boundaries of evolution. Front Ecol Evol 7:242
Kerkhoff AJ, Enquist BJ, Elser JJ, Fagan WF (2005) Plant allometry, stoichiometry and the temperature-dependence of primary productivity. Global Ecol Biogeogr 14(6):585–598
Kim H, Smith HB, Mathis C, Raymond J, Walker SI (2019) Universal scaling across biochemical networks on Earth. Sci Adv 5(1):eaau0149
Klausmeier CA, Litchman E, Levin SA (2004a) Phytoplankton growth and stoichiometry under multiple nutrient limitation. Limnol Oceanogr 49(4):1463–1470
Klausmeier CA, Litchman E, Daufresne T, Levin SA (2004b) Optimal nitrogen-to-phosphorus stoichiometry of phytoplankton. Nature 429(6988):171–174
Klausmeier C, Litchman E, Levin SA (2007) A model of flexible uptake of two essential resources. J Theor Biol 246(2):278–289
Klausmeier CA, Litchman E, Daufresne T, Levin SA (2008) Phytoplankton stoichiometry. Ecol Res 23(3):479–485
Klausmeier CA, Tilman, D (2002) Spatial models of competition. In: Competition and coexistence, Eds. Sommer, Ulrich, Worm, Boris. Chap. 3, pp. 43–78. Springer, Berlin, Heidelberg
Kremer CT, Klausmeier CA (2013) Coexistence in a variable environment: eco-evolutionary perspectives. J Theor Biol 339:14–25
Legović T, Cruzado A (1997) A model of phytoplankton growth on multiple nutrients based on the Michaelis–Menten–Monod uptake, Droop's growth and Liebig's law. Ecol Model 99(1):19–31
Levin SA (1970) Community equilibria and stability, and an extension of the competitive exclusion principle. Am Nat 104(939):413–423
Levins R, Culver D (1971) Regional coexistence of species and competition between rare species. Proceed Nat Acad Sci 68(6):1246–1248
Liefer JD, Garg A, Fyfe MH, Irwin AJ, Benner I, Brown CM, Follows MJ, Omta AW, Finkel ZV (2019) The macromolecular basis of phytoplankton C:N: P under nitrogen starvation. Front Microbiol 10:763
Lineweaver CH, Chopra A (2012) What Can Life on Earth Tell Us About Life in the Universe?. In: Seckbach J (ed) Genesis–In The Beginning. Precursors of Life, Chemical Models and Early Biological Evolution, pp: 799–815. Springer, Dordrecht
Litchman E, Klausmeier CA, Schofield OM, Falkowski PG (2007) The role of functional traits and trade-offs in structuring phytoplankton communities: scaling from cellular to ecosystem level. Ecol Lett 10(12):1170–1181
Loladze I, Elser JJ (2011) The origins of the Redfield nitrogen–to–phosphorus ratio are in a homoeostatic protein–to–rRNA ratio. Ecol Lett 14(3):244–250
Lϕvdal T, Skjoldal E, Heldal M, Norland S, Thingstad T (2008) Changes in morphology and elemental composition of *Vibrio splendidus* along a gradient from carbon-limited o phosphate-limited growth. Microb Ecol 55(1):152–161
McKay CP (2008) An approach to searching for life on Mars, Europa, and Enceladus. In: Botta O, Bada J, Gómez Elvira J, Javaux E, Selsis F, Summons R (eds) Strategies of Life Detection, pp. 49–54. Springer Science & Business
Meroueh SO, Bencze KZ, Hesek D, Lee M, Fisher JF, Stemmler TL, Mobashery S (2006) Three-dimensional structure of the bacterial cell wall peptidoglycan. Proceed Nat Acad Sci 103(12):4404–4409
Neveu M, Poret-Peterson A, Anbar A, Elser J (2016) Ordinary stoichiometry of extraordinary microorganisms. Geobiology 14(1):33–53

Nichols JW, Deamer DW (1980) Net proton-hydroxyl permeability of large unilamellar liposomes measured by an acid-base titration technique. Proceed Nat Acad Sci 77(4):2038–2042
Redfield AC (1958) The biological control of chemical factors in the environment. Am Sci 46(3):205–221
Rhee G–Y (1973) A continuous culture study of phosphate uptake, growth rate and polyphosphate in *Scenedesmus* sp. J Phycol 9(4):495–506
Rhee G–Y (1978) Effects of N: P atomic ratios and nitrate limitation on algal growth, cell composition, and nitrate uptake. Limnol Oceanograp 23(1):10–25
Savage V, Gillooly J, Woodruff W, West G, Allen A, Enquist B, Brown J (2004) The predominance of quarter-power scaling in biology. Func Ecol 18:257–282
Savage V, Gillooly J, Brown J, West G, Charnov E (2004) Effects of body size and temperature on population growth. Am Nat 163(3):429–441
Sheldon R, Parsons T (1967) A continuous size spectrum for particulate matter in the sea. J Fisheries Board Canada 24(5):909–915
Shuler M, Leung S, Dick C (1979) A mathematical model for the growth of a single bacterial cell. Ann New York Acad Sci 326(1):35–52
Sterner RW, Andersen T, Elser JJ, Hessen DO, Hood JM, McCauley E, Urabe J (2008) Scale-dependent carbon: nitrogen:phosphorus seston stoichiometry in marine and freshwaters. Limnol Oceanogr 53(3):1169–1180
Szalontai B, Nishiyama Y, Gombos Z, Murata N (2000) Membrane dynamics as seen by Fourier transform infrared spectroscopy in a cyanobacterium, *Synechocystis* PCC 6803: the effects of lipid unsaturation and the protein–to–lipid ratio. Biochim Biophys Acta Biomembr BBA-Biomembranes 1509(1–2):409–419
Tang EP (1995) The allometry of algal growth rates. J Plankton Res 17(6):1325–1335
Taniguchi DA, Franks PJ, Poulin FJ (2014) Planktonic biomass size spectra: an emergent property of size-dependent physiological rates, food web dynamics, and nutrient regimes. Mar Ecol Progr Series 514:13–33
Tilman D (1982) Resource competition and community structure. Princeton University Press, Princeton, USA
Verdy A, Follows M, Flierl G (2009) Optimal phytoplankton cell size in an allometric model. Mar Ecol Progr Series 379:1–12
Volterra V (1926) Variazioni e fluttuazioni del numero d'individui in specie animali conviventi. Memoria della Reale Accademia Nazionale dei Lincei 2: 31–113
Volterra V (1931) Leçons sur la Théorie Mathématique de la Lutte pour la Vie, Gauthier–Villars, Paris
Vrede T, Dobberfuhl DR, Kooijman S, Elser JJ (2004) Fundamental connections among organism C:N: P stoichiometry, macromolecular composition, and growth. Ecology 85(5):1217–1229
Wang HS, Lineweaver CH, Ireland TR (2018) The elemental abundances (with uncertainties) of the most Earth–like planet. Icarus 299:460–474
Ward BA, Dutkiewicz S, Jahn O, Follows MJ (2012) A size-structured food-web model for the global ocean. Limnol Oceanograp 57(6):1877–1891
West G, Brown J (2005) The origin of allometric scaling laws in biology from genomes to ecosystems: towards a quantitative unifying theory of biological structure and organization. J Exp Biol 208(9):1575–1592
Young PA, Desch SJ, Anbar AD, Barnes R, Hinkel NR, Kopparapu R, Madhusudhan N, Monga N, Pagano MD, Riner MA et al (2014) Astrobiological stoichiometry. Astrobiology 14(7):603–626

Publisher's Note Springer Nature remains neutral with regard to jurisdictional claims in published maps and institutional affiliations.

Bulletin of Mathematical Biology (2021) 83:67
https://doi.org/10.1007/s11538-021-00892-6

Society for Mathematical Biology

SPECIAL ISSUE: CELEBRATING J. D. MURRAY

The Effect of Covert and Overt Infections on Disease Dynamics in Honey-Bee Colonies

Nicholas F. Britton[1] · K. A. Jane White[1]

Received: 15 April 2020 / Accepted: 16 March 2021 / Published online: 6 May 2021

Abstract

Viral diseases of honey bees are important economically and ecologically and have been widely modelled. The models reflect the fact that, in contrast to the typical case for vertebrates, invertebrates cannot acquire immunity to a viral disease, so they are of SIS or (more often) SI type. Very often, these diseases may be transmitted vertically as well as horizontally, by vectors as well as directly, and through the environment, although models do not generally reflect all these transmission mechanisms. Here, we shall consider an important additional complication the consequences of which have yet to be fully explored in a model, namely that both infected honey bees and their vectors may best be described using more than one infection class. For honey bees, we consider three infection classes. Covert infections occur when bees have the virus under control, such that they do not display symptoms of the disease, and are minimally or not at all affected by it. Acutely overtly infected bees often exhibit severe symptoms and have a greatly curtailed lifespan. Chronically overtly infected bees typically have milder symptoms and a moderately shortened lifespan. For the vector, we consider just two infection classes which are covert infected and overt infected as has been observed in deformed-wing virus (DWV) vectored by varroa mites. Using this structure, we explore the impact of spontaneous transition of both mites and bees from a covertly to an overtly infected state, which is also a novel element in modelling viral diseases of honey bees made possible by including the different infected classes. The dynamics of these diseases are unsurprisingly rather different from the dynamics of a standard SI or SIS disease. In this paper, we highlight how our compartmental structure for infection in honey bees and their vectors impact the disease dynamics observed, concentrating in particular on DWV vectored by varroa mites. If there is no spontaneous transition,

Nick Britton very sadly passed away in December 2020, while this paper was still in review stage. On his request, Jane White, his long-time colleague and friend at the University of Bath, revised and finalised the manuscript in response to reviewer comments.

✉ K. A. Jane White
k.a.j.white@bath.ac.uk

1 Department of Mathematical Sciences and Centre for Mathematical Biology, University of Bath, Bath, UK

then a basic reproduction number R_0 exists. We derive a condition for $R_0 > 1$ that reflects the complexities of the system, with components for vertical and for direct and vector-mediated horizontal transmission, using the directed graph of the next-generation matrix of the system. Such a condition has never previously been derived for a honey-bee–mite–virus system. When spontaneous transitions do occur, then R_0 no longer exists, but we introduce a modification of the analysis that allows us to determine whether (i) the disease remains largely covert or (ii) a substantial outbreak of overt disease occurs.

Keywords *Apis mellifera* · Deformed-wing virus · Varroa mites · Covert and overt infections · Spontaneous transition · Directed graph

1 Introduction

Colonies of the western honey bee *Apis mellifera* are almost always infected by viruses, with infection transmitted via various routes, including horizontally, vertically (from queen to egg), venereally, by physical or biological vectors, and through the environment (particularly stored food resources). One of the most common is deformed-wing virus (DWV), found in more than 80% of colonies in the latest USDA-APHIS survey in the USA. Until recently DWV was considered a minor problem (Rosenkranz et al. 2010), with infections usually without obvious pathology, called *covert* (Evans and Schwarz 2011). This is no longer the case, at least in temperate climates. Overt infection, either acute or chronic, is now common (Martin et al. 2013). Acute overt infection (with deformed wings and early death) may be seen in honey bees infected as pupae, while those infected as adults may exhibit chronic overt infection (with some cognitive deficit and possible reduced longevity). Three infectious classes of honey bees should therefore be distinguished, (i) covert, (ii) acutely overt, and (iii) chronically overt. Note that alternative terms for these different classes of infection are in wide use in the literature, but we follow the usage recommended by De Miranda and Genersch (2010) in their definitive review of DWV. A table in that paper (adapted from Hails et al. (2008)) contains a helpful summary of DWV transmission routes and outcomes.

Wilfert et al. (2016) describe the recent spread of DWV as a global epidemic. Schroeder and Martin (2012) and Martin et al. (2013) state that DWV is 'the most likely candidate responsible for the majority of the colony losses that have occurred across the world over the last 50 years', and 'the key pathogen involved in colony collapse', a conclusion backed up by other studies (Highfield et al. 2009; Genersch et al. 2010). The transformation of the disease from predominantly covert to substantially overt has been crucial. This transformation may have been exacerbated by the use of neonicotinoid pesticides (Di Prisco et al. 2013), but a likely more fundamental cause is the parasitic varroa mite *Varroa destructor* (Highfield et al. 2009; Genersch et al. 2010). These mites were originally found only in colonies of the eastern honey bee *Apis cerana*, but invaded western honey bee *Apis mellifera* populations from the middle of the 20th century onwards, probably as a consequence of commercial transportation of western honey bees to the natural range of the eastern honey bee. Colonies of western honey bees worldwide (except in Australia) are now typically infested. Compared to *Apis cerana*, which employs hygienic methods to defend itself effectively against infestations, *Apis mellifera* are badly affected, and significant infestations often lead to the death of the colony (Dietemann et al. 2012). The mites have two life stages, *phoretic* and *reproductive*. At the phoretic stage, they attach themselves to adult bees and feed on their haemolymph, occasionally moving from one host bee to the next. At the reproductive stage, they move to the brood cells of the colony where they reproduce and feed on larval bees.

Modelling the effect of pathogens on the population dynamics of invertebrates has a long history (Anderson and May 1981) and includes previous work specifically in the context of honey bees, mites and/or virus. Sumpter and Martin (2004) created a model to consider DWV assuming a fixed mite population size. Eberl's group has concentrated on modelling acute bee paralysis virus (ABPV) (Eberl et al. 2010; Ratti et al. 2012, 2015, 2017). Others have not been specific about the virus concerned (Kang et al. 2016; Bernardi and Venturino 2016; Dénes and Ibrahim 2019), although DWV and ABPV are usually considered as examples. Dénes and Ibrahim (2019) take a different approach from others in modelling honey bees according to whether they are infested by mites, and if so whether those mites are infected by virus or not.

We shall consider the effect of the varroa mite and DWV together on the population dynamics of the western honey bee. To do this, we require some insight into the interaction between mites and DWV. There is strong evidence that the virus replicates within the mite (Kevan et al. 2006; Gisder et al. 2009). So the virus may be ingested by a mite at the phoretic stage (in the haemolymph of a covertly or overtly infected adult bee), may replicate within the mite, and may be passed on at high levels to a larval bee in a brood cell when the mite is at the reproductive stage (Yue and Genersch 2005). Typically, the larval bee then shows acute overt symptoms of DWV at the adult stage, with characteristically deformed wings, and dies within 2 or 3 days of emergence (Gisder et al. 2009). The mite acts not simply as a physical vector but as a biological vector for the virus (Kevan et al. 2006), and amplifies the effects of the pathogen from covert to overt. The mites themselves may be infected with DWV at a low or at a high level, depending on whether replication has occurred or not. This determines the effect that they have on their honey-bee hosts, and it is necessary in a model to distinguish these two infectious classes. We shall call infections at a low-level covert, and at a high-level overt, although the mite does not seem to suffer symptoms even from high-level infections.

The most important quantity for an infectious disease is R_0, the basic reproduction number, which determines whether and how widely disease will spread if introduced into an initially disease-free population. For the first time, we shall derive expressions to determine whether $R_0 > 1$ in a dynamically varying honey-bee–mite–virus system, when a disease-free steady state exists. When there is no such steady state we shall introduce a new analysis that determines whether a disease remains predominantly covert (as DWV did before varroa mites became established) or breaks out and becomes a substantially overt disease, leading in the case of DWV to widespread colony losses. To do this, we build our mathematical model sequentially. In the following section, we propose a model to describe the interaction between honey bees and mites. Having established the key dynamic properties of that system, we extend our model to incorporate viral infection which allows us to derive expressions for R_0 as described above. In the conclusions, we discuss the significance of the work presented, both the approach to calculating R_0 and the results in the context of understanding the importance of overt and covert infections and associated spontaneous transitions from covert to overt infections within mite and honey-bee populations.

2 Modelling Interactions Between Honey Bees and Mites

In the absence of viral infection, we define $N(t)$ and $M(t)$ to be the number of honey bees in a colony and mites in that colony at time t. Following previous published work, we make the following model choices and assumptions:

- Bee production depends on the number of workers in the colony since they are necessary to care for the brood and to gather resources for the colony. Consequently, we assume that production $h(N)$ is a saturating function of colony size following (Eberl et al. 2010; Khoury et al. 2011, 2013; Kang et al. 2016) and choose the functional form to follow Eberl's group (Ratti et al. 2012, 2015, 2017):

$$h(N) = \frac{N^2}{A^2 + N^2}$$

  where A^2 is a positive constant.
- We assume that the death rate of bees in a colony due to parasitism by mites is directly proportional to the number of mites in the environment. This differs from previous authors who have used a mass action assumption (Ratti et al. 2012, 2015, 2017; Kang et al. 2016). It was chosen such that the per capita honey-bee death rate due to parasitism is proportional to the number of mites per honey bee which we interpret as a measure of stress on the bee that leads to its increased death rate.
- Mites physically attach themselves to their hosts and so we follow Eberl's group and use a Leslie–Gower approach (Pielou 1977) by assuming that mites grow logistically during the summer months with a carrying capacity that is proportional to the size of the host colony (Ratti et al. 2012, 2015, 2017). In the winter, we assume that mites die at a constant per capita rate.

Using these assumptions, our model for the honey-bee–mite interactions is:

$$\begin{aligned} \frac{dN}{dt} &= f(N, M) = \alpha h(N) - \mu N - \gamma M, \\ \frac{dM}{dt} &= g(N, M) = \begin{cases} rM(1 - M/(kN)) & \text{if } r > 0, \\ -sM & \text{if } r = 0, \end{cases} \end{aligned} \tag{1}$$

with $\alpha > 0$ and $r > 0$ in the growing season, $\alpha = r = 0$ in the winter. The remaining model parameters μ, γ, k and s are all positive constants which take the following meaning: μ is the per capita natural death rate of honey bees; γM is the parasite-related death rate of the bees; r is the intrinsic growth rate of the mites which grow logistically with a carrying capacity kN; the parameter s denotes the per capita death

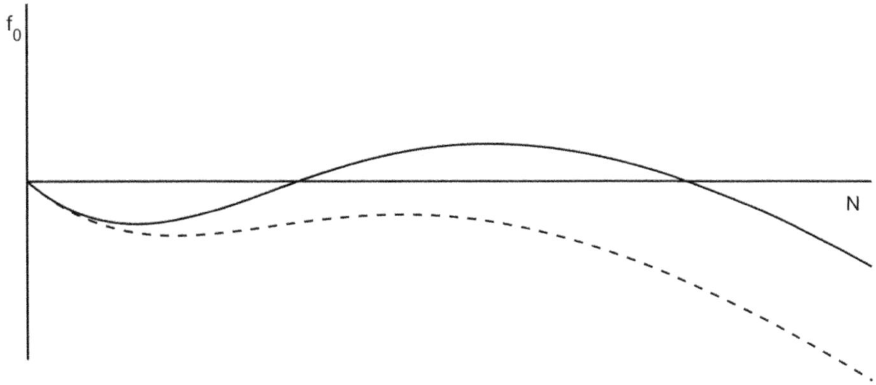

Fig. 1 The function f_0, describing the growth rate of honey bees within a colony in the absence of mites as given in (2). The solid line corresponds to the case $\alpha > 2\mu A$ while the dashed line is for $\alpha < 2\mu A$

rate of mites in the winter period. The parameters vary with time in a temperate climate, even within the growing season, however for the work presented here, we assume that the parameters are constant and focus on the summer period.

2.1 Honey-Bee Dynamics in the Absence of Mites

If $M(0) = 0$, then $M(t) = 0$ for all t, so the N equation becomes

$$\begin{aligned}\frac{\mathrm{d}N}{\mathrm{d}t} &= f_0(N) = f(N, 0) = \alpha h(N) - \mu N \\ &= -N\frac{\mu N^2 - \alpha N + \mu A^2}{N^2 + A^2}.\end{aligned} \tag{2}$$

The function f_0 is as shown in Fig. 1, for $\alpha < 2\mu A$ and $\alpha > 2\mu A$.

The bifurcation structure for this system is easy to analyse. It has a stable steady state at $N = 0$. Two other steady states, $0 < \hat{N}_1 < \hat{N}_2$, the first unstable and the second stable, appear by a saddle–node bifurcation for $\alpha > 2\mu A$. We may write $\hat{N}_i = An_i(\beta)$ for $i = 1, 2$, where $\beta = \alpha/(\mu A)$, and

$$n_1(\beta) = \frac{1}{2}(\beta - \sqrt{\beta^2 - 4}), \qquad n_2(\beta) = \frac{1}{2}(\beta + \sqrt{\beta^2 - 4}), \tag{3}$$

real and positive for $\beta > 2$, or $\alpha > 2\mu A$. The bifurcation diagram is as shown in Fig. 2.

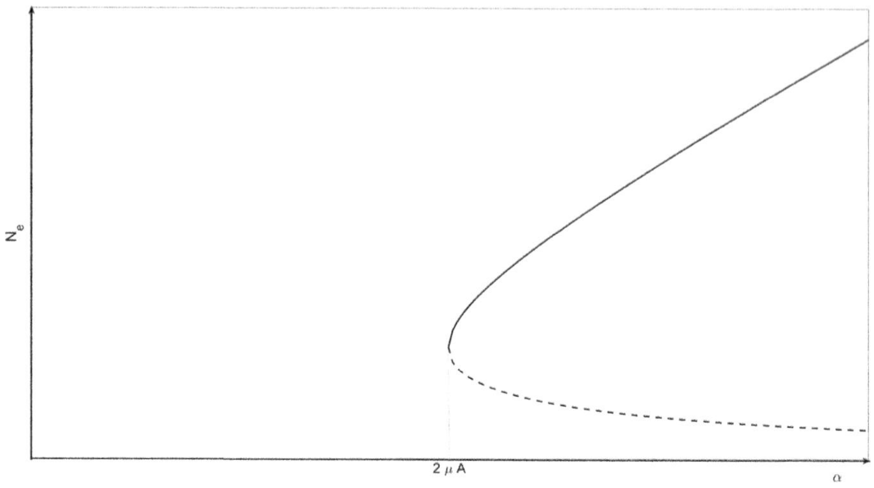

Fig. 2 Bifurcation diagram showing how the equilibrium values, N_e, of (2) and their stability vary as the parameter α (corresponding to the growth in colony size) increases. For small α the colony cannot be maintained; once the critical threshold $\alpha = 2\mu A$ is exceeded, the colony exhibits bistable dynamics. The solid line corresponds to locally stable equilibria, $N_e = 0$ and $N_e = \hat{N}_2$ while the dotted line represents an unstable equilibrium, $N_e = \hat{N}_1$

2.2 Analysis of the Full System

During the growing season, our model system (1) can be written as,

$$\begin{aligned}\frac{dN}{dt} &= f(N, M) = \alpha h(N) - \mu N - \gamma M,\\ \frac{dM}{dt} &= g(N, M) = rM\left(1 - \frac{M}{kN}\right).\end{aligned} \tag{4}$$

The system has a singularity at $N = 0$. But the transformation to N, $\varphi = M/N$ leads to

$$\begin{aligned}\frac{dN}{dt} &= \alpha h(N) - \mu N - \gamma N\varphi,\\ \frac{d\varphi}{dt} &= r\varphi(1 - \varphi/k) - \alpha\varphi h(N)/N + \mu\varphi - \gamma\varphi^2,\end{aligned}$$

which has no singularities and is in Kolmogorov form, so that the positive quadrant of (N, φ) space is positively invariant. It follows that the positive quadrant of (N, M) space is positively invariant, despite the singularity in g and the $-\gamma M$ term in the N equation.

The nullclines and steady states for this system are as follows. For $g(N, M) = 0$, either $M = \hat{G}(N) = 0$ or $M = G^*(N) = kN$. For $f(N, M) = 0$, $M = F(N) = (1/\gamma) f_0(N)$, where f_0 is as in (2). The nullclines $M = F(N)$ and $M = \hat{G}(N) = 0$

intersect where

$$\hat{Q}(N) = \mu N^2 - \alpha N + \mu A^2 = 0.$$

The roots of the quadratic $\hat{Q}(N) = 0$ are given as before by $\hat{N}_i = An_i(\beta)$, where $\beta = \alpha/(\mu A)$ and the functions n_i are defined in (3), real and positive if $\beta > 2$, $\alpha > 2\mu A$. The nullclines $M = F(N)$ and $M = G^*(N) = kN$ intersect where

$$Q^*(N) = (\mu + k\gamma)N^2 - \alpha N + (\mu + k\gamma)A^2 = 0.$$

This quadratic is the same as $\hat{Q}$ above but with μ replaced by $\mu + k\gamma$. Its roots are given by $N_i^* = An_i(\beta)$, where the functions n_i are again as in (3) but now $\beta = \alpha/((\mu + k\gamma)A)$, and are real and positive if $\beta > 2$, $\alpha > 2(\mu + k\gamma)A$.

There are therefore three cases.

(i) $\alpha < 2\mu A$: the curve $M = F(N)$ never enters the positive quadrant.
(ii) $2\mu A < \alpha < 2(\mu + k\gamma)A$: the curve $M = F(N)$ is in the positive quadrant between $\hat{N}_1$ and $\hat{N}_2$, but never intersects the straight line $M = G^*(N) = kN$.
(iii) $\alpha > 2(\mu + k\gamma)A$: the curve $M = F(N)$ is in the positive quadrant between $\hat{N}_1$ and $\hat{N}_2$, and intersects the straight line $M = G^*(N)$ at N_1^* and N_2^*. It is clear that $\hat{N}_1 < N_1^* < N_2^* < \hat{N}_2$.

We shall consider each case in turn.

Case (i), $\alpha < 2\mu A$.

The set D given by

$$D = \{(N, M) \mid 0 < N < C,\ 0 < M < kC\}$$

is positively invariant for any positive constant C, so that the origin is globally asymptotically stable. The honey-bee production rate is not under any circumstances sufficient to outweigh the death rate. Henceforth, we shall assume that $\alpha > 2\mu A$.

Case (ii), $2\mu A < \alpha < 2(\mu + k\gamma)A$

There are no periodic solutions in $\mathbb{R}_+^2$ (since there are no steady states there). No trajectories approach $(\hat{N}_1, 0)$ and the only trajectories to approach $(\hat{N}_2, 0)$ do so along the N axis. The set D given by

$$D = \{(N, M) \mid 0 < N < C,\ 0 < M < kC\}$$

is positively invariant for any constant $C > \hat{N}_2$ (and for $0 < C < \hat{N}_1$). By the Poincaré–Bendixson theorem, the origin is globally asymptotically stable (in the strictly positive quadrant). This case is shown graphically in Fig. 3.

Case (iii), $\alpha > 2(\mu + k\gamma)A$.

There are semi-trivial (mite-free) steady states at $(\hat{N}_1, 0)$ and $(\hat{N}_2, 0)$, and non-trivial steady states at (N_1^*, M_1^*) and (N_2^*, M_2^*), where $M_1^* = kN_1^*$ and $M_2^* = kN_2^*$.

The character of each steady state is determined by the Jacobian matrix J, where

$$J(N, M) = \begin{pmatrix} \alpha h'(N) - \mu & -\gamma \\ (rM^2)/(kN^2) & r\,(1 - (2M)/(kN)) \end{pmatrix}.$$

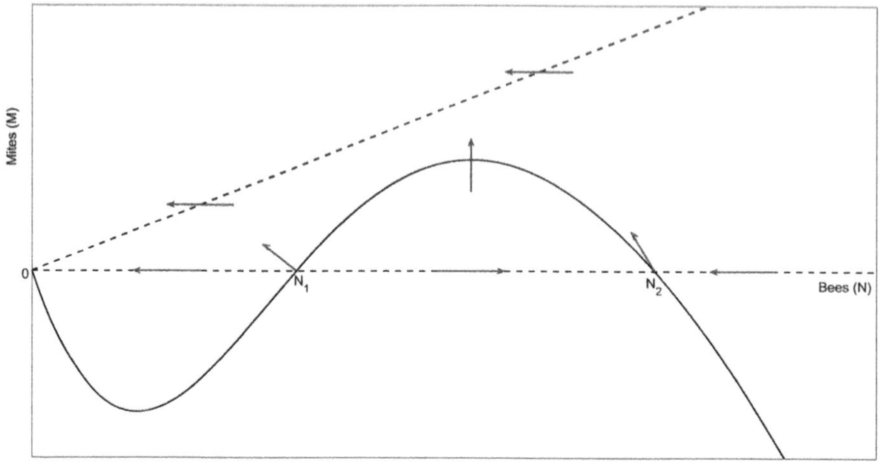

Fig. 3 The phase plane for Case (ii) with medium honey-bee production rate, $2\mu A < \alpha < 2(\mu + k\gamma)A$. The dashed lines represent the M nullclines along which $g(N, M) = 0$ and the solid line represents the N nullcline along which $f(N, M) = 0$. The red arrows show the direction of solution trajectories as time increases. The origin is globally asymptotically stable in the strictly positive quadrant

For $(\hat{N}_1, 0)$ and $(\hat{N}_2, 0)$,

$$J(\hat{N}, 0) = \begin{pmatrix} \alpha h'(\hat{N}) - \mu & -\gamma \\ 0 & r \end{pmatrix},$$

with eigenvalues $\alpha h'(\hat{N}) - \mu$ and r. The slope of the curve $M = F(N)$ is $F'(N) = (\alpha h'(N) - \mu)/\gamma$. Hence, $\alpha h'(\hat{N}_1) - \mu > 0$, $(\hat{N}_1, 0)$ is an unstable node. And $\alpha h'(\hat{N}_2) - \mu < 0$, $(\hat{N}_2, 0)$ is a saddle point.

For (N_1^*, M_1^*) and (N_2^*, M_2^*),

$$J^* = J(N^*, M^*) = \begin{pmatrix} \alpha h'(N^*) - \mu & -\gamma \\ rk & -r \end{pmatrix}.$$

At (N_1^*, M_1^*), the slope of the curve $M = F(N)$ is greater than k, $\alpha h'(N_1^*) - \mu > k\gamma$, so

$$\det J_1^* = \det J(N_1^*, M_1^*) = -r(\alpha h'(N_1^*) - \mu) + rk\gamma < 0,$$

and (N_1^*, M_1^*) is a saddle point.

Similarly $\alpha h'(N_2^*) - \mu < k\gamma$, so $\det J_2^* = \det J(N_2^*, M_2^*) > 0$, and (N_2^*, M_2^*) is a stable or unstable node or focus. Also, $\operatorname{tr} J_2^* = \alpha h'(N_2^*) - \mu - r < k\gamma - r$, so $\operatorname{tr} J_2^* < 0$ and (N_2^*, M_2^*) is stable as a solution of the disease-free system if r is sufficiently large, and in particular if

$$r > k\gamma. \tag{5}$$

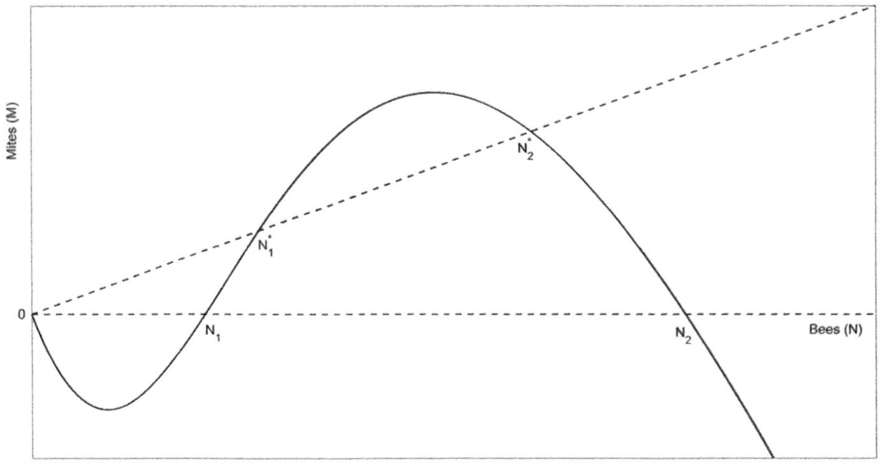

Fig. 4 The phase plane for Case (iii) with high honey-bee production rate, $2(\mu + k\gamma)A < \alpha$. As above, the dashed lines represent the M nullclines along which $g(N, M) = 0$ and the solid line represents the N nullcline along which $f(N, M) = 0$. The phase plane exhibits bistable properties such that there are two locally stable equilibrium points, the origin and N_2^*, separated by an unstable equilibrium N_1^*

It is not globally stable because of the Allee effect built into the model. In fact, the analysis of case (ii) restricted to the positively invariant set D given by

$$D = \{(N, M) \mid 0 < N < C,\ 0 < M < kC\}$$

with $0 < C < \hat{N}_1$ shows that the origin is still asymptotically stable (but of course no longer globally asymptotically stable). This case is shown graphically in Fig. 4.

3 Modelling Viral Infection within the Honey-Bee and Mite Ecosystem

Central to our assumption that infection classes in the bee populations should be compartmentalised we divide the bee colony according to infection status using the empirical evidence that very few bees in a virus-infected colony are uninfected (Yue and Genersch 2005). At time t, individual bees may be in one of three states: covertly infected $X(t)$, chronically overtly infected $Y(t)$ or acutely infected $Z(t)$ such that

$$X + Y + Z = N.$$

Similarly, very few mites in a DWV-infected colony are virus-free (Anguiano-Baez et al. 2016), and we neglect them. The virus titre in infected mites varies from around 10^8 particles to 10^{10} or as much as 10^{12} viral genome equivalents (Gisder et al. 2009), depending on whether or not virus replication has taken place in the mite. Therefore, we create two compartments for the mite population at time t: covert infected $U(t)$

and overt infected $V(t)$ such that

$$U + V = M.$$

Once overtly infected, bees and mites remain overtly infected throughout their life. Figure 5a and b shows the transfer between these classes separately for the bees and the mites and should be used as an aide memoire as we now describe the demographic and infection processes that we combine together to describe viral infection within the honey-bee and mite ecosystem.

Honey-bee production The infection status of adult bees emerging from brood cells will clearly impact the model dynamics. Several different approaches have been taken in the literature each arising from different underlying assumptions (Sumpter and Martin 2004; Eberl et al. 2010; Ratti et al. 2012, 2017; Bowen-Walker et al. 1999; Bernardi and Venturino 2016; Kang et al. 2016). We follow an approach that is closest to Sumpter and Martin (2004) who model the infection status of adult bees emerging from brood cells phenomenologically, using data suggesting that phoretic mites entering the reproductive stage (whose infection status is part of the model) are Poisson distributed among brood cells (Martin 1995; Salvy et al. 1999). This phenomenon directly impacts the class into which newborn bees emerge: with probability e^{-cV}, where c is a positive constant, a newborn is covertly infected and contributes to population X and with probability $1 - e^{-cV}$, they are acutely infected and contribute to population Z. In the case of acute infection, there is an additional chance of mortality and so the birth rate of acutely infected bees is reduced by a factor p, $0 \leq p < 1$, compared with the covertly infected newborns.

We also assume that infection status of the queen and workers impacts their ability to produce viable eggs in the following ways:

- The queen bee is covertly infected, and transmits the covert but not the overt disease to her offspring;
- Acutely overtly infected bees do not contribute to production;
- Chronically overtly infected bees make a reduced contribution with parameter κ (Hails et al. 2008; Sumpter and Martin 2004).

Honey-bee mortality The natural per capita death rates μ, ν and ζ of X, Y and Z bees depend on the bee's infection status, such that $\mu < \nu < \zeta$ (Hails et al. 2008). Mite-related honey-bee death is assumed to be independent of infection status of the bee with rate parameter γ and assuming a frequency-dependent functional form.

Mite production and mortality There is no evidence in the literature that mites are affected by DWV and so we assume that there is no negative impact on mites that have the virus. Consequently, mite dynamics are assumed to follow those described in Sect. 2 with an intrinsic growth rate r, carrying capacity kN and winter mortality rate s, independent of infection status. Overt disease may be transmitted vertically and so we assume that a fraction θ of new infections from overtly infected mites produce overtly infected mites; the remainder are covertly infected.

Rates of infection transmission for adult bees and mites Transmission of infection for both adult bees and mites results in movement between covert and overt infected status both due to interactions between individuals and as a result of spontaneous

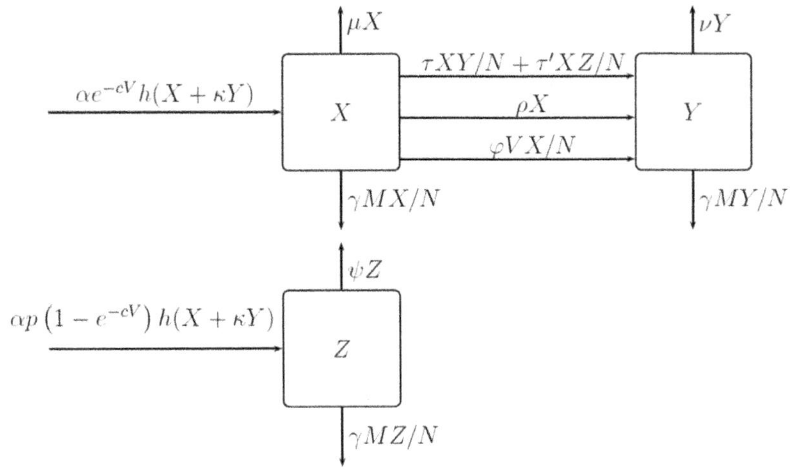

(a) Transfer diagram for honey-bees between the three classes covertly infected, chronically overtly infected and acutely overtly infected classes.

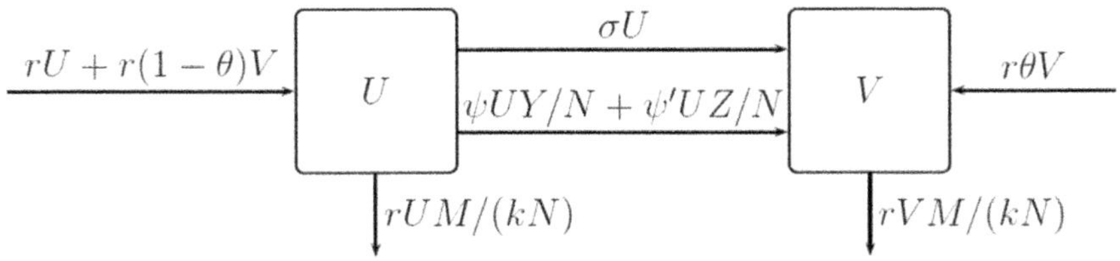

(b) The transfer diagram for mite between the two classes of covert and overt infection.

Fig. 5 Transfer diagrams showing the flow between infection classes separately for the honey bees, shown in (**a**), and mites, shown in (**b**). Justification of the different components and their functional forms is given in the text

transition where growth in viral load within an individual results in a change in the status. We assume the following transmission routes:

- Horizontal transmission of chronic overt infection from one adult bee to another does occur (Hails et al. 2008). Disease may be transmitted in its chronic overt form from a chronically overtly or an acutely overtly infectious bee to a covertly infectious bee, with frequency-dependent infectious contact parameter τ and τ', respectively. This may be a minor route of transmission.

- Disease may be transmitted from mites to bees and vice versa, with φ the infectious contact parameter from mites to bees, ψ (respectively, ψ' for acutely overtly infected bees) that from bees to mites, with frequency-dependent transmission.
- Spontaneous transition from covert to chronic overt disease may occur in an adult bee (Sumpter and Martin 2004; Nazzi et al. 2012; Di Prisco et al. 2013), by viral replication. We assume this happens at a per capita rate ρ. Such transitions may be rare. We assume that spontaneous transitions from covert to acute overt disease in larval bees do not occur, as acutely overtly infected bees are not observed in mite-free colonies.
- Spontaneous transition from covert to overt infection has been observed in mites (Kevan et al. 2006; Yue and Genersch 2005; Gisder et al. 2009). We include them in our model assuming a per capita rate of transition σ. Autonomous spontaneous transitions in the opposite direction may possibly occur but the evidence for them is less clear and we have not included them. We note that spontaneous transition in bees in particular may be rare.

Combining these model components and assumptions, we present our model system (for the spring, summer and autumn periods) takes the form:

$$\begin{aligned}
\frac{\mathrm{d}X}{\mathrm{d}t} =& \alpha e^{-cV} h(X+\kappa Y) - \mu X - \gamma M\frac{X}{N} - \rho X - \tau X\frac{Y}{N} - \tau' X\frac{Z}{N} - \varphi\frac{X}{N}V, \\
\frac{\mathrm{d}Y}{\mathrm{d}t} =& \rho X + \tau X\frac{Y}{N} + \tau' X\frac{Z}{N} - \nu Y - \gamma M\frac{Y}{N} + \varphi\frac{X}{N}V, \\
\frac{\mathrm{d}Z}{\mathrm{d}t} =& \alpha p(1-e^{-cV})h(X+\kappa Y) - \zeta Z - \gamma M\frac{Z}{N}, \\
\frac{\mathrm{d}U}{\mathrm{d}t} =& rU + r(1-\theta)V - rU\frac{M}{kN} - \sigma U - \psi U\frac{Y}{N} - \psi' U\frac{Z}{N}, \\
\frac{\mathrm{d}V}{\mathrm{d}t} =& r\theta V - rV\frac{M}{kN} + \sigma U + \psi U\frac{Y}{N} + \psi' U\frac{Z}{N},
\end{aligned} \tag{6}$$

where $h(N) = N^2/(A^2+N^2)$ and $\mu < \nu < \zeta$.

In winter, there is no brood, so $\alpha = 0$, $r = 0$, and the equations for U and V are replaced by

$$\frac{\mathrm{d}U}{\mathrm{d}t} = -sU, \quad \frac{\mathrm{d}V}{\mathrm{d}t} = -sV.$$

3.1 Analysis of the Mite-Free Model for Bees and Disease

Acutely overtly infected (Z) bees are only produced by vector-borne transmission, so we take $Z = 0$. The X and Y equations are given by

$$\begin{aligned}\frac{\mathrm{d}X}{\mathrm{d}t} &= \alpha h(X + \kappa Y) - \mu X - \rho X - \tau X\frac{Y}{N},\\ \frac{\mathrm{d}Y}{\mathrm{d}t} &= -\nu Y + \rho X + \tau X\frac{Y}{N}.\end{aligned} \tag{7}$$

We seek steady states (X^*, Y^*), with $X^* + Y^* = N^*$, and $x^* = X^*/N^*$, $y^* = Y^*/N^*$, and $x^* + y^* = 1$. Then, from the Y equation,

$$0 = -\nu y^* + \rho(1 - y^*) + \tau y^*(1 - y^*), \tag{8}$$

a quadratic equation with one negative root and one root between 0 and 1. From now on, let y^* denote the root between 0 and 1 and let $x^* = 1 - y^*$, also between 0 and 1. The sum of the equations in (7) at steady state then gives

$$\alpha h(\xi N^*) = \mu\eta N^*, \tag{9}$$

where $\xi = x^* + \kappa y^* < 1$, $\mu\eta = \mu x^* + \nu y^* > \mu$, or $\eta > 1$, since $\kappa < 1$, $\nu > \mu$. (Note that the expressions ξ and η may be given explicitly in terms of the parameters of the system.) This is just a rescaled version of the standard equation $\alpha h(N) = \mu N$, from (2), with solutions given by (3). We can therefore immediately give the solutions as

$$N_i^* = \frac{A}{\xi} n_i\left(\frac{\alpha\xi}{\mu\eta A}\right), \tag{10}$$

positive and realistic for $\alpha\xi > 2\mu\eta A$. The expression simplifies if Y bees are not dysfunctional compared to X bees, $\kappa = 1$ and $\nu = \mu$, since then $\xi = 1$ and $\eta = 1$. Unsurprisingly, the more dysfunctional Y bees are compared to X bees, in other words the smaller κ is and/or the larger ν is, the larger the production rate α has to be to prevent the colony collapsing to zero. We shall now consider three special cases.

3.1.1 Case (i): No Horizontal Transmission from Bee to Bee, $\tau = 0$

The assumption that $\tau = 0$ is an assumption that all previous models except Kang et al. (2016) have made. The resulting Y equation is given by

$$\frac{\mathrm{d}Y}{\mathrm{d}t} = \rho X - \nu Y,$$

so that $X^* = \nu N^*/(\nu + \rho)$, $Y^* = \rho N^*/(\nu + \rho)$. There is no Y-free steady state, except $(0, 0)$. Every colony with $\rho > 0$ contains bees with overt disease, although if ρ

is small (see cases (ii) and (iii) below) there are very few of them, and if $\rho = 0$ there are none.

3.1.2 Case (ii): No Spontaneous Transition in Bees, $\rho = 0$

All previous models have made this assumption, usually with $\tau = 0$ as well. It is likely to be at least a good approximation, unless the bees' immune system has been compromised by mites or neonicotinoids (Di Prisco et al. 2013). The transition from X to Y occurs by contact instead of spontaneously, and the equations are

$$\begin{aligned}\frac{\mathrm{d}X}{\mathrm{d}t} &= \alpha h(X+\kappa Y) - \mu X - \tau X\frac{Y}{N} = f(X,Y),\\ \frac{\mathrm{d}Y}{\mathrm{d}t} &= -\nu Y + \tau X\frac{Y}{N} = g(X,Y).\end{aligned} \tag{11}$$

Seeking steady states $(X^*, Y^*) = (N^*x^*, N^*y^*)$, as before, Eq. (8) becomes $-\nu y^* + \tau y^*(1-y^*) = 0$, so (a) $y^* = 0$ or (b) $x^* = 1-y^* = \nu/\tau$, $y^* = 1-\nu/\tau$, realistic if and only if $\nu < \tau$. For alternative (b), the quantity N^* may then be calculated from the sum of the X and the Y equation in the usual way, leading to Eq. (9) with $\xi = \kappa+(1-\kappa)\nu/\tau$, $\mu\eta = \mu\nu/\tau + \nu(1-\nu/\tau)$, and two solutions given by Eq. (10), positive and realistic for $\alpha\xi > 2\mu\eta A$. Let us now consider alternative (a), with $Y^* = y^* = 0$. The steady-state X equation then becomes very familiar, $\alpha h(X^*) = \mu X^*$, with two solutions $X^* = N^* = \hat{N}_i = An_i(\alpha/(\mu A))$ for $i = 1, 2$, realistic for $\alpha > 2\mu A$.

One overt-disease-free steady state always exists, the trivial steady state (0, 0), and there are two more given by $(\hat{N}_1, 0)$ and $(\hat{N}_2, 0)$ as long as $\alpha > 2\mu A$. There are also two steady states with overt disease, $(X_1^*, Y_1^*) = (N_1^*x^*, N_1^*y^*)$ and $(X_2^*, Y_2^*) = (N_2^*x^*, N_2^*y^*)$, described above, as long as both $\nu < \tau$ and $\alpha\xi > 2\mu\eta A$.

The overt-disease-free steady states appear by a saddle–node bifurcation at $\alpha = 2\mu A$, and we know from bifurcation theory (as in Sect. 2.1) that $(\hat{N}_1, 0)$ is unstable and $(\hat{N}_2, 0)$ stable as solutions of the Y-free system. We wish to test whether $(\hat{N}_2, 0)$ is stable as a solution of the full system (11). So consider introducing an overtly infected Y bee (a *primary*) into the system at the steady state $(\hat{N}_2, 0)$. While the primary is in the Y compartment it makes infectious contacts at rate $\tau X/N = \tau$ at the steady state. Bees in the Y compartment leave it at per capita rate ν, so they spend time $1/\nu$ in the compartment on average. Hence, the primary makes an expected $R_0^- = \tau/\nu$ infectious contacts. R_0^- is called the *basic reproduction number*. There is a threshold $R_0^- = 1$ for spread of overt disease: it spreads if $R_0^- > 1$ but not if $R_0^- < 1$. So the disease spreads if $\tau > \nu$, but not if $\tau < \nu$. If $\tau < \nu$ there is no steady state with overt disease. If $\tau > \nu$ and $\alpha\xi < 2\mu\eta A$, there is still no steady state with overt disease. The trajectory starting with a perturbation from $(\hat{N}_2, 0)$ must tend to (0, 0). If $\tau > \nu$ and $\alpha\xi > 2\mu\eta A$ there are two steady states with overt disease. The trajectory starting with a perturbation from $(\hat{N}_2, 0)$ tends either to (0, 0) or to (X_2^*, Y_2^*), unless there is a Hopf bifurcation allowing periodic solutions about (X_2^*, Y_2^*). This system is essentially an SI model, and its bifurcation behaviour is quite different from the system with spontaneous transition, $\rho > 0$. In particular, there is a threshold value $R_0^- = \tau/\nu = 1$ below which overt disease cannot exist. The notation R_0^- emphasises that this is the basic reproduction number *in the absence of mites*. The difference between this and the corresponding basic reproduction number R_0 with mites present that we shall discuss in the next section is the basis for the different behaviours of infected bee colonies with and without mites, and for an explanation of the grievous effect of mites on bee colonies.

3.1.3 Case (iii): Very Little Spontaneous Transition in Bees, ρ Small

This is likely to be a good assumption unless the bees' immune systems are compromised, which could be because of high levels of infestation by mites (Nazzi et al. 2012) or because of high levels of neonicotinoid pesticides in the environment (Di Prisco et al. 2013). Equation (8) still holds, with solutions $y^* = \rho/(\nu - \tau) + O(\rho^2)$, $y^* = 1 - \nu/\tau + O(\rho)$. If $\nu > \tau$ the first of these is realistic, and the corresponding steady states have low ($O(\rho)$) prevalence of overt disease, while if $\nu < \tau$ the second is, and the steady states have $O(1)$ prevalence. Strictly speaking there is no basic reproduction number, but $R_0 = \tau/\nu$ still has a role to play: overt disease is maintained at a low level for $R_0^- < 1$ but not for $R_0^- > 1$.

3.2 The Complete System

We shall now analyse the complete system (6), with honey bees, varroa mites, and DWV. We recall that in spring, summer and autumn, the model equations are given by

$$\begin{aligned}
\frac{\mathrm{d}X}{\mathrm{d}t} &= \alpha e^{-cV} h(X + \kappa Y) - \mu X - \gamma M\frac{X}{N} - \rho X - \tau X\frac{Y}{N} - \tau' X\frac{Z}{N} - \varphi\frac{X}{N}V, \\
\frac{\mathrm{d}Y}{\mathrm{d}t} &= \rho X + \tau X\frac{Y}{N} + \tau' X\frac{Z}{N} - \nu Y - \gamma M\frac{Y}{N} + \varphi\frac{X}{N}V, \\
\frac{\mathrm{d}Z}{\mathrm{d}t} &= \alpha p(1 - e^{-cV})h(X + \kappa Y) - \zeta Z - \gamma M\frac{Z}{N}, \\
\frac{\mathrm{d}U}{\mathrm{d}t} &= rU + r(1-\theta)V - rU\frac{M}{kN} - \sigma U - \psi U\frac{Y}{N} - \psi' U\frac{Z}{N}, \\
\frac{\mathrm{d}V}{\mathrm{d}t} &= r\theta V - rV\frac{M}{kN} + \sigma U + \psi U\frac{Y}{N} + \psi' U\frac{Z}{N}.
\end{aligned}$$

3.2.1 Case (i): No Spontaneous Transition in Bees and Mites, $\rho = \sigma = 0$

Motivated by our analysis of the mite-free system, we shall start by assuming that there is no spontaneous transition to overt disease, $\rho = \sigma = 0$. Let $\alpha > 2A(\mu + k\gamma)$ (so N_2^* exists) and let r be so large that (N_2^*, M_2^*) is stable as a solution of the disease-free system (e.g. $r > k\gamma$, see Eq. (5)).

We shall analyse this system using the next-generation matrix method (Diekmann et al. 1990; Van den Driessche and Watmough 2008). The next-generation matrix K is a generalisation of the basic reproduction number R_0. In this case, it is 3×3, with rows and columns related to the three overt disease classes Y, Z, and V, which are zero in an overt-disease-free steady state. It is given by $K = FD^{-1}$, where F is the matrix whose component F_{ij} gives the rate at which individuals in overtly infected class i arise through infection by those in class j near the steady state, and D_{ii} denotes the rate at which those in class i leaves that class through death. (In general, we would have to consider those that entered a particular infected class by transition from another infected class, and those that left an infected class other than through death, but there are no such processes in this system.) Then, K_{ij} denotes the number of disease offspring a primary of class j produces in class i throughout the life-time of its disease. It may be shown that the basic reproduction number R_0 for the system at the steady state is given by the largest eigenvalue of K. Here, the next-generation matrix K about the overt-disease-free steady state S_2^* is given by

$$\begin{aligned} K &= \begin{pmatrix} K_{YY} & K_{YZ} & K_{YV} \\ K_{ZY} & K_{ZZ} & K_{ZV} \\ K_{VY} & K_{VZ} & K_{VV} \end{pmatrix} \\ &= \begin{pmatrix} \tau & \tau' & \varphi \\ 0 & 0 & \alpha p c h_2^* \\ k\psi & 0 & r\theta \end{pmatrix} \begin{pmatrix} 1/(\nu + k\gamma) & 0 & 0 \\ 0 & 1/(\zeta + k\gamma) & 0 \\ 0 & 0 & 1/r \end{pmatrix} \\ &= \begin{pmatrix} \tau/(\nu + k\gamma) & \tau'/(\zeta + k\gamma) & \varphi/r \\ 0 & 0 & \alpha p c h_2^*/r \\ k\psi/(\nu + k\gamma) & k\psi'/(\zeta + k\gamma) & \theta \end{pmatrix}, \end{aligned} \tag{12}$$

where we have written $h_2^* = h(N_2^*)$. The characteristic polynomial P of K is given by

$$\begin{aligned} P(\lambda) =& -\lambda(\lambda - \theta)\left(\lambda - \frac{\tau}{\nu + k\gamma}\right) + \left(\frac{k\varphi\psi}{r(\nu + k\gamma)} + \frac{\alpha p c h_2^* k\psi'}{r(\zeta + k\gamma)}\right)\lambda \\ &+ \frac{\alpha p c h_2^*}{r}\left(\frac{k\psi\tau' - k\psi'\tau}{(\nu + k\gamma)(\zeta + k\gamma)}\right). \end{aligned} \tag{13}$$

The roots of the characteristic equation $P(\lambda) = 0$ are the eigenvalues of K. This is a cubic with a negative λ^3 coefficient, so it has at least one root greater than 1 (and hence $R_0 > 1$) if $P(1) > 0$. Hence, $P(1) > 0$ is sufficient for $R_0 > 1$, but it is not necessary: a cubic P with $P(1) < 0$ may have two roots greater than 1. In that case, it may be easier to check whether $R_0 > 1$ by considering circuits of transmission of overt infection in a directed graph.

Arcs of transmission are shown in Fig. 6. We shall define a (single-generation) circuit of transmission as a sequence of arcs starting from node i and ending at node i but not otherwise visiting node i. The simplest circuits of transmission are arcs direct from a node to itself: mites V infecting other mites (which then enter V) vertically, V→V; and bees Y infecting other bees (which then enter Y) horizontally, Y→Y. We shall denote the partial basic reproduction numbers for these processes as

$$R_0^V = K_{VV} = \theta$$

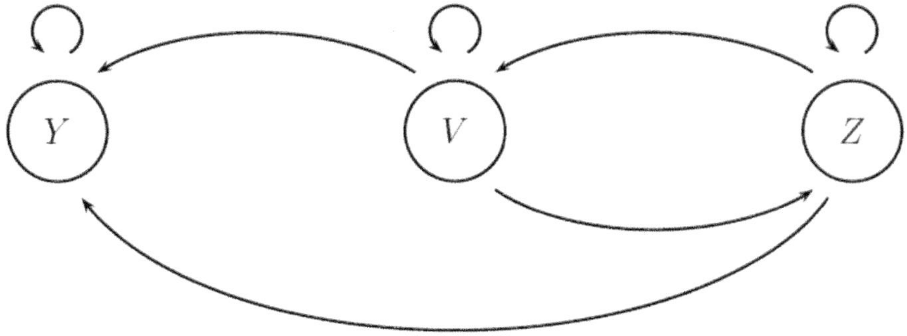

Fig. 6 Arcs of transmission for overt infection. There is an arc (or directed edge) from i to j if the component K_{ij} in the next-generation matrix K is positive. The arc weights (not shown in the diagram) are the components K_{ij}

and

$$R_0^Y = K_{YY} = \tau/(\nu + k\gamma).$$

The circuits of length 2 are Y bees infecting mites (which then enter V) infecting other bees (which then enter Y), Y→V→Y, and Z bees infecting mites (which then enter V) infecting other bees (which then enter Z), Z→V→Z. We shall denote the partial basic reproduction numbers for these processes as R_0^{YV} and R_0^{VZ}, where

$$(R_0^{YV})^2 = K_{YV}K_{VY} = (k\varphi\psi)/(r(\nu + k\gamma))$$

and

$$(R_0^{VZ})^2 = K_{VZ}K_{ZV} = (\alpha pch_2^* k\psi')/(r(\zeta + k\gamma)).$$

Finally, there is a circuit of length 3, mites V infecting bees (which then enter Z) infecting other bees (which then enter Y) infecting mites (which then enter V), V→Z→Y→V. We shall denote the partial basic reproduction number for this process as R_0^{VZY}, where

$$(R_0^{YVZ})^3 = K_{YV}K_{VZ}K_{ZY} = (\alpha pch_2^* k\psi\tau')/(r(\nu + k\gamma)(\zeta + k\gamma)).$$

How many secondaries in class Y does a primary in class Y produce? We need to count the circuits in the graph in Fig. 6 that start at Y and end at Y. First, there is the circuit Y→Y, with partial basic reproduction number R_0^Y. Then, there is the circuit Y→V→Y, but before returning from V to Y one may traverse the circuits V→V and V→Z→V an indefinite number of times and in any order. The partial basic reproduction number for all these circuits is

$$(R_0^{YV})^2 \left(1 + (R_0^V + (R_0^{VZ})^2) + (R_0^V + (R_0^{VZ})^2)^2 + \cdots\right).$$

If $R_0^V + (R_0^{VZ})^2 > 1$, then this is unbounded and the disease invades whatever the other parameters of the system, while if $R_0^V + (R_0^{VZ})^2 < 1$ we may write this as $(R_0^{YV})^2/(1 - R_0^V - (R_0^{VZ})^2)$. Finally, there is the circuit Y→V→Z→Y, where from V one may again traverse the circuits V→V and V→Z→V an indefinite number of times and in any order, to obtain invasion if $R_0^V + (R_0^{VZ})^2 > 1$ and $(R_0^{YVZ})^3/(1 - R_0^V - (R_0^{VZ})^2)$ if $R_0^V + (R_0^{VZ})^2 < 1$. The condition for growth, that there are more secondaries than primaries, should be that either (i)

$$R_0^V + (R_0^{VZ})^2 > 1 \tag{14}$$

or (ii) $R_0^V + (R_0^{VZ})^2 < 1$ and

$$R_0^Y + \frac{(R_0^{YV})^2 + (R_0^{YVZ})^3}{1 - R_0^V - (R_0^{VZ})^2} > 1. \tag{15}$$

Note that this second inequality is always satisfied if $R_0^Y > 1$. The expression on the left-hand side is not R_0, which is defined as the leading eigenvalue of a linear operator, but is the number of secondaries in class Y produced by a primary in class Y, and hence gives the same condition for growth that R_0 does. Indeed, the characteristic equation $P(\lambda) = 0$ reduces to

$$\begin{aligned} P(\lambda) = &-\lambda(\lambda - R_0^V)(\lambda - R_0^Y) \\ &+ ((R_0^{YV})^2 + (R_0^{VZ})^2)\lambda + ((R_0^{YVZ})^3 - R_0^Y(R_0^{VZ})^2) = 0, \end{aligned}$$

and the condition $P(1) > 0$, equivalent to $R_0 > 1$, reduces to

$$(R_0^Y - 1)(1 - R_0^V - (R_0^{VZ})^2) + (R_0^{YV})^2 + (R_0^{YVZ})^3 > 0,$$

or (15) if $R_0^V + (R_0^{YV})^2 < 1$.

Case (i)(a): $\rho = \sigma = 0$, $\tau = \tau' = 0$.

The parameters τ and τ' for horizontal transmission from bee to bee are small (so small that such transmission has not been included in most previous models). Let us assume that they are negligible, $\tau = \tau' = 0$, while maintaining the assumption that there is no spontaneous transition, $\rho = \sigma = 0$. The edges Y→Y and Z→Y in the graph disappear (so that $R_0^Y = (R_0^{YVZ})^3 = 0$), and it is easier to count the circuits from V back to V rather than from Y back to Y. They are V→V, V→Y→V and V→Z→V. The condition that there are more secondaries than primaries, equivalent to the condition that $R_0 > 1$, is therefore

$$R_0^V + (R_0^{VZ})^2 + (R_0^{YV})^2 > 1.$$

This is the inequality (15) with $R_0^Y = (R_0^{YVZ})^3 = 0$. All the terms are mite-related, and it is therefore the mites that drive overt disease in honey-bee colonies in this case, as we might have inferred from Sect. 3.1.1.

Case(i)(b): $\rho = \sigma = 0$, ζ large.

Here, we exploit the fact that the parameter ζ is large (since Z bees only survive 2 or 3 days compared to more than 20 days for X and Y bees). Then, the characteristic equation (13) has a root close to zero, $\lambda_1 = \alpha pch_2^*(\psi\tau' - \psi'\tau)/(\varphi\psi(\zeta + k\gamma))$. To leading order, the other two roots are the solutions of the quadratic

$$Q(\lambda) = (\lambda - \theta)\left(\lambda - \frac{\tau}{\nu + k\gamma}\right) - \frac{k\varphi\psi}{r(\nu = k\gamma)},$$

which are $\lambda_2 < \min\{\theta, \tau/(\nu + k\gamma)\}$ and $\lambda_3 = R_0 > \max\{\theta, \tau/(\nu + k\gamma)\}$. The characteristic equation has a single root greater than 1 if and only if $Q(1) < 0$, or, in terms of partial basic reproduction numbers,

$$R_0^Y + \frac{(R_0^{YV})^2}{1 - R_0^V} > 1. \tag{16}$$

3.2.2 Case (ii): ρ and σ Small, ζ Large

Now let us relax the assumption that $\rho = \sigma = 0$, so spontaneous transition to overt disease does occur in bees and mites, as is realistic. To simplify the algebra, we shall retain the realistic assumption that ζ is large, so that we can neglect the Z class and the associated equation. Then, seeking steady states (X^*, Y^*, U^*, V^*) of the X, Y, U and V equations in (6), we obtain $M = kN$ as before, and from the V equation

$$(r(1 - \theta) + \sigma + \psi y^*)v^* = \sigma + \psi y^*,$$

where $y^* = Y^*/N^*$, $v^* = V^*/M^*$, as usual. Then, from the Y equation, using $x^* + y^* = 1$,

$$\rho(1 - y^*) - \nu y^* - k\gamma y^* + k\varphi(1 - y^*)v^* + \tau y^*(1 - y^*) = 0.$$

Eliminating v^* between these two equations, we obtain $C(y^*) = 0$, where

$$\begin{aligned} C(y) = &(r(1 - \theta) + (\sigma + \psi y))\left(-\tau y(1 - y) - \rho(1 - y) + (\nu + k\gamma)y\right) \\ &- k\varphi(1 - y)(\sigma + \psi y) = 0. \end{aligned}$$

This is a cubic with $C(0) = -(r(1 - \theta) + \sigma)\rho - k\varphi\sigma < 0$, $C(1) = (r(1 - \theta) + \sigma + \psi))(\nu + k\gamma) > 0$, and there is a root y^* of $C(y) = 0$ in $(0, 1)$. If ρ and σ are small, then so is $C(0)$, so there is a root near zero, given by

$$y = \frac{r(1 - \theta)\rho + k\varphi\sigma}{r(1 - \theta)(-\tau + \nu + k\gamma) - k\varphi\psi},$$

to leading order in ρ and σ. If this expression is positive, it is the root in $(0, 1)$, and is $O(\rho, \sigma)$, so the steady state has low prevalence of overt infection Y (and hence V). If it is negative on the other hand it is not the root in $(0, 1)$, the root in $(0, 1)$ is not small, and there is an outbreak, high levels of overt infection Y and V. The condition for an outbreak is therefore

$$r(1-\theta)(-\tau+\nu+k\gamma)-k\varphi\psi<0,$$

or $(1-\theta)(1-\tau/(\nu+k\gamma))-k\varphi\psi/(r(\nu+k\gamma))<0$, or, in terms of partial basic reproduction numbers after dividing by $1-\theta$,

$$R_0^Y+\frac{(R_0^{YV})^2}{1-R_0^V}>1,$$

exactly the condition for $R_0>1$ derived in (16). There is no standard basic reproduction number for this system, since there is no overt-disease-free steady state. However, the basic reproduction number R_0 for the system with $\rho=\sigma=0$ still has a role to play: there is an outbreak of overt infection if $R_0>1$. The condition for an outbreak without mites is the unrealistic condition $R_0^Y>1$, and overt disease is again mite driven.

4 Conclusions

There are many differences in detail between our model and previously published ones. However there is one key difference driven by the biology that allows us new insight into the bee-mite-virus system. This is our distinction between covert and overt infection in bees and between low- and high-level infections in mites together with the associated possibility of spontaneous transitions between infected classes caused by replication of the virus within the bee or the mite population. These transitions have been widely reported in the literature, and the transition in mites in particular is recognised as crucial to the recent epidemiology of DWV. For DWV, it is also necessary to distinguish between those bees that gained their overt infection in brood cells and those that gained it in the hive, which are acutely and chronically overtly infected, respectively.

We have focussed our analysis on insights gained from exploring the dependence of R_0, the basic reproduction number, on model parameters and the origin of its constituent components. This presented particular challenges when the model system included spontaneous transition in bees or mites, but we addressed that by considering our model as a perturbation from a baseline system with no spontaneous transition, for which R_0 could be calculated.

We found it easier to determine the size of R_0 by analysing the weighted directed graph associated with K, the next-generation matrix. From this graph, we were able to extract expressions for the number of secondaries of class i produced by a primary of class i, and hence determine conditions for growth of the epidemic after a perturbation from the steady state, equivalent to $R_0>1$. These conditions are given in terms of the weights of the arcs in the directed graph, or equivalently in terms of partial basic reproduction numbers for circuits in the graph.

Further exploration of the conditions suggests that simple transmission of overt virus between adult bees could theoretically be sufficient to maintain overt infection in a colony (if $R_0^Y > 1$), but this is unlikely with realistic parameters. Alternatively, it could be maintained solely by vertical transmission in mites coupled with their interactions with acutely overtly infected bees (if $R_0^V + R_0^{VZ})^2 > 1$), but this is also unlikely given that vertical transmission R_0^V is a probability and therefore less than 1, and acutely infected adult bees die so quickly that their role in infecting mites is unlikely to be important. Therefore, it seems that circuits of transmission involving mites and chronically overtly infected bees $(R_0^{YV})^2 + (R_0^{YVZ})^3)$ must be involved.

If there are no mites, the condition for overt infection reduces to $R_0^Y > 1$, which is unrealistic, in agreement with observations of DWV in colonies before the arrival of varroa.

In the perturbed system with spontaneous transitions (from covert to overt infection for bees and mites), if these transitions are rare, and making the realistic assumption for simplicity that acutely overtly infected bees are also rare, then the basic reproduction number R_0 for the unperturbed system still has a role to play. If $R_0 < 1$, then overt infection is rare, while if $R_0 > 1$ then there is an outbreak of overt infection. As in the unperturbed system, the condition for an outbreak without mites is the unrealistic condition $R_0^Y > 1$, and overt disease is again mite driven.

Finally, if spontaneous transitions lead to high rates of overt infection in either bees or mites or both, then it is clear that overt disease will be prevalent independently of other processes. Spontaneous transitions in mites may be reasonably common. Spontaneous transitions in bees are in general rare, but neonicotinoids can lead to compromised immune systems and hence an outbreak of overt disease, even in the absence of mites.

Our work demonstrates the impact of distinct infection classes for both honey bees and their infection vector in maintaining viral infections within a honey-bee colony. It also highlights the importance of spontaneous transition between infection classes in both populations. As such, it provides an important theoretical contribution to inform future studies both theoretical and practical as we strive to find new approaches to preserve honey-bee populations worldwide.

References

Attila D, Ibrahim Mahmoud A (2019) Global dynamics of a mathematical model for a honeybee colony infested by virus-carrying *Varroa* mites. J Appl Math Comput 61(1):349–371

Bowen-Walker PL, Martin SJ, Gunn A (1999) The transmission of deformed wing virus between honeybees (*Apis mellifera* L.) by the ectoparasitic mite *Varroa jacobsoni* oud. J Invertebr Pathol 73(1):101–106
Constanze Y, Elke G (2005) RT-PCR analysis of deformed wing virus in honeybees (*Apis mellifera*) and mites (*Varroa destructor*). J Gen Virol 86(12):3419–3424
De Miranda JR, Elke G (2010) Deformed wing virus. J Invertebr Pathol 103:S48–S61
Eberl Hermann J, Frederick Mallory R, Kevan Peter G (2010) Importance of brood maintenance terms in simple models of the honeybee-varroa destructor-acute bee paralysis virus complex. Electron J Differ Equ (EJDE) 2010:85–98
Elke G, Der Ohe V, Werner KH, Annette S, Christoph O, Ralph B, Stefan B, Wolfgang R, Werner M, Sebastian G et al (2010) The German bee monitoring project: a long term study to understand periodically high winter losses of honey bee colonies. Apidologie 41(3):332–352
Evans Jay D, Schwarz Ryan S (2011) Bees brought to their knees: microbes affecting honey bee health. Trends Microbiol 19(12):614–620
Francesco N, Brown Sam P, Desiderato A, Fabio DP, Gennaro DP, Paola V, Giorgio DV, Federica C, Emilio C, Francesco P (2012) Synergistic parasite-pathogen interactions mediated by host immunity can drive the collapse of honeybee colonies. PLoS Pathog 8(6):e1002735
Gennaro DP, Valeria C, Desiderato A, Paola V, Emilio C, Francesco N, Giuseppe G, Francesco P (2013) Neonicotinoid clothianidin adversely affects insect immunity and promotes replication of a viral pathogen in honey bees. Proc Nat Acad Sci 110(46):18466–18471
Hails Rosemary S, Ball Brenda V, Genersch E (2008) Infection strategies of insect viruses. In: Aubert, Michel; Ball, Brenda; Fries, Ingemar; Moritz, Robin; Milani, Norberto; Bernardinelli, Iris, (eds) Virology and the honey bee. Office for Official Publications of the European Communities, Luxembourg 255–276
Highfield Andrea C, Aliya EN, Mackinder Luke CM, Noël LM-LJ, Hall Matthew J, Martin Stephen J, Schroeder Declan C (2009) Deformed wing virus implicated in overwintering honeybee colony losses. Appl Environ Microbiol 75(22):7212–7220
Kevan Peter G, Hannan MA, Ostiguy N, Guzman-Novoa Ernesto (2006) A summary of the varroa-virus disease complex in honey bees. Am Bee J 694–697
Khoury David S, Myerscough Mary R, Barron Andrew B (2011) A quantitative model of honey bee colony population dynamics. PLoS ONE 6(4):e18491
Khoury David S, Barron Andrew B, Myerscough Mary R (2013) Modelling food and population dynamics in honey bee colonies. PLoS ONE 8(5):e59084
Malcolm AR, Mccredie MR (1981) The population dynamics of microparasites and their invertebrate hosts. Philos Trans R Soc Lond B Biol Sci 291:451–524
Martin Stephen J, Ball Brenda V, Carreck Norman L (2013) The role of deformed wing virus in the initial collapse of varroa infested honey bee colonies in the UK. J Apic Res 52(5):251–258
Martin SJ (1995) Reproduction of *Varroa jacobsoni* in cells of *Apis mellifera* containing one or more mother mites and the distribution of these cells. J Apic Res 34(4):187–196
Odo D, Peter HJA, Metz Johan AJ, Johan AJ (1990) On the definition and the computation of the basic reproduction ratio r 0 in models for infectious diseases in heterogeneous populations. J Math Biol 28(4):365–382
Peter R, Pia A, Bettina Z (2010) Biology and control of *Varroa destructor*. J Invertebr Pathol 103:S96–S119
Pielou EC (1977) Mathematical ecology. Number 574.50151. Wiley
Ratti V, Kevan PG, Eberl HJ (2012) A mathematical model for population dynamics in honeybee colonies infested with varroa destructor and the acute bee paralysis virus. Can Appl Math Q 1–27
Ricardo A-B, Ernesto G-N, Hamiduzzaman MM, Espinosa-Montano Laura G, Adriana C-B (2016) *Varroa destructor* (mesostigmata: Varroidae) parasitism and climate differentially influence the prevalence, levels, and overt infections of deformed wing virus in honey bees (hymenoptera: Apidae). J Insect Sci 16(1):44
Salvy M, Capowiez Y, Le Conte Y, Clément J-L (1999) Does the spatial distribution of the parasitic mite *Varroa jacobsoni* Oud. (mesostigmata: Varroidae) in worker brood of honey bee *Apis mellifera* L. (hymenoptera: Apidae) rely on an aggregative process? Naturwissenschaften 86(11):540–543
Sara B, Ezio V (2016) Viral epidemiology of the adult *Apis mellifera* infested by the varroa destructor mite. Heliyon 2(5):e00101
Schroeder Declan C, Martin Stephen J (2012) Deformed wing virus: the main suspect in unexplained honeybee deaths worldwide. Virulence 3(7):589–591

Sebastian G, Pia A, Elke G (2009) Deformed wing virus: replication and viral load in mites (*Varroa* destructor). J Gen Virol 90(2):463–467

Sumpter David JT, Martin Stephen J (2004) The dynamics of virus epidemics in *Varroa*-infested honey bee colonies. J Anim Ecol 73(1):51–63

Van den Driessche P, Watmough J (2008) Further notes on the basic reproduction number. In: Mathematical epidemiology. Springer, pp 159–178

Vardayani R, Kevan Peter G, Eberl Hermann J (2015) A mathematical model of the honeybee–*Varroa destructor*–acute bee paralysis virus system with seasonal effects. Bull Math Biol 77(8):1493–1520

Vardayani R, Kevan Peter G, Eberl Hermann J (2017) A mathematical model of forager loss in honeybee colonies infested with *Varroa destructor* and the acute bee paralysis virus. Bull Math Biol 79(6):1218–1253

Vincent D, Jochen P, Denis A, Jean-Daniel C, Nor C, Benjamin D, de Miranda J, Delaplane K, Dillier F-X, Fuch S et al (2012) Varroa destructor: research avenues towards sustainable control. J Apic Res 51(1):125–132

Wilfert L, Long G, Leggett HC, Schmid-Hempel P, Butlin R, Martin SJM, Boots M (2016) Deformed wing virus is a recent global epidemic in honeybees driven by *Varroa* mites. Science 351(6273):594–597

Yun K, Krystal B, Talia D, Ying W, Gloria DG-H (2016) Disease dynamics of honeybees with varroa destructor as parasite and virus vector. Math Biosci 275:71–92

Bulletin of Mathematical Biology (2021) 83:65
https://doi.org/10.1007/s11538-021-00899-z

Society for Mathematical Biology

SPECIAL ISSUE: CELEBRATING J. D. MURRAY

The Signature of Endemic Populations in the Spread of Mountain Pine Beetle Outbreaks

Dean Koch[1] · Mark A. Lewis[1] · Subhash Lele[1]

Received: 3 August 2020 / Accepted: 30 March 2021 / Published online: 1 May 2021

Abstract

The mountain pine beetle (MPB) is among the most destructive eruptive forest pests in North America. A recent increase in the frequency and severity of outbreaks, combined with an eastward range expansion towards untouched boreal pine forests, has spurred a great interest by government, industry and academia into the population ecology of this tree-killing bark beetle. Modern approaches to studying the spread of the MPB often involve the analysis of large-scale, high-resolution datasets on landscape-level damage to pine forests. This creates a need for new modelling tools to handle the unique challenges associated with large sample sizes and spatial effects. In two companion papers (Koch et al. in Environ Ecol Stat. https://doi.org/10.1007/s10651-020-00456-2, 2020a; J R Soc Interface 17(170):20200434, 2020b), we explain how the computational challenges of dispersal and spatial autocorrelation can be addressed using separable kernels. In this paper, we use these ideas to capture nonstationary patterns in the dispersal flights of MPB. This facilitates a landscape-level inference of subtle properties of MPB attack behaviour based on aerial surveys of killed pine. Using this model, we estimate the size of the cryptic endemic MPB population, which formerly has been measurable only by means of costly and time-intensive ground surveys.

The authors thank the Lewis Research Group for providing expert advice and feedback, as well as Victor Shegelski and the Sperling Lab for providing flight mill data. MAL is also grateful for support through NSERC and the Canada Research Chair Program.

✉ Dean Koch
dkoch@ualberta.ca

Mark A. Lewis
mlewis@ualberta.ca; mark.lewis@ualberta.ca

Subhash Lele
slele@ualberta.ca

[1] Department of Mathematical and Statistical Sciences, University of Alberta (UofA), 11324 89 Ave NW, Edmonton, AB T6G 2G1, Canada

Keywords Mountain pine beetle · Endemic · Nonstationary · Redistribution · Kernel · Modelling

1 Introduction

The mountain pine beetle (MPB) *Dendroctonus ponderosae* Hopkins (Coleoptera Curculionidae) is a tree-killing species of bark beetle native to pine forests of Western North America. Each year for a short period in summer, adult MPBs seek to complete their life cycle by attacking a suitable living host pine. During attacks, MPBs bore into the bark, introducing fungal pathogens in the process, and ultimately girdle the tree (Taylor et al. 2006). Death follows swiftly for a pine whose defence systems fail to eject these attackers. When an attack succeeds, MPBs use the host to feed and reproduce, laying eggs in galleries excavated underneath its bark.

A successful attack often leads to a MPB outbreak, in which local populations rise dramatically and large numbers of healthy pine are attacked over a period of several years. These outbreaks are major disturbance agents of pine forests, so mathematical models to explain their origin and how they spread across the landscape are of great interest to forest ecologists.

With few exceptions, the adult MPBs die after reproduction, and their progeny emerge as teneral adults the following summer to begin a new life cycle. This semelparity, and the approximately linear relationship between reproductive success and host death, are mathematically convenient properties when constructing models to track year-to-year changes in MPB populations. For example, if a total of B beetles have attacked a stand containing H susceptible pines, killing a fraction $\phi(B)$ of them, then a rough estimate of the MPB population emerging in the next year is $\lambda\phi(B)H$, where $\lambda > 0$ represents average per-stem productivity.

Modellers sometimes exploit this relationship to project outbreak dynamics ahead to future years (Heavilin and Powell 2008), and to explore ecological factors in population growth (Aukema et al. 2008). However, two major complications in MPB dynamics become apparent when attempting to link the host mortality fraction $\phi(B)$ with the underlying beetle population (Nelson et al. 2008). First, any spatially explicit model for $\phi(B)$ must account for dispersal flights, which allow localized outbreaks to spread into nearby areas. Second, $\phi(B)$ must reflect the eruptive and nonlinear nature of MPB population growth. We review these aspects briefly below, before introducing a new mathematical approach to the modelling problem.

1.1 Dispersal Flights

In modelling the evolution of an outbreak over multiple years, it is often convenient to track the beetle population in discrete time, where B_t is the value of B in year t. Such models cannot easily relate B_{t+1} and $\lambda\phi(B_t)H$ without incorporating dispersal. Flights of the MPB allow it to escape depleted stands, spark outbreaks in neighbouring areas, and expand its range (de la Giroday et al. 2012). By modelling B_t as the outcome of a spatially explicit dispersal event, we are better equipped to capture these interesting and important ecological phenomena, and achieve a higher precision in fitting $\phi(B)$ to data.

A variety of MPB dispersal models can be found in the literature (e.g. Goodsman et al. 2016; Preisler et al. 2012; Aukema et al. 2008; Heavilin and Powell 2008), but most make two simplifying assumptions out of mathematical convenience: that movements occur in all directions with equal probability (isotropy); and that patterns of redistribution do not vary with spatial location (stationarity).

The main novelty in the methods presented here is that we drop both of these assumptions. Our dispersal model has the flexibility to capture directed and location-dependent (anisotropic and nonstationary) events. It is meant as a phenomenological alternative to dynamical systems-based approaches to the same problem (e.g. Garlick et al. 2011; Powell and Bentz 2014; Powell et al. 2018), but with a simpler mathematical representation that borrows computationally efficient methods from spatial statistics. Our mathematical approach is based on ideas presented in two companion papers; on covariance structure (Koch et al. 2020a) and redistribution kernels (Koch et al. 2020b).

1.2 Colonization Curves

The colonization curve, function $\phi(B)$, should be highly nonlinear to accommodate the distinct behaviours exhibited in different phases of MPB populations (Berryman 1978). During the incipient-epidemic phase, attacks occur at densities low enough to be defended by hosts, so cooperative efforts in overcoming these defences lead to a positive density dependence (Allee effect) in $\phi(B)$ (Boone et al. 2011). However, as the number of attacking individuals rises, the MPB enters epidemic and post-epidemic phases, in which the density dependence turns negative as a result of scramble competition (Woodell and Peters 1992).

Empirical data on $\phi(B)$ therefore reveal an S-shaped, or sigmoid relationship (Raffa and Berryman 1983). This form is reminiscent of the familiar type-III functional responses for parasitism behaviour (Holling 1959), and indeed many aspects of MPB population dynamics are well described by this parasitoid–prey systems theory (Goodsman et al. 2016). In Sect. 2, we show how these functions can be adapted to model MPB population growth, generalizing the models of Heavilin and Powell (2008) and Koch et al. (2020b).

1.3 Aerial Overview Surveys

The extent to which a model can be complexified is limited by the amount and quality of data available for fitting and validation. Thus, while the ideas outlined above lead to sophisticated population models, they also demand an unusually large spatial dataset for parameter inference. For this reason, we demonstrate our methods in an analysis of data from the Aerial Overview Survey (AOS) in British Columbia (BC).

The AOS maintains an annual record of the spatial patterns of insect damage to forests in BC. Operators fly in fixed-wing aircraft over most of the forested land in the province each summer, logging the locations of damage (and the presumed cause) as polygon and point data on maps, which are then digitized and published online.

These data are sometimes dismissed as too imprecise for detailed population modelling, since the process of visual observation and manual delineation of damaged areas is prone to human error (e.g. Kautz 2014; Wulder et al. 2006). Nevertheless, because the AOS covers such an impressively large extent and timeline of forest damage patterns in BC, a considerable body of landscape-level MPB research draws from the AOS and its predecessor, the Forest Insect and Disease Survey (e.g. Aukema et al. 2006; Robertson et al. 2009; Chen and Walton 2011; Reyes et al. 2012; Sambaraju et al. 2012; Chen et al. 2015).

1.4 Paper Outline

We show in Sect. 2 how a generalization of the Heavilin and Powell (2008) model allows us to relate data from the AOS with ground surveys of MPB activity. We use these ideas in Sect. 3 to demonstrate the remarkable amount of information that can be extracted—with the right modelling tools—from the AOS alone.

In particular, our model accurately estimates the size of the cryptic, low-density endemic MPB population using only spatial data on outbreaks. This is remarkable given that pine mortality caused by the endemic phase happens at levels far below the operational detection threshold of the AOS (Cooke and Carroll 2017). Studies of endemic MPB more typically rely on intensive ground surveys of attacked pine (e.g. Boone et al. 2011; Bleiker et al. 2014). Our model estimates the rate of endemic attacks using only AOS data on outbreak-level pine mortality.

This is important because, in comparison with more reliable ground survey methods, aerial survey programs such as the AOS are a far less expensive and time-consuming means of monitoring MPB activity over large geographical areas. Note that similar datasets are available for the neighbouring province of Alberta (AB), in which a highly consequential MPB range expansion is currently underway.

Section 2.1 introduces the model by reviewing a popular mathematical representation for the colonization curve $\phi(B)$, before introducing several refinements in Sects. 2.2, 2.3 and 2.4. Our representation of dispersal flight is then introduced in Sect. 2.5, and an error model suitable for the AOS dataset is proposed in Sect. 2.6. We demonstrate the model in Sect. 3 by fitting to data on outbreaks of the MPB in BC during the years 2006–2008.

2 Methods

Our case study covers a pine-rich region of roughly 10,000 km^2, centred over the Merritt Timber Supply Area (TSA) of Southern BC (Fig. 1). We divided this into a 1 hectare (ha) resolution grid (*sensu* Aukema et al. 2006) to form a 1000×1000 lattice of cells, with matching layers provided by the province (http://www.hectaresbc.org) on wildfire, cutblocks and topography.

As we are interested in how dispersal patterns are related to outbreak development, we analysed the attack years 2006–2008, in which a large number of pine-leading stands would see transitions from endemic to epidemic behaviour (the incipient-

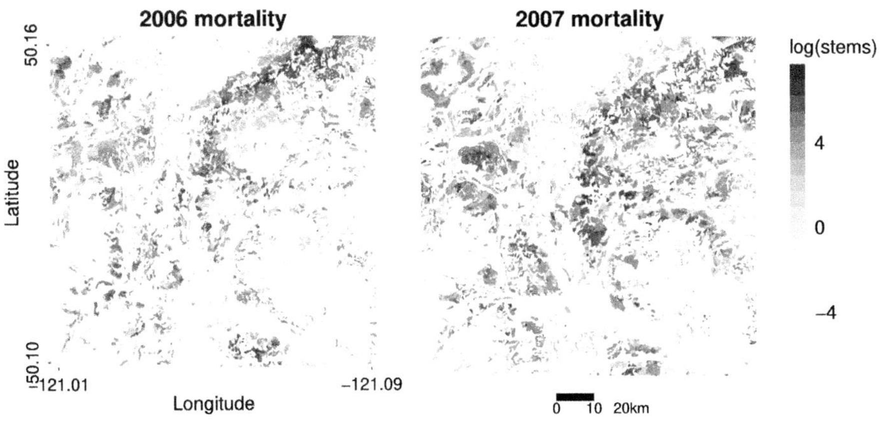

Fig. 1 (Colour figure online) Hosts killed by MPB ($\phi_{i,t} H_{i,t}$, in stems/ha) in the summers of 2006–2007. AOS data on damage severity were rasterized to approximate susceptible host mortality ($\phi_{i,t}$). Host density $H_{i,t}$ was derived from pine volume estimates in Beaudoin et al. (2014), as described in Appendix 1.1

Table 1 Notation for state variables in the MPB attack dynamics model

Location i	Vectorized	Definition	Units	Type
$H_{i,t}$	$\boldsymbol{H}_t$	Pre-attack susceptible pine density	stems/ha	Observed
$\phi_{i,t}$	$\boldsymbol{\phi}_t$	Proportion of $H_{i,t}$ killed by MPB	unitless	
$\tilde{B}_{i,t}$	$\tilde{\boldsymbol{B}}_t$	Emerging MPB density (pre-dispersal)	females/ha	Latent
$B_{i,t}$	$\boldsymbol{B}_t$	MPB attack density (post-dispersal)		

Indexing is by year t and location i, and boldface denotes the vector of all n locations, e.g. $\boldsymbol{\phi}_t = (\phi_{1,t}, \phi_{2,t}, \ldots, \phi_{n,t})'$

epidemic population phase). This period captures the peak of an epidemic in the Merritt TSA (in terms of basal area damaged) at a time when around one out of four cells in the area exhibited crown-fade due to MPB activity.

Our analysis tracks four state variables, indexed by year (t) and location (i): Only two of them are measured in practice: pine mortality ($\phi_{i,t}$) and host density ($H_{i,t}$, in stems/ha) (Appendices 1.1 and 1.2); The others, MPB density *pre*-dispersal $\tilde{B}_{i,t}$ and *post*-dispersal $B_{i,t}$ (in females/ha), are latent variables, inferred by the model but never directly observed (Table 1).

In the remainder of Sect. 2, we construct a model connecting these four state variables. We start in Sect. 2.1 by extending the red-top model of Heavilin and Powell (2008), interpreting its attack parameters in a new light. We then link this attack model to novel submodels describing four important components of MPB population dynamics (Nelson et al. 2008): stand susceptibility, endemic populations, reproduction and dispersal (Sects. 2.2, 2.3, 2.4 and 2.5, respectively). Finally, in Sect. 2.6 we describe the data and statistical methodology used for fitting the full model, before presenting our results and connecting them to empirical findings from the MPB literature in Sects. 3 and 4.

2.1 Attack Dynamics (ϕ)

Our equation for pine mortality $\phi_{i,t}(B_{i,t})$ generalizes the red-top model of Heavilin and Powell (2008) to better match the types of colonization curves fitted in Cooke and Carroll (2017). The red-top model is best introduced by focusing at first on a particular location and year; so for notational convenience we omit the subscripts i and t until they are needed again in Sect. 2.4. Thus, for i and t fixed, we relate the attack density B (females/ha) to pine mortality ϕ by:

$$\text{proportion of } H \text{ killed} = \phi(B) = \frac{B^\kappa}{a^\kappa + B^\kappa} \quad \text{where } a > 0,\ \kappa > 0. \tag{1}$$

Parameter a is the half-saturation value, or attack density (in females/ha) at which 50% mortality occurs, and κ is a shape parameter controlling the density dependence. The special case $\kappa = 2$ recovers the red top model of Heavilin and Powell (2008) (after multiplying both sides by H). Other κ values reflect alternative regimes of density dependence. Larger κ and/or a values coincide with a stronger defensive response by pines. When $\kappa \leq 1$, the Allee effect vanishes, reflecting compromised defences, as might occur, for example, during a drought.

Parameter estimation becomes simpler if Eq. (1) can be made linear in its parameters. Observing that the odds ratio of pine mortality $\phi/(1-\phi)$ is $(B/a)^\kappa$, we can take logarithms to get a linear equation on the logit-log scale:

$$\text{logit}(\phi) = -\kappa \log(a) + \kappa \log(B), \tag{2}$$

where $\log(B)$ is the logarithm of attack density, and $\log(a)$ the density (on the log scale) at which one half of susceptible hosts are expected to be colonized.

This also happens to be the mathematical form of the colonization curve fitted in Cooke and Carroll (2017) to the data reported in Boone et al. (2011) on attacked pines in our study area. Their analysis estimated $\hat{\kappa} = 1.66$ for the 2 years leading up to 2006. In years prior, a much lower value (0.56) was estimated, suggesting that environmental stressors on pine may have relaxed the Allee effect and bolstered endemic populations to spark the large-scale outbreaks of 2006–2008 (Fig. 1).

Once started, outbreaks are not easily stopped. Irruptions in MPB populations are accompanied by behavioural changes in which host-preference switches from stressed to healthy pine (Carroll et al. 2006). This allows population growth to continue even after pine vigour recovers from a period of stress. Above a certain density threshold, B_T, the MPBs have sufficient numbers to cooperatively attack a healthy pine (a mass attack), releasing them from the ordinary pressures of the Allee effect and marking the beginning of the incipient-epidemic phase.

The nature of this density dependence is reflected by (equivalent) Eqs. (1) and (2). The case $\kappa > 1$, corresponding to attacks on healthy pine, is illustrated graphically in

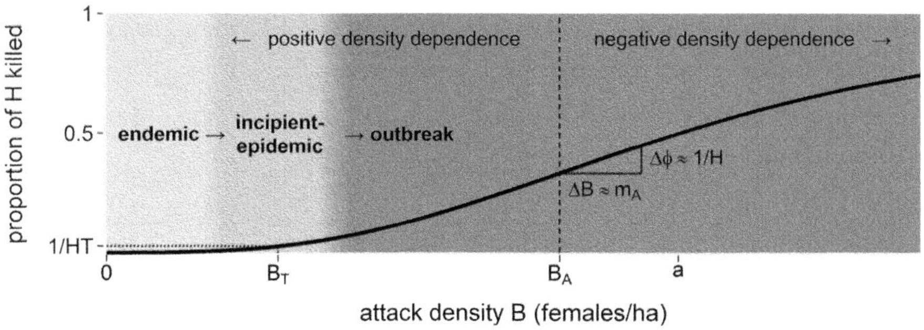

Fig. 2 (Colour figure online) Host mortality as a function (Eq. 1) of MPB attack density B for $\kappa = 3$. Above the inflection point (B_A) is a regime of negative density dependence. When $B \approx 0$, the endemic population is too small to mass-attack healthy pine. When B rises to the incipient-epidemic transition point B_T, mass attacks become feasible and the MPBs are released from the endemic phase. At moderate densities, each attacked pine accounts for $\approx m_A$ beetles. At higher densities, intraspecific competition leads to diminishing returns and negative density dependence

Fig. 2. We discuss two features of this curve below, the incipient-epidemic transition point (B_T) and mass attack number (m_A). We report on their estimates in Sect. 3.

2.1.1 The Incipient-Epidemic Transition (B_T)

Empirical data from our study area suggest that a density of $\hat{B}_T \approx 300$–600 females/ha is sufficient to initiate the incipient-epidemic phase in an area of $T = 15.3$ ha (Carroll et al. 2006). B_T is of course scale dependent; Amman (1984) estimated a quite different transition point at the $T = 40.5$ ha scale.

However, given any scale of interest, and given the values of κ and a, B_T can be estimated via equation (1). This is done by setting $\phi(B_T) = 1/HT$ (i.e. assuming a single host death in the specified area, T) and inverting (1) to get:

$$B_T = \frac{a}{\sqrt[\kappa]{HT - 1}}. \tag{3}$$

2.1.2 The Mass Attack Number (m_A)

The number of MPB aggregating during attacks is carefully moderated by pheromones. This allows the beetle to optimize its reproductive success in mass attacks, by attacking in numbers high enough to overcome tree defences, but low enough to avoid crowd competition (Taylor et al. 2006). For example, Raffa and Berryman (1983) reported an optimum of around 61 attackers/m^2, implying that a tree with 5.5 m^2 of bark available for attack (typical of the pine-leading stands in our study area) would have an optimum around 340 females/stem.

Under ideal conditions for MPB attack, this optimal density will presumably match the average attack density per attacked tree, which we call the *mass attack number* m_A (in females/stem). This average is approximated by the slope of $\phi(B)$ near its inflection point $B_A = a\sqrt[\kappa]{(\kappa - 1)/(\kappa + 1)}$ (where $\phi''(B_A) = 0$), since, at this intermediate density, $\phi(B)$ is nearly linear, and increases with B at rate $\phi'(B_A) \approx 1/(m_A H)$. From (1), we can therefore compute the approximation:

$$m_A \approx 1/\left(\phi'(B_A)H\right) = \frac{4B_A\kappa}{H(\kappa^2 - 1)}. \tag{4}$$

2.2 Stand Susceptibility (*a*)

Although (3) and (4) both depend nonlinearly on κ, both equations scale linearly with the half-saturation value a. Stands that are highly susceptible to MPB attack have lower values (requiring fewer attacking beetles to initiate an outbreak), and vice versa. We can therefore interpret a as a simple measure of susceptibility to attack.

One can expect a to vary with environmental factors, such as weather, and stand characteristics, such as pine density. These factors have a complex and nuanced relationship with susceptibility (Preisler et al. 2012), and a clear biology-based model for this relationship is lacking. To avoid overcomplicating our model, we simply take the best linear approximation on the logit-log scale, writing $\boldsymbol{x\beta} = -\kappa \log(a)$ for a set of unknown regression parameters $\boldsymbol{\beta} = (\beta_1, \ldots \beta_{n_\beta})$ and covariates $\boldsymbol{x} = (x_1, \ldots x_{n_\beta})$.

Thus, in equation (2) for the mortality log-odds, logit(ϕ), we swap out the intercept term with a linear predictor. Similar regression models, such as in Aukema et al. (2008) and Preisler et al. (2012), have been useful for identifying environmental factors that have a significant ($\beta_k \neq 0$) effect on outbreak occurrence. For our purposes, $\boldsymbol{\beta}$ simply serves as a (location-wise) correction of a through which to estimate MPB population sizes, so we do not focus on the β_k or their effect sizes in our analysis. However, interested readers will find the full set of linear regression covariates listed in Appendix 1.2.

2.3 Endemic Populations (ϵ)

The (aspatial) red-top model of Heavilin and Powell (2008) has no endemic equilibrium: low-density populations are viewed as unstable, tending to extinction and occurring only by means of immigrations from a reservoir of distant outbreaks appearing stochastically across the landscape. However, empirical data (e.g. Boone et al. 2011; Bleiker et al. 2014) suggest that resident endemic populations are widespread and persistent. These low-density populations subsist on a small number of defensively compromised pines and an assemblage of secondary bark beetle species that assist in the colonization of weakened trees.

We introduce a stable endemic equilibrium into the red-top model (Eq. 1) by adding a small positive term $\epsilon > 0$ (in females/ha) to the post-dispersal MPB population (B) in the red-top model at all sites/years prior to attack. Specifically, with the addition of endemic beetles the attack function $\phi(B)$ becomes:

$$\text{proportion of } H \text{ killed} = \phi(B + \epsilon) = \frac{(B + \epsilon)^\kappa}{a^\kappa + (B + \epsilon)^\kappa} \quad \text{where } \epsilon > 0, \tag{5}$$

which, on the log scale, produces $\text{logit}(\phi) = -\kappa \log(a) + \kappa \log(B + \epsilon)$.

This constant introduces a spatially uniform background level of MPB. Should an in-flight from a neighbouring outbreak occur, its density is added to the endemic cohort ϵ, and the combined population attacks pines according to Eq. (1). The effect of ϵ is therefore to boost the effective size of spreading populations, increasing the likelihood that an incipient-epidemic transition will succeed in sparking a local outbreak.

In the absence of immigrating MPB, the endemic population is too small to attack healthy pines, so it instead seeks out defensively weakened trees. Because this pool of suitable hosts is ephemeral and extremely small compared to H, these endemic MPBs incur a much higher flight-establishment mortality cost than do outbreaking populations: Taylor et al. (2006) estimate the generation mortality of endemic MPB at 97.5%. Assuming most of this loss can be attributed to the search flight, the rate of attack on defensively weakened hosts under this model would be $(1 - 0.975)\epsilon$ females/ha (or slightly above), with the healthy pine population variable H unaffected.

However, if an endemic population joins with a cohort of immigrating outbreak-level MPB, suitable hosts suddenly become abundant, and the flight-establishment losses should drop accordingly. The generation mortality in populations capable of mass-attacks is thought to lie in the range 80–98.6% (Taylor et al. 2006; Amman 1984). We assume that these losses mostly occur as a result of tree defences and crowd-competition. Unlike search flight losses, the latter are subsumed into $\phi(B)$ under the model (1). Therefore, we estimate the total number of attacking beetles at a given site as the sum of ϵ and any MPB (local or immigrant) originating from mass-attacked trees.

2.4 Reproduction (λ)

Reproduction connects subsequent years, so we must now make the dependence of our model variables on time and location explicit. In the red-top model, reproduction is summarized by $\tilde{B}_{i,t} = \lambda_{t-1}\phi_{i,t-1}H_{i,t-1}$. This expresses that $\tilde{B}_{i,t}$, the density of (non-endemic) mature MPB emerging in year t at location i, is proportional to the number of mass-attacked stems in year $t-1$.

The productivity parameter λ_{t-1} specifies the average number of female MPB brood to emerge from each attacked tree in the year following an attack. This counts only those individuals that hatch, survive to maturity and engage in search flights for new hosts. Note that by aggregating demographic information to the level of the tree (and year), we forego some precision. However, this formulation simplifies the model considerably, summarizing in a single constant the many MPB within-tree growth and development processes that cannot be observed in aerial surveys (Berryman 1974).

Under this model, productivity λ_t is not identifiable from data on $\phi_{i,t}$ and $H_{i,t}$ without knowledge of $a_{i,t}$. So we instead fixed the value of $\lambda_t = \lambda$ in all years to a plug-in estimate of $\lambda = (2/3)(250) = 166.7$ (females/stem) suggested by empirical productivity data for epidemic phase MPB (Cole and Amman 1969, Fig. 9), and assuming a 1:2 male–female sex ratio (Reid 1962). This productivity value is consistent with a 90% generation mortality rate, calculated using the brood production regression

Table 2 Parameters of the generalized red-top model ($\boldsymbol{\theta}_t$), organized into categories of attack ($\boldsymbol{\theta}_{\phi_t}$), dispersal ($\boldsymbol{\theta}_{D_t}$) and error ($\boldsymbol{\theta}_{V_t}$)

Submodel	Vector	Contents	Definition	Units
Attack	$\boldsymbol{\theta}_{\phi_t}$	κ_t	Density dependence shape value	unitless
		λ	Beetle production per attacked host	females/stem
		ϵ_t	Emerging endemic MPB population level	females/ha
		$a_{i,t}$	Half-saturation / susceptibility value	females/ha
		$\boldsymbol{\beta}_t$	Linear regression coefficients for $a_{i,t}$	–
dispersal	$\boldsymbol{\theta}_{D_t}$	$\boldsymbol{\Delta}_{k,t}$	pWMY kernel: angle, shape and range	–
Error	$\boldsymbol{\theta}_{V_t}$	σ_t^2	Marginal variance	unitless
		$\boldsymbol{\rho}_t$	Gaussian autocorrelation range (x and y)	km

All except for λ are fitted to data separately by year (t). For dispersal, a 5-parameter product-WMY (pWMY) kernel (Appendix 2) is assigned to each of $m = 625$ data blocks, indexed by $k = 1, \ldots m$. A vector of 44 regression coefficients ($\boldsymbol{\beta}_t$) defines stand susceptibility through the linear model $\kappa_t \log(a_{i,t}) = \boldsymbol{x}_{i,t}\boldsymbol{\beta}_t$ for local covariates $\boldsymbol{x}_{i,t}$ (Appendix 1.3), where i indexes location

in (Safranyik 1988, eq. 14) on the mean diameters (Carroll et al. 2006) and heights (Safranyik and Linton 1991) of pine in our study area.

Although a time (and space) dependent λ would be more realistic, it would complicate the model considerably. We do however allow all other process model parameters to vary with time (e.g. ϵ_t, κ_t, $\boldsymbol{\beta}_t$, and the parameters of D_t), estimating them separately for each year in our analysis. Variations in productivity are therefore reflected in changing stand susceptibility $a_{i,t}$, which varies both spatially and temporally through $\boldsymbol{\beta}_t$ and the local covariates $\boldsymbol{x}_{i,t}$ (Table 2).

2.5 Dispersal ($\tilde{B} \rightarrow B$)

Dispersal can be represented in population models using redistribution kernels (Neubert et al. 1995). These are functions, D_t, specifying a probability distribution for location following movement events. If the emerging MPB population $\tilde{B}_{t,i}$ is observed at n spatial locations, D_t specifies an $n \times n$ matrix ($\boldsymbol{D}_t$) whose i, jth entry $[D_t]_{ij}$ is the expected proportion of the population $\tilde{B}_{t,j}$ that will move to cell i in the course of dispersal (Appendix 2). Thus, after adding the endemic MPB, the expected attack density is $\mathbb{E}\left(B_{i,t}\right) = \epsilon_t + \sum_j \left([D_t]_{ij}\tilde{B}_{t-1,j}\right)$. The equivalent matrix-vector equation

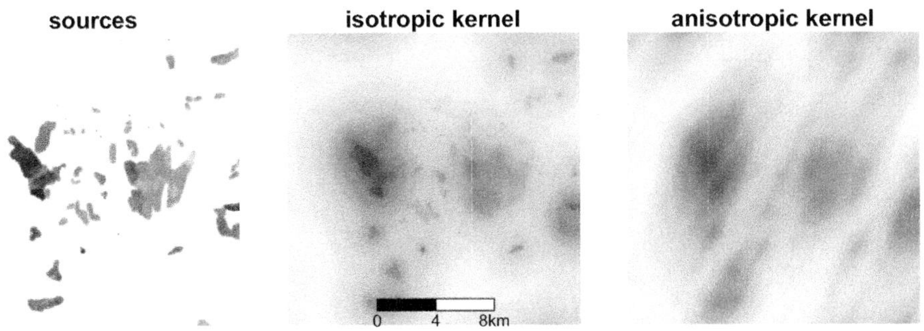

Fig. 3 (Colour figure online) MPB density pre-dispersal (left) and post-dispersal (middle and right) for two models of MPB flight patterns: an isotropic Bessel kernel (middle) with parameters from Goodsman et al. (2016), and an anisotropic pWMY kernel (right) parametrized to resemble the Bessel, but with the addition of a northeast-facing directionality

$\mathbb{E}(\boldsymbol{B}_t) = \epsilon_t \boldsymbol{I} + \boldsymbol{D}_t \tilde{\boldsymbol{B}}_t$ allows us to drop the cumbersome location indices (i), so we will use this simplified notation whenever possible.

The ith entry of $\mathbb{E}(\boldsymbol{B}_t)$ estimates the density of attackers, sometimes called beetle pressure, at stand location i in year t. Beetle pressure is a common feature of MPB outbreak risk models (e.g. Wulder et al. 2006; Preisler et al. 2012), where it expresses proximity to infestations by a weighted sum of severity values or presence/absence indicators in a neighbourhood of the target stand. The weights in this calculation are provided by the kernel function D_t.

The choice of D_t therefore reflects assumptions about how MPBs redistribute in search of new hosts. Ad-hoc assignments of weights to $[D_t]_{ij}$ often suffice in simpler predictive models (e.g. Kärvemo et al. 2014; Kunegel-Lion et al. 2019) but, whenever possible, it is desirable to choose a kernel derived from models of the physical flight process (Nelson et al. 2008).

Our flight model approximates the Whittle–Matérn–Yasuda (WMY) kernel family (Yasuda 1975), which describes diffusive movements through complex habitat (Koch et al. 2020b). Included in this family are a number of distinct isotropic kernels that have been advocated in previous studies of similar datasets (e.g. Turchin and Thoeny 1993; Heavilin and Powell 2008; Goodsman et al. 2016). Figure 3 (middle) is one example, arising from diffusion with constant settling.

We calculate the $[D_t]_{ij}$ values using pWMY kernels, which in addition to closely approximating the WMY, easily incorporate anisotropic (directed) movement patterns (Fig. 3, right) as might be expected from the effect of local winds (Ainslie and Jackson 2011) and patchy habitat (Powell et al. 2018).

Writing $\boldsymbol{\delta}_{ij} = \left(x_i - x_j, y_i - y_j\right)'$ for the vector difference between the x-y coordinates at locations i and j, the equation of the anisotropic pWMY kernel is:

$$D_t\left(\boldsymbol{\delta}_{ij}; \alpha, \rho_x, \nu_x, \rho_y, \nu_y\right) = c D_W\left(d_{ij}^x; \rho_x, \nu_x\right) D_W\left(d_{ij}^y; \rho_y, \nu_y\right),$$
$$\text{where } (d_{ij}^x, d_{ij}^y)' = \boldsymbol{R}_\alpha \boldsymbol{\delta}_{ij}, \text{ and } D_W(d; \rho, \nu) = (d/\rho)^\nu K_\nu(d/\rho), \tag{6}$$

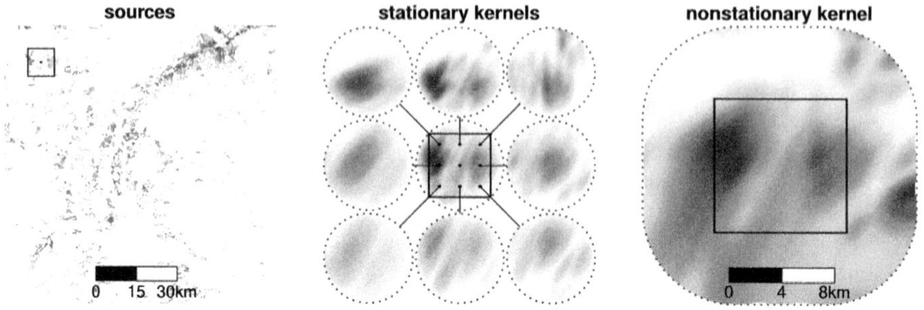

Fig. 4 (Colour figure online) A nonstationary flight pattern estimation scheme: stationary kernels are separately fitted to small overlapping blocks of data (at left, a block and its centroid). Expected beetle pressure (detail, at right) is computed as the distance-weighted average of nearby kernel predictions. The middle panel shows the nearest 9 block centroids and their kernel predictions before averaging

where K_ν denotes the νth order modified Bessel function of the second kind, $\boldsymbol{R}_\alpha$ is the standard 2D rotation matrix for angle α, and c is the kernel normalization constant. The parameters of this kernel are explained in detail in Appendix 2.

Importantly, the pWMY can be computed far more quickly than the WMY. Computational simplicity allows different dispersal patterns to be quickly fitted at different sites within a dataset. In our study area, this revealed a complex pattern of directionality (nonstationarity) that varies depending on the position of the source population.

Nonstationarity in dispersal patterns over a large geographic area is unsurprising in light of work by Powell and Bentz (2014), whose differential equation-based movement model connects environmental cues to direction and motility in MPB flights. Recognizing the importance of this nonstationarity, but lacking high-resolution data on its cues, we opted for a novel phenomenological model that combines multiple stationary (pWMY) kernels to form a nonstationary one.

We fitted each pWMY kernel to a relatively small square geographical area (a block) before combining them by computing a weighted average of their fitted values, with weights inversely related to distance from the block centroid (Fig. 4). The effective contribution of each kernel to beetle pressure $\mathbb{E}(\boldsymbol{B}_t)$ is therefore restricted to a neighbourhood (dashed outer line in Fig. 4, right) of the block over which it was fitted (solid line).

The resulting nonstationary dispersal model is itself a redistribution kernel, so we refer to it as D_t (with associated matrix $\boldsymbol{D}_t$). Its explicit mathematical form is derived in Appendix 2.

The virtue of this approach is that it captures complex (nonstationary) dispersal patterns by means of simpler stationary kernels, whose parameters can be fitted rapidly by well-established techniques over small neighbourhoods within which a stationarity assumption is reasonable. Moreover, there is no requirement for detailed environmental data, such as the stand density values used by Powell and Bentz (2014). Movement patterns are instead estimated directly from the available attack data.

Our construction of D_t used a total of 625 pWMY kernels in a 25×25 grid arrangement of blocks, each of size 10×10 km. Since each pWMY kernel captures only the local flight patterns within its respective block, we chose a distance-weighting function (Appendix 2) that assigns zero weight beyond the centroid-to-corner distance within a block (7.1 km). This scheme tracks movements up to 14.2 km, a reasonable upper bound on self-powered dispersal given laboratory studies suggesting fewer than 10% of MPBs are capable of flight beyond this distance (Shegelski et al. 2019).

To avoid overparameterizing an already complicated model—and lacking data on wind patterns—we assumed that atmospherically driven flight events (such as those documented by Jackson et al. 2008) were rare enough to ignore. Furthermore, although both block size and the number of blocks can be viewed as tuning parameters for the dispersal model, we assigned them ad-hoc values in this case to (roughly) coincide with the aforementioned self-powered dispersal limitations of MPB, rather than attempting to optimize them via model selection.

Some edge effects are unavoidable with this modelling strategy. For example, $[D_t]_{ij}$ values for a location coinciding with a block centroid will be determined almost entirely by the data within that single block, whereas for a location halfway between block centroids, the $[D_t]_{ij}$ values are influenced by data from two (or more) overlapping blocks—a much larger geographical extent. We believe, however, that this type of inconsistency pales in comparison with the roughcast assumption of stationary and isotropic dispersal patterns.

2.6 Model-Fitting

2.6.1 Data

Pine density $\boldsymbol{H}_t$ was estimated using the model output of Beaudoin et al. (2014) for the year 2001, after adjusting for losses due to wildfire, logging and pest damage incurred during the intervening years (Appendix 1.1). For simplicity, we did not attempt to model regeneration, but rather assume that changes in density due to growth were small enough to ignore over the period 2001–2008.

Pine mortality data are drawn from the AOS of the Merritt TSA (Fig. 1) for the attack years 2006–2008. These were rasterized by standard methods (Appendix 1.2) to produce a 1000×1000 grid of sample locations at a 1 ha resolution, matching the geometry of the pine density dataset. To avoid edge effects in dispersal calculations, we excluded a $\approx$10km buffer at the edge of this grid from the response data, forming the (logit-transformed) vector $\boldsymbol{\phi}_t$ from the subgrid of dimensions 893×893 centred on this region (a within-year sample size of 797,449 points).

2.6.2 Statistical Model

A redistribution kernel is a probabilistic model—it connects MPB damage patterns to the *expected* density of attackers arriving next year at each location $\mathbb{E}(\boldsymbol{B}_t)$. Variations of $\boldsymbol{B}_t$ about this mean should therefore be modelled as error. Investigations into ecological dispersal by Preston (1948) and Limpert et al. (2001) inform us these errors are likely to be lognormally distributed. Assuming, $\left(\mathbb{E}(B_{i,t}) - B_{i,t}\right) \overset{iid}{\sim}$ lognormal$(0, \tilde{\sigma}_t^2)$, we can summarize Sects. 2.1, 2.2, 2.3, 2.4 and 2.5 in the equation:

$$\underbrace{\operatorname{logit}(\boldsymbol{\phi}_t)}_{\text{pine mortality log-odds}} = \underbrace{\boldsymbol{X}_t\boldsymbol{\beta}_t}_{\text{susceptibility}} + \underbrace{\kappa_t \log(\epsilon_t \boldsymbol{I} + \lambda \boldsymbol{D}_t (\boldsymbol{\phi}_{t-1} \odot \boldsymbol{H}_{t-1}))}_{\text{beetle pressure}} + \underbrace{\boldsymbol{Z}_t}_{\text{error}}, \quad (7)$$

where $\boldsymbol{X}_t = (\boldsymbol{x}'_{1,t}, \ldots \boldsymbol{x}'_{n,t})'$ is the (covariate) data matrix for year t, and $\boldsymbol{Z}_t$ is the vector of process errors arising from $\boldsymbol{B}_t$. The logit and log functions are applied element-wise, and the symbol $\odot$ denotes element-wise multiplication. This slight abuse of notation allows us to suppress the location indices i and write the complete model (7) in terms of length-n vector operations.

Under the lognormal assumption, $\boldsymbol{Z}_t$ is mean-zero multivariate normal (MVN), with a variance $\kappa_t \tilde{\sigma}_t^2$ that scales with the strength of the density dependence in $\phi(B)$. We assume that measurement error introduces an additional mean-zero MVN random vector appearing additively on the logit scale of (7). Since these errors are presumably independent of $\boldsymbol{B}_t$, their effect (by standard MVN theory) is to simply increase the variance of $\boldsymbol{Z}_t$. Thus, ignoring any autocorrelation (for now), we could write $\boldsymbol{Z}_t \sim \text{MVN}\left(\boldsymbol{0}, \sigma_t^2 \boldsymbol{I}\right)$, where σ_t^2 is the sum of the variances from process and measurement error.

For simplicity, we ignored temporal autocorrelation by treating each year of data in the analysis as independent, as is commonly done in large-scale MPB outbreak analyses (e.g. Heavilin and Powell 2008; Goodsman et al. 2016). While this is not ideal, it avoids the difficulties associated with aligning consecutive years of raster data containing a large number of slight positional errors (Wulder et al. 2009), while simplifying the error model both mathematically and computationally.

Spatial autocorrelation, on the other hand, is more easily corrected using covariograms (Chilès and Delfiner 2012). For computational efficiency, we used the Gaussian covariogram, which generates a covariance matrix $\boldsymbol{V}_t$ (to replace $\sigma_t^2 \boldsymbol{I}$ above) based on σ_t^2 and a pair of correlation range parameters, $\boldsymbol{\rho}_t$. In this model, the logarithm of the likelihood function for observations of $\boldsymbol{\phi}_t$, given $\boldsymbol{\phi}_{t-1}$ and $\boldsymbol{X}_t$ is proportional to:

$$\mathcal{L}(\boldsymbol{\theta}_t \mid \boldsymbol{Z}_t) = -\log(\det(\boldsymbol{V}_t)) - \boldsymbol{Z}_t' \boldsymbol{V}_t^{-1} \boldsymbol{Z}_t \quad \text{where } \boldsymbol{\theta}_t = \left(\boldsymbol{\theta}_{\phi_t}, \boldsymbol{\theta}_{D_t}, \boldsymbol{\theta}_{V_t}\right) \quad (8)$$

with $\boldsymbol{Z}_t$ as defined in (7), and model parameters $\boldsymbol{\theta}_t$ organized into components of attack dynamics, $\boldsymbol{\theta}_{\phi_t} = (\kappa_t, \lambda, \epsilon_t, \beta_t)$; error, $\boldsymbol{\theta}_{V_t} = (\sigma_t^2, \boldsymbol{\rho}_t)$, and dispersal $\boldsymbol{\theta}_{D_t} = (\boldsymbol{\Delta}_{1,t}, \ldots \boldsymbol{\Delta}_{625,t})$; as in Table 2. The model can now be fitted to data by maximum likelihood estimation (MLE), which finds the maximizer of (8), called $\hat{\boldsymbol{\theta}}_t = (\hat{\boldsymbol{\theta}}_{\phi_t}, \hat{\boldsymbol{\theta}}_{D_t}, \hat{\boldsymbol{\theta}}_{V_t})$.

Our estimation method for $\boldsymbol{\theta}_t$ is based on the 2-step algorithm described in Crujeiras and Van Keilegom (2010), but with a block-wise approach to approximating the large number of parameters in $\boldsymbol{\theta}_{D_t}$. Each of the 625 pWMY kernels is fitted independently to the data in its block, before being combined to form the nonstationary kernel matrix $\hat{\boldsymbol{D}}_t$. By assuming $\boldsymbol{D}_t \approx \hat{\boldsymbol{D}}_t$, estimation of the remaining parameters $\boldsymbol{\theta}_{\phi_t}$ and $\boldsymbol{\theta}_{V_t}$ then becomes straightforward using generalized least squares (GLS) based methods (Chilès and Delfiner 2012). Simulations indicated that our approach yields unbiased and reasonably precise estimates of $\boldsymbol{\theta}_t$ (Appendix 3).

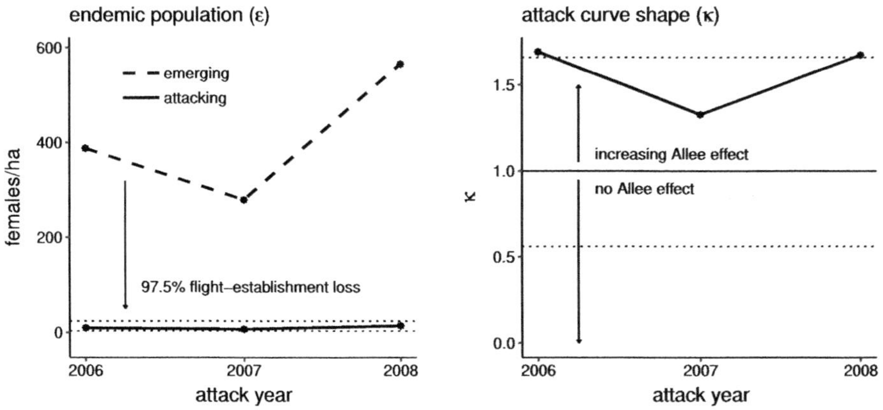

Fig. 5 Fitted attack parameters. At left, estimates of the endemic population and expected attack rates lying within the range (dotted lines) reported in Boone et al. (2011). At right, estimates of the attack curve shape compared with reference levels from Cooke and Carroll (2017) (dotted lines)

3 Results

The estimated endemic densities and attack curve shapes in all 3 years (Fig. 5) matched closely with ground surveys of our study area during the period 2001–2005. We estimated ϵ_t, the endemic contribution, at 388, 279 and 566 (females/ha), respectively, for the years 2006–2008. Note that these densities are well above what is considered normal for the endemic phase (Safranyik and Carroll 2006), as they represent populations before flight-establishment loss. After correcting for this loss (97.5%), our estimates suggest a range of 7–14 attackers/ha in endemic-only populations, similar to the ranges reported in Boone et al. (2011) and Bleiker et al. (2014).

A density dependence in attack was detected in all years, with κ estimated at 1.69, 1.32 and 1.67. Note that the estimates in 2006 and 2008 very nearly matched the value of 1.66 reported by Cooke and Carroll (2017) for pooled colonization curve data from the preceding years 2002–2003 and 2005 (Fig. 5). This indicates that not only is density dependence detectable from stand-level AOS data (in the absence of failed attack counts)—supporting the findings of Goodsman et al. (2016) on Allee effects—but also that the precise shape of the attack curve in (9) can be estimated from aerial data on $\boldsymbol{\phi}_t$ and $\boldsymbol{H}_t$ alone. This includes both the Allee and compensatory (crowd competition) effects highlighted in Fig. 2.

Estimates of stand susceptibility $a_{i,t}$ varied across the landscape, being spatially dependent on $\boldsymbol{x}_{i,t}$. Locations unsuitable to MPB (such as unforested areas) tended to assume extremely large $a_{i,t}$ values whereas areas with optimal habitat for MPB assumed much smaller ones.

Restricting our attention to optimal stands only, i.e. those having a density of 800–1500 stems/ha and aged > 80 yrs (Carroll et al. 2006), representing around 150,000 locations—the observed distribution of susceptibility values can be compared to empirical data from similar outbreaks. For example, the modes of the estimated $m_{A_{i,t}}$ values over these optimal stands were centred at 336, 932 and 480 females/stem, for the

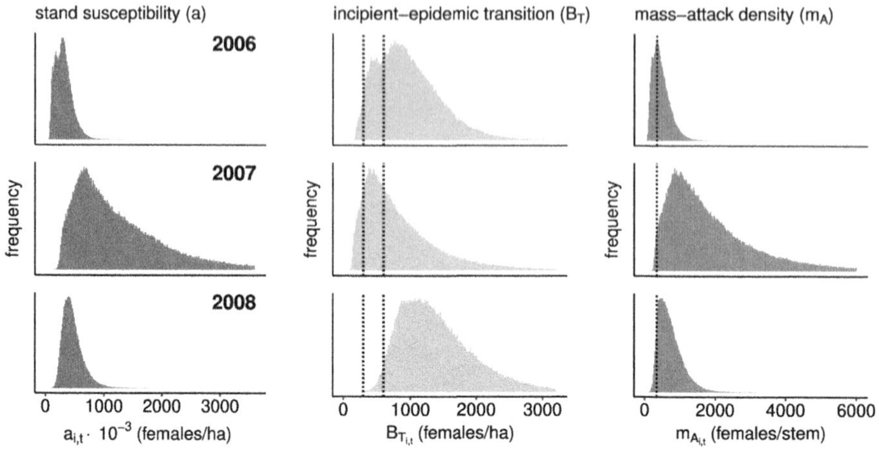

Fig. 6 (Colour figure online) Histograms of estimated susceptibility ($a_{i,t}$ left) in stands optimal for MPB in the years 2006–2008, and two associated quantities: (middle) the beetle pressure required for one mass attack per 15 ha, with dotted lines indicating an empirical range (Cooke and Carroll 2017); and the mass attack number (right), with a dotted line indicating the optimum of Raffa and Berryman (1983)

years 2006–2008, respectively (Fig. 6). This is reasonably consistent with the 300–617 females/stem range observed in our study area by Safranyik and Linton (1991) during a previous outbreak in 1984.

Using data on average diameters and attack heights for these optimal stands (23 cm, Carroll et al. 2006; and 11.36 m, Safranyik and Linton 1991, respectively), we estimated a typical bark area of 5.5 m^2/stem (Safranyik 1988, eq. 6). Our typical per-m^2 observed attack density ($m_{A_{i,t}}/5.5$) therefore lay in the range of 61–170 females/m^2. Note that the lower end of this range (observed in 2006) coincides exactly with the optimal attack density measured by Raffa and Berryman (1983) (Fig. 6, right).

This shows that 2006 was a year of strong population growth for MPB, with a relatively low threshold for outbreak emergence (B_T), and mass attack numbers (m_A) centred at or near the optimum for brood production. Populations continued to expand through the next 2 years, with a large number of incipient-epidemic transition events, followed by a collapse. Our model indicates that in optimal habitat, these events typically happened when MPB attack densities increased through the range 427–1114 of females/ha (the modes of the estimated $B_{T_{i,t}}$ by year; Fig. 6, middle). This agrees with empirical observations by (Cooke and Carroll 2017) of a transition point in the 300–600 range during the 5 years leading up to 2006, and indicates that B_T values spiked as the epidemic neared collapse in 2008.

On dividing the $B_{T_{i,t}}$ values in Fig. 6 by our estimates for ϵ_t, and taking medians, we find that a factor of 2.5–3.2 increase in the endemic population was typically sufficient to initiate an outbreak. These findings support the observation of Carroll et al. (2006) that the incipient-epidemic transition point seems to occur at a level slightly above the density required to mass-attack a single pine. Our model expresses this quantity by the ratio $B_{T_{i,t}}/m_{A_{i,t}}$, whose median values (in optimal MPB habitat) were 2.2, 0.5 and 2.0 in the years 2006–2008, respectively.

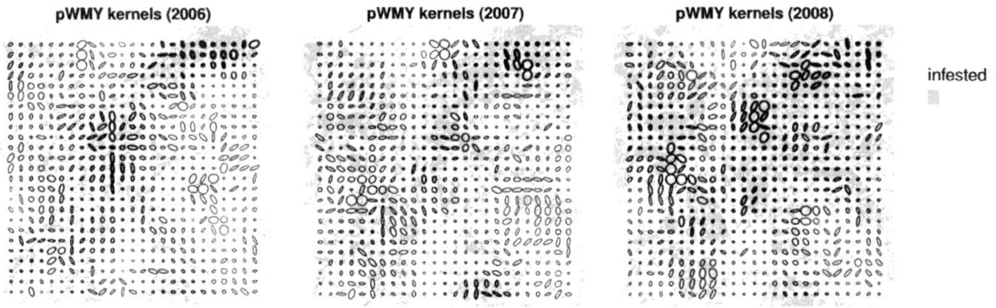

Fig. 7 (Colour figure online) Diffusion ellipses summarizing the angle and effective range corresponding to each of the 625 fitted pWMY parameter sets used to construct $\hat{\boldsymbol{D}}_t$ for each year. Each ellipse inscribes a contour of constant density for dispersal from its centre. Line thickness is scaled to match the estimated number of MPB displaced, emphasizing major outbreak centres. Infestations from the previous year are shaded to indicate the spatial distribution of source populations. Uninfested areas tended to produce small ellipses—these should be viewed as uninformative, as the model had no data from which estimate flight patterns in those blocks

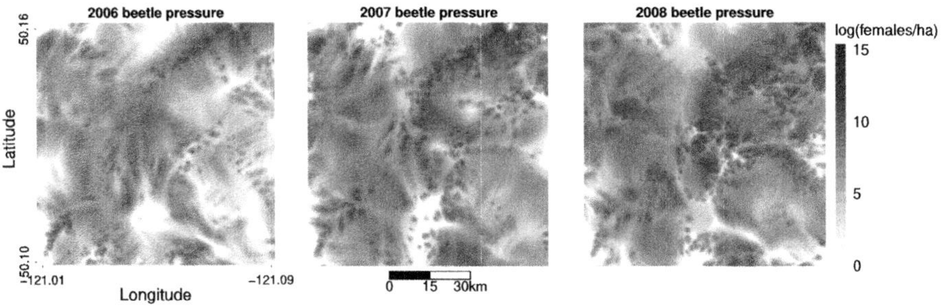

Fig. 8 (Colour figure online) Heatmaps of $\log(\lambda \hat{\boldsymbol{D}}_t\,(\boldsymbol{\phi}_{t-1} \odot \boldsymbol{H}_{t-1}))$, the fitted beetle pressure values arising from flight events in the years 2006–2008 (excluding endemic MPB). $\hat{\boldsymbol{D}}_t$ is the moving average of predictions from a 25 × 25 grid of local stationary models, each fitted to a local subset of the data

Flight events under the fitted model are summarized by the block-wise redistribution kernel estimates. Our pWMY kernels identified a large number of highly directed (anisotropic) dispersal events in all years. The grid of fitted dispersal kernel parameters ($\hat{\boldsymbol{\theta}}_{D_{k,t}}$) that generate $\hat{D}_t$ (Fig. 7) resembles a smooth vector field, raising some interesting questions as to the driving forces behind these patterns.

The combination of these stationary fitted kernels to form the nonstationary kernel ($\hat{D}_t$) brings into focus a complex landscape of MPB movement patterns (Fig. 8), illustrating how detailed information on beetle pressure can be recovered from AOS data by rethinking the usual modelling assumptions about dispersal.

Note that our model was constructed for parameter inference, rather than predictions of future outbreak locations. However, our methodology for estimating beetle pressure could easily be adapted to serve a forecasting role. We illustrate the idea in Fig. 9, where the empirical value of $B_T = 450$ (the midpoint of the range reported in Cooke and Carroll 2017) is used as threshold for outbreak development. The plot shows how our model delineates infested areas under two different scenarios; the first with no endemic population, and the second with ϵ_t set to its estimated value from 2006. Notice that neither $a_{i,t}$, κ_t nor $\boldsymbol{\theta}_{V_t}$ is needed for this classification.

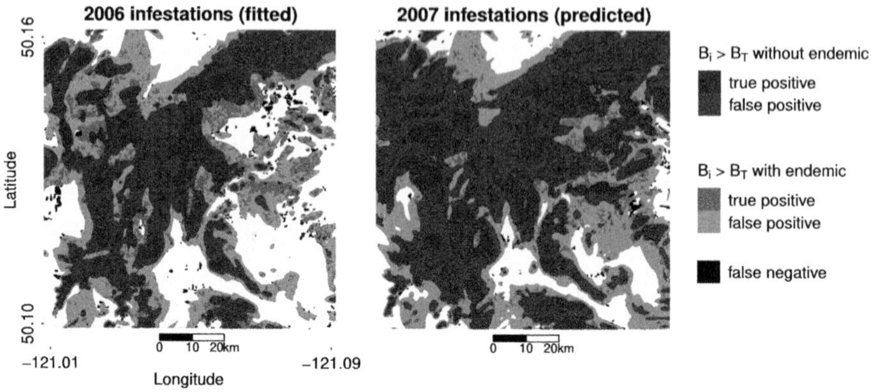

Fig. 9 (Colour figure online) Infested locations and next-year forecasts using the equation $\hat{\boldsymbol{B}}_t = \hat{\epsilon}\boldsymbol{I} + \hat{\boldsymbol{D}}_t \tilde{\boldsymbol{B}}_t$ from Sect. 2.5. Using the fitted values of $\hat{\epsilon}_t$ and $\hat{\boldsymbol{D}}_t$ from the training year 2006 (left), locations were classified as infested (shaded) if the predicted beetle pressure exceeded $B_T = 450$. Using these same parameters along with the observed attack damage and pine density in 2006, we then predicted infestations in 2007 (right). For comparison, an endemic-free estimate is also plotted (darker shaded regions) by replacing $B_{i,t}$ with $B_{i,t} - \epsilon_t$. The effect is to withdraw the contours of infested areas inwards, considerably limiting the estimated spread

The true positive rate in the training year 2006 was 93.5%, and in the forecast for 2007, it improved to 98.0%. By including the endemic population in our beetle pressure estimates, the contours of the infestation predictions broadened, sometimes by several kilometres. This improved detection rates substantially (true positive rate in 2006 without the endemic component: 71%; and in 2007: 84%).

4 Discussion

The S-shaped colonization curves that characterize the nonlinearity of MPB attack dynamics (e.g. Raffa and Berryman 1983; Boone et al. 2011) are usually fitted to field data on individual attacked trees, so they relate attack density to the mortality among pines undergoing attack. This is a conditional probability model. For example, the model of Cooke and Carroll (2017) has the form:

$$\text{logit}\,(\Pr(\text{pine mortality} \mid \text{attack})) = A + \kappa \log(N_a) = (A - \kappa \log(c)) + \kappa \log(B) \quad (9)$$

where A is a dimensionless intercept; and N_a is the number of stems attacked within the study plot, which we expect to scale according to $cN_a \approx B$ with the attack density B (in females/ha).

Our model, however, is based on aerial data, from which failed attacks cannot be resolved. In (2), we therefore related B to the *unconditional* probability of stand-level mortality $\Pr(\text{pine mortality} \mid \text{attack}) \Pr(\text{attack})$, which we called ϕ. Notice that when $\Pr(\text{attack}) = 1$, both the red-top model of Heavilin and Powell (2008) and our generalization (1) coincide exactly with (9). In reality, attack rates will be much lower, so in the high-level description (1) we assumed that the logit-linear relationship (9) remains after aggregating mortality data at the 1 hectare scale. Our results supported this assumption, with estimates of κ in close agreement with the field data reported by Cooke and Carroll (2017).

In Sect. 2.1, we showed how, via stand-susceptibility (a), this κ value is mathematically linked to the mass attack number (m_A) and the incipient-epidemic transition point (B_T). Our comparison of point estimates for these parameters with empirical data from previous years showed reasonably good agreement, supporting the theory behind formulae (3) and (4). This illustrates one way in which our model can be used to study ecological questions about MPB attack dynamics at the level of the individual tree, while using only (stand-level) AOS data for parameter fitting.

For instance, the observed increase in B_T in 2008, along with the elevated m_A levels in 2007–2008, can be attributed to host depletion, as MBP tend to select pine of a certain phloem, size and vigour class for colonization (Shrimpton and Thomson 1985; Cole and McGregor 1983; Raffa and Berryman 1983). As the preferred hosts become scarce, MPBs are thought to balance increasing fitness costs by first intensifying mass attacks on the few that remain (Lewis et al. 2010), thus effectively increasing m_A above its optimal level. Similarly, a scarcity of suitable mass-attack targets can be expected to make spontaneous eruptions from the endemic phase less likely.

Furthermore, our results on ϵ_t shed a mathematical light on how outbreaks might sporadically arise across the landscape—if environmental conditions were to double or triple the number of injured/weakened pines available to the endemic population, this could allow it to grow to the point of exceeding B_T in the absence immigrating MPB—in accordance with the theory of Berryman (1978), and the explanation of Cooke and Carroll (2017) as to the origin of the outbreaks analysed in Sect. 3.

In-flights of MPB are equally important to understanding MPB outbreak dynamics. This is clear from the large number of spatial regression studies pointing to beetle pressure as the single most significant factor in outbreak development (e.g. Aukema et al. 2008; Preisler et al. 2012; Sambaraju et al. 2012). However, there remains little consensus in the modelling literature on how best to represent beetle pressure mathematically.

As we explained in Sect. 2.5, beetle pressure simply expresses our modelling assumptions about MPB dispersal; Different modelling approaches handle this problem in different ways. With few exceptions (such as Powell and Bentz 2014; Powell et al. 2018), forecasting models tend to reconstruct beetle pressure in a heuristic way, by defining infestation indicator variables that are summed over local spatial neighbourhoods (see, e.g. Shore et al. 2000; Aukema et al. 2008; Robertson et al. 2009; Kunegel-Lion et al. 2019). Many attack dynamics regression models also employ this trick (e.g. Zhu et al. 2010; Preisler et al. 2012; Sambaraju et al. 2012; Kärvemo et al. 2014), and indeed a stationary and isotropic kernel-based representation (as in Heavilin and Powell 2008; Goodsman et al. 2016) is simply a refinement that finds a biology-based shape (and range) for the filter. Our method refined this idea further, in a novel way, by introducing directedness and location-dependence by means of a weighted combination of stationary kernels.

As we observed in a previous study (Koch et al. 2020b), the precision gained through the use of anisotropic kernels appears to far outweigh the drawbacks associated with the introduction of additional dispersal parameters. Moreover, we believe our refined flight model shows promise not only in formulating beetle pressure (as we do here), but as a tool for studying nonstationary dispersal processes more generally. Future work might look for connections between $\boldsymbol{\theta}_{D_{k,t}}$ and environmental drivers such as prevailing wind direction, as a means of studying the dispersal process itself. For example, one could analyse whether patterns of directionality might arise in reaction to population density, both of beetles and hosts, similar to work by Powell and Bentz (2014) (we thank an anonymous reviewer for these suggestions).

Though we did not analyse the kernel parameters ($\boldsymbol{\theta}_{D_t}$) in detail, it is worth remarking that in most of the pWMY kernels a leptokurtic pattern of dispersal was favoured over the simpler Gaussian model of bio-diffusion. This highlights the versatility of the pWMY in modelling different flight mechanisms (Koch et al. 2020b), and suggests that a wide range of MPB flight behaviours are realized across the landscape: including both the fat-tailed patterns, proposed by Goodsman et al. (2016) and Turchin and Thoeny (1993); and the Gaussian, suggested by Heavilin and Powell (2008).

Note that the model-fitting procedure of Sect. 2.6 was constructed to study attack dynamics (at least) 1 year after they occur, not to predict them in future summers, nor is our estimate of stand susceptibility $a_{i,t}$ (as a log-linear function of local covariates) intended for extrapolation. A more judicious choice of covariates whose values can be projected in time (combined with a significance-based covariate selection) would be needed in a predictive risk model for MPB damage. Nonetheless, we think the framework in (7)—and in particular the nonstationary approach to dispersal—will be helpful in building model-based solutions to management and forecasting problems.

We illustrated this briefly in the next-year classification example of Fig. 9. Note that our high detection rates lie near the level mentioned in Fettig et al. (2014) for stabilizing outbreaks by mitigation measures (such as cut and burn). However, with high recall comes a high false positive rate (low precision); Moreover, the 2007 prediction required information on pre-dispersal density that is typically not available until *after* the attack summer being predicted—recall that $\tilde{\boldsymbol{B}}_t$ is derived from crown fade data with a 1-year lag. One possible solution would be to iterate equation (7) with simulated error to produce a suite of multi-year forecasts under various scenarios of stand susceptibility and process error, an idea we plan to explore in future work.

Figure 9 illustrates an important consequence of the ubiquity of endemic MPB in their natural range: it increases the potential for outbreaks to spread. The potential for range expansion may be therefore be underestimated if the endemic contribution to MPB outbreaks is ignored. This will be of particular relevance in contemporary areas of concern, such as the Boreal forest in Alberta (Safranyik et al. 2010). The establishment of endemic populations in these areas should be closely monitored, as our results show that they have the potential to accelerate the spread of outbreaks, and thus speed the range expansion of the MPB.

Supplementary Information The online version contains supplementary material available at https://doi.org/10.1007/s11538-021-00899-z.

Author Contributions All authors contributed to the model conception, design and application. The first draft of the manuscript was written by Dean Koch and all authors commented on previous versions of the manuscript. All authors read and approved the final manuscript.

Funding This research was supported by a grant to MAL from the Natural Science and Engineering Research Council of Canada (Grant No. NET GP 434810-12) to the TRIA Network.

Availability of data and materials All data used in this paper are available in a zip archive as electronic supplementary material (Online resource 2).

Declarations

Conflict of interest TRIA Network was supported by contributions from Alberta Agriculture and Forestry, Foothills Research Institute, Manitoba Conservation and Water Stewardship, Natural Resources Canada-Canadian Forest Service, Northwest Territories Environment and Natural Resources, Ontario Ministry of Natural Resources and Forestry, Saskatchewan Ministry of Environment, West Fraser and Weyerhaeuser.

Ethics approval Not applicable.

Consent to participate Not applicable.

Consent for publication Not applicable.

Code availability The data analysis results reported in this paper can be reconstructed using the R script files included in the electronic supplementary material (Online resource 3).

References

Ainslie B, Jackson PL (2011) Investigation into mountain pine beetle above-canopy dispersion using weather radar and an atmospheric dispersion model. Aerobiologia (Bologna) 27(1):51–65. https://doi.org/10.1007/s10453-010-9176-9

Amman GD (1984) Mountain pine beetle (Coleoptera: Scolytidae) mortality in three types of infestations. Environ Entomol 13(1):184–191. https://doi.org/10.1093/ee/13.1.184

Aukema BH, Carroll AL, Zhu J, Raffa KF, Sickley TA, Taylor SW (2006) Landscape level analysis of mountain pine beetle in British Columbia, Canada: spatiotemporal development and spatial synchrony within the present outbreak. Ecography (Cop) 29(3):427–441. https://doi.org/10.1111/j.2006.0906-7590.04445.x

Aukema BH, Carroll AL, Zheng Y, Zhu J, Raffa KF, Dan Moore R, Stahl K, Taylor SW (2008) Movement of outbreak populations of mountain pine beetle: influences of spatiotemporal patterns and climate. Ecography (Cop) 31(3):348–358. https://doi.org/10.1111/j.0906-7590.2007.05453.x

Beaudoin A, Bernier PY, Guindon L, Villemaire P, Guo XJ, Stinson G, Bergeron T, Magnussen S, Hall RJ (2014) Mapping attributes of Canada's forests at moderate resolution through kNN and MODIS imagery. Can J For Res 44(5):521–532. https://doi.org/10.1139/cjfr-2013-0401

Berryman AA (1974) Dynamics of bark beetle populations: towards a general productivity model 1. Environ Entomol 3(4):579–585. https://doi.org/10.1093/ee/3.4.579

Berryman AA (1978) Towards a theory of insect epidemiology. Res Popul Ecol (Kyoto) 19(2):181–196. https://doi.org/10.1007/BF02518826

Bleiker KP, O'Brien MR, Smith GD, Carroll AL (2014) Characterisation of attacks made by the mountain pine beetle (Coleoptera: Curculionidae) during its endemic population phase. Can Entomol 146(3):271–284. https://doi.org/10.4039/tce.2013.71

Boone CK, Aukema BH, Bohlmann J, Carroll AL, Raffa KF (2011) Efficacy of tree defense physiology varies with bark beetle population density: a basis for positive feedback in eruptive species. Can J For Res 41(6):1174–1188. https://doi.org/10.1139/x11-041

Carroll AL, Aukema BH, Raffa KF, Linton DA, Smith GD, Lindgren BS (2006) Mountain pine beetle outbreak development: the endemic—incipient epidemic transition. Can For Serv Mt Pine Beetle Initiat Proj 1:27

Chen H, Walton A (2011) Mountain pine beetle dispersal: spatiotemporal patterns and role in the spread and expansion of the present outbreak. Ecosphere 2(6):art66. https://doi.org/10.1890/es10-00172.1

Chen H, Jackson PL, Ott PK, Spittlehouse DL (2015a) A spatiotemporal pattern analysis of potential mountain pine beetle emergence in British Columbia, Canada. For Ecol Manag 337(4):11–19. https://doi.org/10.1016/j.foreco.2014.10.034
Chilès JP, Delfiner P (2012) Geostatistics: modeling spatial uncertainty, vol 2, 2nd edn. Wiley, Hoboken. https://doi.org/10.1002/9781118136188
Cole, W and Amman G (1969) Mountain pine beetle infestations in relation to lodgepole pine diameters, vol 95. US Department of Agriculture, Forest Service, Intermountain Forest & Range
Cole WE, McGregor MD (1983) Estimating the rate and amount of tree loss from mountain pine beetle infestations. Technical Report INT-318, US Department of Agriculture, Forest Service, Intermountain Forest and Range..., https://doi.org/10.5962/bhl.title.68709
Cooke BJ, Carroll AL (2017) Predicting the risk of mountain pine beetle spread to eastern pine forests: considering uncertainty in uncertain times. For Ecol Manag 396:11–25. https://doi.org/10.1016/j.foreco.2017.04.008
Crujeiras RM, Van Keilegom I (2010) Least squares estimation of nonlinear spatial trends. Comput Stat Data Anal 54(2):452–465. https://doi.org/10.1016/j.csda.2009.09.014
de la Giroday HMC, Carroll AL, Aukema BH (2012) Breach of the northern Rocky Mountain geoclimatic barrier: initiation of range expansion by the mountain pine beetle. J Biogeogr 39(6):1112–1123. https://doi.org/10.1111/j.1365-2699.2011.02673.x
Fettig CJ, Gibson KE, Munson AS, Negrón JF (2014) A comment on Management for mountain pine beetle outbreak suppression: does relevant science support current policy? Forests 5(4):822–826. https://doi.org/10.3390/f5040822
Garlick MJ, Powell JA, Hooten MB, McFarlane LR (2011) Homogenization of large-scale movement models in ecology. Bull Math Biol 73(9):2088–2108. https://doi.org/10.1007/s11538-010-9612-6
Goodsman DW, Koch D, Whitehouse C, Evenden ML, Cooke BJ, Lewis MA (2016) Aggregation and a strong Allee effect in a cooperative outbreak insect. Ecol Appl 26(8):2621–2634. https://doi.org/10.1002/eap.1404
Heavilin J, Powell J (2008) A novel method of fitting spatio-temporal models to data, with applications to the dynamics of mountain pine beetles. Nat Resour Model 21(4):489–524. https://doi.org/10.1111/j.1939-7445.2008.00021.x
Holling CS (1959) Some characteristics of simple types of predation and parasitism. Can Entomol 91(7):385–398. https://doi.org/10.4039/Ent91385-7
Jackson PL, Straussfogel D, Lindgren BS, Mitchell S, Murphy B (2008) Radar observation and aerial capture of mountain pine beetle, Dendroctonus ponderosae Hopk. (Coleoptera: Scolytidae) in flight above the forest canopy. Can J For Res 38(8):2313–2327. https://doi.org/10.1139/X08-066
Kärvemo S, Van Boeckel TP, Gilbert M, Grégoire JC, Schroeder M (2014) Large-scale risk mapping of an eruptive bark beetle—importance of forest susceptibility and beetle pressure. For Ecol Manag 318:158–166. https://doi.org/10.1016/j.foreco.2014.01.025
Kautz M (2014) On correcting the time-lag bias in aerial-surveyed bark beetle infestation data. For Ecol Manag 326:157–162. https://doi.org/10.1016/j.foreco.2014.04.010
Koch D, Lele S, Lewis M (2020a) Computationally simple anisotropic lattice covariograms. Environ Ecol Stat. https://doi.org/10.1007/s10651-020-00456-2
Koch DC, Lewis MA, Lele SR (2020b) A unifying theory for two-dimensional spatial redistribution kernels with applications in population spread modelling. J R Soc Interface 17(170):20200434
Kunegel-Lion M, McIntosh RL, Lewis MA (2019) Management assessment of mountain pine beetle infestation in cypress hills, SK. Can J For Res 49(2):154–163. https://doi.org/10.1139/cjfr-2018-0301
Lewis MA, Nelson W, Xu C (2010) A structured threshold model for mountain pine beetle outbreak. Bull Math Biol 72(3):565–589. https://doi.org/10.1007/s11538-009-9461-3
Limpert E, Stahel WA, Abbt M (2001) Log-normal distributions across the sciences: keys and clues on the charms of statistics, and how mechanical models resembling gambling machines offer a link to a handy way to characterize log-normal distributions, which can provide deeper insight into va. Bioscience 51(5):341–352
Nelson WA, Potapov A, Lewis MA, Hundsdörfer AE, He F (2008) Balancing ecological complexity in predictive models: a reassessment of risk models in the mountain pine beetle system. J Appl Ecol 45(1):248–257. https://doi.org/10.1111/j.1365-2664.2007.01374.x
Neubert MG, Kot M, Lewis MA (1995) Dispersal and pattern formation in a discrete-time predator–prey model. Theor Popul Biol 48(1):7–43. https://doi.org/10.1006/tpbi.1995.1020

Powell JA, Bentz BJ (2014) Phenology and density-dependent dispersal predict patterns of mountain pine beetle (Dendroctonus ponderosae) impact. Ecol Model 273:173–185. https://doi.org/10.1016/j.ecolmodel.2013.10.034

Powell JA, Garlick MJ, Bentz BJ, Friedenberg N (2018) Differential dispersal and the Allee effect create power-law behaviour: distribution of spot infestations during mountain pine beetle outbreaks. J Anim Ecol 87(1):73–86. https://doi.org/10.1111/1365-2656.12700

Preisler HK, Hicke JA, Ager AA, Hayes JL (2012) Climate and weather influences on spatial temporal patterns of mountain pine beetle populations in Washington and Oregon. Ecology 93(11):2421–2434. https://doi.org/10.1890/11-1412.1

Preston FW (1948) The commonness, and rarity, of species. Ecology 29:254–283. https://doi.org/10.2307/1930989

Raffa KF, Berryman AA (1983) The role of host plant resistance in the colonization behavior and ecology of bark beetles (Coleoptera: Scolytidae). Ecol Monogr 53(1):27–49. https://doi.org/10.2307/1942586

Reid RW (1962) Biology of the mountain pine beetle, Dendroctonus monticolae Hopkins, in the East Kootenay Region of British Columbia I. Life Cycle, Brood Development, and Flight Periods1. Can Entomol 94(5):531–538. https://doi.org/10.4039/Ent94531-5

Reyes PE, Zhu J, Aukema BH (2012) Selection of spatial–temporal lattice models: assessing the impact of climate conditions on a mountain pine beetle outbreak. J Agric Biol Environ Stat 17(3):508–525. https://doi.org/10.1007/s13253-012-0103-0

Robertson C, Farmer CJ, Nelson TA, MacKenzie IK, Wulder MA, White JC (2009) Determination of the compositional change (1999–2006) in the pine forests of British Columbia due to mountain pine beetle infestation. Environ Monit Assess 158(1–4):593–608. https://doi.org/10.1007/s10661-008-0607-9

Safranyik L (1988) Estimating attack and brood totals and densities of the mountain pine beetle in individual lodgepole pine trees. Can Entomol 120(4):323–331. https://doi.org/10.4039/Ent120323-4

Safranyik L, Linton DA (1991) Unseasonably low fall and winter temperatures affecting mountain pine beetle and pine engraver beetle populations and damage in the British Columbia Chilcotin region. J Entomol Soc B C 88:17–21

Safranyik L, Carroll AL (2006) The biology and epidemiology of the mountain pine beetle in lodgepole pine forests. In: The mountain pine beetle: a synthesis of biology, management and impacts on lodgepole pine, vol 11, pp 3–66. https://doi.org/10.1673/031.011.12701

Safranyik L, Carroll AL, Régnière J, Langor DW, Riel WG, Shore TL, Peter B, Cooke BJ, Nealis VG, Taylor SW (2010) Potential for range expansion of mountain pine beetle into the boreal forest of North America. Can Entomol 142(5):415–442. https://doi.org/10.4039/n08-CPA01

Sambaraju KR, Carroll AL, Zhu J, Stahl K, Moore RD, Aukema BH (2012) Climate change could alter the distribution of mountain pine beetle outbreaks in western Canada. Ecography (Cop) 35(3):211–223. https://doi.org/10.1111/j.1600-0587.2011.06847.x

Shegelski VA, Evenden ML, Sperling FA (2019) Morphological variation associated with dispersal capacity in a tree-killing bark beetle Dendroctonus ponderosae Hopkins. Agric For Entomol 21(1):79–87. https://doi.org/10.1111/afe.12305

Shore TL, Safranyik L, Lemieux JP (2000) Susceptibility of lodgepole pine stands to the mountain pine beetle: testing of a rating system. Can J For Res 30(1):44–49. https://doi.org/10.1139/x99-182

Shrimpton DM, Thomson AJ (1985) Relationship between phloem thickness and lodgepole pine growth characteristics. Can J For Res 15(5):1004–1008. https://doi.org/10.1139/x85-161

Taylor SW, Carroll AL, Alfaro RI, Safranyik L, Safranyik L, Wilson B (2006) The mountain pine beetle: a synthesis of biology. Canadian Forest Service, Victoria

Turchin P, Thoeny WT (1993) Quantifying dispersal of southern pine beetles with mark-recapture experiments and a diffusion model. Ecol Appl 3(1):187–198. https://doi.org/10.2307/1941801

Woodell SRJ, Peters RH (1992) A critique for ecology, vol 80. Cambridge University Press, Cambridge. https://doi.org/10.2307/2261026

Wulder MA, Dymond CC, White JC, Leckie DG, Carroll AL (2006) Surveying mountain pine beetle damage of forests: a review of remote sensing opportunities. For Ecol Manag 221(1–3):27–41. https://doi.org/10.1016/j.foreco.2005.09.021

Wulder MA, White JC, Grills D, Nelson T, Coops NC, Ebata T (2009) Aerial overview survey of the mountain pine beetle epidemic in British Columbia: communication of impacts. BC J Ecosyst Manag 10(1):45–58

Yasuda N (1975) The random walk model of human migration. Theor Popul Biol 7(2):156–167. https://doi.org/10.1016/0040-5809(75)90011-8

Zhu J, Huang HC, Reyes PE (2010) On selection of spatial linear models for lattice data. J R Stat Soc Ser B Stat Methodol 72(3):389–402. https://doi.org/10.1111/j.1467-9868.2010.00739.x

Publisher's Note Springer Nature remains neutral with regard to jurisdictional claims in published maps and institutional affiliations.

Bulletin of Mathematical Biology (2021) 83:55
https://doi.org/10.1007/s11538-021-00891-7

Society for Mathematical Biology

SPECIAL ISSUE: CELEBRATING J. D. MURRAY

Modeling the Effect of HIV/AIDS Stigma on HIV Infection Dynamics in Kenya

Ben Levy[1] · Hannah E. Correia[2,3] · Farainumashe Chirove[5] · Marilyn Ronoh[4] · Ash Abebe[5] · Moatlhodi Kgosimore[6] · Obias Chimbola[7] · M. Hellen Machingauta[7] · Suzanne Lenhart[8] · K. A. Jane White[9]

Received: 30 September 2020 / Accepted: 15 March 2021 / Published online: 5 April 2021

Abstract

Stigma toward people living with HIV/AIDS (PLWHA) has impeded the response to the disease across the world. Widespread stigma leads to poor adherence of preventative measures while also causing PLWHA to avoid testing and care, delaying important treatment. Stigma is clearly a hugely complex construct. However, it can be broken down into components which include internalized stigma (how people with the trait feel about themselves) and enacted stigma (how a community reacts to an individual with the trait). Levels of HIV/AIDS-related stigma are particularly high in sub-Saharan Africa, which contributed to a surge in cases in Kenya during the late twentieth century. Since the early twenty-first century, the United Nations and governments around the world have worked to eliminate stigma from society and resulting public health education campaigns have improved the perception of PLWHA over time, but HIV/AIDS remains a significant problem, particularly in Kenya. We take a data-driven approach to create a time-dependent stigma function that captures both the level of internalized and enacted stigma in the population. We embed this within

✉ K. A. Jane White
maskajw@bath.ac.uk

1 Department of Mathematics, Fitchburg State University, Fitchburg, MA, USA

2 Harvard Data Science Initiative, Harvard University, Cambridge, MA, USA

3 Department of Biostatistics, Harvard University, Boston, MA, USA

4 School of Mathematics, University of Nairobi, Nairobi, Kenya

5 Department of Mathematics and Statistics, Auburn University, Auburn, AL, USA

6 Department of Biometry and Mathematics, Botswana University of Agriculture and Natural Resources, Gaborone, Botswana

7 Mathematics and Statistical Sciences, Botswana International University of Science and Technology, Palapye, Botswana

8 Mathematics Department, University of Tennessee, Knoxville, TN, USA

9 Department of Mathematical Sciences, University of Bath, Bath, UK

a compartmental model for HIV dynamics. Since 2000, the population in Kenya has been growing almost exponentially and so we rescale our model system to create a coupled system for HIV prevalence and fraction of individuals that are infected that seek treatment. This allows us to estimate model parameters from published data. We use the model to explore a range of scenarios in which either internalized or enacted stigma levels vary from those predicted by the data. This analysis allows us to understand the potential impact of different public health interventions on key HIV metrics such as prevalence and disease-related death and to see how close Kenya will get to achieving UN goals for these HIV and stigma metrics by 2030.

Keywords HIV · Stigma · Kenya · Mathematical model · UN goals

1 Introduction

HIV/AIDS-related stigma and discrimination continue to impede the progress of *responses to* HIV/AIDS across the world (Chesney and Smith 1999). While the percentage of people expressing discriminatory attitudes toward people living with HIV/AIDS has decreased over time, on average more than half of adults in 36 countries across the globe still express discriminatory attitudes (ICF 2018). People living with HIV/AIDS (PLWHA) who experience high levels of HIV/AIDS-related stigma avoid testing and delay initiating HIV/AIDS care and treatment (Golub and Gamarel 2013; Price et al. 2019; Remien et al. 2015; Ti et al. 2013; Treves-Kagan et al. 2017). Further, individuals living with HIV/AIDS *avoid* frequenting hospitals for treatment or collecting antiretroviral therapy (ART) drugs for fear of health workers disclosing their HIV/AIDS status to the communities (Kagee et al. 2011; Mills et al. 2006). Available data across 19 countries confirm that one in four PLWHA face discrimination in health care (Global Network of People with HIV/AIDS and International Community of Women living with HIV/AIDS 2017), and one in five avoid healthcare treatment due to fear of discrimination (King et al. 2013; Nyblade et al. 2017). Approximately one in every eight PLWHA are denied health care due to stigma regarding their status, and women living with HIV/AIDS face greater discrimination in health care than their male counterparts (Global Network of People with HIV/AIDS and International Community of Women living with HIV/AIDS 2017). Stigma or fear of stigma results in poor adherence to pre-exposure prophylaxis and antiretroviral therapy, leading to high HIV/AIDS viral loads (Buregyeya et al. 2017; Croome et al. 2017; Katz et al. 2013; Patel et al. 2016). Stigmatized PLWHA are also less likely to disclose their HIV/AIDS status to their sex partner(s) (McKay and Mutchler 2011).

In sub-Saharan Africa, rates of HIV/AIDS-related stigma remain particularly high, and so do infection levels. In Kenya, a peak in new HIV infections in 1995 was followed by a peak in deaths attributed to HIV/AIDS in 2004, and although numbers of new HIV infections are falling, the decrease has been no more than 1000 individuals per year since 2010 (Global Burden of Disease Collaborative Network 2018).

At the United Nations (UN) General Assembly Special Session on HIV/AIDS in 2001, African governments agreed to combat all forms of discrimination against PLWHA and subsequently the UN released *the "Getting to Zero*" initiative in 2011.

The goals of this initiative were to get new infections, discrimination, and deaths from HIV/AIDS to zero by 2030, clearly recognizing the importance of reductions in both infection and stigma levels in order to achieve the ambitious goal. However, HIV/AIDS-related stigma and discrimination are difficult to overcome solely through top-down initiatives and messaging campaigns (Campbell and Cornish 2010; Johnny and Mitchell 2006; Parkhurst 2014) and while there has been progress, it seems unlikely that the zero goals will be achieved.

Many researchers have formulated mathematical models to understand the dynamics of HIV/AIDS. We are aware of work on epidemiological models for HIV infection levels and spread in Africa (Nyabadza et al. 2011; Simwa and Pokhariyal 2003), including some models that consider interventions such as treatment, use of condoms, and contact tracing (Hyman et al. 2003; Moghadas et al. 2003; Granich et al. 2009). Some models include features representing information that causes changes in the behavior of individuals living in a society with strong HIV prevalence (Joshi et al. 2008; Ronoh et al. 2020). However, there are very few examples that include stigma explicitly within infectious disease dynamic models. We call attention to a system of four ODEs used for showing dynamics and game theoretical results illustrating interactions of stigmatization and prevalence in a generic infectious disease (Reluga et al. 2019). Two recent papers used structural equation modeling and cohort scenario analysis to examine the effects of stigma on African women with HIV (Logie et al. 2016; Prudden et al. 2017).

Here, we seek to investigate the effects of HIV/AIDS-related stigma on the dynamics of an HIV infection model which includes a class of infected individuals that are receiving treatment. We specifically focus on understanding the effects of stigma on HIV/AIDS dynamics in Kenya which has some of the highest estimated prevalence of HIV/AIDS in the world (UNAIDS 2018). Our model approach has two strands. First we use survey data to create a time-varying measure of stigma in the adult population of Kenya; we use that to build a model for stigma which feeds into our compartmental model for HIV infection dynamics.

We begin in Sect. 2, by establishing a model for population stigma, parameterizing it using data from Kenya Demographic and Health surveys (CBS et al. 2004; Kenya National Bureau of Statistics (KNBS) and ICF Macro 2010, 2014). This feeds into a compartmental model for HIV infection in Kenya and the associated parameter estimation in Sect. 3. Our results, presented in Sect. 4, explore how baseline stigma parameters impact infection prevalence and HIV-related deaths. We modify the parameter estimates to undertake a numerical exploration focussed on understanding how changes to internalized versus enacted stigma would have impacted HIV infection measures; we complement this with a simple steady-state analysis to gain insight into how the infection dynamics evolve. In the final section, we discuss our results in the context of the impact of stigma on HIV dynamics in Kenya highlighting the urgent need to gather more data on stigma and its associated impact on HIV dynamics.

2 Modeling Population Stigma

Stigma is a socially devalued attribute that gives rise to social inequality in the form of labeling, stereotyping, devaluation, status loss, or discrimination arising from the social judgment applied to a person or group who possesses the devalued attribute (Earnshaw and Chaudoir 2009; Van Brakel 2006). It keeps those with a socially devalued attribute in a position of relative subordination to those without the devalued attribute (Link and Phelan 2001; Parker and Aggleton 2003).

One of the main approaches to measuring HIV/AIDS-related stigma is the assessment of discriminatory attitudes, including measures calculated from questions regarding a person's potential actions toward a PLWHA (Van Brakel 2006; Earnshaw and Chaudoir 2009). Select studies have also measured stigma through interviews with PLWHA asking how many times or how often they have experienced various forms of discrimination over the past year (Neuman and Obermeyer 2013). Indices for HIV/AIDS-related stigma have been developed previously, however most were intended for use in the USA and few have been broadly deployed (Van Brakel 2006).

PLWHA experience stigma through three mechanisms (Earnshaw and Chaudoir 2009; Van Brakel 2006):

- enacted or experienced stigma;
- anticipated or perceived stigma; and
- internalized stigma.

There are two stages to our modeling activity. Firstly, we estimate population-level stigma in Kenya using data from the Kenya Demographic and Health Surveys (KDHS) from 2003, 2009, and 2014. This results in only three data points which are insufficient to make accurate predictions. However, the points allow us to predict parameters of our dynamic model for stigma. In the second stage, we create a simple linear model to describe the change in stigma over time using mechanistic principles and guided by Occam's Razor.

2.1 Obtaining Data Points for Stigma in Kenya

The KDHS from 2003, 2009, and 2014 provide data on HIV/AIDS knowledge, relevant behavior, and attitudes toward PLWHA captured at the national level and for demographically homogeneous subpopulations (CBS et al. 2004; Kenya National Bureau of Statistics (KNBS) and ICF Macro 2010, 2014). The questionnaire module on attitudes toward PLWHA asks survey respondents familiar with AIDS the following four questions:

1. Would you buy fresh vegetables from a shopkeeper or vendor if you knew that this person had HIV?
2. If a member of your family became sick with AIDS, would you be willing to care for her or him in your own household?
3. In your opinion, if a female teacher has the AIDS virus, but is not sick, should she be allowed to continue teaching in the school?
4. If a member of your family got infected with the AIDS virus, would you want it to remain a secret or not?

The first three questions capture the stigma mechanism of social distancing from PLWHA, while the fourth question aims to measure perceived or anticipated stigma enacted by others should the respondent be associated with HIV/AIDS (Chan and Tsai 2016). Question 2 in the DHS has been found to be interpreted very differently by men and women and so is unreliable for inclusion in our estimation of stigma (Cordes et al. 2017). Stigmatizing responses for the three remaining questions were as follows:

1A. "No, I would not buy fresh vegetables from a shopkeeper or vendor if I knew that this person had HIV."
3A. "No, a female teacher who has the AIDS virus, but is not sick, should not be allowed to continue teaching in the school."
4A. "Yes, I would want my family member's AIDS virus infection to remain a secret."

Women exhibit higher levels of internalized and enacted stigma than men in Sub-Saharan Africa (Geary et al. 2014; Mugoya and Ernst 2014) and are considered critical pathways to reducing community-level stigma (Kelly et al. 2017). Additionally, stigma is likely to be underestimated by surveys (Kalichman et al. 2019; Maughan-Brown 2010). We therefore constructed a measure of stigma as the proportion of female respondents across Kenya who answered at least two questions of the remaining three (Questions 1, 3, and 4) in a stigmatizing manner, resulting in the time-ordered pairs of data:

$$(2003, 0.3622), (2008, 0.2654), (2014, 0.2654).$$

We interpret these values as the fraction of Kenyans that have a stigmatizing view of HIV/AIDS irrespective of infection status and consider this to be a measure of population-wide stigma. Raw data used to calculate these values are given in "Appendix 1."

We acknowledge the difficulty in accurately measuring stigma through surveys and the limitations of the KDHS questions, including bias from respondents indicating they do not engage in stigmatizing behaviors and concerns over how some questions may be understood by respondents (Cordes et al. 2017; USAIDS 2005; Yoder and Nyblade 2004). However, the KDHS questionnaire is the only study gathering standardized, national-level information on attitudes toward PLWHA for many countries at regular intervals over time and therefore allows us to consider the effects of stigma on a national population of PLWHA. Our estimates of stigma were validated by comparing our estimated values with findings of smaller studies within Kenya between 2003 and 2014 using more comprehensive instruments for measuring stigma (National Empowerment Network of People Living With HIV and AIDS in Kenya (NEPHAK) et al. 2011; Neuman and Obermeyer 2013).

2.2 Creating a Mechanistic Model for Stigma in Kenya

We let $\sigma(t)$ represent population-level stigma as defined above and make the following model assumptions:

- There is a lower bound σ_i for $\sigma(t)$ which corresponds to population levels of internalized stigma at equilibrium;

- The rate at which stigma changes in the population is directly proportional to the difference between current levels of stigma and the lower bound.

Combining these assumptions gives rise to the model equation:

$$\frac{d\sigma}{dt} = \nu(\sigma_i - \sigma) \tag{1}$$

where ν and σ_i are positive constants and $\sigma(0)$ is specified. While simple in structure, this model still provides a caricature of the three components of stigma—internalized, enacted, and perceived—and allows us to determine the impact of interventions on each of these components:

- At equilibrium, $\sigma = \sigma_i$. Therefore, we interpret σ_i as the population-level *internalized stigma;*
- The rate ν at which stigma changes over time represents how *enacted and perceived stigma* change due to external drivers such as advertising campaigns for HIV treatment.

Taking our starting point to correspond to the year 2000, we use the Curve Fitting Toolbox in MATLAB to obtain parameter estimates $\nu = 0.24$, $\sigma_i = 0.23$ and $\sigma(0) = 0.5$. Using these values, we find the solution to (1),

$$\sigma(t) = 0.27e^{-0.24(t-2000)} + 0.23. \tag{2}$$

See Fig. 1 for the fit of this function to the three data points from the KDHS. We consider this function of $\sigma(t)$ to be the primary analytical scenario representative of the observed levels of stigma in Kenya during 2003–2014. In the Results section, we use this fit as our baseline from which to consider alternative parameter values corresponding to different levels of internalized or enacted/perceived stigma.

3 Modeling Infection Dynamics

Since 2000, the adult population in Kenya (16–64-year olds) has been growing just over 3% per annum and so we cannot make the commonly used assumption for infectious disease modeling that there is a constant population. Rather, we assume that the adult population is growing exponentially (a good fit with the data as shown in the following section) and let $N(t)$ denote the adult population in Kenya at time t (t measured in years). We use a compartmental structure for the population, assuming that there are two infected classes:

- $I_1(t)$ denotes individuals who are infected with HIV that are seeking treatment (individuals experiencing little or no impact from population-level stigma); and
- $I_2(t)$ denotes individuals with HIV that are not seeking treatment because they experience and are impacted by the population level stigma.

With this structure, we note that the number of individuals in the population that are not infected with HIV, the susceptibles $S(t)$ can be calculated using the simple relation:

$$S(t) = N(t) - I_1(t) - I_2(t).$$

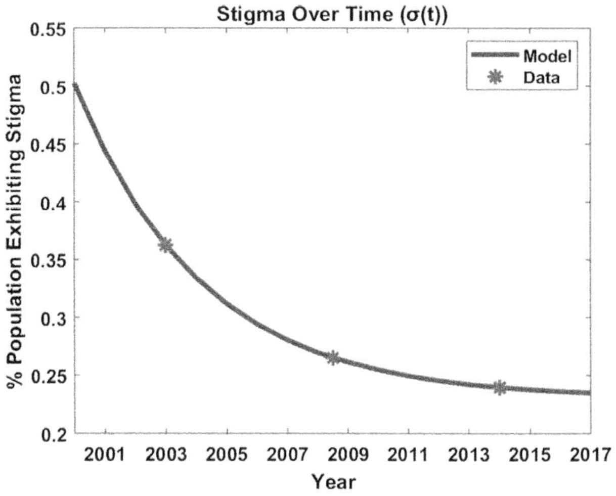

Fig. 1 Fitted curve for $\sigma(t)$ (blue line) shown with data points (orange stars) from the three Kenya Demographic and Health Surveys. The starting point was chosen to correspond to the year 2000

Following standard practice, we assume a frequency-dependent infection rate with individuals joining class I_1 or I_2 depending on the level of population stigma σ. Individuals may move between the two infected classes at rates dependent on σ and may die from natural and/or disease related causes. Using these simple assumptions, together with (1), we obtain the model system of ODEs, shown also in the schematic presented in Fig. 2:

$$\frac{\mathrm{d}N}{\mathrm{d}t} = rN \tag{3a}$$

$$\frac{\mathrm{d}I_1}{\mathrm{d}t} = \left(1 - \frac{\sigma}{\sigma_{\max}}\right)(\beta_1 I_1 + \beta_2 I_2)\frac{(N - I_1 - I_2)}{N} - \gamma_1(\sigma)I_1 + \gamma_2(\sigma)I_2 - \mu_1 I_1, \tag{3b}$$

$$\frac{\mathrm{d}I_2}{\mathrm{d}t} = \frac{\sigma}{\sigma_{\max}}(\beta_1 I_1 + \beta_2 I_2)\frac{(N - I_1 - I_2)}{N} + \gamma_1(\sigma)I_1 - \gamma_2(\sigma)I_2 - \mu_2 I_2, \tag{3c}$$

$$\frac{\mathrm{d}\sigma}{\mathrm{d}t} = \nu(\sigma_i - \sigma). \tag{3d}$$

with associated positive initial conditions for each variable. The parameter r represents the intrinsic growth rate of the population. The parameters β_i, $i = 1, 2$ denote the transmission rates from individuals in compartment I_i, and μ_i is the corresponding death rate from those compartments (due to natural and disease-related causes). The parameter $\sigma_{\max}$ denotes the maximum impact of stigma on newly infected individuals, i.e., if $\sigma = \sigma_{\max}$ then all newly infected individuals will move into the I_2 class and will not seek treatment. Our model system is positively invariant and so our solution set will remain positive throughout, given the positivity of the initial conditions.

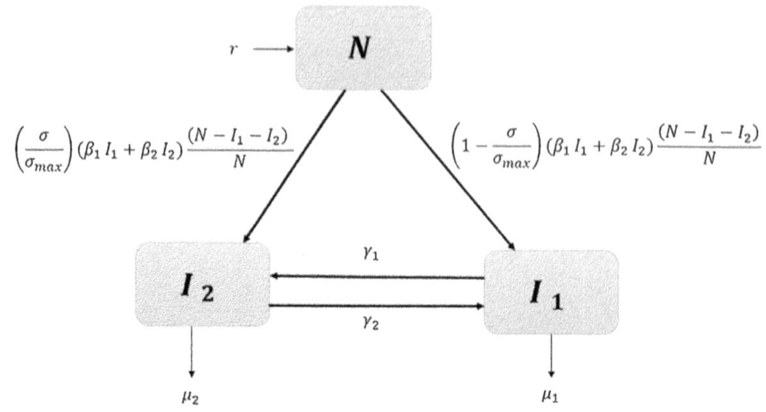

Fig. 2 Schematics of the compartmental model

Movement between the two infected classes is represented by the rate functions γ_i which satisfy the following properties:

- γ_1 is a convex increasing function of σ such that as stigma increases in the population, the rate at which individuals move from the treated I_1 class to the untreated I_2 class increases;
- γ_2 is a convex, decreasing function of σ such that as stigma increases in the population, the rate at which individuals move from untreated I_2 to treated I_1 decreases.

These choices were made in the absence of empirical data as parsimonious, based on the principle that the more prevalent stigma is within a population then the more likely individuals are to try and avoid seeking treatment for HIV and that this behavior would become more pronounced the higher the level of stigma (hence the assumption of convexity).

Our data-driven choice of exponential growth for the total population means that the model system (3) does not admit any non-trivial steady-state solutions. Moreover, the infection classes are not easily linked to infection data from Kenya and they do not correspond to standard measures of infection, such as infection prevalence. With this in mind, we chose to transform the model system to consider infection prevalence $P(t)$ and the fraction of infected individuals seeking treatment $V(t)$ using the relations:

$$P = \frac{I_1+I_2}{N}, \quad V = \frac{I_1}{I_1+I_2}.$$

This transformation gives rise to the transformed (P, V) model system:

$$\frac{dP}{dt} = P[(1-P)(\beta_1 V + \beta_2(1-V)) - \mu_1 V - \mu_2(1-V) - r], \tag{4a}$$

$$\frac{dV}{dP} = \left(1 - \frac{\sigma}{\sigma_{\max}} - V\right)(\beta_1 V + \beta_2(1-V))(1-P) - \gamma_1 V + \gamma_2(1-V)$$
$$+ (\mu_2 - \mu_1)V(1-V) \tag{4b}$$

$$\frac{d\sigma}{dt} = \nu(\sigma_i - \sigma) \tag{4c}$$

which we use in our model analysis, and simulations. Note that positivity of the values of I_1 and I_2 guarantees that V is well defined. We simulate our system using the parameters estimated in the next section, taking our starting point to correspond to the year 2004.

3.1 Parameter Estimation

Data indicating that individuals did not begin to take up treatment for HIV/AIDS in Kenya until 2004 motivated us to set $t = 0$ corresponding to the year 2004. We used published data (The World Bank, World Development Indicators 2019; Global Burden of Disease Collaborative Network 2018; The World Bank 2019), given in "Appendix A.1," to estimate initial conditions:

$$N(0) = 19{,}881{,}691,\ P(0) = \frac{1{,}556{,}539}{19{,}881{,}691} \approx 0.08,\ V(0) = 2,$$

and the average yearly growth rate for the adult population in Kenya over this period,

$$r \approx 0.032.$$

Since there is evidence that antiretroviral therapy can reduce transmission of HIV by up to 96% (Cohen et al. 2013, 2016), we take

$$\beta_1 = 0.1\beta_2.$$

Studies also agree that antiretroviral treatment reduces the likelihood of death due to the disease by more than 50% (Kasamba et al. 2012; Violari et al. 2008), and so we impose the constraint

$$\mu_1 \leq 0.5\mu_2.$$

Movement between I_1 and I_2 depends on the level of stigma that exists in society and therefore also change over time. As a result, we assume that $\gamma_1(t)$ and $\gamma_2(t)$ are both functions of $\sigma(t)$. For the purpose of parameter estimation and simulations, we chose

$$\gamma_1(t) = b\sigma(t)^2 \quad \text{and} \quad \gamma_2(t) = c(1 - \sqrt{\sigma(t)})$$

both of which satisfy the qualitative characteristics described in Sect. 3. A second, distinct pair of functions was also used to validate model results; details can be found in "Appendix A.3.1."

Including the parameters embedded within the $\gamma_i(t)$ functions ($i = 1, 2$), we have 6 unknown values to estimate: $\beta_2, \mu_1, \mu_2, \sigma_{\max}, b$, and c. To do this, we used data from

Table 1 Parameters used in our model. We obtained the estimate for $r \approx 0.032$ and the relationship $\beta_1 \approx 0.1\beta_2$ from the literature (Gapminder 2016; United Nations, Department of Economic and Social Affairs, Population Division 2019; Cohen et al. 2013). In cases where a parameter was estimated from data, we have provided the bounds used in the optimization problem

Parameter	Bounds	Estimated Value	Units
r		0.032	Years^{-1}
β_1		0.0082	Years^{-1}
β_2	[0 0.2]	0.082	Years^{-1}
μ_1	[0.021 0.2]	0.021	Years^{-1}
μ_2	[0.021 0.2]	0.068	Years^{-1}
$\sigma_{\max}$	[0 0.5]	0.50	None
b	[0 100]	2.09×10^{-7}	Years^{-1}
c	[0 50]	0.133	Years^{-1}

Kenya (2004–2017) given in "Appendix A.1" (The World Bank, World Development Indicators 2019; Global Burden of Disease Collaborative Network 2018; The World Bank 2019).

Details of the fitting algorithm and its goodness of fit are provided in "Appendix A.2." The resulting parameter estimates (together with those obtained directly from the literature) are given in Table 1.

From the estimated parameters in Table 1, we note that our estimated death rates satisfy

$$\mu_1 \approx 0.31\mu_2,$$

which agrees with findings that antiretroviral treatment reduces death in adults by around 34% (Kasamba et al. 2012).

The parameter b was estimated as 2.09×10^{-7} (Table 1), resulting in very little flow from the treated class (I_1) to the untreated class (I_2).

4 Results

Figure 3 presents our model output using the estimated parameters together with the epidemiological data from Kenya. We explore how sensitive our model is to changes in "Appendix A.3," which includes model simulation using alternative functional forms for $\gamma_1(t)$ and $\gamma_2(t)$. That work gives us confidence in the values of our disease-related parameters and confirms that the behaviors seen in the baseline case are qualitatively similar for alternative functional choices where we have no evidence or data on which to make our choices. Using our parameter estimates, the number of treated individuals (I_1) surpasses the number of non-treated individuals (I_2) in 2014, which agrees with the data. As time progresses, $\sigma(t)$ decreases allowing for an increasing percentage of new infections to begin treatment immediately. This accurately reflects what took place historically as the obtained data indicate that individuals in Kenya did not seek treatment for HIV/AIDS prior to 2000.

There are three distinct components of this results section. Firstly, we compare model outputs when parameter estimates associated with $\sigma(t)$ are varied from those

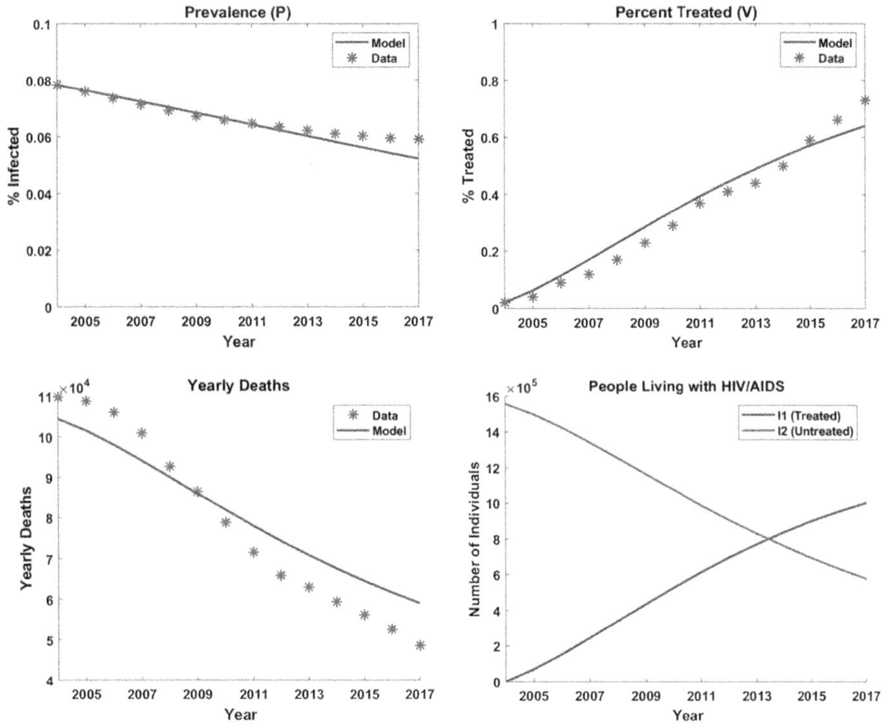

Fig. 3 Model output compared to data after parameter estimation process

described in Sect. 3.1. This allows us to understand better how stigma has impacted the HIV dynamics observed in Kenya over the period of interest. Next, we use model predictions to see how close Kenya might be to the UN "Getting to Zero" goal in 2030 and finally we undertake a standard steady-state analysis to explore the impact of long-term stigma on the fraction of PLWHA who seek treatment.

4.1 Alternative Stigma Scenarios

We compare model predictions for the number of new cases each year, the total number of people being treated for HIV and deaths of people with HIV in the time interval 2004–2017, and we determine the time at which the models predict more infected people are seeking treatment than those who are not. We consider four scenarios that maintain σ at a constant value; this can be interpreted as ignoring enacted stigma ($\nu = 0$) but allowing population-wide internalized stigma to assume different levels, including the best-case scenario of no stigma. We consider one case in which ν is increased above the value estimated in Sect. 3.1 to explore the potential for a greater impact on reducing enacted stigma, and finally we compare these results to the case in which there is no internalized stigma assumed in the population $\sigma_i = 0$. Results from these numerical explorations are presented in Fig. 4 and Table 2.

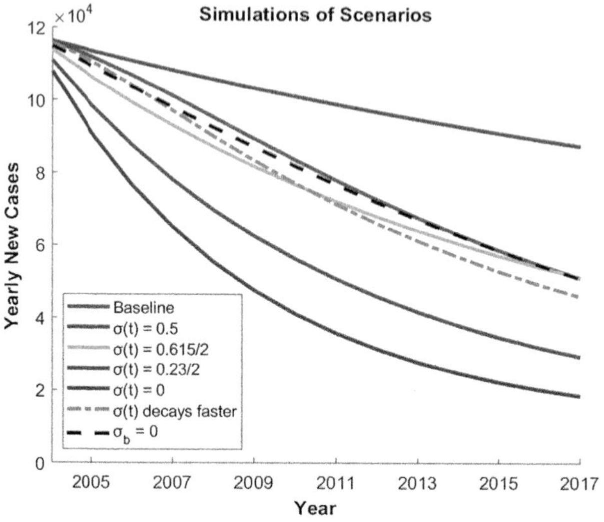

Fig. 4 New cases 2004–2017 from the 7 scenarios described in Sect. 4. The baseline simulation uses our estimated parameters. We also simulated four scenarios where $\sigma(t)$ is constant over time, a case where $\sigma(t)$ decays twice as fast ($\nu = 0.48$), and a scenario where we refit our $\sigma(t)$ function fixing internalized stigma as zero ($\sigma_i = 0$ and $\sigma(t) = 0.4e^{-0.04t}$)

Table 2 Results of various scenarios all using the initial conditions from Sect. 3.1

Scenario	Switching month	Total deaths(millions)	Total treated(millions)
Baseline	113	1.13	1.04
$\sigma(t) = \sigma_{\max} = 1/2$	298	1.31	0.60
$\sigma(t) = 0.615/2$	104	1.09	1.02
$\sigma(t) = \sigma_{\min} = 0.23/2$	60	0.91	1.25
$\sigma(t) = 0$	41	0.79	1.34
$\sigma(t)$ decays faster ($\nu = 0.48$)	99	1.08	1.09
$\sigma_i = 0$ ($\sigma(t) = .4e^{-.04t}$)	113	1.12	1.04

Simulations are from 2004 through 2017. The switching month is defined as the month after January 2004 in which the percent of individuals that are being treated surpasses the percent of those not receiving treatment ($I_1(t) > I_2(t)$). Columns 3 and 4 display total values at the end of each simulation

4.1.1 Constant Stigma

Scenario 1 fixes the stigma level at its level in 2000 by setting

$$\sigma = \sigma_{\max} = 1/2.$$

This scenario considers what would occur if the National AIDS Control Council (NACC) was never formed in Kenya, resulting in a sustained stigma toward PLWHA.

In the second scenario, we let

$$\sigma = \sigma_{\min} = 0.23/2.$$

This corresponds to the stigma level starting and remaining at its lowest level, i.e., at its internalized stigma level.

The third scenario we considered corresponds to the midpoint of the previous two scenarios,

$$\sigma = 0.615/2.$$

In this case, $\frac{1}{\sigma_{\max}} * 0.615/2 = 0.615$. According to our estimate of $\sigma(t)$, this midpoint occurred in 2005 and represents a scenario where the NACC formed, but did not effectively reduce stigma in Kenya resulting in 61.5% of new cases flowing into the untreated class for the duration of the simulation.

Lastly, we simulate the case where $\sigma = 0$, representing a scenario where all stigma toward PLWHA is eliminated from society in Kenya so that all new cases immediately begin treatment. This hypothetical scenario may not be feasible, but it represents a "best case."

There are clear differences in the results from the scenarios where stigma is constant over time. Note that in the case where $\sigma = \sigma_{\max}$, even though all new cases flowing into I_2 there is still movement from the untreated class (I_2) into the treated class (I_1). This simulation represents a "worst case" scenario where it takes nearly 300 months for the number of treated to surpass the number of untreated and over 1.31 million deaths occur after the 14-year simulation. In the simulations where $\sigma = 0.615/2$, model output remains most similar to the baseline scenario of $\sigma(t)$ given in (2). In the cases where $\sigma = 0.23$ and $\sigma = 0$, we see significant decreases in the switching month and total deaths, as well as noticeable increases in the number of treated.

4.1.2 Stigma Decays Faster

Here, we simulate a scenario in which the rate at which stigma decays is twice that predicted using the stigma data from Kenya. Specifically, we set

$$\nu = 2 \times 0.24 = 0.48.$$

This represents a situation where public health education was more effective allowing for the perception of PLWHA to improve at a more rapid rate and can be thought of as a reduction in enacted stigma.

Although there was not a dramatic difference in most metrics from this scenario compared to the baseline simulation, allowing stigma to decay at a faster rate does result in slightly improved metrics across the board. We also note how the decrease in total deaths in this scenario is the same value as the increase in total treated. Thus, even though the number of yearly cases does not see a significant decline in this scenario (see Fig. 4), this highlights how a more rapid improvement in the perception of PLWHA (i.e., a reduction in enacted stigma) can save lives through more individuals seeking antiretroviral treatment.

4.1.3 No Internalized Stigma

Finally, we simulate a scenario where there is no internalized stigma, $\sigma_i = 0$. In this case, we must first re-fit the function $\sigma(t)$ to the KDHS survey data using $\sigma_i = 0$ because the solution trajectory that best fits the three data points we have estimated but for which $\sigma_i = 0$, cannot be derived from our existing solution (2). The best fit solution gives

$$\sigma(t) = 0.4e^{-0.04t}.$$

This case has similar output to the baseline scenario shown in Fig. 4, producing the same switching month and a slight reduction in total deaths. The reason for the similarity is that even though $\sigma(t) \to 0$ as $t \to \infty$, the baseline $\sigma(t)$ function (2) is not dramatically different during the simulated time frame. Having said that, the level of stigma at $t = 0$ for this function is lower than that predicted with our baseline, data-driven estimate for $\sigma(t)$ and so more infected individuals move into the treated class early in the simulation resulting in a reduction in total cases and therefore also total deaths. Considering the clear importance of receiving treatment, allowing stigma to entirely dissipate from society would undoubtedly have a more significant impact when considering extended time frames.

4.2 Meeting UN Goals

As described in the introduction, the UN initiative "Getting to Zero" aimed to reduce the number of new infections, the level of discrimination, and deaths from HIV/AIDS to zero by 2030. Our model predictions for these three measures are given in Table 3, where row 1 shows model output where an internalized level of stigma is assumed while row 2 assumes $\sigma_i = 0$.

It should be noted that our model continues to assume exponential growth for the whole population until 2030 which certainly over-estimates the likely population in Kenya in 2030. That not with-standing, it is clear that there is likely to be a shortfall in achieving these goals. This is supported by our baseline model as output suggests that in 2030 about 23% of the population will stigmatize PLWHA, resulting in over 24,000 new cases and over 38,000 deaths that year. We obtain similar, though slightly lower, estimates in the case where the $\sigma_i = 0$. Having said that, this is over 70% reduction since 2003 and the reductions in the number of new cases and deaths by 2030 are considerable.

4.3 Understanding the Dynamics

Figure 5 shows the long term predictions of our model system using the parameter set fitted to Kenyan data. Our simulation predicts that as prevalence decays toward

Table 3 Model output in 2030

Stigma level	Yearly new cases	Yearly deaths
0.231	24,427	38,448
0.120	21,500	36,364

Row 1 displays output from our baseline model while row 2 considers the case where $\sigma_i = 0$ and $\sigma(t) = 0.4e^{-0.04t}$

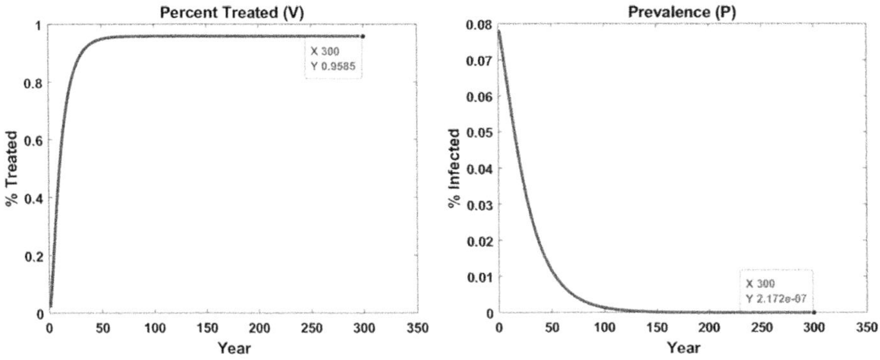

Fig. 5 Simulation showing the long-term model predictions for (P, V) demonstrating the monotonic decline in prevalence to zero with around 96% of PLWHA seeking treatment

zero, the fraction of PLWHA seeking treatment stabilizes to a nonzero level. We use steady-state analysis as a proxy to explore this observation and find that it predicts that an infection-free equilibrium (zero prevalence) may arise under a range of conditions. At steady state, $\sigma^* = \sigma_i$. When $P^* = 0$, the corresponding equilibrium V^* solves the quadratic equation:

$$AV^2 + BV + C = 0$$

where

$$\begin{aligned}
A &= \beta_2 - \beta_1 + \mu_1 - \mu_2 \\
B &= \beta_1\left(1 - \frac{\sigma_i}{\sigma_{\max}}\right) - \beta_2\left(2 - \frac{\sigma_i}{\sigma_{\max}}\right) + \mu_2 - \mu_1 - \gamma_1 - \gamma_2 \\
C &= \beta_2\left(1 - \frac{\sigma_i}{\sigma_{\max}}\right) + \gamma_2.
\end{aligned}$$

Of course, it is entirely unrealistic to assume that model parameters would remain unchanged over an extended period. What is interesting to glean from this analysis is a model estimate for the fraction of individuals infected with HIV that will seek out treatment (around 96 % in our particular parameter set for V^* from that quadratic equation, and also matching with numerical predictions shown in Fig. 5). More details of our brief analysis are given in "Appendix A.4."

5 Discussion and Conclusions

Our motivation in undertaking the work presented here was to understand the impact of stigma on HIV prevalence in Kenya by developing and analyzing a mathematical model that could be adapted easily to explore HIV dynamics in other countries. We were careful to invoke Occam's Razor such that the model parameters could be estimated using data readily available in the literature and where we did employ model assumptions, for example in determining how the population measure of stigma affected the movement of infected individuals between the treated and untreated classes, we validated the model outcomes by checking our results were independent of the particular functional form (provided it satisfied our baseline assumption).

Since the population in Kenya is growing at around 3% per annum, one of our first challenges was to think about how to understand infection dynamics within a growing population. We addressed this by transforming the model system into one in which the state variables measured infection prevalence and the fraction of infected individuals seeking treatment. The next was to consider how stigma should impact the dynamics. With a paucity of data available to validate assumptions, we chose a parsimonious approach and decoupled the time evolution of stigma from the infection dynamics. Although the resulting model was simple in form, we were able to identify three critical components of stigma—internalized stigma and enacted plus perceived stigma—within the model. We used this to good effect in our analysis when we explored how changes to these two elements would have changed the amount of HIV infection in Kenya under a range of different scenarios. What emerged from that, as shown in Fig. 4, is that a reduction in internalized stigma (measured in the model by σ_i) would not have had a big impact on reducing incidence of infection in Kenya in the period 2004–2017; by contrast, reductions in enacted stigma would have reduced incidence by around 8%. Figure 4 also shows that if there was no stigma associated with HIV infections, then incidence of HIV infection would be lowest. This is hardly surprising. However, even in that scenario, the UN goal of Zero in 2030 would not be achieved according to our model predictions.

The model prediction that reducing enacted stigma might have more impact on reducing incidence of HIV than reducing internalized stigma provides a potential recommendation to those working in public health in Kenya. With limited resources available to tackle stigma, our model suggests that activities that target enacted stigma might be of greater benefit in the current situation than those targeting internalized stigma. This may be a welcome message—persuading communities to alter their view of HIV infection may provide a more tangible target for public health campaigns than initiatives that focus on individuals within those community.

It is clear that our model representation of stigma is simplistic. That was intentional for two reasons. Firstly, we did not find data-driven evidence in the literature that would link stigma dynamically to HIV infection dynamics. This meant that we could have chosen stigma to depend on infection prevalence, on HIV-related death, infection prevalence, and/or some combination of all of these. Our results may have been interesting but they may not have been relevant to Kenya. Secondly, we wanted to highlight the potential importance of incorporating stigma into our HIV model in order to support the argument that a focus on gathering data on sociological processes that impact infectious disease dynamics should start to take priority. While we move toward that goal, Kenya may not achieve the UN goal "Getting to Zero in 2030," but it is certainly moving in the right direction.

Acknowledgements The research was based upon work supported in part by the National Science Foundation under Grant No. NSF-DMS-1343651 (US-Africa Masamu Advanced Study Institute (MASI) and Workshop Series in Mathematical Sciences). HEC was supported by a U.S. National Science Foundation predoctoral fellowship Grant No. NSF-DGE-1414475. SL was partially supported by the National Institute for Mathematical and Biological Synthesis, supported by the National Science Foundation through NSF Award DBI-1300426, with additional support from The University of Tennessee, Knoxville. BL received support from Fitchburg State University. KAJW was supported by two LMS Scheme 5 grants (51607, 51710) and also supported financially from the International Research Initiator Scheme at the University of Bath (2016).

A Appendix

A.1 Data for Kenya

Data used to estimate key parameters β_2, μ_1, μ_2, $\sigma_{\max}$,b, and c include the fractions of Kenyans with a stigmatizing view of HIV/AIDS, i.e., the number of female respondents answering at least two questions in a stigmatizing manner divided by the total number of responses to all three questions (Table 4; CBS et al. 2004; Kenya National Bureau of Statistics (KNBS) and ICF Macro 2010, 2014), along with the total adult population, total number of HIV/AIDS cases, number of yearly HIV/AIDS-related deaths, and percent of PLWHA that are receiving antiretroviral therapy treatment for HIV/AIDS for Kenya from 2004 through 2017 (Table 5).

Table 4 Data for stigma used to estimate parameters in our model

Year	TFR for all three questions	SR for at least two questions
2003	8037	2911
2008	8349	2216
2014	14,633	3509

TFR total number of female responses, *SR* stigmatized responses

Table 5 Data for adult population (The World Bank, World Development Indicators 2019), total cases (Global Burden of Disease Collaborative Network 2018), yearly deaths (Global Burden of Disease Collaborative Network 2018), and percent treated (The World Bank 2019) used to estimate parameters in our model

Year	Adult population	Total cases	Yearly deaths	Percent treated
2004	19,881,691	1,556,539	109,769	2
2005	20,500,063	1,555,886	108,881	4
2006	21,134,459	1,556,425	106,011	9
2007	21,756,530	1,552,905	100,971	12
2008	22,384,120	1,549,630	92,753	17
2009	23,050,075	1,551,573	86,497	23
2010	23,771,983	1,564,451	78,894	29
2011	24,525,795	1,585,198	71,528	37
2012	25,345,229	1,607,145	65,842	41
2013	26,216,858	1,630,997	62,910	44
2014	27,117,617	1,657,895	59,355	50
2015	28,036,000	1,689,424	56,188	59
2016	28,988,590	1,727,026	52,482	66
2017	29,957,102	1,772,350	48,502	73

A.2 Parameter Estimation Methodology

The total population together with the total cases gives the percentage of the population that is PLWHA. Using these data, we fit our model to the percent of the population that are a PLWHA (PI^*), the percent of the population that is a PLWHA and is receiving antiretroviral treatment (PT^*), and the number of yearly deaths caused by HIV/AIDS (YD^*). The asterisk represents data points while we use the equivalent notation without an asterisk to represent model output.
We estimate parameters by minimizing the objective function:

$$\text{minimize } J(\mathbf{x}) = \frac{||PI(\mathbf{x}) - PI^*||_2}{||PI^*||_2} + \frac{||PT(\mathbf{x}) - PT^*||_2}{||PT^*||_2} + \frac{||YD(\mathbf{x}) - YD^*||_2}{||YD^*||_2},$$

where $\mathbf{x}^T = [\beta_2, \mu_1, \mu_2, \sigma_{\max}, b, c]$ is the vector of unknown parameters from Table 1 (which also provides parameter bounds where these are known). Each vector comprises 14 components corresponding to the years, 2004–2017. We divide by the magnitude of the data to normalize each term.

The optimization problem was implemented in MATLAB using the fmincon function in the Optimization Toolbox. Parameter bounds and constraints that were imposed are detailed in the main text. Since fmincon is a local solver, we used MATLAB *multistart* to choose 200 starting points to fully explore the parameter space. All starting points converged to one of two local minimums, with the optimal set of parameters resulting in an objective function output of $J = 0.24$ and the parameter values shown in Table 1. The second local minimum produced an objective function value twice as

Table 6 Model sensitivity analysis results for three outcomes of interest: Total Cases after the 14-year simulation (TC), Yearly Cases in final year of the simulation (YC), and Yearly Deaths in final year of the simulation (YD)

	TC	YC	YD
β_2	0.991*	0.981*	0.956*
μ_1	− 0.043	0.044	0.842*
μ_2	− 0.880*	− 0.893*	0.152
$\sigma_{\max}$	0.833*	0.809*	0.835*
b	− 0.050	− 0.053	0.117
c	− 0.727*	− 0.821*	− 0.866*

Each entry represents the corresponding partial rank correlation coefficient (PRCC) and statistically significant values ($p < 0.0001$) have an asterisk

large at $J = 0.48$ and resulted in the same parameter values as the optimal set except that $b \approx 40$ rather than $b = 2.09 \times 10^{-7}$ and $c \approx 5$ rather than $c = 0.133$. These values of b and c at the second minimum are unreasonably large, especially $b \approx 40$. More specifically, since $b\sigma(t)^2$ determines the percent of individuals moving from I_1 to I_2 per unit time, our model requires that $0 \geq b * \sigma(t)^2 \leq 1$ where $0 \geq \sigma(t)^2 \leq 1$. Thus, if $b \approx 40$ then $40\sigma(t)^2 > 1$ for $\sigma(t) > 1/\sqrt{40} \approx 0.16$, which is problematic. Thus, the optimal set of parameters values are those given in Table 1.

A.3 Sensitivity Analysis

We performed a global sensitivity analysis for our model using Latin Hypercube Sampling (LHS) to sample the parameter space and Partial Rank Correlation Coefficients (PRCC) to evaluate the sensitivity outcome variables of interest to changes in our 6 estimated parameter values. We created intervals for each parameters to sample from by extending 50% above and below the estimated values shown in Table 1. Uniform probability distributions were used for each parameter interval and we drew 100 samples from each interval. PRCC provides us with a way to evaluate the monotonicity of relationship of a parameter with each outcome variable of interest while holding all the remaining parameters constant, even when the relationship is not linear. We chose to evaluate the sensitivity of three outcome variables to changes in our parameters: Total Cases after the 14-year simulation, Yearly Cases in final year of the simulation, and Yearly Deaths in final year of the simulation (Blower and Dowlatabadi 1994; Marino et al. 2008). Results from this process are given in Table 6.

Regardless of the variable of interest, our model is sensitive to changes in β_2, $\sigma_{\max}$, and c. While β_2 will determine the number of new infections per unit time, $\sigma_{\max}$ and c control the number of individuals that are in the treated (I_1) and untreated (I_2) classes. Death rate μ_1 has a statistically significant impact on yearly deaths while μ_2 is statistically significant with respect to the two outcome variables associated with number of cases.

Table 7 Parameters used to test how sensitive our parameter estimation results are to the functional forms of $\gamma_1(t)$ and $\gamma_2(t)$

Parameter	Bounds	Estimated Value	Units
r		0.032	Years^{-1}
β_1		0.0082	Years^{-1}
β_2	[0 0.2]	0.082	Years^{-1}
μ_1	[0.021 0.2]	0.021	Years^{-1}
μ_2	[0.021 0.2]	0.068	Years^{-1}
σ_{max}	[0 0.5]	0.50	None
b	[0 100]	2.09×10^{-7}	Years^{-1}
c	[0 50]	9.22	Years^{-1}
d	[0 500]	200	Years^{-1}

We obtained the estimate for $r \approx 0.032$ and the relationship $\beta_1 \approx 0.1\beta_2$ from the literature (Gapminder 2016; United Nations, Department of Economic and Social Affairs, Population Division 2019; Cohen et al. 2013). In cases where a parameter was estimated from data, we have provided the bounds used in the optimization problem

A.3.1 Fitting Alternate Forms of $\gamma_1(t)$ and $\gamma_2(t)$

To determine how sensitive our parameter estimation results are to the functional forms of $\gamma_1(t)$ and $\gamma_2(t)$, we refit the model with the following functions that also satisfy the qualitative requirements stated in Sect. 3.1:

$$\gamma_1(t) = b\sigma(t),$$
$$\gamma_2(t) = \frac{c}{1 + d\sigma(t)}.$$

We followed exactly the process described in Sect. 3.1 and obtained the same parameter estimates as shown in Table 7 for the parameters not in the γ_1 and γ_2 functions. The corresponding simulation plots are depicted in Fig. 6. Additionally, we performed a sensitivity analysis on these parameters as described in "Appendix A.3." The magnitude of the resulting PRCC values, their signs, and statistical significance closely aligned with those depicted in Table 6. The consistency of these results with those using the initial pair of functions suggests that we have reasonable estimates for our model parameters.

A.4 Steady State and Stability Calculations

At steady state, $\sigma^* = \sigma_i$, and solutions of the transformed (P, V) system (4) satisfy the coupled algebraic equations:

$$P[(1 - P)(\beta_1 V + \beta_2(1 - V)) - \mu_1 V - \mu_2(1 - V) - r] = 0 \tag{5a}$$

$$\left(1 - \frac{\sigma_i}{\sigma_{max}} - V\right)(\beta_1 V + \beta_2(1 - V))(1 - P) - \gamma_1 V + \gamma_2(1 - V)$$

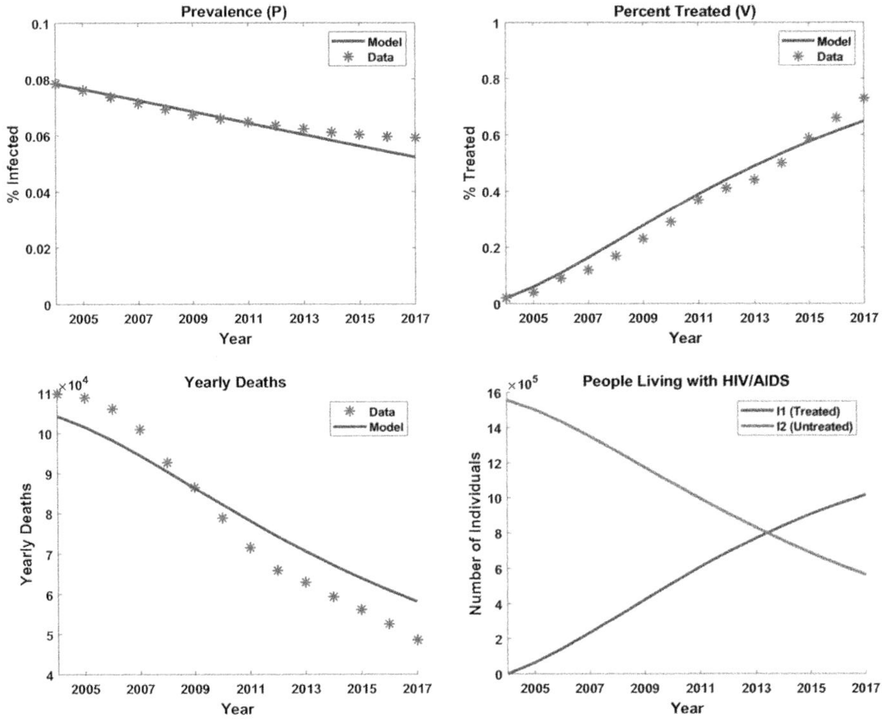

Fig. 6 Model output compared to data after parameter estimation process using alternate forms for $\gamma_1(t)$ and $\gamma_2(t)$

$$+ (\mu_2 - \mu_1)V(1 - V) = 0. \tag{5b}$$

Solving (5) gives either $P^* = 0$ or

$$P^* = 1 - \frac{\mu_1 V^* + \mu_2(1 - V^*) + r}{(\beta_1 V^* + \beta_2(1 - V^*))}. \tag{6}$$

We are interested in the equilibrium with $P^* = 0$. This satisfies the first equation in (5); substituting into the second, we find that V must satisfy the quadratic equation:

$$AV^2 + BV + C = 0 \tag{7}$$

where

$$\begin{aligned} A &= \beta_2 - \beta_1 + \mu_1 - \mu_2 \\ B &= \beta_1\left(1 - \frac{\sigma_i}{\sigma_{\max}}\right) - \beta_2\left(2 - \frac{\sigma_i}{\sigma_{\max}}\right) + \mu_2 - \mu_1 - \gamma_1 - \gamma_2 \\ C &= \beta_2\left(1 - \frac{\sigma_i}{\sigma_{\max}}\right) + \gamma_2. \end{aligned} \tag{8}$$

Solving (7) gives the standard quadratic solutions

$$V_1^* = \frac{-B + \sqrt{B^2 - 4AC}}{2A},$$
$$V_2^* = \frac{-B - \sqrt{B^2 - 4AC}}{2A}.$$

Since $C > 0$, the existence of equilibrium points is determined by the sign of A. Local stability of this infection-free equilibrium is determined by analysis of the Jacobian matrix

$$J = \begin{bmatrix} a_{11} & 0 \\ a_{21} & a_{22} \end{bmatrix} \tag{9}$$

where

$$a_{11} = \beta_1 V^* + \beta_2(1 - V^*) - \mu_1 V^* - \mu_2(1 - V^*) - r$$
$$a_{21} = -\left(1 - \frac{\sigma_i}{\sigma_{\max}} - V^*\right)(\beta_1 V^* + \beta_2(1 - V^*))$$
$$a_{22} = -(\beta_1 V^* + \beta_2(1 - V^*)) + \left(1 - \frac{\sigma_i}{\sigma_{\max}} - V^*\right)(\beta_1 - \beta_2) - \gamma_1 - \gamma_2$$
$$+(\mu_2 - \mu_1)(1 - 2V^*).$$

Local stability is determined by the sign of the trace and determinant of J. Since this matrix is lower triangular, the stability conditions simplify to

$$a_{11} < 0, \quad \text{and} \quad a_{22} < 0.$$

For the parameter estimates, we have for Kenya, $A = \beta_2 - \beta_1 + \mu_1 - \mu_2 > 0$. The scenario $A > 0$ gives two possible positive roots for V^*. We can show, by contradiction, that $V_1^* > 1$ as follows. Assuming that $V_1^* \leq 1$, then

$$-B + \sqrt{B^2 - 4AC} < 2A$$

and hence

$$B^2 - 4AC < 4A^2 + 4AB + B^2.$$

Substituting for A, B and C into this inequality gives

$$\frac{\sigma_i}{\sigma_{\max}}\beta_1 + \gamma_1 \leq 0$$

which is a contradiction.

A direct manipulation on V_2^* shows that $0 < V_2^* < 1$, which gives the meaningful root for our model to get an equilibrium point $(0, V_2^*)$. Using the parameters values in Table 7, we obtain $V_2^* = 0.9588$ with $a_{11} < 0$ and $a_{22} < 0$ and so we conclude that this equilibrium is locally stable.

Remark 1 The case for $A < 0$ leads to existence of one positive root of equation (5). An interior equilibrium point also exists when $P^* \neq 0$ and given by (6). Details for existence and stability of these equilibria have intentionally been omitted as we focus on the relevant equilibrium point predicted by fitting the model to data.

References

Blower SM, Dowlatabadi H (1994) Sensitivity and uncertainty analysis of complex models of disease transmission: an HIV model, as an example. Int Stat Rev/Rev Int Stat 229–243

Buregyeya E, Naigino R, Mukose A, Makumbi F, Esiru G, Arinaitwe J, Musinguzi J, Wanyenze RK (2017) Facilitators and barriers to uptake and adherence to lifelong antiretroviral therapy among HIV infected pregnant women in Uganda: a qualitative study. BMC Pregnancy Childbirth 17(1):94

Campbell C, Cornish F (2010) Towards a "fourth generation" of approaches to HIV/AIDS management: creating contexts for effective community mobilisation. AIDS Care 22(sup2):1569–1579

Central Bureau of Statistics (CBS) [Kenya], Ministry of Health (MOH) [Kenya], and ORC Macro (2004) Kenya Demographic and Health Survey 2003. CBS, MOH and ORC Macro, Calverton

Chan BT, Tsai AC (2016) HIV stigma trends in the general population during antiretroviral treatment expansion: analysis of 31 countries in sub-Saharan Africa, 2003–2013. J Acquir Immune Defic Syndr 72(5):558–564

Chesney MA, Smith AW (1999) Critical delays in HIV testing and care: the potential role of stigma. Am Behav Sci 42(7):1162–1174

Cohen MS, Chen YQ, McCauley M, Gamble T, Hosseinipour MC, Kumarasamy N, Hakim JG, Kumwenda J, Grinsztejn B, Pilotto JH et al (2016) Antiretroviral therapy for the prevention of HIV-1 transmission. N Engl J Med 375(9):830–839

Cohen MS, Smith MK, Muessig KE, Hallett TB, Powers KA, Kashuba AD (2013) Antiretroviral treatment of HIV-1 prevents transmission of HIV-1: where do we go from here? Lancet 382(9903):1515–1524

Cordes JL, Stangl A, Krishnaratne S, Hoddinott G, Mathema H, Bond V, Seeley J, Hargreaves JR (2017) Trends in responses to DHS questions should not be interpreted as reflecting an increase in "anticipated stigma" in Africa. JAIDS 75(1)

Croome N, Ahluwalia M, Hughes LD, Abas M (2017) Patient-reported barriers and facilitators to antiretroviral adherence in sub-Saharan Africa. AIDS 31(7):995–1007

Earnshaw VA, Chaudoir SR (2009) From conceptualizing to measuring HIV stigma: a review of HIV stigma mechanism measures. AIDS Behav 13(6):1160

Gapminder (2016) HYDE Database—Total population of Kenya. Published online at OurWorldInData.org. https://ourworldindata.org/grapher/projected-population-by-country?country=~KEN [Online Resource; Accessed May 2020]

Geary C, Parker W, Rogers S, Haney E, Njihia C, Haile A, Walakira E (2014) Gender differences in HIV disclosure, stigma, and perceptions of health. AIDS Care 26(11):1419–1425

Global Burden of Disease Collaborative Network (2018) Global Burden of Disease Study 2017 (GBD 2017) Results. Published online at OurWorldInData.org. https://ourworldindata.org/global-rise-of-education [Online Resource; Accessed May 2020]

Global Network of People with HIV/AIDS and International Community of Women living with HIV/AIDS (2017). People Living with HIV Stigma Index. [Online]. http://www.stigmaindex.org/

Golub SA, Gamarel KE (2013) The impact of anticipated HIV stigma on delays in HIV testing behaviors: findings from a community-based sample of men who have sex with men and transgender women in New York City. AIDS Patient Care STDS 27(11):621–627

Granich RM, Gilks CF, Dye C, De Cock KM, Williams BG (2009) Universal voluntary HIV testing with immediate antiretroviral therapy as a strategy for elimination of HIV transmission: a mathematical model. Lancet 373:48–57

Hyman J, Li J, Stanley E (2003) Modelling the impact of screening and contact tracing in reducing the spread of HIV. Math Biosci 181:17–54
ICF (2018) The Demographic and Health Surveys Program STATcompiler. [Online; Accessed April 2020]. http://www.statcompiler.com
Johnny L, Mitchell C (2006) "Live and Let Live": an analysis of HIV/AIDS-related stigma and discrimination in international campaign posters. J Health Commun 11(8):755–767
Joshi H, Lenhart S, Albright K, Gipson K (2008) Modelling the effect of information campaigns on the HIV epidemic in Uganda. Math Biosci Eng 5(AIMS Press):757–770
Kagee A, Remien R, Berkman A, Hoffman S, Campos L, Swartz L (2011) Structural barriers to ART adherence in Southern Africa: challenges and potential ways forward. Glob Public Health 6(1):83–97
Kalichman SC, Mathews C, Banas E, Kalichman MO (2019) Treatment adherence in HIV stigmatized environments in South Africa: stigma avoidance and medication management. Int J STD AIDS 30(4):362–370
Kasamba I, Baisley K, Mayanja BN, Maher D, Grosskurth H (2012) The impact of antiretroviral treatment on mortality trends of HIV-positive adults in rural Uganda: a longitudinal population-based study, 1999–2009. Trop Med Int Health 17(8):e66–e73
Katz IT, Ryu AE, Onuegbu AG, Psaros C, Weiser SD, Bangsberg DR, Tsai AC (2013) Impact of HIV-related stigma on treatment adherence: systematic review and meta-synthesis. J Int AIDS Soc 16(3 Suppl 2):18640
Kelly J, Reid MJ, Lahiff M, Tsai AC, Weiser SD (2017) Community-level HIV stigma as a driver for HIV transmission risk behaviors and sexually transmitted diseases in Sierra Leone: a population-based study. JAIDS 75(4)
Kenya National Bureau of Statistics (KNBS) and ICF Macro (2010) Kenya Demographic and Health Survey 2008–09. KNBS and ICF Macro, Calverton
Kenya National Bureau of Statistics (KNBS) and ICF Macro (2014) Kenya Demographic and Health Survey 2014. KNBS and ICF Macro, Calverton
King EJ, Maman S, Bowling JM, Moracco KE, Dudina V (2013) The influence of stigma and discrimination on female sex workers' access to HIV services in St. Petersburg. Russia. AIDS Behav 17(8):2597–2603
Link BG, Phelan JC (2001) Conceptualizing stigma. Ann Rev Sociol 27(1):363–385
Logie CH, Jenkinson JIR, Earnshaw V, Tharao W, Loutfy M (2016) A structural equation model of HIV-related stigma, racial discrimination, housing insecurity and wellbeing among African and Caribbean Black women living with HIV in Ontario, Canada. PLoS One 9:1–20
Marino S, Hogue IB, Ray CJ, Kirschner DE (2008) A methodology for performing global uncertainty and sensitivity analysis in systems biology. J Theor Biol 254(1):178–196
Maughan-Brown B (2010) Stigma rises despite antiretroviral roll-out: a longitudinal analysis in South Africa. Soc Sci Med 70(3):368–374
McKay T, Mutchler MG (2011) The effect of partner sex: nondisclosure of HIV status to male and female partners among men who have sex with men and women (MSMW). AIDS Behav 15(6):1140–1152
Mills EJ, Nachega JB, Buchan I, Orbinski J, Attaran A, Singh S, Rachlis B, Wu P, Cooper C, Thabane L, Wilson K, Guyatt GH, Bangsberg DR (2006) Adherence to antiretroviral therapy in sub-Saharan Africa and North America: a meta-analysis. JAMA 296(6):679–690
Moghadas M, Gumel A, Mcleod R, Gordon R (2003) Could condoms stop the AIDS epidemic? J Theor Med 5:171–181
Mugoya GC, Ernst K (2014) Gender differences in HIV-related stigma in Kenya. AIDS Care 26(2):206–213
National Empowerment Network of People Living With HIV and AIDS in Kenya (NEPHAK), Global Network of People with HIV/AIDS, and International Community of Women living with HIV/AIDS (2011) PLHIV Stigma Index Kenyan Country Assessment
Neuman M, Obermeyer CM (2013) Experiences of stigma, discrimination, care and support among people living with HIV: a four country study. AIDS Behav 17(5):1796–1808
Nyabadza F, Mukandavire Z, Hove-Musekwa S (2011) Modelling the HIV/AIDS epidemic trends in South Africa: insights from a simple mathematical model. Nonlinear Anal Real World Appl 12:2091–2104
Nyblade L, Reddy A, Mbote D, Kraemer J, Stockton M, Kemunto C, Krotki K, Morla J, Njuguna S, Dutta A, Barker C (2017) The relationship between health worker stigma and uptake of HIV counseling and testing and utilization of non-HIV health services: the experience of male and female sex workers in Kenya. AIDS Care 29(11):1364–1372
Parker R, Aggleton P (2003) HIV and AIDS-related stigma and discrimination: a conceptual framework and implications for action. Soc Sci Med 57(1):13–24

Parkhurst JO (2014) Structural approaches for prevention of sexually transmitted HIV in general populations: definitions and an operational approach. J Int AIDS Soc 17(1):19052
Patel RC, Stanford-Moore G, Odoyo J, Pyra M, Wakhungu I, Anand K, Bukusi EA, Baeten JM, Brown JM (2016) "Since both of us are using antiretrovirals, we have been supportive to each other": facilitators and barriers of pre-exposure prophylaxis use in heterosexual HIV serodiscordant couples in Kisumu. Kenya. J Int AIDS Soc 19(1):21134
Price DM, Howell JL, Gesselman AN, Finneran S, Quinn DM, Eaton LA (2019) Psychological threat avoidance as a barrier to HIV testing in gay/bisexual men. J Behav Med 42(3):534–544
Prudden HJ, Hamilton M, Foss AM, Adams ND, Stockton M, Black V, Nyblade L (2017) Can mother-to-child transmission of HIV be eliminated without addressing the issue of stigma? Modeling the case for a setting in South Africa. PLoS One 12:1–19
Reluga T, Smith RA, Hughes DP (2019) Dynamic and game theory of infectious disease stigma. J Theor Biol 476:95–107
Remien RH, Bauman LJ, Mantell JE, Tsoi B, Lopez-Rios J, Chhabra R, DiCarlo A, Watnick D, Rivera A, Teitelman N, Cutler B, Warne P (2015) Barriers and facilitators to engagement of vulnerable populations in HIV primary care in New York City. J Acquir Immune Defic Syndr 69(Suppl 1):S16-24
Ronoh M, Chirove F, Wairim J, Ogama W (2020) Modeling disproportional effects of educating infected Kenyan youth on HIV/AIDS. J Biol Syst 28(02):311–349
Simwa R, Pokhariyal G (2003) A dynamical model for stage-specific HIV incidences with application to Sub-Saharan Africa. Appl Math Comupt 146(1):93–104
The World Bank (2019) Antiretroviral therapy coverage (% of people living with HIV)—Kenya. [Data File; Accessed May 2020]. https://data.worldbank.org/indicator/SH.HIV.ARTC.ZS?end=2018&locations=KE&start=2004&view=chart&year=2004
The World Bank, World Development Indicators (2019) Population, total—Kenya. [Data File; Accessed May 2020]. https://data.worldbank.org/indicator/SP.POP.TOTL?locations=KE
Ti L, Hayashi K, Kaplan K, Suwannawong P, Wood E, Montaner J, Kerr T (2013) HIV test avoidance among people who inject drugs in Thailand. AIDS Behav 17(7):2474–2478
Treves-Kagan S, El Ayadi AM, Pettifor A, MacPhail C, Twine R, Maman S, Peacock D, Kahn K, Lippman SA (2017) Gender, HIV testing and stigma: the association of HIV testing behaviors and community-level and individual-level stigma in rural South Africa differ for men and women. AIDS Behav 21(9):2579–2588
UNAIDS (2018) Kenya 2018 profile. [Online; Accessed November 2019]. https://www.unaids.org/en/regionscountries/countries/kenya
United Nations, Department of Economic and Social Affairs, Population Division (2019) World Population Prospects: the 2019 Revision. Published online at OurWorldInData.org. https://ourworldindata.org/grapher/projected-population-by-country?country=~KEN' [Online Resource; Accessed May 2020]
USAIDS (2005) Working report measuring HIV stigma: Results of a field test in Tanzania. [Online]. https://www.icrw.org/wp-content/uploads/2016/10/Working-Report-Measuring-HIV-Stigma-Results-of-a-Field-Test-in-Tanzania.pdf
Van Brakel WH (2006) Measuring health-related stigma—a literature review. Psychol Health Med 11(3):307–334
Violari A, Cotton MF, Gibb DM, Babiker AG, Steyn J, Madhi SA, Jean-Philippe P, McIntyre JA (2008) Early antiretroviral therapy and mortality among HIV-infected infants. N Engl J Med 359(21):2233–2244
Yoder PS, Nyblade, L (2004) Comprehension of questions in the Tanzania AIDS indicator survey. DHS Qualitative Research Study No. 10. ORC Macro, Calverton

Bulletin of Mathematical Biology (2021) 83:41
https://doi.org/10.1007/s11538-021-00870-y

Society for Mathematical Biology

SPECIAL ISSUE: CELEBRATING J. D. MURRAY

Bespoke Turing Systems

Thomas E. Woolley[1] · Andrew L. Krause[2] · Eamonn A. Gaffney[2]

Received: 7 September 2020 / Accepted: 11 February 2021 / Published online: 19 March 2021

Abstract

Reaction–diffusion systems are an intensively studied form of partial differential equation, frequently used to produce spatially heterogeneous patterned states from homogeneous symmetry breaking via the Turing instability. Although there are many prototypical "Turing systems" available, determining their parameters, functional forms, and general appropriateness for a given application is often difficult. Here, we consider the reverse problem. Namely, suppose we know the parameter region associated with the reaction kinetics in which patterning is required—we present a constructive framework for identifying systems that will exhibit the Turing instability within this region, whilst in addition often allowing selection of desired patterning features, such as spots, or stripes. In particular, we show how to build a system of two populations governed by polynomial morphogen kinetics such that the: patterning parameter domain (in any spatial dimension), morphogen phases (in any spatial dimension), and even type of resulting pattern (in up to two spatial dimensions) can all be determined. Finally, by employing spatial and temporal heterogeneity, we demonstrate that mixed mode patterns (spots, stripes, and complex prepatterns) are also possible, allowing one to build arbitrarily complicated patterning landscapes. Such a framework can be employed pedagogically, or in a variety of contemporary applications in designing synthetic chemical and biological patterning systems. We also discuss the implications that this freedom of design has on using reaction–diffusion systems in biological modelling and suggest that stronger constraints are needed when linking theory and experiment, as many simple patterns can be easily generated given freedom to choose reaction kinetics.

Keywords Turing patterns · Identifiability

Thomas E. Woolley and Andrew L. Krause equal first authors.

✉ Thomas E. Woolley
woolleyt1@cardiff.ac.uk

1 Cardiff School of Mathematics, Cardiff University, Senghennydd Road, Cardiff CF24 4AG, UK

2 Mathematical Institute, University of Oxford, Andrew Wiles Building, Radcliffe Observatory Quarter, Woodstock Road, Oxford OX2 6GG, UK

1 Introduction

Alan Turing's chemical theory of morphogenesis Turing (1952) is a remarkable model of spatial pattern formation, providing a mechanistic, predictive framework through which many biological developmental systems can be understood (Woolley et al. 2017; Maini et al. 2012, 2016). Specifically, the theory demonstrates that two distinct diffusible populations (known as morphogens) can produce stationary heterogeneous spatial patterns, if the interactions of the two populations satisfy specific criteria. It has been applied widely to understand biological patterns across numerous scales and taxa (Marcon and Sharpe 2012).

Turing-type patterning is also studied in a variety of other settings, such as in nonlinear optics (Oppo 2009; Ardizzone et al. 2013; Chembo et al. 2017), geochemistry (Baurmann et al. 2004; McBride and Picard 2004), astrophysics (Smolin 1996), reaction–advection–diffusion systems (Klika et al. 2018; Krause et al. 2018a; Van Gorder et al. 2019), network-organised media (Nakao and Mikhailov 2010; Asllani et al. 2014), spatial ecology (Sherratt 2012; Hata et al. 2014; Taylor et al. 2020), and social dynamics (Wakano et al. 2009). Whilst it has been known for decades that chemical systems can be engineered to produce specific Turing patterns (Vanag and Epstein 2001), recently Tan et al. (2018) employed such designed Turing systems to manufacture a porous filter for use in water purification. Experimental efforts are also presently underway to engineer Turing-type patterns in synthetic biological experiments using genetically engineered bacteria (Grant et al. 2016; Boehm et al. 2018; Karig et al. 2018). Alongside a growing theoretical literature on the control of Turing patterns (Pismen 1994; Li and Ji 2004; Kashima et al. 2015), these experimental endeavours demonstrate valuable applications of using engineering design principles to create specific patterns in a variety of settings.

However, despite enormous theoretical and experimental advances, the theory has, so far, had some fundamental obstructions in explaining biological patterns (Woolley et al. 2017). One specific criticism that is often levelled at Turing's theory is that the parameter region over which patterning can occur is typically relatively small and, thus, fine tuning may become an issue (Murray 1982; Scholes et al. 2019), especially when patterning processes in development occur robustly across a large variety of scales and environments. A number of solutions to this problem have been suggested such as by considering realistic noise in the system, which enlarges the pertinent parameter domain (Woolley et al. 2011; Schumacher et al. 2013; Biancalani et al. 2010), by increasing the number of species that are being modelled (Marcon et al. 2016; Diego et al. 2018), or by considering domain growth, which is ubiquitous in developing systems (Crampin et al. 1999). In contrast to these endeavours, this article provides a constructive user's guide to building a Turing system that is unstable in a specifically chosen, arbitrarily large parameter region associated with polynomial kinetics (though in principle one can apply this approach to more general self-organisation systems). Consistent with convention, we will call such parameter spaces "Turing spaces".

This construction highlights several of the inherent difficulties of model selection and parameter identifiability issues that are important across modelling in general, and especially mathematical biology, where universal fundamental principles are rare (Warne et al. 2019; Clermont and Zenker 2015; Maclaren and Nicholson 2019). Specifically, without careful experimental design and rigorous first-principles-based models, the best one can often do is offer a phenomenological interpretation that is consistent with observed data and can be used to hypothesise underlying causal mechanisms. Further experimental work is then required to test any suggested mechanism via theoretically derived predictions. Moreover, the modelling choices that enable consistency of simulation and experiment are frequently motivated by parsimony and Occam's razor, but this rarely prohibits arbitrarily many possible variations that can consistently match predictions. Of course, phenomenological models can still be useful in providing causal hypotheses, and guiding intuition (Seul and Andelman 1995; Gelfert 2018). However, even if a model is consistent with the underlying mechanisms, it may be impractical, or impossible, to generate data that will specify all parameters uniquely (Anguelova et al. 2007). This paper demonstrates the extreme freedom a modeller has when faced with an observed pattern and a lack of biological constraint within a reaction–diffusion system, whilst highlighting that this freedom results in problems regarding meaningful interpretation. This makes a stronger case for constraining models not only by consistency with particular experimental results but also to known biological pathways and interactions.

Recent methods have been developed for parameter inference of reaction–diffusion models, subject to a given set of reaction kinetics (Garvie et al. 2010; Krämer et al. 2013; Campillo-Funollet et al. 2019; Kazarnikov and Haario 2020). Such inference methods have been proposed to quantify aspects of signalling dynamics from observed morphogen concentrations in different taxa (Dewar et al. 2010), though again often using phenomenological reaction kinetics that are undoubtedly a caricature of the real morphogen signalling dynamics. Similar ideas have been used to quantify and classify the role of different aspects of agent-based models of pattern formation (such as stochasticity and interactions) in models of zebrafish pigmentation (McGuirl et al. 2020). If the underlying morphogen interactions are known, and hence, the nonlinearities are given, then these techniques provide a powerful way for using observations to determine key features of the morphogen dynamics. However, in practice the choice of the nonlinear kinetics is poorly constrained a priori and one has instead readily observable patterning features. In particular, common classifications of patterns are mainly qualitative. For example, we often seek to specify whether a heterogeneity is that of spots, or stripes, and whether the peaks and troughs of the morphogens are aligned (in-phase patterning) or whether the peaks of one population correspond to troughs of another (out of phase patterning) (see Fig. 1). However, we will show that such qualitative requirements on observable patterns can mean little in constraining the nonlinear kinetics. Hence, a major theme of our results will be advocating for more mechanism-based constraints of the reaction kinetics in reaction–diffusion models of pattern formation.

We will demonstrate that if one has freedom in the choice of reaction kinetics, as is often the case in practice, then model selection and parameter identifiability are theoretically impossible when one simply is considering the presence, or absence, of simple one- and two-dimensional simple patterns (e.g. peaks, troughs, spots, stripes, and labyrinths). In particular, we will demonstrate how to construct polynomial morphogen kinetics such that the patterning parameter domain, morphogen phases, and aspects of the resulting pattern can all be determined at will. We demonstrate examples of these constructions numerically, giving examples of non-contiguous parameter domains where patterning occurs, as well as parameter regimes where the resulting patterns are qualitatively different in specified ways. Such bespoke Turing systems may be valuable for designing chemical, or biological systems, as they will suggest general qualitative features that either have to be (i) discovered, thereby directing future experimental design, or (ii) designed, in the case of synthetic patterning construction. Whilst we will focus on designing a specific chemical system able to match any desired parameter space, the ideas we present should be valuable for the analysis and design of other self-organising systems, and to help illuminate issues of robustness in Turing systems.

Note that the results of this work can either be viewed in a positive or negative light. The positive view commends Turing's theory for being general enough to mimic a huge range of biological complexity, whilst the negative view suggests that such modelling does not allow us to distinguish Turing's mechanism from alternative hypotheses without further constraints from biology. This has been discussed widely before, such as by Oster (1988); "So what good are models? The foregoing discussion may seem a bit depressing to theoretical biologists: all manner of models produce the same spatial patterns. Thus it is not generally possible to distinguish between models solely from the patterns they generate, and the question I raised at the beginning begs an answer: what good are models of pattern formation?" Despite some progress, and use of modelling by experimental biologists (Economou and Green 2014), these fundamental issues have not really been overcome. Our work further sheds light on this by demonstrating how easily reaction–diffusion systems can be made to fit arbitrarily complex patterning scenarios. We discuss this further in the conclusion, offering a more nuanced opinion which suggests that this work be viewed as a celebration of Turing's theory for its beauty, both theoretically and numerically, whilst offering a word of caution as to its over-zealous application to understand any and all pattern formation.

2 Framework

In the following, we will consider two continuous populations, u and v, termed "morphogens", which is a standard and general name for the populations whose interactions may generate patterning. Critically, we are intrinsically assuming that each population is made up of a large number of individuals, so as to allow a continuum description, although there has been extensive work done on low-copy number reaction–diffusion systems (Adamer et al. 2020; Woolley et al. 2012).

We use the term morphogen to indicate that the populations represent interacting biological/chemical species that are able to generate spatiotemporal complexity. We do not specify the populations' make up in terms of whether they represent cellular, or chemical populations (for example) (Kondo and Miura 2010). In many cases, these species represent physical, observable quantities and, thus, must be positive to be physically meaningful, though this is not always the case in reaction–diffusion models used for spatial patterning, such as those involving cellular transmembrane potential differences (Sánchez-Garduno et al. 2019). For concreteness, we will require positive solutions throughout, but note that relaxing this requirement makes the analysis below easier, in general.

Our goal is to show that we can construct a set of interaction kinetics, which produce spotted, or striped/labyrinthine, Turing patterns in a specified parameter region, $\Omega \in \mathbb{R}^n$. In Sect. 4, we exploit rapid changes in parameter values to demonstrate that we can produce both spots and stripes across different spatial and temporal regions of the same system, effectively sampling from different parameter regimes. Precise details of the resulting patterns will of course depend on geometry and nonlinearity, as well as the heterogeneity employed, but we show that hybrid patterns can be constructed by piecing together different regions in parameter space.

We illustrate a summary of our results in Fig. 1 in the form of a flowchart. We remark that this choice of kinetics, given in the equations in Fig. 1, contains all of the parameter dependencies in the level set function $S(\boldsymbol{p})$, where $\boldsymbol{p} = (p_1, \ldots, p_n)$ are the parameters, which can be adjusted to construct the desired pattering properties as detailed throughout this paper. Thus, this illustrates how we can design parameter spaces for a desired pattern, rather than be constrained by a specific reaction mechanism.

Furthermore, the basic design principles should be applicable more generally if one has sufficient control over how parameters enter into the system (e.g. through the use of catalysis, reactions on differing timescales, etc.). Additionally, the design framework shown in this toy model has implications for the use of Turing systems in other frameworks as we discuss later.

2.1 Linear Analysis

We first briefly review the basic linear instability analysis for reaction–diffusion systems, in order to motivate our choice of nonlinear kinetics. Let $B \subset \mathbb{R}^m$ be a bounded spatial domain with boundary ∂B, such that for all $\boldsymbol{x} \in B$, $u(\boldsymbol{x}, t)$ and $v(\boldsymbol{x}, t)$ are defined to be population densities satisfying the equations:

$$\frac{\partial u}{\partial t} = D_u \nabla^2 u + f(u, v), \tag{1}$$

$$\frac{\partial v}{\partial t} = D_v \nabla^2 v + g(u, v), \tag{2}$$

where $f(u, v)$ and $g(u, v)$ are functions describing the reaction kinetics. Moreover, D_u and D_v are constant (positive) diffusion coefficients. For simplicity, we assume u and v satisfy zero-flux boundary conditions on $\boldsymbol{x} \in \partial B$.

In the rest of the paper, we restrict the dimension of B to be two, or less, for simplicity of presentation. Our results regarding the existence of Turing patterns and their phase are independent of dimension, whereas the specific results and simulations regarding stripes or spot selection, discussed in Sect. 3.3, are strictly a two-dimensional phenomenon. Although the resulting dynamics can be much richer in more complex cases (for example, through: higher domain dimensions; advective transport; and increased numbers of species (Klika et al. 2012; Marcon et al. 2016; Arcuri and Murray 1986; Aragón et al. 2012; Woolley 2014, 2017; Van Gorder et al. 2019)), the proposed

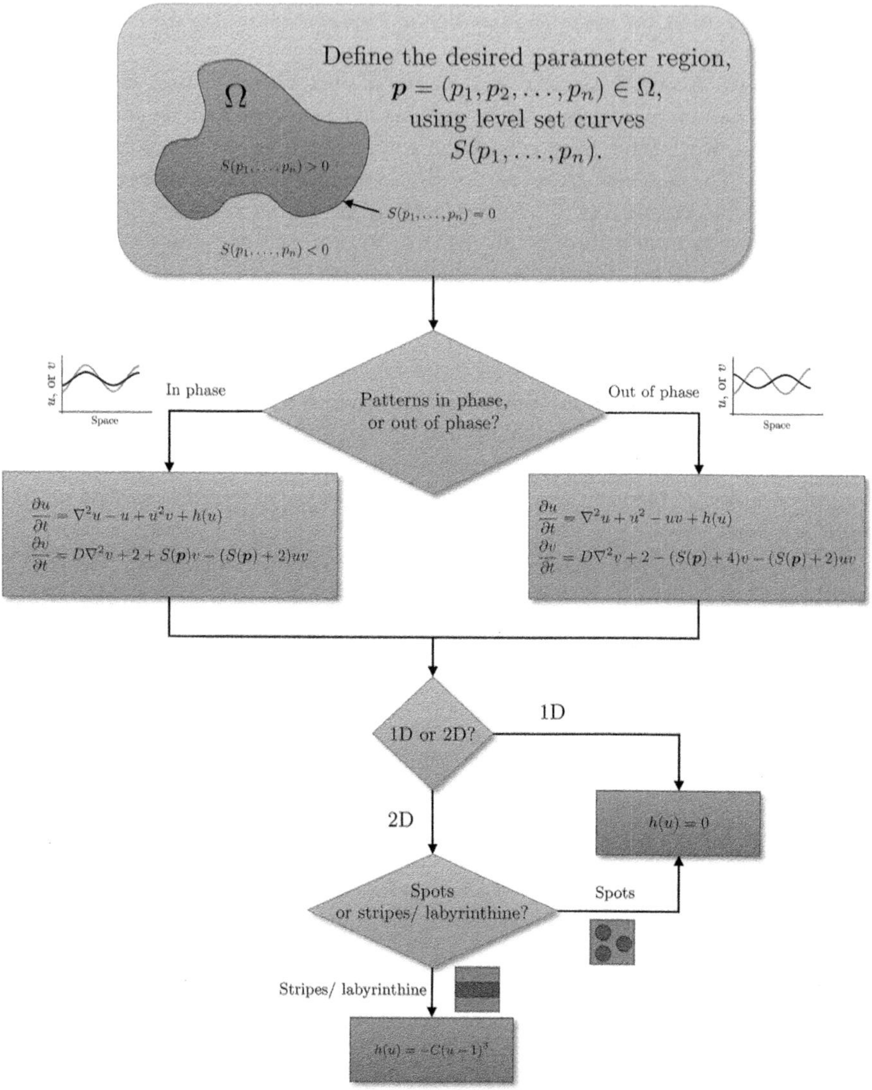

Fig. 1 Flowchart for constructing a Turing system that will work in any specified parameter region and provide a desired pattern. Note that D, C, and the solution domain must be "sufficiently large". See text for details (Color figure online)

framework can be generalised to these cases. Indeed, with cases of higher complexity there are often even more degrees of freedom to design kinetics that match observed patterns. Here, we consider the simplest case through which we can construct specific kinetic forms of f and g, in order to satisfy the necessary conditions over a given parameter region for patterns, of any kind, to exist.

For Eqs. (1) and (2) to present a Turing instability, or diffusion-driven instability, the system must satisfy two specific conditions. Firstly, there must be a homogeneous steady state (here taken to be positive), which is stable in the absence of diffusion. Secondly, the steady state must become unstable when diffusion is considered. Hence, we assume that Eqs. (1) and (2) admit at least one positive, constant, homogeneous steady state (u_c, v_c) such that $f(u_c, v_c) = g(u_c, v_c) = 0$. We will neglect pattern formation due to stabilised fronts and other phenomena that can occur in multistable systems (Vastano et al. 1987), focusing solely on patterning due to Turing instability of a single homogeneous steady state. To examine the linear stability of this steady state, we consider a small perturbation,

$$\begin{pmatrix} \hat{u} \\ \hat{v} \end{pmatrix} = \begin{pmatrix} u_c \\ v_c \end{pmatrix} + \begin{pmatrix} \epsilon_u(\boldsymbol{x}, t) \\ \epsilon_v(\boldsymbol{x}, t) \end{pmatrix}. \tag{3}$$

We will consider a one-dimensional interval for the linear derivation, i.e. $B = [0, L]$. Then, the perturbations are assumed to have the following form $(\epsilon_u, \epsilon_v) = (\epsilon_{u0}, \epsilon_{v0}) \exp(\lambda t) \cos(kx)$, where $0 < |\epsilon_{u0}| \ll 1$ and $0 < |\epsilon_{v0}| \ll 1$ are constants and the cosine form of the perturbation is chosen to satisfy the boundary conditions. We consider a single such Fourier term of wave mode k (by considering orthogonality and completeness of such functions). Hence, instability of a particular mode depends on its associated wavenumber, k. For this choice of domain, we have that $k = n\pi/L$, for some integer n, which is nonnegative, without loss of generality.

Substituting Eq. (3) into Eqs. (1) and (2) and linearising f and g in the usual way (e.g. see (Murray 2003a)), we can deduce conditions depending on the Jacobian of the kinetics, the diffusion parameters, and the Laplacian spectrum, here given by admissible k. The usual criteria for Eqs. (1) and (2) to admit a Turing instability are the following inequalities

$$f_u + g_v < 0, \tag{4}$$

$$f_u g_v - f_v g_u > 0, \tag{5}$$

$$D_v f_u + D_u g_v > 2\sqrt{D_u D_v (f_u g_v - f_v g_u)} > 0, \tag{6}$$

$$k_-^2 < \left(\frac{n\pi}{L}\right)^2 < k_+^2 \text{ where}$$

$$k_\pm^2 = \frac{D_v f_u + D_u g_v \pm \sqrt{(D_v f_u + D_u g_v)^2 - 4 D_u D_v (f_u g_v - f_v g_u)}}{2 D_u D_v}, \tag{7}$$

where subscripts of f and g denote partial differentiation with respect to the indexed variable and all partial derivatives are evaluated at the steady state. The first two inequalities (4) and (5) enforce that the steady state is stable in the absence of diffusion. Inequality (6) then ensures that this steady state can be driven to instability when diffusion is included. Finally, inequality (7) states that the patterning domain has to admit an unstable mode and can generally be satisfied for a suitable choice of the integer, n, by making the domain sufficiently large, since $k_- < k_+$ if there is a linear instability. Critically, the patterns have an intrinsic wavelength and the domain has to be larger than this wavelength in order for the pattern to appear, at least before nonlinear waveform selection dynamics manifest as patterning develops. We will focus on the first three inequalities (4)–(6), since inequality (7) can always be satisfied by a sufficiently large domain. Whether this size is biologically relevant and/or reasonable depends on the scales involved with the problem and, thus, can only be considered on a case by case basis, in which at least some of the parameters are known to within orders of magnitude.

From considering inequalities (4) and (5), we find that the Turing conditions impose specific sign structures on the partial derivatives (Dillon et al. 1994). Noting $f_u = g_v = 0$ is not consistent with relations (4)–(6) and that at least one f_u, and g_v must be positive because of inequality (6) then, without loss of generality, we take $f_u > 0$, whence the Jacobian of first-order partial derivatives,

$$\boldsymbol{J} = \begin{pmatrix} f_u & f_v \\ g_u & g_v \end{pmatrix}, \tag{8}$$

must have one of the following sign structures

$$\boldsymbol{J}_p = \begin{pmatrix} + & - \\ + & - \end{pmatrix}, \text{ or } \boldsymbol{J}_c = \begin{pmatrix} + & + \\ - & - \end{pmatrix}, \tag{9}$$

and also $D_v > D_u$.

Kinetics with the $\boldsymbol{J}_p$ sign structure are known as pure kinetics, and those with the $\boldsymbol{J}_c$ sign structure are known as cross-kinetics. Although, generally, the same patterns are available in each case, the sign structure does influence how these patterns appear in the two morphogen populations, e.g. the peaks and troughs of u and v will be out of phase in the cross-kinetics case (i.e. peaks of u will correspond to troughs of v, and vice versa) and in phase in the pure kinetics case, at least for linearised solutions, though, in practice, this relation is typically inherited even when nonlinear dynamics eventually feature.

We will use the sign structure of $\boldsymbol{J}_c$ throughout most of the paper, though note that the following derivation works *mutatis mutandis* in the case that $\boldsymbol{J}_p$ is chosen. We proceed noting that the linear stability of the system is defined completely by the action of the perturbed equations

$$\frac{\partial \epsilon_u}{\partial t} = D_u \frac{\partial^2 \epsilon_u}{\partial x^2} + f_u \epsilon_u + f_v \epsilon_v, \tag{10}$$

$$\frac{\partial \epsilon_v}{\partial t} = D_v \frac{\partial^2 \epsilon_v}{\partial x^2} + g_u \epsilon_u + g_v \epsilon_v, \tag{11}$$

where $f_u, f_v > 0$ and $g_u, g_v < 0$.

We specify new variables

$$x = [x]x', t = [t]t', \epsilon_u = [\epsilon_u]\epsilon_u', \epsilon_v = [\epsilon_v]\epsilon_v', \tag{12}$$

where in each case the bracketed term is a constant dimensional scale and the primed term is the new non-dimensionalised variable. Further, we specify the dimensional scales as

$$[t] = \frac{1}{f_u}, \quad [x] = \sqrt{\frac{D_u}{f_u}}, \tag{13}$$

and specify a required equality,

$$f_u[\epsilon_u] = f_v[\epsilon_v], \tag{14}$$

which leaves one of the scales, ($[\epsilon_u]$, or $[\epsilon_v]$) as a free parameter. Using this scaling, we have the simpler system

$$\frac{\partial \epsilon_u}{\partial t} = \frac{\partial^2 \epsilon_u}{\partial x^2} + \epsilon_u + \epsilon_v, \tag{15}$$

$$\frac{\partial \epsilon_v}{\partial t} = D\frac{\partial^2 \epsilon_v}{\partial x^2} - F\epsilon_u - G\epsilon_v, \tag{16}$$

where the primes on the variables have been omitted, and we define $F = |g_u| f_v / f_u^2$, $G = |g_v|/f_u$, and $D = D_v/D_u > 1$, with all three parameter groupings strictly positive.

Under this transformation, the Turing instability criteria simplify to

$$F > G > 1, \tag{17}$$

$$D - G > 2\sqrt{D(F-G)} > 0. \tag{18}$$

By inequality (17) and the positivity of the parameters, we know that $\sqrt{D(F-G)} > 0$ is guaranteed; thus, we only need to satisfy $D - G > 2\sqrt{D(F-G)}$. Moreover, because $D - G$ grows linearly with increasing D, whilst $\sqrt{D(F-G)}$ grows sub-linearly, we are guaranteed to be able to satisfy inequality (18) if we choose D large enough. Namely, the minimum possible diffusion is

$$D_c = \left(\sqrt{F-G} + \sqrt{F}\right)^2, \tag{19}$$

which is well defined, because $F > G$ by inequality (17). Although inequality (18) is satisfied for any $D > D_c$, whether D (the ratio of the two diffusion coefficients) can be as high as required in a particular application can, once again, only be determined on a case by case basis, through applying the known data to the equations and deriving the appropriate scales.

If we further define $F' = F - 1$, $G' = G - 1$, then the only criterion that has to be satisfied is $F' > G' > 0$. Other than this, we can always choose D and L large enough to ensure that a system is able to pattern at a suitable wave number. This criterion is not only simple, but also highlights the required relative strengths of the activation and inhibition effects of the two populations. Namely, under the signs of $\boldsymbol{J}_c$ if we identify u to be the (self-)activator and v to be the (self-)inhibitor, then we see that the strength of inhibition of u on v, i.e. F, has to be stronger than the self-inhibition, i.e. G. Equally, $G' > 0$ suggests that G has to be greater than unity so that self-inhibition is also stronger than self-activation.

Under the signs of $\boldsymbol{J}_p$, although the perturbed system would have the form

$$\frac{\partial \epsilon_u}{\partial t} = \frac{\partial^2 \epsilon_u}{\partial x^2} + \epsilon_u - \epsilon_v, \tag{20}$$

$$\frac{\partial \epsilon_v}{\partial t} = D\frac{\partial^2 \epsilon_v}{\partial x^2} + F\epsilon_u - G\epsilon_v, \tag{21}$$

we have the same criteria for instability. Once again we see that the activation of u on v has to be stronger than the self-inhibition of v. Thus, in the non-dimensionalised case, where the influence of u and v on u is taken to be the same relative strength, then, regardless of the sign structure, the influence of the self-activator on the self-inhibitor has to be stronger than the self-inhibition. In turn, this influence is stronger than the influence of either morphogen on the activator, which is a useful constraint to place on biological systems that are suggested to act through a Turing instability. Whilst these conclusions are only strictly true in the two morphogen case, analogous results can be drawn for systems with more populations, typically resulting in restrictions to the kinds of interactions between groups of morphogens that can support pattern formation (Satnoianu et al. 2000; Marcon et al. 2016).

We note that all of the parameters from the full nonlinear model that influence linear stability are embedded in the parameters F and G, corresponding to elements of the Jacobian at the steady state. Suppose our desired parameter domain for a Turing instability, $\Omega \in \mathbb{R}^n$, is bounded by the level set curve $S(\boldsymbol{p}) = 0$, where $\boldsymbol{p} = (p_1, p_2, \ldots, p_n)$ are parameters that influence the reaction kinetics. For orientation purposes let $S(\boldsymbol{p}) > 0$ for all $\boldsymbol{p} \in \Omega$ and $S(\boldsymbol{p}) < 0$ otherwise. If we can decompose S into the difference of two positive functions, then we can find a system that patterns as required, since $S(\boldsymbol{p}) = F' - G'$ satisfies our constraints. We can choose, for instance,

$$F' = S(\boldsymbol{p}) + \eta, G' = \eta, \tag{22}$$

for any $\eta > 0$, say $\eta = 1$. Hence, $F' > G' > 0$ if $\boldsymbol{p} \in \Omega$ and, so, we have constructed a linear system that will be Turing unstable in any parameter domain that we choose. Outside of this domain, we have not constrained the system. Thus, patterns may exist, though by construction, the homogeneous steady state will be unstable to spatially constant modes for $S(\boldsymbol{p}) < 0$. Extending these ideas so that $S(\boldsymbol{p}) = 0$ gives the boundary of the Turing space, $S(\boldsymbol{p}) > 0$, whereas the region $S(\boldsymbol{p}) < 0$ requires a more detailed analysis and is the subject of future work.

2.2 Defining a Full Set of Kinetics

We now specify a set of nonlinear kinetics f and g. Nonlinearities are needed in the system to bound the otherwise exponential growth of any instability exhibited by the linear system of Eqs. (15) and (16). Further, under our current assumptions, we must also constrain the system to ensure that f and g never drive the system to negative values of u and v or suffer from finite time blow up. For simplicity, we consider polynomial kinetics of the following forms:

$$f = a_1 + b_1 u + c_1 u^\alpha v, \quad g = a_2 + b_2 v + c_2 uv, \tag{23}$$

where the α, a_i, b_i and c_i are all constants. To maintain positivity near the origin, we enforce the condition that a_1 and a_2 are both nonnegative, whilst b_i and c_i are free to take any sign. To constrain the six parameters further, we require a positive steady state. Without loss of generality, we choose $(1, 1)$ to be the critical point; hence, we have to satisfy the equations $f(1, 1) = g(1, 1) = 0$. Further, we require that $f_u(1, 1) = 1 = f_v(1, 1)$, $g_u(1, 1) = -F$ and $g_v(1, 1) = -G$. Solving these equations simultaneously results in the following requirements

$$a_1 = -2+\alpha, \quad b_1 = -\alpha+1, \quad c_1 = 1, \quad a_2 = G, \quad b_2 = F-G, \quad c_2 = -F. \tag{24}$$

Hence, α is the only free parameter and because we require $a_1 \geq 0$, then we observe that $\alpha \geq 2$. So for simplicity, we can choose $\alpha = 2$. Thus,

$$\frac{\partial u}{\partial t} = \frac{\partial^2 u}{\partial x^2} - u + u^2 v, \tag{25}$$

$$\frac{\partial v}{\partial t} = D\frac{\partial^2 v}{\partial x^2} + G' + 1 + (F' - G')v - (F' + 1)uv, \tag{26}$$

is a pure kinetic Turing system that presents a diffusion-driven instability whenever $F' > G' > 0$, given a diffusion constant and domain that are large enough. Consequently, using Eq. (22), with $\eta = 1$

$$\frac{\partial u}{\partial t} = \frac{\partial^2 u}{\partial x^2} - u + u^2 v, \tag{27}$$

$$\frac{\partial v}{\partial t} = D\frac{\partial^2 v}{\partial x^2} + 2 + S(\boldsymbol{p})v - (S(\boldsymbol{p}) + 2)uv, \tag{28}$$

presents a diffusion-driven instability whenever $\boldsymbol{p} \in \Omega$.

The results derived for a_2 in Eq. (24) demonstrate why we took $\eta = 1$, so that $F' = S(\boldsymbol{p}) + 1$ and $G' = 1$. Specifically, if the parameter space defines a closed bounded domain, $\overline{\Omega} = \{\boldsymbol{p} \in \mathbb{R}^n : S(\boldsymbol{p}) \geq 0\}$, we could have defined $G' = S(\boldsymbol{p})$ and $F' = \sup_{\boldsymbol{p}\in\overline{\Omega}} S(\boldsymbol{p})$, which would also ensure that the Turing instability criterion is only fulfilled when $\boldsymbol{p} \in \Omega$. However, $S(\boldsymbol{p}) = G' < 0$ outside of Ω and, so, the positivity of $G' + 1$ would not be guaranteed. Consequently, solution trajectories that remain positive for all time (and, thus, physically feasible) are not guaranteed.

Another benefit from defining the desired parameter space using level sets is that since F' and G' are bounded, and able to attain their bounds, we can specify a maximum value for D_c. Namely, a Turing pattern is possible for all $\boldsymbol{p} \in \Omega$ if

$$D > \sup_{\boldsymbol{p}\in\overline{\Omega}} \left(\sqrt{F(\boldsymbol{p}) - G(\boldsymbol{p})} + \sqrt{F(\boldsymbol{p})}\right)^2 = \sup_{\boldsymbol{p}\in\overline{\Omega}} \left(\sqrt{S(\boldsymbol{p})} + \sqrt{S(\boldsymbol{p}) + 2}\right)^2. \tag{29}$$

As an even more simple and conservative approach, we could define $D = 4\sup_{\boldsymbol{p}\in\overline{\Omega}} F(\boldsymbol{p}) = 4\left(2 + \sup_{\boldsymbol{p}\in\overline{\Omega}} S(\boldsymbol{p})\right)$, which always satisfies inequality (29). Further, upon fixing D we can calculate the minimum domain, L_c, that is required through Eq. (7) and noting that the minimum wave number occurs when $n = 1$, namely,

$$L_c^2 > \frac{2\pi^2 D}{D - G + \sqrt{(D-G)^2 + 4D(F-G)}}. \tag{30}$$

We now comment on the $\boldsymbol{J}_p$ sign case as Eqs. (25) and (26), or Eqs. (27) and (28), are only able to produce the $\boldsymbol{J}_c$ sign system. However, the same process as above can be followed except that we use

$$f = a_1 + b_1 u^\alpha + c_1 uv, \quad g = a_2 + b_2 v + c_2 uv, \tag{31}$$

and require $f_u(1,1) = 1, f_v(1,1) = -1, g_u(1,1) = F, g_v(1,1) = -G$. Consequently, with $\alpha = 2$, we are able to create the following pure kinetic system that is unstable whenever $F' > G' > 0$

$$\frac{\partial u}{\partial t} = \frac{\partial^2 u}{\partial x^2} + u^2 - uv, \tag{32}$$

$$\frac{\partial v}{\partial t} = D\frac{\partial^2 v}{\partial x^2} + G' + 1 - (F' + G')v - (F' + 1)uv, \tag{33}$$

or whenever $\boldsymbol{p} \in \Omega$

$$\frac{\partial u}{\partial t} = \frac{\partial^2 u}{\partial x^2} + u^2 - uv, \tag{34}$$

$$\frac{\partial v}{\partial t} = D\frac{\partial^2 v}{\partial x^2} + 2 - (S(\boldsymbol{p}) + 4)v - (S(\boldsymbol{p}) + 2)uv, \tag{35}$$

for large enough L and large enough D (see Eqs. (29) and (30)). Further systems are generated for $\alpha > 2$.

Finally, we note that Eqs. (25) and (26) and Eqs. (32) and (33) are highly non-unique. Indeed, we may add any arbitrary terms of the form $(u-1)^\beta (v-1)^\gamma H(u, v)$ where H is any smooth function and either $\beta > 1, \gamma \geq 0$, or $\gamma > 1$ and $\beta \geq 0$. However, we must ensure that these additional terms do not cause the system to present a finite time blow up, which is a weak constraint in practice. In the current case, of considering (u, v) to be morphogens, we also specify that the choice of H should not violate the positivity of the trajectories. However, in more general cases, e.g. one of (u, v) measuring transmembrane potential differences, this condition can also be relaxed.

Critically, although these extra nonlinear terms will not influence the stability characteristics of the homogeneous steady state at (1,1), they may create new steady states that have different parameter regions of existence and stability. Equally, higher-order terms are able to influence the observed patterns (Ermentrout 1991), which we will see in Sect. 3.3. For simplicity, we are defining our Turing space Ω in terms of parameters $\boldsymbol{p}$ subject to $D > D_c$, and for L sufficiently large. For a bounded patterning space Ω, we can define a largest value of D_c by the supremum over $\boldsymbol{p} \in \Omega$ as in (29). Hence, for all $D > D_c$, Ω will coincide exactly with the set of kinetic parameters which admit Turing instabilities. This allows us to focus the discussion solely on the kinetic parameters, though in general the domain length scale and ratio of diffusion coefficients should be included in the definition the Turing space. Additionally, if we choose Ω to be unbounded, then D_c may no longer be a finite number (as one may need larger diffusion ratios to satisfy (18)), and hence, there will still be a dependence of Ω on D. We will now give a few examples of this framework to design parameter spaces, before discussing nonlinear selection effects.

3 Examples

Numerical simulations of the following reaction–diffusion systems were run using the finite element software COMSOL Multiphysics 5.3 using a backward differentiation formula scheme in time. The initial conditions are uniformly randomly generated about the steady state 1, with range 0.1. Simulations were run until the supremum of the time derivative over all grid points was less than 10^{-3} and until simulation time was greater than or equal to 10^4. One-dimensional domains were chosen to have a discretisation of at least 10^3 elements. Two-dimensional domains were chosen to have a discretisation of at least 2.5×10^4 triangular elements. In each case, representative simulations were rerun with halved mesh sizes to ensure that the observed patterns did not change with discretisation.

It should be noted that our stopping tolerances (time greater than 10^4 and time derivative less than 10^{-3}) were simply chosen to ensure that the system was close to a steady state. The values themselves have no specific meaning as we are working within a non-dimensionalised setting. However, such thresholds are a critical point for a modeller to consider in applications, as without well-defined scales for the kinetic rates, spatial resolution, and time duration of pattern formation, we cannot conclude, for example, whether 10^4 time units is a sensible length of time (Woolley et al. 2010).

3.1 Circular Parameter Region

Suppose we want our system to be a pure kinetic system and to pattern within the circular parameter region $\Omega = \{(a, b) \in \mathbb{R} | S(a, b) = 1 - a^2 - b^2 > 0\}$. Note that we do not make any claims as to the system dynamics outside of Ω. We can define $F' = 1 - a^2$ and $G' = b^2$ and consider

$$\frac{\partial u}{\partial t} = \nabla^2 u + u^2 - uv, \tag{36}$$

$$\frac{\partial v}{\partial t} = D\nabla^2 v + 1 + b^2 - (1 - a^2 - b^2)v - (2 - a^2)uv. \tag{37}$$

By considering the maximum of Eq. (19), plotted in Fig. 2a, we derive that any diffusion rate larger than $D_c = (1 + \sqrt{2})^2 \approx 5.8$ will cause the system to present a Turing instability for all $(a, b) \in \Omega$. For a conservative approach, we choose $D = \sup_{\overline{\Omega}} 4(1 + F') = 8$. Further, we also need to ensure that the domain size is larger than

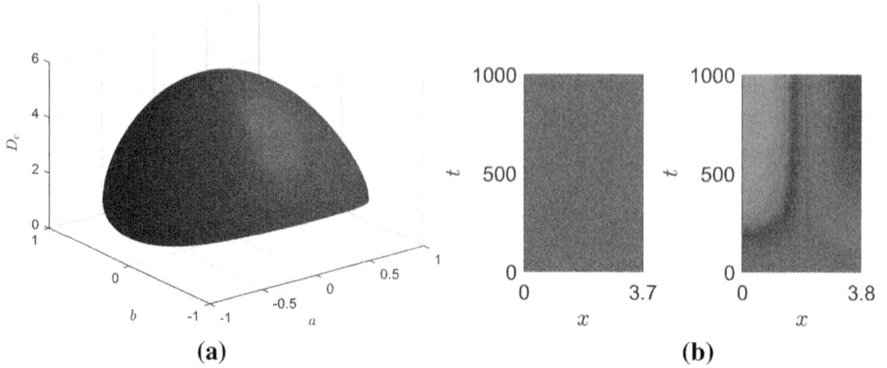

Fig. 2 **a** For the case that we require Turing instabilities to occur with the parameter region $\Omega = \{(a, b) \in \mathbb{R} | S(a, b) = 1 - a^2 - b^2 > 0\}$ we illustrate the critical values of the diffusion rate, given by Eq. (19), namely $D_c = \left(\sqrt{1-a^2-b^2} + \sqrt{2-a^2}\right)^2$. The maximum value of $D_c \approx 5.8$ occurs at the origin. **b** Space-time simulations of u from solving Eqs. (36) and (37) with $L = 3.7$ and $L = 3.8$ for the left and right plots, respectively. The simulations demonstrate that although the system may present a Turing instability, the domain has to be large enough for the instability to be realised and the results agree with the theoretical prediction that $L_c \approx 3.77$. Parameters $D = 8$, $a = 0 = b$. The colour axis in both figures runs from blue at $u = 0.9$ to yellow at $u = 1.1$. Because we have chosen cross-kinetics, the variable v (not shown) also presents a single boundary peak that is out of phase with u (Color figure online)

$$L_c = \sup_{\overline{\Omega}} \pi \left(\frac{D}{D - G' - 1 + \sqrt{(D - G' - 1)^2 - 4D(F' - G')}} \right)^{1/2}$$
$$= \frac{4\pi}{\sqrt{7 + \sqrt{17}}} \approx 3.77. \tag{38}$$

Figure 2b illustrates that even though the bound derived in Eq. (38) is only a sufficient condition, it still serves as quite a sharp bifurcation point, at least in this case. Specifically, for the parameters defined in the caption of Fig. 2b a mode one Turing pattern appears when $L = 3.8 > L_c$, but not for slightly smaller domains.

All of the analytical results and bounds still hold in two spatial dimensions, with an increase in the number of potential patterning structures. Specifically, whereas in one dimension we are only able to generate peaks and troughs, two dimensions provide the option of stripes, spots and labyrinthine patterns, since there is now a vector of wave modes, $\boldsymbol{k} = (k_x, k_y)$ that can be unstable. Patterning structures become even more complex in three spatial dimensions and beyond (Leppänen et al. 2002, Callahan and Knobloch 1999). Critically, a simple encompassing framework to categorise such higher spatial dimensional patterns does not currently exist, as the distinction relies on nonlinear dynamics (Ermentrout 1991).

We now numerically evidence that for the chosen kinetics, all patterns in the circular parameter domain correspond to spot patterns. In order to sweep over Ω, we transform the parameters (a, b) into their polar forms $a = r\cos(\theta)$, $b = r\sin(\theta)$ and consider multiple values for r and θ. Note that simulations were completed over the (r, θ) domain of $[0, 2] \times [0, 2\pi)$, using 40 equally spaced values of r and 10

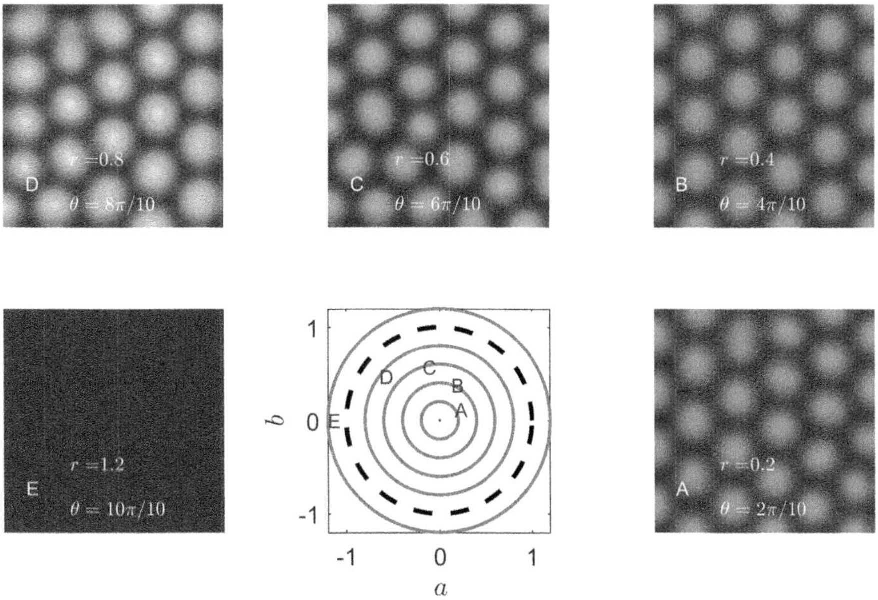

Fig. 3 Representative two-dimensional simulations of Eqs. (36) and (37) demonstrating that Turing patterns exist if and only if $(a, b) \in \Omega$, the unit circle, as required. The central image on the bottom row illustrates the parameter region. The red solid lines indicate circles of radii 0.2, 0.4, 0.6, 0.8, and 1.2, respectively, and the black dashed curve represent the circle of radius 1, which is the bifurcation line. The letters **a**–**e** on this subfigure represent the parameter values chosen for each of the surrounding subfigures **a**–**e**. The specific values of radius, r, and angle θ are noted on each subfigure. Each simulation occurs on a square domain of side length 50. The colour axis is the same for all simulations and runs from blue at a value of 0.5 to yellow at a value of 5 (Color figure online)

equally spaced values of θ. Figure 3 illustrates five representative final steady states of these 400 simulations for different values of (r, θ). Although pattern alignment and extremal values vary, the patterns within the Ω parameter region are all spot patterns. By construction, the patterns only exist when $0 \leq r < 1$ (subfigures A–D) and not for $r > 1$ (subfigure E).

3.2 Non-contiguous Parameter Regions

We now give an example of a reaction–diffusion system that is Turing unstable on an unbounded and non-contiguous parameter region, assuming a large enough spatial domain and diffusion ratio, D, can be prescribed. To take the complexity one step further, we construct a system that takes a pure kinetic form on one of the regions, whilst taking a cross-kinetic form on the other.

We begin by specifying the disconnected domain,

$$\Omega = \{(a, b) \in \mathbb{R} | a, b > 0 \text{ and } (a - 3)^2 - b^2 \geq 1\}, \tag{39}$$

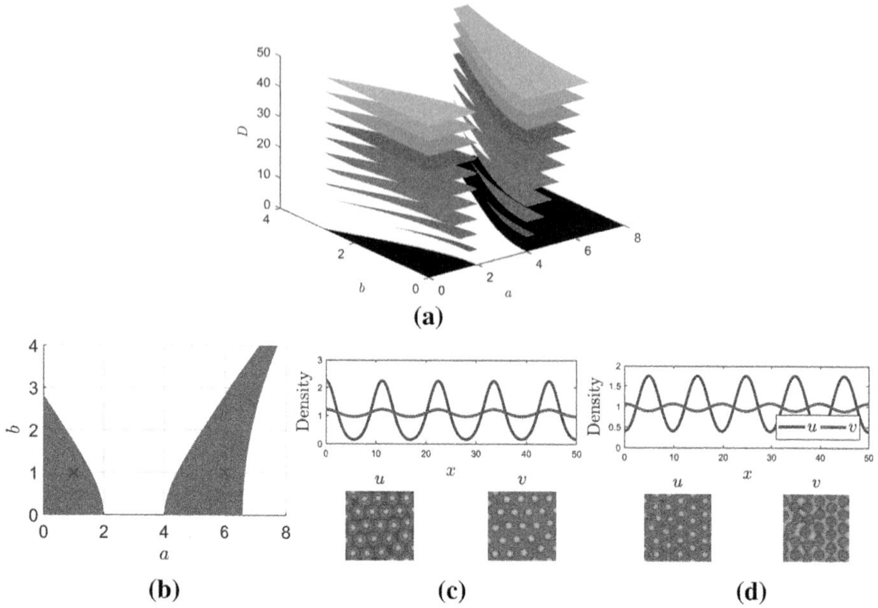

Fig. 4 **a** Slices of the (a, b, D) parameter space. The black area is the region $(a - 3)^2 - b^2 \geq 1$ within the range of (a, b) shown. The coloured slices depict the region of Turing space for given values of D. **b** Fixing $D = 50$ specifies a suitable non-contiguous parameter space in the (a, b) parameter plane. The two crosses specify the parameters used for simulations in (**c**) and (**d**), respectively. **c** and **d** are simulations of Eqs. (45) and (46) and illustrate the available patterns in the left and right regions, respectively. The top one-dimensional simulation demonstrates that the pattern is in phase in the left-hand region and out of phase in the right-hand region, as desired. The two-dimensional simulations illustrate both populations, u and v, demonstrating that the outcome is once again spots, which are either in phase, or out of phase. Parameters $D = 50$, **c** $(a, b) = (1, 1)$ and **d** $(a, b) = (6, 1)$. Each two-dimensional simulation occurs on a square domain of side length 50. The colour axes range from blue at the lowest to yellow at the highest and, from left to right, are: (0.1,3.3), (0.9,1.3), (0.2,2.4) and (0.8,1.1) (Color figure online)

illustrated in Fig. 4a, which specifies the two non-contiguous regions that we will be working with. If we simply wanted to specify that the Turing pattern was active on only both these regions, we could simply use the kinetics derived in Sect. 2.2 using $F = (a - 3)^2$ and $G = b^2 + 1 > 1$. However, we note that because the space is unbounded, we can no longer specify a single value for D_c for all parameter space. For any given parameter set, $(a, b) \in \Omega$, we are still able to use the inequalities derived in Sect. 2 to create a valid diffusion constant and domain size; however, this has to be computed parametrically. Here, we consider the case that $0 \leq a \leq 6$ and, so, $D = 50$ is sufficient (see Fig. 4b).

Critically, further refinement of the kinetics is required to address the constraint that the kinetics are of different types in the different regions. To enforce this, we return to the linearised perturbation system of Eqs. (15) and (16)

$$\frac{\partial \epsilon_u}{\partial t} = \frac{\partial^2 \epsilon_u}{\partial x^2} + \epsilon_u + \epsilon_v, \tag{40}$$

$$\frac{\partial \epsilon_v}{\partial t} = D\frac{\partial^2 \epsilon_v}{\partial x^2} - (a-3)^2\epsilon_u - (b^2+1)\epsilon_v, \tag{41}$$

and redefine the perturbation $\epsilon_u = \epsilon_u'/(a-3)$; thus, the perturbation changes sign depending on whether a is greater than or less than 3. The perturbed system is now of the form

$$\frac{\partial \epsilon_u'}{\partial t} = \frac{\partial^2 \epsilon_u'}{\partial x^2} + \epsilon_u' + (a-3)\epsilon_v, \tag{42}$$

$$\frac{\partial \epsilon_v}{\partial t} = D\frac{\partial^2 \epsilon_v}{\partial x^2} - (a-3)\epsilon_u' - (b^2+1)\epsilon_v, \tag{43}$$

and, thus, has a cross-kinetic Jacobian sign structure if $a > 3$, i.e. (a, b) is in the right-hand region of Ω, whilst it has a pure kinetic Jacobian sign structure if $a < 3$, i.e. (a, b) is in the left-hand region of Ω. Thus, knowing the structure of the Jacobian that we require, i.e.

$$\boldsymbol{J} = \begin{pmatrix} 1 & a-3 \\ a-3 & -b^2-1 \end{pmatrix}, \tag{44}$$

we define generic kinetic forms, akin to Eq. (31), and determine the constants in these kinetics analogously to derivations in Sect. 2.2. A suitable set of kinetics is found to be

$$\frac{\partial u}{\partial t} = \nabla^2 u - u + (4-a)u^2 + (a-3)u^2 v, \tag{45}$$

$$\frac{\partial v}{\partial t} = D\nabla^2 v + b^2 + 1 + (3-a)uv + (a-4-b^2)v. \tag{46}$$

It should be noted here that system (45)–(46) highlights a critical point made earlier that the presented theory guarantees that our patterning requirements are met within the defined parameter region. Outside of this region, we have not provided any constraints. Indeed, in the case of $a = 3$ (45) suffers from finite time blow up. However, $a = 3$ is outside of the parameter region of interest of $(a-3)^2 - b^2 \geq 1$. There are many ways of fixing such problems, e.g. by replacing the term $(4-\alpha)u^2$ with $(4-\alpha)u^2/(1+u^2\epsilon^2)$ for sufficiently small ϵ. However, in this paper we focus on the dynamics within Ω, where $S(\boldsymbol{p}) > 0$ and leave further subtleties to future work.

Simulations of Eqs. (45) and (46) in the two different regions of Ω can be seen in Fig. 4c, d. As desired, we observe that, in the both one- and two-dimensional simulations, the constructed equations do, indeed, provide in phase Turing patterns in the left-hand region of Ω and out of phase patterns in the right-hand region of Ω (see Fig. 4b).

3.3 Spots to stripes

Thus far, we have demonstrated how to create a set of kinetics that can be tailored to suit any parameter region in either case of wanting the populations in phase or out of phase. Although a greater complexity of pattern is possible in spatial dimensions higher than one, all two-dimensional simulations so far have only demonstrated spot patterns. Hence, in this section we demonstrate that the choice of pattern can similarly be specified as required in the two-dimensional case. Namely, the pattern kinetics can be tuned using a new parameter to provide either spots or labyrinthine patterns depending on the desired application.

Achieving such specificity is an application of a previously proven result, where Ermentrout (1991) used weakly nonlinear perturbation theory to demonstrate that the choice of pattern in a two-dimensional Turing system depends on the nonlinear competition between quadratic and cubic terms in the kinetics. Specifically, if the cubic term dominates, then stripe/labyrinthine patterns are more likely, whereas spots appear in the case that the quadratic term dominates. Critically, the result demonstrated by Ermentrout (1991) assumes a square-based pattern template of spots. This provides the further result that spots and stripes are never stable within the same parameter region. More recent literature (Bozzini et al. 2015; Ma et al. 2019) considers a wider range of stripe and spot arrangements (e.g. rhombi, mixed modes, super-squares, hexagonal, etc.) and demonstrates that although multiple pattern types can be stable at the same time, their existence and stability parameter regions still depend on the interactions of the cubic and quadratic terms. However, it should be noted that domain shape, curvature, and boundary conditions can also influence this bifurcation (Krause et al. 2019, 2020a). Away from boundaries on sufficiently large domains (compared to the pattern wavelength), we anticipate that the results of Ermentrout (1991) hold, as these were derived for periodic square domains and our results demonstrate the anticipated patterning control.

The analysis of the most often observed case, hexagonally arranged spots, is complicated by the fact that its wave mode template produces resonant secular terms in the weakly nonlinear expansion at quadratic order that cannot be completely removed. Rather, a Fredholm solvability criterion has to be applied and assumptions about quadratic terms remaining small are required. Thus, we must remember that all of the reported results only hold in the weakly nonlinear regime and should be taken only as approximate results formally valid only near the boundaries of the Turing regime (where pattern amplitudes are small).

To investigate the influence of cubic terms, we need to extend the basic system (32)–(33). As mentioned in Sect. 2.2, we are able to add any term of the form $(u-1)^\beta(v-1)^\gamma$ with $\beta > 1, \gamma \geq 0$ or $\gamma > 1, \beta \geq 0$ to Eqs. (32) and (33) and the linear stability analysis stays the same. This suggests that subtracting a term of the form $C(u-1)^3$ from any of the kinetic forms derived here should lead to stripe patterns as C is increased, whilst not influencing the desired patterning parameter region. Note that we subtract the cubic term to maintain positivity near $(u = 0, v = 0)$ and to ensure the cubic term does not risk finite time blow up. Thus, we consider the following equations

$$\frac{\partial u}{\partial t} = \nabla^2 u - u + u^2 v - C(u-1)^3, \tag{47}$$

$$\frac{\partial v}{\partial t} = D\nabla^2 v + G' + 1 + (F' - G')v - (F' + 1)uv. \tag{48}$$

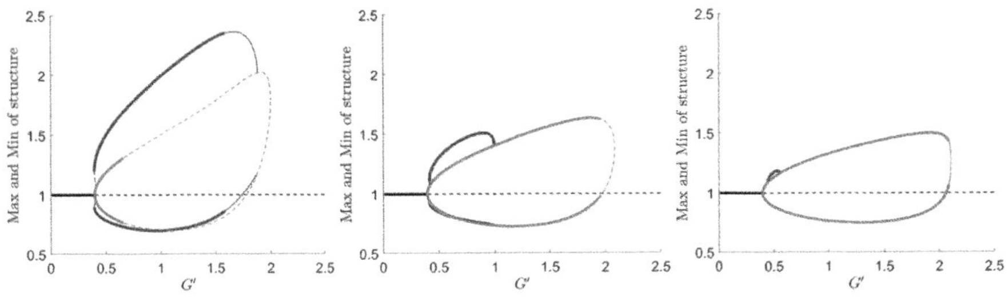

(a) $C = 0$ (b) $C = 0.2$ (c) $C = 0.5$

Fig. 5 Bifurcation diagram of the steady states of Eqs. (47) and (48) on a square domain of side length 10 presenting the maximum and minimum of the patterns in the u population. The lines indicate the parameter dependence of the homogeneous solution (black), stripe solution (red), and hexagonal spot solution (blue). The thick solid lines represent the Max and Min values of stable solutions, whilst the thin dashed lines represent the Max and Min values of unstable solutions, and the graphs illustrate that stripe solutions dominate at larger C. Parameters: $D = 9$, $F' = 2$ and C are shown below each figure

As discussed above, the cubic term does not influence the linear analysis. Thus, patterns are able to form whenever $F' > G'$, assuming D and the domain size, $[0, L] \times [0, L]$, are large enough.

Having subjected Eqs. (47) and (48) to a weakly nonlinear expansion technique (not shown), it appears that the dynamics we are interested in (namely the ability of the C parameter to control pattern type) is outside the scope of the weakly nonlinear regime. Specifically, although for small C we are able to track the start of bifurcation from the stable homogeneous state to stable heterogeneous patterns, the point at which the patterns cease to be stable is defined at a higher order. Further, as we will see in the upcoming simulations, as C increases the spot pattern begins to bifurcate from the stripe solution, rather than from homogeneity, a result that is completely outside weakly nonlinear analysis. Thus, to understand the bifurcation properties of system (47)–(48) in more detail, we turn to simulation. Namely, we use the MATLAB numerical continuation software package `pde2path`, which is designed to derive PDE bifurcation structures (Uecker et al. 2014; Dohnal et al. 2014; Engelnkemper et al. 2019).

Figure 5 illustrates that the numerical continuation matches the expected results from the literature (Ermentrout 1991; Bozzini et al. 2015; Ma et al. 2019), namely that as the cubic term increases, the system tends to present labyrinthine patterns. Specifically, in Fig. 5a, we observe that the spot solution branch bifurcates subcritically and as C increases, the region of stable spot pattern existence shrinks. Further, although stripe patterns are available when $C = 0$, we see that their region of stability increases dramatically as C increases (compare the red curves in Fig. 5a, b). Finally, by the time $C = 0.5$ the spot solution barely exists (see Fig. 5c).

Using these ideas, we now extend the non-contiguous parameter region kinetics, Eqs. (45) and (46), to possess either stripe or spot patterns depending on parameter choices. Taking the system from Sect. 3.2, we subtract a cubic term,

$$\frac{\partial u}{\partial t} = \nabla^2 u - u + (4 - a)u^2 + (a - 3)u^2 v - C(u - 1)^3, \tag{49}$$

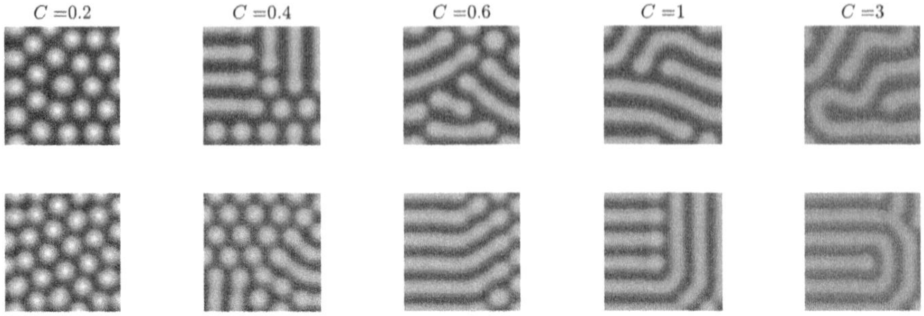

Fig. 6 The action of increasing the influence of a cubic term in Eqs. (45) and (46). Each two-dimensional simulation illustrates u on a square domain of side length 50. The colour axes range from blue at the lowest to yellow at the highest, and the top row is plotted on a consistent range of (0.2, 2.6), whilst the bottom row is plotted on a consistent range of (0.4, 2). The top row parameters are ($a = 1, b = 1$), and thus, v will be in phase. The bottom row parameters are ($a = 6, b = 1$), and thus, v will be out of phase

$$\frac{\partial v}{\partial t} = D\nabla^2 v + b^2 + 1 + (3 - a)uv + (a - 4 - b^2)v. \tag{50}$$

Figure 6 illustrates the influence of increasing C on the pattern structure for both pure (top row) and cross (bottom row) kinetics. As expected, we observe in both kinetic types that as the value of C increases left to right, the pattern transitions from spots to labyrinthine structures. We also note the amplitude of the pattern is suppressed with increasing C due to the cubic term having an inhibitory role.

4 Applications Beyond Homogeneous Systems

As a final example, we extend the model given by Eqs. (49) and (50) to have spatial and temporal heterogeneity, to obtain an enormous range of patterns. Using such heterogeneities, one can in principle design quite elaborate patterning fields that can change across space and time. It should be noted that including such spatiotemporal heterogeneities is not simply a matter of mathematical interest, as many biological systems demonstrate such complexity (Maini and Woolley 2019). For example, Malayan tapirs are born with dappled white stripes and spots on a brown coat. Critically, these patterns do not follow Turing's simple theory relating spatial size and pattern complexity, namely, thinner/ smaller domains should have simpler patterns than larger domains, whereas tapirs have stripes on their bodies breaking up to spots on its legs. Moreover, these patterns disappear during the first few months after birth and transition into a large-scale striped pattern of dark head and front legs, light body, and dark hind legs (Othmer et al. 2009) (see Fig. 7). Furthermore, transitions between solid spots and "rosette" patterns can be observed in leopards and jaguars as they mature (Liu et al. 2006; Werdelin and Olsson 1997). We do not claim that these evolving patterns are necessarily due to reaction and diffusion, only that there is an interest in extending such pattern forming systems to include spatial and temporal complexity.

There are many technical difficulties in generalising the linear theory explored in earlier sections, though some progress has been made; see, for instance, Krause et al.

(a) (b)

Fig. 7 Pattern evolution on a Malayan Tapir going from **a** broken stripes and spots on the baby to **b** large banded colouring on the adult. Any patterning theory for such an animal would have to account for this spatiotemporal complexity

(2020b), Kozák et al. (2019) for spatial heterogeneity and Klika and Gaffney (2017), Madzvamuse et al. (2010), Van Gorder (2020) for temporal forcing. We also remark that a large number of papers have explored these concepts, especially in terms of precursor patterns or environmental modulation as cited within the preceding references. Broadly summarising this earlier published work, whilst such heterogeneity can induce novel effects, if it is either sharply separated, or accompanied by sufficiently small diffusion, then the stability and instability criteria of a linear analysis hold locally, with only minor generalisation in the analogous result for slow temporal forcing (Madzvamuse et al. 2010). Thus, we anticipate the validity of linear stability and instability criteria applied locally for the stripe and spot transitions, considered here.

We focus on demonstrating our results through simulation, leaving a detailed mathematical justification for future work. Whilst our intuition for the "localization" of patterning above can be justified near a steady state, we will show that the predictions of the linear theory, developed in the papers cited above, appear to hold even when transitioning between patterned states which are far from equilibrium. Justifying such evolving states is beyond the scope of what we aim to do here, which is merely to demonstrate the ability of a modeller to design sophisticated patterns in space and time.

To proceed, we replace C and a with functions of space and time, which we will specify for our examples. We also assume that the linear and weakly nonlinear results of the previous sections hold, at least for well-chosen functions, and give two examples where spatial and temporal heterogeneity interact with the patterning process in interesting ways. In both examples, we will consider a domain of size 120×60 where half of the domain is in a stripe parameter regime, and the other half in a spot regime. We first demonstrate switching between these in time by changing the local parameters in each to oscillate between stripe and spot parameters, all by a simple choice of the function C. In the second example, we leave both spatial regions fixed, but instead exploit the parameter a to globally change between pure and cross-kinetic systems, so that u and v change phase.

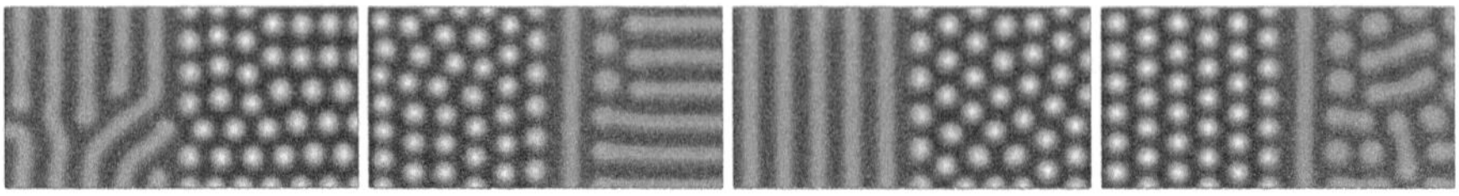

(a) $t = 4400$ **(b)** $t = 7400$ **(c)** $t = 10400$ **(d)** $t = 13400$

Fig. 8 Simulations of Eqs. (49) and (50) with Eq. (51) on a domain of size 120×60 with $a = 1$, $b = 1$, and $D = 50$ at different points in time, specified under each figure. The plots show simulations of u. The v pattern (not shown) is in phase and is, thus, qualitatively similar to u. The times correspond to 100 time units before the next transition. u ranges from 0.08 to 3.3

For the first example, we set $a = 1$ and

$$C = 1 - H\left(\cos\left(\frac{2\pi t}{6000}\right)\right) + \left(2H\left(\cos\left(\frac{2\pi t}{6000}\right)\right) - 1\right)\left(\frac{1 + \tanh(x - 60)}{2}\right), \tag{51}$$

where H is a sharp Heaviside step function. Hence, C rapidly transitions between $C = 0$ and $C = 1$, both in time and across the domain, giving a spatiotemporal transition from spots to stripes as shown for constant C in Fig. 6. The spatial heterogeneity is sharp across the midpoint of the domain (though it is smoothed slightly to avoid numerical errors in the finite element discretization across this interface). The temporal switching changes the kinetic parameter at this interface every $t = 3000$ units of time, which is sufficient time for patterns to get close to an equilibrium state. As a technical aside, we note that we increased the number of triangular finite elements for these simulations to $\sim 50,000$ and restricted the maximum timestep of the solver to 10 in order to resolve the temporal switching.

We plot example simulations of this system in Fig. 8, demonstrating fully formed patterns across the domain, which oscillate in space and time between stripes and spots. We note that the "stripe" parameter regime does not always give labyrinthine structures as it sometimes degenerates into spots, but the maximal value of C can be increased such that this does not happen. (Multiplying the tanh in (51) by 5, for instance, is sufficient to do this, but at the cost of making the amplitude of the stripes lower and hence harder to see against higher-amplitude spots.) We note that whilst the linear analysis is not valid here, as we are really transitioning from a completely patterned stated to another with the temporal oscillation, the emergent bifurcation structures predicted by the linear and weakly nonlinear theories seem to be captured in these simulations.

We next give an example of changing the parameter a in order to oscillate between in and out of phase kinetics in time, again on a spatial domain, which itself is composed of stripe and spot regions. We take,

$$C = 1 + \tanh(x - 60), \quad a = 1 + 5H\left(\cos\left(\frac{2\pi t}{6000}\right)\right), \tag{52}$$

with H again being the Heaviside, and the same considerations as above regarding smoothed spatial steps and sharp temporal ones. Here, C transitions from 0 to 2 over the spatial domain, whilst a oscillates in time from 1 to 6, as in the static examples

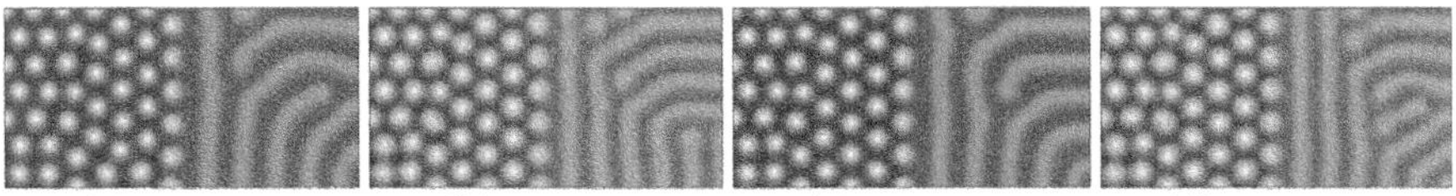

(a) $t = 4400$ (b) $t = 7400$ (c) $t = 10400$ (d) $t = 13400$

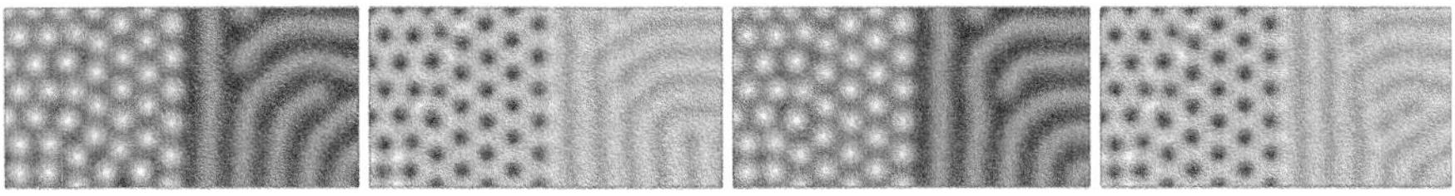

(e) $t = 4400$ (f) $t = 7400$ (g) $t = 10400$ (h) $t = 13400$

Fig. 9 Simulations of Eqs. (49) and (50) with Eq. (52) on a domain of size 120×60 with $b = 1$, and $D = 50$ at different points in time. The top row shows simulations of u, and the bottom row simulations of v at the given times, which correspond to 100 time units before the next transition. u (top) ranges from 0.08 to 3.3, and v (bottom) ranges from 0.9 to 1.3

given in Fig. 6. As before, we use the same slow period of oscillation so that the kinetics stay approximately constant in time for 3000 units of time.

Snapshots of simulating Eqs. (49) and (50) with parameters (52) are illustrated in Fig. 9, where we observe that even though both u and v are driven by temporally evolving kinetics, it is primarily v that toggles its phase. Further, the pattern structures remain broadly intact (even when carried out for several more oscillations), though we do see some changes in the striped half of the domain, as these structures are generally less stable to perturbation than spots. Simulations of only spot dynamics (not shown) confirm this as we can indefinitely oscillate between in and out of phase kinetics without modifying the number or organization of spot patterns, as long as sufficient time is given for the pattern to relax between such oscillations.

4.1 Patterns on Complex Prepatterns

Thus far, we have only demonstrated simple binary heterogeneities in space and time. In real biological systems, complex heterogeneity is more often the rule, with these simple ideal cases the exception (Weber et al. 2019; Warmflash et al. 2014; Woolley et al. 2014). If the patterning wavelength is sufficiently smaller than the length scale of any background or exogenous heterogeneity, then these observations about localised patterns can be shown more formally for sufficiently small diffusion, regardless of the diffusion ratio, D, and such scales often apply in development, for instance (Krause et al. 2020b). Hence, even extremely complicated underlying prepatterning systems can often lead to simple combinations of spot and stripe structures, and one can use this intuition to engineer hybrid patterns in a piecewise fashion.

As a first demonstration, we construct a system that presents distinct regions of patterning behaviour due to a complex heterogeneity. The heterogeneity is taken from a photograph of James Murray, author of (Murray 2003b, a), in Fig. 10a, and is reduced to a simpler segmented image (Fig. 10c) based on intensity in Fig. 10b. Step function heterogeneities are again used to vary C and a in simulations shown in Fig. 10d, which demonstrates that spots, labyrinthine, and inverse spot/labyrinthine patterns can all

Fig. 10 Imaging pipeline from initial photograph to final Turing pattern with a complex heterogeneity. **a** Initial photograph. **b** Photograph has been converted to greyscale. The photograph also has enhanced contrast to emphasise the different regions. **c** Segmented image using intensity as a segmentation parameter. This is used as a parameter prepattern for Eqs. (49) and (50). Namely, $C = 1$ and $a = 1$ within the black region, $C = 0$ and $a = 1$ in the white region and $C = 0.4$ and $a = 6$ in the grey region. In terms of the Turing pattern in the concentration of v in **d**, this provides regions of: spots in the black regions; stripes in the white region and inverse spots and stripes in the grey region. Other parameter values: $D = 50$, $a = 1$ and the patterning domain size is equal to the pixels in the photograph, namely $(0, 558) \times (0, 734)$. The colour axis goes from 0.8 (light) to 1.5 (dark)

exist together in the same simulation, with boundaries between regions conforming to the underlying heterogeneity.

Finally, we consider a system where we want to preserve a prepattern in some parts of the domain, but undergo a specified Turing-type patterning in other parts of the domain. To proceed, we reconsider the kinetics of Eq. (31), but rather than enforcing the steady state to be $(u, v) = (1, 1)$ we consider a steady state at $(u, v) = (U, V)$. We therefore consider the system,

$$\frac{\partial u}{\partial t} = \nabla^2 u + a_1 + b_1 u^2 + c_1 uv - C(u - U)^3, \tag{53}$$

$$\frac{\partial v}{\partial t} = D\nabla^2 v + a_2 + b_2 v + c_2 uv. \tag{54}$$

Following the same procedure as before (but using conditions around an arbitrary steady state) and setting $F' = S + 1$ and $G' = 1$, we find

$$a_1 = V - U, \quad a_2 = 2V, \quad b_1 = \frac{U - 2V}{U}, \quad b_2 = \frac{SU + 2U - 2V}{V},$$
$$c_1 = U^{-2}, \quad c_2 = -\frac{S + 2}{V}. \tag{55}$$

Generalising this result so that U, or V, depends on the spatial variables entails a solution (U, V) that is no longer a steady solution to the reaction–diffusion system (as the Laplacian term will not vanish), but can instead be thought of as a steady state to the equations in the absence of transport. If the diffusion is sufficiently small, or equivalently the domain is sufficiently large compared to gradients in heterogeneity, then this can be justified asymptotically (Krause et al. 2020b). Asymptotically such a

(a) (b)

Fig. 11 **a** Modified version of Fig. 10**a** to distinguish three regions. The outer background is white, an inner background boundary is light grey, and a central complex inner image is shades of dark grey. **b** Values of u from simulations of Eqs. (53)–(55). We take $D = 20$ and use a rectangular domain of size 400 by 503. We take a heterogeneous steady state as $(U, V) = (U(x), 1)$ and vary the first component in space based on the inverse intensity of the image in (**a**), scaled from 1 in the white region to 11 in the dark region. We take $S = 1$ in the outer white and grey regions, and $S = -1$ in the inner region, and $C = 0$ in the white region, $C = 2$ in the grey region, and $C = 20$ in the inner region. In the innermost region, a heterogeneous steady state value is stable to perturbations, and hence, the concentration of u can essentially read out the complex spatial heterogeneity

heterogeneous steady state can satisfy only heterogeneous kinetics in the absence of diffusion, but can be driven locally unstable to a more patterned state.

We introduce the heterogeneity similarly to Fig. 10 where distinct regions will have different patterning features, but we want to preserve complex details in the central region of the original image. For this reason, we darken the central region, as shown in Fig. 11a, so that the lightest portion of the central region is darker than the outer part of the image. We also introduce a second grey area around this so that there are three distinct regions: a white outer border, a grey inner border, and darker detailed centre region. In this centre area, we want to stabilise a heterogeneous steady state and so also make S depend on space so that the system is only Turing unstable in the outer regions. We will also increase the value of C in this region to ensure stripes and labyrinthine patterns in this region. See Fig. 11a for the image used as a heterogeneity, and Fig. 11b for results from our simulation.

By design, Fig. 11b thus shows a transition from spots in the outermost region, to a labyrinthine boundary surrounding a detailed pattern matching the original heterogeneity. Given the size of the domain, and the value of C used, the solution in these regions closely mirrors the input pattern, effectively preserving all of the spatial structure inherent in this region. Both modulation of Turing patterns and a morphogen conforming to a complex prepattern directly have been implicated in reconciling Turing-type patterning with other theories of spatial patterning such as positional information (Green and Sharpe 2015). This example shows that matching such predictions can be implemented in a straightforward and systematic manner by suitably defining kinetics without any parameter fitting.

5 Discussion

We have presented a relatively straightforward way to construct a rudimentary reaction–diffusion system, which can be designed to have specific simple output patterns, within arbitrarily elaborate parameter regimes. Specifically, we showed how this system can be used to create unbounded, or non-contiguous parameter regimes, as well as how the phase and two-dimensional pattern type can be specified within each region. Such an approach is useful pedagogically and might be employed to design Turing-type systems for use in both applied mathematical research, and plausibly industrial applications, as described in the introduction. Beyond this, our results show reaction–diffusion systems may be designed to match complicated parameter dependencies and simple patterning properties. This suggests that consistency with the patterning of a reaction–diffusion model with freedom in the reaction kinetics may be a relatively weak test of the hypothesis of a Turing patterning mechanism. For instance, when reaction–diffusion systems are evoked to understand heterogeneous biological patterns, the exact kinetic pathways behind a physical application are unknown, since the nature of the morphogens is themselves unspecified. Two populations are presumed to interact through an activator–inhibitor feedback, and the experimental outcome being matched is typically some form of heterogeneous signal, usually taking the form of high and low concentrations as a one-dimensional signal, or as spot, or stripe patterns in two dimensions (Woolley et al. 2014; Cho et al. 2011). Due to this lack of mechanistic detail, there is freedom to suggest any form of kinetics that have the correct interaction form (pure, or cross, kinetics) to produce the final heterogeneous solutions that matches the desired pattern. Usually, these kinetics are taken from a small pool of well understood examples (Gierer and Meinhardt 1972; Schnakenberg 1979; Barrio et al. 1999).

Here, we have simplified the standard Turing conditions down to an inequality on two variables (inequality (17)), assuming that the diffusion ratio and spatial domain can be taken as large as required. This simplification allows us to define a set of cross and pure kinetics that will present a Turing instability in any desired parameter region. The kinetic forms are particularly simple in the case that the parameter region can be specified by level set curves. Finally, we demonstrated that any two-dimensional reaction–diffusion system can be further tailored to produce either a spot or labyrinthine pattern in two dimensions, as required, with mixed patterns also possible when sharp spatial heterogeneities are included in the parameter values.

This suggests that if we are unable to constrain our models, through domain and diffusion scales or kinetic terms, then the reaction–diffusion framework alone is able to produce essentially any simple desired qualitative pattern and is thus automatically consistent with any simple stripe or spot pattern and even patterns with sharp transitions between spots and stripes. Moreover, recent work into larger interaction networks has demonstrated that the Turing instability is even less constrained when more populations are considered. Work by Marcon et al. (2016) shows that there are Jacobian sign structures (dictating local interaction types as activatory, or inhibitory) that are able to pattern without constraining the diffusion ratio parameter, though with the inclusion of immobile species. Further, their analysis demonstrates that Jacobian sign structures specify all the different in and out of phase relationships between the morphogens.

Thus, unless detailed knowledge of all species' interactions is known, there is once again flexibility in the choice of potential kinetics.

Notably, even though the assumptions of large diffusion ratios and large domains are generally necessary, there are many ways these assumptions can be weakened, such as by using substrate binding to slow one of the populations down, creating the large diffusion ratio required for a two species model (Klika et al. 2012). In fact, such mechanisms were exploited in the well-known Lengyel–Epstein system (Lengyel and Epstein 1991), which modelled chemical examples of Turing instabilities. Because of the enormous complexity of biological systems, difficulties with these assumptions are often neglected or presented with a brief mention of including more species (Dougoud et al. 2019; Liu et al. 2006; Cho et al. 2011). Additionally, if theoreticians are able to introduce prepatterns into their simulations then (as seen in Sects. 4 and 4.1), there is scope to combine simpler patterns to generate almost arbitrarily complicated patterns.

Furthermore, modelling certain experimental perturbations is also feasible. Usually, experimental data are focused on generating pattern transitions. Namely, a chemical factor is added into the system, which changes the regulation of an interaction (e.g. upregulation of the influence of the activator on the inhibitor), generating an observable pattern transition between spots to stripes, or, vice versa. Such outcomes can be encompassed in the presented framework. Namely, if we want a pattern to transition between spots and stripes, we can introduce a feedback between the experiment and a cubic term like that seen in Sect. 3.3.

Currently, we have not presented a means of guaranteeing that other mechanisms of patterning will not be feasible outside of Ω, which is equivalent to $S(\boldsymbol{p}) > 0$. Thus, we cannot strictly claim a method of eradicating self-organisation external to Ω using the kinetic parameters. This will be the subject of future work. However, even without the use of kinetic parameters, if there is freedom to assume experimental perturbations influence diffusion rates, then shrinking the ratio, D, will cause patterns to be eradicated, even with parameters such that $S(\boldsymbol{p}) > 0$.

Nonetheless, such subtleties do not ameliorate the overarching observations that, with free reign, it is feasible to construct kinetics that generate any simple homogeneous, or sharply stratified Turing pattern within any specified region of parameter space. Hence, we conclude that critical consideration is required when reaction–diffusion models are compared with experimental data; especially with regard to the extent, the reaction kinetics have been constrained. In particular, the interpretation of such comparisons is only as informative as the constraints that are placed upon the model, noting that mechanism-based restrictions on the kinetics will reduce such freedoms.

Although the universality of these Turing systems potentially limits the insights that we are able to generate with minimal biological constraints, they are nonetheless useful (Gelfert 2018). For example, recent work with experimentalists on generating patterns with decreasing wavelengths has shown that homogeneous Turing systems are unable to generate sufficient wavelength modulation across a spatial domain. Thus, extra gradients in the kinetics are needed, which then lead to the investigation of whether such gradients exist in the experimental system, as well as an ensuing theory of Turing patterns in heterogeneous domains (Krause et al. 2018b, 2020b).

Equally, experimentalists are constantly working within the freedom of Turing's theory to try and further constrain possible mechanisms. Recent work by Economou et al. (2020) investigated the biochemical interactions of proteins FGF, Hh, Wnt, and BMP and their ability to generate transverse ridges in the mouths of mice. Critically, by identifying activation patterns and the response of these patterns to multiple inhibitory perturbations they were able to substantially constrain the number of possible kinetics to 154 potential interaction networks. Thus, although we are not quite at the stage where we can uniquely define a set of biological interactions for a given patterning mechanism, we are optimistic that future collaborations could achieve this goal.

Finally, the fundamental idea behind the Turing instability is extremely valuable across the spectrum of experimental, modelling, and theoretical studies. Namely, the Turing instability demonstrates that a system can be more than the sum of its components. A diffusive system is thought to generally dissipate patterns and connecting this to a stable set of kinetics should, intuitively, not create sustained heterogeneity. Thus, philosophically, the Turing pattern demonstrates the need for mathematical modelling of biological ideas, where verbal models and intuition can only take us so far (Maini et al. 2016, 2012; Woolley et al. 2017).

In summary, the richness of pattern exhibited by reaction–diffusion equations and the frequent freedom in the choice of kinetics entails that many putative Turing systems are subject to severe difficulties of model selection and parameter identifiability. This is highlighted here by an explicit and constructive demonstration of how an extremely diverse range of patterns, within any specified region of parameter space, can be reproduced using a non-unique constructive algorithm, as summarised in Fig. 1.

As such pattern matching given freedom in the choice of kinetics is a poor discriminator of mechanism on comparing theoretical and experimental studies. Further understanding the detailed mechanisms of biological self-organisation is required for testing Turing's hypothesis against experimental studies, and more generally elucidating its place within larger developmental schema, for example. From the experimental side, we would encourage more focus on spatial modulation of patterning features such as wavelength and explicit constraints on signalling networks (e.g. the nonlinear interactions between activator and inhibitor networks). Future, theoretical studies should also address limitations of Turing's theory, such as the tendency of a limited variation in the wavelength or determining fundamental principles valid beyond two-species models.

Acknowledgements This research emerged from ideas that developed during BBSRC grant BB/N006097/1. In particular, we would like to thank Dr Denis Headon (The Roslin Institute, University of Edinburgh) for motivating this paper by asking the question of whether a non-contiguous parameter domain was possible. TEW would also like to thank the Bremen-Cardiff Alliance fund and in particular Prof. Jens Rademacher (University of Bremen) for discussion on bifurcation simulations and theory. Finally, TEW would like to thank the playwright Alan Harris for taking an interest in this work and using it as a basis for his play "Gather Itself Up". The tapir photographs in Fig. 7 are from the wikimedia photograph library and are used under a creative commons license 3.0. The baby tapir photograph was taken by Sasha Kopf, and the adult tapir photograph was taken by Fred Hsu.

References

Adamer MF, Harrington HA, Gaffney EA, Woolley TE (2020) Coloured noise from stochastic inflows in reaction-diffusion systems. Bull Math Biol 82(4):44. https://doi.org/10.1007/s11538-020-00719-w ISSN 0092-8240

Anguelova M, Cedersund G, Johansson M, Franzen CJ, Wennberg B (2007) Conservation laws and unidentifiability of rate expressions in biochemical models. IET Syst Biol 1(4):230–237

Aragón JL, Barrio RA, Woolley TE, Baker RE, Maini PK (2012) Nonlinear effects on turing patterns: time oscillations and chaos. Phys Rev E 86(2):026201

Arcuri P, Murray JD (1986) Pattern sensitivity to boundary and initial conditions in reaction-diffusion models. J Math Biol 24(2):141–165

Ardizzone V, Lewandowski P, Luk M-H, Tse Y-C, Kwong N-H, Lücke A, Abbarchi M, Baudin E, Galopin E, Bloch J, Baudin E, Galopin E, Lemaitre A, Tsang CY, Chan KP, Leung PT, Roussignol PH, Binder R, Tignon J, Schumacher S (2013) Formation and control of turing patterns in a coherent quantum fluid. Sci Rep 3:3016

Asllani M, Busiello DM, Carletti T, Fanelli D, Planchon G (2014) Turing patterns in multiplex networks. Phys Rev E 90(4):042814

Barrio RA, Varea C, Aragón JL, Maini PK (1999) A two-dimensional numerical study of spatial pattern formation in interacting Turing systems. Bull Math Biol 61(3):483–505

Baurmann M, Ebenhöh W, Feudel U (2004) Turing instabilities and pattern formation in a benthic nutrient-microorganism system. Math Biosci Eng 1(1):111

Biancalani T, Fanelli D, Di Patti F (2010) Stochastic Turing patterns in the Brusselator model. Phys Rev E 81(4):046215. https://doi.org/10.1103/PhysRevE.81.046215

Boehm CR, Grant PK, Haseloff J (2018) Programmed hierarchical patterning of bacterial populations. Nat Commun 9:776. https://doi.org/10.1038/s41467-018-03069-3

Bozzini B, Gambino G, Lacitignola D, Lupo S, Sammartino M, Sgura I (2015) Weakly nonlinear analysis of turing patterns in a morphochemical model for metal growth. Comput Math Appl 70(8):1948–1969

Callahan TK, Knobloch E (1999) Pattern formation in three-dimensional reaction-diffusion systems. Phys D 132(3):339–362

Campillo-Funollet E, Venkataraman C, Madzvamuse A (2019) Bayesian parameter identification for turing systems on stationary and evolving domains. Bull Math Biol 81(1):81–104

Chembo YK, Gomila D, Tlidi M, Menyuk CR (2017) Theory and applications of the lugiato-lefever equation. Eur Phys J D 71:299

Cho SW, Kwak S, Woolley TE, Lee MJ, Kim EJ, Baker RE, Kim HJ, Shin JS, Tickle C, Maini PK, Jung HS (2011) Interactions between shh, sostdc1 and wnt signaling and a new feedback loop for spatial patterning of the teeth. Development 138:1807–1816 ISSN 0950-1991

Clermont G, Zenker S (2015) The inverse problem in mathematical biology. Math Biosci 260:11–15

Crampin EJ, Gaffney EA, Maini PK (1999) Reaction and diffusion on growing domains: scenarios for robust pattern formation. Bull Math Biol 61(6):1093–1120

Dewar MA, Kadirkamanathan V, Opper M, Sanguinetti G (2010) Parameter estimation and inference for stochastic reaction-diffusion systems: application to morphogenesis in d melanogaster. BMC Syst Biol 4(1):21

Diego X, Marcon L, Müller P, Sharpe J (2018) Key features of turing systems are determined purely by network topology. Phys Rev X 8(2):021071

Dillon R, Maini PK, Othmer HG (1994) Pattern formation in generalized Turing systems. J Math Biol 32(4):345–393

Dohnal T, Rademacher JDM, Uecker H, Wetzel D (2014) pde2path 2.0: multi-parameter continuation and periodic domains. In: Ecker H, Steindl A, Jakubek S (eds.), Proceedings of 8th European nonlinear dynamics conference

Dougoud M, Mazza C, Schwaller B, Pecze L (2019) Extending the mathematical palette for developmental pattern formation: Piebaldism. Bull Math Biol 81(5):1461–1478
Economou Andrew D, Monk Nicholas AM, Green Jeremy BA (2020) Perturbation analysis of a multi-morphogen turing reaction-diffusion stripe patterning system reveals key regulatory interactions. Development 147(20): ISSN 0950-1991. https://doi.org/10.1242/dev.190553. URL https://dev.biologists.org/content/147/20/dev190553
Economou AD, Green JBA (2014) Modelling from the experimental developmental biologists viewpoint. Seminars in cell and developmental biology, vol 35. Elsevier, Amsterdam, pp 58–65
Engelnkemper S, Gurevich SV, Uecker H, Wetzel D, Thiele U (2019) Continuation for thin film hydrodynamics and related scalar problems. Computational modelling of bifurcations and instabilities in fluid dynamics. Springer, New York, pp 459–501
Ermentrout B (1991) Stripes or spots? Nonlinear effects in bifurcation of reaction-diffusion equations on the square. Proc Math Phys Sci, 434(1891): 413–417, ISSN 09628444. URL http://www.jstor.org/stable/51838
Garvie MR, Maini PK, Trenchea C (2010) An efficient and robust numerical algorithm for estimating parameters in turing systems. J Comput Phys 229(19):7058–7071
Gelfert A (2018) Models in search of targets: exploratory modelling and the case of turing patterns. Philosophy of science. Springer, New York, pp 245–269
Gierer A, Meinhardt H (1972) A theory of biological pattern formation. Biol Cybern 12(1):30–39
Grant PK, Dalchau N, Brown JR, Federici F, Rudge TJ, Yordanov B, Patange O, Phillips A, Haseloff J (2016) Orthogonal intercellular signaling for programmed spatial behavior. Mol Syst Biol 12(1)
Green JBA, Sharpe J (2015) Positional information and reaction-diffusion: two big ideas in developmental biology combine. Development 142(7):1203–1211
Hata S, Nakao H, Mikhailov AS (2014) Dispersal-induced destabilization of metapopulations and oscillatory turing patterns in ecological networks. Sci Rep 4:3585
Karig D, Martini KM, Lu T, DeLateur NA, Goldenfeld N, Weiss R (2018) Stochastic turing patterns in a synthetic bacterial population. Proc Natl Acad Sci 115(26):6572–6577
Kashima K, Ogawa T, Sakurai T (2015) Selective pattern formation control: spatial spectrum consensus and turing instability approach. Automatica 56:25–35
Kazarnikov A, Haario H (2020) Statistical approach for parameter identification by turing patterns. J Theor Biol 110319
Klika V, Gaffney EA (2017) History dependence and the continuum approximation breakdown: the impact of domain growth on turing's instability. Proc R Soc A 473(2199):20160744
Klika V, Baker RE, Headon D, Gaffney EA (2012) The influence of receptor-mediated interactions on reaction-diffusion mechanisms of cellular self-organisation. B Math Biol 74(4):935–957
Klika V, Kozák M, Gaffney EA (2018) Domain size driven instability: self-organization in systems with advection. SIAM J Appl Math 78(5):2298–2322
Kondo S, Miura T (2010) Reaction-diffusion model as a framework for understanding biological pattern formation. Science 329(5999):1616–1620. https://doi.org/10.1126/science.1179047, URL http://www.sciencemag.org/cgi/content/abstract/329/5999/1616
Kozák M, Gaffney EA, Klika V (2019) Pattern formation in reaction-diffusion systems with piecewise kinetic modulation: an example study of heterogeneous kinetics. Phys Rev E 100(4):042220
Krämer S, Laflorencie N, Stern R, Horvatić M, Berthier C, Nakamura H, Kimura T, Mila F (2013) Spatially resolved magnetization in the bose-einstein condensed state of bacusi 2 o 6: evidence for imperfect frustration. Phys Rev B 87(18):180405
Krause AL, Klika V, Maini PK, Headon D, Gaffney EA (2020a) Isolating patterns in open reaction-diffusion systems. arXiv:2009.13114
Krause AL, Klika V, Woolley TE, Gaffney EA (2018b) Heterogeneity induces spatiotemporal oscillations in reaction-diffusion systems. Phys Rev E 97(5)
Krause AL, Burton AM, Fadai NT, Van Gorder RA (2018a) Emergent structures in reaction-advection-diffusion systems on a sphere. Phys Rev E 97:042215. https://doi.org/10.1103/PhysRevE.97.042215
Krause AL, Ellis MA, Van Gorder RA (2019) Influence of curvature, growth, and anisotropy on the evolution of turing patterns on growing manifolds. Bull Math Biol 81(3):759–799
Krause AL, Klika V, Woolley TE, Gaffney EA (2020b) From one pattern into another: analysis of turing patterns in heterogeneous domains via wkbj. J R Soc Interf 17(162):20190621
Lengyel I, Epstein IR (1991) Modeling of Turing structures in the chlorite-iodide-malonic acid-starch reaction system. Science 251(4994):650–652

Leppänen T, Karttunen M, Kaski K, Barrio RA, Zhang L (2002) A new dimension to turing patterns. Phys D 168:35–44
Li QS, Ji L (2004) Control of turing pattern formation by delayed feedback. Phys Rev E 69(4):046205
Liu RT, Liaw SS, Maini PK (2006) Two-stage turing model for generating pigment patterns on the leopard and the jaguar. Phys Rev E 74(1):011914
Ma M, Gao M, Carretero-González R (2019) Pattern formation for a two-dimensional reaction-diffusion model with chemotaxis. J Math Anal Appl 475(2):1883–1909
Maclaren OJ, Nicholson R (2019) What can be estimated? Identifiability, estimability, causal inference and ill-posed inverse problems. arXiv preprint arXiv:1904.02826
Madzvamuse A, Gaffney EA, Maini PK (2010) Stability analysis of non-autonomous reaction-diffusion systems: the effects of growing domains. J Math Biol 61(1):133–164
Maini PK, Woolley TE (2019) The turing model for biological pattern formation. Springer, New York, pp 189–204
Maini PK, Woolley TE, Baker RE, Gaffney EA, Lee SS (2012) Turing's model for biological pattern formation and the robustness problem. Interf Focus 2(4):487–496
Maini PK, Woolley TE, Gaffney EA, Baker RE (2016) The once and future turing, chapter 15: biological pattern formation. Cambridge University Press, Cambridge
Marcon L, Sharpe J (2012) Turing patterns in development: what about the horse part? Curr Opin Genet Dev 22(6):578–584
Marcon L, Diego X, Sharpe J, Müller P (2016) High-throughput mathematical analysis identifies turing networks for patterning with equally diffusing signals. eLife 5:e14022
McBride EF, Picard MD (2004) Origin of honeycombs and related weathering forms in oligocene macigno sandstone, Tuscan coast near Livorno, Italy. Earth Surf Process Landf 29(6):713–735
McGuirl MR, Volkening A, Sandstede B (2020) Topological data analysis of zebrafish patterns. Proc Nat Acad Sci 117(10):5113–5124
Murray JD (1982) Parameter space for Turing instability in reaction diffusion mechanisms: a comparison of models. J Theor Biol 98(1):143
Murray JD (2003a) Mathematical biology II: spatial models and biomedical applications, vol 2, 3rd edn. Springer, New York
Murray JD (2003b) Mathematical biology I: an introduction, vol 1, 3rd edn. Springer, New York
Nakao H, Mikhailov AS (2010) Turing patterns in network-organized activator-inhibitor systems. Nat Phys 6(7):544–550
Oppo G-L (2009) Formation and control of turing patterns and phase fronts in photonics and chemistry. J Math Chem 45(1):95
Oster GF (1988) Lateral inhibition models of developmental processes. Math Biosci 90(1–2):265–286
Othmer HG, Painter K, Umulis D, Xue C (2009) The intersection of theory and application in elucidating pattern formation in developmental biology. Math Model Nat Phenom 4(4):3
Pismen LM (1994) Turing patterns and solitary structures under global control. J Chem Phys 101(4):3135–3146
Sánchez-Garduno F, Krause AL, Castillo JA, Padilla P (2019) Turing-hopf patterns on growing domains: the torus and the sphere. J Theor Biol 481:136–150
Satnoianu RA, Menzinger M, Maini PK (2000) Turing instabilities in general systems. J Math Biol 41(6):493–512
Schnakenberg J (1979) Simple chemical reaction systems with limit cycle behaviour. J Theor Biol 81(3):389–400
Scholes NS, Schnoerr D, Isalan M, Stumpf MPH (2019) A comprehensive network atlas reveals that turing patterns are common but not robust. Cell Syst 9(3):243–257
Schumacher LJ, Woolley TE, Baker RE (2013) Noise-induced temporal dynamics in turing systems. Phys Rev E 87(4):042719
Seul M, Andelman D (1995) Domain shapes and patterns: the phenomenology of modulated phases. Science 267(5197):476–483
Sherratt JA (2012) Turing patterns in deserts. Conference on Computability in Europe. Springer, New York, pp 667–674
Smolin L (1996) Galactic disks as reaction-diffusion systems. arXiv preprint arXiv:astro-ph/9612033 [astro-ph]
Tan Z, Chen S, Peng X, Zhang L, Gao C (2018) Polyamide membranes with nanoscale turing structures for water purification. Science 360(6388):518–521

Taylor NP, Kim H, Krause AL, Van Gorder RA (2020) A non-local cross-diffusion model of population dynamics I: emergent spatial and spatiotemporal patterns. Bull Math Biol, In Press
Turing AM (1952) The chemical basis of morphogenesis. Phil Trans R Soc Lond B 237:37–72
Uecker H, Wetzel D, Rademacher JDM (2014) pde2path: a matlab package for continuation and bifurcation in 2D elliptic systems. Numer Math Theory Meth Appl 7(1):58–106
Van Gorder RA, Klika V, Krause AL Turing conditions for pattern forming systems on evolving manifolds. arXiv:1904.09683 [nlin.PS]
Van Gorder RA (2020) Turing and benjamin-feir instability mechanisms in non-autonomous systems. Proc R Soc A 476(2238):20200003
Van Gorder RA, Kim H, Krause AL (2019) Diffusive instabilities and spatial patterning from the coupling of reaction-diffusion processes with stokes flow in complex domains. J Fluid Mech 877:759–823
Vanag VK, Epstein IR (2001) Pattern formation in a tunable medium: the belousov-zhabotinsky reaction in an aerosol ot microemulsion. Phys Rev Lett 87(22):228301
Vastano JA, Pearson JE, Horsthemke W, Swinney HL (1987) Chemical pattern formation with equal diffusion coefficients. Phys Lett A 124(6–7):320–324
Wakano JY, Nowak MA, Hauert C (2009) Spatial dynamics of ecological public goods. Proc Nat Acad Sci 106(19):7910–7914
Warmflash A, Sorre B, Etoc F, Siggia ED, Brivanlou AH (2014) A method to recapitulate early embryonic spatial patterning in human embryonic stem cells. Nat Methods 11(8):847–854
Warne DJ, Baker RE, Simpson MJ (2019) Using experimental data and information criteria to guide model selection for reaction-diffusion problems in mathematical biology. Bull Math Biol 81(6):1760–1804
Weber EL, Woolley TE, Yeh C-Y, Ou K-L, Maini PK, Chuong C-M (2019) Self-organizing hair peg-like structures from dissociated skin progenitor cells: new insights for human hair follicle organoid engineering and turing patterning in an asymmetric morphogenetic field. Exp Dermatol 28:355–366. https://doi.org/10.1111/exd.13891 ISSN 0906-6705
Werdelin L, Olsson L (1997) How the leopard got its spots: a phylogenetic view of the evolution of felid coat patterns. Biol J Linn Soc 62(3):383–400
Woolley TE (2014) 50 visions of mathematics, chapter 48: mighty morphogenesis. Oxford Univ, Press, Oxford
Woolley TE (2017) Pattern production through a chiral chasing mechanism. Phys Rev E 96(3):032401
Woolley TE, Baker RE, Maini PK, Aragón JL, Barrio RA (2010) Analysis of stationary droplets in a generic Turing reaction-diffusion system. Phys Rev E 82(5):051929. https://doi.org/10.1103/PhysRevE.82.051929
Woolley TE, Baker RE, Gaffney EA, Maini PK (2011) Stochastic reaction and diffusion on growing domains: understanding the breakdown of robust pattern formation. Phys Rev E 84(4):046216. https://doi.org/10.1103/PhysRevE.84.046216
Woolley TE, Baker RE, Gaffney EA, Maini PK, Seirin-Lee S (2012) Effects of intrinsic stochasticity on delayed reaction-diffusion patterning systems. Phys Rev E 85(5):051914
Woolley TE, Baker RE, Tickle C, Maini PK, Towers M (2014) Mathematical modelling of digit specification by a sonic hedgehog gradient. Dev Dyn 243(2):290–298
Woolley TE, Baker RE, Maini PK (2017) The turing guide, chapter 35: turing's theory of morphogenesis. Oxford Univ, Press, Oxford

Bulletin of Mathematical Biology (2021) 83:31
https://doi.org/10.1007/s11538-020-00841-9

Society for Mathematical Biology

SPECIAL ISSUE: CELEBRATING J. D. MURRAY

Calcium Dynamics and Water Transport in Salivary Acinar Cells

James Sneyd[1] · Elias Vera-Sigüenza[1] · John Rugis[1] · Nathan Pages[1] · David I. Yule[2]

Received: 30 September 2020 / Accepted: 25 November 2020 / Published online: 17 February 2021

Abstract
Saliva is secreted from the acinar cells of the salivary glands, using mechanisms that are similar to other types of water-transporting epithelial cells. Using a combination of theoretical and experimental techniques, over the past 20 years we have continually developed and modified a quantitative model of saliva secretion, and how it is controlled by the dynamics of intracellular calcium. However, over approximately the past 5 years there have been significant developments both in our understanding of the underlying mechanisms and in the way these mechanisms should best be modelled. Here, we review the traditional understanding of how saliva is secreted, and describe how our work has suggested important modifications to this traditional view. We end with a brief description of the most recent data from living animals and discuss how this is now contributing to yet another iteration of model construction and experimental investigation.

Keywords Saliva secretion · Calcium oscillations · Water transport · Mathematical modelling · Three-dimensional finite element computations

1 Introduction

Secretion of saliva from the major salivary glands provides both fluid that hydrates and lubricates the oral cavity and proteins that begin to digest food. In addition, factors are also present in saliva that protect the oral cavity and upper gastrointestinal tract from bacterial and fungal assault. Malfunction of the salivary glands resulting in a

✉ James Sneyd
sneyd@math.auckland.ac.nz

1 Department of Mathematics, The University of Auckland, Level 2, Building 303, 38 Princes Street, Auckland, New Zealand

2 School of Medicine and Dentistry, University of Rochester Medical Center, 601 Elmwood Ave, Box 711, Rochester, NY, USA

reduction in fluid flow (a condition called *xerostomia*) results in difficulty swallowing and chewing food, a marked increase in dental carries and susceptibility to oral candidiasis, and leads to a severe deterioration in the quality of life. Xerostomia is associated with the auto-immune disease, Sjögren's syndrome, and is also commonly seen after radiotherapy for head and neck cancers (Ship et al. 1991; Melvin 1991; Fox and Maruyama 1997; Daniels and Wu 2000).

The most important early studies of saliva secretion were those of Lundberg (1956, 1957a, b) who was the first to propose that the principal underlying mechanism was the transport of Cl^- through the cell. A specific qualitative model was proposed by Silva et al. (1977); it is these studies that have provided the foundation for practically all subsequent work, including ours.

For almost 20 years, our group has used a combination of experimental and theoretical approaches to study the mechanisms of saliva secretion (Sneyd et al. 2003; Gin et al. 2007; Palk et al. 2010, 2012), and in 2014 we wrote a review of our major results (Sneyd et al. 2014). However, so much has changed since that time that another review is not only timely, but necessary. In particular, it is slowly becoming clearer that the traditional understanding of how saliva secretion works is not entirely accurate. The broad outlines remain the same, but the details have changed significantly.

The study of saliva secretion is a (fairly small) part of a much larger body of work on water transport by epithelial cells (Diamond and Bossert 1967; Swanson 1977; Weinstein and Stephenson 1979, 1981b, a; Reuss 2002; Dawson 1992; Mathias and Wang 2005; Chara and Brusch 2015). Water transport by epithelia is important in a number of body parts; the lung epithelium maintains a surface layer of water which is critical for mucociliary clearance (with sample models by Novotny and Jakobsson 1996b, a; Warren et al. 2009; Garcia et al. 2013; Sandefur et al. 2017; Wu et al. 2018), the gut epithelium absorbs water (sample models by Larsen et al. 2000; Larsen 2002), while water absorption is a crucial function of epithelial cells in the kidney (sample models by Weinstein 1994, 1999; Layton and Layton 2005a, b; Layton 2011a, b; Weinstein 2020). In a manner similar to salivary secretion, tear and sweat glands also secrete water. Each of these areas relies on underlying models of water transport, a field which remains controversial, with a number of competing theories that we shall not discuss here. The interested reader is referred to Hill (2008), an excellent and readable review of the pros and cons of five major theories of physiological water transport.

2 Saliva Secretion: The Traditional View

The traditional explanation of saliva secretion goes as follows (Foskett and Melvin 1989; Martinez 1990; Nauntofte 1992; Cook et al. 1994; Melvin 1999; Turner and Sugiya 2002; Melvin et al. 2005; Ambudkar 2012; Lee et al. 2012). In a salivary gland, there are two major types of cells; firstly, the acinar cells, which are arranged in groups like a bunch of grapes or a blackberry (hence the name, from the Latin *acinus*, meaning a berry) and, secondly, the duct cells. Primary saliva is produced by the acinar cells and is collected in the acinar lumen (analogous to the stem of a bunch of grapes) whence it flows through the ducts to the mouth, emerging as secondary saliva (usually just called saliva).

2.1 Production of Primary Saliva

A salivary gland acinar cell is a polarised epithelial cell; it has a basal surface which faces to the interstitium and an apical surface which faces to the acinar lumen. At rest, the entry of Na^+, Cl^- and K^+ through a $Na^+/K^+/2Cl^-$ cotransporter (NKCC1) in the basal membrane is balanced by the exit of Cl^- through the apical membrane into the lumen, the exit of K^+ across the basal membrane through a number of different types of K^+ channels, and the removal of Na^+ by the Na^+/K^+ ATPase (NaK) which is ubiquitous in all cells. This achieves internal ionic balance, control of cell volume and a very low flow of water through the cell, driven by slight differences in osmolarity; the water comes in across the basal membrane and exits the cell across the apical membrane into the lumen.

At rest, most cells have a cytoplasmic $[Cl^-]$ that minimises the electrochemical potential of Cl^- from inside to out. Indeed, many models of cell volume control—for example, the classic model of Tosteson and Hoffman (1960), see Keener and Sneyd (2008) for an introduction to models of cell volume control—incorporate this as a fundamental assumption. However, secretory epithelial cells such as salivary gland acinar cells accumulate Cl^- to much higher concentrations; they do this by the combined action of the basal NKCC1 cotransporter and the basal HCO_3^-/Cl^- exchangers (AE2 and AE4). The NKCC1 cotransporter flux is, in turn, maintained by the $[Na^+]$ gradient created by the NaK ATPase.

It follows that, at rest, a salivary acinar cell is primed for Cl^- release into the lumen and thus primed for water transport. However, flow at rest is kept small as the apical Cl^- channels are mostly closed, thus allowing for only a small Cl^- flux into the lumen.

Production of much greater quantities of primary saliva is initiated by extracellular hormones or neurotransmitter (such as acetylcholine or cholecystokinin) that bind to G-protein-coupled receptors on the basal membrane, stimulating the production of inositol trisphosphate and the subsequent release of Ca^{2+} from the endoplasmic reticulum (Berridge et al. 2000; Berridge 2009; Ambudkar 2012, 2014; Dupont et al. 2016). The increase in cytosolic $[Ca^{2+}]$ activates two ion channels in particular; in the apical membrane Ca^{2+}-activated Cl^- channels (ClCa) release Cl^- into the lumen and depolarise the apical membrane. At the same time, Ca^{2+}-activated K^+ channels (KCa) in the basal membrane repolarise the membrane, allowing for a continued flow of Cl^- from the cell into the lumen. Water follows by osmosis, while charge balance in the lumen is maintained by a paracellular Na^+ current through the tight junctions into the lumen. The primary saliva that collects in the lumen is thus high in Na^+ and Cl^-.

In summary, the underlying driving force for saliva secretion is the transport of Cl^- from the interstitium, through the acinar cell, into the lumen. This sets up small osmotic gradients down which water flows from the interstitium, into the cell, and thence into the lumen. This is illustrated in Fig. 1.

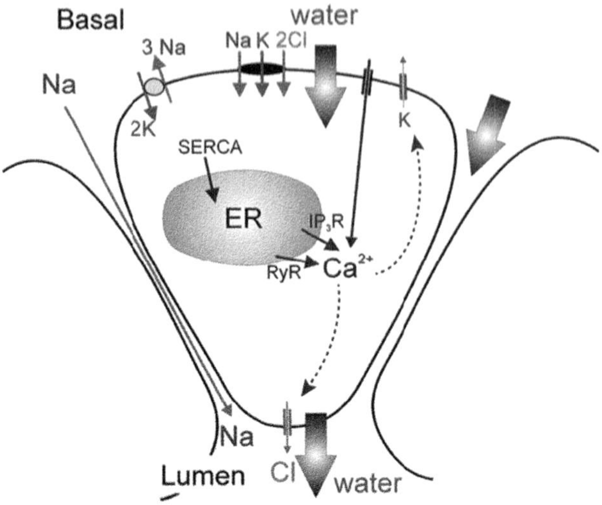

Fig. 1 Traditional view of saliva production (Melvin et al. 2005; Lee et al. 2012). At rest, uptake of Cl^- by a NKCC1 cotransporter maintains a high intracellular $[Cl^-]$. Agonist stimulation results in the release of Ca^{2+} from the endoplasmic reticulum (ER) through inositol trisphosphate receptors (IP_3R) and (possibly) ryanodine receptors (RyR). Calcium is resequestered into the ER by ATPase Ca^{2+} pumps (SERCA). Ca^{2+} activates Cl^- channels on the apical membrane and K^+ channels on the basal membrane (with possibly some on the apical membrane also). This results in the flow of Cl^- into the lumen, whereupon water follows by osmosis. Hence, in summary, Cl^- transport across the cell provides an osmotic gradient down which water flows, from the interstitium, through the cell, into the lumen. Charge balance is maintained by a paracellular Na^+ current

2.2 The Role of Ducts

As the primary saliva flows through the ducts, its ionic composition is changed by the duct cells that line the duct. Primary saliva has a high $[Na^+]$ and $[Cl^-]$ but a low $[K^+]$ and $[HCO_3^-]$ content. The duct cells replace the Na^+ and Cl^- with K^+ and HCO_3^-, so that secondary saliva has a high $[K^+]$ and $[HCO_3^-]$ but low $[Na^+]$ and $[Cl^-]$. However, the duct cells are impermeable to water.

Ionic balance is not at steady state by the time the saliva emerges into the mouth. We know this because the ionic composition of the secondary saliva depends on the rate of saliva secretion, with higher rates of secretion causing a decrease in $[K^+]$, and an increase in $[Na^+]$ in the secondary saliva (Mangos et al. 1973a, b).

2.3 Calcium Dynamics

Because saliva secretion is initiated by a rise in cytosolic $[Ca^{2+}]$, it is important to understand in detail the mechanisms that control $[Ca^{2+}]$, mechanisms that are common to almost all cell types. At steady state, the cytosolic $[Ca^{2+}]$ is low; Ca^{2+} influx into the cytoplasm is tightly controlled, while ATPase Ca^{2+} pumps (SERCA pumps) actively pump Ca^{2+} from the cytoplasm into the endoplasmic reticulum (ER), resulting in a resting $[Ca^{2+}]$ of around 50 nM, but a much higher $[Ca^{2+}]$ in the ER. Outside the

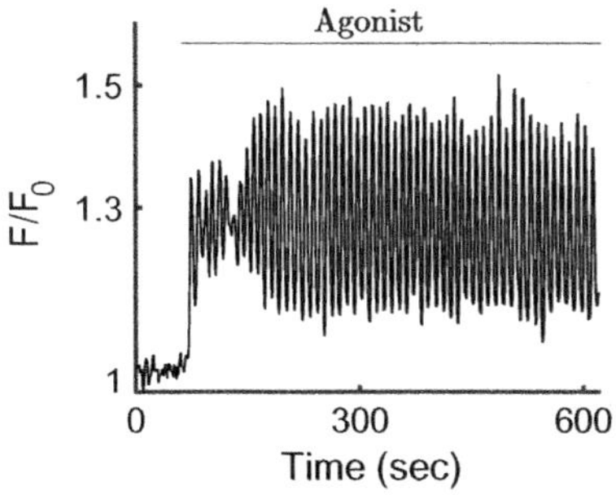

Fig. 2 Time series of the mean fluorescence of the Ca^{2+} indicator Fluo-4, averaged over a whole acinar cell in an intact mouse parotid gland (Pages et al. 2019). These oscillations were stimulated by the agonist carbachol, which binds to acetylcholine receptors. The fluorescence is closely related to $[Ca^{2+}]$, although the relationship is not linear. It is a good working assumption that the fluorescence is an analogue of $[Ca^{2+}]$, with oscillations with the same qualitative properties

cell, $[Ca^{2+}]$ is of the order of a few mM. In addition, the cytoplasm has many Ca^{2+}-binding proteins, or Ca^{2+} buffers, which contribute to the maintenance of a low resting cytosolic $[Ca^{2+}]$.

Hence, the cell cytoplasm is under enormous Ca^{2+} pressure. If Ca^{2+} channels in the cell membrane or the ER are opened, then Ca^{2+} will flow into the cytoplasm, quickly raising the cytosolic $[Ca^{2+}]$. Once these channels shut again, Ca^{2+} is removed from the cytoplasm by ATPase pumps, whereupon the influx channels reopen, leading to a second rise in $[Ca^{2+}]$. Repetition of this process results in repetitive spikes of increased $[Ca^{2+}]$, often called Ca^{2+} oscillations.

One particularly important concept for understanding (and modelling) Ca^{2+} oscillations is the idea of the *calcium toolbox*. There are many types of Ca^{2+} currents and fluxes that can let Ca^{2+} into the cytoplasm or remove it; these are the components of the toolbox. Membrane Ca^{2+} channels allow Ca^{2+} to flow from the interstitium into the cytoplasm, while ER Ca^{2+} channels (such as the inositol trisphosphate receptor, IPR, or the ryanodine receptor, RyR) allow Ca^{2+} to flow from the ER into the cytoplasm. Similarly, there are Ca^{2+} channels in other organelles such as the mitochondria and lysosomes. Ca^{2+} is removed from the cytoplasm by a range of different mechanisms, principally ATPases in the ER membrane and the cell membrane.

Each of the toolbox components has its own model, many of which can be found in Dupont et al. (2016). By using various combinations of these models, it is possible to construct a wide range of different models of Ca^{2+} oscillations, each adapted to a particular cell type.

The toolbox components also have well-defined spatial distributions, which are different for every cell type. By including these spatial distributions, it is possible to construct spatially distributed models tailored for different cell types. In this way, the study of Ca^{2+} dynamics combines limited variety with almost infinite flexibility. There are only a relatively small number of toolbox components, but they may be combined in essentially an infinite number of ways, leading to an endless variety of outcome.

Typical Ca^{2+} oscillations in an acinar cell from an intact mouse parotid gland are shown in Fig. 2. The overall shape is characteristic and seen in all responding cells in the gland, although the frequencies and amplitudes show considerable variety. The oscillations have a period of around 9 s, and are superimposed on a raised plateau.

2.4 Earlier Models

In our earlier work, we combined the above mechanisms into a unified spatially homogeneous model of saliva secretion, which was presented in detail in our previous review (Sneyd et al. 2014).

1. Agonist stimulation leads to the production of IP_3 which binds to IP_3 receptors (IPR) on the ER membrane, releasing Ca^{2+} from the ER into the cytosol. The IPR are restricted to the apical region (Kasai et al. 1993; Thorn et al. 1993; Thorn 1996; Nathanson et al. 1994; Lee et al. 1997), but IP_3 diffuses quickly through the cell. Hence, the IP_3 (which is produced at the basal membrane where the agonist receptors are located) is able quickly to activate the IPR at the other end of the cell.
2. Modulation of the production and degradation of IP_3 by Ca^{2+} leads to a feedback loop that results in an oscillation of cytosolic $[Ca^{2+}]$.
3. These Ca^{2+} oscillations activate Cl^- channels in the apical membrane and K^+ channels in the basal membrane. Cl^- flows into the lumen, while cytosolic Cl^- is maintained by the influx of Cl^- through the NKCC1 cotransporter in the basal membrane.
4. The osmotic gradient set up by these Cl^- fluxes drive the flow of water through the cell, from the interstitium to the lumen.

3 Recent Developments in Modelling Salivary Acinar Cells

Not surprisingly, the model as described above has had to be modified extensively as new experimental data have appeared, often in direct response to questions posed by the modelling.

3.1 Calcium Dynamics

One important parameter in the model is the exact spatial distribution of the IPR. It has been known for some years that the IPR are situated in the apical region of the acinar cell (Kasai et al. 1993; Thorn et al. 1993; Thorn 1996; Nathanson et al. 1994; Lee et al. 1997), but limited resolution in the data meant that it was only possible to say that the IPR were within approximately 200 nm of the apical ClCa channels. This supported the construction of models in which Ca^{2+} was able to modulate the rate of production and/or degradation of IP_3, leading to positive or negative feedback that could result in oscillations. Such models are called Class II models (Dupont et al. 2016); we had already successfully tested predictions from a Class II model in a pancreatic acinar

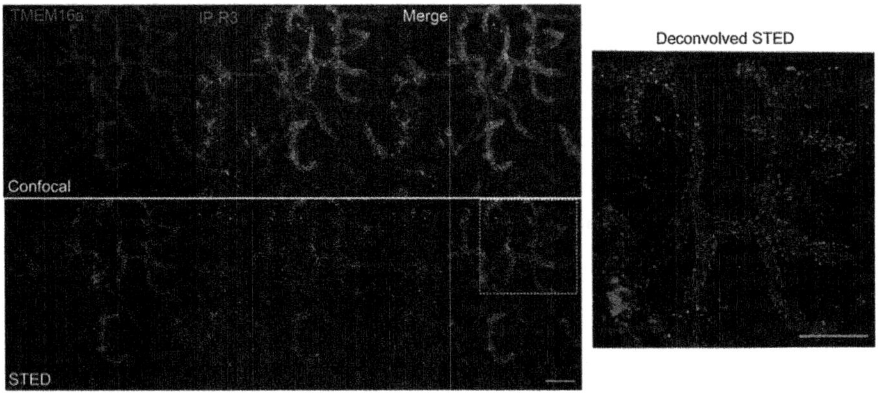

Fig. 3 Upper left panel: maximum projection of a confocal z stack (vertical resolution of 7 μm) showing ClCa channels (TMEM16a) and type 3 IPR. The entire section was 80 μm thick. Lower left panel: maximum projection of a stimulated emission depletion (STED) image. The STED image clearly shows that the proteins have distinct localisation. Right panel: Huygens deconvolution of the boxed region of the STED merged image. Scale bars = 5 μm

cell (Sneyd et al. 2006) which was additional motivation for using a Class II model for a parotid acinar cell, which is similar to a pancreatic acinar cell in many ways.

However, using Leica STED superresolution microscopy (Pages et al. 2019) we were able to show that the IPR and the ClCa channels are nearly always located not further than 50 nm from each other, and, furthermore, appear to be organised in puncta rather than being spread homogeneously through the apical region (Fig. 3). The spatial distribution of the IPR is critical, as they are the site of Ca^{2+} release. In a Class II model, the Ca^{2+} released from the IPR must be able to stimulate further production of IP_3, which can occur only at the basolateral membrane (as this is where PLC is located). This means that the sites of Ca^{2+} release and IP_3 production cannot be too far apart. The old Class II model, which always struggled to cope with the requirement that the IPR were within 200 nm of the apical membrane, was completely beaten by the new figure of 50 nm, and it became clear that our hypothesis for the mechanism underlying the Ca^{2+} oscillations was incorrect.

Our new model of Ca^{2+} oscillations in parotid acinar cells is a so-called Class I model, in which oscillations appear via modulation by Ca^{2+} of the IPR open probability (Dupont et al. 2016). Although in many Class I models this modulation takes the form of fast activation by Ca^{2+} followed by slower inactivation by Ca^{2+}, our latest IPR model (Cao et al. 2013; Sneyd et al. 2017) has a different mechanism. The IPR is still activated quickly by Ca^{2+}, but the inactivation of the IPR is now a result of the time-dependent rate at which the IPR responds to changes in $[Ca^{2+}]$. Although the overall behaviour is very similar to that of older Class I models, the underlying physiological assumptions are quite different.

With this new model, the apical region of the cell essentially turns into a pacemaking region. Ca^{2+} oscillations are generated, entirely within the apical region, by the interactions between Ca^{2+} and the IPR, and these oscillations are propagated as waves throughout the cell by a secondary excitable mechanism, based either on ryanodine receptors or a different type of IPR (Pages et al. 2019). The Ca^{2+} oscillations in the apical region activate the ClCa channels, while the propagated Ca^{2+} waves activate the basal KCa channels.

3.2 Chloride Reuptake

Another critical component of the saliva secretion mechanism is Cl^- reuptake. The Cl^- that is lost to the lumen across the apical membrane must be replenished by Cl^- entry across the basal membrane. Approximately 70% of this reuptake is accomplished by the NKCC1 cotransporter which brings in 1 Na^+, 1 K^+, and 2 Cl^- ions per cycle (Evans et al. 2000). However, the remaining 30% of Cl^- reuptake is HCO_3^- dependent, and involves two anion exchangers, AE2 (Slc4a2) and AE4 (Slc4a4) (Peña-Münzenmayer et al. 2015). Both AE2 and AE4 bring Cl^- into the cell and take HCO_3^- out, but AE4 anion transport is also linked to Na^+ exit.

The AE2 exchangers are paired with the NHE1 Na^+/H^+ exchangers, in a complex that allows for the control of intracellular pH, the removal of HCO_3^- and Na^+, and reuptake of Cl^-.

Although both AE2 and AE4 are functionally expressed in salivary acinar cells and thus potentially involved in Cl^- reuptake, Peña-Münzenmayer et al. (2015) showed that, surprisingly, knockout of AE2 had no effect on saliva secretion, while knockout of AE4 decreased saliva secretion by approximately 35%. The most likely explanation for this discrepancy was the cotransport of Na^+ by the AE4, but it was not clear why such cotransport would make such a large difference.

This puzzle was explained, at least in part, by Vera-Sigüenza et al. (2018) who showed quantitatively how increased AE4 activity can compensate for the loss of AE2 activity. However, the loss of Na^+ transport when the AE4 are knocked out turns out to be critical; since the AE2 do not exchange Na^+, they are unable to compensate for the loss of AE4 activity, and thus, Cl^- reuptake and saliva secretion are impaired upon AE4 knockout.

3.3 Potassium Channels and Pumping

In the traditional view of saliva secretion, KCa channels on the basolateral membrane are responsible for maintaining cell depolarisation during the efflux of Cl^- into the lumen. Since there is very little K^+ in primary saliva, it was thought that there could be few, if any, K^+ channels in the apical membrane, as otherwise it was thought that there would necessarily be significant levels of K^+ in the primary saliva. Furthermore, there was no evidence for the presence either of functional K^+ channels or NaK ATPases in the apical membrane.

This belief was challenged by Almássy et al. (2012) who used flash photolysis of caged Ca^{2+} to demonstrate the presence of functional KCa channels in the apical membrane of parotid acinar cells. Although modelling studies demonstrated that the presence of some K^+ channels in the apical membrane could increase the total amount of saliva produced (Palk et al. 2012), there remained no explanation of how this could result in primary saliva with very low $[K^+]$.

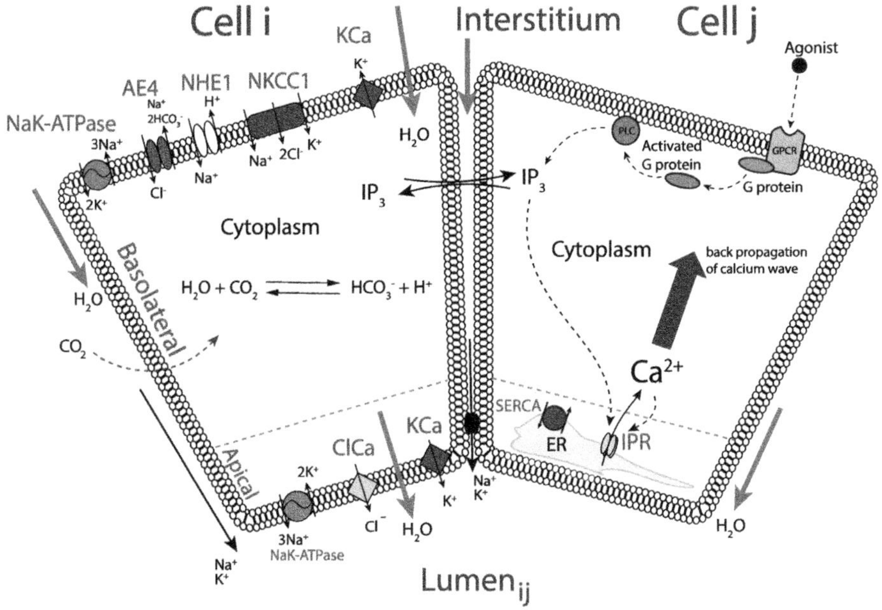

Fig. 4 New model of saliva secretion is similar in most respects to the traditional view discussed above, but with some important differences. Ca^{2+} is accumulated by the NKCC1 cotransporter and the AE4 anion exchanger, both of which are in the basal membrane. Upon agonist stimulation, IP_3 diffuses through the cell to the apical membrane where it binds to IPR located within 50 nm of the apical membrane. The apical region acts as a pacemaker, generating Ca^{2+} oscillations locally via Ca^{2+} modulations of the IPR open probability, and these oscillations are propagated across the cell by a secondary active mechanism, which could be either RyR or a different type of IPR. K^+ channels in the apical and basal regions maintain membrane depolarisation as Ca^{2+} flows into the lumen, and lumenal $[K^+]$ is kept low by apical NaK ATPases

This discrepancy was partially resolved by Almássy et al. (2018) who showed that the apical membrane of parotid acinar cells contains both K^+ channels and NaK ATPase pumps, thus providing a mechanism for transport of K^+ into and out of the lumen. In the same paper, a mathematical model of this new mechanism showed how the lumenal $[K^+]$ could be decreased approximately threefold by the NaK ATPase pumps in the apical membrane, while maintaining approximately 40% of the total cellular K^+ conductance in the apical membrane.

3.4 Summary of New Model

Our new model, incorporating the changes described above, is shown in Fig. 4. The major differences from the traditional model discussed above are, firstly, the presence of a substantial KCa current, as well as NaK ATPases, in the apical membrane, and, secondly, the tight clustering of the IPR with the apical ClCa channels.

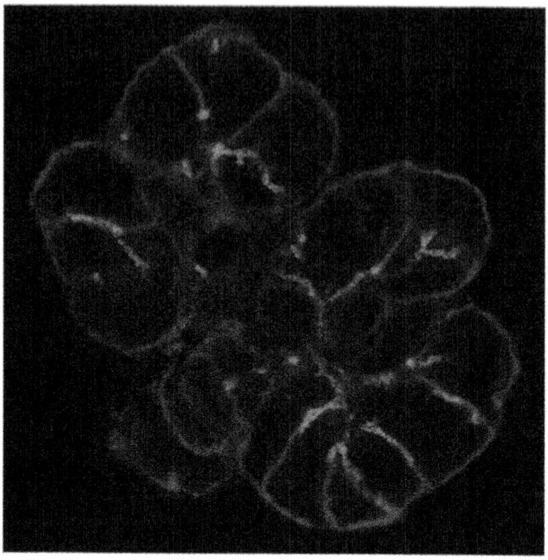

Fig. 5 A typical frame from a z stack of confocal slices. The NaK ATPases are shown in red (giving the position of the basal membrane), while the apical ClCa channels are shown in green (giving the position of the apical membranes). By stacking multiple such frames a three-dimensional picture of the group of cells, together with a branched lumen, can be reconstructed

4 Solving the Acinus Model

An important feature of our model was the construction of a three-dimensional finite element mesh upon which the model equations were to be solved. The mesh was based on z stacks of confocal slices with the apical and basal membranes labelled (Pages et al. 2019). The basal membranes were determined by labelling the NaK ATPase, while the apical membranes were determined by labelling the ClCa channels.

A typical frame from the z stack is shown in Fig. 5, with the NaK ATPase shown in red and the ClCa channels shown in green. Each membrane component can be traced (this was initially done manually) and then combined into a three-dimensional mesh. Each cell has well-defined apical and basal regions, and the boundaries between cells are conformal, in that the same set of boundary mesh points are in both the neighbouring cells. This allows for the computational solution of models with intercellular fluxes; thus, the model incorporates the intercellular diffusion of both Ca^{2+} and IP_3, so that the cells are coupled.

The final grid (with an idealised lumenal structure) is shown in Fig. 6. In each cell, the apical region has a finger-like structure and sits between two different cells, while the overall structure of the lumen is similar to that of a tree. When considering this reconstruction it is important to keep in mind a number of things. Firstly, it is an idealisation only, based on a semi-manual reconstruction from the experimental z stack. Thus, one would necessarily expect there to be deviations from the structure of the actual cells. Secondly, the lumen is not a structure in its own right, it is instead simply a gap between cells. Here, the gap is drawn as a cylinder, but that is merely a simplification for the sake of clarity. In an alternative reconstruction the lumen has a varying width. Thirdly, important cell types, such as the myoepithelial cells and the duct cells, are entirely missing. Finally, the IPR are clustered within 50 nm at most of the apical membrane and thus display the same fingerlike structure as the lumen.

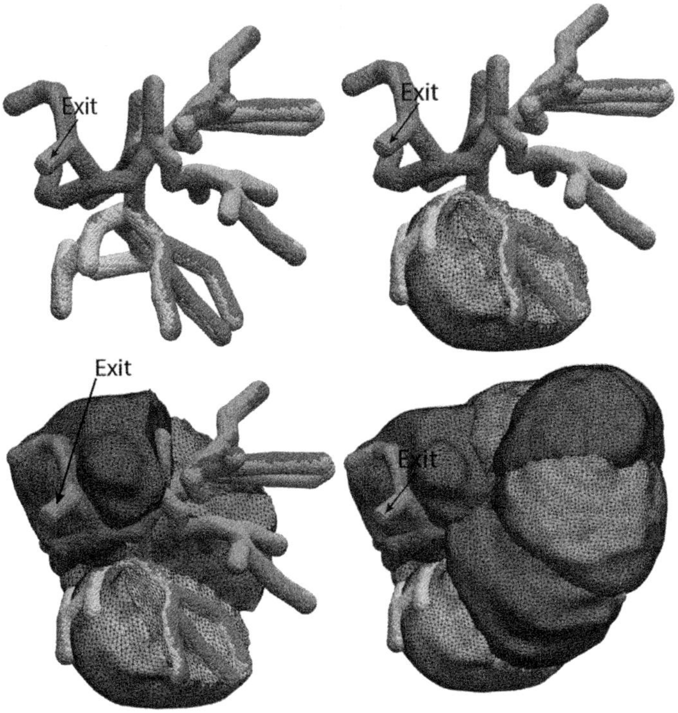

Fig. 6 Three-dimensional grid of seven acinar cells, together with an idealised reconstruction of the lumenal structure. The lumen is colour-coded according to the colours of the cells that border each lumenal segment. It is important to keep in mind that the lumen is not a stand-alone structure; it is simply the gaps between cells. The top left panel shows the isolated lumen; the top right panel shows how a single cell fits between the branches of the lumen; the bottom left panel shows how three cells fits in the lumenal branches; the bottom right panel shows a view of all seven cells, with the lumen now almost entirely obscured

However, other receptors and channels are distributed uniformly over the appropriate membrane.

The model equations (given in full detail in Pages et al. 2019; Vera-Sigüenza et al. 2019, 2020) are solved on the 7-cell structure shown in Fig. 6. Ca^{2+} and IP_3 obey reaction–diffusion equations inside each cell and can move between cells at a rate that is proportional to the local concentration difference. All the cytosolic ions, with the exception of Ca^{2+}, are assumed to diffuse so fast that they can be treated as spatially homogeneous, but the apical and basal membrane potentials are modelled separately. Because of buffering, the effective diffusion coefficient of Ca^{2+} in the cytoplasm is low, allowing for significant intracellular $[Ca^{2+}]$ gradients. Fluid flow inside the lumen is highly simplified, being described simply by a total flow in response to changes in inflow and outflow. Pressure in the lumen is not included in the model.

The volume of the cell is computed also, and incorporated in the reaction–diffusion equations. However, our modelling of changes in volume is highly simplified; we do not recompute the mesh for each change in volume, we simply scale the diffusion coefficients and transport rates as needed. Since we have no information on how the

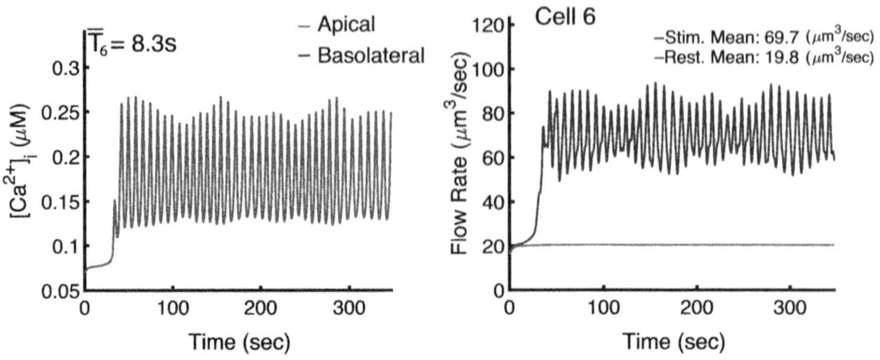

Fig. 7 Typical model results from cell 6 (Vera-Sigüenza et al. 2020). Results from the other cells are qualitatively similar. Left panel: time series of $[Ca^{2+}]$, spatially averaged over the apical and basal regions. At this resolution is it difficult to tell the difference between the apical and basal responses, but they are not identical. Right panel: the resting and stimulated rates of saliva secretion. Full details of the model output, including the apical and basal membrane potentials, and the cytoplasmic and lumenal concentration of Na^+, K^+ and Cl^-, can be found in Vera-Sigüenza et al. (2020)

shape of the acinar cell and the lumen changes as the cell shrinks, this is the simplest approach, requiring the fewest assumptions.

5 What Have We Learned from the Model?

Solution of the model equations gives Ca^{2+} responses and fluid flow that closely resemble data. A typical result (from cell 6) is shown in Fig. 7. In the apical region $[Ca^{2+}]$ oscillates with a period of about 8 s and each oscillation initiates a travelling wave of Ca^{2+} (not shown in the figure) which travels to the basal region, giving basal $[Ca^{2+}]$ oscillations with the same period and almost identical amplitude. The oscillations are superimposed on a raised baseline. Slight variations in amplitude are caused by intercellular coupling of nonidentical oscillators. Because each cell has a different shape and thus makes a different amount of IP_3 and has different numbers of IPR, each cell has a different oscillation period and amplitude. Since the cells are coupled by the intercellular diffusion of both IP_3 and Ca^{2+} (although the intercellular diffusion of Ca^{2+} has no discernible effect) these differences in oscillation period cause slight variations in the amplitude and period of the oscillations inside each cell, as expected from the general theory (Neu 1979; Ermentrout and Kopell 1984; Kopell and Ermentrout 1986, 1990; Mirollo and Strogatz 1990; Kuramoto 2003).

5.1 Oscillation Period is Unimportant

The usual theory of Ca^{2+} oscillations holds that the oscillations are a frequency-encoded signal; since a sustained rise in $[Ca^{2+}]$ is toxic to cells, the best way for a cell to use Ca^{2+} as an intracellular messenger is to carry the signal encoded in the frequency (Berridge and Galione 1988; Cuthbertson 1989; Berridge 1990; Thul et al.

2008; Dupont et al. 2011). There is considerable evidence from multiple cell types that this is indeed the case for many signalling pathways (Gu and Spitzer 1995; Dolmetsch et al. 1998; Li et al. 1998; Watt et al. 2000; Nash et al. 2002).

However, this does not appear to be the case for salivary acinar cells. The model predicts that, for a given average $[Ca^{2+}]$, the rate of saliva secretion is almost entirely independent of oscillation frequency (Vera-Sigüenza et al. 2020). Because it is difficult to measure accurately the water flow through a single cell, this prediction yet remains untested. However, the prediction is supported by preliminary intravital measurements (i.e. measurements in a living mouse, as discussed later) from mouse submandibular acinar cells which suggest strongly that the average $[Ca^{2+}]$ increases with increased stimulation, with little change in frequency. Thus, the rate of saliva secretion, which increases with increased stimulation, appears to depend on the average $[Ca^{2+}]$ rather than the Ca^{2+} oscillation frequency.

5.2 Do We Need a Multiscale Model?

Our multiscale model incorporates many known features of saliva secretion (particularly in the parotid gland), ranging from properties of the IPR measured at the single-channel level (Siekmann et al. 2012; Cao et al. 2013) through to responses across multiple cells. It includes a range of membrane ion transport mechanisms that control intracellular ion concentrations and pH, and is based on an anatomical reconstruction so that the model is solved in a realistic domain.

Although these might seem like model advantages, this is not necessarily so. The full model is complex to program and takes a long time to run (a full run can take well over a day), which is potentially a serious problem. We wish to use the model to investigate how saliva secretion depends on a range of parameters, and to predict the outcome from a range of pathological situations. In addition, ideally we want to predict the fluid flow from an entire salivary gland, not just from the 7 cells chosen in this reconstruction. However, since it takes many hours to solve the model on only 7 cells, solving the model on 10,000 cells is not possible. Neither is it realistic to perform many hundreds of simulations, testing various parameter values and interventions.

On the other hand, we can have some faith in the predictions from this model. After all, predictions from a simple, but inaccurate model are easily obtained, but useless.

We attempt to resolve this conundrum by determining how far the full model can be simplified while still retaining the essential model behaviour. We emphasise that by "essential behaviour" we mean the total amount of fluid secreted by the group of cells. If a simplification of the model causes significant changes to the responses of individual cells, but no significant change to the average amount of fluid secreted by the acinus, then we claim that the essential behaviour is preserved by the simplification.

It turns out that the vast majority of the model complexity can be omitted without a significant change to the model predictions.

5.2.1 The Lumenal Structure is Unimportant

We have previously shown that, in the limit as the membrane permeability to water gets large, and the osmotic gradient gets small, cells that are coupled by a common lumen act essentially as a single super-cell (Maclaren et al. 2012). In this case, the total amount of fluid secreted by N coupled cells is just the sum of the fluid secreted by each cell acting independently. Although this is theoretically true in the limit, it was unclear how accurate this would be in a more realistic situation. It turns out that, in our model, the membrane permeability to water is high enough, and the osmotic gradient is low enough, that this situation holds in most cases (Vera-Sigüenza et al. 2020). Simulations of a simplified version of the model, in which each cell is independent of the others (and not coupled via the common branching lumen shown in Fig. 6), show that total fluid flow (mostly) remains practically unchanged.

This result is not an artefact of our particular lumenal topology. When the model is computed for an arbitrary selection of different lumenal topologies (i.e. with the cells connected by a common lumen in a different way), the total fluid flow is the same in each case (Vera-Sigüenza et al. 2020), and in each case is equal to the flow that would be generated by 7 cells acting independently.

The exception to this general rule is for model cells that demonstrate spontaneous Ca^{2+} oscillations at rest. In the model this can happen when the cell is small, but this is an artefact of parameter selection, as we did not parameterise the model differently for every cell, thus avoiding any hint that our results depend on careful cell-dependent parameter selection. However, in an actual salivary gland we have not seen any evidence that any cells exhibit spontaneous Ca^{2+} oscillations at rest.

The reason for this exception is that, if an isolated cell has a spontaneous Ca^{2+} oscillation at rest (and thus significant fluid flow at rest), this oscillation can disappear when the cell is connected to others via a common lumen. In this case, the coupling via a common lumen has a significant effect on the resting fluid flow, although no significant effect on the fluid flow when all the cells are stimulated.

5.2.2 Spatial Structure and Intercellular Communication are Unimportant

If the internal spatial structure of each cell is ignored and the Ca^{2+} dynamics modelled with an ODE rather than a PDE, there is very little difference in total fluid flow. There are two slightly different ways of constructing an approximate ODE; firstly, one can simply model the cell as a homogeneous region, making no distinction between apical and basal regions, or, secondly, one can model the cell as two coupled spatially homogeneous regions, corresponding to the apical and basal regions. In either case, the total fluid flow of the acinus is almost unchanged (Vera-Sigüenza et al. 2020).

More surprising, perhaps, is the result that the intercellular diffusion of Ca^{2+} and IP_3 makes no significant difference to the total fluid flow. Intercellular diffusion of Ca^{2+} makes little difference to anything at all; this is because any Ca^{2+} that gets through a gap junction (which are assumed to be homogeneously distributed on the lateral surfaces of each cell) is quickly bound by Ca^{2+} buffers.

However, intercellular diffusion of IP_3 has a significant effect on the intracellular Ca^{2+} dynamics of each cell, particularly at low coupling strength. This is because each

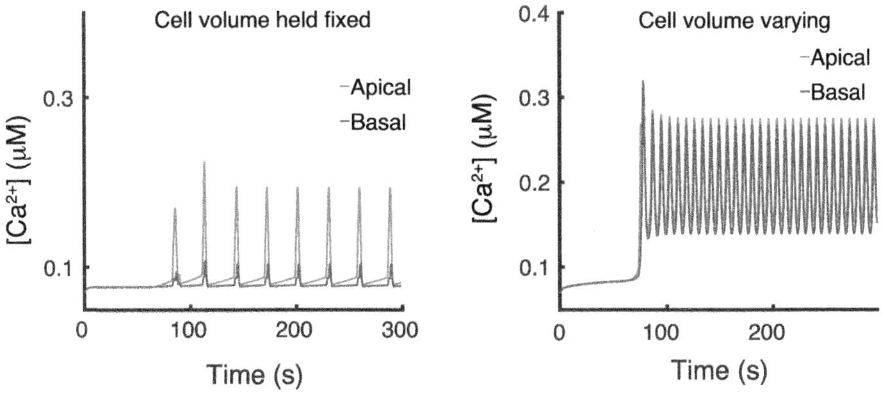

Fig. 8 Model oscillations from a single independent cell (Pages et al. 2019), showing the effect of incorporating changes in cell volume. If the volume of the cell is artificially held constant, the model oscillation is slower and not superimposed on a raised plateau. The concentrations are averaged over the apical and basal regions

cell is an autonomous oscillator with a period different to that of its neighbours. When non-identical oscillators are coupled, we know from the general theory that when the coupling strength is high the oscillators can synchronise, but when the coupling strength is lower (but not zero) much more complex behaviour can appear (Torre 1975).

Although this general result remains true in our model, the overall fluid flow simply doesn't care. The details of the oscillations in each cell might change, but the average [Ca^{2+}] doesn't change greatly, and that is all that matters for the fluid flow. Thus, even at high intercellular coupling strength, when the individual cellular oscillators are significantly perturbed by their neighbours, total fluid flow remains unaffected. This is true for all stimulation levels.

5.3 The Change in Volume is Critical

As it transports water, each cell shrinks in volume by approximately 30% (Foskett and Melvin 1989), and this effect is incorporated into our model. Indeed, it's not possible to construct a model of water transport without taking the volume change into account. In our model, the change in volume is included both in the water transport equations and in the Ca^{2+} dynamics equations. Surprisingly, it turns out that the change in volume upon stimulation plays a significant role in the Ca^{2+} dynamics. When volume changes are not incorporated in the Ca^{2+} dynamics model, the Ca^{2+} oscillations have a much larger period and are not superimposed on a raised baseline (Fig. 8). However, incorporation of volume changes into the Ca^{2+} model changes the oscillations into the form seen experimentally.

It follows that the Ca^{2+} dynamics of a salivary acinar cell cannot be understood independently of an understanding of water transport; these two processes are inextricably linked.

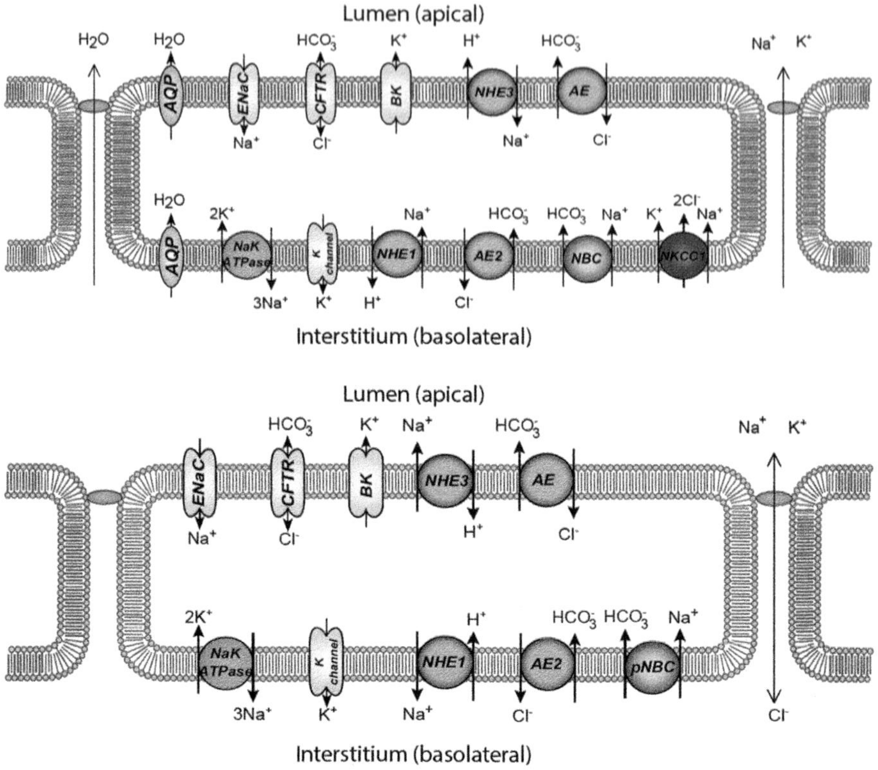

Fig. 9 Schematic diagrams of two different kinds of duct cells (Fong et al. 2017). The upper panel describes a duct cell that allows for transcellular and paracellular water transport, while the lower panel describes a duct cell that allows no water transport. ENaC is the epithelial Na^+ channel; CFTR is the cystic fibrosis transmembrane regulator anion channel that allows both Cl^- and HCO_3^- current; BK is the usual K^+ channel (i.e. not the Ca^{2+}-dependent version seen in acinar cells); NHE1 and NHE3 are Na^+/H^+ exchangers; NBC is a Na^+/HCO_3^- cotransporter; K channel is a generic unspecified K^+ channel

6 Recent Developments in Modelling Salivary Duct Cells

Less is known about salivary duct cells than about salivary acinar cells. The traditional view of them is that they mostly just transport Na^+ and Cl^- out of the duct, and K^+ and HCO_3^- into the duct, while remaining impermeable to water. There are two major types of duct cells; intercalated duct cells, which are closer to the acinus, and striated duct cells, which are further downstream. There is evidence that intercalated duct cells transport water (Lee et al. 2012; Hong et al. 2014), although less than acinar cells do, but there remains general agreement that striated duct cells are impermeable to water.

The secondary saliva that flows out of the ducts and into the mouth is hypotonic to the interstitium, which explains why much of the duct must be impermeable to water. If it wasn't, then water would be drawn out of the duct by osmosis, leading to a decrease in saliva secretion.

Recent developments in gene therapy to treat xerostomia arising from irradiation for head and neck cancers have motivated a more detailed understanding of how duct cells work. Briefly, in the 1990s it was conjectured that transferral of the gene for the human hAQP1 aquaporin (water channel) to salivary ducts which had been damaged by irradiation might be able to restore salivary function, at least partially (Baum et al. 2006, 2010, 2017). This was first shown to work well in rats and minipigs, and around 2012 a clinical trial showed that this therapy would also work in humans (Gao et al. 2010; Baum et al. 2015).

However, given current understanding of how the ducts work, this outcome was rather a puzzle. It was thought that the aquaporins were mostly ending up in the duct cells, thus making them permeable to water, but it was not clear why this didn't result in water leaving the ducts, rather than entering them.

An initial model by Patterson et al. (2012) was later modified and extended by Fong et al. (2017) (Fig. 9), based largely on the qualitative model of Lee et al. (2012). They concluded that, firstly, the hAQP had to be transfected only into the intercalated duct cells, and that transfection of the gene also changed the expression of other ionic transport mechanisms. For example, the model predicted that, after transfection, the nonsecretory duct cells would have fewer ENaC and CFTR channels, but more apical NHE exchangers. These model predictions have not yet been tested, and neither have they been rigorously tested for robustness. Hence, it is fair to say that our current understanding of how ducts work, either before or after transfection, remains uncertain.

7 The Next Generation of Models

Most recently (in early 2020), exciting new data have appeared that threaten to overturn yet again our ideas of how salivary acinar cells work.

Developments in genetic manipulation and microscopy have now made it possible to measure, at high spatial and temporal resolution, Ca^{2+} responses in the submandibular gland of a living mouse, in response to direct neural stimulation. Since all previous results were from isolated cells or glands, and the responses were stimulated by direct application of agonist to the cells, this is the first time it has been possible to determine salivary gland Ca^{2+} responses in a realistic physiological setting.

Preliminary results are surprising in two particular ways.

1. It appears that, in response to neural stimulation ranging from 2 to 10 Hz, the $[Ca^{2+}]$ in submandibular acinar cells oscillates with a period of around 1 s, much faster than the oscillations seen in isolated cells or glands (in response to direct application of agonist). These fast oscillations still occur on a raised baseline. Preliminary results are shown in Fig. 10.
2. The fast oscillations described above occur principally in the apical regions of the cell, and do not appear to be propagated to the basal regions to any significant extent. Preliminary results are shown in Fig. 11.

Clearly, these new data paint a very different picture than the older data from isolated cells and glands. The Ca^{2+} oscillations seem no longer to be periodic intracellular waves, being far more spatially restricted, and their frequency is much higher.

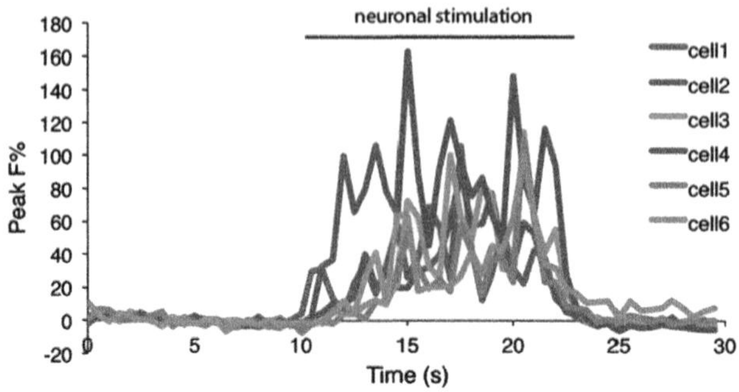

Fig. 10 Preliminary data from intravital measurements of Ca^{2+} oscillations in acinar cells from the submandibular gland acinar cell of a living mouse. (These data are a low-resolution version of more extensive high-resolution data that are submitted for publication.) The oscillations were initiated by neuronal stimulation, which results in the release of acetylcholine. Neuronal stimulation was 5 mA at 5 Hz, for 12 s, and responses were measured at 2 fps. The 6 cells are all from different acini. These intravital oscillations were stimulated by activation of the same receptor as those from an intact gland, as shown in Fig. 2, but their frequency is significantly higher

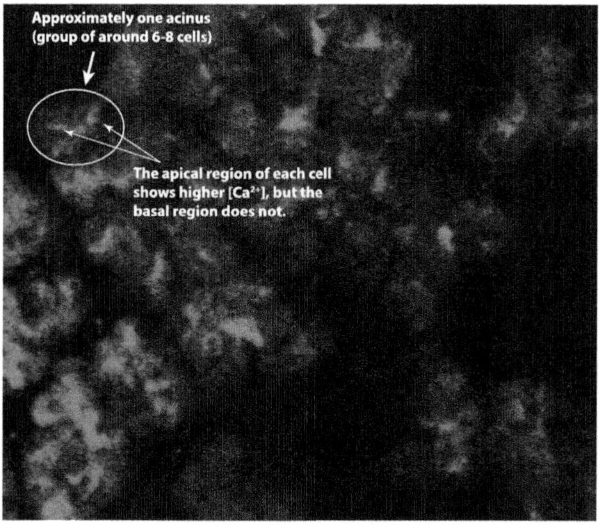

Fig. 11 A still image from a video of intravital Ca^{2+} oscillations in acinar cells from the submandibular gland of a living mouse, in response to neuronal stimulation. (These data are a low-resolution version of more extensive high-resolution data that are submitted for publication.) The majority of the cells demonstrate Ca^{2+} oscillations only in their apical regions

We must, of course, keep in mind that the new data are from the submandibular gland, not the parotid gland, and thus, it is possible that Ca^{2+} oscillations in parotid gland acinar cells (in living animals) are similar to those observed in isolated parotid acinar cells and glands. However, because of the extensive similarities between parotid and submandibular acinar cells, this is not the most plausible way to resolve the discrepancies between the old data and the new. It is more likely that the discrepancies are caused by the fact that, in the living animal, the cells are stimulated directly by the nerves, as would occur physiologically.

7.1 The Proposed New Paradigm

It is particularly interesting that our new observations of Ca^{2+} signalling in submandibular acinar cells are consistent with previous work on the spatial positioning of the KCa channels and the NaK ATPase pumps.

As we discussed above, there is mounting evidence that the apical membrane of acinar cells contains KCa channels and NaK ATPase pumps, in addition to the ClCa channels. Our new intravital data suggest why this should be so. First, we note that, since the ClCa current moves the membrane potential towards the Nernst potential of Cl^- (in which case the Cl^- current would stop), the cell requires the activation of K^+ channels to counterbalance the Cl^- current. However, in the absence of a propagated Ca^{2+} wave across the cell, KCa channels can only be activated if they are in a region where the $[Ca^{2+}]$ increases, i.e. in the apical region. Thus, our new data suggest that the cell will secrete water properly only if the apical region has a substantial K^+ current. This apical K^+ current, rather than being a non-critical byproduct of the modelling studies, now appears to have morphed into a critical feature.

In addition, since we still know that the primary saliva is low in K^+, the K^+ that flows into the lumen needs to be removed, and this appears to be the function of the apical NaK ATPase pumps.

Thus, all these results are converging to a new version of saliva secretion in salivary gland acinar cells, as illustrated in Fig. 12. In this new model the regulation of water transport occurs almost exclusively in the apical region; because of their spatial proximity, the Ca^{2+} released from the IPR has enhanced access to the two critical Ca^{2+}-dependent ion channels that control the membrane potential and the Cl^- current.

There is, as yet, no quantitative description of this proposed new model, so it remains unclear how well this new approach will deal with the harsh requirements of a stable cell volume and electroneutrality. Although it is also unclear whether or not our previous conclusions about the unimportance of the spatial structure of the acinar lumen will remain accurate, it is a reasonable first guess that they will. Finally, it is plausible that intracellular spatial variation of $[Ca^{2+}]$ will be even less relevant in a new model of an acinar cell; we thus anticipate that an ODE model of the apical region of the cell, coupled to the water transport mechanisms as before, will be an accurate and predictive model.

Many spatial questions remain, also. For example, what exactly is the spatial pattern of ACh released by neuronal stimulation? Is each acinus innervated equally, and does each acinar cell receive the same level of agonist stimulation? Does it matter if they do, or don't?

7.2 What About Duct Cells?

So far we have concentrated our attention on the next generation of models of the salivary acinar cells, simply because more is known about them. However, new data

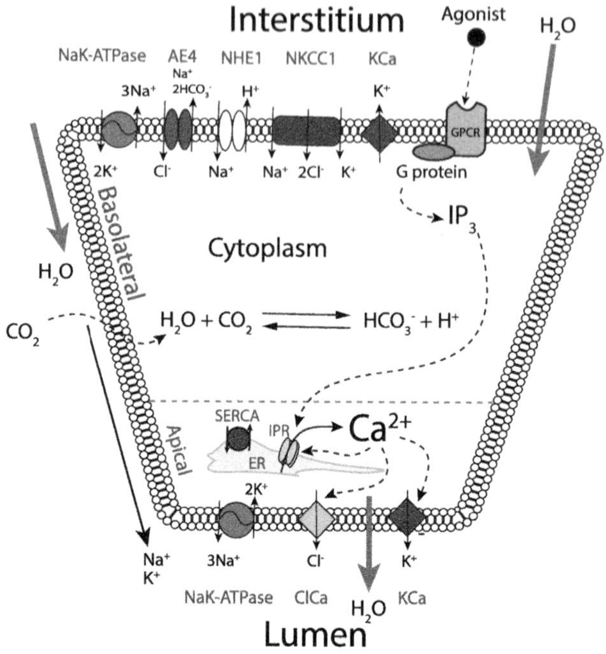

Fig. 12 A proposed new hypothesis for saliva secretion in salivary acinar cells. The only difference from the model shown in Fig. 4 is that there is now no Ca^{2+} wave propagated across the cell, from apical to basal, and thus, all the necessary Ca^{2+}-modulated ion channels now reside in the apical region. They can also exist in the basal membrane but do not need to be activated to a significant extent. No quantitative description of this proposed model yet exists

are beginning to appear from duct cells now also, again in the submandibular gland. These new data take time to collect as they rely on mouse genetic constructs which take time to breed and grow, but preliminary data are already good enough for us to be able to start the construction of three-dimensional models of acini and ducts together.

Given that one of our major conclusions is that the spatial structure of the lumen is not important, and given also that our new understanding of how an acinar cell works suggests that the intracellular structure is also unimportant, why would we bother at all with constructing three-dimensional models of a combined acinus and duct?

There are two principal answers to this question. Firstly, we still do not know if our previous conclusions will stand the test of the new model, based on our new data. Secondly, we also do not know how spatial structure affects the function of the duct. Although we might suspect it has little effect, we will not know for sure unless we test it.

On the other hand, our previous mesh construction methods were partially manual and very time-consuming; it would not seem to be a good use of time to continue with such slow (although accurate) approaches if our expectation is that the spatial structure will have little importance. We have thus developed a method for the fully-automated construction of a three-dimensional mesh of a combined acinus and duct with the correct spatial statistics on average, as determined from the data. Any particular mesh will not be an accurate description of any particular group of cells, but it will have the correct average properties. This work remains ongoing.

One of the first questions to be addressed by this new generation of duct models is whether or not our predictions from the simplified model (Fong et al. 2017) of hAQP transfection remain valid in a more accurate model. In addition, the robustness of these predictions with respect to spatial structure needs to assessed carefully.

8 Conclusions

Over the past 15 years or so, our combined theoretical/experimental work has taught us a great deal about how saliva secretion works, and we are on the verge of the next generation of results and models. Even more importantly, we are on the verge of discovering much more about how saliva secretion works in living animals, in response to physiological patterns of neuronal stimulation.

From the modelling point of view, one of our most important results has a rather negative flavour. We have strong evidence that a multiscale model is not necessary if the goal is to understand how much saliva is secreted. The Ca^{2+} dynamics in individual cells are dependent on a host of factors, including intercellular permeability to IP_3, and the size and structure of the apical and basal regions. Nevertheless, it appears that each acinar cell simply averages out the $[Ca^{2+}]$ in the apical region of the cell, and pays almost no attention to the fine structure of the Ca^{2+} oscillations. The oscillation period, for example, plays no significant role in determining the rate of saliva secretion. In addition, the fine spatial structure of the acinar lumen appears to play no significant role in determining the rate of saliva secretion.

In short, if one cares only about calculating how much primary saliva is going to be produced, one might as well simply write down an ODE for the apical region of each cell, calculate the fluid flow for each cell independently, and then add them all up.

It is somewhat ironic that we had to construct a multiscale three-dimensional model to show that we didn't need one.

Acknowledgements Since this review appears in a special volume to mark the 90th birthday of Jim Murray, it would be remiss of me (J.S.) not to acknowledge the enormous debt I owe to Jim. Jim has been, for my entire career, an inspiration to me. Not just to me, of course; Jim has been a giant of mathematical biology for decades now, and I am only one of the many who has read, and reread, his book "Mathematical Biology" for both duty and pleasure. It is absolutely fitting that this review, dedicated as it is to a series of results in which it is not always easy to decide whether the modelling stimulated the experiment, or vice versa, appears in a volume dedicated to Jim. I suspect (and hope) he will like this approach, similar as it is to his own. He is, after all, the shoulders I stand on. And, even then, I'm not entirely certain I see any further than he does. This work was supported by NIH Grant 2R01DE019245, and by the Marsden Fund of the Royal Society of New Zealand.

Compliance with Ethical Standards

Conflict of interest The authors declare that they have no conflict of interest.

References

Almássy J, Won JH, Begenisich TB, Yule DI (2012) Apical Ca^{2+}-activated potassium channels in mouse parotid acinar cells. J Gen Physiol 139(2):121–33

Almássy J, Siguenza E, Skaliczki M, Matesz K, Sneyd J, Yule DI, Nánási PP (2018) New saliva secretion model based on the expression of Na^+-K^+ pump and K^+ channels in the apical membrane of parotid acinar cells. Pflugers Arch 470(4):613–621. https://doi.org/10.1007/s00424-018-2109-0

Ambudkar IS (2012) Polarization of calcium signaling and fluid secretion in salivary gland cells. Curr Med Chem 19(34):5774–81

Ambudkar IS (2014) Ca^{2+} signaling and regulation of fluid secretion in salivary gland acinar cells. Cell Calcium 55(6):297–305. https://doi.org/10.1016/j.ceca.2014.02.009

Baum BJ, Zheng C, Cotrim AP, Goldsmith CM, Atkinson JC, Brahim JS, Chiorini JA, Voutetakis A, Leakan RA, Van Waes C, Mitchell JB, Delporte C, Wang S, Kaminsky SM, Illei GG (2006) Transfer of the AQP1 cDNA for the correction of radiation-induced salivary hypofunction. Biochim Biophys Acta 1758(8):1071–7. https://doi.org/10.1016/j.bbamem.2005.11.006

Baum BJ, Adriaansen J, Cotrim AP, Goldsmith CM, Perez P, Qi S, Rowzee AM, Zheng C (2010) Gene therapy of salivary diseases. Methods Mol Biol 666:3–20. https://doi.org/10.1007/978-1-60761-820-1

Baum BJ, Alevizos I, Chiorini JA, Cotrim AP, Zheng C (2015) Advances in salivary gland gene therapy—oral and systemic implications. Expert Opin Biol Ther 15(10):1443–54. https://doi.org/10.1517/14712598.2015.1064894

Baum BJ, Afione S, Chiorini JA, Cotrim AP, Goldsmith CM, Zheng C (2017) Gene therapy of salivary diseases. Methods Mol Biol 1537:107–123. https://doi.org/10.1007/978-1-4939-6685-1

Berridge MJ (1990) Calcium oscillations. J Biol Chem 265:9583–86

Berridge MJ (2009) Inositol trisphosphate and calcium signalling mechanisms. Biochim Biophys Acta 1793(6):933–40

Berridge MJ, Galione A (1988) Cytosolic calcium oscillators. FASEB J 2:3074–3082

Berridge MJ, Lipp P, Bootman MD (2000) The versatility and universality of calcium signalling. Nat Rev Mol Cell Biol 1(1):11–21

Cao P, Donovan G, Falcke M, Sneyd J (2013) A stochastic model of calcium puffs based on single-channel data. Biophys J 105(5):1133–42. https://doi.org/10.1016/j.bpj.2013.07.034

Chara O, Brusch L (2015) Mathematical modelling of fluid transport and its regulation at multiple scales. Biosystems 130:1–10. https://doi.org/10.1016/j.biosystems.2015.02.004

Cook D, Van Lennep E, Ml R, Ja Y (1994) Secretion by the major salivary glands. In: Johnson L (ed) Physiology of the gastrointestinal tract, 3rd edn. Raven Press, New York, pp 1061–2017

Cuthbertson KSR (1989) Intracellular calcium oscillators. In: Goldbeter A (ed) Cell to cell signalling: from experiments to theoretical models. Academic Press, London, pp 435–447

Daniels TE, Wu AJ (2000) Xerostomia—clinical evaluation and treatment in general practice. J Calif Dent Assoc 28(12):933–41

Dawson D (1992) Water transport principles and perspectives. In: Seldin D, Giebisch G (eds) The kidney physiology and pathophysiology. Raven Press, New York, pp 301–316

Diamond JM, Bossert WH (1967) Standing-gradient osmotic flow. A mechanism for coupling of water and solute transport in epithelia. J Gen Physiol 50(8):2061–83

Dolmetsch RE, Xu K, Lewis RS (1998) Calcium oscillations increase the efficiency and specificity of gene expression. Nature 392(6679):933–6. https://doi.org/10.1038/31960

Dupont G, Combettes L, Bird GS, Putney JW (2011) Calcium oscillations. Cold Spring Harb Perspect Biol 3(3):a004226. https://doi.org/10.1101/cshperspect.a004226

Dupont G, Falcke M, Kirk V, Sneyd J (2016) Models of calcium signalling, interdisciplinary applied mathematics, vol 43. Springer, Berlin. https://doi.org/10.1007/978-3-319-29647-0

Ermentrout G, Kopell N (1984) Frequency plateaus in a chain of weakly coupled oscillators. SIAM J Math Anal 15:215–37

Evans RL, Park K, Turner RJ, Watson GE, Nguyen HV, Dennett MR, Hand AR, Flagella M, Shull GE, Melvin JE (2000) Severe impairment of salivation in Na^+/K^+/2Cl^- cotransporter (NKCC1)-deficient mice. J Biol Chem 275(35):26720–26726. https://doi.org/10.1074/jbc.M003753200

Fong S, Chiorini JA, Sneyd J, Suresh V (2017) Computational modeling of epithelial fluid and ion transport in the parotid duct after transfection of human aquaporin-1. Am J Physiol Gastrointest Liver Physiol 312(2):G153–G163. https://doi.org/10.1152/ajpgi.00374.2016

Foskett JK, Melvin JE (1989) Activation of salivary secretion: coupling of cell volume and $[Ca^{2+}]_i$ in single cells. Science 244(4912):1582–5

Fox RI, Maruyama T (1997) Pathogenesis and treatment of Sjögren's syndrome. Curr Opin Rheumatol 9(5):393–9

Gao R, Yan X, Zheng C, Goldsmith CM, Afione S, Hai B, Xu J, Zhou J, Zhang C, Chiorini JA, Baum BJ, Wang S (2010) AAV2-mediated transfer of the human aquaporin-1 cDNA restores fluid secretion from irradiated miniature pig parotid glands. Gene Ther 18(1):38–42

Garcia GJM, Boucher RC, Elston TC (2013) Biophysical model of ion transport across human respiratory epithelia allows quantification of ion permeabilities. Biophys J 104(3):716–26. https://doi.org/10.1016/j.bpj.2012.12.040

Gin E, Crampin EJ, Brown DA, Shuttleworth TJ, Yule DI, Sneyd J (2007) A mathematical model of fluid secretion from a parotid acinar cell. J Theor Biol 248(1):64–80

Gu X, Spitzer NC (1995) Distinct aspects of neuronal differentiation encoded by frequency of spontaneous Ca^{2+} transients. Nature 375(6534):784–7. https://doi.org/10.1038/375784a0

Hill AE (2008) Fluid transport: a guide for the perplexed. J Membr Biol 223(1):1–11. https://doi.org/10.1007/s00232-007-9085-1

Hong JH, Park S, Shcheynikov N, Muallem S (2014) Mechanism and synergism in epithelial fluid and electrolyte secretion. Pflugers Arch 466(8):1487–99. https://doi.org/10.1007/s00424-013-1390-1

Kasai H, Li YX, Miyashita Y (1993) Subcellular distribution of Ca^{2+} release channels underlying Ca^{2+} waves and oscillations in exocrine pancreas. Cell 74:669–77

Keener J, Sneyd J (2008) Mathematical physiology, 2nd edn. Springer, New York

Kopell N, Ermentrout G (1986) Symmetry and phaselocking in chains of weakly coupled oscillators. Commun Pure Appl Math 39(5):623–60

Kopell N, Ermentrout G (1990) Phase transitions and other phenomena in chains of coupled oscillators. SIAM J Appl Math 50(4):1014–52

Kuramoto Y (2003) Chemical oscillations, waves, and turbulence. Courier Corporation, New York

Larsen EH (2002) Analysis of the sodium recirculation theory of solute-coupled water transport in small intestine. J Physiol 542(1):33–50

Larsen EH, Sørensen JB, Sørensen JN (2000) A mathematical model of solute coupled water transport in toad intestine incorporating recirculation of the actively transported solute. J Gen Physiol 116(2):101–24. https://doi.org/10.1085/jgp.116.2.101

Layton AT (2011a) A mathematical model of the urine concentrating mechanism in the rat renal medulla. I. Formulation and base-case results. Am J Physiol Renal Physiol 300(2):F356–F371. https://doi.org/10.1152/ajprenal.00203.2010

Layton AT (2011b) A mathematical model of the urine concentrating mechanism in the rat renal medulla. II. Functional implications of three-dimensional architecture. Am J Physiol Renal Physiol 300(2):F372–F384. https://doi.org/10.1152/ajprenal.00204.2010

Layton AT, Layton HE (2005) A region-based mathematical model of the urine concentrating mechanism in the rat outer medulla. I. Formulation and base-case results. Am J Physiol Renal Physiol 289(6):F1346–F1366. https://doi.org/10.1152/ajprenal.00346.2003

Layton AT, Layton HE (2005) A region-based mathematical model of the urine concentrating mechanism in the rat outer medulla. II. Parameter sensitivity and tubular inhomogeneity. Am J Physiol Renal Physiol 289(6):F1367–F1381. https://doi.org/10.1152/ajprenal.00347.2003

Lee MG, Xu X, Zeng W, Diaz J, Wojcikiewicz RJ, Kuo TH, Wuytack F, Racymaekers L, Muallem S (1997) Polarized expression of Ca^{2+} channels in pancreatic and salivary gland cells Correlation with initiation and propagation of $[Ca^{2+}]_i$ waves. J Biol Chem 272(25):15765–15770

Lee MG, Ohana E, Park HW, Yang D, Muallem S (2012) Molecular mechanism of pancreatic and salivary gland fluid and HCO_3 secretion. Physiol Rev 92(1):39–74. https://doi.org/10.1152/physrev.00011.2011

Li W, Llopis J, Whitney M, Zlokarnik G, Tsien RY (1998) Cell-permeant caged $InsP_3$ ester shows that Ca^{2+} spike frequency can optimize gene expression. Nature 392(6679):936–41. https://doi.org/10.1038/31965

Lundberg A (1956) Secretory potentials and secretion in the sublingual gland of the cat. Nature 177(4519):1080–1. https://doi.org/10.1038/1771080a0

Lundberg A (1957a) The mechanism of establishment of secretory potentials in sublingual gland cells. Acta Physiol Scand 40(1):35–58. https://doi.org/10.1111/j.1748-1716.1957.tb01476.x

Lundberg A (1957b) Secretory potentials in the sublingual gland of the cat. Acta Physiol Scand 40(1):21–34. https://doi.org/10.1111/j.1748-1716.1957.tb01475.x

Maclaren OJ, Sneyd J, Crampin EJ (2012) Efficiency of primary saliva secretion: an analysis of parameter dependence in dynamic single-cell and acinus models, with application to aquaporin knockout studies. J Membr Biol 245(1):29–50. https://doi.org/10.1007/s00232-011-9413-3

Mangos JA, McSherry NR, Irwin K, Hong R (1973a) Handling of water and electrolytes by rabbit parotid and submaxillary glands. Am J Physiol 225(2):450–5. https://doi.org/10.1152/ajplegacy.1973.225.2.450

Mangos JA, McSherry NR, Nousia-Arvanitakis S, Irwin K (1973b) Secretion and transductal fluxes of ions in exocrine glands of the mouse. Am J Physiol 225(1):18–24. https://doi.org/10.1152/ajplegacy.1973.225.1.18

Martinez JR (1990) Cellular mechanisms underlying the production of primary secretory fluid in salivary glands. Crit Rev Oral Biol Med 1(1):67–78

Mathias RT, Wang H (2005) Local osmosis and isotonic transport. J Membr Biol 208(1):39–53

Melvin JE (1991) Saliva and dental diseases. Curr Opin Dent 1(6):795–801

Melvin JE (1999) Chloride channels and salivary gland function. Crit Rev Oral Biol Med 10(2):199–209

Melvin JE, Yule D, Shuttleworth T, Begenisich T (2005) Regulation of fluid and electrolyte secretion in salivary gland acinar cells. Annu Rev Physiol 67(1):445–469

Mirollo RE, Strogatz SH (1990) Synchronization of pulse-coupled biological oscillators. SIAM J Appl Math 50(6):1645–1662

Nash MS, Schell MJ, Atkinson PJ, Johnston NR, Nahorski SR, Challiss RAJ (2002) Determinants of metabotropic glutamate receptor-5-mediated Ca^{2+} and inositol 1,4,5-trisphosphate oscillation frequency. Receptor density versus agonist concentration. J Biol Chem 277(39):35947–35960. https://doi.org/10.1074/jbc.M205622200

Nathanson MH, Fallon MB, Padfield PJ, Maranto AR (1994) Localization of the type 3 inositol 1,4,5-trisphosphate receptor in the Ca^{2+} wave trigger zone of pancreatic acinar cells. J Biol Chem 269(7):4693–6

Nauntofte B (1992) Regulation of electrolyte and fluid secretion in salivary acinar cells. Am J Physiol 263(6 Pt 1):G823–37

Neu J (1979) Chemical waves and the diffusive coupling of limit cycle oscillators. SIAM J Appl Math 36:509–515

Novotny JA, Jakobsson E (1996a) Computational studies of ion-water flux coupling in the airway epithelium. I. Construction of model. Am J Physiol 270(6 Pt 1):C1751–C1763

Novotny JA, Jakobsson E (1996b) Computational studies of ion-water flux coupling in the airway epithelium. II. Role of specific transport mechanisms. Am J Physiol 270(6 Pt 1):C1764–C1772

Pages N, Vera-Sigüenza E, Rugis J, Kirk V, Yule DI, Sneyd J (2019) A model of Ca^{2+} dynamics in an accurate reconstruction of parotid acinar cells. Bull Math Biol 81(5):1394–1426. https://doi.org/10.1007/s11538-018-00563-z

Palk L, Sneyd J, Shuttleworth TJ, Yule DI, Crampin EJ (2010) A dynamic model of saliva secretion. J Theor Biol 266(4):625–40

Palk L, Sneyd J, Patterson K, Shuttleworth TJ, Yule DI, Maclaren O, Crampin EJ (2012) Modelling the effects of calcium waves and oscillations on saliva secretion. J Theor Biol 305:45–53

Patterson K, Catalán MA, Melvin JE, Yule DI, Crampin EJ, Sneyd J (2012) A quantitative analysis of electrolyte exchange in the salivary duct. Am J Physiol Gastrointest Liver Physiol 303(10):G1153–63. https://doi.org/10.1152/ajpgi.00364.2011

Peña-Münzenmayer G, Catalán MA, Kondo Y, Jaramillo Y, Liu F, Shull GE, Melvin JE (2015) Ae4 (Slc4a9) anion exchanger drives Cl-uptake-dependent fluid secretion by mouse submandibular gland acinar cells. J Biol Chem 290(17):10677–10688. https://doi.org/10.1074/jbc.M114.612895

Reuss L (2002) Water transport controversies—an overview. J Physiol 542(1):1–2

Sandefur CI, Boucher RC, Elston TC (2017) Mathematical model reveals role of nucleotide signaling in airway surface liquid homeostasis and its dysregulation in cystic fibrosis. Proc Natl Acad Sci USA 114(35):E7272–E7281. https://doi.org/10.1073/pnas.1617383114

Ship JA, Fox PC, Baum BJ (1991) How much saliva is enough? “Normal” function defined. J Am Dent Assoc 122(3):63–9

Siekmann I, Wagner LE, Yule D, Crampin EJ, Sneyd J (2012) A kinetic model for type I and II IP_3R accounting for mode changes. Biophys J 103(4):658–68

Silva P, Stoff J, Field M, Fine L, Forrest JN, Epstein FH (1977) Mechanism of active chloride secretion by shark rectal gland: role of Na-K-ATPase in chloride transport. Am J Physiol 233(4):F298–306. https://doi.org/10.1152/ajprenal.1977.233.4.F298

Sneyd J, Tsaneva-Atanasova K, Bruce JIE, Straub SV, Giovannucci DR, Yule DI (2003) A model of calcium waves in pancreatic and parotid acinar cells. Biophys J 85(3):1392–405. https://doi.org/10.1016/S0006-3495(03)74572-X

Sneyd J, Tsaneva-Atanasova K, Reznikov V, Bai Y, Sanderson MJ, Yule DI (2006) A method for determining the dependence of calcium oscillations on inositol trisphosphate oscillations. Proc Natl Acad Sci USA 103(6):1675–80. https://doi.org/10.1073/pnas.0506135103

Sneyd J, Crampin E, Yule D (2014) Multiscale modelling of saliva secretion. Math Biosci 257:69–79. https://doi.org/10.1016/j.mbs.2014.06.017

Sneyd J, Han JM, Wang L, Chen J, Yang X, Tanimura A, Sanderson MJ, Kirk V, Yule DI (2017) On the dynamical structure of calcium oscillations. Proc Natl Acad Sci USA 114(7):1456–1461. https://doi.org/10.1073/pnas.1614613114

Swanson CH (1977) Isotonic water transport in secretory epithelia. Yale J Biol Med 50(2):153–63

Thorn P (1996) Spatial domains of Ca^{2+} signaling in secretory epithelial cells. Cell Calcium 20:203–214

Thorn P, Lawrie AM, Smith PM, Gallacher DV, Petersen OH (1993) Local and global cytosolic Ca^{2+} oscillations in exocrine cells evoked by agonists and inositol trisphosphate. Cell 74(4):661–8

Thul R, Bellamy TC, Roderick HL, Bootman MD, Coombes S (2008) Calcium oscillations. Adv Exp Med Biol 641:1–27

Torre V (1975) Synchronization of non-linear biochemical oscillators coupled by diffusion. Biol Cybern 17:137–144

Tosteson D, Hoffman J (1960) Regulation of cell volume by active cation transport in high and low potassium sheep red cells. J Gen Physiol 44:169–94

Turner RJ, Sugiya H (2002) Understanding salivary fluid and protein secretion. Oral Dis 8(1):3–11

Vera-Sigüenza E, Catalán MA, Peña-Münzenmayer G, Melvin JE, Sneyd J (2018) A mathematical model supports a key role for Ae4 (Slc4a9) in salivary gland secretion. Bull Math Biol 80(2):255–282. https://doi.org/10.1007/s11538-017-0370-6

Vera-Sigüenza E, Pages N, Rugis J, Yule DI, Sneyd J (2019) A mathematical model of fluid transport in an accurate reconstruction of parotid acinar cells. Bull Math Biol 81(3):699–721. https://doi.org/10.1007/s11538-018-0534-z

Vera-Sigüenza E, Pages N, Rugis J, Yule DI, Sneyd J (2020) A multicellular model of primary saliva secretion in the parotid gland. Bull Math Biol 82(3):38. https://doi.org/10.1007/s11538-020-00712-3

Warren NJ, Tawhai MH, Crampin EJ (2009) A mathematical model of calcium-induced fluid secretion in airway epithelium. J Theor Biol 259(4):837–849

Watt SD, Gu X, Smith RD, Spitzer NC (2000) Specific frequencies of spontaneous Ca^{2+} transients upregulate GAD 67 transcripts in embryonic spinal neurons. Mol Cell Neurosci 16(4):376–87. https://doi.org/10.1006/mcne.2000.0871

Weinstein A (1994) Mathematical models of tubular transport. Ann Rev Physiol 56:691–709

Weinstein AM (1999) Modeling epithelial cell homeostasis: steady-state analysis. Bull Math Biol 61(6):1065–1091

Weinstein AM (2020) A mathematical model of the rat kidney. Antidiuresis. Am J Physiol Renal Physiol II. https://doi.org/10.1152/ajprenal.00046.2020

Weinstein AM, Stephenson JL (1979) Electrolyte transport across a simple epithelium. Steady-state and transient analysis. Biophys J 27(2):165–86. https://doi.org/10.1016/S0006-3495(79)85209-1

Weinstein AM, Stephenson JL (1981a) Coupled water transport in standing gradient models of the lateral intercellular space. Biophys J 35(1):167–191

Weinstein AM, Stephenson JL (1981b) Models of coupled salt and water transport across leaky epithelia. J Membr Biol 60(1):1–20

Wu D, Boucher RC, Button B, Elston T, Lin CL (2018) An integrated mathematical epithelial cell model for airway surface liquid regulation by mechanical forces. J Theor Biol 438:34–45. https://doi.org/10.1016/j.jtbi.2017.11.010

Bull Math Biol (2021) 83:26
https://doi.org/10.1007/s11538-021-00859-7

Society for Mathematical Biology

SPECIAL ISSUE: CELEBRATING J. D. MURRAY

Modelling Cell Invasion: A Review of What JD Murray and the Embryo Can Teach Us

Paul M. Kulesa[1,2] · Jennifer C. Kasemeier-Kulesa[1] · Jason A. Morrison[1] · Rebecca McLennan[1] · Mary Cathleen McKinney[1] · Caleb Bailey[3]

Received: 10 November 2020 / Accepted: 8 January 2021 / Published online: 17 February 2021

Abstract

Cell invasion and cell plasticity are critical to human development but are also striking features of cancer metastasis. By distributing a multipotent cell type from a place of birth to distal locations, the vertebrate embryo builds organs. In comparison, metastatic tumor cells often acquire a de-differentiated phenotype and migrate away from a primary site to inhabit new microenvironments, disrupting normal organ function. Countless observations of both embryonic cell migration and tumor metastasis have demonstrated complex cell signaling and interactive behaviors that have long confounded scientist and clinician alike. James D. Murray realized the important role of mathematics in biology and developed a unique strategy to address complex biological questions such as these. His work offers a practical template for constructing clear, logical, direct and verifiable models that help to explain complex cell behaviors and direct new experiments. His pioneering work at the interface of development and cancer made significant contributions to glioblastoma cancer and embryonic pattern formation using often simple models with tremendous predictive potential. Here, we provide a brief overview of advances in cell invasion and cell plasticity using the embryonic neural crest and its ancestral relationship to aggressive cancers that put into current context the timeless aspects of his work.

Keywords Cell invasion · Plasticity · Modeling · Neural crest · Melanoma · Neuroblastoma

✉ Paul M. Kulesa
pmk@stowers.org

[1] Stowers Institute for Medical Research, Kansas City, MO 64110, USA

[2] Department of Anatomy and Cell Biology, School of Medicine, University of Kansas, Kansas City, KS 66160, USA

[3] Department of Biology, Brigham Young University-Idaho, Rexburg, ID 83460, USA

1 Introduction

The neural crest (NC) is a multipotent and highly invasive stem cell population that travels along stereotypical pathways in discrete streams throughout the vertebrate embryo (Fig. 1). Neural crest cells give rise to diverse derivatives that include bone, cartilage, neurons, and melanocytes (LeDouarin and Kalcheim 1999). Pathologies resulting from abnormal neural crest patterning are called neurocristopathies, of which there are currently 66 different classifications ranging from craniofacial deformations to life-threatening heart deformities (Vega-Lopez et al. 2018). The predictability of neural crest cell trajectories and differentiation into specific cell types, together with accessibility to in vivo imaging, surgical and molecular manipulation, and gene profiling make the neural crest an ideal model system to study questions in cell invasion and plasticity. Further, since the neural crest are the ancestral cell

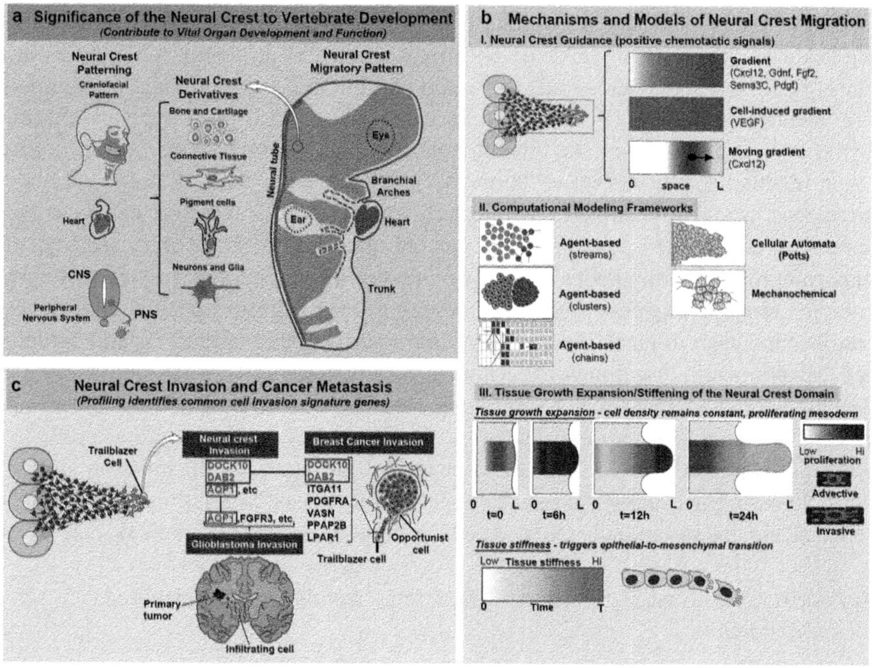

Fig. 1 The neural crest as an experimental and theoretical model for cell invasion in development and cancer. **a** The significance of the neural crest to vertebrate development includes the diverse set of derivatives to which multipotent cells may contribute during organogenesis. The neural crest migrate extensively throughout the embryo to nearly every organ and contribute to the craniofacial pattern (by invasion into the branchial arches), heart formation, and the peripheral nervous system. **b** Neural crest cells are guided by attraction (shown) and inhibitory cues that are present within the microenvironments through which cells travel. These cues may be in different forms, including (I) chemotactic signals ingradient, cell-induced gradient, and moving gradient profiles and this complexity has spurred the creation of (II) multiple computational modeling frameworks. These modeling frameworks must be coordinated with emerging empirical measurements of (III) tissue growth expansion (via cell proliferation and cell dynamics) and tissue stiffening. **c** Advances in isolation and gene profiling of the most invasive cells throughout embryonic and cancer cell invasion are revealing common genes that are shared between the neural crest, breast cancer and glioblastoma

type of aggressive pediatric neuroblastoma and melanoma cancers, this model system is appropriate to consider questions at the interface of development and cancer. In this article, we provide an overview of invasion and plasticity of the neural crest and neural crest-derived cancers by summarizing experimental data and modelling. In particular, we highlight the main forces driving neural crest invasion, how this has influenced models of cell invasion, and speculate how emerging single cell gene expression of the most invasive neural crest relate to a JDM-derived proliferation-invasion (PI) model(s) for glioblastoma. We then present the strengths of the embryonic neural crest microenvironment to study human neural crest-derived cancer cell invasion, its ability to regulate tumor cell plasticity, and relate this to current and emerging mathematical models of neuroblastoma and melanoma.

2 Neural Crest Cell Guidance

2.1 Chemoattractants in Space and Time

Neural crest cells emerge from the dorsal neural tube in a head-to-trunk manner and must travel through dense extracellular matrix and other cell types to reach peripheral targets (Fig. 1a). A number of guidance factors have emerged as neural crest cell chemoatttractants (Theveneau and Mayor 2012; Kulesa and McLennan 2015). These factors include vascular endothelial growth factor (VEGF) (McLennan et al. 2010) and platelet-derived growth factor (PDGF) (Eberhart et al. 2008; He and Soriano 2013) in the head, fibroblast growth factor (FGF) in the head, cardiac and trunk regions (Kubota and Ito 2000; Creuzet et al. 2006; Sato et al. 2011; Dunkel et al. 2020), glial cell-derived neurotrophic factor (GDNF) described in the gut (Lake and Heuckeroth 2013), stromal cell-derived factor 1 (Cxcl12) (Olesnicky-Killian et al. 2009; Kasemeier-Kulesa et al. 2010; Saito et al. 2012; Theveneau et al. 2013) in the head and trunk, and brain-derived neurotrophic factor (BDNF) present in the trunk (Kasemeier-Kulesa et al. 2015). These chemotactic factors may be present at a peripheral target, such as in the case of Cxcr4-positive chick trunk neural crest cells that migrate from the dorsal neural tube to the ventral-positioned dorsal aorta, the source of Cxcl12 (Kasemeier-Kulesa et al. 2010; Saito et al. 2012). Neural crest cells also chase a moving tissue, as in the example of Cxcr4-positive Xenopus neural crest that are attracted to Cxcl12 secreted from the epibranchial placode (Theveneau et al. 2013). Chemical signals may also be distributed uniformly throughout the neural crest migratory domain, as is the case with VEGF in the chick head (McLennan et al. 2010). Lastly, spinal cord-derived axons secrete BDNF, which is then sensed by TrkB-expressing trunk neural crest cells, promoting an interaction between the axons and the neural crest to establish the connection between the central and peripheral nervous systems (Kasemeier-Kulesa et al. 2015). Identification of the tissue sources of these chemotactic factors, present spatially in various tissues within the neural crest microenvironments and their tight regulation in time, adds to the complexity of the neural crest cell migration problem (Fig. 1b).

2.2 Combining Modelling and Experiment to Better Understand Collective Cell Migration

How do neural crest cells respond to an attraction signal that is secreted by a moving target? This question has driven experimentalists to better understand neural crest cell behavior at the individual and collective levels (Fig. 1b). Using a rule-based Cellular Potts framework, Szabo et al (2016) showed that contact inhibition of locomotion (CIL) and local cell–cell co-attraction (CoA) are sufficient for neural crest cells to move as a collective group within a simplified two-dimensional domain flanked by inhibitory signals. CIL and CoA cooperate to produce persistence of polarity of a collective group by suppressing random protrusions, as modeled by the integration of chemical RhoGTPase activity with mechanical cell membrane dynamics (Merchant et al. 2018).

In the absence of a graded chemoattractant, collective migration persists within a two-dimensional domain of specific width to ensure that CoA is functional; for example, a narrow domain limits the strength of CoA (cells remain too close) resulting in the lack of suppression of new protrusions and spreading out of the cell cluster (Merchant et al. 2018). Further, in the presence of a graded chemoattractant and without flanking inhibitory zones, a collective group of cells may move toward the source and more persistently as the size of the group increases (Merchant and Feng 2020). Similarly, the introduction of a moving chemoattractant source into their Cellular Potts model framework (Szabo et al. 2019) reproduces the observed chase and run behavior of the neural crest and epibranchial placode (Theveneau et al. 2013). The strength and duration of the CIL response for each cell can be modelled by an explicit dynamical equation (Colombi et al. 2020), with these models setting the stage for future studies of how the chase phase is determined by the diffusive and reaction kinetics of the Cxcl12 attractant.

In contrast, the mechanisms underlying persistence of neural crest cell polarity are not well understood, especially when cells are presented with a chemical signal uniformly distributed throughout the migratory domain (Fig. 1b). Chick neural crest cells at the invasive stream front display filopodial protrusions in many distinct directions (Teddy and Kulesa 2004). Through careful measurements, local bursts of filopodia precede the generation of larger protrusions that are spatially biased and correlate with the direction of cell guidance (Genuth et al. 2018), confirming a polarized morphology of neural crest cells distinct from CIL. Differences in leader versus follower cell behaviors and readout of guidance cues were predicted by an agent-based model of neural crest cell migration (McLennan et al. 2012; Schumacher et al. 2016). In this model, the uniform distribution of VEGF signal in the chick cranial neural crest microenvironment is consumed by all cells and diluted by empirically measured tissue growth that together establish a cell-induced gradient that drives leader persistence of polarity (McLennan et al. 2012, 2015). Interestingly, models in cancer metastasis report robust self-induced gradients since cells maintain the chemoattractant at optimal concentrations (Tweedy et al. 2016; Tweedy and Insall 2020), suggesting a common theme for further exploration at the interface of development and cancer.

2.3 In Vivo Measurements of Tissue Growth Expansion and Stiffening Drive Modeling

Innovative methods in tissue growth and dynamics measurements are advancing our understanding of the role of the tissues through which neural crest cells migrate collectively and both complement and complicate our understanding of cell chemotaxis mechanisms. In vivo measurements of cell proliferation and cell density within the chick cranial neural crest migratory domain show non-uniform mesoderm growth in space and time (McKinney et al. 2020). Further, the mesoderm stiffens over time and this stiffening appears to initiate neural crest epithelial-to-mesenchymal transition and triggers collective cell migration (Barriga et al. 2018). Modeling efforts to explore how the mechanics of tissue dynamics drive collective neural crest cell migration are underway (Giniunaite et al. 2019) and benefit from these recent in vivo data. These new modelling efforts may draw on sophisticated mechanochemical models of morphogenesis put forth by Murray and Oster (Murray et al. 1983; Murray and Maini 1986). Mechanochemical models captured key in vitro observations of the interactions between the mechanical forces generated by cells and their extracellular matrix and led to explanations of developmental processes such as limb cartilage formation and its evolution (Oster et al. 1988). With the expanding medical role of the neural crest as a means of cartilage repair, there will be an important synergy between modeling and experiment that combine the knowledge of cell migration, cell differentiation and in vivo tissue mechanics that leverage the strengths of the earlier work of JDM and colleagues.

3 A Novel Gene Signature of the Invasive Neural Crest Cells

Neural crest cells encounter complex developing tissues during migration. These migratory routes are replete with mesoderm, extracellular matrix, and endothelial cells forming vasculature. Given the complex cell migratory behaviors and diverse derivatives of the neural crest that we have described above, a natural question to ask is how travel through different microenvironments alters gene expression within the neural crest. That is, cells at the leading edge of the invasive front see a virgin territory and their gene expression changes may offer hints on cell invasion signaling pathways. Recent technical advances have reduced the cost and technical barriers to performing whole transcriptome profiling by bulk and single cell RNA-seq. This has led to multiple studies of fate restriction and lineage choices of the cranial (Williams et al. 2019; Soldatov et al. 2019; Lumb et al. 2017) and trunk (Simoes-Costa and Bronner 2016) neural crest along with target and neighboring tissues (Plouhinec et al. 2017; Xu et al. 2019). However, what remains unclear is the extent of molecular heterogeneities within the migratory neural crest cell streams throughout migration and more specifically, what are the differences in expression between the leader and follower cells.

The capacity of a neural crest cell to become invasive requires a change in gene expression such that the cell migrates and interacts with different microenvironments, but what are these genes and are they functionally significant to invasion?

Lead cells remain within the migratory front with small changes in cell neighbor relationships and no widespread leap frogging or cell rearrangements (Kulesa et al. 2008; Ridenour et al. 2015, 2016). We mentioned earlier that a cell-induced gradient model of neural crest migration predicted leader and follower strategy with clear functional differences (McLennan et al. 2012). Isolation of the leading edge neural crest cells and subpopulations within a migratory stream in the head revealed molecular heterogeneities that depend on space (McLennan et al. 2012, 2015), found using single cell polymerase chain reaction (PCR) technology to amplify small segments of DNA (McLennan et al. 2012) and single-cell (sc) RNA-seq (McLennan et al. 2015). Tissue transplantation of the followers into the lead position show cranial neural crest cells rapidly change gene expression to match the lead invasive signature, demonstrating the gene expression pattern is driven by the local microenvironment and not hard-wired in premigratory neural crest cells (McLennan et al. 2012).

Deeper analysis of the chick cranial neural crest stream recently revealed there is a novel transcriptional signature of the most invasive neural crest cells that is conserved in time and contains a subset of genes related to cell polarity, cell motility, and cell contractility (Morrison et al. 2017). Perturbation of a subset of these genes, including the water channel protein Aquaporin-1 (McLennan et al. 2020), leads to reduced cell invasion. Interestingly, AQP-1 has been detected in several aggressive human cancers and its expression correlates with poor disease prognosis (Tomita et al. 2017; De Ieso and Yool 2018), suggesting the exciting possibility that common properties of cell invasion may underlie development and cancer.

3.1 The Neural Crest Invasion Signature and Its Relationship to Cancer Metastasis

The vast majority of all cancer-related deaths can be ascribed to the metastatic dissemination of cells from the primary tumor. The high rate of misdiagnosis of metastatic cancer is, in part, due to a lack of definitive molecular biomarkers of disease onset and progression. Deciphering the molecular signature of invasive cancers, including the ability to predict metastatic potential and the preferred site of metastatic dissemination, would improve diagnosis and prognosticate personalized treatment strategies.

Melanoma is initiated by the neoplastic transformation of neural crest-derived melanocytes. This has led to the hypothesis that transformed melanocytes may be intrinsically predisposed to increased metastatic potential due to the inherent invasive abilities of the neural crest (Gupta et al. 2005; Meier et al. 1998; Strizzi et al. 2011). Advances in cell isolation and scRNA-seq have proven invaluable for elucidating cell-type specific transcriptional signatures, often defined by an ensemble of genes rather than high expression of any single gene (Dueck et al. 2016; Zeisel et al. 2016), during single time-point events in melanoma (Tirosh et al. 2016; Song et al. 2020). These patient sample-based databases have been mined using regression analysis to establish a 12-gene signature predictive of patient survival rates in melanoma (Song et al. 2020). Of these genes, DOCK10 (Dedicator of Cytokinesis 10)-mediated Cdc42 activation has been shown to be associated with amoeboid invasion of melanoma cells (Gadea et al. 2008). Further, DOCK10 was identified as part of a 7-gene invasion signature derived from analyzing gene expression differences in trailblazers (lead invasive cells) versus opportunists (followers) and measurements of reduction in invasive properties using human breast cancer cell lines (Fig. 1c) grown in an in vitro ECM invasion assay (Westcott et al. 2015). Dock10 is present in the embryonic neural crest cell invasion signature (Morrison et al. 2017), demonstrating links between the embryonic neural crest, neural crest-derived and other human cancers.

If the neural crest invasion signature shares some genes with human metastatic melanoma and breast cancer, may it share genes and signaling pathways with invasive glioblastoma (Fig. 1c)? Would this information help to make current glioblastoma models more accurate and/or offer potential clinical targets that better define the invasive front in patient specimens or inhibit diffusion, respectively? Shortly after arriving at the University of Washington, JDM pioneered the development of simple yet elegant linear partial differential equation (PDE) mathematical models of glioblastoma after discussions on the subject with the prominent neurosurgeon, Charles Alvord (described in his autobiography; Murray 2018). In their model, the invasive nature of glioblastoma is defined by cell proliferation and diffusion (Tracqui et al. 1995; Woodward et al. 1996). Using patient in vivo parameters obtained from two successive brain scans, the model estimates survival times and response to resection or radiation treatment, often predicting that the tumor invasive front was far more invasive than MRI detection limits would permit (Burgess et al. 1997). Revisions of the model have incorporated differences in cell motility through white versus grey matter (Swanson et al. 2000).

It is intriguing to consider how the application of recent information from theoretical leader/follower neural crest cell migration models (Schumacher et al. 2016; Giniunaite et al. 2019) and experimentally derived gene signatures of glioblastoma patients (Darmanis et al. 2017) may advance recent brain cancer modeling (Jackson et al. 2015; Bruno et al. 2020). For example, scRNAseq of infiltrating cells isolated from four glioblastoma patients and compared with tumor core cells found a consistent gene signature between patients (Darmanis et al. 2017), suggesting a common mechanism of invasion. Analysis of gene expression of infiltrating cells in glioblastoma revealed significant enrichment of categories relevant to tumor migration (Demuth and Berens 2004), including cell–cell adhesion and AQP-1 anion transport (in common with the invasive embryonic neural crest as shown in Fig. 1c). Aquaporins 1 and 4 are both strongly elevated in glioblastoma and correlate with poor chance of survival (Yool and Ramesh 2020). Inhibition of AQP-1 has shown an enhanced blocking effect on colon cancer cell motility (De Ieso et al. 2019). Thus, factors derived from the embryonic neural crest invasion signature and glioblastoma patient data may inform further glioblastoma models that consider the function of invading cells and provide the medical community with novel clinical targets and markers of invasion that enhance detection of the invasive front.

4 Using the Embryonic Neural Crest Microenvironment to Predict and Control Cancer Cell Plasticity and Invasion

Cancer cells must regulate plasticity and invasion in order to metastasize. Accordingly, the role of tumor cell–microenvironment interactions are of critical importance and at the forefront of metastatic disease progression (Hinshaw and Shevde 2019). The embryonic neural crest offers a unique model system in which to study cell-microenvironment interactions in vivo, since cell emigration from the dorsal neural tube and migration along stereotypical pathways are thought to be sculpted by both intrinsic and extrinsic guidance cues found within the neural tube and surrounding microenvironments (LeDouarin and Kalcheim 1999; Trainor et al. 2002; Kulesa et al. 2004).

Do signals within the embryonic neural crest cell microenvironment that regulate normal development also control the plasticity and invasive ability of neural crest-derived cancers? To address this, a number of embryo transplantation models have been developed with over a 100-year history in the chick embryo (reviewed in Karnofsky et al. 1952) and more recently the zebrafish (Lee et al. 2005; Taylor and Zon 2009; Veinotte et al. 2014) that permit single cell interrogation of changes in human melanoma and neuroblastoma gene expression and cell dynamics after transplantation into the head or trunk neural crest cell migratory pathways (Kulesa et al. 2006; Bailey and Kulesa 2014; Delloye-Bourgeois et al. 2017). These studies have shown that aggressive human metastatic melanoma or neuroblastoma cells respond to both permissive and inhibitory neural crest microenvironmental signals that regulate the timing and direction of invasion to coincide with the neural crest migration pattern (Bailey and Kulesa 2014; Delloye-Bourgeois et al. 2017). Further, manipulation of Ephrin signaling in melanoma (Bailey and Kulesa 2014) or Semaphorins in neuroblastoma (Delloye-Bourgeois et al. 2017) alters the invasive cell behaviors and may be targets for therapeutics.

The embryonic neural crest microenvironment is also capable of regulating human neural crest-derived cancer cell plasticity (Kasemeier-Kulesa et al. 2018a; Delloye-Bourgeois and Castellani 2019). When human metastatic melanoma cells are exposed to the chick embryonic neural crest microenvironment, a subset of cells re-express MART-1 (a marker for melanosome formation) which is absent in nearly 70% of melanoma patients (Boiko et al. 2010) and induced through TrkA and CD271 receptors to the ligand nerve growth factor (NGF) (Kasemeier-Kulesa et al. 2018b). Importantly, CD271 expression correlates with human melanoma invasiveness (Boiko et al. 2010; Radke et al. 2017) and is thought to underlie phenotype switching between human melanoma proliferation and invasion (Restivo et al. 2017). Mathematical models of melanoma (thoroughly reviewed in Albrecht et al. 2020; Morais et al. 2017) that incorporate molecular heterogeneity may directly benefit from signals/pathways identified from the embryo that represent potential mechanisms to control tumor stemness properties.

In contrast to the rich history of mathematical models of melanoma, predictive modeling of pediatric neuroblastoma cancer has received less attention and is slowly emerging (Panetta et al. 2008; He et al. 2018). This is primarily due to a lack of fundamental knowledge of how signaling mechanisms that underlie normal sympathetic nervous system (SNS) development are corrupted to lead to neuroblastoma initiation and disease progression (Johnsen et al. 2019). The SNS arises from a stereotyped multi-step migration of neural crest cells that form a primary sympathetic ganglion and then move collectively to establish a neural circuit with axons projecting from the ventral spinal cord. Gene profiling of human patient samples with known unfavorable disease outcome has identified a developmental signaling receptor, TrkB, as a marker of high-risk neuroblastoma and target for therapy (Tetri et al. 2020; MacFarland et al. 2020).

Discovery of TrkB and its ligand, brain-derived neurotrophic factor (BDNF) play a critical role during normal SNS development, and this has led to the postulate that TrkB/BDNF and an associated developmental signaling network underlie neuroblastoma initiation (Kasemeier-Kulesa et al. 2015). Through modeling discussions, this has led to the development of a predictive algorithm of neuroblastoma based on Boolean modeling (Kasemeier-Kulesa et al. 2018), similar to what has been successfully deployed in colon cancer prediction (Wynn et al. 2014). Together, these studies highlight the success of an interdisciplinary approach to neuroblastoma as a blend of developmental and cancer cell biology signaling that may be integrated within a clear, logical, direct and verifiable modeling framework championed by JDM.

5 Discussion

One need only to read the first several lines of any mathematical biology paper written by JDM to learn that he and his colleagues intended to present a significant problem then offer a clear, logical, direct, and verifiable model solution. In their models applied to medical problems, such as glioblastoma and wound healing, JDM and colleagues endeavored to present a modeling solution that predicted cell growth and invasion and thus could be of accurate predictive use for treatment strategy. The success of his approach relied on a close, sometimes intimate, dialogue with experimentalists that would often lay bare their confidences in some data versus other, but the end result was a trusted, sustainable partnership that produced informative models. This strategy sounds simplistic, but applied mathematicians and experimentalists often continue to work separately such that models are translationally uninformative to human disease or predict the obvious. On the other hand, experimentalists are often driven to distraction by technical advances that promise enormous data collection at increasingly small scales of minutia that lead further away from the mechanistic basis of their problem. As we move forward to tackle ever-present questions in cancer and birth defects and train the next generation of interdisciplinary scientists, we should be reminded of JDM's contributions and most importantly his approach to the side-by-side education and collaboration between multiple fields and investigators as a model for success.

Acknowledgements PMK and JCK would like to thank the generous funding from the St. Baldrick's Foundation, and the Stowers Institute for Medical Research.

References

Albrecht M, Lucarelli P, Kulms D, Sauter T (2020) Computational models of melanoma. Theor Biol Med Model 17(1):8. https://doi.org/10.1186/s12976-020-00126-7

Bailey CM, Kulesa PM (2014) Dynamic interactions between cancer cells and the embryonic microenvironment regulate cell invasion and reveal EphB6 as a metastasis suppressor. Mol Cancer Res 2(9):1303–1313

Barriga Eh, Franze K, Charras G, Mayor R (2018) Tissue stiffening coordinates morphogenesis by triggering collective cell migration in vivo. Nature 554(7693):523–527

Boiko AD et al (2010) Human melanoma-initiating cells express neural crest nerve growth factor receptor CD271. Nature 466(7302):133–137

Bruno et al (2020) Progress and opportunities to advance clinical cancer therapeutics using tumor dynamic models. Clin Cancer Res 26(8):1787–1795

Burgess PK, Kulesa PM, Murray JD, Alvord EC (1997) The interactions of growth rates and diffusion coefficients in a three-dimensional mathematical model of gliomas. J Neuropathol Exp Neurol 56(6):704–713

Colombi A, Scianna M, Painter KJ, Preziosi L (2020) Modelling chase-and-run migration in heterogeneous populations. J Math Biol 80(1–2):423–456

Creuzet SE, Martinez S, LeDouarin NM (2006) The cephalic neural crest exerts a critical effect on forebrain and midbrain development. PNAS 103(38):14033–14038

Darmanis S et al (2017) Single-Cell RNA-seq analysis of infiltrating neoplastic cells at the migrating front of human glioblastoma. Cell Rep 21:1399–1410

De Ieso ML, Yool AJ (2018) Mechanisms of aquaporin-facilitated cancer invasion and metastasis. Front Chem 6:135

De Ieso ML et al (2019) Combined pharmacological administration of AQP1 ion channel blocker AqB011 and water channel blocker Bacopaside II amplifies inhibition of colon cancer cell migration. Sci Rep 9:12635

Delloye-Bourgeois C, Castellani V (2019) Hijacking of embryonic programs by neural crest-derived neuroblastoma: from physiological migration to metastatic dissemination. Front MolNeurosci 12:52

Delloye-Bourgeois C et al (2017) Microenvironment-driven shift of cohesion/detachment balance within tumors induces a switch toward metastasis in neuroblastoma. Cancer Cell 32(4):427–443

Demuth T, Berens ME (2004) Molecular mechanisms of glioma cell migration and invasion. J Neurooncol 70(2):217–228

Dueck HR et al (2016) Assessing characteristics of RNA amplification methods for single cell RNA sequencing. BMC Genom 17(1):966

Dunkel H, Chaverra M, Bradley R, Lefcort F (2020) FGF signaling is required for chemokinesis and ventral migration of trunk neural crest cells. Dev Dyn 249(9):1077–1097

Eberhart JK et al (2008) MicroRNA Mirn140 modulates Pdgf signaling during palatogenesis. Nat Genet 40(3):290–298

Gadea G, Sanz-Moreno V, Self A, Godi A, Marshall CJ (2008) DOCK10-mediated Cdc42 activation is necessary for amoeboid invasion of melanoma cells. Curr Biol 18(19):1456–1465

Genuth MA, Allen C, Mikawa T, Weiner OD (2018) Chick cranial neural crest cells use progressive polarity refinement, not contact inhibition of locomotion, to guide their migration. Dev Biol 444:S252–S261

Giniunaite R, Baker RE, Kulesa PM, Maini PK (2019) Modelling collective cell migration: neural crest as a model paradigm. J Math Biol 80(1–2):481–504

Gupta PB et al (2005) The melanocyte differentiation program predisposes to metastasis after neoplastic transformation. Nat Genet 37(10):1047–1054

He F, Soriano P (2013) A critical role for PDGFR-alpha signaling in medial nasal process development. PLoS Genet 9(9):e1003851. https://doi.org/10.1371/journal.pgen.1003851

He Y, Kodali A, Wallace DI (2018) Predictive modeling of neuroblastoma growth dynamics in xenograft model after bevacizumab anti-VEGF treatment. Bull Math Biol 80(8):2026–2048

Hinshaw DC, Shevde LA (2019) The tumor microenvironment innately modulates cancer progression. Cancer Res 79(18):4557–4566

Jackson PR, Juliano J, Hawkins-Daarud A, Rockne RC, Swanson KR (2015) Patient-specific mathematical neuro-oncology: using a simple proliferation and invasion tumor model to inform clinical practice. Bull Math Biol 77(5):846–856

Johnsen JI, Dyberg C, Wickstrom M (2019) Neuroblastoma—a neural crest derived embryonal malignancy. Front MolNeurosci Rev 12:9. https://doi.org/10.3389/fnmol.2019.00009.eCollection
Karnofsky DA, Ridgway LP, Patterson PA (1952) Tumor transplantation to the chick embryo. Ann NY Acad Sci 55(2):313–329
Kasemeier-Kulesa JC, Kulesa PM (2018) The convergent roles of CD271/p75 in neural crest-derived melanoma plasticity. Dev Biol 444:S352-355
Kasemeier-Kulesa JC, McLennan R, Romine MH, Kulesa PM, Lefcort F (2010) CXCR4 controls ventral migration of sympathetic precursor cells. J Neurosci 30(39):13078–13088
Kasemeier-Kulesa JC, Morrison JA, Lefcort F, Kulesa PM (2015) TrkB/BDNF signalling patterns the sympathetic nervous system. Nat Commun 6:8281
Kasemeier-Kulesa JC, Romine MH, Morrison JA, Bailey CM, Welch DR, Kulesa PM (2018a) NGF reprograms metastatic melanoma to a bipotent glial-melanocyte neural crest-like precursor. Biol Open 7(1):bio030817. https://doi.org/10.1242/bio.030817
Kasemeier-Kulesa JC, Schnell S, Woolley T, Spengler JA, Morrison JA, McKinney MC, Pushel I, Wolfe LA, Kulesa PM (2018b) Predicting neuroblastoma using developmental signals and a logic-based model. Biophys Chem 238:30–38
Kubota Y, Ito K (2000) Chemotactic migration of mesencephalic neural crest cells in the mouse. Dev Dyn 217(2):170–179
Kulesa PM, McLennan R (2015) Neural crest migration: trailblazing ahead. F1000 Prime Rep 7:02
Kulesa PM, Ellies DL, Trainor PA (2004) Comparative analysis of neural crest cell death, migration, and function during vertebrate embryogenesis. Dev Dyn 229:14–29
Kulesa PM, Kasemeier-Kulesa JC, Teddy JM, Margaryan NV, Seftor EA, Seftor RE, Hendrix MJ (2006) Reprogramming metastatic melanoma cells to assume a neural crest cell-like phenotype in an embryonic microenvironment. PNAS 103(10):3752–3757
Kulesa PM, Teddy JM, Stark DA, Smith SE, McLennan R (2008) Neural crest invasion is a spatially-ordered progression into the head with higher cell proliferation at the migratory front as revealed by the photoactivatable protein. KikGR Dev Biol 316(2):275–287
Lake JI, Heuckeroth RO (2013) Enteric nervous system development: migration, differentiation, and disease. J PhysiolGastrointest Liver Physiol 305(1):G1–G24
LeDouarin NM, Kalcheim C (1999) The neural crest, 2nd edn. Cambridge University Press, Cambridge
Lee LMJ, Seftor EA, Bonde G, Cornell RA, Hendrix MJC (2005) The fate of human malignant melanoma cells transplanted into zebrafish embryos: assessment of migration and cell division in the absence of tumor formation. Dev Dyn 233(4):1560–1570
Lumb R, Buckberry S, Secker G, Lawrence D, Schwarz Q (2017) Transcriptomic profiling reveals expression signatures of cranial neural crest cells arising from different axial levels. BMC Dev Biol 17(1):5
MacFarland SP, Naraparaju K, Iyer R, Guan P, Kolla V, Hu Y, Tan K, Brodeur GM (2020) Mechanisms of entrectinib resistance in a neuroblastoma xenograft model. Cancer Ther 19(3):920–926
McKinney MC, McLennan R, Giniunaite R, Baker RE, Maini PK, Othmer HG, Kulesa PM (2020) Visualizing mesoderm and neural crest cell dynamics during chick head morphogenesis. Dev Biol 461(2):184–196
McLennan R, Teddy JM, Kasemeier-Kulesa JC, Romine MH, Kulesa PM (2010) Vascular endothelial growth factor (VEGF) regulates cranial neural crest migration in vivo. Dev Biol 339(1):114–125
McLennan R, Dyson L, Prather KW, Morrison JA, Baker RE, Maini PK, Kulesa PM (2012) Multi-scale mechanisms of cell migration during development: theory and experiment. Development 139(16):2935–2944
McLennan R, Schumacher LJ, Morrison JA, Teddy JM, Ridenour DA, Box AC, Semerad CL, Li H, McDowell W, Kay D, Maini PK, Baker RE, Kulesa PM (2015) Neural crest migration is driven by a few trailblazer cells with a unique molecular signature narrowly confined to the invasive front. Development 142(11):2014–2025
McLennan R, McKinney MC, Teddy JM, Morrison JA, Kasemeier-Kulesa JC, Ridenour DA, Manthe CA, Giniunaite R, Robinson M, Baker RE, Maini PK (2020) Neural crest cells bulldoze through the microenvironment using Aquaporin 1 to stabilize filopodia. Development 147(1):dev185231. https://doi.org/10.1242/dev.185231
Meier F, Satyamoorthy K, Nesbit M, Jsu MY, Schittek B, Garbe C, Herlyn M (1998) Molecular events in melanoma development and progression. Front Biosci 3:D1005–D10010. https://doi.org/10.2741/a341

Merchant B, Feng JJ (2020) A Rho-GTPase based model explains group advantage in collective chemotaxis of neural crest cells. PhysBiol 17(3):036002. https://doi.org/10.1088/1478-3975/ab71f1
Merchant B, Edelstein-Keshet L, Feng JJ (2018) A Rho-GTPase based model explains spontaneous collective migration of neural crest cell clusters. Dev Biol 444:S262-273
Morais MCC et al (2017) Stochastic model of contact inhibition and the proliferation of melanoma in situ. Sci Rep 7(1):8026
Morrison JA, McLennan R, Wolfe JA, Gogol MM, Meier S, McKinney MC, Teddy JM, Holmes L, Semerad CL, Box AC, Li H, Hall KE, Perera AG, Kulesa PM (2017) Single-cell transcriptome analysis of avian neural crest migration reveals signatures of invasion and molecular transitions. Elife. 6:e28415. https://doi.org/10.7554/eLife.28415
Murray JDM (2018) My gift of polio-an unexpected life from Scotland's rustic hills to Oxford's hallowed halls and beyond. Chauntecleer Press, Toronto
Murray JD, Maini PK (1986) A new approach to the generation of pattern and form in embryology. Sci Prog 70(280):539–553
Murray JD, Oster GF, Harris AK (1983) A mechanical model for mesenchymal morphogenesis. J Math Biol 17(1):125–129
Olesnicky-Killian EC, Birkholz DA, Artinger KB (2009) A role for chemokine signaling in neural crest cell migration and craniofacial development. Dev Biol 333(1):161–172
Oster GF, Shubin N, Murray JD, Alberch P (1988) Evolution and morphogenetic rules: the shape of the vertebrate limb in ontogeny and phylogeny. Evolution 42(5):862–884
Panetta JC, Schaiquevich SVM, Stewart CF (2008) Using pharmacokinetic and pharmacodynamic modeling and simulation to evaluate importance of schedule in topotecan therapy for pediatric neuroblastoma. Clin Cancer Res 14(1):318–325
Plouhinec JL et al (2017) A molecular atlas of the developing ectoderm defines neural, neural crest, placode, and nonneural progenitor identity in vertebrates. PLoS Biol. 15(10):e2004045. https://doi.org/10.1371/journal.pbio.2004045
Radke J, Rossner F, Redmer T (2017) CD271 determines migratory properties of melanoma cells. Sci Rep 7:9834
Restivo G, Diener J, Cheng PF, Kiowski G, Bonalli M, Biedermann T et al (2017) Low neurotrophin receptor CD271 regulates phenotype switching in melanoma. Nat Commun 8:1988
Richardson J et al (2016) Leader cells define directionality of trunk, but not cranial, neural crest cell migration. Cell Rep 15:2076–2088
Ridenour DA, McLennan R, Teddy JM, Semerad CL, Haug JS, Kulesa PM et al (2015) The neural crest cell cycle is related to phases of migration in the head. Development 141:1095–1103
Saito D, Takase Y, Murai H, Takahashi Y (2012) The dorsal aorta initiates a molecular cascade that instructs sympatho-adrenal specification. Science 336:1578–1581
Sato A et al (2011) FGF8 signaling is chemotactic for cardiac neural crest cells. Dev Biol 354(1):18–30
Schumacher L, Kulesa PM, McLennan R, Baker RE, Maini PK (2016) Multidisciplinary approaches to understanding collective cell migration in developmental biology. Open Biol 6:160056. https://doi.org/10.1098/rsob.160056
Simoes-Costa M, Bronner ME (2016) Reprogramming of avian neural crest axial identity and cell fate. Science 352(6293):1570–1573
Soldatov R et al. (2019) Spatiotemporal structure of cell fate decisions in murine neural crest. Science 364(6444):eaas9536. https://doi.org/10.1126/science.aas9536
Song LB et al (2020) A twelve-gene signature for survival prediction in malignant melanoma patients. Ann Transl Med 8(6):312
Strizzi L, Hardy KM, Kirsammer GT, Gerami P, Hendrix MJC (2011) Embryonic signaling in melanoma: potential for diagnosis and therapy. Lab Invest 91(6):819–824
Swanson KR, Alvord EC, Murray JD (2000) A quantitative model for differential motility of gliomas in grey and white matter. Cell Prolif 33:317–329
Szabo A et al (2016) In vivo confinement promotes collective migration of neural crest cells. J Cell Biol 213(5):543–555
Szabo A et al (2019) Neural crest streaming as an emergent property of tissue interactions during morphogenesis. PLoS Comp Biol 15(4):e1007002. https://doi.org/10.1371/journal.pcbi.1007002
Taylor AM, Zon LI (2009) Zebrafish tumor assays: the state of transplantation. Zebrafish 6:339–346
Teddy JM, Kulesa PM (2004) In vivo evidence for short- and long-range cell communication in cranial neural crest cells. Development 131:6141–6151

Tetri LH et al (2020) RET receptor expression and interaction with TRK receptors in neuroblastomas. Oncol Rep 44(1):263–272

Theveneau E, Mayor R (2012) Neural crest delamination and migration: from epithelium-to-mesenchyme transition to collective cell migration. Dev Biol 366(1):34–54

Theveneau E, Steventon B, Scarpa E, Garcia S, Trepat X, Streit A, Mayor R (2013) Chase-and-run between adjacent cell populations promotes directional collective migration. Nat Cell Biol 15(7):763–772

Tirosh et al (2016) Dissecting the multicellular ecosystem of metastatic melanoma by single-cell RNA-seq. Science 352(6282):189–196

Tomita Y et al (2017) Role of aquaporin 1 signalling in cancer development and progression. Int J MolSci 18(2):299

Tracqui P, Cruywagen GC, Woodward DE, Bartoo GT, Murray JD, Alvord EC (1995) A mathematical model of glioma growth: the effect of chemotherapy on spatio-temporal growth. Cell Prolif 28(1):17–31

Trainor PA, Sobieszczuk D, Wilkinson D, Krumlauf R (2002) Signalling between the hindbrain and paraxial tissues dictates neural crest migration pathways. Development (Cambridge, England) 129:433–442

Tweedy L, Insall RH (2020) Self-generated gradients yield exceptionally robust steering cues. Front Cell Dev Biol. 8:133. https://doi.org/10.3389/fcell.2020.00133

Tweedy L et al (2016) Self-generated chemoattractant gradients: attractant depletion extends the range and robustness of chemotaxis. PLoS Biol. 42:46–51. https://doi.org/10.1016/j.ceb.2016.04.003

Vega-Lopez GA, Cerrizuela S, Tribulo C, Aybar MJ (2018) Neurocristopathies: new insights 150 years after the neural crest discovery. Dev Biol 444:S110-143

Veinotte CJ, Dellaire G, Berman JN (2014) Hooking the big one: the potential of zebrafish xenotransplantation to reform cancer drug screening in the genomic era. Dis Model Mech 7(7):745–754

Westcott JM et al (2015) An epigenetically distinct breast cancer cell subpopulation promotes collective invasion. J Clin Invest 125(5):1927–1943

Williams RM, Candido-Ferreira I, Repapi E, Gavriouchkina D, Senanayake U, Ling ITC, Telenius J, Taylor S, Hughes J, Sauka-Spengler T (2019) Reconstruction of the global neural crest gene regulatory network in vivo. Dev Cell 51(2):255–276

Woodward DE, Cook J, Tracqui P, Cruywagen GC, Murray JD, Alvord EC (1996) A mathematical model of glioma growth: the effect of extent of surgical resection. Cell Prolif 29(6):269–288

Wynn ML, Consul N, Merajver SD, Schnell S (2014) Inferring the effects of honokiol on the notch signaling pathway in SW480 colon cancer cells. Cancer Inform 13:1–12

Xu J, Liu Y, Adam M, Clouthier DE, Potter S, Jiang R (2019) Hedgehog signaling patterns the oral-aboral axis of the mandibular arch. Elife 14:8

Yool AJ, Ramesh S (2020) Molecular targets for combined therapeutic strategies to limit glioblastoma cell migration and invasion. FrontiPharmacol 11:358. https://doi.org/10.3389/fphar.2020.00358

Zeisel A et al (2016) Cell types in the mouse cortex and hippocampus revealed by single-cell RNA-seq. Science 347(6226):1138–1142

Bulletin of Mathematical Biology (2021) 83:32
https://doi.org/10.1007/s11538-021-00861-z

Society for Mathematical Biology

SPECIAL ISSUE: CELEBRATING J. D. MURRAY

Controlling the Spatial Spread of a Xylella Epidemic

Sebastian Aniţa[1,2] · **Vincenzo Capasso**[3] · **Simone Scacchi**[4]

Received: 31 August 2020 / Accepted: 31 December 2020 / Published online: 17 February 2021

Abstract

In a recent paper by one of the authors and collaborators, motivated by the Olive Quick Decline Syndrome (OQDS) outbreak, which has been ongoing in Southern Italy since 2013, a simple epidemiological model describing this epidemic was presented. Beside the bacterium *Xylella fastidiosa*, the main players considered in the model are its insect vectors, *Philaenus spumarius*, and the host plants (olive trees and weeds) of the insects and of the bacterium. The model was based on a system of ordinary differential equations, the analysis of which provided interesting results about possible equilibria of the epidemic system and guidelines for its numerical simulations. Although the model presented there was mathematically rather simplified, its analysis has highlighted threshold parameters that could be the target of control strategies within an integrated pest management framework, not requiring the removal of the productive resource represented by the olive trees. Indeed, numerical simulations support the outcomes of the mathematical analysis, according to which the removal of a suitable amount of weed biomass (reservoir of *Xylella fastidiosa*) from olive orchards and surrounding areas resulted in the most efficient strategy to control the spread of the OQDS. In addition, as expected, the adoption of more resistant olive tree cultivars has been shown to be a good strategy, though less cost-effective, in controlling the pathogen. In this paper for a more realistic description and a clearer interpretation of the proposed control measures, a spatial structure of the epidemic system has been included, but, in order to keep mathematical technicalities to a minimum, only two players have been described in a dynamical way, trees and insects, while the weed biomass is taken to be a given quantity. The control measures have been introduced only on a subregion of the whole habitat, in order to contain costs of intervention. We show that such a practice can lead to the eradication of an epidemic outbreak. Numerical simulations confirm both the results of the previous paper and the theoretical results of the model with a spatial structure, though subject to regional control only.

Keywords *Xylella fastidiosa* · Olive trees · Epidemics · Mathematical model · Reaction–diffusion models · Numerical simulations · Control strategies · Regional control

---

Extended author information available on the last page of the article

Mathematics Subject Classification 35-XX · 35B40 · 37N25 · 92C80 · 92D30 · 92D40 · 93B99

1 Introduction

The etiological agent of olive tree disease known as olive quick decline syndrome (OQDS) is the plant pathogenic bacterium *Xylella fastidiosa*, which is a vector-borne bacterium.

The main vector of *Xylella fastidiosa* in Southern Italy has been identified in the so-called meadow spittlebug, i.e., the Philaenus spumarius, a xylem sap-feeding specialist. Their juvenile form (nymphs) develops on weeds or ornamental plants, confined in a foam produced by themselves for protection from predators and temperature, while their adult form moves to olive tree canopies. Experiments have shown a larger infection prevalence of adults on olive trees than on weeds; this fact may lead to the assumption of infection of adults from infected olive trees more than from weeds.

X. fastidiosa transmission is the result of four events [see e. g. Almeida et al. (2005), Redak et al. (2004) and references cited therein]:

(a) acquisition from a source plant;

(b) attachment and retention to the vector's foregut;

(c) inoculation into a new host;

(d) development of the infection in a plant after inoculation.

A successful transmission also includes bacterial multiplication.

Once a plant is infected, bacteria multiply within the xylem vessels inducing the production of a gel in the plant xylem, which occludes the xylem vessels, thus inhibiting the flux of water through the lymph vessels eventually blocking the nutrition of the plant. Typical symptoms are leaf scorch, dieback of twigs, branches and even of the whole plant (Carlucci et al. 2013).

Sanitation of infected olive trees is unfeasible; the scope of our research is the mathematical modeling of the population dynamics of the ecosystem in the presence of infection. The availability of a sound mathematical model may lead to predictive analysis of the relevant populations, so as to suggest possible eradication strategies, or at least possible optimal control strategies.

In a previous paper (Brunetti et al. 2020), based on the outbreak of OQDS in Southern Italy, a model describing the epidemic was presented. It consists of a system of ordinary differential equations (ODEs) describing the evolution of the main three players, i.e., the insect vector, *Philaenus spumarius*, the olive trees, and weeds. A preliminary mathematical analysis and related numerical simulations have shown that "the removal of a significant amount of weeds (acting as a reservoir for juvenile insects) from olive orchards and surrounding areas has resulted in the most efficient strategy to control the spread of the OQDS. In addition, as expected, the adoption of more resistant olive tree cultivars has been shown to be a good strategy, though less cost effective, in controlling the pathogen."

The scope of the present paper is to extend the above results to a spatially structured model and to show in a more rigorous way that the spatial expansion of an OQDS outbreak can indeed be stopped by acting either on the weed biomass or on the choice of the olive cultivar. We have adopted a deterministic reaction–diffusion model. We recall that reaction–diffusion models can be interpreted as mean-field approximations of individual based models, which are more appropriate at the micro scale. Of course, in our case individual behaviors and possible randomness are lost. On the other hand, our approach allows a mathematical "qualitative" analysis of the system, including the derivation of eradicability theorems. Such a "qualitative" analysis has driven on the one side, our numerical experiments and, on the other side, anticipates future investigations on optimal control problems in a variety of scenarios.

In order to keep mathematical technicalities to a minimum, here the weed biomass is taken as a given field parameter and as in Brunetti et al. (2020), the insect life cycle has not been taken into account [more information about this last aspect can be found in Rossini et al. (2020)]. As possible control actions, insect traps, weed cut, choice of the cultivar, and reduction of contact rates have been taken into account. In the numerical simulations, particular attention has been paid to weed cut and choice of the cultivar, the other control measures being more difficult to implement. It is worth mentioning that, in recent investigations presented in Schneider et al. (2020), the authors, by means of a cellular automaton simulator, have confirmed the relevance of the olive cultivar as a possible effective control strategy.

A relevant contribution of our approach consists of the restriction of measures of intervention (control) only to a subregion of the whole habitat of interest. Due to diffusion, any point of the domain Ω is strongly connected to any other point of Ω, so that any action taken in a subregion $\omega \subset \Omega$ will eventually propagate to the whole habitat. This is the "leit motiv" of our proposal concerning regional control: "think globally, act locally" [see Aniţa and Capasso (2009)]. This practice may contribute in a significant way to improve the ratio cost/effectiveness of real control strategies. Mathematical analysis and numerical simulations have been carried out showing that it is indeed possible to eradicate the disease by such local action. Our aim is to analyze optimal control problems in future investigations, which may possibly lead to the identification of an optimal choice of the subregion of intervention.

The paper is organized as follows. In Sect. 2, the mathematical model is presented. In Sect. 3, possible control strategies are proposed, based on which an eradicability result is shown. Finally, in Sect. 4, numerical simulations are presented which confirm the analytical results. In the numerical simulations, the relevant parameters have been taken from Brunetti et al. (2020).

2 Our Model

Since the feeding behavior and metabolic processes are qualitatively similar for both nymphs and adults (Janse and Obradovic 2010), we will consider only one stage of active vectors of the infection. We will consider a spatially structured model which includes the population of vector insects and the population of infected olive trees.

We will denote by $C_I(x,t)$ the spatial density of the total population of insects, by $s_1(x,t)$ the spatial density of susceptible insects, and by $i_1(x,t)$ the spatial density of infected insects:

$$C_I(x,t) = s_1(x,t) + i_1(x,t).$$

$C_T(x,t)$ will denote the spatial density of the total biomass of olive trees, $s_2(x,t)$ the spatial density of the biomass of healthy trees, $i_2(x,t)$ the spatial density of the biomass of infected trees:

$$C_T(x,t) = s_2(x,t) + i_2(x,t).$$

All the parameters in the model are nonnegative quantities.

- INSECTS

Due to the fact that bacteria reside only in the foregut of an adult insect, the latter generate only healthy offspring [see, e.g., Almeida et al. (2005), Redak et al. (2004), and references cited therein]. The parameter r denotes the birth rate of new insects.

Reproduction, however, can occur only if bushes and other plants, whether healthy or infected, are present, explaining the presence of a "carrying capacity" $M(x)$ in the logistic term, describing the biomass of all such plants, which we simply call *weeds*.

Both healthy and infected insects experience natural mortality at a rate n; χ is a scale parameter.

Finally, we assume that insects may diffuse in the habitat (with constant diffusion coefficient to avoid purely technical complications).

As far as the local incidence of infection for insects, at point $x \in \overline{\Omega}$, and time $t \geq 0$, as in previous models (Capasso 1984; Aniţa and Capasso 2009), we assume that it arises due to biting of infected olive trees at any point $x' \in \Omega$ of the habitat, within a spatial neighborhood of x represented by a suitable probability kernel $k(x,x')$, depending on the specific structure of the local ecosystem [see also Shcherbacheva et al. (2018)]; as a trivial simplification, one may assume $k(x,\cdot)$ as a Gaussian density centered at x; hence, the "local incidence rate (i.r.)" for insects, at point $x \in \Omega$, and time $t \geq 0$, is taken as

$$(i.r.)_I(x,t) = \beta s_1(x,t) \int_\Omega k(x,x') i_2(x',t) \mathrm{d}x'.$$

Hence, the spatial dynamics of susceptible insects is expressed by the following equation

$$\begin{aligned}\frac{\partial s_1}{\partial t}(x,t) &= d\Delta s_1(x,t) + rC_I(x,t)[M(x) - \chi s_1(x,t)] - ns_1(x,t)\\ &\quad - \beta s_1(x,t) \int_\Omega k(x,x') i_2(x',t) \mathrm{d}x', \end{aligned} \tag{1}$$

while the spatial dynamics of infective insects is expressed by the following equation

$$\begin{aligned}\frac{\partial i_1}{\partial t}(x,t) &= d\Delta i_1(x,t) - ni_1(x,t) - r\chi i_1(x,t) C_I(x,t)\\ &\quad + \beta s_1(x,t) \int_\Omega k(x,x') i_2(x',t) \mathrm{d}x'. \end{aligned} \tag{2}$$

Table 1 Model parameters

Symbol	Description
r	Insect birth rate
χ	Insect intraspecific competition rate
n	Insect mortality rate
q	Healthy tree (canopy) regrowth rate
C	Tree carrying capacity parameter
ℓ	Elimination rate of trees by pruning or logging
b	Infection rate of trees by infected tools
μ	Infected tree mortality rate
α	Infected tree recovery rate
β	Insect infection rate by infected trees
ζ	Tree infection rate by infected insects

Both Eqs. (1) and (2) act in a spatial domain $\Omega \subset \mathbb{R}^2$, at times $t \in (0, +\infty)$. They are subject to suitable initial conditions, and we assume homogeneous Neumann boundary conditions.

- OLIVE TREES

For the olive trees, it is better to refer to their canopies, so that we may consider pruning and regrowth. Healthy trees (canopy) are produced at constant net regrowth rate q, get infected by contact with infected insects at rate ζ or by human activities, such as budding and grafting, at rate b. For trees, in view of their long survival, we can neglect natural mortality. Infected trees experience disease-related mortality μ and human-induced mortality ℓ due to pruning and logging.

Hence, the spatial dynamics of trees is expressed by the following equations

$$\begin{aligned}\frac{\partial s_2}{\partial t}(x,t) &= (q-\ell)s_2(x,t) - s_2(x,t)\frac{C_T(x,t)}{C} \\ &\quad - (\zeta i_1(x,t) + b\ell i_2(x,t))s_2(x,t) + \alpha i_2(x,t), \qquad (3)\\ \frac{\partial i_2}{\partial t}(x,t) &= -\mu i_2(x,t) - \ell i_2(x,t) - i_2(x,t)\frac{C_T(x,t)}{C} \\ &\quad + (\zeta i_1(x,t) + b\ell i_2(x,t))s_2(x,t) - \alpha i_2(x,t). \qquad (4)\end{aligned}$$

Both Eqs. (3) and (4) act in the spatial domain $\Omega \subset \mathbb{R}^2$, at times $t \in (0, +\infty)$. They are subject to suitable initial conditions.

Assumptions:

- $\Omega \subset \mathbb{R}^2$ is a bounded domain with a smooth boundary;
- $M \in L^\infty(\Omega)$, $M(x) \geq 0$ a.e. in Ω;
- $k \in L^\infty(\Omega \times \Omega)$, $k(x, x') \geq 0$ a.e. in $\Omega \times \Omega$; normalized to β;
- $d,\ r,\ \chi, n,\ q,\ \ell,\ C,\ \mu,\ \zeta,\ b,\ \alpha,\ \beta$ are positive constants.

3 Control and Eradicability

By summing Eqs. (1) and (2), we obtain an equation for the total insect population $C_I(x, t)$

$$\frac{\partial C_I}{\partial t}(x,t) = d\Delta C_I(x,t) + rC_I(x,t)[M(x) - \chi C_I(x,t)] - nC_I(x,t).$$

For M constant, pure Neumann boundary conditions allow constant in space and time solutions, satisfying the following equilibrium equation

$$rC_I^*(M - \chi C_I^*) - nC_I^* = 0,$$

that we may rewrite as

$$[r(M - \chi C_I^*) - n]C_I^* = 0.$$

This equation may admit two solutions

$$C_I^* = 0,$$

and the solution of

$$r(M - \chi C_I^{**}) - n = 0.$$

This shows that for $M \leq \dfrac{n}{r}$, no other nonnegative solutions are feasible, but the trivial one. Otherwise, another equilibrium is feasible

$$C_I^{**} = \frac{1}{\chi}\left(M - \frac{n}{r}\right), \tag{5}$$

provided

$$M > \frac{n}{r}.$$

From the above simple reasoning, we may conjecture that a way to eradicate the disease is to eliminate the insect population by a significant reduction of the carrying capacity M, which may be obtained by eliminating weeds in the relevant olive orchards. Equation (5) shows the quantitative role of the scale parameter χ; a smaller value of χ allows a larger value of C_I^{**}, so that we may say that $\dfrac{M}{\chi}$ plays the role of an "effective carrying capacity" of insects. Numerical simulations, reported in Sect. 4, illustrate these facts.

A more accurate analysis follows. Assume that certain constant controls are considered in a non-empty open subset $\omega \subset \Omega$.

Possible regional controls (acting in ω) are the following:

C1: Traps : $n \to n + \gamma_1$; increase insect death rate by insecticides and/or treated nets.

C2: Weed cut : $M \to M(1-\gamma_{21})$; decrease carrying capacity of insects by eliminating weeds in the olive orchards.
C3: Choice of the cultivar: $\zeta \to \zeta(1-\gamma_{22})$; decrease infection rate of trees, acting on the cultivar.
C4: Reduce contacts: $\beta \to \beta(1-\gamma_{23})$; decrease contact rate of trees with insects by installing treated nets.
C5: Clean tools: $b \to b(1-\gamma_3)$; decrease infection rate from tools to trees by disinfection.

For various reasons, including cost reduction, we may consider the possibility of implementing the proposed control measures only on a suitable subregion $\omega \subset \Omega$ of the whole habitat, including areas already affected by the epidemic, augmented by a preventive confinement area.

Let $\mathbb{I}_\omega$ denote the characteristic function of ω. Then, the controlled system is

$$\frac{\partial s_1}{\partial t}(x,t) = d\Delta s_1(x,t) + rC_I(x,t)[M(x)(1-\gamma_{21}\mathbb{I}_\omega(x)) - \chi s_1(x,t)] - (n+\gamma_1\mathbb{I}_\omega(x))s_1(x,t) - \beta(1-\gamma_{23}\mathbb{I}_\omega(x))s_1(x,t)\int_\Omega k(x,x')i_2(x',t)\mathrm{d}x', \tag{6}$$

$$\frac{\partial i_1}{\partial t}(x,t) = d\Delta i_1(x,t) - (n+\gamma_1\mathbb{I}_\omega(x))i_1(x,t) - r\chi i_1(x,t)C_I(x,t) + \beta(1-\gamma_{23}\mathbb{I}_\omega(x))s_1(x,t)\int_\Omega k(x,x')i_2(x',t)\mathrm{d}x', \tag{7}$$

$$\frac{\partial s_2}{\partial t}(x,t) = (q-\ell)s_2(x,t) - s_2(x,t)\frac{C_T(x,t)}{C} - (\zeta(1-\gamma_{22}\mathbb{I}_\omega(x))i_1(x,t) + b(1-\gamma_3\mathbb{I}_\omega(x))\ell i_2(x,t))s_2(x,t) + \alpha i_2(x,t), \tag{8}$$

$$\frac{\partial i_2}{\partial t}(x,t) = -\mu i_2(x,t) - \ell i_2(x,t) - i_2(x,t)\frac{C_T(x,t)}{C} + (\zeta(1-\gamma_{22}\mathbb{I}_\omega(x))i_1(x,t) + b(1-\gamma_3\mathbb{I}_\omega(x))\ell i_2(x,t))s_2(x,t) - \alpha i_2(x,t). \tag{9}$$

We assume that there is no flux of insects through the boundary of Ω (the domain is isolated):

$$\frac{\partial s_1}{\partial \nu}(x,t) = \frac{\partial i_1}{\partial \nu}(x,t) = 0, \quad x \in \partial\Omega,\ t > 0, \tag{10}$$

and that the following initial conditions are satisfied

$$s_j(x,0) = s_{j0}(x),\ i_j(x,0) = i_{j0}(x), \quad x \in \Omega,\ j \in \{1,2\}. \tag{11}$$

Assumptions:

- $s_{10},\ i_{10} \in L^\infty(\Omega),\quad s_{10}(x),\ i_{10}(x) \geq 0,$ a.e. $x \in \Omega$;
- $s_{20},\ i_{20} \in L^\infty(\Omega),\quad s_{20}(x),\ i_{20}(x) \geq 0,$ a.e. $x \in \Omega$;
- $\gamma_1 \geq 0, \gamma_{21}, \gamma_{22}, \gamma_{23} \in [0, 1], \gamma_3 \in [0, 1]$ are constants.

Let (s_1, i_1, s_2, i_2) be the solution to (6)–(11), satisfying

$$s_j,\ i_j \in L^\infty_{loc}(\overline{\Omega} \times [0, +\infty)),$$
$$s_j(x,t),\ i_j(x,t) \geq 0,\ \text{a.e. } (x,t) \in \Omega \times (0, +\infty),$$

$j \in \{1, 2\}$. We postpone, for the time being, the proof of the existence and uniqueness of such a solution.

As far as the trees are concerned, we may easily obtain an upper bound for the total canopy.

If we now add Eqs. (8) and (9) and take into account the initial conditions, we find that $C_T = s_2 + i_2$ satisfies

$$\begin{cases} \dfrac{\partial C_T}{\partial t} = (q - \ell)C_T - \dfrac{1}{C}C_T^2 - (\mu + q)i_2, & x \in \Omega,\ t > 0 \\ C_T(x, 0) = C_{T0}(x) = s_{20}(x) + i_{20}(x), & x \in \Omega. \end{cases} \tag{12}$$

If we denote by $\tilde{C}_T$ the solution to

$$\begin{cases} \dfrac{d\tilde{C}_T}{dt} = (q - \ell)\tilde{C}_T - \dfrac{1}{C}\tilde{C}_T^2, & t > 0 \\ \tilde{C}_T(0) = \|C_{T0}\|_\infty + 1, \end{cases} \tag{13}$$

we may state that

$$C_T(x,t) \leq \tilde{C}_T(t),\quad \text{a.e. } (x,t) \in \Omega \times (0, +\infty) \tag{14}$$

(here, and throughout this paper, $\|\cdot\|_p$ denotes the usual norm in $L^p(\Omega)$); it is obvious that $\tilde{C}_T$ is space independent.

To prove this inequality, we set $y(x,t) := \tilde{C}_T(t) - C_T(x,t)$ and note that y is the solution to

$$\begin{cases} \dfrac{\partial y}{\partial t}(x,t) = \theta(x,t)y(x,t) + g(x,t), & x \in \Omega,\ t > 0 \\ y(x, 0) = y_0(x), & x \in \Omega, \end{cases}$$

where $y_0(x) = \tilde{C}_T(0) - C_T(x, 0)$, $\theta(x,t) = q - \ell - \frac{1}{C}(\tilde{C}_T(t) + C_T(x,t))$, $g(x,t) = (\mu + q)i_2(x,t)$, if $x \in \Omega$, $t > 0$. Since y_0 and g are nonnegative, we have that $y(x,t) \geq 0$, a.e. $x \in \Omega$, $\forall t \geq 0$, and the inequality (14) follows.

As a consequence of (13)

(j) if $q \leq \ell$, then $\tilde{C}_T(t) \to 0$, as $t \to +\infty$; hence, $C_T(x,t) \to 0$, a.e. $x \in \Omega$, as $t \to +\infty$;

(jj) if $q > \ell$, then $\tilde{C}_T(t) \to C(q-\ell)$, as $t \to +\infty$.

We note that the quantity $C(q-\ell)^+$ is the carrying capacity of the tree population. We may now also note that if we denote by $\overline{C}_T$ the solution to

$$\begin{cases} \dfrac{d\overline{C}_T}{dt} = (q-\ell)\overline{C}_T, & t > 0 \\ \overline{C}_T(0) = \|C_{T0}\|_\infty + 1, \end{cases}$$

by similar arguments as for the previous inequality, we may also state that

$$\tilde{C}_T(t) \leq \overline{C}_T(t), \quad t > 0.$$

It is evident that $\overline{C}_T$ is space independent too. Altogether we may state

$$C_T(x,t) \leq \tilde{C}_T(t) \leq \overline{C}_T(t), \quad \text{a.e. } (x,t) \in \Omega \times (0,+\infty),$$

For the insect population, we may add Eqs. (6) and (7) and take into account the boundary and initial conditions, to obtain that $C_I = s_1 + i_1$ satisfies

$$\begin{cases} \dfrac{\partial C_I}{\partial t} = d\Delta C_I + r[M(x)(1-\gamma_{21}\mathbb{I}_\omega(x)) - \chi C_I]C_I - (n+\gamma_1\mathbb{I}_\omega(x))C_I, & x \in \Omega,\ t > 0 \\ \dfrac{\partial C_I}{\partial \nu}(x,t) = 0, & x \in \partial\Omega,\ t > 0 \\ C_I(x,0) = C_{I0}(x) = s_{10}(x) + i_{10}(x), & x \in \Omega. \end{cases} \tag{15}$$

By Banach's fixed point theorem and using the comparison result for the solutions to the linear parabolic equations [see Friedman (1964) and Protter and Weinberger (1994)], the existence and uniqueness of a nonnegative solution to (15) follow. Moreover, the solution satisfies

$$C_I(x,t) \leq \overline{C}_I(x,t), \quad \text{a.e. } (x,t) \in \Omega \times (0,+\infty),$$

where $\overline{C}_I$ is the solution to the linear parabolic equation

$$\begin{cases} \dfrac{\partial \overline{C}_I}{\partial t} = d\Delta \overline{C}_I + rM(x)(1-\gamma_{21}\mathbb{I}_\omega(x))\overline{C}_I - (n+\gamma_1\mathbb{I}_\omega(x))\overline{C}_I, & x \in \Omega,\ t > 0 \\ \dfrac{\partial \overline{C}_I}{\partial \nu}(x,t) = 0, & x \in \partial\Omega,\ t > 0 \\ \overline{C}_I(x,0) = C_{I0}(x) = s_{10}(x) + i_{10}(x), & x \in \Omega. \end{cases}$$

Turning back to (6)–(11), it follows via Banach's fixed point theorem (and using the comparison of the solutions to the linear parabolic equations and the comparison of the solutions to the linear first order ODEs) that there exists a unique solution (s_1, i_1, s_2, i_2) such that

$$s_j,\ i_j \in L^\infty_{loc}(\overline{\Omega} \times [0,+\infty)),$$

$$0 \leq s_1(x,t),\ i_1(x,t) \leq \overline{C}_I(x,t),\ \text{a.e. } (x,t) \in \Omega \times (0,+\infty),$$
$$0 \leq s_2(x,t),\ i_2(x,t) \leq \overline{C}_T(t),\ \text{a.e. } (x,t) \in \Omega \times (0,+\infty),$$

$j \in \{1, 2\}$ [for such an argument see Anița and Capasso (2009, 2012) and Anița et al. (2009)].

Let C_I^* be the maximal nonnegative solution to

$$\begin{cases} -d\Delta C_I - [rM(x)(1-\gamma_{21}\mathbb{I}_\omega(x)) - (n+\gamma_1\mathbb{I}_\omega(x))]C_I + r\chi C_I^2 = 0, & x \in \Omega \\ \dfrac{\partial C_I}{\partial \nu}(x) = 0, & x \in \partial\Omega, \end{cases} \tag{16}$$

and let $\tilde{\lambda}_1$ be the principal eigenvalue for the problem

$$\begin{cases} -d\Delta\varphi - [rM(x)(1-\gamma_{21}\mathbb{I}_\omega(x)) - (n+\gamma_1\mathbb{I}_\omega(x))]\varphi = \lambda\varphi, & x \in \Omega \\ \dfrac{\partial \varphi}{\partial \nu}(x) = 0, & x \in \partial\Omega. \end{cases} \tag{17}$$

It is obvious that $\tilde{\lambda}_1$ depends on Ω, ω, and the controls γ_1 and γ_{21}, and is an increasing function of γ_1, γ_{21}, and ω—by inclusion—via Rayleigh's principle.

By the same methods as in Anița et al. (2009), the following proposition can be shown to hold.

Proposition 1 *Under the above assumptions*

(i) *If* $\tilde{\lambda}_1 \geq 0$, *then* (16) *admits only the trivial solution.*
Moreover, for any initial condition, the solution to (15) *satisfies*

$$C_I(\cdot, t) \to 0 \quad \text{in } L^\infty(\Omega),$$

as $t \to +\infty$.

(ii) *If* $\tilde{\lambda}_1 < 0$, *then* (16) *has two nonnegative solutions: the trivial one and* $C_I^* > 0$.
Moreover, if C_{I0} *is not identically* 0, *then the solution to* (15) *satisfies*

$$C_I(\cdot, t) \to C_I^* \quad \text{in } L^\infty(\Omega),$$

as $t \to +\infty$.

We note that Proposition 1 confirms for the full reaction–diffusion system, the outcomes of the preliminary analysis presented at the beginning of Sect. 3. In particular, it is of interest to observe that, for small values of the weed distribution, $M(x)$, or a large value of the control parameter, γ_{21}, or a large domain of intervention, ω, $\tilde{\lambda}_1$ may become greater than or equal to zero, so that the only nonnegative solution of (16) is the trivial one, and the whole insect population eventually goes extinct.

The case $\tilde{\lambda}_1 < 0$ requires further investigation, since an additional nontrivial value of C_I^* is possible; we may then require suitable threshold conditions for the eventual extinction of the epidemic.

Now, let λ_1 be the principal eigenvalue for the problem

$$\begin{cases} -d\Delta\varphi + (n + \gamma_1 \mathbb{I}_\omega(x))\varphi + r\chi C_I^*(x)\varphi = \lambda\varphi, & x \in \Omega \\ \dfrac{\partial\varphi}{\partial\nu} = 0, & x \in \partial\Omega. \end{cases} \tag{18}$$

Notice that $\lambda_1 \geq n$ and that if $\gamma_1 \nearrow +\infty$, then $\lambda_1 \nearrow \lambda_1^*$, where λ_1^* is the principal eigenvalue for

$$\begin{cases} -d\Delta\varphi + n\varphi + r\chi C_I^*(x)\varphi = \lambda\varphi, & x \in \Omega \setminus \overline{\omega} \\ \varphi = 0, & x \in \partial\omega \\ \dfrac{\partial\varphi}{\partial\nu} = 0, & x \in \partial\Omega. \end{cases}$$

Notice that λ_1^* may be as large as we wish if γ_1 is sufficiently large and/or if ω is sufficiently large.

The following result concerns the eradicability of the disease in the most interesting situation when $\tilde{\lambda}_1 < 0, q > \ell$ and $C_{I0} \neq 0_{L^\infty(\Omega)}$.

Theorem 1 *If* $\tilde{\lambda}_1 < 0, q > \ell$, $C_{I0} \neq 0_{L^\infty(\Omega)}$ *and*

$$[\|C_I^*\|_\infty \|k\|_2 + \zeta C(q-\ell)^+]^2 < \lambda_1[\mu + l + \alpha - b\ell C(q-\ell)^+], \tag{19}$$

then

$$i_1(\cdot, t) \longrightarrow 0, \quad i_2(\cdot, t) \longrightarrow 0 \quad \text{in } L^1(\Omega),$$

as $t \to +\infty$.

Notice that condition (19) holds if, for example, γ_1 is sufficiently large and/or the subset ω is sufficiently large.

Proof Since $C_{I0} \neq 0_{L^\infty(\Omega)}$, then $\forall\varepsilon > 0, \exists t(\varepsilon) \geq 0, \forall t, t \geq t(\varepsilon)$:

$$C_I^*(x) - \varepsilon < C_I(x,t) < C_I^*(x) + \varepsilon, \quad \text{a.e. } x \in \Omega,\ t \geq t(\varepsilon),$$

and

$$0 \leq C_T(x,t) < C(q-\ell)^+ + \varepsilon, \quad \text{a.e. } x \in \Omega,\ t \geq t(\varepsilon).$$

Let $(\tilde{\imath}_1, \tilde{\imath}_2)$ be the solution to

$$\begin{cases} \dfrac{\partial \tilde{\imath}_1}{\partial t} = d\Delta \tilde{\imath}_1 - (n + \gamma_1 \mathbb{I}_\omega(x))\tilde{\imath}_1 - r\chi(C_I^*(x) - \varepsilon)\tilde{\imath}_1 \\ \qquad + (C_I^*(x) + \varepsilon)(1 - \gamma_{23}\mathbb{I}_\omega(x)) \displaystyle\int_\Omega k(x, x')\tilde{\imath}_2(x', t)\mathrm{d}x', & x \in \Omega,\ t > t(\varepsilon) \\ \dfrac{\partial \tilde{\imath}_1}{\partial \nu}(x, t) = 0, & x \in \partial\Omega,\ t > t(\varepsilon) \\ \dfrac{\partial \tilde{\imath}_2}{\partial t} = -(\mu + \ell)\tilde{\imath}_2 - \alpha\tilde{\imath}_2 \\ \qquad + [\zeta(1 - \gamma_{22}\mathbb{I}_\omega(x))\tilde{\imath}_1 + b(1 - \gamma_3\mathbb{I}_\omega(x))\ell\tilde{\imath}_2] \cdot [C(q - \ell)^+ + \varepsilon], & x \in \Omega,\ t > t(\varepsilon) \\ \tilde{\imath}_1(x, t(\varepsilon)) = i_1(x, t(\varepsilon)),\ \tilde{\imath}_2(x, t(\varepsilon)) = i_2(x, t(\varepsilon)), & x \in \Omega \end{cases} \tag{20}$$

(the existence, uniqueness, and nonnegativity of the solution follows via a fixed point argument). We have that

$$0 \le i_j(x, t) \le \tilde{\imath}_j(x, t) \quad \text{a.e. } x \in \Omega, \forall t \ge t(\varepsilon),$$

$j \in \{1, 2\}$. This follows from considering the system satisfied by $(\tilde{\imath}_1 - i_1, \tilde{\imath}_2 - i_2)$ and showing in a standard manner that

$$(\tilde{\imath}_1 - i_1)^-(x, t) = 0,\ (\tilde{\imath}_1 - i_1)^-(x, t),\ x \in \Omega,\ t \ge t(\varepsilon).$$

If we multiply the first equation in (20) by $\tilde{\imath}_1$ and integrate over Ω, we obtain that

$$\begin{aligned} \frac{1}{2}\frac{\mathrm{d}}{\mathrm{d}t}\left(\int_\Omega |\tilde{\imath}_1(x, t)|^2 \mathrm{d}x\right) \le & -d\int_\Omega |\nabla\tilde{\imath}_1(x, t)|^2\mathrm{d}x - \int_\Omega (n + \gamma_1\mathbb{I}_\omega(x))|\tilde{\imath}_1(x, t)|^2\mathrm{d}x \\ & - r\chi\int_\Omega (C_I^*(x) - \varepsilon)|\tilde{\imath}_1(x, t)|^2\mathrm{d}x \\ & + \int_\Omega (C_I^*(x) + \varepsilon)\tilde{\imath}_1(x, t)\int_\Omega k(x, x')\tilde{\imath}_2(x', t)\mathrm{d}x'\mathrm{d}x \\ \le & -\lambda_1\|\tilde{\imath}_1(t)\|_2^2 + r\chi\varepsilon\|\tilde{\imath}_1(t)\|_2^2 + (\|C_I^*\|_\infty + \varepsilon)\|k\|_2\|\tilde{\imath}_1(t)\|_2\|\tilde{\imath}_2(t)\|_2 \end{aligned}$$

(by Rayleigh's principle), $\forall t \ge t(\varepsilon)$.

If we multiply the second PDE in (20) by $\tilde{\imath}_2$ and integrate over Ω, we obtain that

$$\begin{aligned} \frac{1}{2}\frac{\mathrm{d}}{\mathrm{d}t}\left(\int_\Omega |\tilde{\imath}_2(x, t)|^2\mathrm{d}x\right) = & -(\mu + \ell)\int_\Omega |\tilde{\imath}_2(x, t)|^2\mathrm{d}x \\ & + b\ell[C(q - \ell)^+ + \varepsilon]\int_\Omega |\tilde{\imath}_2(x, t)|^2\mathrm{d}x - \alpha\int_\Omega |\tilde{\imath}_2(x, t)|^2\mathrm{d}x \\ & + \zeta[C(q - \ell)^+ + \varepsilon]\int_\Omega \tilde{\imath}_1(x, t)\tilde{\imath}_2(x, t)\mathrm{d}x \end{aligned}$$

$$\leq -[\mu + \ell + \alpha - bl(C(q-\ell)^+ + \varepsilon)]\|\tilde{\imath}_2(t)\|_2^2 \\ + \zeta[C(q-\ell)^+ + \varepsilon]\|\tilde{\imath}_1(t)\|_2\|\tilde{\imath}_2(t)\|_2, \quad \forall t \geq t(\varepsilon).$$

We may infer that

$$\begin{aligned}\frac{1}{2}\frac{\mathrm{d}}{\mathrm{d}t}(\|\tilde{\imath}_1(t)\|_2^2 + \|\tilde{\imath}_2(t)\|_2^2) &\leq -(\lambda_1 - r\chi\varepsilon)\|\tilde{\imath}_1(t)\|_2^2 \\ &\quad - [\mu + \ell + \alpha - b\ell(C(q-\ell)^+ + \varepsilon)]\|\tilde{\imath}_2(t)\|_2^2 \\ &\quad + [(\|C_I^*\|_\infty + \varepsilon)\|k\|_2 + \zeta(C(q-\ell)^+ + \varepsilon)]\|\tilde{\imath}_1(t)\|_2\|\tilde{\imath}_2(t)\|_2 \\ &\leq -\frac{1}{2}\min\{\lambda_1 - r\chi\varepsilon, \mu + \ell + \alpha - b\ell(C(q-\ell)^+ + \varepsilon)\}(\|\tilde{\imath}_1(t)\|_2^2 + \|\tilde{\imath}_2(t)\|_2^2),\end{aligned}$$

$\forall t \geq t(\varepsilon)$, if

$$[(\|C_I^*\|_\infty + \varepsilon)\|k\|_2 + \zeta(C(q-\ell)^+ + \varepsilon)]^2 \\ < (\lambda_1 - r\chi\varepsilon)[\mu + \ell + \alpha - b\ell(C(q-\ell)^+ + \varepsilon)].$$

This condition holds for any sufficiently small $\varepsilon > 0$ [because (19) holds]. On the other hand, since condition (19) holds, it follows that for any $\varepsilon > 0$ sufficiently small, we have that

$$\frac{1}{2}\min\{\lambda_1 - r\chi\varepsilon, \mu + \ell + \alpha - b\ell(C(q-\ell)^+ + \varepsilon)\} = a > 0,$$

and we conclude that $\|\tilde{\imath}_1(t)\|_2^2 + \|\tilde{\imath}_2(t)\|_2^2$ converges to 0 as $t \to +\infty$, at the rate of $\exp\{-2at\}$. Since Ω is bounded, it follows that

$$\tilde{\imath}_1(\cdot, t) \to 0, \quad \tilde{\imath}_2(\cdot, t) \to 0 \quad \text{in } L^1(\Omega),$$

and that

$$i_1(\cdot, t) \to 0, \quad i_2(\cdot, t) \to 0 \quad \text{in } L^1(\Omega),$$

as $t \to +\infty$, at least as fast as $\exp\{-at\}$ (i.e., the total number of infected insects and the total number of infected trees tend to 0, exponentially). It means that the disease is eradicable. □

Remark 1 Recall that if the diffusion coefficient is strictly positive, $d > 0$, then any point of the domain Ω is strongly connected to any other point of Ω, so that actions taken in a subregion $\omega \subset \Omega$ will eventually propagate to the whole habitat.

On the other hand, due to the lack of diffusion for i_2, it is reasonable to expect the control to be more effective (leading to a faster decay of i_1 and i_2) if γ_{22}, γ_{23}, and γ_3 act on the whole domain Ω. In this case, the controlled system becomes

$$\frac{\partial s_1}{\partial t}(x,t) = d\Delta s_1(x,t) + rC_I(x,t)[M(x)(1-\gamma_{21}\mathbb{I}_\omega(x)) - \chi s_1(x,t)]$$
$$- (n+\gamma_1\mathbb{I}_\omega(x))s_1(x,t) - \beta(1-\gamma_{23})s_1(x,t)\int_\Omega k(x,x')i_2(x',t)\mathrm{d}x', \tag{21}$$

$$\frac{\partial i_1}{\partial t}(x,t) = d\Delta i_1(x,t) - (n+\gamma_1\mathbb{I}_\omega(x))i_1(x,t) - r\chi i_1(x,t)C_I(x,t)$$
$$+ \beta(1-\gamma_{23})s_1(x,t)\int_\Omega k(x,x')i_2(x',t)\mathrm{d}x', \tag{22}$$

$$\frac{\partial s_2}{\partial t}(x,t) = (q-\ell)s_2(x,t) - s_2(x,t)\frac{C_T(x,t)}{C}$$
$$- (\zeta(1-\gamma_{22})i_1(x,t) + b(1-\gamma_3)\ell i_2(x,t))s_2(x,t) + \alpha i_2(x,t), \tag{23}$$

$$\frac{\partial i_2}{\partial t}(x,t) = -\mu i_2(x,t) - \ell i_2(x,t) - i_2(x,t)\frac{C_T(x,t)}{C}$$
$$+ (\zeta(1-\gamma_{22})i_1(x,t) + b(1-\gamma_3)\ell i_2(x,t))s_2(x,t) - \alpha i_2(x,t). \tag{24}$$

The boundary conditions and the initial conditions are as before.

There exists a unique solution (s_1, i_1, s_2, i_2) to (21)–(24) with the above mentioned boundary, and initial conditions, satisfying

$$s_j,\ i_j \in L^\infty_{loc}(\overline{\Omega}\times[0,+\infty)),$$
$$s_j(x,t),\ i_j(x,t) \geq 0,\ \text{a.e. } (x,t)\in\Omega\times(0,+\infty),$$

$j \in \{1,2\}$. This follows from using a similar argument as for (6)–(11).

In this special case, the following result holds:

Theorem 2 *If $\tilde{\lambda}_1 < 0, q > \ell$, $C_{I0} \neq 0_{L^\infty(\Omega)}$ and*

$$[(1-\gamma_{23})\|C_I^*\|_\infty\|k\|_2 + \zeta(1-\gamma_{22})C(q-\ell)^+]^2$$
$$< \lambda_1[\mu + l + \alpha - b(1-\gamma_3)\ell C(q-\ell)^+] \tag{25}$$

then

$$i_1(\cdot,t) \longrightarrow 0,\quad i_2(\cdot,t)\longrightarrow 0 \quad \text{in } L^1(\Omega),$$

as $t \to +\infty$.

Notice that (25) is a weaker assumption than (19).

Proof Using a comparison result for (i_1, i_2), we obtain that

$$0 \leq i_j(x,t) \leq \tilde{\imath}_j(x,t) \quad \text{a.e. } x\in\Omega, \forall t \geq t(\varepsilon),$$

$j \in \{1,2\}$, where $(\tilde{\imath}_1, \tilde{\imath}_2)$ is the solution to

$$\begin{cases} \dfrac{\partial \tilde{\imath}_1}{\partial t} = d\Delta \tilde{\imath}_1 - (n + \gamma_1 \mathbb{I}_\omega(x))\tilde{\imath}_1 - r\chi(C_I^*(x) - \varepsilon)\tilde{\imath}_1 \\ \qquad + (C_I^*(x) + \varepsilon)(1 - \gamma_{23}) \displaystyle\int_\Omega k(x, x')\tilde{\imath}_2(x', t)\mathrm{d}x', & x \in \Omega,\ t > t(\varepsilon) \\ \dfrac{\partial \tilde{\imath}_1}{\partial \nu}(x, t) = 0, & x \in \partial\Omega,\ t > t(\varepsilon) \\ \dfrac{\partial \tilde{\imath}_2}{\partial t} = -(\mu + \ell)\tilde{\imath}_2 - \alpha \tilde{\imath}_2 \\ \qquad + [\zeta(1 - \gamma_{22})\tilde{\imath}_1 + b(1 - \gamma_3)\ell \tilde{\imath}_2] \cdot [C(q - \ell)^+ + \varepsilon], & x \in \Omega,\ t > t(\varepsilon) \\ \tilde{\imath}_1(x, t(\varepsilon)) = i_1(x, t(\varepsilon)),\ \tilde{\imath}_2(x, t(\varepsilon)) = i_2(x, t(\varepsilon)), & x \in \Omega. \end{cases} \tag{26}$$

If we multiply the first equation in (26) by $\tilde{\imath}_1$ and integrate over Ω, we obtain that

$$\begin{aligned} \frac{1}{2}\frac{\mathrm{d}}{\mathrm{d}t}\left(\int_\Omega |\tilde{\imath}_1(x,t)|^2 \mathrm{d}x\right) &\le -d\int_\Omega |\nabla \tilde{\imath}_1(x,t)|^2 \mathrm{d}x - \int_\Omega (n + \gamma_1 \mathbb{I}_\omega(x))|\tilde{\imath}_1(x,t)|^2 \mathrm{d}x \\ &\quad - r\chi \int_\Omega (C_I^*(x) - \varepsilon)|\tilde{\imath}_1(x,t)|^2 \mathrm{d}x \\ &\quad + (1 - \gamma_{23}) \int_\Omega (C_I^*(x) + \varepsilon)\tilde{\imath}_1(x,t) \int_\Omega k(x,x')\tilde{\imath}_2(x',t)\mathrm{d}x'\mathrm{d}x \\ &\le -\lambda_1 \|\tilde{\imath}_1(t)\|_2^2 + r\chi\varepsilon \|\tilde{\imath}_1(t)\|_2^2 \\ &\quad + (1 - \gamma_{23})(\|C_I^*\|_\infty + \varepsilon)\|k\|_2 \|\tilde{\imath}_1(t)\|_2 \|\tilde{\imath}_2(t)\|_2 \end{aligned}$$

(by Rayleigh's principle).

If we multiply the second PDE in (26) by $\tilde{\imath}_2$ and integrate over Ω, we obtain that

$$\begin{aligned} \frac{1}{2}\frac{\mathrm{d}}{\mathrm{d}t}\left(\int_\Omega |\tilde{\imath}_2(x,t)|^2 \mathrm{d}x\right) &= -(\mu + \ell)\int_\Omega |\tilde{\imath}_2(x,t)|^2 \mathrm{d}x \\ &\quad + b(1 - \gamma_3)\ell[C(q - \ell)^+ + \varepsilon] \int_\Omega |\tilde{\imath}_2(x,t)|^2 \mathrm{d}x \\ &\quad - \alpha \int_\Omega |\tilde{\imath}_2(x,t)|^2 \mathrm{d}x \\ &\quad + \zeta(1 - \gamma_{22})[C(q - \ell)^+ + \varepsilon] \int_\Omega \tilde{\imath}_1(x,t)\tilde{\imath}_2(x,t)\mathrm{d}x \\ &\le -[\mu + \ell + \alpha - b(1 - \gamma_3))l(C(q - \ell)^+ + \varepsilon)]\|\tilde{\imath}_2(t)\|_2^2 \\ &\quad + \zeta(1 - \gamma_{22})[C(q - \ell)^+ + \varepsilon]\|\tilde{\imath}_1(t)\|_2 \|\tilde{\imath}_2(t)\|_2, \quad \forall t \ge t(\varepsilon). \end{aligned}$$

We may infer that

$$\begin{aligned} \frac{1}{2}\frac{\mathrm{d}}{\mathrm{d}t}(\|\tilde{\imath}_1(t)\|_2^2 + \|\tilde{\imath}_2(t)\|_2^2) &\le -(\lambda_1 - r\chi\varepsilon)\|\tilde{\imath}_1(t)\|_2^2 \\ &\quad - [\mu + \ell + \alpha - b(1 - \gamma_3)\ell(C(q - \ell)^+ + \varepsilon)]\|\tilde{\imath}_2(t)\|_2^2 \end{aligned}$$

$$
\begin{aligned}
&+[(1-\gamma_{23})(\|C_I^*\|_\infty+\varepsilon)\|k\|_2+\zeta(1-\gamma_{22})(C(q-\ell)^+ +\varepsilon)]\\
&\times\|\tilde{\imath}_1(t)\|_2\|\tilde{\imath}_2(t)\|_2\\
&\le -\frac{1}{2}\min\{\lambda_1-r\chi\varepsilon,\mu+\ell+\alpha-b(1-\gamma_3)\ell(C(q-\ell)^+ +\varepsilon)\}\\
&\times(\|\tilde{\imath}_1(t)\|_2^2+\|\tilde{\imath}_2(t)\|_2^2),
\end{aligned}
$$

$\forall t \ge t(\varepsilon)$, if

$$
\begin{aligned}
&[((1-\gamma_{23})\|C_I^*\|_\infty+\varepsilon)\|k\|_2+\zeta(1-\gamma_{22})(C(q-\ell)^+ +\varepsilon)]^2\\
&<(\lambda_1-r\chi\varepsilon)[\mu+\ell+\alpha-b(1-\gamma_3)\ell(C(q-\ell)^+ +\varepsilon)].
\end{aligned}
$$

This condition holds for $\varepsilon > 0$ sufficiently small (because (25) holds). On the other hand, since condition (25) is satisfied, then for $\varepsilon > 0$ sufficiently small, we have that

$$
\frac{1}{2}\min\{\lambda_1-r\chi\varepsilon,\mu+\ell+\alpha-b(1-\gamma_3)\ell(C(q-\ell)^+ +\varepsilon)\}=\tilde{a}>0,
$$

and we conclude that $\|\tilde{\imath}_1(t)\|_2^2+\|\tilde{\imath}_2(t)\|_2^2$ converges to 0 as $t \to +\infty$, at the rate of $\exp\{-2\tilde{a}t\}$. Since Ω is bounded, it follows that

$$
\tilde{\imath}_1(\cdot,t)\to 0,\quad \tilde{\imath}_2(\cdot,t)\to 0 \quad \text{in } L^1(\Omega),
$$

and that

$$
i_1(\cdot,t)\to 0,\quad i_2(\cdot,t)\to 0 \quad \text{in } L^1(\Omega),
$$

as $t \to +\infty$, at least as fast as $\exp\{-\tilde{a}t\}$ (i.e., the total number of infected insects and the total number of infected trees tend to 0, exponentially). This means that the disease is eradicable. Notice that $\tilde{a} \ge a$. □

4 Numerical Simulations

The numerical strategy adopted to approximate the controlled system (Equations (6)–(9)) consists of the finite element method for space discretization and the finite difference method for time discretization. This procedure is the state of the art for the solution of parabolic partial differential equations (PDEs); see, e.g., Quarteroni and Valli (1994).

Space discretization We first apply a standard Galerkin procedure to the weak formulations of the controlled system [Eqs. (6)–(9)]. For the computational domain Ω, we have taken a rectangle of size $400 \times 80\,\text{km}^2$, which mimics, for example, the whole region of Apulia in Southern Italy, from South (right-hand side of the domain) to North (left-hand side of the domain). The rectangular domain has been discretized by a uniform grid of 200×40 bilinear finite elements ($Q1$), yielding a total number of 8241 discretization nodes. The stiffness matrix is computed exactly, whereas the mass matrix is obtained by applying the mass lumping technique.

Time discretization After the spatial discretization, we obtain a semi-discrete problem that consists of a system of ordinary differential equations (ODEs). We solve these ODEs by employing a first order semi-implicit finite difference scheme, where the linear diffusion terms are approximated by Backward Euler, whereas the nonlinear reaction terms are approximated by forward Euler. As a result, at each time step we must find the solution of a linear system of algebraic equations of dimension $4 \times 8241 = 32964$ degrees of freedom. The system is solved by Gaussian elimination with the built-in function in MATLAB. The time step size is $\Delta t = 0.0002$ years. For further details on the numerical discretization of parabolic problems, we refer the reader to Quarteroni and Valli (1994).

4.1 Parameter Calibration

The values of the simulation parameters are listed in Table 2. The diffusion coefficient in the PDEs is set to $d = 1e - 4\,\text{km}^2/\text{year}$. The total simulation time is $T = 10$ years. The initial distributions, used in all next numerical simulations, are the following:

$$\begin{cases} s_1(x,0) = 100 \\ i_1(x,0) = 20\exp(-100(x_1 - 380)^2 - 100(x_2 - 40)^2) \\ s_2(x,0) = 50 \\ i_2(x,0) = 0 \end{cases}$$

That is, we assume constant initial distributions for healthy insects and trees, and we localize the initial presence of infected insects on the right-hand side of the domain.

4.2 Numerical Results

A couple of numerical experiments are reported here to show the efficacy of the controls on the weed biomass and the olive cultivar.

In Figs. 1 and 2, we have taken $M = 1$ and $\chi = 0.03$, while in Figs. 3 and 4, $M = 1$ and $\chi = 0.01$. The colorbar indicates the scaled values of the distribution corresponding to the colors adopted in the map: It goes from blue, associated with low values, to yellow, associated with high values.

In both computational experiments, control measures have been applied only in the right-hand section of the habitat, simulating the subregion of the Apulian region already affected by the xylella epidemic, augmented by a "containment band," i.e., the subregion

$$\omega = \{x = (x_1, x_2) \in \Omega : x_1 > 250\,\text{km}\}.$$

Experiment 1 $\chi = 0.03$, $\zeta = 0.8\,\text{year}^{-1}$: we first run the model without control, and thus, the control parameters $(\gamma_{21}, \gamma_1, \gamma_{23}, \gamma_{22}, \gamma_3)$ are set to zero everywhere. The xylella epidemic starts spreading as a travelling wave from the right portion of the domain, where the initial condition for the infected insects is positive; see Fig. 1.

Table 2 Values of the model parameters

Symbol	Unit	Value range	Simulation value	References
r	year^{-1}	[37,400]	200	Silva et al. (2015) and Yurtsever (2000)
χ	–	[0.001, 0.05]	0.03	Brunetti et al. (2020)
n	year^{-1}	[0.95, 0.99]	0.98	Brunetti et al. (2020)
q	year^{-1}	[0.2, 0.7]	0.5	Villalobos et al. (2006)
C	–	100	100	Brunetti et al. (2020)
ℓ	year^{-1}	0.01	0.01	Brunetti et al. (2020)
b	–	0.05	0.05	Brunetti et al. (2020)
μ	year^{-1}	[0.8, 1]	0.9	Saponari et al. (2017, 2018)
α	year^{-1}	[0.1, 0.5]	0.3	Brunetti et al. (2020)
β	year^{-1}	[0.5, 1]	1	Cornara et al. (2017)
ζ	year^{-1}	[0.2, 0.8]	0.8	Boscia et al. (2017) and Fierro et al. (2019)

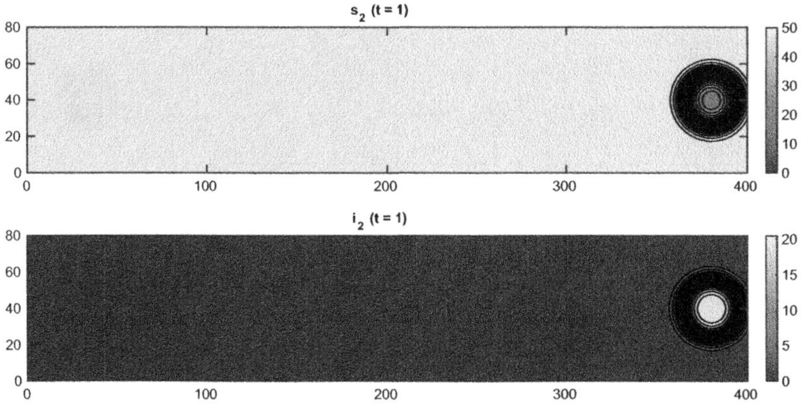

(a) $t = 1$ *year*. Top figure: healthy trees (s_2); bottom figure: infective trees (i_2).

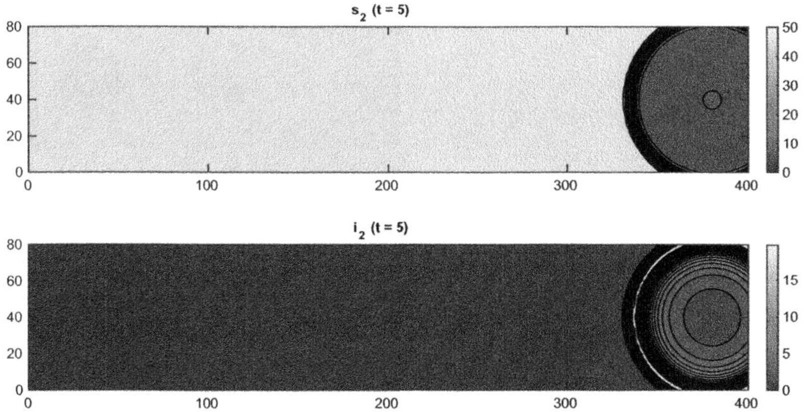

(b) $t = 5$ *years*. Same format as panel (a).

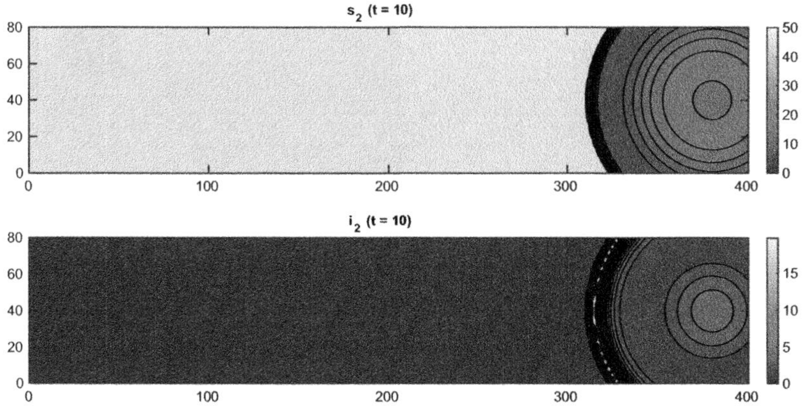

(c) $t = 10$ *years*. Same format as panel (a).

Fig. 1 Experiment 1. Spatial distributions of olive trees without control: $\chi = 0.03$, $d = 1\text{e}{-4}\,\text{km}^2/\text{year}$ (Color figure online)

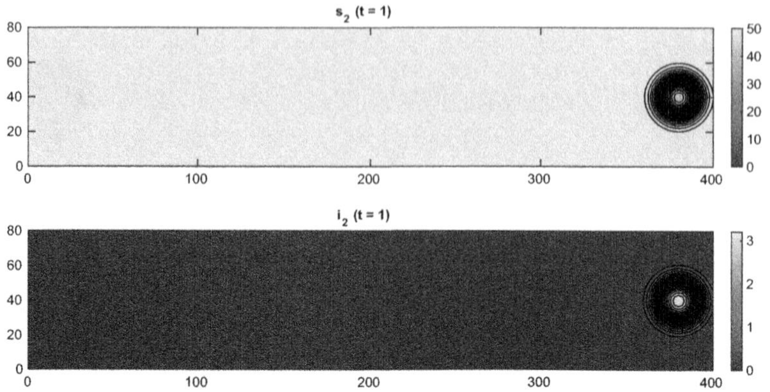

(a) $t = 1$ *year*. Top figure: healthy trees (s_2); bottom figure: infective trees (i_2).

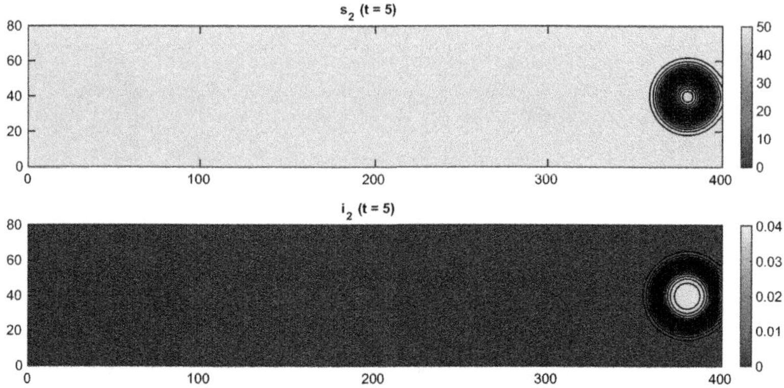

(b) $t = 5$ *years*. Same format as panel (a).

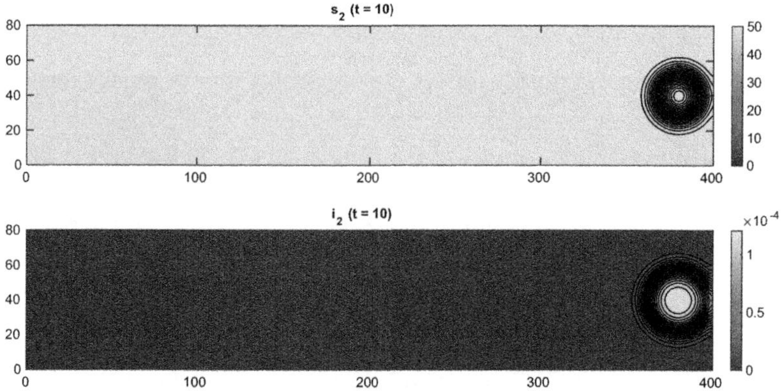

(c) $t = 10$ *years*. Same format as panel (a).

Fig. 2 Experiment 1. Spatial distributions of olive trees with regional control. $\chi = 0.03$, $d = 1e - 4\,\text{km}^2/\text{year}$, $\{x = (x_1, x_2) \in \Omega : x_1 > 250\,\text{km}\}$. $(\gamma_{21}, \gamma_1, \gamma_{23}, \gamma_{22}, \gamma_3) = (0.999, 0.2, 0.2, 0.2, 0.2)$ (Color figure online)

When we apply regional control, setting the control parameters

$$(\gamma_{21}, \gamma_1, \gamma_{23}, \gamma_{22}, \gamma_3) = (0.999, 0.2, 0.2, 0.2, 0.2),$$

the xylella epidemic is stopped at the origin and no travelling front occurs; see Fig. 2. This result confirms our conjecture that a significant weed cut in the olive orchards strongly decreases the carrying capacity of insects, yielding a successful blocking of the xylella epidemic spread.

In order to investigate how the diffusion coefficient d influences the results, we have re-run both tests with $d = 2\text{e}{-4},\ 5\text{e}{-4},\ 1\text{e}{-3}$ (all in km^2/year). We do not observe any significant qualitative difference.

Experiment 2. $\chi = 0.01, \zeta = 0.2\ year^{-1}$: as above, we first run the model without control, and thus, the control parameters $(\gamma_{21}, \gamma_1, \gamma_{23}, \gamma_{22}, \gamma_3)$ are set to zero everywhere. The xylella epidemic starts spreading as a travelling wave from the right portion of the domain, where the initial condition of the infected insects is positive; see Fig. 3.

When we apply a regional control, setting the control parameters

$$(\gamma_{21}, \gamma_1, \gamma_{23}, \gamma_{22}, \gamma_3) = (0.6, 0.2, 0.2, 0.6, 0.2),$$

the xylella epidemic is stopped at the origin and no travelling front occurs; see Fig. 4. This result clearly indicates that for more resistant olive cultivar, we may still block an epidemic with a lower level of weed cut in the olive orchards, even though, by using $\chi = 0.001$, we have a larger effective insect carrying capacity.

As in the previous experiment, we have also re-run both tests with $d = 2\text{e}{-4},\ 5\text{e}{-4},\ 1\text{e}{-3}$ (all in km^2/year). Even in this case, we do not observe any significant qualitative difference.

5 Concluding Remarks

The results reported in this paper confirm that the most promising target for an effective and cost-efficient control of the *X. fastidiosa* epidemic is represented by agricultural management practices consisting of the removal of the weeds in the whole relevant habitat of the olive orchards. A further interesting strategy, as expected, is represented by the use of more resistant cultivar (Brunetti et al. 2020) [see also Schneider et al. (2020) and references cited therein].

Here, we have extended the ordinary differential equation (ODE) model introduced in Brunetti et al. (2020) to an integro-partial differential system which takes into account the spatial structure of the relevant epidemic system, subject to possible control strategies, including weed cut, insect traps, treated nets, more resistant cultivar, etc. As emphasized in Remark 1, theoretical results show that eradication of an epidemic outbreak is possible by the implementation of the above mentioned control measures only on a suitable subregion of the whole habitat, including the area already affected by the outbreak, possibly augmented by a suitable "containment band."

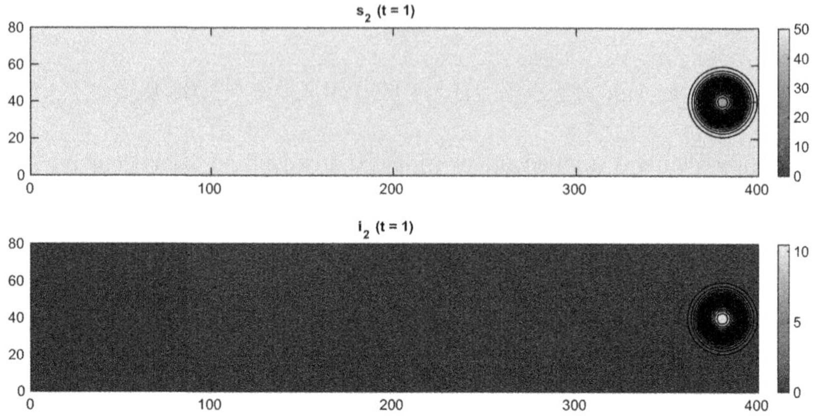

(a) $t = 1$ *year*. Top figure: healthy trees (s_2); bottom figure: infective trees (i_2).

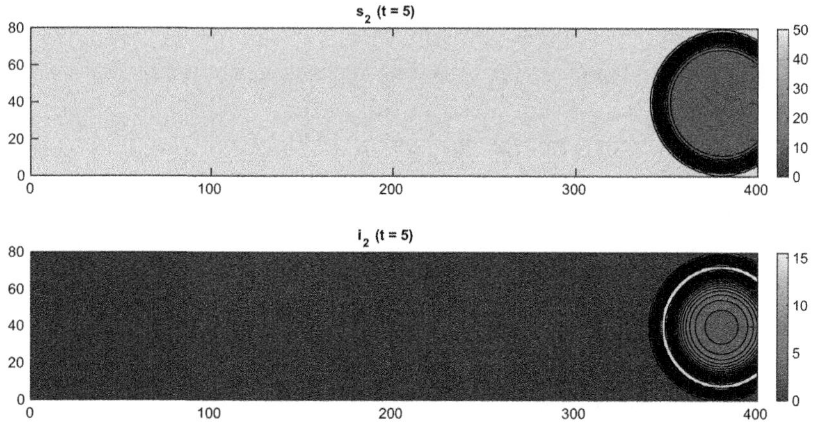

(b) $t = 5$ *years*. Same format as panel (a).

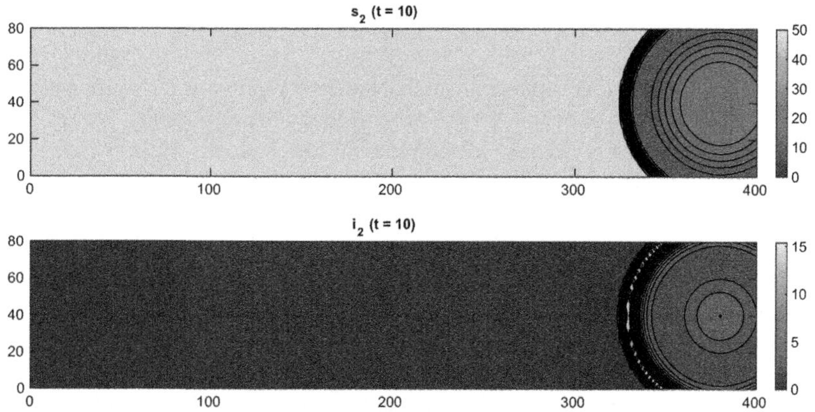

(c) $t = 10$ *years*. Same format as panel (a).

Fig. 3 Experiment 2. Spatial distributions of olive trees without control: $\chi = 0.01$, $d = 1e - 4\,\mathrm{km}^2/\mathrm{year}$ (Color figure online)

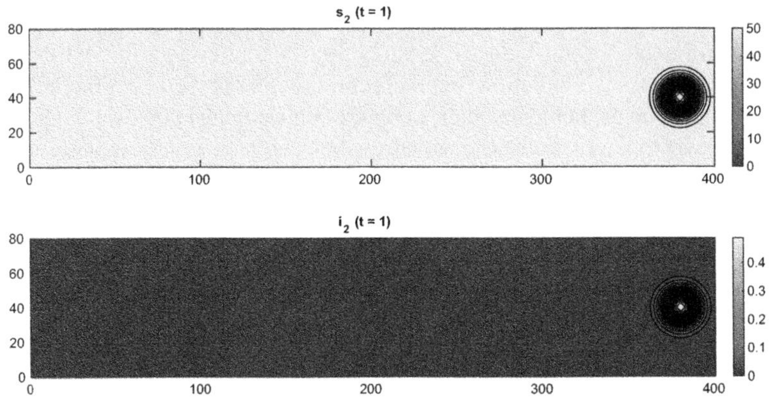

(a) $t = 1$ *year*. Top figure: healthy trees (s_2); bottom figure: infective trees (i_2).

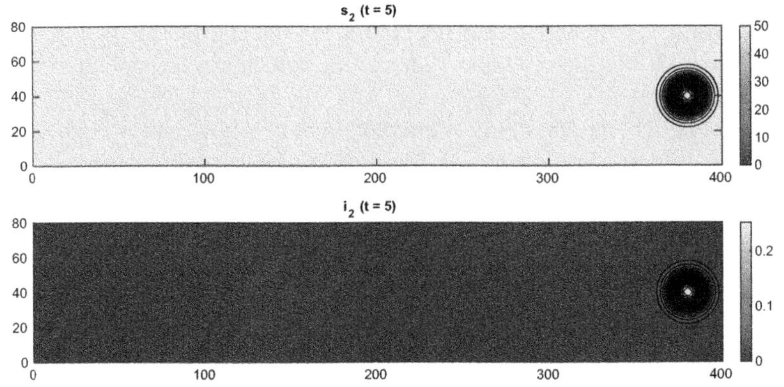

(b) $t = 5$ *years*. Same format as panel (a).

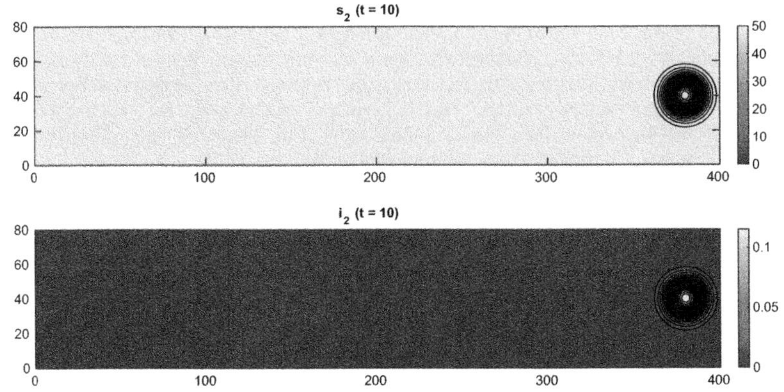

(c) $t = 10$ *years*. Same format as panel (a).

Fig. 4 Experiment 2. Spatial distributions of olive trees with regional control. $\chi = 0.01$, $d = 1e - 4\,\text{km}^2/\text{year}$, $\{x = (x_1, x_2) \in \Omega : x_1 > 250\,\text{km}\}$. $(\gamma_{21}, \gamma_1, \gamma_{23}, \gamma_{22}, \gamma_3) = (0.6, 0.2, 0.2, 0.6, 0.2)$ (Color figure online)

Unfortunately in practice, it might be unfeasible to act on the diffusion parameters; they are related to the ecological structure of the relevant habitat, including the behavior of the insect population. Intervention strategies should then include possible modification of insect behavior, such as by the use of treated nets.

Future investigations will be devoted to the search of an optimal set of control parameters, γ's, together with an optimal subregion of intervention ω. It is clear that, for a realistic optimal control problem, relevant participating costs need to be included: e.g., production, losses, and management costs.

Once again, validation of the model proposed here represents a key issue: although we have tried to make explicit the assumptions underlying our model, they have not yet been validated by comparison with experimental data. Therefore, we caution that our results are far from being conclusive for *X. fastidiosa* subsp. *pauca* - *Philaenus spumarius* olive tree epidemics. However, it is desirable that, with additional features that make it more realistic, our model might provide the foundations for designing optimal control strategies by public authorities.

It is worth reporting here a statement (abridged) taken from the very recent paper (Matricardi et al. 2020) on COVID-19 modeling, which can be applied to the role of any model:

"a model is only an approximate interpretation of reality and it is always wrong in some small or relevant elements. The destiny of the model presented here is to be rapidly improved thanks to novel knowledge coming from new observations and better assumptions. The Authors hope that many and more brilliant minds will read the present pages, will identify and highlight putative mistakes, will get inspiration for their research, and will produce better, more complete, and useful models........ If the speculations presented here on implications for surveillance, control, and therapy of [Xylella] will contribute, even only minimally, to save ..... [olive trees]........., then the Authors have accomplished their small mission. "

Acknowledgements The authors are indebted to Professor Matteo Montagna and Dr. Matteo Brunetti of the Department of Agricultural and Environmental Sciences—Production, Landscape, Agroenergy of the Università degli Studi di Milano *La Statale* for relevant discussions on the biological aspects of this research. The role of Professor Ezio Venturino of the Department of Mathematics "Giuseppe Peano" of the Università di Torino has been significant in the identification of the relevant ingredients of the epidemic model. Thanks are also due to the Anonymous Reviewers and the Editors of the journal for their accurate reading and comments, which have led to a significant improvement of the presentation.

Funding Open access funding provided by Università degli Studi di Milano.

Compliance with Ethical Standards

Conflict of interest The authors declare that they have no conflict of interest.

References

Almeida RPP, Blua MJ, Lopes JRS, Purcell AH (2005) Vector transmission of *Xylella fastidiosa*: applying fundamental knowledge to generate disease management strategies. Ann Entomol Soc Am 98:775–786

Aniţa S, Capasso V (2009) A stabilization strategy for a reaction-diffusion system modelling a class of spatially structured epidemic systems (think globally, act locally). Nonlinear Anal Real World Appl 10:2026–2035

Aniţa S, Capasso V (2012) Stabilization of a reaction-diffusion system modelling a class of spatially structured epidemic systems via feedback control. Nonlinear Anal Real World Appl 13:725–735

Aniţa S, Fitzgibbon W, Langlais M (2009) Global existence and internal stabilization for a class of predator-prey systems posed on non coincident spatial domains. Discrete Cont Dyn Syst B 11:805–822

Boscia D, Altamura G, Saponari M, Tavano D, Zicca S, Pollastro P, Silletti MR, Savino VN, Martelli GP, Delle Donne A, Mazzotta S, Signore PP, Troisi M, Drazza P, Conte P, D'Ostuni V, Merico S, Perrone G, Specchia F, Stanca A, Tanieli M (2017) Incidenza di Xylella in oliveti con disseccamento rapido. Informatore Agrario 27:47–50

Brunetti M, Capasso V, Montagna M, Venturino E (2020) A mathematical model for *Xylella fastidiosa* epidemics in the Mediterranean regions Promoting good agronomic practices for their effective control. Ecol Model 432:109204. https://doi.org/10.1016/j.ecolmodel.2020.109204

Capasso V (1984) Asymptotic stability for an integro-differential reaction–diffusion system. J Math Anal Appl 103:575–588

Carlucci A, Lops F, Marchi G, Mugna L, Has Surico G (2013) Xylella fastidiosa "chosen" olive trees to establish in the Mediterranean basin? Phytopathol Mediterranea 52:541–544

Cornara D, Cavalieri V, Dongiovanni C, Altamura G, Palmisano F, Bosco D, Porcelli F, Almeida RPP, Saponari M (2017) Transmission of *Xylella fastidiosa* by naturally infected Philaenus spumarius (Hemiptera, Aphrophoridae) to different host plants. J Appl Entomol 141:80–87

Fierro A, Liccardo A, Porcelli F (2019) A lattice model to manage the vector and the infection of the Xylella fastidiosa on olive trees. Sci Rep 9:8723

Friedman A (1964) Partial differential equations of parabolic type. Prentice-Hall Inc, Upper Saddle River

Janse JD, Obradovic A (2010) *Xylella fastidiosa*: its biology, diagnosis, control and risks (Minireview). J Plant Pathol 92:S13.5-S14.8

Matricardi PM, Dal Negro RW, Nisini R (2020) The first holistic immunological model of COVID-19: implications for prevention, diagnosis, and public health mesaures. Pedriatr Allergy Immunol 31:454–470

Protter MH, Weinberger HE (1994) Maximum principles in differential equations. Springer, Berlin

Quarteroni A, Valli A (1994) Numerical approximation of partial differential equations. Springer, Berlin

Redak RA, Purcell AH, Lopes JRS, Blua MJ, Mizell RF III, Andersen PC (2004) The biology of xylem fluid-feeding insect vectors of *Xylella fastidiosa* and their relation to disease epidemiology, applying fundamental knowledge to generate disease management. Annu Rev Entomol 49:243–270

Rossini L, Severini M, Contarini M, Speranza S (2020) A novel modelling approach to describe an insect life cycle vis-á-vis plant protection: description and application in the case study of the *Tuta absoluta*. Ecol Model 409:108778. https://doi.org/10.1016/j.ecolmodel.2019.108778

Saponari M, Boscia D, Altamura G, Loconsole G, Zicca S, D'Attoma G, Morelli M, Palmisano F, Saponari A, Tavano D, Savino VN, Dongiovanni C, Martelli GP (2017) Isolation and pathogenicity of *Xylella fastidiosa* associated to the olive quick decline syndrome in southern Italy. Sci Rep 7:17723

Saponari M, Giampetruzzi A, Loconsole G, Boscia D, Saldarelli P (2018) *Xylella fastidiosa* in olive in Apulia: where we stand. Phytopathology 109(2):175–186

Schneider K, van der Werf W, Cendoya M, Maurits M, Navas-Cortes JA (2020) Impact of *Xylella fastidiosa* subspecies pauca in European olives. PNAS 117:9250–9259

Shcherbacheva A, Haario H, Killeen GF (2018) Modeling host-seeking behavior of African malaria vector mosquitoes in the presence of long-lasting insecticidal nets. Math Biosci 295:36–47

Silva SE, Rodrigues ASB, Marabuto E, Yurtsever S, Borges PAV, Quartau JA, Paulo OS, Seabra SG (2015) Differential survival and reproduction in colour forms of *Philaenus spumarius* give new insights to the study of its balanced polymorphism. Ecol Entomol 40:759–766
Villalobos FJ, Testi L, Hidalgo J, Pastor M, Orgaz F (2006) Modelling potential growth and yield of olive (*Olea europaea* L.) canopies. Eur J Agron 24:296–303
Yurtsever S (2000) On the polymorphic meadow spittlebug, *Philaenus spumarius* (L.) (Homoptera: Cercopidae). Turk J Zool 24:447–460

Affiliations

Sebastian Aniţa[1,2] · Vincenzo Capasso[3] · Simone Scacchi[4]

✉ Vincenzo Capasso
vincenzo.capasso@unimi.it

Sebastian Aniţa
sanita@uaic.ro

Simone Scacchi
simone.scacchi@unimi.it

1 Faculty of Mathematics, "Alexandru Ioan Cuza" University of Iaşi, 700506 Iaşi, Romania

2 "Octav Mayer" Institute of Mathematics of the Romanian Academy, 700506 Iaşi, Romania

3 ADAMSS (Centre for Advanced Applied Mathematical and Statistical Sciences), Universitá degli Studi di Milano "La Statale", 20133 Milan, Italy

4 Department of Mathematics, Universitá degli Studi di Milano "La Statale", 20133 Milan, Italy

Bulletin of Mathematical Biology (2021) 83:29
https://doi.org/10.1007/s11538-021-00860-0

Society for Mathematical Biology

SPECIAL ISSUE: CELEBRATING J. D. MURRAY

The Role of Cytoplasmic MEX-5/6 Polarity in Asymmetric Cell Division

Sungrim Seirin-Lee[1]

Received: 16 August 2020 / Accepted: 14 January 2021 / Published online: 17 February 2021

Abstract
In the process of asymmetric cell division, the mother cell induces polarity in both the membrane and the cytosol by distributing substrates and components asymmetrically. Such polarity formation results from the harmonization of the upstream and downstream polarities between the cell membrane and the cytosol. MEX-5/6 is a well-investigated downstream cytoplasmic protein, which is deeply involved in the membrane polarity of the upstream transmembrane protein PAR in the *Caenorhabditis elegans* embryo. In contrast to the extensive exploration of membrane PAR polarity, cytoplasmic polarity is poorly understood, and the precise contribution of cytoplasmic polarity to the membrane PAR polarity remains largely unknown. In this study, we explored the interplay between the cytoplasmic MEX-5/6 polarity and the membrane PAR polarity by developing a mathematical model that integrates the dynamics of PAR and MEX-5/6 and reflects the cell geometry. Our investigations show that the downstream cytoplasmic protein MEX-5/6 plays an indispensable role in causing a robust upstream PAR polarity, and the integrated understanding of their interplay, including the effect of the cell geometry, is essential for the study of polarity formation in asymmetric cell division.

Keywords Pattern formation · Cell polarity

1 Introduction

Asymmetric cell division is an elegant developmental process that creates cell diversity (Campanale et al. 2017; Knoblich 2008; Gönczy 2005). A mother cell distributes

Supplementary Information The online version contains supplementary material available at https://doi.org/10.1007/s11538-021-00860-0.

✉ Sungrim Seirin-Lee
seirin.lee@gmail.com ; seirin@hiroshima-u.ac.jp

[1] Department of Mathematics, Department of Mathematical and Life Sciences, Graduate School of Integrated Science for Life, Hiroshima University, Kagamiyama 1-3-1, Hiroshima 700-0046, Japan

substrates and components asymmetrically before cell division and transfers them to two daughter cells, asymmetrically. Ultimately, this leads to two daughter cells with different functions and sizes. One representative experimental model of asymmetric cell division is the fertilized egg cell of *Caenorhabditis elegans* (Cuenca et al. 2002; Gönczy 2005; Motegi and Seydoux 2013). With the entry of sperm, a fertilized egg cell undergoes symmetry breaking in the posterior pole site (Fig. 1a). Concurrent with the symmetry breaking, the acto-myosin network in the cell cortex begins contracting from the site of symmetry breaking and stops contracting in the middle of the cell (Nishikawa et al. 2017; Niwayama et al. 2011). It is known that acto-myosin contraction causes cortical flow directed from the posterior to the anterior side of the cell, and cytoplasmic flow directed from the anterior to the posterior side in the center of the cell but directed from the posterior to the anterior side in the periphery of the cell membrane (Gönczy 2005; Goehring et al. 2011b; Niwayama et al. 2011) (Fig. 1a, blue arrows).

Initially, PAR-6, PAR-3, and PKC-3, a group known as anterior proteins (aPAR), are homogeneously distributed in the membrane and cytosol, while PAR-2 and PAR-1, a group known as posterior proteins (pPAR), are homogeneously distributed in the cytosol. However, once symmetry breaking occurs, these protein groups begin to form exclusive polarity domains in the membrane (Fig. 1a). pPAR relocates to the site of symmetry breaking, and aPAR relocates to the opposite site. The location of the polarity domain of these protein groups determines the anterior–posterior axis of the mother cell, and the boundary of the two exclusive polarity domains in the membrane is maintained for approximately 16 min (Gönczy 2005) after the establishment phase of the polarity, which is observed to be approximately 6–8 min (Cowan and Hyman 2004).

PAR polarity in the membrane is considered to play the central role in regulating the entire process of asymmetric cell division; therefore, both experimental and theoretical approaches to elucidate the mechanism of PAR polarity formation have been extensively studied (Cortes et al. 2018; Hoege and Hyman 2013; Lang and Munro 2017; Motegi and Seydoux 2013; Rappel and Levine 2017; Seirin-Lee 2020; Seirin-Lee et al. 2020a; Small and Dawes 2017; Zonies et al. 2010). The formation of an exclusive domain is underlined by the mutual inhibition dynamics between the anterior and posterior protein groups in which the aPAR/pPAR protein transmits pPAR/aPAR protein from the membrane to the cytosol. Theoretically, it has been demonstrated that bi-stability, due to mutual inhibition dynamics, and mass conservation are the basic mechanisms of polarity formation (Kuwamura et al. 2018; Seirin-Lee et al. 2020b; Trong et al. 2014).

Interestingly, similar polarity dynamics are also observed for cytoplasmic proteins (Cuenca et al. 2002; Daniels et al. 2010). The cytoplasmic MEX-5/6 protein simultaneously creates a polarity in the cytosol with PAR polarity formation in the membrane (Fig. 1a). The cytoplasmic MEX-5/6 protein, distributed homogeneously in the cytosol before symmetry breaking, becomes polarized to the anterior side, and the boundary of the MEX-5/6 polarity domain is observed in a location similar to the boundary of the anterior and posterior polarity domains (Cuenca et al. 2002; Schubert et al. 2000). Unlike the mechanism of PAR polarity formation in the membrane, it was found that MEX-5/6 has two different diffusive types: slow-diffusing and fast-diffusing, and it was suggested that MEX-5/6 creates polarity using the conversion dynamics of mobility. In the early stages of MEX-5/6 polarity studies, it had been hypothesized that the conversion dynamics of mobility is regulated by the phosphorylation and dephosphorylation cycle directly controlled by the membrane pPAR and aPAR proteins (Daniels et al. 2010). However, Griffin et al. (2011) suggested that pPAR (PAR-1) plays a key role and promotes the conversion from slow-diffusing MEX-5/6 to fast-diffusing MEX-5/6, but that aPAR does not play a direct role in the conversion of MEX-5/6 diffusivity. Furthermore, they hypothesized that the phosphatase PP2A antagonizes PAR-1-dependent phosphorylation of MEX-5, returning MEX-5 to the slow-diffusing state. However, aPAR (PKC-3) has been found to be significantly involved in regulating the conversion dynamics of the fast diffusive type of MEX-5/6 to the slow diffusive type (Wu et al. 2018), though it was supposed that the regulation of the conversion dynamics is likely to be indirectly regulated by aPAR proteins (Griffin et al. 2011).

While the mechanism underlying cytoplasmic polarity MEX-5/6 has been well investigated experimentally at a molecular level, a theoretical approach that integrates experimental observations is lacking. Moreover, it is not clear how MEX-5/6 polarity formation is related to the dynamics of the PARs. In particular, there has not been a study that explores how the cortical and cytoplasmic flows interact with MEX-5/6 when realistic cell geometry is included. Thus, in this study, we focus on three issues: Firstly, we formulate the MEX-5/6 model by combining it with the upstream PAR dynamics. Secondly, we explore how MEX-5/6 polarity in the cytosol affects the spatial and temporal dynamics of membrane PAR polarity. Finally, we investigate the effect of the flow dynamics and cell geometry. We explore how these two factors affect the dynamics of the cytoplasmic proteins and, consequently, the formation of membrane PAR polarity. In this study, we also introduce a general method, using phase-field modeling, to combine cell geometry with a convection–reaction–diffusion system. This method will present an easy numerical technique to simulate convection–reaction–diffusion equations on a higher-dimensional bulk-surface domain of various cell shapes.

This study suggests that it is not only the upstream polarity of PARs that dominates the downstream polarity of MEX-5/6, but also that the downstream polarity of MEX-5/6 can critically affect both the spatial and temporal dynamics of PAR polarity, and that the interaction between membrane PAR polarity and cytoplasmic MEX-5/6 polarity is vital for inducing robust cell polarity during asymmetric cell division.

2 Model Development

2.1 PARs Model

Mathematical models for the polarity formation of PAR dynamics have been proposed in several studies (Dawes and Munro 2011; Goehring et al. 2011b; Seirin-Lee and Shibata 2015; Tostevin and Howard 2008; Trong et al. 2014). All these models suggested similar mathematical structures based on bi-stability and mass conservation for the creation of polarity. Therefore, we adopt the standard model for PAR dynamics suggested by Seirin-Lee and Shibata (2015), and extend it to a higher-dimensional bulk-surface model. Let us define the cytosol by $\Omega \subset \mathbb{R}^N$ and the membrane by $\partial\Omega$ ($\equiv \Gamma$), where

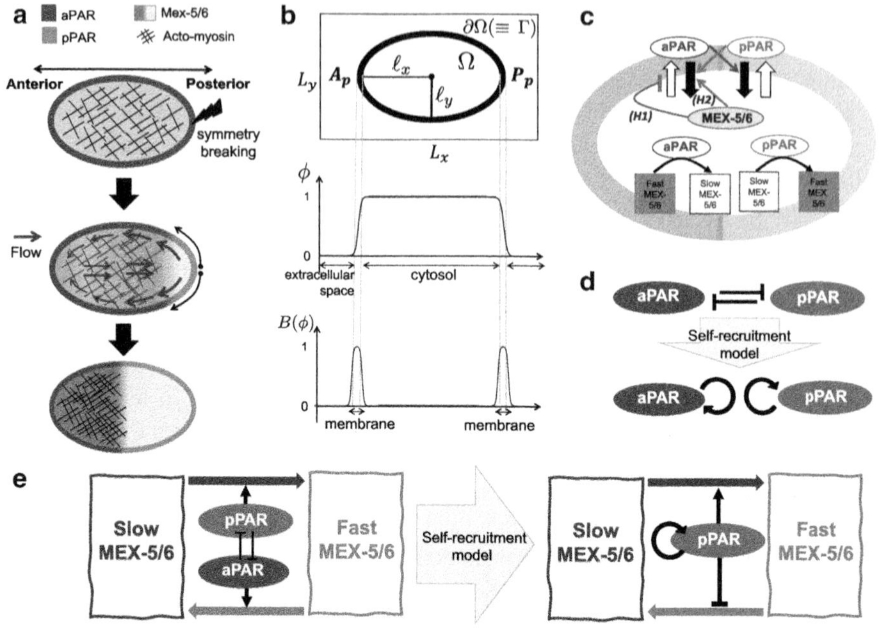

Fig. 1 Schematic diagrams for polarity formation in the *C. elegans* embryo and mathematical model. **a** Dynamics of polarity formation in the *C. elegans* embryo. **b** The description of a cell using the phase-field function (ϕ). $\Omega \cup \partial\Omega$ is a cell region, $L_x \times L_y$ is the simulation domain, and ℓ_x and ℓ_y are radii of the long axis and short axis of the cell, respectively. A_p and P_p are polar points at the anterior and posterior sides, respectively. **c** Diagram of the dynamics of aPAR, pPAR, and MEX-5/6. The black arrows indicate transportation and conversion, while the colored arrows and the inhibition symbol indicate interactions between the proteins. (H1) and (H2) indicate each assumption for how MEX-5/6 regulates aPAR. **d**, **e** Diagram of the reduction to the self-recruitment model. The black arrows and inhibition symbol indicate interaction between the proteins. The colored arrows indicate conversion of the slow and fast diffusion types of MEX-5/6

Ω is an open subset of $\mathbb{R}^N$ such that $\bar{\Omega}$ represents a cell (Fig. 1b). We also define the concentrations of anterior proteins (aPAR) in the membrane and cytosol by $[A_m](\mathbf{x}, t)$ and $[A_c](\mathbf{x}, t)$, respectively, and the concentrations of posterior proteins (pPAR) in the membrane and cytosol by $[P_m](\mathbf{x}, t)$ and $[P_c](\mathbf{x}, t)$, respectively, where $\mathbf{x} \in \mathbb{R}^N$ and $t \in [0, \infty)$. Then, the PAR polarity model is given by

$$
\begin{aligned}
&\frac{\partial[A_m]}{\partial t} + \nabla_\Gamma \cdot (\mathbf{v}_m(\mathbf{x}, t)[A_m]) \\
&\quad = D_m^A \nabla_\Gamma^2 [A_m] + F_{\text{on}}^A(\mathbf{x}, t)[A_c] - F_{\text{off}}^A(\mathbf{x}, t)[A_m] \quad \text{on } \mathbf{x} \in \partial\Omega, \\
&\frac{\partial[A_c]}{\partial t} + \nabla \cdot (\mathbf{v}_c(\mathbf{x}, t)[A_c]) = D_c^A \nabla^2 [A_c] \quad \text{on } \mathbf{x} \in \Omega, \\
&D_c^A \frac{\partial[A_c]}{\partial \mathbf{n}} - \mathbf{v}_c(\mathbf{x}, t)[A_c] = -F_{\text{on}}^A(\mathbf{x}, t)[A_c] + F_{\text{off}}^A(\mathbf{x}, t)[A_m] \quad \text{on } \mathbf{x} \in \partial\Omega, \\
&\frac{\partial[P_m]}{\partial t} + \nabla_\Gamma \cdot (\mathbf{v}_m(\mathbf{x}, t)[P_m])
\end{aligned}
$$

$$
\begin{aligned}
&= D_m^P \nabla_\Gamma^2 [P_m] + F_{\text{on}}^P(\mathbf{x},t)[P_c] - F_{\text{off}}^P(\mathbf{x},t)[P_m] \quad \text{on } \mathbf{x} \in \partial\Omega, \\
&\frac{\partial [P_c]}{\partial t} + \nabla \cdot (\mathbf{v}_c(\mathbf{x},t)[P_c]) = D_c^P \nabla^2 [P_m] \quad \text{on } \mathbf{x} \in \Omega, \\
&D_c^P \frac{\partial [P_c]}{\partial \mathbf{n}} - \mathbf{v}_c(\mathbf{x},t)[P_c] = -F_{\text{on}}^P(\mathbf{x},t)[P_c] + F_{\text{off}}^P(\mathbf{x},t)[P_m] \quad \text{on } \mathbf{x} \in \partial\Omega, \quad (1)
\end{aligned}
$$

where $\mathbf{n}$ is the inner normal vector on $\partial\Omega$. Here, $\mathbf{v}_m$ and $\mathbf{v}_c$ are cortical and cytoplasmic flow velocity functions, respectively, D_m^A and D_m^P are the diffusion rates of aPAR and pPAR in the membrane, respectively, and D_c^A and D_c^P are the diffusion rates of aPAR and pPAR in the cytosol, respectively. F_{on}^A and F_{on}^P are the on-rate functions of aPAR and pPAR from the cytosol to the membrane, respectively, and F_{off}^A and F_{off}^P are the off-rate functions of aPAR and pPAR, respectively. Note that $D_c^A > D_m^A$ and $D_c^P > D_m^P$ because diffusion in the cytosol is faster than that in the membrane (Kuhn et al. 2011; Goehring et al. 2011b). We define the detailed form of the flow velocity functions in Sect. 2.5.

The off-rate functions reflect the effect of the mutual inhibition of aPAR and pPAR. aPAR/pPAR transports pPAR/aPAR from the membrane to the cytosol (Fig. 1c), and we select the functional forms suggested in Seirin-Lee and Shibata (2015):

$$
F_{\text{off}}^A(\mathbf{x},t) = \alpha_1 + \frac{K_1[P_m]^n}{K_2 + \overline{K}_2[P_m]^n}, \qquad F_{\text{off}}^P(\mathbf{x},t) = \alpha_2 + \frac{K_3[A_m]^n}{K_4 + \overline{K}_4[A_m]^n},
$$

where α_1 and α_2 are basal off-rates, and $K_1, K_2, \overline{K}_2, K_3, K_4$ and $\overline{K}_4$ are positive constants determining the off-rates. $n(> 1)$ is the Hill coefficient, and we select $n = 2$ for the simulations. The on-rate functions are given by

$$
F_{\text{on}}^A(\mathbf{x},t) = \gamma_1, \qquad F_{\text{on}}^P(\mathbf{x},t) = \gamma_2, \qquad (2)
$$

where γ_1 and γ_2 are the on-rates of aPAR and pPAR, respectively.

2.2 MEX-5/6 Model

It is well known that MEX-5/6 has both slow and fast diffusion types in the cytosol, and that the conversion of one diffusion type to the other is regulated by PAR proteins (Daniels et al. 2010; Griffin et al. 2011; Wu et al. 2018). To develop a MEX-5/6 model combined with the PARs dynamics, we first formulate a general conversion model of MEX-5/6 diffusion. Defining the concentrations of the fast diffusive type of MEX-5/6 and slow diffusive type of MEX-5/6 by $[M_f](\mathbf{x},t)$ and $[M_s](\mathbf{x},t)$, respectively, the general MEX-5/6 conversion model is given by

$$
\begin{aligned}
&\frac{\partial [M_s]}{\partial t} + \nabla \cdot (\mathbf{v}_c(\mathbf{x},t)[M_s]) \\
&= D_s \nabla^2 [M_s] + G_{F\to S}^C(\mathbf{x},t)[M_f] - G_{S\to F}^C(\mathbf{x},t)[M_s], \quad \text{on } \mathbf{x} \in \Omega, \\
&D_s \frac{\partial [M_s]}{\partial \mathbf{n}} - \mathbf{v}_c(\mathbf{x},t)[M_s]
\end{aligned}
$$

$$
\begin{aligned}
&= G^M_{F\to S}(\mathbf{x},t)[M_f] - G^M_{S\to F}(\mathbf{x},t)[M_s], \quad \text{on } \mathbf{x}\in\partial\Omega,\\
&\frac{\partial[M_f]}{\partial t} + \nabla\cdot(\mathbf{v}_c(\mathbf{x},t)[M_f])\\
&= D_f\nabla^2[M_f] + G^C_{S\to F}(\mathbf{x},t)[M_s] - G^C_{F\to S}(\mathbf{x},t)[M_f], \quad \text{on } \mathbf{x}\in\Omega,\\
&D_f\frac{\partial[M_f]}{\partial\mathbf{n}} - \mathbf{v}_c(\mathbf{x},t)[M_f]\\
&= G^M_{S\to F}(\mathbf{x},t)[M_s] - G^M_{F\to S}(\mathbf{x},t)[M_f], \quad \text{on } \mathbf{x}\in\partial\Omega, \qquad (3)
\end{aligned}
$$

where D_s and D_f are the diffusion coefficients for slow and fast diffusive types of MEX-5/6, respectively, with $D_f > D_s$. $G^{\ell}_{S\to F}(\mathbf{x},t)$ and $G^{\ell}_{F\to S}(\mathbf{x},t)$ are conversion functions from the slow diffusive type to the fast diffusive type, and from the fast diffusive type to the slow diffusive type, respectively, where ℓ denotes either cytosol (C) or membrane (M).

Next, we derive forms for $G^{\ell}_{S\to F}(\mathbf{x},t)$ and $G^{\ell}_{F\to S}(\mathbf{x},t)$. In the wild type of *C. elegans*, the diffusion rate of MEX-5/6 in the posterior side is notably higher than in the anterior side. The studies by Daniels et al. (2010) and Griffin et al. (2011) suggest that pPAR (PAR-1) regulates the slow type of MEX-5/6, and the diffusion rate of MEX-5/6 of the posterior side in the PAR-1 mutant cell is significantly decreased compared to the wild type. This result indicates that PAR-1 promotes the conversion dynamics of the slow type of MEX-5/6 to the fast type. Thus, we suppose that the conversion rate from the slow type to the fast type has a positive correlation with pPAR concentration, and we propose that

$$G^M_{S\to F}(\mathbf{x},t) = \mu_1[P_m](\mathbf{x},t), \quad G^C_{S\to F}(\mathbf{x},t) = \mu_3[P_c](\mathbf{x},t),$$

where μ_1 and μ_3 are positive correlation constants reflecting the effective strength of pPAR on the conversion rate around the cell membrane and within the bulk cytosol, respectively.

The study by Wu et al. (2018) suggests that the diffusion rate of MEX-5/6 in the anterior side of the aPAR(PKC-3) mutant cell is significantly increased compared to the wild type. This result suggests two hypotheses: either PKC-3 promotes the conversion dynamics of the fast type of MEX-5/6 to the slow type, or PKC-3 plays a role in inhibiting the conversion dynamics from the slow type to the fast type of MEX-5/6. However, the diffusion rates of both PKC-3 and PAR-1 mutant cells did not show a notable difference from that of only PKC-3 mutant cell. This indicates that PKC-3 does not play the latter role, and it may promote the conversion of fast type to slow type of MEX-5/6. Thus, we suppose that the conversion rate from the fast type to the slow type has a positive correlation with aPAR concentration, and this leads us to define

$$G^M_{F\to S}(\mathbf{x},t) = \mu_2[A_m](\mathbf{x},t), \quad G^C_{F\to S}(\mathbf{x},t) = \mu_4[A_c](\mathbf{x},t), \qquad (4)$$

where μ_2 and μ_4 are positive correlation constants reflecting the effective strength of aPAR on the conversion rate in the cell membrane and bulk cytosol, respectively.

2.3 MEX-5/6-Combined-PARs Model

Experimental observations of the *C. elegans* embryo suggest that MEX-5/6 regulates the expansion of the pPAR domain by helping to exclude the aPAR domain, rather than directly promoting pPAR localization (Cuenca et al. 2002; Schubert et al. 2000), which suggests the possibility that MEX-5/6 may directly regulate the translocation dynamics of aPAR between the membrane and the cytosol. On the other hand, the detailed molecular mechanism of the interaction between MEX-5/6 and the PARs is still unclear. Thus, we consider two possible assumptions. We suppose that the experimental observation of Cuenca et al. (2002) is related to either the on-rate or the off-rate of aPAR in the model (1). Thus, we assume that in one model, MEX-5/6 inhibits the recruitment of aPAR from the cytosol to the membrane (i.e., the on-rate of aPAR), and in the other model, we assume that MEX-5/6 promotes the transport of aPAR from the membrane to the cytosol (i.e., the off-rate of aPAR). We call these models *H1* and *H2*, respectively (Fig. 1c). We formulate the simplest type of model as follows:

$$\begin{aligned} H1: &\quad F_{\text{on}}^{A,H_1}(\mathbf{x},t) = \frac{F_{\text{on}}^{A}(\mathbf{x},t)}{1+\mu_0[M](\mathbf{x},t)}, \\ H2: &\quad F_{\text{off}}^{A,H_2}(\mathbf{x},t) = F_{\text{off}}^{A}(\mathbf{x},t)(1+\mu_0[M](\mathbf{x},t)), \end{aligned} \tag{5}$$

where $[M](\mathbf{x},t) = [M_f](\mathbf{x},t) + [M_s](\mathbf{x},t)$, and μ_0 is either the inhibition rate of aPAR recruitment or the promotion rate of aPAR transport from the membrane to cytosol by MEX-5/6. Note that $\mu_0 = 0$ recovers the original model (1) without the effect of MEX-5/6. We name the combination of model (1) including (5), with model (3), the *MEX-5/6-combined-PARs Model* (Fig. 1c).

2.4 MEX-5/6-Combined-Self-Recruitment pPAR Model

The self-recruitment model of PAR dynamics was first suggested in Seirin-Lee and Shibata (2015), in which the aPAR-pPAR model is reduced to either an aPAR alone or a pPAR alone model, and the effect of the off-rate (by the mutual inhibition) is replaced by a self-recruitment effect resulting from either aPAR or pPAR itself. By applying this reduction to the self-recruitment model, we can study the interaction of PAR polarity dynamics with pPAR alone. This gives us more precise information on how pPAR is directly, or indirectly, involved in MEX-5/6 dynamics (Fig. 1d, e). Thus, we here reduce the MEX-5/6-combined-PARs Model (1)–(3) to the self-recruitment form and show that the conversion model of MEX-5/6 by aPAR and pPAR given in (3) is essentially equivalent to the conversion model through pPAR alone.

The polarity of the PARs is formed only in the membrane. We find that the interface between the aPAR and pPAR domains in the membrane is sufficiently narrow and that the concentration of pPAR in the membrane is very low where the concentration of aPAR is high (Fig. 3a, middle panel). Thus, we approximate the effect of the off-rate of aPAR (F_{off}^{A}) under the condition that $[P_m] \ll 1$ by Taylor expansion :

$$F_{\text{off}}^A(\mathbf{x}, t) = \alpha_1 + \frac{K_1[P_m]^n}{K_2 + \overline{K}_2[P_m]^n} \approx \alpha_1 + F'(0)[P_m] + \frac{F''(0)}{2}[P_m]^2 + O([P_m]^3).$$

We can easily calculate $F'(0) = 0$ for $n \geq 2$, $F''(0) = 2K_1K_2^{-1}$ for $n = 2$ and $F''(0) = 0$ for $n \geq 3$. Assuming $n = 2$, we obtain

$$F_{\text{off}}^A(\mathbf{x}, t) \approx \alpha_1 + \frac{K_1}{K_2}[P_m]^2 + O([P_m]^3). \tag{6}$$

We assume that the fast diffusion of aPAR in the cytosol leads to a well-mixed state and that the concentration of aPAR in the cytosol quickly approaches an equilibrium state, namely $A_c^* = (1/\Omega)\int_\Omega [A_c](\mathbf{x}, t)d\mathbf{x}$. This leads to

$$-F_{\text{on}}^{A,H_1}(\mathbf{x}, t)A_c^* + F_{\text{off}}^A(\mathbf{x}, t)[A_m] \approx 0 \ \text{ for } H1,$$
$$-F_{\text{on}}^{A}(\mathbf{x}, t)A_c^* + F_{\text{off}}^{A,H_2}(\mathbf{x}, t)[A_m] \approx 0 \ \text{ for } H2.$$

With the approximation (6) and the on-rate function F_{on}^A (2), we have the same approximation for both *H1* and *H2*, such that

$$[A_m] \approx \frac{F_{\text{on}}^A(\mathbf{x}, t)A_c^*}{(1 + \mu_0[M])F_{\text{off}}^A(\mathbf{x}, t)} = \frac{\delta_2}{(1 + \mu_0[M])(1 + \delta_1[P_m]^2)} \tag{7}$$

where $\delta_1 = K_1/(K_2\alpha_1)$ and $\delta_2 = \gamma_1 A_c^*/\alpha_1$. Thus, both the *H1* and *H2* models are essentially the same, and the effect of MEX-5/6 on aPAR dynamics is likely to decrease the concentration of aPAR in the membrane.

Substituting the approximate equation (7) for $[A_m]$ into the off-rate function F_{off}^P of pPAR, (2), we obtain

$$F_{\text{off}}^P(\mathbf{x}, t) = \alpha_2 + \frac{\beta_4}{\beta_1 + (1 + \mu_0[M])^2(\beta_2 + \beta_3[P_m]^2) + O([P_m]^4)}, \tag{8}$$

where $\beta_1 = \overline{K}_4\delta_2^2$, $\beta_2 = K_4$, $\beta_3 = 2K_4\delta_1$, and $\beta_4 = K_3\delta_2^2$. Finally, $G_{F\to S}^C(\mathbf{x}, t)$ and $G_{F\to S}^M(\mathbf{x}, t)$, given in (4), are transformed to

$$G_{F\to S}^C(\mathbf{x}, t) = \mu_4 A_c^*, \quad G_{F\to S}^M(\mathbf{x}, t) = \mu_2 \frac{\delta_2}{(1 + \mu_0[M])(1 + \delta_1[P_m]^2)} \tag{9}$$

from equation (7).

Combining Eqs. (6) and (9) with the pPAR equations of the model (1) and the MEX-5/6 model (3), we obtain the *MEX-5/6-combined-self-recruitment pPAR Model* as follows:

$$\frac{\partial[P_m]}{\partial t} + \nabla_\Gamma \cdot (\mathbf{v}_m[P_m]) = D_m^P \nabla_\Gamma^2 [P_m]$$

$$
\begin{aligned}
&+\gamma_2[P_c] - \left\{\alpha_2 + \frac{\beta_4}{\beta_1 + (1+\mu_0[M])^2(\beta_2 + \beta_3[P_m]^2)}\right\}[P_m], \text{ on } \mathbf{x} \in \partial\Omega, \\
&\frac{\partial [P_c]}{\partial t} + \nabla \cdot (\mathbf{v}_c[P_c]) = D_c^P \nabla^2 [P_m], \text{ on } \mathbf{x} \in \Omega, \\
&D_c^P \frac{\partial [P_c]}{\partial \mathbf{n}} - \mathbf{v}_c(\mathbf{x}, t)[P_c] \\
&= -\gamma_2[P_c] + \left\{\alpha_2 + \frac{\beta_4}{\beta_1 + (1+\mu_0[M])^2(\beta_2 + \beta_3[P_m]^2)}\right\}[P_m], \text{ on } \mathbf{x} \in \partial\Omega,
\end{aligned} \tag{10}
$$

and

$$
\begin{aligned}
&\frac{\partial [M_s]}{\partial t} + \nabla \cdot (\mathbf{v}_c[M_s]) \\
&= D_s \nabla^2 [M_s] + \mu_4\delta_3[M_f] - \mu_3[P_c][M_s], \quad \text{on } \mathbf{x} \in \Omega, \\
&D_s \frac{\partial [M_s]}{\partial \mathbf{n}} - \mathbf{v}_c[M_s] \\
&= \mu_2 \frac{\delta_2}{(1+\mu_0[M])(1+\delta_1[P_m]^2)}[M_f] - \mu_1[P_m][M_s], \quad \text{on } \mathbf{x} \in \partial\Omega, \\
&\frac{\partial [M_f]}{\partial t} + \nabla \cdot (\mathbf{v}_c[M_f]) \\
&= D_f \nabla^2 [M_f] + \mu_3[P_c][M_s] - \mu_4\delta_3[M_f], \quad \text{on } \mathbf{x} \in \Omega, \\
&D_f \frac{\partial [M_f]}{\partial \mathbf{n}} - \mathbf{v}_c[M_f] \\
&= \mu_1[P_m][M_s] - \mu_2 \frac{\delta_2}{(1+\mu_0[M])(1+\delta_1[P_m]^2)}[M_f], \quad \text{on } \mathbf{x} \in \partial\Omega,
\end{aligned} \tag{11}
$$

where we have replaced A_c^* by a positive parameter, δ_3, without loss of generality. The model is a conservation system and the total mass of pPAR and MEX-5/6 is conserved.

From a direct comparison between the MEX-5/6-combined-PARs Model (1)–(3) and MEX-5/6-combined-self-recruitment pPAR Model (10)–(11), we find that the regulation network involving aPAR, pPAR, and MEX-5/6 can be reduced to the network of pPAR-alone conversion control in the MEX-5/6 dynamics (Fig. 1e). We can interpret the model (11) such that conversion from the fast diffusive type to slow diffusive type of MEX-5/6 is promoted constantly (the terms $\mu_4\delta_3$ and $\mu_2\delta_2$) by a substrate and it is simultaneously down-regulated by pPAR in the membrane. The substrate may be considered to be phosphatase PP2A, and the inhibition by pPAR on the membrane may be considered as an indirect role of aPAR on the conversion of MEX-5/6, as suggested in Griffin et al. (2011) and Wu et al. (2018). Our model reduction suggests that the direct conversion model involving aPAR and pPAR is essentially the same as the pPAR-alone conversion model. Indeed, we confirm that the two models essentially show similar dynamics (see Fig. 3). In this paper, we explore our results on the MEX-5/6-combined-self-recruitment pPAR Model (10)–(11).

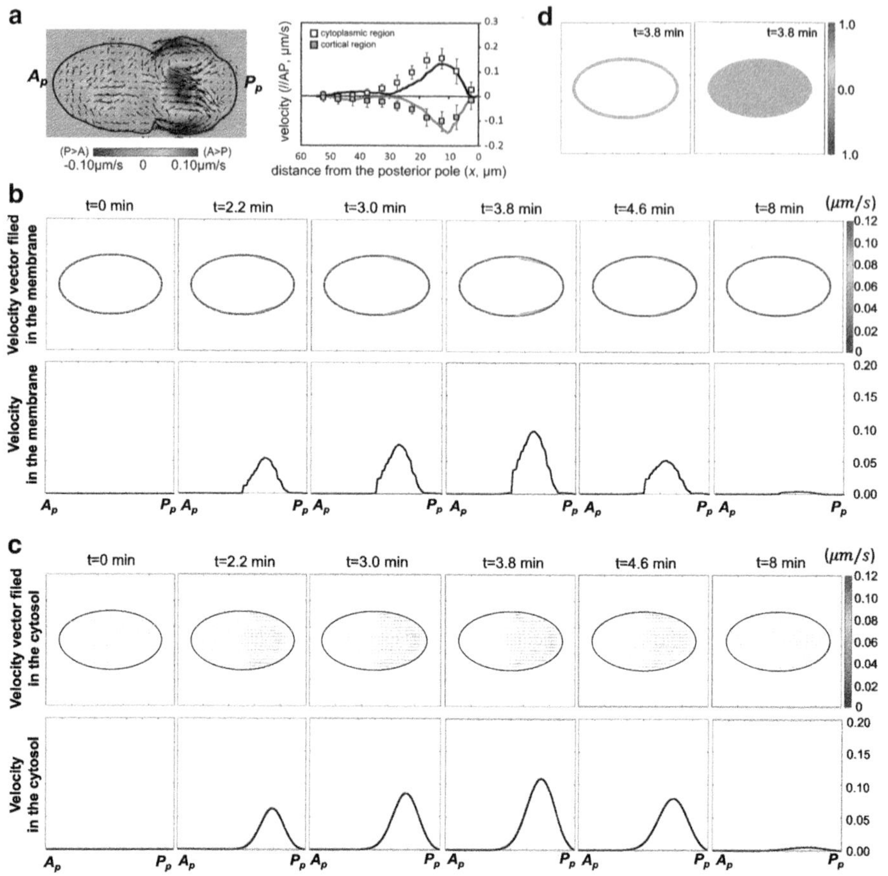

Fig. 2 Flow velocity function. **a** Experimental data of flow velocity (adapted from Niwayama et al. (2011)). Right panel shows the velocity distribution of cytoplasmic streaming. Lines in the left panel show the velocity distribution along the AP axis in vivo reconstructed with the moving particle simulation. Boxes in the left panel show the experimental data (gray: cortical flow, white: cytoplasmic flow). **b, c** Representative example of the flow velocity function $\mathbf{v}(\mathbf{x}, t) = (v^x, v^y)$ given by (12). The vectors indicate the direction, and the color map shows the speed of flow velocities. The one-dimensional velocity data ($|\mathbf{v}|$) in the membrane have been plotted on the cell circumference, the data in the cytosol have been plotted on $(x, y) = (x, L_y/2)$. **d** A representative simulation of incompressibility at maximal flow velocity. The numerical simulations show that $\nabla \cdot \mathbf{v} \approx O(2 \times 10^{-2})$ during the flow

2.5 Flow Velocity Function

After sperm entry, the acto-myosin network located in the cell cortex begins contracting toward the anterior side from the posterior side (Gönczy 2005) (Fig. 1a), causing a cortical flow in the same direction as the acto-myosin contraction, and a cytoplasmic flow in the opposite direction of the center of the cytosol and in the same direction as the cortical flow in the periphery of the membrane (Niwayama et al. 2011) (Fig. 2a). The velocity of the flows has been investigated in detail experimentally, revealing that the maximal velocity (approximately 0.156 ± 0.044 $\mu m/s$) is reached during the establishment phase and the velocity goes to zero around the time that the establishment phase terminates (Goehring et al. 2011b; Niwayama et al. 2011). In this study, we explicitly formulate the flow velocity function, $\mathbf{v}(\mathbf{x}, t) = (v^x(\mathbf{x}, t), v^y(\mathbf{x}, t))$, defined for the entire cell region ($\bar{\Omega}$) based on the experimental data in Niwayama et al. (2011).

The flow velocity function is given by

$$
\begin{aligned}
v^x(\mathbf{x}, t) &= \frac{c_3[x - (L_x/2 - \ell_x)]^2[(L_x/2 + \ell_x) - x]\cos(c_4\pi(y - L_y/2))}{c_1 + \exp[c_2(x - L_x)^2/(1 + c_5 t)]} \times \tau(t), \\
v^y(\mathbf{x}, t) &= \frac{c_8[x - (L_x/2 + \ell_x/2)]\sin(c_7\pi(y - L_y/2))}{c_1 + \exp[c_2(x - (L_x/2 + \ell_x/2))^2/(1 + c_5 t)]} \times \tau(t),
\end{aligned}
\tag{12}
$$

where

$$
\tau(t) = \begin{cases} \frac{t}{T_0}\{c_6(T_0 - t) + 1\} & t < T_0 \\ \frac{1}{1 + c_9(t - T_0)^2} & t \geq T_0 \end{cases}
$$

and $c_i (i = 1 \ldots 9)$ are positive constants. T_0 is the temporal point at which the velocity is maximum. We select the parameter values so that the flow velocity function approaches the maximal velocity of about 0.12 μ m/s at 3.8 min, and the flow ceases around 8 min (Fig. 2b, c, Movie S1). The cortical flow velocity function, $\mathbf{v}_m(\mathbf{x}, t)$, is given by the value of $(v^x(\mathbf{x}, t), v^y(\mathbf{x}, t))$ on the domain $\partial\Omega$, and the cytoplasmic flow velocity function, $\mathbf{v}_c(\mathbf{x}, t)$, is given by the value of $(v^x(\mathbf{x}, t), v^y(\mathbf{x}, t))$ on the domain Ω. Detailed temporal and spatial data for the cortical and cytoplasmic flows used in our simulations are shown in Fig. 2b, c. Note that the flow functions in a cell satisfy incompressibility almost everywhere (Fig. 2d and Fig. S1)

2.6 Model Incorporated with the Bulk-Surface Cellular Geometry

Here, we introduce a method to combine the phase-field function with the bulk-surface system (10)–(11). The method allows for a simple numerical technique to solve a convection–reaction–diffusion model on any high-dimensional cellular shape, and we can also simply include the flow dynamics in the model system. Let us express a fixed cell domain using a phase-field function for some time t^*, namely $\phi(\mathbf{x}, t^*)$, as follows (Fig. 1b):

$$
\begin{aligned}
&\text{Cytosol} \equiv \{\mathbf{x} | \phi(\mathbf{x}, t^*) = 1\}, \quad \text{Membrane} \equiv \{\mathbf{x} | 0 < \phi(\mathbf{x}, t^*) < 1\}, \\
&\text{Extracellular region} \equiv \{\mathbf{x} | \phi(\mathbf{x}, t^*) = 0\}.
\end{aligned}
$$

We now explain a general method to make a phase-field function of a cell. Let us first define the free energy function, E_0, of Ginzburg–Landau type for a cell such that

$$
E_0 = \int_A \frac{\varepsilon^2}{2} |\nabla\phi|^2 + g(\phi) \mathrm{d}x
$$

where A denotes the area of the system in which ϕ is defined, $\varepsilon(>0)$ is a sufficiently small constant that defines the thickness of the cell membrane, and $g(\phi)=\frac{1}{4}\phi^2(1-\phi)^2$. Here, the symmetric potential $g(\phi)$ is used for setting the local minima at $\phi=0$ and $\phi=1$.

Next, we define the energy function which determines the volume of the cell such that

$$E_1=\alpha\left(\int_A h(\phi)\mathrm{d}\mathbf{x}-\overline{V}\right)^2,$$

where $\alpha(>0)$ is the intensity constant of the energy for cell volume, $\overline{V}$ is the target volume of the cell, and $h(\phi)=\phi^3(10-15\phi+6\phi^2)$, which is used for the induction of an energetic asymmetry between $\phi=0$ and $\phi=1$, driving the interface while keeping $\phi=0$ and $\phi=1$ as local minima of the energy function (see Appendix of Seirin-Lee et al. (2017) for more detail). Finally, we define a time evolution equation for the total energy of the cell, satisfying

$$\mu^{-1}\frac{\partial\phi}{\partial t}=-\frac{\delta(E_0+E_1)}{\delta\phi},$$

where $\mu(>0)$ is the constant defining the mobility of the interface. Substituting for E_0 and E_1 into the above equation, we arrive at the equation

$$\mu^{-1}\frac{\partial\phi}{\partial t}=\varepsilon^2\nabla^2\phi+\phi(1-\phi)\left\{\phi-\frac{1}{2}-60\alpha\phi(1-\phi)\left(V(t)-\overline{V}\right)\right\}. \tag{13}$$

By providing the target cell volume ($\bar{V}$) and initial conditions, we can readily generate cells that have different shapes and sizes. We generate the *C. elegans* embryo by setting $\overline{V}$ as the actual size (the area in two-dimensional simulations) of the embryo and the initial condition to be an ellipse with the embryo scale of short and long axes.

With the cell phase-field function ϕ, we rewrite the MEX-5/6-combined-self-recruitment pPAR model (10)–(11) in a form in which the cell geometry is reflected (Seirin-Lee 2016; Teigen et al. 2009; Wang et al. 2017). The model system, combined with the phase-field function that we used for numerical simulations, is given by

$$\begin{aligned}
&\frac{\partial B(\phi)[P_m]}{\partial t}+\nabla\cdot(B(\phi)\mathbf{v}_m[P_m])=D_m^P\nabla\cdot(B(\phi)\nabla[P_m])\\
&\quad+B(\phi)\left[\gamma_2[P_c]-\left\{\alpha_2+\frac{\beta_4}{\beta_1+(1+\mu_0[M])^2(\beta_2+\beta_3[P_m]^2)}\right\}[P_m]\right],\\
&\frac{\partial\phi[P_c]}{\partial t}+\nabla\cdot(\phi\mathbf{v}_c[P_c])=D_c^P\nabla\cdot(\phi\nabla[P_c])\\
&\quad+|\nabla\phi|\left[-\gamma_2[P_c]+\left\{\alpha_2+\frac{\beta_4}{\beta_1+(1+\mu_0[M])^2(\beta_2+\beta_3[P_m]^2)}\right\}[P_m]\right],\\
&\frac{\partial\phi[M_s]}{\partial t}+\nabla\cdot(\phi\mathbf{v}_c[M_s])=D_s\nabla\cdot(\phi\nabla[M_s])+\phi\{\mu_4\delta_3[M_f]-\mu_3[P_c][M_s]\}
\end{aligned}$$

$$+|\nabla\phi|\left\{\mu_2\frac{\delta_2}{(1+\mu_0[M])(1+\delta_1[P_m]^2)}[M_f]-\mu_1[P_m][M_s]\right\},$$

$$\frac{\partial\phi[M_f]}{\partial t}+\nabla\cdot(\phi\mathbf{v}_c[M_f])=D_f\nabla\cdot(\phi\nabla[M_f])+\phi\{\mu_3[P_c][M_s]-\mu_4\delta_3[M_f]\}$$

$$+|\nabla\phi|\left\{\mu_1[P_m][M_s]-\mu_2\frac{\delta_2}{(1+\mu_0[M])(1+\delta_1[P_m]^2)}[M_f]\right\}, \tag{14}$$

for $\mathbf{x}\in A(\equiv[0,L_x]\times[0,L_y])$, where $B(\phi)=\nu\phi^2(1-\phi)^2$ ($\nu>0$), a function defining the membrane region (Fig. 1b). The cortical flow velocity function, $\mathbf{v}_m(\mathbf{x},t)$, is given by $\mathbf{v}_m(\mathbf{x},t)=B(\phi(\mathbf{x},t))\mathbf{v}(\mathbf{x},t)$, and the cytoplasmic flow velocity function, $\mathbf{v}_c(\mathbf{x},t)$, is given by $\mathbf{v}_c(\mathbf{x},t)=\phi(\mathbf{x},t)\mathbf{v}(\mathbf{x},t)$.

One can confirm that a sharp interface limit recovers the boundary conditions in the cytosol equations of the phase-field combined model (see "Appendix A" for more detail). Note that we can numerically solve the bulk-surface model (10)–(11) using a standard finite difference method on a square. The details of the initial conditions and parameter values are given in "Appendix B."

3 Results

3.1 Regeneration of PAR and MEX-5/6 Polarities

We first confirm that the MEX-5/6-combined-PARs model (1)–(3) and the MEX-5/6-combined-self-recruitment pPAR model (10)–(11) are essentially the same (Fig. 3), and there are no qualitative differences in the model dynamics, suggesting that the two different conversion dynamics suggested by Daniels et al. (2010) and Griffin et al. (2011) can be reconsidered by our mathematical models, and they have essentially the same mathematical structure. This implies that our model is integrating all relevant molecular dynamics observed in the previous experiments and is thus a general model to capture the dynamics of both MEX-5/6 and PAR, simultaneously.

In our model, we confirmed that PAR polarity in the membrane and MEX-5/6 polarity in the cytosol are simultaneously generated (Fig. 3b, Movie S2, S3), as observed experimentally (Cuenca et al. 2002). The simulations showed that the establishment phase finishes at approximately 6–7 min, and the boundary of the pPAR domain stops at around the middle of the cell, for a representative parameter set. With a small initial stimulus at the posterior polar site (Fig. 3b, $t=0$ panel), the pPAR domain begins to emerge toward the anterior from the posterior, and simultaneously MEX-5/6 generates polarity in the same direction on the emergence of pPAR. The domain boundary of MEX-5/6 is always determined in a location similar to the domain boundary of pPAR in the periphery of the membrane (Fig. 3c, upper panel). This result suggests that polarity formation of MEX-5/6 and PAR is very interactive, both temporally and spatially.

On the other hand, the MEX-5/6 concentration profile in the bulk region of the cytosol shows that the distribution of MEX-5/6 is not clearly distinguished by the two domains of different concentration levels (Fig. 3c, lower panel). This is likely to be

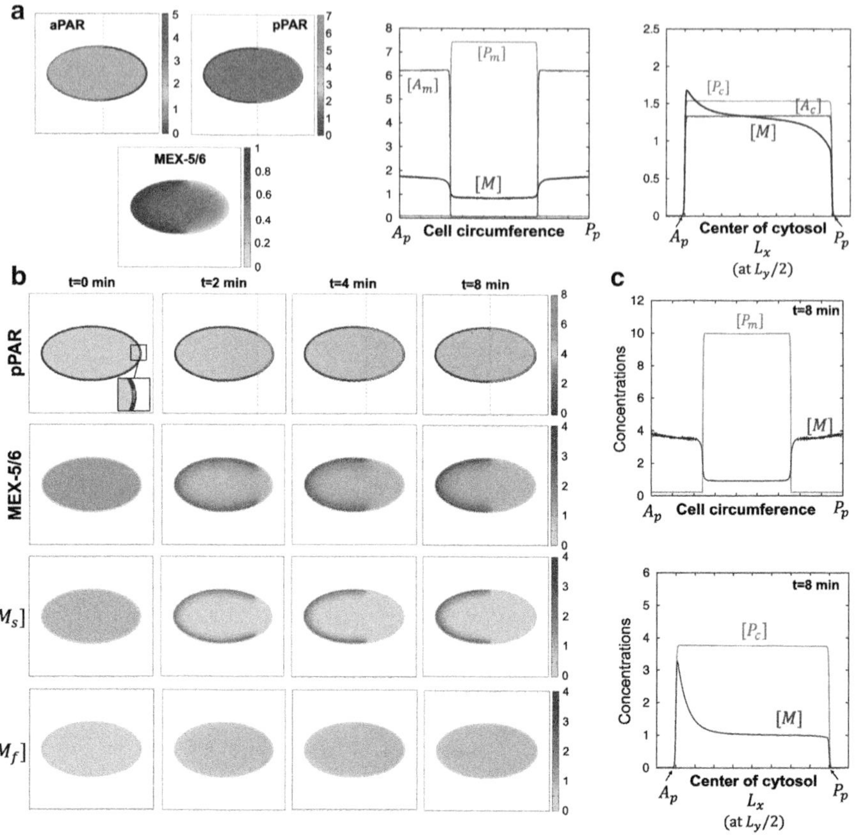

Fig. 3 Dynamics of PARs and MEX-5/6 polarity formation. **a** Representative simulation results of the MEX-5/6-combined-PAR model (1)–(3) without flow. Left panels show the polarities of aPAR, pPAR, and MEX-5/6. The concentrations in the cell circumference are plotted in the middle panel and the concentrations in the cytosol are plotted in the right panel. **b**, **c** Representative simulation results of the MEX-5/6-combined-self-recruitment pPAR model (10)–(11) without flow. The gray dotted line indicates the boundary location of the polarity domains. The detailed parameter values are given in "Appendix B"

a consequence of the homogeneity of the pPAR concentration in the cytosol (Seirin-Lee et al. 2020b). We further found that the slow diffusive MEX-5/6 has a similar distribution shape to that of the total MEX-5/6, whereas the fast diffusive MEX-5/6 is almost homogeneous. This indicates that the conversion of MEX-5/6 to a slow diffusive type is essential to create the MEX-5/6 polarity, and that the inhibition/activation role of pPAR on the conversion of MEX-5/6 to a slow/fast diffusive type is critical. Our bulk-surface model proposes that the polarity of MEX-5/6 is mainly formed in the periphery of the cell membrane rather than in the bulk space of the cytosol, and that the heterogeneity of pPAR polarity in the membrane plays an important role in generating MEX-5/6 polarity.

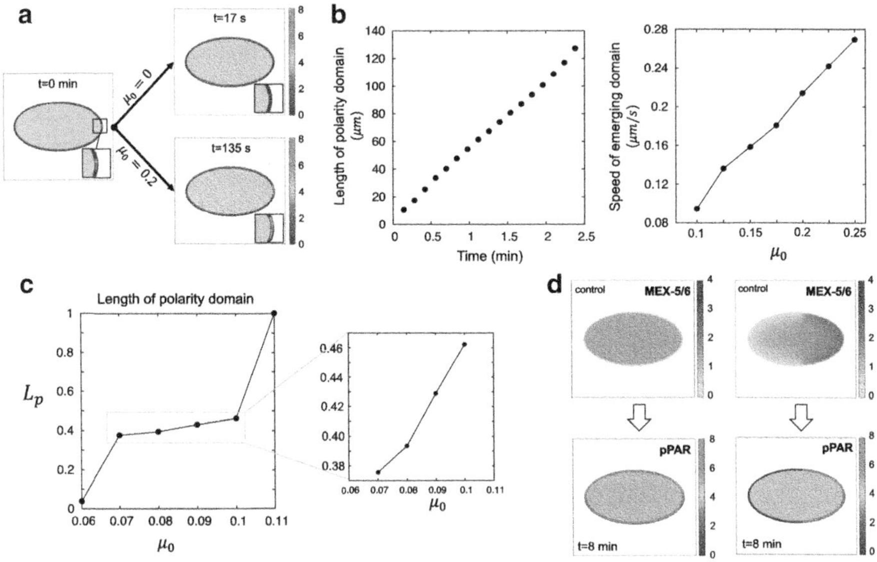

Fig. 4 Role of MEX-5/6 polarity on PAR polarity formation. **a** The effect of MEX-5/6 on the symmetry breaking phase. **b** The effect of MEX-5/6 on the establishment phase. The left panel shows the change in length of the pPAR domain and the right panel shows the effect of MEX-5/6 on the emerging speed of the pPAR polarity domain. The data were measured using the average speed over the interval [0.5 min, 1 min] for each simulation. **c** The effect of MEX-5/6 on the maintenance phase. The effect of MEX-5/6 on the length of the pPAR polarity domain is shown. **d** The effect of MEX-5/6 polarity on pPAR polarity. The upper panels show how the polarity of MEX-5/6 was numerically controlled and the lower panels show the resultant pPAR polarity for each case. The detailed parameter values are given in "Appendix B"

3.2 Role of MEX-5/6 Polarity on PAR Polarity Formation

How the upstream PAR proteins, and their polarity, influence formation of MEX-5/6 have been investigated in detail, both experimentally and mathematically (Wu et al. 2018; Seirin-Lee et al. 2020b). However, it is still unknown how the downstream MEX-5/6 polarity influences the upstream PAR polarity. Thus, we explore here the biochemical roles of MEX-5/6 on PAR polarity formation by investigating how MEX-5/6 affects PAR polarity formation, spatially and temporally, with respect to the symmetry breaking, establishment, and maintenance phases, without flow effects. To see the influence on the symmetry breaking phase, we investigated whether the polarity pattern can emerge for the cases when the effect of MEX-5/6 is absent ($\mu_0 = 0$) or present ($\mu_0 > 0$). We found that MEX-5/6 can promote symmetry breaking (Fig. 4a), implying that MEX-5/6 supports pPAR invasion into the membrane by suppressing aPAR. To see how MEX-5/6 controls pPAR recruitment, we analyzed how the property of the bi-stability of pPAR dynamics can be affected by the parameter μ_0 (see "Appendix D"). The analysis showed that μ_0 leads to a wider parameter region for the bi-stability of pPAR dynamics (Fig. S2), suggesting that MEX-5/6 plays an important supporting role in the formation of pPAR polarity.

Next, we investigated the emerging speed of polarity pattern in the establishment phase (Fig. 4b). We found that the speed is almost constant (Fig. 4b, left panel) before it enters the maintenance phase, in which the speed should be slower in order to halt the speed of establishment polarity pattern. Thus, we explored how the emerging speed is affected by MEX-5/6, namely μ_0, during the early stage of the establishment phase (Fig. 4b, right panel). The results show that the emerging speed is highly affected by changes in μ_0. The emerging speed increased by more than double its value when μ_0 doubled in magnitude, indicating that MEX-5/6 plays a critical role in regulating the temporal dynamics of PAR polarity formation.

Finally, to see the effect of MEX-5/6 in the maintenance phase, we focused on two phenomena: one is the length scale of the pPAR polarity domain (L_p = [Length of polarity domain]/[Length of cell circumference]) and the other is the location of pPAR polarity. We first found that the length scale of the pPAR domain can be affected by MEX-5/6 (Fig. 4c). As μ_0 increased, the length of the pPAR domain increased. The total mass of proteins is conserved; therefore, it is likely that the MEX-5/6 redistributed the pPAR protein between the membrane and cytosol through helping pPAR stay in the membrane. However, the parameter range of μ_0 where pPAR forms a stationary polarity pattern was approximately within 8% variation of the length scale, and there existed two threshold values of μ_0 at which pPAR either fails to invade, or spreads throughout the whole cell membrane. This indicates that the effect of MEX-5/6 on the length scale of the pPAR domain may be negligible, but the maintenance of pPAR polarity may be tightly regulated by MEX-5/6.

We also found that the asymmetry of MEX-5/6 is critical to maintain the pPAR domain. To test this, we set $D_f = D_s$ and controlled MEX-5/6 to be spatially homogeneous (Fig. 4d, left panels). The result showed that pPAR fails to maintain polarity and the polarity domain spread throughout the whole cell membrane. On the other hand, we found that the location of MEX-5/6 polarity does not affect pPAR polarity (Fig. 4d, right panels). To see how the location of MEX-5/6 polarity affects pPAR polarity, we switched the conversion roles of MEX-5/6 artificially in the MEX-5/6 model (11) such that

$$
\begin{aligned}
\frac{\partial [M_s]}{\partial t} &= D_s \nabla^2 [M_s] - \mu_4 \delta_3 [M_s] + \mu_3 [P_c][M_f], && \text{on } \mathbf{x} \in \Omega, \\
D_s \frac{\partial [M_s]}{\partial \mathbf{n}} &= -\mu_2 \frac{\delta_2}{(1+\mu_0[M])(1+\delta_1[P_m]^2)} [M_s] + \mu_1 [P_m][M_f], && \text{on } \mathbf{x} \in \partial\Omega, \\
\frac{\partial [M_f]}{\partial t} &= D_f \nabla^2 [M_f] - \mu_3 [P_c][M_f] + \mu_4 \delta_3 [M_s], && \text{on } \mathbf{x} \in \Omega, \\
D_f \frac{\partial [M_f]}{\partial \mathbf{n}} &= -\mu_1 [P_m][M_f] + \mu_2 \frac{\delta_2}{(1+\mu_0[M])(1+\delta_1[P_m]^2)} [M_s], && \text{on } \mathbf{x} \in \partial\Omega.
\end{aligned}
$$

Using this model, we controlled the location of MEX-5/6 polarity so that it had a high concentration in the posterior side. Unexpectedly, the simulation result showed that pPAR polarity is maintained robustly, even though the MEX-5/6 polarity is formed on the opposite site to the wild-type case. Taking these results together, we conclude that the location of MEX-5/6 polarity is not essential, but the asymmetry of MEX-5/6 distribution is indispensable for PAR polarity maintenance.

3.3 Interplay with the Flows and Cell Geometry

Cortical and cytoplasmic flows induced by acto-myosin contraction have been considered as critical factors in the patterning phase of PAR polarity formation (Goehring et al. 2011b). Nevertheless, it has not yet been explored how these flows can affect cytoplasmic polarity. To see how the cytoplasmic protein MEX-5/6 interacts with the flow dynamics and, consequently, affects the polarity dynamics, we have explored three cases: wild-type case, flow absent case, and flow present case only for MEX-5/6, in which the advection terms in the pPAR model (10) are removed. We first compared the temporal dynamics of polarity pattern for the aforementioned three cases (Fig. 5A). We found that the flows around the membrane can speed up the patterning time of pPAR, although the cytoplasmic flow in the bulk cytosol space has the opposite direction to that of the cortical flow (black dots and white dots in Fig. 5A). This indicates that the temporal dynamics of pPAR are affected strongly by the cortical flow, rather than the cytoplasmic flow.

On the other hand, we found that the interplay of MEX-5/6 and flows can slow down pattern emergence and negatively affect the temporal dynamics of patterning (red dots and white dots in Fig. 5A). This supposes that the cytoplasmic flows around the membrane transport the slow diffusive type of MEX-5/6 from the posterior pole to the anterior pole, resulting in lower MEX-5/6 concentration, and a weakening of the positive effect of MEX-5/6 for the pPAR to stay on the membrane (namely, either the inhibition effect on aPAR recruitment, or the activation effect on aPAR transmembrane off-rate). This result indicates that the flow dynamics do not always play a role in promoting PAR polarity, but can affect it negatively via the interplay with MEX-5/6. Nevertheless, such a negative effect is likely to be eliminated by the positive effect of the cortical flow on pPAR.

Next, we explored how the directions of flow interplay with MEX-5/6 dynamics and, consequently, influence the dynamics of PAR polarity. For this, we artificially imposed an opposite direction for the flows to the wild type only in MEX-5/6, with no flow in pPAR (Fig. 5B(b1),(b2)). We found that the flow directions greatly affected the dynamics of MEX-5/6, leading to completely different PAR polarity patterns. In contrast, there was no difference in the final polarity pattern (stationary steady state) in the case that both pPAR and MEX-5/6 are simultaneously affected by oppositely directed flows, although the temporal dynamics of patterning was affected when the velocity of flows was increased (Fig. 5B(b3)). This result indicates that when the flows affect pPAR and MEX-5/6 simultaneously, the influence of flow direction on the spatial dynamics of polarity can be negligible. To confirm this, we investigated the effect of the spatial position of symmetry breaking (Fig. 5C). We set the symmetry breaking position of pPAR polarity to be in a perpendicular location to that for the wild-type case but with the wild-type flow dynamics. This setting gives us the situation in which both pPAR and MEX-5/6 are strongly perturbed by flow. Nevertheless, we found that the final polarity domain is formed robustly, in the middle portion of the cell, even

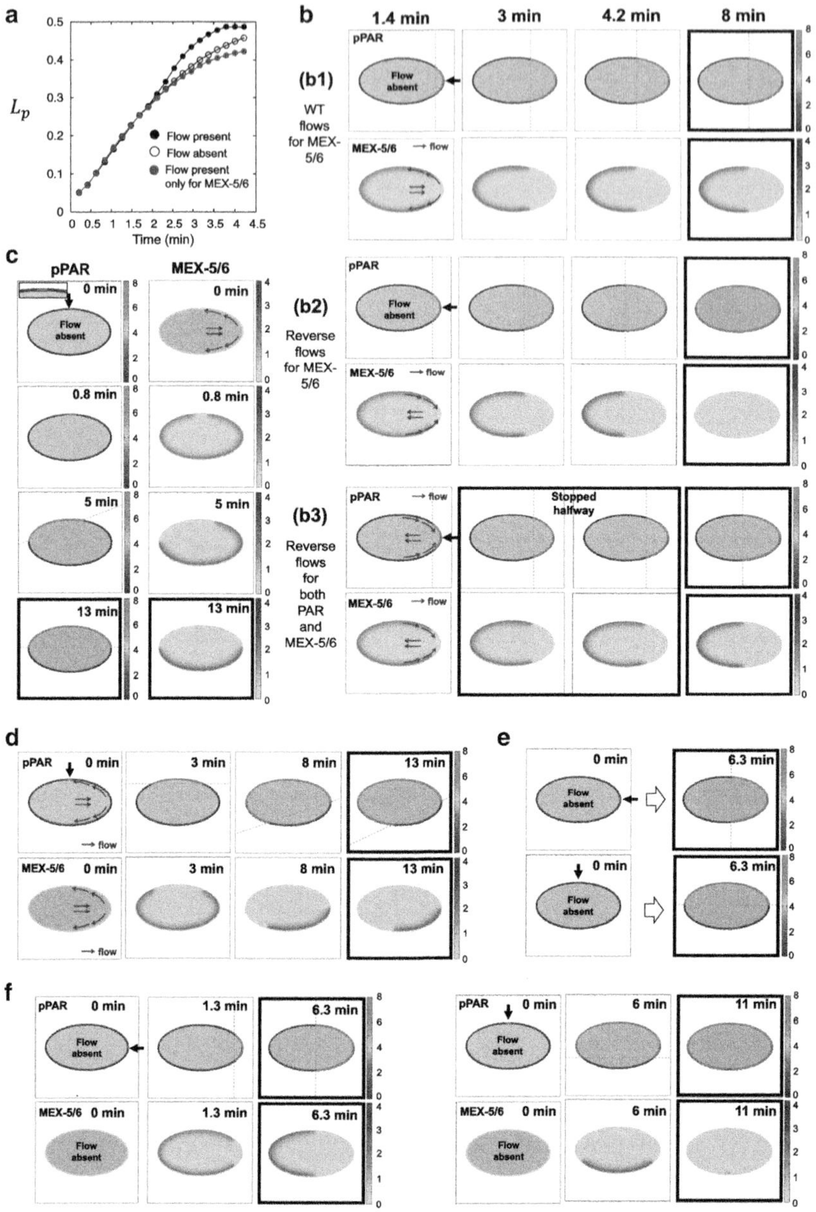

Fig. 5 Interplay with the flows and cell geometry. **a** The effect of flow on the emerging speed. **b** The effect of flow direction on polarity patterning. The reverse flows are given by replacing c_3 and c_8 by $-c_3$ and $-c_8$, respectively, in the flow functions (12). **c** The effect of a different location of the symmetry breaking point without flow effects on pPAR dynamics. **d** The effect of a different location of the symmetry breaking point when wild-type flows are included. **e** pPAR polarity dynamics with a different location of symmetry breaking point when MEX-5/6 does not affect pPAR ($\mu_0 = 0$ in the model (10)–(11)) and flows are absent. **f** Comparison of polarity dynamics with different symmetry breaking points without flow effects. The figures surrounded by a bold square are the steady-state patterns. The gray dotted line indicates the boundary location of polarity domains. The detailed parameter values are given in "Appendix B"

though the patterning phase is strongly affected by the flow directions (Fig. 5C). This result suggests that the MEX-5/6 is affected by the direction of flows, but its influence on spatial patterning can be greatly restricted by interaction with the PAR dynamics.

On the other hand, we found a very intriguing result by comparing Fig. 5B(b1) and D. In Fig. 5D, we considered only the flow effect for MEX-5/6 to be the same as Fig. 5B(b1) but with a different symmetry breaking position. That is, in these two numerical experiments, MEX-5/6 and pPAR are undergoing the same biochemical interactions under the same effect of cytoplasmic flows. However, the final polarity patterns are very different. Thus, we hypothesize that cell geometry may affect the state of biochemical interaction of cytoplasmic proteins and, consequently, results in different patterns. To confirm this, we investigated the effect of different symmetric breaking positions without flows (Fig. 5E, F). We found that this cell geometry effect is not seen in the PAR alone model, namely when we removed the effect of MEX-5/6 with $\mu_0 = 0$ (Fig. 5E), but the polarity pattern dynamics were dramatically changed, both temporally and spatially with respect to the cell geometry, when the effect of MEX-5/6 was included (Fig. 5F). This result proposes that the dynamics of cytoplasmic protein can be strongly affected by cell geometry and, consequently, the interplay between MEX-5/6 and cell geometry can lead to a dramatical change in PAR polarity dynamics. Summarizing the results, the effect of the interplay of MEX-5/6 with flows on the spatial dynamics of PAR polarity may be negligible, but the effect of cell geometry can be critical.

4 Discussion

For the last ten years, the polarity phenomenon in asymmetric cell division has been extensively studied by both experimental and theoretical approaches. In particular, PAR polarity in the membrane has been found to be the most upstream regulator which controls the dynamics of cytoplasmic polarity. However, how downstream polarity in the cytosol affects PAR polarity, and how they interact, is still unknown. In this study, we revisited the question of polarity formation of the cytoplasmic protein MEX-5/6 by combining the dynamics of the upstream protein PAR and explored how the cytoplasmic polarity of MEX-5/6 interacts with the membrane PAR polarity in the high-dimensional bulk-surface model.

The mechanism for generating MEX-5/6 polarity has been studied in MEX-5/6 alone models (Daniels et al. 2010; Wu et al. 2018), which assumed that the conversion rate of MEX-5/6 diffusion is spatially heterogeneous, and did not include PAR dynamics. By contrast, we have developed a conversion model in which the conversion rate functions of the two diffusion types of MEX-5/6 are determined by PAR proteins in a concentration-dependent manner, so that the model formulation directly includes the dynamics of PAR proteins. In our model, we supposed that the conversion rate function from fast type to slow type depends on the concentration of aPAR. However, this term was converted to a suppression term by pPAR concentration in the self-recruitment model (Sect. 2.4). The formulation of our self-recruitment model implicitly includes the dynamics of PAR-1, which acts repressively on the phosphorylated substance PP2A, converting the fast type of MEX-5/6 to the slow type (Griffin et al. 2011). In fact, for the conversion dynamics of MEX-5/6 diffusion type, Daniels et al. (2010) assumed a direct conversion from fast type to slow type by aPAR. In contrast, Griffin et al. (2011) proposed that pPAR represses the conversion of MEX-5/6 from fast diffusive type to the slow type. However, our model formulation and analysis showed that such apparently contrasting propositions are essentially the same. This suggests that the conversion assumptions of our model capture an essential mechanism of interplay between cytoplasmic protein MEX-5/6 and PAR proteins, which may suggest a general mathematical structure for pattern formation in cytoplasmic proteins.

With our MEX-5/6-combined-pPAR model, we explored the specific role of MEX-5/6 on PAR polarity formation. We found that MEX-5/6 can play a critical role in inducing the symmetry breaking. Motegi et al. (2011) showed that the recruitment of PAR-2 around the posterior pole by microtubule transport may promote symmetry breaking. Our study suggests that symmetry breaking can also be promoted by the repression of aPAR recruitment around the posterior pole by MEX-5/6, implying that the self-recruitment of pPAR is indirectly promoted. These results propose that the first stage of symmetry breaking in asymmetric division may be regulated by the synergistic effect of multiple positive feedbacks of pPAR recruitment from cytosol to membrane. We also found that the length of the pPAR domain tends to be shorter as the regulation effect (μ_0) of MEX-5/6 on PAR decreases. This is consistent with the dynamics observed in previous experiments with the *mex-5(RNAi)* and *mex-6(RNAi)* embryos of *C. elegans*, where the length of the PAR-2 domain was shorter than that of the wild type (Cuenca et al. 2002; Schubert et al. 2000). However, we also found that the length scale of the pPAR polarity domain is not sensitive to the regulation effect of MEX-5/6, indicating that the length of the PAR domain may be robustly regulated by other factors, such as the total mass of PAR proteins (Seirin-Lee and Shibata 2015; Goehring et al. 2011b). Our study suggests that the upstream polarity of the PARs, and the downstream polarity of MEX-5/6, significantly regulate each other with respect to both spatial and temporal dynamics in polarity formation. Even if cytoplasmic polarity serves as a downstream regulator, cytoplasmic polarity can play a critical role in upstream PAR polarity, and the balance of their bi-directional regulation is important for generating robust polarity.

In the history of the study of pattern formation, the effect of domain geometry has been considered as an important factor that can regulate spatial patterning (Crampin et al. 1999; Dawes and Iron 2013; Murray 1993; Seirin-Lee 2017), and there is biological evidence supporting the hypothesis that the shape, or size, of domain is likely to play a critical role in determining cell function, via regulation of pattern formation (Kondo and Asai 1995; Seirin-Lee et al. 2019). In this study, we found that cell geometry may play an important role in the dynamics of cytoplasmic protein in polarity formation. Many cell polarity studies using mathematical models have been focused on the dynamics of membrane polarity in simplified one-dimensional domains, neglecting cell geometry. In general, the fast diffusion in cytosol, and the homogeneity of cytosol concentration, have validated this model simplification. However, our study suggests that the effect of cell geometry on the cytoplasmic protein, which creates a spatial heterogeneity in the bulk cytosol space, can play a critical role in the dynamics of polarity patterning in both membrane and cytosol, and that cell geometry should not be neglected. Furthermore, the flow dynamics is likely to be affected by cell geometry (Mittasch et al. 2018), which may consequently affect PAR polarity formation.

In this study, we presented simulation results for representative parameter sets. However, a rigorous mathematical analysis of the high-dimensional bulk-surface model of PARs alone proves that the polarity pattern exists within a large parameter range (Morita and Seirin-Lee 2020). Furthermore, our bi-stability analysis shows that the parameter region can be extended as the effect of MEX-5/6 increases (Fig. S2B), implying that we could have a polarity pattern in the MEX-5/6-combined-PAR model which is robust to the values of the kinetic parameters. However, it is a mathematical challenge to analyze the bifurcation structure, existence of the polarity solution, and the details of polarity dynamics, in the high-dimensional bulk-surface MEX-5/6-combined-PARs system.

Finally, our study proposes that to understand the whole process of cell polarity in asymmetric cell division, it is vital to integrate biochemical interaction, biophysical dynamics, and cell geometry.

Acknowledgements This work was supported by JSPS KAKENHI Grant Numbers JP19H01805 and JP17KK0094, and by the JSPS A3 Foresight Program.

Appendix A Sharp Interface Limit and Zero-Flux Boundary Condition

We show that a sharp interface limit recovers the boundary conditions in the cytosol equations of the phase-field combined model (14). The general form of the cytoplasmic protein dynamics in our model is given by

$$\frac{\partial u}{\partial t} + \nabla \cdot (\mathbf{v}u) = D\nabla^2 u \text{ on } \mathbf{x} \in \Omega, \tag{15}$$

$$D\frac{\partial u}{\partial \mathbf{n}} - \mathbf{v}u = -F(u) \text{ on } \mathbf{x} \in \Gamma, \tag{16}$$

where u is the concentration of cytoplasmic protein, D is the diffusion coefficient, $\mathbf{v}$ is the flow velocity, and $\mathbf{n}$ is the normal vector on $\Gamma(\equiv \partial\Omega)$. $F(u)$ is a function describing a reaction on Γ and it is not necessarily only a function of u. The cytoplasmic model equation incorporating cell shape is given by

$$\phi\frac{\partial u}{\partial t} + \nabla \cdot (\phi \mathbf{v}u) = D\nabla \cdot (\phi \nabla u) + \nabla\phi F(u) \text{ on } \mathbf{x} \in \mathbb{R}^N, \tag{17}$$

where ϕ is the phase-field function at a fixed time, *i.e.*, $\phi = \phi(\mathbf{x})$, defined by

$$\phi(\mathbf{x}) = 1 \text{ on cytosol}, \quad 0 < \phi(\mathbf{x}) < 1 \text{ on cell membrane},$$
$$\phi(\mathbf{x}) = 0 \text{ on extracellular space}.$$

In what follows, we show that Eq. (17) recovers the boundary condition (16) in the sharp interface. Let us define T_ξ to be the interface region of ϕ with thickness ξ. That is,

$$T_\xi = \{\mathbf{x} \in \mathbb{R}^N | 0 < \phi(\mathbf{x}) < 1\} \quad \text{and} \quad \lim_{\xi \to 0} T_\xi = \Gamma.$$

Integrating Eq. (17) over the interface yields

$$\begin{aligned} \int_{T_\xi} \phi \frac{\partial u}{\partial t} \, \mathrm{d}V &= \int_{T_\xi} \nabla \cdot (D\phi \nabla u - \phi \mathbf{v} u) \, \mathrm{d}V + \int_{T_\xi} \nabla \phi F(u) \, \mathrm{d}V, \\ &= \int_{\partial T_\xi} (D\phi \nabla u - \phi \mathbf{v} u) \cdot \mathbf{n} \, \mathrm{d}S + \int_{T_\xi} \nabla \phi F(u) \, \mathrm{d}V, \\ &= \int_{\partial T_\xi} \phi (D \nabla u - \mathbf{v} u) \cdot \mathbf{n} \, \mathrm{d}S + \int_{T_\xi} \nabla \phi F(u) \, \mathrm{d}V. \end{aligned} \tag{18}$$

On the other hand, substituting the original Eq. (15) into the left-hand side of equation (18) gives

$$\begin{aligned} \int_{T_\xi} \phi \frac{\partial u}{\partial t} \, \mathrm{d}V &= \int_{T_\xi} \phi \nabla \cdot (D \nabla u - \mathbf{v} u) \, \mathrm{d}V \\ &= \int_{\partial T_\xi} \phi (D \nabla u - \mathbf{v} u) \cdot \mathbf{n} \, \mathrm{d}S - \int_{T_\xi} \nabla \phi \cdot (D \nabla u - \mathbf{v} u) \, \mathrm{d}V. \end{aligned} \tag{19}$$

Thus, from (18) and (19), we obtain

$$\int_{T_\xi} \nabla \phi \cdot [D \nabla u - \mathbf{v} u + F(u)] \, \mathrm{d}V = 0.$$

Since $\nabla \phi \neq 0$ as $\xi \to 0$,

$$-D \frac{\partial u}{\partial \mathbf{n}} + \mathbf{v} u = F(u) \quad \text{as } \xi \to 0.$$

should hold.

Appendix B Initial Conditions and Parameter Values

Before fertilization, aPAR is homogeneously distributed in the membrane, and pPAR and MEX-5/6 are homogeneously distributed in the cytosol. The core mechanism for inducing symmetry breaking remains unclear, as does the reason why the pPAR domain begins around the posterior polar site. On the other hand, it is known that some signals from the centrosome and its microtubule asters help recruit pPAR into the membrane around the posterior polar site (Rose and Gönczy 2014; Motegi et al. 2011; Motegi and Seydoux 2013). Thus, we define the initial conditions such that all proteins are at homogeneous steady state, and some small stimulus of pPAR is added in a small region of the membrane.

The detailed form of the initial conditions are :

$$
\begin{aligned}
&[A_m](\mathbf{X},0)c[A_m]_0(1+\delta\psi(\mathbf{X})) \text{ on } \mathbf{X}\in\partial\Omega,\\
&[A_c](\mathbf{X},0)=[A_c]_0(1+\delta\psi(\mathbf{X})) \text{ on } \mathbf{X}\in\Omega,\\
&[P_m](\mathbf{X},0)=[P_m]_0+\sigma \text{ on } \mathbf{X}\in\omega\subset\partial\Omega,\\
&[P_m](\mathbf{X},0)=[P_m]_0(1+\delta\psi(\mathbf{X})) \text{ on } \mathbf{X}\in\partial\Omega\setminus\omega,\\
&[P_c](\mathbf{X},0)=[P_c]_0(1+\delta\psi(\mathbf{X})) \text{ on } \mathbf{X}\in\Omega,\\
&[M_f](\mathbf{X},0)=[M_f]_0(1+\delta\psi(\mathbf{X})) \text{ on } \mathbf{X}\in\Omega,\\
&[M_s](\mathbf{X},0)=[M_s]_0(1+\delta\psi(\mathbf{X})) \text{ on } \mathbf{X}\in\Omega,
\end{aligned}
$$

where σ is the strength of the signal, and ω is the sufficiently small region in which the signal is imposed. δ is a positive constant and has a very small value. Typically, we set $\delta = 0.02$. $\psi(\mathbf{X})$ is a random function with uniform distribution, which takes values in the range $[-0.5, 0.5]$. $[A_m]_0$, $[A_c]_0$, $[P_m]_0$, $[P_c]_0$, $[M_f]_0$ and $[M_s]_0$ are equilibrium concentrations. We set $[A_m]_0$, $[P_m]_0$, $[M_s]_0$ and $[M_f]_0$, directly, where $[A_c]_0$ and $[P_c]_0$ are calculated as the equilibrium values in model (1)–(3), or (10)–(11) when $\mu_0 = 0$.

We selected the parameter values based on experimental data for the *C. elegans* embryo. A fertilized *C. elegans* egg is usually elliptical, and the radii of the long and short axes are observed typically in the range of 27.0 ± 1.7 μm and 14.8 ± 1.0 μm, respectively (Goehring et al. 2011a). The diffusion rates of aPAR and pPAR in the membrane were selected based on the diffusion rate of PAR-6 and PAR-2, respectively (Goehring et al. 2011a). The cytoplasmic diffusion rates of aPAR and pPAR were chosen from the data in Kuhn et al. (2011). For the diffusion rates of MEX-5/6, we used data from Daniels et al. (2010) and Wu et al. (2018), and for the flow velocity, we used data from Goehring et al. (2011a) and Niwayama et al. (2011). We selected the time scale to coincide with the quantitative data and the qualitative dynamics of PAR polarity. The unit time in the non-dimensionalized system corresponds to the dimensional time, 0.3375 s. In the simulations, we used the non-dimensionalized model, combined with the phase-field functions, and solved the equations on the non-dimensional region 1×1, where the dimensional space $L_x \times L_y$ has been scaled by $L_x = L_y = 75$ μm (Fig. 1a). Since the phase-field function is time-independent and is used to define the cell region, we assume the parameters used in the phase-field function as dimensionless quantities. The parameter values are listed in Table 1, and the detailed parameter values used in the flow functions, phase-field function, and the figures are given in the following. For the other figures, the parameter values in Table 1 were used.

Table 1 Representative parameter set. $\mathcal{M}$ has the scale of protein number used to non-dimensionalize the model

Parameter	Dimensional value	Dimensionless value
$^{\dagger}\ell_x$	30.0 (μm)	0.4
$^{\dagger}\ell_y$	16.875 (μm)	0.225
t	0.3375 (s)	1.0
$^{\dagger}D_m^P$	0.120 (μm^2/s)	7.2×10^{-6}
$^{\ddagger}D_c^P$	6.0 (μm^2/s)	3.6×10^{-4}
$^{*}D_f$	8.5 (μm^2/s)	5.134×10^{-4}
$^{*}D_s$	0.59 (μm^2/s)	3.54×10^{-5}
γ_2	5.9 (s^{-1})	2.0
α_2	1.8 (s^{-1})	0.6
β_1	0.59 (s^{-1})	0.2
β_2	0.15 (s^{-1})	0.05
β_3	2.07 (s$^{-1}\mathcal{M}^{-2}$)	0.7
β_4	100.93 (s^{-2})	12.45
μ_0	0.1 ($\mathcal{M}^{-1}$)	0.1
μ_1	0.18 (s^{-1})	0.05
μ_2	0.59 (s^{-1})	0.2
μ_3	0.015 (s^{-1})	0.005
μ_4	0.015 (s^{-1})	0.005
δ_1	4.5 ($\mathcal{M}^{-2}$)	4.5
δ_2	5.0 ($\mathcal{M}$)	5.0
δ_3	0.5 ($\mathcal{M}$)	0.5

†Goehring et al. (2011a),‡Kuhn et al. (2011), * Wu et al. (2018)

Flow functions, (12): $c_1 = 0.001,\ c_2 = 400.0,\ c_3 = 0.02,\ c_4 = 4.2,\ c_5 = 0.01,\ c_6 = 0.00001,\ c_7 = 3.5,\ c_8 = 0.0021,\ c_9 = 0.00005,\ T_0 = 3.3375$ min
Phase-field function, (13): $\alpha = 120,\ \overline{V} = 0.2828,\ \mu = 1.0,\ \varepsilon = 2.0\times10^{-3},\ \nu = 16.0$
Figure 1a: $D_m^A = 0.000001652,\ D_c^A = 0.00036,\ D_m^P = 0.00000072,\ D_c^P = 0.00036,\ \gamma_1 = \gamma_2 = 0.3,\ \alpha_1 = \alpha_2 = 0.06,\ K_1 = K_3 = 0.4,\ K_2 = K_4 = 1.0,\ \overline{K}_2 = \overline{K}_4 = 0.05,\ \mu_0 = 0.01,\ \mu_1 = 0.05,\ \mu_2 = 0.01,\ \mu_3 = \mu_4 = 0.005$.
Figure 4a: $\beta_3 = 0.5$, $\mu_0 = 0$ or 0.2, and the other parameters are given in Table 1.
Figure 4b: $\mu_0 = 0.2$ and the other parameters are given in Table 1.

Appendix C Incompressibility

The detailed figure of incompressibility is shown in Fig. S1.

Appendix D Supporting Role of MEX-5/6 on pPAR Patterning

Here, we show that MEX-5/6 can play a supporting role in inducing pPAR polarity patterning by analyzing a parameter space where bi-stability is possible. Let us assume that the concentration of MEX-5/6 is given by a constant, namely, M^*. Then, from Eq. (10), the equilibrium state of pPAR can be written as

$$\gamma_2 P_c^* - \left\{\alpha_2 + \frac{\beta_4}{\beta_1 + (1+\mu_0 M^*)^2(\beta_2 + \beta_3 (P_m^*)^2)}\right\} P_m^* = 0, \tag{20}$$

where P_c^* and P_m^* are the equilibrium concentrations of pPAR in the cytosol and the membrane, respectively. Note that

$$P_c^* = \frac{1}{|\Omega|}\int_\Omega P_c^* = \frac{1}{|\Omega|}\left(P_{\text{tot}} - \int_{\partial\Omega} P_m^*\right) = P_{\text{tot}}^* - AP_m^*, \tag{21}$$

where $P_{\text{tot}} = \int_\Omega P_c + \int_{\partial\Omega} P_m$, $P_{\text{tot}}^* = P_{\text{tot}}/|\Omega|$, and $A = |\partial\Omega|/|\Omega|$. Substituting Eq. (21) into Eq. (20), we obtain

$$\begin{aligned} G(P_m^*) &= -(\alpha_2 + A\gamma_2)(P_m^*)^3 + P_{\text{tot}}^*\gamma_2(P_m^*)^2 \\ &\quad - \frac{(\alpha_2 + A\gamma_2)(\beta_1 + \beta_2(1+\mu_0 M^*)^2 + \beta_4)}{\beta_3(1+\mu_0 M^*)^2} P_m^* \\ &\quad + \gamma_2 P_{\text{tot}}^* \frac{\beta_1 + \beta_2(1+\mu_0 M^*)^2}{\beta_3(1+\mu_0 M^*)^2} = 0. \end{aligned}$$

For two positive stable states, the equation $G'(p_m^*) = 0$ should have two positive solutions (Fig. S2A). Denoting the two solutions by P_1 and P_2, the following property is satisfied :

$$P_1 + P_2 > 0, \qquad P_1 P_2 > 0, \tag{22}$$

and

$$\beta_4 < \left[\frac{\beta_3(P_{\text{tot}}^*\gamma_2)^2}{3(\alpha_2 + A\gamma_2)} - (\alpha_2 + A\gamma_2)\beta_2\right](1+\mu_0 M)^2 - (\alpha_2 + A\gamma_2)\beta_1.$$

We can easily check that the conditions (22) are satisfied for all positive parameter values. Thus, we compare the parameter region of γ_2 (on-rate) and β_4 (magnitude of off-rate) for bi-stability with respect to the cases $\mu_0 = 0$ and $\mu_0 = 0.2 > 0$. The result shows that the case of $\mu_0 > 0$ gives a wider parameter space for bi-stability than the case of $\mu_0 = 0$.

References

Campanale JP, Sun TY, Montell DJ (2017) Development and dynamics of cell polarity at a glance. J Cell Sci 130:1201–1207

Cortes DB, Dawes A, Liu J, Nickaeen M, Strychalski W, Maddox AS (2018) Unite to divide-how models and biological experimentation have come together to reveal mechanisms of cytokinesis. J Cell Sci 131:1–10

Cowan CR, Hyman AA (2004) Asymmetric cell division in *C. elegans*: cortical polarity and spindle positioning. Annu Rev Cell Dev Biol 20:427–453

Crampin EJ, Gaffney EA, Maini PK (1999) Reaction and diffusion on growing domains: scenarios for robust pattern formation. Bull Math Biol 61:1093–1120

Cuenca AA, Schetter A, Aceto D, Kemphues K, Seydoux G (2002) Polarization of the *C. elegans* zygote proceeds via distinct establishment and maintenance phases. Development 130:1255–1265

Daniels BR, Dobrowsky TM, Perkins EM, Sun SX, Wirtz D (2010) Mex-5 enrichment in the *C. elegans* early embryo mediated by differential diffusion. Development 137:2579–2585

Dawes AT, Iron D (2013) Cortical geometry may influence placement of interface between par protein domains in early *Caenorhabditis elegans* embryos. J Theor Biol 333:27–37

Dawes AT, Munro EM (2011) PAR-3 oligomerization may provide an actin-independent mechanism to maintain distinct Par protein domains in the early *Caenorhabditis elegans* embryo. Biophys J 101:1412–1422

Goehring NW, Hoege C, Grill SW, Hyman AA (2011a) PAR proteins diffuse freely across the anterior-posterior boundary in polarized *C. elegans* embryos. J Cell Biol 193(3):583–594

Goehring NW, Trong PK, Bois JS, Chowdhury D, Nicola EM, Hyman AA, Grill SW (2011b) Polarization of PAR proteins by advective triggering of a pattern-forming system. Science 334(6059):1137–1141

Gönczy P (2005) Asymmetric cell division and axis formation in the embryo. WormBook.org. https://doi.org/10.1895/wormbook.1.30.1

Griffin EE, Odde DJ, Seydoux G (2011) Regulation of the MEX-5 gradient by a spatially segregated kinase/phosphatase cycle. Cell 146:955–968

Hoege C, Hyman AA (2013) Principles of PAR polarity in *Caenorhabditis elegans* embryos. Mol Cell Biol 14:315–322

Knoblich JA (2008) Mechanisms of asymmetric stem cell division. Cell 132:583–597

Kondo S, Asai R (1995) A reaction-diffusion wave on the skin of the marine angelfish Pomacanthus. Nature 376:765–768

Kuhn T, Ihalainen TO, Hyvaluoma J, Dross N, Willman SF, Langowski J, Vihinen-Ranta M, Timonen J (2011) Protein diffusion in mammalian cell cytoplasm. PLoS ONE 6(8):e22962

Kuwamura M, Seirin-Lee S, Ei S-I (2018) Dynamics of localized unimodal patterns in reaction-diffusion systems related to cell polarization by extracellular signaling. SIAM J Appl Math 78(6):3238–3257

Lang CF, Munro E (2017) The PAR proteins: from molecular circuits to dynamic self-stabilizing cell polarity. Development 144:3405–3416

Mittasch M, Gross P, Nestler M, Fritsch AW, Iserman C, Kar M, Munder M, Voigt A, Alberti S, Grill SW, Kreysing M (2018) Non-invasive perturbations of intracellular flow reveal physical principles of cell organization. Nat Cell Biol 20:344–351

Morita Y, Seirin-Lee S (2020) Long time behaviour and stable pattern in the systems of cell polarity model. Preprint

Motegi F, Seydoux G (2013) The PAR network: redundancy and robustness in a symmetry-breaking system. Philos Trans R Soc B 368:20130010

Motegi F, Zonies S, Hao Y, Cuenca AA, Griffin E, Seydoux G (2011) Microtubules induce self-organization of polarized PAR domains in *Caenorhabditis elegans* zygotes. Nat Cell Biol 13(11):1361–1367

Murray JD (1993) Mathematical biology II: spatial models and biomedical applications, 3rd edn. Springer, Berlin

Nishikawa M, Naganathan SR, Jülicher F, Grill SW (2017) Symmetry breaking in a bulk surface reaction diffusion model for signalling networks. eLife 6:e19595

Niwayama R, Shinohara K, Kimura A (2011) Hydrodynamic property of the cytoplasm is sufficient to mediate cytoplasmic streaming in the *Caenorhabiditis elegans* embryo. PNAS 108(29):11900–11905

Rappel W-J, Levine H (2017) Mechanisms of cell polarization. Curr Opin Syst Biol 3:43–53

Rose LS, Gönczy P (2014) Polarity establishment, asymmetric division and segregation of fate determinants in early *C. elegans* embryo. WormBook

Schubert CM, Lin R, de Vries CJ, Plasterk RHA, Priess JR (2000) MEX-5 and MEX-6 unction to establish soma/germline asymmetry in early *C. elegans* embryo. Mol Cell 5:671–682
Seirin-Lee S (2016) Lateral inhibition-induced pattern formation controlled by the size and geometry of the cell. J Theor Biol 404:51–65
Seirin-Lee S (2017) The role of domain in pattern formation. Dev Growth Differ 59:396–404
Seirin-Lee S (2020) From a cell to cells in asymmetric cell division and polarity formation?: Shape, length, and location of par polarity. Dev Growth Differ 62:188–195
Seirin-Lee S, Shibata T (2015) Self-organization and advective transport in the cell polarity formation for asymmetric cell division. J Theor Biol 382:1–14
Seirin-Lee S, Tashiro S, Awazu A, Kobayashi R (2017) A new application of the phase-field method for understanding the reorganization mechanisms of nuclear architecture. J Math Biol 74:333–354
Seirin-Lee S, Osakada F, Takeda J, Tashiro S, Kobayashi R, Yamamoto T, Ochiai H (2019) Role of dynamic nuclear deformation on genomic architecture reorganization. PLOS Comput Biol 15(8):e1007289
Seirin-Lee S, Gaffney EA, Dawes AT (2020a) CDC-42 interactions with Par proteins are critical for proper patterning in polarization. Cells 9:2036
Seirin-Lee S, Sukekawa T, Nakahara T, Ishii H, Ei S-I (2020b) Transitions to slow or fast diffusions provide a general property for in-phase or anti-phase polarity in a cell. J Math Biol 80:1885–1917
Small LE, Dawes AT (2017) PAR proteins regulate maintenance-phase myosin dynamics during *Caenorhabditis elegans* zygote polarization. Mol Biol Cell 28:2220–2231
Teigen KE, Li X, Lowengrub J, Wang F, Voigt A (2009) A diffuse-interface approach for modeling transport, diffusion and adsorption/desorption of material quantities on a deformable interface. Commun Math Sci 4(7):1009–1037
Tostevin F, Howard M (2008) Modeling the establishment of PAR protein polarity in the one-cell *C. elegans* embryo. Biophys J 95:4512–4522
Trong PK, Nicola EM, Goehring NW, Kumar KV, Grill SW (2014) Parameter-space topology of models for cell polarity. New J Phys 16:065009
Wang W, Tao K, Wang J, Yang G, Ouyang Q, Wang Y, Zhang L, Liu F (2017) Exploring the inhibitory effect of membrane tension on cell polarization. PLOS Comput Biol 13(1):e1005354
Wu Y, Han B, Li Y, Munro E, Odde DJ, Griffin EE (2018) Rapid diffusion-state switching underlies stable cytoplasmic gradients in *Caenorhabditis elegans* zygote. PNAS 115(36):E8440–E8449
Zonies S, Motegi F, Hao Y, Seydoux G (2010) Symmetry breaking and polarization of the *C. elegans* zygote by the polarity protein PAR-2. Development 137:1669–1677

Bulletin of Mathematical Biology (2020) 82:143
https://doi.org/10.1007/s11538-020-00814-y

Society for Mathematical Biology

SPECIAL ISSUE: CELEBRATING J. D. MURRAY

A Mechanistic Investigation into Ischemia-Driven Distal Recurrence of Glioblastoma

Lee Curtin[1] · **Andrea Hawkins-Daarud**[1] · **Alyx B. Porter**[2] · **Kristoffer G. van der Zee**[3] · **Markus R. Owen**[3] · **Kristin R. Swanson**[1]

Received: 1 April 2020 / Accepted: 25 September 2020 / Published online: 7 November 2020

Abstract
Glioblastoma (GBM) is the most aggressive primary brain tumor with a short median survival. Tumor recurrence is a clinical expectation of this disease and usually occurs along the resection cavity wall. However, previous clinical observations have suggested that in cases of ischemia following surgery, tumors are more likely to recur distally. Through the use of a previously established mechanistic model of GBM, the Proliferation Invasion Hypoxia Necrosis Angiogenesis (PIHNA) model, we explore the phenotypic drivers of this observed behavior. We have extended the PIHNA model to include a new nutrient-based vascular efficiency term that encodes the ability of local vasculature to provide nutrients to the simulated tumor. The extended model suggests sensitivity to a hypoxic microenvironment and the inherent migration and proliferation rates of the tumor cells are key factors that drive distal recurrence.

Keywords Glioblastoma · Hypoxia · PIHNA · Ischemia · Tumor growth

1 Introduction

Glioblastoma (GBM) is the most aggressive primary brain tumor (Louis et al. 2016). It is uniformly fatal with a median survival from diagnosis of only 15 months with standard of care treatment, consisting of a combination of resection, chemotherapy and radiotherapy (Stupp et al. 2009). Due to the sensitive location of the tumor, there is a reliance on clinical imaging to assess tumor treatment response and progression. Enhancement on T1-weighted magnetic resonance imaging (T1Gd MRI) with

✉ Lee Curtin
curtin.lee@mayo.edu

1 Mathematical NeuroOncology Lab, Precision Neurotherapeutics Innovation Program, Mayo Clinic, Phoenix, AZ 85054, USA

2 Division of Neuro-Oncology, Mayo Clinic, Phoenix, AZ 85054, USA

3 School of Mathematical Sciences, University of Nottingham, Nottingham, UK

gadolinium contrast shows regions where gadolinium has leaked through disrupted vasculature. T2-weighted MRI (T2 MRI) shows infiltrative edema, fluid that has leaked from vasculature. Abnormalities on T1Gd MRI spatially correlate with the bulk of the tumor mass with central dark regions typically showing necrosis, whereas edema visible on T2 MRI corresponds to regions of lower tumor cell density.

An unfortunate clinical expectation following surgical resection is tumor recurrence, which usually presents on the edge of the resection cavity (Chamberlain 2011); this is known as a local recurrence. The recurrent tumor will occasionally appear as enhancement on T1Gd MRI in a different region of the brain, away from the primary site, which is known as a distant recurrence (Chamberlain 2011). In some cases, the T2 MRI abnormality will become much larger relative to the enhancement on T1Gd MRI, and these cases are known as diffuse recurrences (Chamberlain 2011). In a retrospective study by Thiepold et al., it was shown that patients with GBM who had also suffered from perioperative ischemia, defined as an inadequate blood supply to a part of the brain following resection, were more likely to have a distantly and/or diffusely recurring GBM (Thiepold et al. 2015). A disruption in normal vasculature can occur following GBM resection and can lead to ischemia, affecting abnormal tissue in the same way it affects the healthy tissue. By reducing available nutrients to the tumor, the tumor is forced toward a hypoxic phenotype and becomes necrotic if the reduction is sustained. Thiepold attributed the observed difference in recurrence patterns to the hypoxic conditions caused by the reduction in vasculature (Thiepold et al. 2015). In retrospective analyses of patient data, Bette *et al.* found further supporting evidence that perioperative ischemia promoted aggressive GBM recurrence patterns (Bette et al. 2016, 2018). Bette *et al.* showed that perioperative infarct volume was positively associated with more multifocal disease and contact to the ventricle, which have both been shown to negatively impact patient survival in a pretreatment setting (Adeberg et al. 2014; Stark et al. 2012).

Spatiotemporal mathematical models have been used extensively to describe the growth of GBM. These models incorporate features of tumor cells such as cell phenotype, migration, proliferation and interactions with other cells to understand how these influence observed growth behavior in GBM. Such models have the ability to provide mechanistic insight into observed tumor growth patterns and treatment effects. Mathematical models have been created to simulate GBM growth on varying spatial scales from clusters of cells (Frieboes et al. 2007; Macklin and Lowengrub 2007; Stein et al. 2007) and murine models (Gallaher et al. 2020; Leder et al. 2014; Rutter et al. 2017) to tissue-level scales seen throughout the presentation of the disease in patients (Hawkins-Daarud et al. 2013; Subramanian et al. 2019; Swan et al. 2018; Swanson et al. 2000, 2003, 2008). Various treatments for GBM have also been modeled, such as resection (Neufeld et al. 2017; Swanson et al. 2008), chemotherapy (Ansarizadeh et al. 2017; Barazzuol et al. 2010; Boujelben et al. 2016) and radiotherapy (Burnet et al. 2006; Leder et al. 2014; Rockne et al. 2015), which are all elements of the current standard of care. Other less widely used and experimental treatments have also been modeled such as anti-angiogenic drugs (Hawkins-Daarud et al. 2013; Scribner et al. 2014) and oncolytic virus therapy (de Rioja et al. 2016).

An example of a tissue-level growth model of GBM is the Proliferation Invasion Hypoxia Necrosis Angiogenesis (PIHNA) model, which has been used to study different mechanisms of tumor development and shows similar growth and progression patterns to those seen in patient tumors (Swanson et al. 2011). Simulated hypoxic events have shown an increase in glioma growth rates in spatiotemporal models of GBM (Martínez-González et al. 2012; Pardo et al. 2016). We have recently found the parameters of the PIHNA model that drive faster outward growth of these simulated tumors and found that those relating to hypoxia were in some cases extremely influential (Curtin et al. 2020).

To the best of our knowledge, the impact of perioperative ischemia on recurrence in GBM has not been mathematically modeled before. We aim to use the PIHNA model to gain insight into the tumor kinetics that may play a role in this behavior.

In this work, we extend a term in the PIHNA model known as the vascular efficiency term, which determines the ability of local vasculature to provide nutrients to the tumor. We carry this out through the inclusion of a nutrient-transport equation parametrized through glucose uptake rates in GBM. We apply this extended PIHNA model to a set of simulated perioperative ischemia cases to determine influential mechanisms in the model that could drive ischemia-induced distal recurrence patterns in GBM. Specifically, we vary migration and proliferation rates of normoxic and hypoxic cell phenotypes, as well as switching rates between these phenotypes. We find that simulated tumors with faster migration and slower proliferation rates are more likely to recur distantly in cases of perioperative ischemia. We see that this can be promoted by changes in switching rates between normoxic and hypoxic cell phenotypes. We have also simulated these same cases with a less intense ischemic event and show that this in turn leads to less distantly recurring tumors. Following an initial exploration of simulated resection and ischemic injury, we present a second case example of simulated perioperative ischemia in which we observe similar recurrence patterns that depend on migration and proliferation rates of the normoxic cells. We also present two example simulations, one with and one without ischemia, to show that ischemia can offset the growth of dense tumor in simulations, which may contribute to diffuse recurrence following perioperative ischemia.

2 Methods

2.1 The PIHNA Model

To simulate glioblastoma growth and spread, we have adapted a previously established tumor growth model—the PIHNA model (Curtin et al. 2020; Swanson et al. 2011). The PIHNA model describes the growth of GBM with the interactions of vasculature in the process of angiogenesis. This model simulates five different species that all depend on space and time and their interactions:

$c(\mathbf{x}, t)$ – the density of normoxic tumor cells,
$h(\mathbf{x}, t)$ – the density of hypoxic tumor cells,
$n(\mathbf{x}, t)$ – the density of necrotic cells,
$v(\mathbf{x}, t)$ – the density of vascular endothelial cells,
$a(\mathbf{x}, t)$ – the concentration of angiogenic factors.

Normoxic cells proliferate with rate ρ and migrate with rate D_c, whereas hypoxic cells do not proliferate and migrate with rate D_h. Cells convert from normoxic to hypoxic phenotypes (with rate β) and from hypoxic to normoxic phenotypes (with rate γ) depending on the ability of the local vascular density to provide nutrients at their location; hypoxic cells in the model become necrotic if they remain in a vasculature-poor region with rate α_h. When any other cell type meets a necrotic cell, they become necrotic with rate α_n, as necrotic cells have been shown to encourage cell death through the creation of an unfavorable microenvironment (Raza et al. 2002; Yang et al. 2013). Angiogenic factors migrate with rate D_a, are created by the presence of normoxic and hypoxic tumor cells (with rates δ_c and δ_h, respectively), decay naturally (λ) and are consumed through the creation and presence of vascular cells. For a more in depth justification of these model parameters, see Curtin et al. (2020); Swanson et al. (2011). The parameter values used in this work, as well as their units, can all be found in Table 1.

Following the literature (Swanson 1999), we have assumed that a high total relative cell density of at least 80% is visible on a T1Gd MRI through the aforementioned imaging abnormalities of enhancement and necrosis present on the image. We have also assumed a total relative density of at least 16% is visible on a T2 MRI, due to the spatial correlation between lower cell densities and edema. In the PIHNA model, this translates to the total cell density $T \geq 0.8$ being visible on a T1Gd MRI and $T \geq 0.16$ being visible on a T2 MRI. By construction, the T1Gd lesion is always less than or equal in size to the T2 lesion, which agrees with patient data (Harpold et al. 2007).

We present a schematic for the PIHNA model in Fig. 1 that indicates the migration and proliferation of individual species as well as the interactions between all model species. The PIHNA model itself is presented in Equations (1)-(5), which we have annotated to give a full description of each of the terms in the model.

$$\overbrace{\frac{\partial c}{\partial t}}^{\substack{\text{Change}\\ \text{of normoxic}\\ \text{cell density}}} = \overbrace{\nabla\cdot(D_c(1-T)\nabla c)}^{\substack{\text{Net diffusion of}\\ \text{normoxic}\\ \text{glioma cells}}} + \overbrace{\rho c(1-T)}^{\substack{\text{Net}\\ \text{proliferation}\\ \text{of normoxic}\\ \text{glioma cells}}} + \overbrace{\gamma h V}^{\substack{\text{Conversion}\\ \text{of hypoxic}\\ \text{to normoxic}}} - \overbrace{\beta c(1-V)}^{\substack{\text{Conversion}\\ \text{of normoxic}\\ \text{to hypoxic}}} - \overbrace{\alpha_n\frac{nc}{K}}^{\substack{\text{Conversion}\\ \text{of normoxic}\\ \text{to necrotic}}} \tag{1}$$

$$\overbrace{\frac{\partial h}{\partial t}}^{\substack{\text{Change}\\ \text{of hypoxic}\\ \text{cell density}}} = \overbrace{\nabla\cdot(D_h(1-T)\nabla h)}^{\substack{\text{Net diffusion of}\\ \text{hypoxic}\\ \text{glioma cells}}} - \overbrace{\gamma h V}^{\substack{\text{Conversion}\\ \text{of hypoxic}\\ \text{to normoxic}}} + \overbrace{\beta c(1-V)}^{\substack{\text{Conversion}\\ \text{of normoxic}\\ \text{to hypoxic}}} - \overbrace{\left(\alpha_h h(1-V) + \alpha_n\frac{nh}{K}\right)}^{\substack{\text{Conversion}\\ \text{of hypoxic}\\ \text{to necrotic}}} \tag{2}$$

$$\overbrace{\frac{\partial n}{\partial t}}^{\substack{\text{Change}\\ \text{of necrotic}\\ \text{cell density}}} = \overbrace{\alpha_h h(1-V)}^{\substack{\text{Conversion}\\ \text{of hypoxic}\\ \text{to necrotic}}} + \overbrace{\alpha_n\frac{n(c+h+v)}{K}}^{\substack{\text{Contact necrosis}\\ \text{of all}\\ \text{living cells}}} \tag{3}$$

Table 1 Parameter definitions and values for the PIHNA model

	Definition	Value/Range	Units	Source
D_c	Diffusion rate of normoxic cells	1–1000	$\frac{\text{mm}^2}{\text{year}}$	Harpold et al. (2007)
D_h	Diffusion rate of hypoxic cells	$(0.1\text{–}100)D_c$	$\frac{\text{mm}^2}{\text{year}}$	Harpold et al. (2007); Martínez-González et al. (2012)*
ρ	Proliferation rate of normoxic cells	10–100	1/year	Harpold et al. (2007)
β	Switching rate from normoxia to hypoxia	$0.1\rho, 0.5\rho$	1/year	Swanson et al. (2011)*
γ	Switching rate from hypoxia to normoxia	0.005, 0.05, 0.5	1/day	Swanson et al. (2011)
α_h	Switching rate from hypoxia to necrosis	0.1β	1/year	Swanson et al. (2011)*
α_n	Rate of contact necrosis	$\log(2)/50$	1/day	Roniotis et al. (2012)
D_v	Diffusion rate of endothelial cells	0.18	$\frac{\text{mm}^2}{\text{year}}$	Swanson et al. (2011)
D_a	Diffusion rate of angiogenic factors	3.15	$\frac{\text{mm}^2}{\text{year}}$	Swanson et al. (2011)
δ_c	Normoxic cell production rate of angiogenic factors	2.77×10^{-13}	$\frac{\mu\text{mol}}{\text{cell}\times\text{year}}$	Swanson et al. (2011)
δ_h	Hypoxic cell production rate of angiogenic factors	5.22×10^{-10}	$\frac{\mu\text{mol}}{\text{cell}\times\text{year}}$	Swanson et al. (2011)
μ	Angiogenesis vasculature production rate	$\log(2)/15$	1/day	Swanson et al. (2011)
q	Consumption of angiogenic factors per cell	1.66	μmol/cell	Swanson et al. (2011)
λ	Natural decay rate of angiogenic factors	15.6	1/day	Swanson et al. (2011)
ω	Rate of removal of angiogenic factors by vasculature	λ/v_0	$\frac{1}{\text{cell}\times\text{day}}$	Swanson et al. (2011)
K	Maximal cell density	2.39×10^5	cells/mm^3	Swanson et al. (2011)
P_c^w	Glucose consumption ratio for normoxic (c)/hypoxic (h) cells in white matter	1.66–4.5	–	Delbeke et al. (1995)*
P_h^w	Glucose consumption ratio for normoxic (c)/hypoxic (h) cells in white matter	1.66–4.5	–	Delbeke et al. (1995)*
P_c^g	Glucose consumption ratio for normoxic (c)/hypoxic (h) cells in gray matter	0.5–2	–	Delbeke et al. (1995)*
P_h^g	Glucose consumption ratio for normoxic (c)/hypoxic (h) cells in gray matter	0.5–2	–	Delbeke et al. (1995)*

*Parameters that we have added/altered from the original publication of the PIHNA model (Swanson et al. 2011)

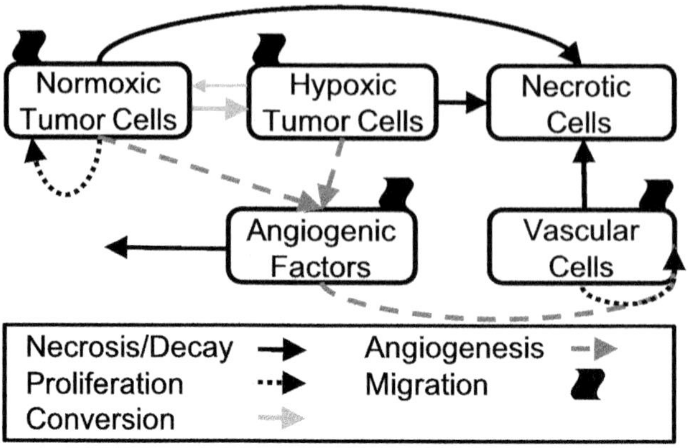

Fig. 1 A schematic for the PIHNA model. Normoxic tumor cells (c) proliferate, migrate, convert toward hypoxia and can become necrotic. Hypoxic tumor cells (h) migrate and can convert back to normoxic cells or to necrotic cells. Necrotic cells (n) accumulate as other cell types die. Angiogenic factors (a) are created in the presence of normoxic and hypoxic cells, migrate, decay and promote the local creation of vasculature. Vascular cells (v) proliferate through the facilitation of angiogenic factors and migrate

$$\overbrace{\frac{\partial v}{\partial t}}^{\substack{\text{Change}\\ \text{of vascular}\\ \text{cell density}}} = \overbrace{\nabla\cdot(D_v(1-T)\nabla v)}^{\substack{\text{Net diffusion of}\\ \text{vasculature}}} + \overbrace{\mu\frac{a}{K_m+a}v(1-T)}^{\substack{\text{Net proliferation}\\ \text{of vasculature}}} - \overbrace{\alpha_n\frac{nv}{K}}^{\substack{\text{Conversion}\\ \text{of vasculature}\\ \text{to necrotic}}} \tag{4}$$

$$\overbrace{\frac{\partial a}{\partial t}}^{\substack{\text{Change}\\ \text{of}\\ \text{angiogenic}\\ \text{factor}\\ \text{conc.}}} = \overbrace{\nabla\cdot(D_a\nabla a)}^{\substack{\text{Net diffusion of}\\ \text{angiogenic factor}}} + \overbrace{\delta_c c+\delta_h h}^{\substack{\text{Net production}\\ \text{of angiogenic}\\ \text{factor}}} \overbrace{-q\mu\frac{a}{K_m+a}v(1-T)-\omega av}^{\substack{\text{Net consumption of}\\ \text{angiogenic factor}}} - \overbrace{\lambda a}^{\text{Decay}} \tag{5}$$

where

$$V(c,h,v) = \frac{v}{v+\frac{\eta_c(D_c,\rho)c+\eta_h(D_h)h}{ps}}. \tag{6}$$

and

$$T = (c+h+n+v)/K. \tag{7}$$

The term V is called the vascular efficiency, and it models the relationship between the vasculature and its effect on the tumor. We let V take values in $[0, 1]$ such that it affects the switching rates between the normoxic (c), hypoxic (h) and necrotic (n) cell populations. When vasculature is abundant relative to other cells, V is close to 1 representing ample nutrient supply. Whereas when vasculature is relatively low, V is close to 0, which represents an unfavorable microenvironment of limited nutrient supply; this promotes conversion toward hypoxic and necrotic cells. In this work, we have extended the vascular efficiency term from previous iterations of the PIHNA model and present the derivation of this term in the next section.

The model equations are run on a two-dimensional slice of a realistic brain geometry from the Brainweb Database (Cocosco et al. 1997; Collins et al. 1998; Kwan et al. 1996, 1999), which spatially differentiates physiological structures such as white matter, gray matter, cerebrospinal fluid (CSF) and anatomical boundaries of the brain. This geometry is an average of multiple MR scans on a single patient to create a brain geometry with 1mm accuracy on and between MR slices. This gives a voxel volume of 1mm^3, which we use to track tumor volume on the two-dimensional brain slice. The simulations are implemented on white and gray matter, with differences in initial vasculature density and nutrient consumption ratios between these tissues, not allowing for growth of the tumor into the CSF or past the boundaries of the brain.

We initiate the simulation with a small normoxic cell population that decreases spatially from a point with coordinates (x_0, y_0)

$$c(\mathbf{x}, 0) = 1000e^{-100R^2}, \tag{8}$$

where $R^2 = (x - x_0)^2 + (y - y_0)^2$. The initial seeding locations are described in Sect. 2.3 and can be seen for the first location in Fig. 2.

The initial vascular cell densities are heterogeneous, set to 3% and 5% of the carrying capacity, K, in white and gray matter, respectively; these values fall within the values for cerebral blood volume found from the literature Yamaguchi et al. (1986). All other spatiotemporal variables are initially set to zero. There are no-flux boundary conditions on the outer boundary of the brain for all variables as well as on the CSF that do not allow growth outside of the brain or into CSF regions.

2.2 Nutrient-Based Vascular Efficiency

The extended vascular efficiency term uses a reaction-transport equation to model the nutrient consumption by the tumor cells. Using this reaction-transport equation for the movement and consumption of nutrient, f,[1] the derivation of the vascular efficiency term goes as follows:

$$\frac{\partial f}{\partial t} = \nabla \cdot (D_f \nabla f) + psv(f_{blood} - f) - \eta_c(D_c, \rho)cf - \eta_h(D_h)hf, \tag{9}$$

where p is the permeability of the blood brain barrier to nutrient, s is the vascular surface area per unit volume, f_{blood} is the concentration of nutrient in the blood which is assumed fixed,[2] η_c is the rate of nutrient consumption by normoxic cells and η_h is the nutrient consumption rate by hypoxic cells. We have let η_c depend on the diffusion (D_c) and proliferation (ρ) rates of normoxic cells, as these processes require energy. A larger D_c and ρ will require more energy as the tumor cells migrate and proliferate

[1] We denote this as f to represent fuel for the cells, to avoid reusing n which is already assigned to necrotic cells.

[2] It is well known that nutrient concentrations in blood (such as glucose concentration) fluctuate throughout a single day; however, we are interested in modeling tumor growth over many days and months, so only consider the average nutrient concentration across these daily fluctuations.

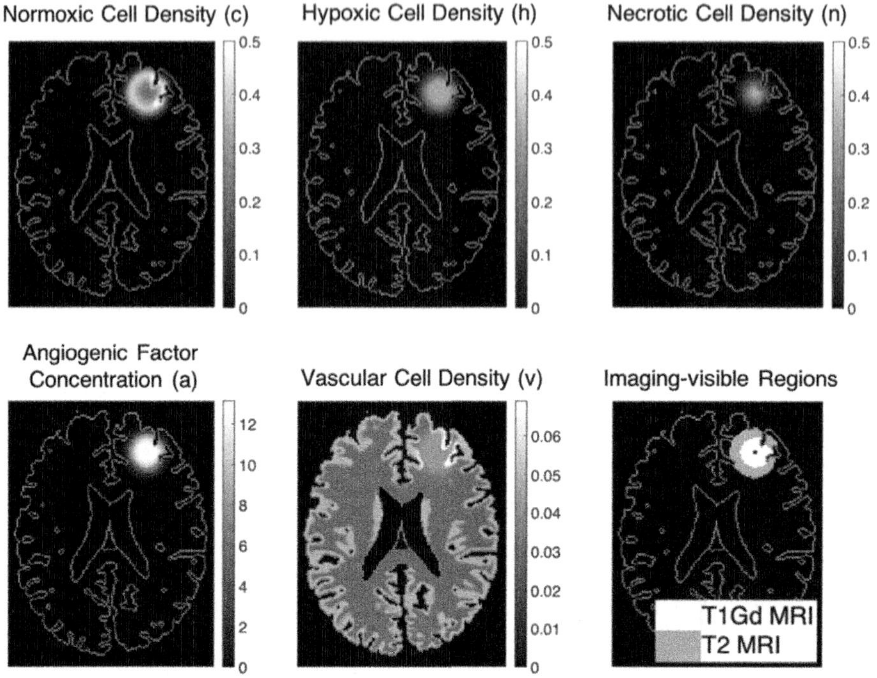

Fig. 2 An example simulation shown at a size equivalent to a circle of 1cm radius on simulated T1Gd MRI. We show all cell densities divided by K and the angiogenic factor concentration divided by K_M. We see how normoxic cells lead the outward growth of the simulated GBM, followed by hypoxic cells and necrotic cells. Angiogenic factors are mostly found in the hypoxic cell region. We also show the regions that are assumed visible on T1Gd MRI ($T \geq 0.8$) and T2 MRI ($T \geq 0.16$) as well as the point where the tumor is initiated (black pixel). In this simulation, $D_h/D_c = 10$, $D_c = 10^{1.5}$mm^2/year, $\rho = 100$/year, $\beta = 0.5\rho$ and $\gamma = 0.05$/day

relatively quickly. Similarly, we have set η_h to depend on the value of D_h, as faster migrating tumor cells require more energy and in turn more nutrient.

Now if we assume that in the timescale of interest, the nutrient concentration rapidly reaches steady state and that the nutrient is consumed much faster than it diffuses, we can eliminate those terms to be left with

$$0 = 0 + psv(f_{blood} - f) - \eta_c(D_c, \rho)cf - \eta_h(D_h)hf \tag{10}$$

and rearrange to get

$$\frac{f}{f_{blood}} = \frac{v}{v + \frac{\eta_c(D_c,\rho)c + \eta_h(D_h)h}{ps}}. \tag{11}$$

We assign this expression as the vascular efficiency term, V, as it corresponds to the ability of the vasculature to provide nutrients to the tumor. This term is similar to that seen in the original formulation of the PIHNA model but now includes the nutrient consumption and extravasation of nutrients from the blood (Swanson et al. 2011).

To estimate the parameters η_c, η_h and ps, we used fludeoxyglucose (FDG) positron emission tomography (PET) data from a paper by Delbeke et al. (1995). FDG is analogous to glucose and can be detected on PET scans. We have chosen glucose as an estimate for our generic nutrient due to the availability of imaging data that we could use to parametrize our vascular efficiency term. Due to the increase in anaerobic respiration of cancer cells compared with normal tissue, known as the Warburg effect (Liberti and Locasale 2016; Warburg 1925), we might expect oxygen uptake ratios to be lower than glucose.

We note that to parametrize our nutrient-based vascular efficiency term, we only need to consider the ratio between η_c/ps and η_h/ps. As both of these expressions are in the same units of $\text{mm}^3/\text{cell}/\text{year}$, their ratio is dimensionless. Delbeke presents the uptake ratios between tumor and healthy tissue within both white and gray matter. To make use of these values, we assume that in a homeostatic healthy brain, the rate of glucose being used by healthy tissue that is not vasculature is equal to the rate of glucose entering from the vasculature. We do not, however, model healthy tissue in the current formulation of the PIHNA model. For the benefit of this section, let us introduce unaffected healthy tissue u_0, with glucose uptake rate η_u, we assume

$$ps\,v_0 = \eta_u(u_0 - v_0), \tag{12}$$

where v_0 is the initial background vascular cell density in the PIHNA model and u_0 is the healthy tissue density. We then have $ps = \eta_u(u_0/v_0 - 1)$, which will always be positive in PIHNA simulations as vasculature takes up a small percentage of brain volume compared to other tissue. We assume that in healthy white matter tissue there is 3% vasculature and in gray there is 5%, so we let $v_0/u_0 = 0.03$ in white matter and $v_0/u_0 = 0.05$ in gray matter; these values fall within realistic values for cerebral blood volume (Yamaguchi et al. 1986). Now the ratios of glucose uptake rates by tumor to the glucose uptake rates by healthy tissue given by Delbeke can be considered as various values of $P_c = \eta_c/\eta_u$ and $P_h = \eta_h/\eta_u$ in the PIHNA model. So Equation 11 is now expressed as

$$V = \frac{v}{v + \frac{P_c\eta_u c + P_h\eta_u h}{\eta_u(u_0/v_0 - 1)}}, \tag{13}$$

and the η_u terms cancel to give

$$V = \frac{v}{v + \frac{P_c c + P_h h}{u_0/v_0 - 1}}. \tag{14}$$

We noted that in the work by Delbeke et al. (1995) there was a spread of relative tumor uptake values for high-grade gliomas within cortical and white matter tissue across 20 patients. As an approximation, we attributed these differences to the nutritional demands of the individual high-grade gliomas. We assign normoxic cells with high (low) D_c and high (low) ρ in the PIHNA model with the higher (lower) glucose uptake rates from the literature, which also vary between white matter and gray matter. We assign hypoxic cells with high (low) D_h high (low) glucose uptake rates in the

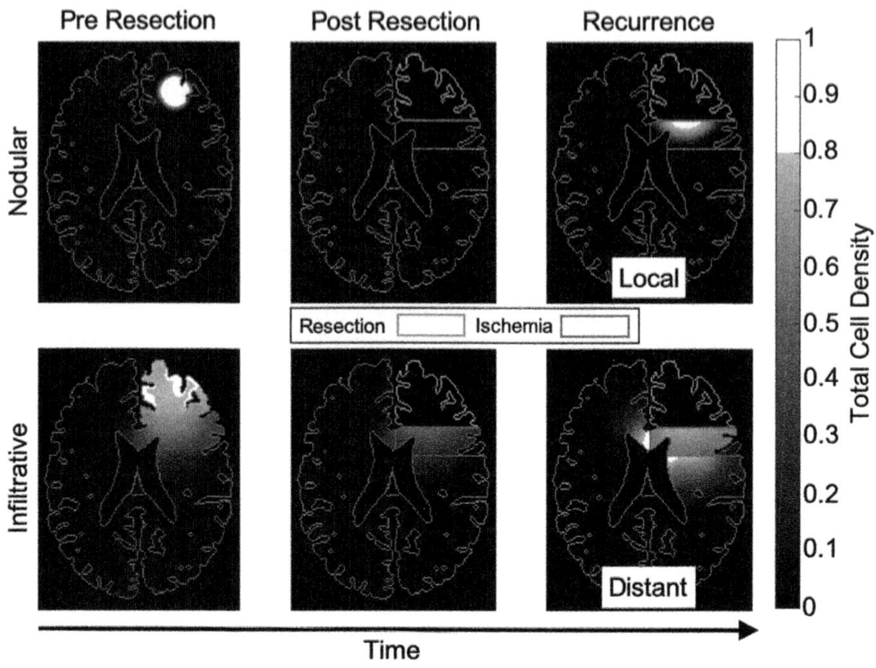

Fig. 3 The tumor undergoes resection that removes the T1Gd imageable tumor cell density at 1cm radius (assumed at 80% of the maximum cell density and shown in white) as well as the surrounding tissue. In these two examples, the nodular tumor (top row) recurs locally, whereas the infiltrative tumor (bottom row) recurs distantly. In these simulations, $\beta = 0.5\rho$, $\gamma = 0.05$/day and $D_h = 10D_c$. For the nodular tumor, $D_c = 10^{0.5}$mm^2/year and $\rho = 10^{1.5}$/year. For the infiltrative tumor, $D_c = 100$mm^2/year and $\rho = 10$/year

same manner as the normoxic cells. The values in between the extremes are assigned using a log linear scale, due to the large range of D_c, D_h and ρ values used in PIHNA simulations. The ratios P_c^w, P_c^g, P_h^w and P_h^g are then given by

$$P_c^{w,g}(D_c, \rho) = G_{max}^{w,g} - (G_{max}^{w,g} - G_{min}^{w,g}) \frac{\log_{10}\left(\frac{D_{c_{max}}}{D_c}\right) + \log_{10}\left(\frac{\rho_{max}}{\rho}\right)}{\log_{10}\left(\frac{D_{c_{max}}}{D_{c_{min}}}\right) + \log_{10}\left(\frac{\rho_{max}}{\rho_{min}}\right)} \tag{15}$$

and

$$P_h^{w,g}(D_h) = G_{max}^{w,g} - (G_{max}^{w,g} - G_{min}^{w,g}) \frac{\log_{10}(D_{h_{max}}/D_h)}{\log_{10}(D_{h_{max}}/D_{h_{min}})}, \tag{16}$$

where we use the extremes of tumor to normal tissue uptake ratios in white matter (taken as $G_{min}^w = 1.66$ and $G_{max}^w = 4.5$) and the extremes of observed uptake ratio in gray matter (taken as $G_{min}^g = 0.5$ and $G_{max}^g = 2$) as the minimum to maximum glucose uptake ratios G_{min} and G_{max}. The maximum and minimum D_c, D_h and ρ values are equal to the maximal and minimal rates that we run in our simulations, see

Table 1. This along with the values of v_0/u_0 give the parametrization of the nutrient-based vascular efficiency term.

2.3 Modeling Resection and Ischemia

Using the PIHNA model, we have simulated a resection that occurs once the tumor has grown to a shape with a volume equivalent to a disk of 1cm radius on simulated T1Gd imaging. The tumor was initiated at $x_0 = 100$, $y_0 = 150$ on the 85th axial slice of the brain geometry. Post-resection, zero-flux boundary conditions are added around the simulated frontal lobectomy (green outline in Fig. 3) so that regrowth into the resection cavity is not possible. This boundary condition was chosen as recurrence into the resection cavity itself is unlikely in GBM. GBM cells have a migratory nature and need structure and nutrients to migrate and grow (Armento et al. 2017). As there is a lack of tissue and vasculature within the resection cavity, it is not an environment conducive to GBM growth. Every resection is the same, in that the same region of brain geometry is removed, which removes all of the enhancing T1Gd region. To incorporate the potential reality that surgery could induce a nearby ischemic event (red outline in Fig. 3), we add subsequent ischemia through a transient reduction in the vasculature term, v, to a region adjacent to the resection cavity wall. We have modeled ischemia as a reduction only at the time point of resection, the vessels then continue to follow the model equations. We reduced the vasculature to 1% of its value at the time of resection thus simulating a near complete ischemic event in the region noted in red in Fig. 3, which is also the value used in all recurrence pattern figures presented in the main text. In further simulations, we have also reduced the vasculature to 10% to see the impact of this less intense simulated ischemia on recurrence locations; results of these simulations are presented in "Appendix" Figs. 12, 13, 14.

We also present a second location with a recurrence that occurs when the tumor has reached a volume of a disk of 0.25cm radius on T1Gd imaging. In this setting, the tumor location was initiated at $x_0 = 44$, $y_0 = 132$ on the 89th coronal slice of the brain geometry. This resection cavity was simulated as a 10 pixel radius around the initial tumor location, with ischemia as a further 5 pixel radius. Zero flux boundary conditions are added to this resection cavity, and perioperative ischemia is simulated in the surrounding tissue.

2.4 Virtual Experiments

We run simulations for different values of normoxic cell migration (D_c with range $1 - 1000\text{mm}^2$/year) and proliferation (ρ with range $10 - 100$/year) and test two values of β (0.1ρ and 0.5ρ), which is the switching rate from the normoxic cell density toward hypoxic cell density, three values of γ (0.005, 0.05, 0.5/day), which is the switching rate back from hypoxic cell density to a normoxic cell density, and the rate of hypoxic to normoxic cell migration, D_h/D_c (1, 10 or 100). We vary the ratio of hypoxic to normoxic cell migration due to evidence that GBM cells migrate faster in hypoxic conditions (Keunen et al. 2011; Zagzag et al. 2006). These parameters were chosen as they represent the tumor's response to hypoxic stress. In previous work, we have observed that all of these parameters (except for β) influence the outward growth rate of PIHNA simulations, which is another consideration of the effects of hypoxia on GBM (Curtin et al. 2020). We note that the varying tumor kinetics (migration rates D_c, D_h and proliferation rate ρ) affect the nodularity of the simulated tumors. Simulations with higher ratios of migration to proliferation will be more infiltrative tumors, whereas those with higher ratios of proliferation to migration will be denser tumor masses with less infiltration and more well-defined tumor cell density boundaries. Examples of this effect can be seen in Fig. 3. We ran further simulations of a second tumor location with $\beta = 0.5\rho$, $\gamma = 0.5$/day and $D_c = 10D_h$, as a validation that the observed dependency of recurrence patterns on migration and proliferation rates were not simply a function of the resection and ischemic regions that we chose in the first location.

2.5 Defining Recurrence Location

Recurrence location of a tumor is classified as the reappearance of the tumor on T1Gd MR scans as is done clinically (Zuniga et al. 2009). If a tumor initially reappears outside of the simulated ischemic region above a certain thresholded size (a disk of radius 2mm on simulated T1Gd MRI) before appearing anywhere else, it is classified as distant. Whereas if it appears within the ischemic region along the cavity wall above the same threshold before anywhere else, it is classed as a local recurrence. Examples of these cases can be seen in Fig. 3. We do not consider the nature of T2 signal in our definition of distant recurrence. We define a mixed recurrence when the tumor appears on simulated T1Gd MRI both inside and outside the ischemic region before the size threshold within either region is reached. This method of defining distant, local and mixed recurrence patterns is applied to both tumor locations presented in this work.

3 Results

3.1 Individual Tumor Kinetics Affect Recurrence Location Following Perioperative Ischemia

Extending on the paper by Thiepold et al., that suggests distant recurrence can occur through ischemia and subsequent hypoxia (Thiepold et al. 2015), the PIHNA model suggests that tumor kinetics also play a role. Figure 3 shows two simulated tumors, one nodular and the other infiltrative, that go through resection and subsequently recur. The recurrence pattern for the nodular tumor is local, whereas the infiltrative tumor recurs distantly. Such distantly recurring tumors remain in lower cell densities within the ischemic region and appear outside on simulated T1Gd imaging as they continue to increase their cell density outside of the ischemic region. The only differences between the two simulations presented in Fig. 3 are the migration and proliferation rates of the tumor cells.

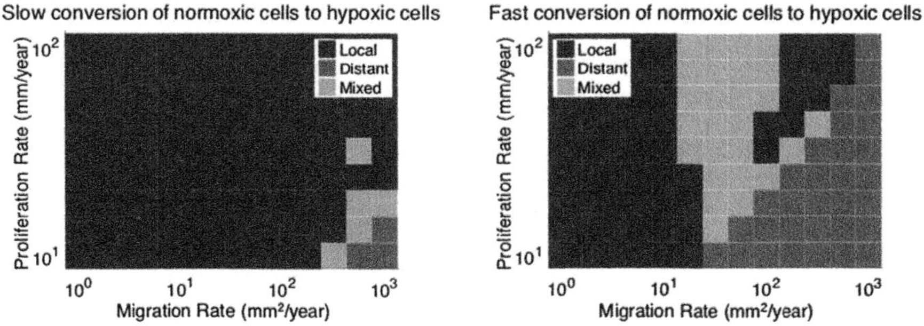

(a) $\beta = 0.1\rho$.

(b) $\beta = 0.5\rho$.

Fig. 4 Recurrence location classified for various D_c, ρ and β for $D_h = 10D_c$ and $\gamma = 0.05$/day. We see that higher values of β (the conversion rate from normoxic to hypoxic cells) lead to a larger proportion of distant recurrences in D_c and ρ parameter space. Higher migration rates, D_c, and lower proliferation rates, ρ, lead to more distantly recurring simulated tumors

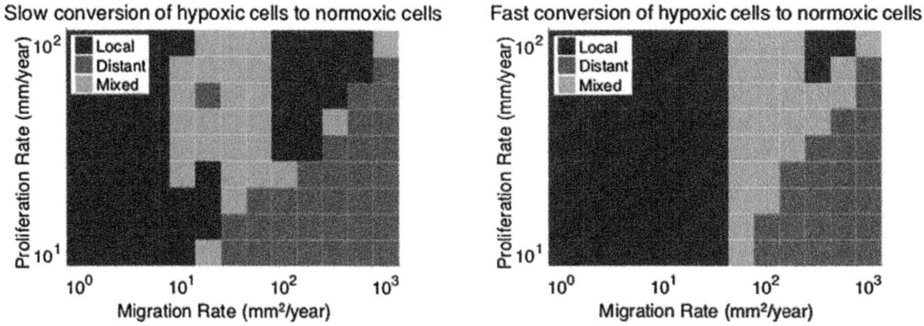

(a) $\gamma = 0.005$/day.

(b) $\gamma = 0.5$/day.

Fig. 5 Recurrence location classified for various D_c, ρ and γ for $D_h = 10D_c$ and $\beta = 0.5\rho$. We see the higher (lower) values of γ lead to a lower (higher) proportion of distant recurrences in D_c and ρ parameter space. Higher migration rates, D_c, and lower proliferation rates, ρ, lead to more distantly recurring simulated tumors

3.2 Tumor Response to Hypoxic Conditions Affects Recurrence Location

By varying individual tumor kinetics (D_c and ρ with $D_h = 10D_c$) and the maximal rate at which tumor cells become hypoxic and in turn necrotic (β), with a fixed vascular ischemia post-resection, we are able to show differing tumor recurrence locations, see Fig. 4. We also varied the maximal rate at which hypoxic tumor cells returned to a normoxic state, γ. Changing this parameter also had an effect on the recurrence patterns, see Fig. 5. A low level of γ promotes more distant recurrence, while a high level of γ promotes local recurrence. Recurrence location is classified as the first reappearance of a tumor on T1Gd MR imaging as described in the previous section.

An increase in β leads to more sensitivity in the tumors to ischemia, which causes them to become more hypoxic and therefore less proliferative within the ischemic

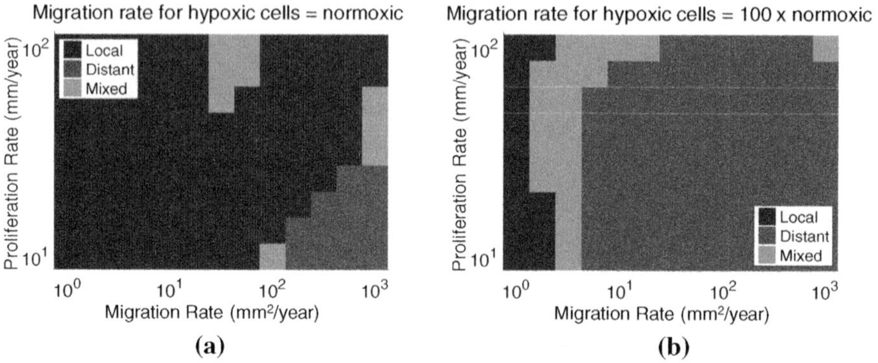

Fig. 6 Recurrence location classified for various D_c, ρ and D_h/D_c for $\beta = 0.5\rho$. We see that the higher (lower) level D_h/D_c leads to a larger (smaller) proportion of distant recurrences in D_c and ρ parameter space. Higher migration rates, D_c, and lower proliferation rates, ρ, lead to more distantly recurring simulated tumors.

region. They are more likely to become more dense and therefore imageable on simulated T1Gd MRI, outside of the ischemic region and be seen as a distant recurrence. Conversely, an increase in γ, the conversion rate from a hypoxic cell phenotype back to normoxic, hinders this effect as it limits the impact of hypoxia on the growth of the simulated tumor. We present other simulation results of varying β and γ in "Appendix" Fig. 10.

3.3 Faster Hypoxic Cell Migration Rates Promote Distant Recurrence

Following this initial analysis, we also varied the hypoxic diffusion rate relative to the normoxic counterpart, D_h/D_c. Along with the simulations where $D_h = 10D_c$ described in the previous section, we have set $D_h/D_c = 1$ and $D_h/D_c = 100$ (see Fig. 6). We see that the higher the D_h/D_c value, the more distantly recurring tumors occur for fixed values of β and γ. The effect of an increase in D_h/D_c is more pronounced for tumors that are more sensitive to the hypoxic environment caused by the ischemia, see "Appendix" Figs. 9, 10, 11. In previous work, we have shown that an increase in D_h/D_c increases the outward growth rate of PIHNA-simulated GBM (Curtin et al. 2020). With faster hypoxic migration rates, the simulated tumor cell densities are able to travel through the hypoxic region faster. These tumors can then reach the region of the brain slice unaffected by ischemia and develop into a dense tumor mass before the tumor develops within the ischemic region.

3.4 Diffuse Recurrence Present Through an Ischemia-Induced Reduction in Dense Tumor

The paper by Thiepold considered distal recurrence as either diffuse or distant (Thiepold et al. 2015). Diffuse recurrence presents as a marked increase in T2 signal without an accompanying increase in T1Gd enhancement. To explore how diffuse

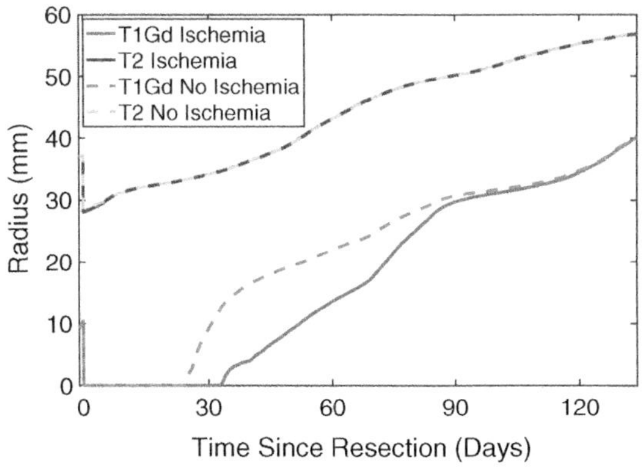

Fig. 7 We present the imageable tumor radii as a function of time for two simulations. The only difference between these simulations is that one has perioperative ischemia and the other does not. Note that the growth of the T1Gd imageable radius is delayed by the ischemia, whereas the T2 imageable radius is minimally affected. This offset of dense tumor growth may be a contributing factor to diffuse recurrence following perioperative ischemia. In these simulations, $D_c = 10^{2.5}$mm^2/year, $\rho = 10^{1.25}$/year, $D_h = 10D_c$, $\beta = 0.5\rho$ and $\gamma = 0.005$/day.

recurrence may occur in the PIHNA model, we present a plot of T1Gd and T2 radii over time for a simulated case of perioperative ischemia, compared with the same simulation without such ischemia. This case recurred distantly and is an example of parameters used in Fig. 5a, with $D_c = 10^{2.5}$mm^2/year, $\rho = 10^{1.25}$/year, $D_h = 10D_c$, $\beta = 0.5\rho$ and $\gamma = 0.005$/day. As can be seen in Fig. 7, we see a reduction in T1Gd signal as a result of the perioperative ischemia for a period of growth, while T2 signal is less affected. If the tumor with perioperative ischemia was observed in this window, the relative T2 signal compared with T1Gd enhancement would appear larger, resulting in a higher probability of the tumor being considered diffuse at recurrence. After some further growth, the T1Gd radii of both simulations meet, as the tumor within the ischemic region eventually grows to a high-enough density to be visible on T1Gd MRI.

3.5 Extent of Vasculature Reduction Impacts Recurrence Patterns

PIHNA model simulations were run with a lower level of reduction in functional vasculature to 10% within the same geometry of perioperative ischemia shown in Fig. 3. This was implemented for all changes in β, γ and D_h presented in the main text. In this setting, all recurrence location results in Figs. 4, 5, 6 showed a shift toward more local recurrence within the values of D_c and ρ that were tested. We present these corresponding recurrence location figures in "Appendix" (Figs. 12, 13, 14).

3.6 Validation in a Second Tumor Location

To explore the simulated recurrence location further, we ran PIHNA simulations for a second tumor location in a coronal view. This tumor was initiated closer to the surface of the brain and resected at a smaller T1Gd imageable tumor volume

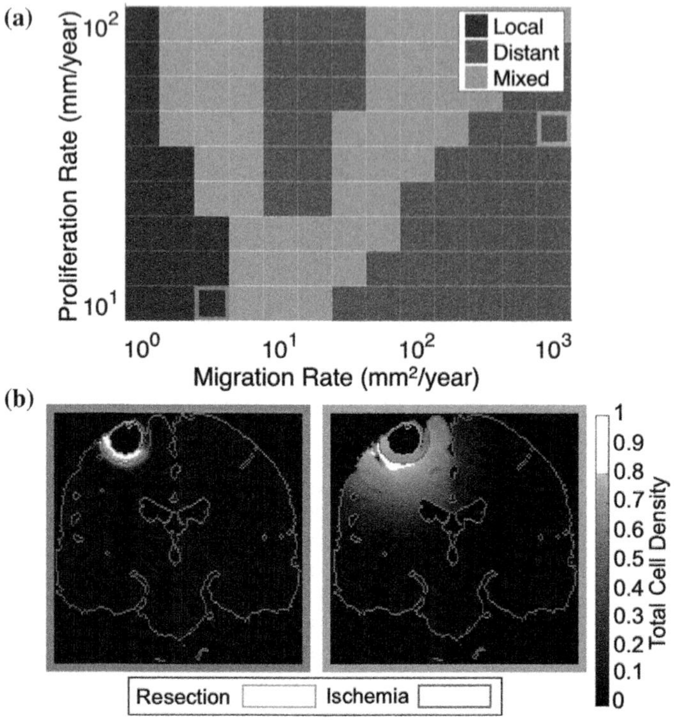

Fig. 8 **a** Recurrence location classified for the second tumor location in coronal view, for various D_c and ρ values. As in the first location, tumors with more diffuse characteristics recur distantly, while those with lower migration rates and faster proliferation rates tended to be more mixed. In these simulations, $\beta = 0.5\rho$, $\gamma = 0.5$/day and $D_h = 10D_c$, parameter values are equivalent to Fig. 5b. **b** Two example simulations, with local recurrence (left) and distant recurrence (right), showing the resection (inner green outline) and ischemic (outer red outline) regions. Migration and proliferation rates of these example simulations are indicated on Subfigure **a**.

(equivalent to a disk of 0.25cm radius). We present recurrence locations and example simulations for $\beta = 0.5\rho$, $\gamma = 0.5$/day and $D_h = 10D_c$ with varying migration and proliferation rates in Fig. 8. Other than the different location and size at resection, these simulations have the same parameters as Fig. 5b. As we observed in the first location, tumors with more diffuse characteristics recurred distantly, whereas those with more nodular characteristics and lower migration rates tended to recur locally or mixed. In this second location, we observe some simulated tumors with faster proliferation rates but mid-range migration rates that also recur distantly, which we also saw some evidence of in the first location, present in Figs. 4b, 5a and 6a.

4 Discussion

Through mathematical modeling, we have found a possible mechanism for distal GBM recurrence in response to ischemia. If the tumor has an invasive phenotype, it can remain unimageable on simulated T1Gd MRI as it travels through the ischemic region (using our assumed threshold of 80% total cell density). Once it reaches healthy intact vasculature, it will return to a normoxic phenotype and proliferate to an imageable density outside of the ischemic region before it does so next to the cavity wall. We see that the switching rate from normoxic cells to hypoxic cells plays a role in this behavior, increasing this rate leads to more distantly recurring tumors within the parameter range of D_c and ρ that we have used (Fig. 4). Conversely, increasing the recovery rate from a hypoxic cell phenotype to a normoxic cell phenotype leads to less distantly recurring tumors (Fig. 5). We also note that an increase in the rate of hypoxic cell migration relative to normoxic cell migration promotes distantly recurring tumors (Fig. 6).

The migratory nature of GBM cells is a key limitation of conventional treatment efficacy and contributes to tumor recurrence (Silbergeld and Chicoine 1997). The dependency of distant recurrence on cell migration shown in these simulations suggests that the use of anti-migratory drugs may reduce the cases of distal recurrence, especially in instances of perioperative ischemia. However, this result is purely theoretical at this stage. These results may also be suggestive of tumor response to hypoxic conditions more generally. Future work may explore patient data to compare preoperative infiltration patterns with distance to recurrence.

We have shown the intensity of the ischemia plays a role in the observed simulated recurrence patterns (see "Appendix" Figs. 12, 13, 14). A reduction in functional vasculature to 10% of its pre-resection value does not promote distant recurrence for as many values of D_c and ρ as the lower value of 1%. As the value of 10% does not promote quite as much hypoxia in the ischemic region, more simulations are able to reach a T1Gd imageable density inside this region and recur locally.

Furthermore, we have shown that similar simulated recurrence patterns occur in a second tumor location that was located closer to the surface of the brain than the first and in a more functionally important region. We also simulated the resection of this second location at a smaller size, yet saw similar patterns of recurrence as the first location. This suggests that the observed model behavior is not simply a function of tumor location or the geometries that we chose for the first resection and ischemic regions. We also ran this simulation in a coronal view of the brain to highlight that this recurrence behavior can occur in any plane. In the future, we can bypass this by moving the model to a more realistic 3D space.

Simulated T1Gd MRI volumes are inhibited within the ischemic region, which may explain why diffuse as well as distant recurrences are observed in patients with perioperative ischemia. If the hypoxic cell phenotype were maintained following exposure to ischemia, the tumor as a whole could remain more diffuse in a clinical sense of a large T2 volume relative to T1Gd. Utilizing the PIHNA model may be a useful tool in our effort to understand patterns of recurrence in GBM and understanding the role of ischemia in recurrence and growth patterns more broadly.

Acknowledgements The authors gratefully acknowledge funding from the National Cancer Institute (R01CA164371, U54CA193489) and the School of Mathematical Sciences at the University of Nottingham.

A Recurrence Results of Other PIHNA Simulations

We present the results of PIHNA simulations that were not shown in the main text. The trends in distant recurrence patterns that we observe in the main text all hold in these simulations, supporting our observations regarding D_h/D_c, β, γ, D_c and ρ (Figs. 9, 10, 11, 12, 13, 14).

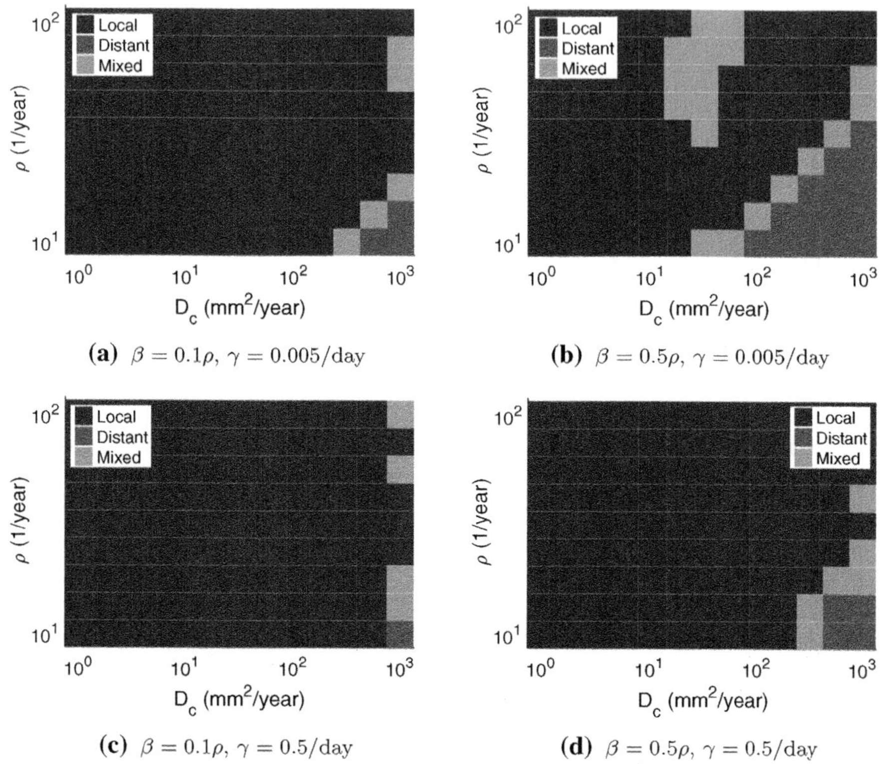

(a) $\beta = 0.1\rho$, $\gamma = 0.005/\text{day}$ (b) $\beta = 0.5\rho$, $\gamma = 0.005/\text{day}$

(c) $\beta = 0.1\rho$, $\gamma = 0.5/\text{day}$ (d) $\beta = 0.5\rho$, $\gamma = 0.5/\text{day}$

Fig. 9 Recurrence location classified for various D_c, ρ, β and levels of ischemia for $D_h = D_c$ for $\gamma = 0.005$/day and $\gamma = 0.5$/day. We see that higher values of β and lower levels of γ lead to a larger proportion of distant recurrences in D_c and ρ parameter space. Higher migration rates, D_c, and lower proliferation rates, ρ, lead to more distantly recurring simulated tumors

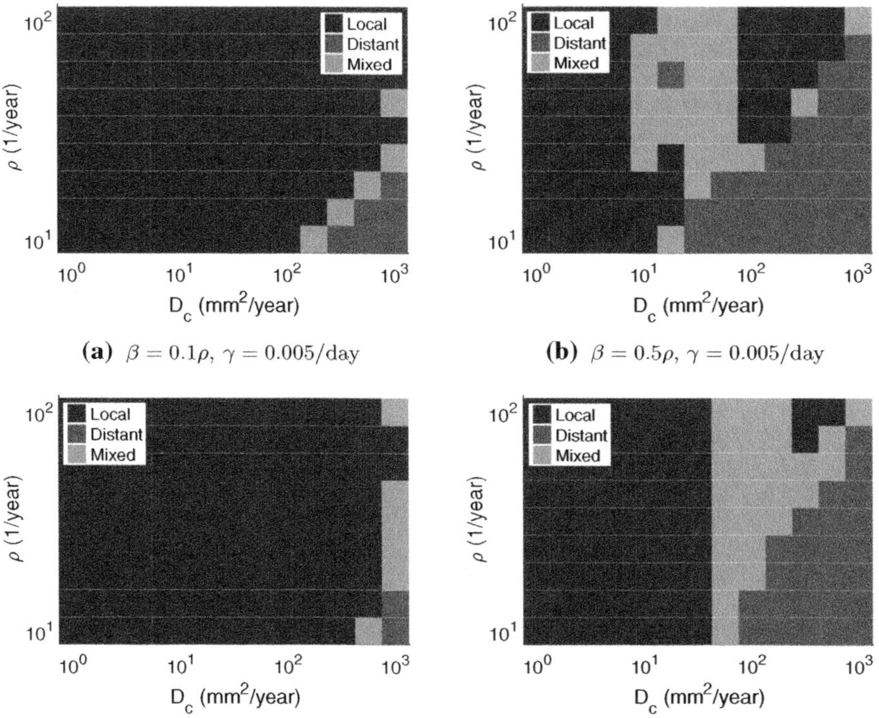

(a) $\beta = 0.1\rho,\ \gamma = 0.005/\text{day}$

(b) $\beta = 0.5\rho,\ \gamma = 0.005/\text{day}$

(c) $\beta = 0.1\rho,\ \gamma = 0.5/\text{day}$

(d) $\beta = 0.5\rho,\ \gamma = 0.5/\text{day}$

Fig. 10 Recurrence location classified for various D_c, ρ, β and levels of ischemia for $D_h = 10D_c$ for $\gamma = 0.005$/day and $\gamma = 0.5$/day. We see that higher values of β and lower levels of γ lead to a larger proportion of distant recurrences in D_c and ρ parameter space. Higher migration rates, D_c, and lower proliferation rates, ρ, lead to more distantly recurring simulated tumors

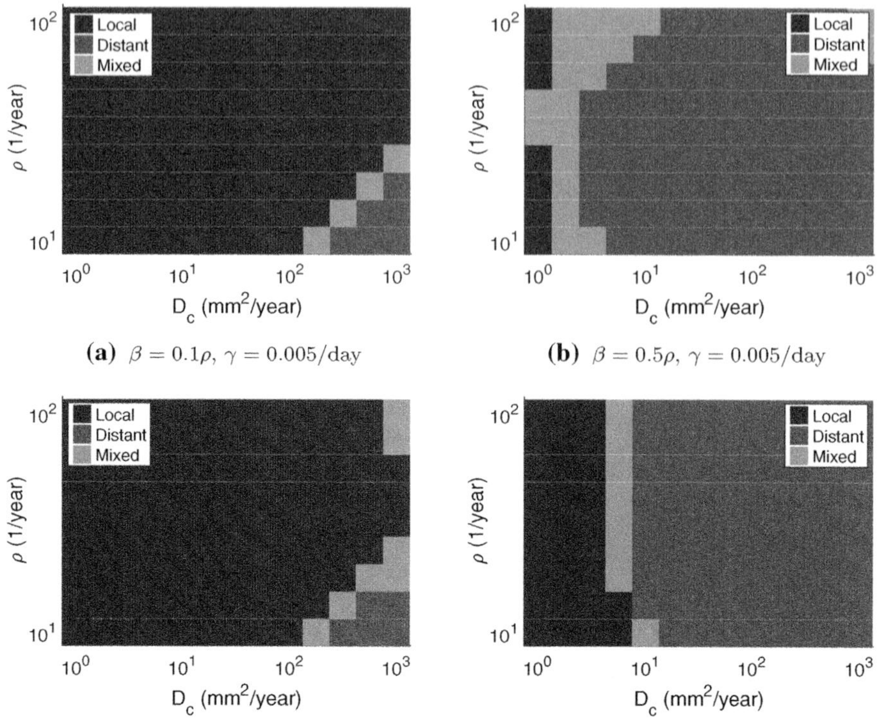

(a) $\beta = 0.1\rho$, $\gamma = 0.005/\text{day}$ **(b)** $\beta = 0.5\rho$, $\gamma = 0.005/\text{day}$

(c) $\beta = 0.1\rho$, $\gamma = 0.5/\text{day}$ **(d)** $\beta = 0.5\rho$, $\gamma = 0.5/\text{day}$

Fig. 11 Recurrence location classified for various D_c, ρ, β and levels of ischemia for $D_h = 100 D_c$ for $\gamma = 0.005$/day and $\gamma = 0.5$/day. We see that higher values of β and lower levels of γ lead to a larger proportion of distant recurrences in D_c and ρ parameter space. Higher migration rates, D_c, and lower proliferation rates, ρ, lead to more distantly recurring simulated tumors

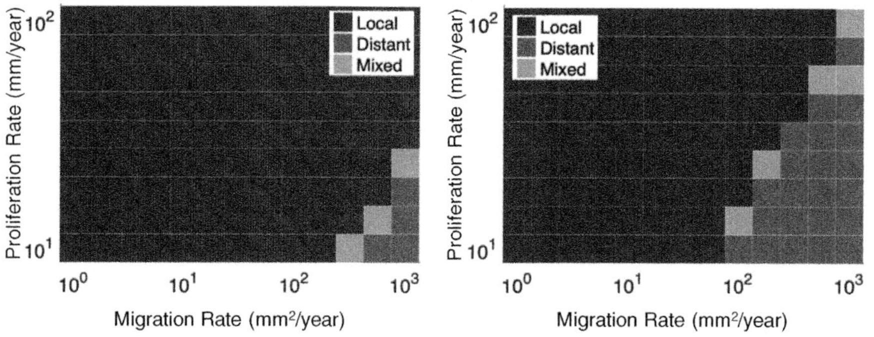

(a) $\beta = 0.1\rho$ **(b)** $\beta = 0.5\rho$

Fig. 12 Recurrence location classified for various D_c, ρ, β and levels of ischemia for $D_h = 10 D_c$ for $\gamma = 0.05$/day. In these simulations, perioperative ischemia was set at 10% of the pre-resection value. We see a larger proportion of local recurrence in these figures compared with those in the main text (see Fig. 4)

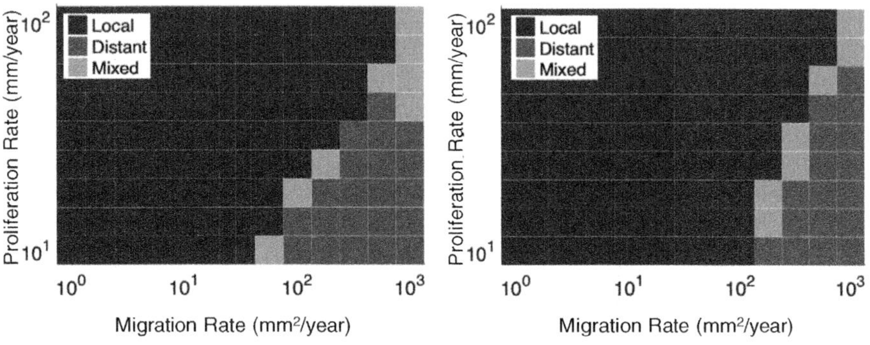

(a) $D_h = D_c$ (b) $D_h = 100D_c$

Fig. 13 Recurrence location classified for various D_c, ρ, β and levels of ischemia for $D_h = 10D_c$ for $\gamma = 0.005$/day. In these simulations, perioperative ischemia was set at 10% of the pre-resection value. We see a larger proportion of local recurrence in these figures compared with those in the main text (see Fig. 5)

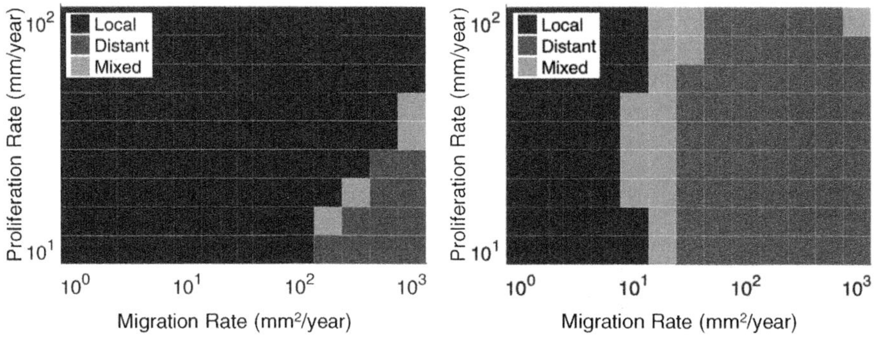

(a) $D_h = D_c$ (b) $D_h = 100D_c$

Fig. 14 Recurrence location classified for various D_c, ρ and D_h levels for $\beta = 0.5\rho$ and $\gamma = 0.05$/day. In these simulations, perioperative ischemia was set at 10% of the pre-resection value. We see a larger proportion of local recurrence in these figures compared with those in the main text (see Fig. 6)

References

Adeberg S, König L, Bostel T, Harrabi S, Welzel T, Debus J, Combs S E (2014) Glioblastoma recurrence patterns after radiation therapy with regard to the subventricular zone. Int J Radiat Oncol Biol Phys 90 4:886–893

Ansarizadeh F, Singh M, Richards D (2017) Modelling of tumor cells regression in response to chemotherapeutic treatment. Appl Math Model 48:96–112

Armento A, Ehlers J, Schötterl S, Naumann U (2017) Molecular mechanisms of glioma cell motility. Exon Publications 73–93

Barazzuol L, Burnet NG, Jena R, Jones B, Jefferies SJ, Kirkby NF (2010) A mathematical model of brain tumour response to radiotherapy and chemotherapy considering radiobiological aspects. J Theor Biol 262(3):553–565

Bette S, Barz M, Huber T, Straube C, Schmidt-Graf F, Combs SE, Delbridge C, Gerhardt J, Zimmer C, Meyer B et al (2018) Retrospective analysis of radiological recurrence patterns in glioblastoma, their prognostic value and association to postoperative infarct volume. Sci Rep 8(1):4561

Bette S, Wiestler B, Kaesmacher J, Huber T, Gerhardt J, Barz M, Delbridge C, Ryang Y-M, Ringel F, Zimmer C et al (2016) Infarct volume after glioblastoma surgery as an independent prognostic factor. Oncotarget 7(38):61945
Boujelben A, Watson M, McDougall S, Yen Y-F, Gerstner ER, Catana C, Deisboeck T, Batchelor TT, Boas D, Rosen B et al (2016) Multimodality imaging and mathematical modelling of drug delivery to glioblastomas. Interface Focus 6(5):20160039
Burnet N, Jena R, Jefferies S, Stenning S, Kirkby N (2006) Mathematical modelling of survival of glioblastoma patients suggests a role for radiotherapy dose escalation and predicts poorer outcome after delay to start treatment. Clin Oncol 18(2):93–103
Chamberlain M (2011) Radiographic patterns of relapse in glioblastoma. J Neuro-oncol 101(2):319–323
Cocosco CA, Kollokian V, Kwan R K-S, Pike GB, Evans AC (1997) Brainweb: Online interface to a 3d mri simulated brain database. In NeuroImage, Citeseer
Collins DL, Zijdenbos AP, Kollokian V, Sled JG, Kabani NJ, Holmes CJ, Evans AC (1998) Design and construction of a realistic digital brain phantom. IEEE Trans Med Imaging 17(3):463–468
Curtin L, Hawkins-Daarud A, Van Der Zee KG, Swanson KR, Owen MR (2020) Speed switch in glioblastoma growth rate due to enhanced hypoxia-induced migration. Bull Math Biol 82(3):1–17
de Rioja VL, Isern N, Fort J (2016) A mathematical approach to virus therapy of glioblastomas. Biol Direct 11(1):1
Delbeke D, Meyerowitz C, Lapidus R, Maciunas R, Jennings M, Moots P, Kessler R (1995) Optimal cutoff levels of f-18 fluorodeoxyglucose uptake in the differentiation of low-grade from high-grade brain tumors with pet. Radiology 195(1):47–52
Frieboes HB, Lowengrub JS, Wise S, Zheng X, Macklin P, Bearer EL, Cristini V (2007) Computer simulation of glioma growth and morphology. Neuroimage 37:S59–S70
Gallaher JA, Massey SC, Hawkins-Daarud A, Noticewala SS, Rockne RC, Johnston SK, Gonzalez-Cuyar L, Juliano J, Gil O, Swanson KR et al (2020) From cells to tissue: How cell scale heterogeneity impacts glioblastoma growth and treatment response. PLoS Comput Biol 16(2):e1007672
Harpold H, Alvord E, Swanson K (2007) The evolution of mathematical modeling of glioma proliferation and invasion. J Neuropathol Exp Neurol 66(1):1–9
Hawkins-Daarud A, Rockne R, Anderson A, Swanson KR (2013) Modeling tumor-associated edema in gliomas during anti-angiogenic therapy and its impact on imageable tumor. Front Oncol 3:66
Keunen O, Johansson M, Oudin A, Sanzey M, Rahim S, Fack F, Thorsen F, Taxt T, Bartos M, Jirik R et al (2011) Anti-vegf treatment reduces blood supply and increases tumor cell invasion in glioblastoma. Proc Natl Acad Sci 108(9):3749–3754
Kwan RK-S, Evans AC, Pike GB (1996) An extensible mri simulator for post-processing evaluation. In: Visualization in biomedical computing, Springer, pp 135–140
Kwan R-S, Evans AC, Pike GB (1999) Mri simulation-based evaluation of image-processing and classification methods. IEEE Trans Med Imaging 18(11):1085–1097
Leder K, Pitter K, LaPlant Q, Hambardzumyan D, Ross BD, Chan TA, Holland EC, Michor F (2014) Mathematical modeling of pdgf-driven glioblastoma reveals optimized radiation dosing schedules. Cell 156(3):603–616
Liberti MV, Locasale JW (2016) The warburg effect: how does it benefit cancer cells? Trends Biochem Sci 41(3):211–218
Louis D, Ohgaki H, Wiestler O, Cavenee W (2016) WHO Classification of Tumours of the Central Nervous System, Revised. Fourth Edition. International Agency for Research on Cancer
Macklin P, Lowengrub J (2007) Nonlinear simulation of the effect of microenvironment on tumor growth. J Theor Biol 245(4):677–704
Martínez-González A, Calvo G, Romasanta L, Pérez-García V (2012) Hypoxic cell waves around necrotic cores in glioblastoma: a biomathematical model and its therapeutic implications. Bull Math Biol 74(12):2875–2896
Neufeld Z, von Witt W, Lakatos D, Wang J, Hegedus B, Czirok A (2017) The role of allee effect in modelling post resection recurrence of glioblastoma. PLoS Comput Biol 13(11):e1005818
Pardo R, Martinez-Gonzalez A, Perez-Garcia VM (2016) Nonlinear ghost waves accelerate the progression of high-grade brain tumors. Commun Nonlinear Sci Numer Simul 39:360–380
Raza SM, Lang FF, Aggarwal BB, Fuller GN, Wildrick DM, Sawaya R (2002) Necrosis and glioblastoma: a friend or a foe? A review and a hypothesis. Neurosurgery 51(1):2–13

Rockne RC, Trister AD, Jacobs J, Hawkins-Daarud AJ, Neal ML, Hendrickson K, Mrugala MM, Rockhill JK, Kinahan P, Krohn KA et al (2015) A patient-specific computational model of hypoxia-modulated radiation resistance in glioblastoma using 18f-fmiso-pet. J R Soc Interface 12(103):20141174

Roniotis A, Sakkalis V, Tzamali E, Tzedakis G, Zervakis, M, Marias K (2012) Solving the pihna model while accounting for radiotherapy. In: Advanced Research Workshop on In Silico Oncology and Cancer Investigation-The TUMOR Project Workshop (IARWISOCI), 2012 5th International, IEEE, pp 1–4

Rutter EM, Stepien TL, Anderies BJ, Plasencia JD, Woolf EC, Scheck AC, Turner GH, Liu Q, Frakes D, Kodibagkar V et al (2017) Mathematical analysis of glioma growth in a murine model. Sci Rep 7(1):1–16

Scribner E, Saut O, Province P, Bag A, Colin T, Fathallah-Shaykh HM (2014) Effects of anti-angiogenesis on glioblastoma growth and migration: model to clinical predictions. PLoS One 9(12):e115018

Silbergeld D, Chicoine M (1997) Isolation and characterization of human malignant glioma cells from histologically normal brain. J Neurosurg 86(3):525–531

Stark AM, van de Bergh J, Hedderich J, Mehdorn HM, Nabavi A (2012) Glioblastoma: clinical characteristics, prognostic factors and survival in 492 patients. Clin Neurol Neurosurg 114(7):840–845

Stein AM, Demuth T, Mobley D, Berens M, Sander LM (2007) A mathematical model of glioblastoma tumor spheroid invasion in a three-dimensional in vitro experiment. Biophys J 92(1):356–365

Stupp R, Hegi M, Mason W, van den Bent M, Taphoorn M, Janzer R, Ludwin S, Allgeier A, Fisher B, Belanger K et al (2009) Effects of radiotherapy with concomitant and adjuvant temozolomide versus radiotherapy alone on survival in glioblastoma in a randomised phase iii study: 5-year analysis of the eortc-ncic trial. Lancet Oncol 10(5):459–466

Subramanian S, Gholami A, Biros G (2019) Simulation of glioblastoma growth using a 3d multispecies tumor model with mass effect. J Math Biol 79(3):941–967

Swan A, Hillen T, Bowman J, Murtha A (2018) A patient-specific anisotropic diffusion model for brain tumour spread. Bull Math Biol 80(5):1259–1291

Swanson K (1999) Mathematical Modeling of the Growth and Control of Tumors. PhD thesis, University of Washington

Swanson KR, Alvord E Jr, Murray J (2000) A quantitative model for differential motility of gliomas in grey and white matter. Cell Prolif 33(5):317–29

Swanson KR, Bridge C, Murray J, Alvord E (2003) Virtual and real brain tumors: using mathematical modeling to quantify glioma growth and invasion. J Neurol Sci 216(1):1–10

Swanson KR, Rockne R, Claridge J, Chaplain M, Alvord E, Anderson A (2011) Quantifying the role of angiogenesis in malignant progression of gliomas: in silico modeling integrates imaging and histology. Cancer Res 71(24):7366–7375

Swanson KR, Rostomily R, Alvord E (2008) A mathematical modelling tool for predicting survival of individual patients following resection of glioblastoma: a proof of principle. Br J Cancer 98(1):113–119

Thiepold A, Luger S, Wagner M, Filmann N, Ronellenfitsch M, Harter P, Braczynski A, Dützmann S, Hattingen E, Steinbach J et al (2015) Perioperative cerebral ischemia promote infiltrative recurrence in glioblastoma. Oncotarget 6(16):14537

Warburg O (1925) The metabolism of carcinoma cells. J Cancer Res 9(1):148–163

Yamaguchi T, Kanno I, Uemura K, Shishido F, Inugami A, Ogawa T, Murakami M, Suzuki K (1986) Reduction in regional cerebral metabolic rate of oxygen during human aging. Stroke 17(6):1220–1228

Yang Y, Hou L, Li Y, Ni J, Liu L (2013) Neuronal necrosis and spreading death in a drosophila genetic model. Cell Death Dis 4(7):e723

Zagzag D, Lukyanov Y, Lan L, Ali MA, Esencay M, Mendez O, Yee H, Voura EB, Newcomb EW (2006) Hypoxia-inducible factor 1 and vegf upregulate cxcr4 in glioblastoma: implications for angiogenesis and glioma cell invasion. Lab Investig 86(12):1221

Zuniga R, Torcuator R, Jain R, Anderson J, Doyle T, Ellika S, Schultz L, Mikkelsen T (2009) Efficacy, safety and patterns of response and recurrence in patients with recurrent high-grade gliomas treated with bevacizumab plus irinotecan. J Neuro-oncol 91(3):329

Bulletin of Mathematical Biology (2020) 82:104
https://doi.org/10.1007/s11538-020-00777-0

Society for Mathematical Biology

SPECIAL ISSUE: CELEBRATING J. D. MURRAY

Characterizing Chemotherapy-Induced Neutropenia and Monocytopenia Through Mathematical Modelling

Tyler Cassidy[1] · Antony R. Humphries[2,3] · Morgan Craig[4,5] · Michael C. Mackey[6,7,8]

Received: 1 April 2020 / Accepted: 11 July 2020 / Published online: 31 July 2020

Abstract
In spite of the recent focus on the development of novel targeted drugs to treat cancer, cytotoxic chemotherapy remains the standard treatment for the vast majority of patients. Unfortunately, chemotherapy is associated with high hematopoietic toxicity that may limit its efficacy. We have previously established potential strategies to mitigate chemotherapy-induced neutropenia (a lack of circulating neutrophils) using a mechanistic model of granulopoiesis to predict the interactions defining the neutrophil response to chemotherapy and to define optimal strategies for concurrent chemotherapy/prophylactic granulocyte colony-stimulating factor (G-CSF). Here, we extend our analyses to include monocyte production by constructing and parameterizing a model of monocytopoiesis. Using data for neutrophil and monocyte concentrations during chemotherapy in a large cohort of childhood acute lymphoblastic leukemia patients, we leveraged our model to determine the relationship between the monocyte and neutrophil nadirs during cyclic chemotherapy. We show that monocytopenia precedes neutropenia by 3 days, and rationalize the use of G-CSF during chemotherapy by establishing that the onset of monocytopenia can be used as a clinical marker for G-CSF dosing post-chemotherapy. This work therefore has important clinical applications as a comprehensive approach to understanding the relationship between monocyte and neutrophils after cyclic chemotherapy with or without G-CSF support.

Keywords Mathematical modeling · Monocytes · Neutrophils · Cyclic chemotherapy · G-CSF · Therapy rationalization

---

Morgan Craig and Michael C. Mackey are co-senior authors.

TC is grateful to the Natural Sciences and Research Council of Canada (NSERC) for support through the PGS-D program. Portions of this work were performed under the auspices of the U.S. Department of Energy under contract 89233218CNA000001 and funded by NIH grants R01-AI116868 and R01-OD011095. ARH is funded by NSERC Discovery Grant RGPIN-2018-05062. MC is funded by NSERC Discovery Grant and Discovery Launch Supplement RGPIN-2018-04546.

Extended author information available on the last page of the article

1 Introduction

In humans, approximately 100,000 hematopoietic stem cells (HSCs) produce 100 billion blood cells a day (Lee-Six et al. 2018). This astonishing hyper-productivity accounts for the hematopoietic system being one of the most intensively studied and best understood stem cell systems (Mackey 2001). All terminal blood cells originate from HSCs that differentiate, proliferate, and mature along many lineages before reaching the circulation to perform their multitude of functions, including oxygenating the body (red blood cells), clotting and wound repair (platelets), and providing immunity from intruders (white blood cells). Neutrophils are important immune regulators and key components of the innate immune response produced by HSCs differentiating into the myeloid lineage, and subsequently undergoing a period of exponential expansion and maturation before transiting into the circulation following sequestration in the bone marrow reservoir (Craig et al. 2016). Monocytes are also myeloid cells that mature within the bone marrow from monoblasts before exiting to the circulation as monocytes, where they remain for several days before marginating in the tissues (Swirski et al. 2014). Tissue-resident monocytes differentiate into macrophages and dendritic cells as a crucial component of an effective immune response (Pittet et al. 2014).

All blood cell production and function is orchestrated through the interaction of a class of small proteins with receptors on the surfaces of HSCs, progenitors, and differentiated cells. These proteins, known as cytokines and chemokines, are produced either by the blood cells themselves or by organs such as the kidneys, liver, and spleen (Roberts 2005; Qian et al. 1998; Athanassakis and Iconomidou 1995). Cytokines act to regulate blood cell production, whereas chemokines direct blood cells through chemotaxis. Though cytokines have overlapping functions to ensure the robustness of the hematopoietic system, many are primary modulators of a specific lineage.

Granulocyte-colony stimulating factor (G-CSF) is the principal cytokine responsible for the regulation of neutrophil production and function (Craig et al. 2015). Its main roles are to modulate the egress of cells from the bone marrow reservoir into the circulation, and control the speed of maturation, proliferation, and differentiation of HSCs into the neutrophil lineage (Craig 2017; Metcalf 1990; Souza et al. 1986; Nagata et al. 1986; Kaushansky 2006; Price et al. 1996; Dale and Mackey 2015; Dale and Welte 2011). G-CSF and neutrophils are inversely related to each other, i.e. when the number of neutrophils in the circulation decreases, G-CSF concentrations rise to stimulate their release and production and vice versa.

This cytokine paradigm (that concentrations of circulating blood cells are reciprocally related to their main cytokine regulator, i.e. a negative feedback) is echoed throughout the hematopoietic system. In particular, monocyte production and function are principally controlled by granulocyte-macrophage colony-stimulating factor (GM-CSF) and macrophage stimulating factor (M-CSF), which act similarly to how G-CSF acts on the neutrophils (Rapoport et al. 1992) to produce monocytes from the HSC, regulate their differentiation into macrophages, and mediate their role in the immune response.

Disrupted cytokine networks are involved in a host of pathologies, including rare oscillatory hematological diseases like cyclic neutropenia (Dale and Hammond 1988;

Dale and Mackey 2015; El Ouriaghli et al. 2003) and cyclic thrombocytopenia (Krieger et al. 2018; Langlois et al. 2017), atherosclerosis (Nahrendorf et al. 2010), and anemia (Mackey 1978). Neutropenia and monocytopenia are characterized by a lack of neutrophils or monocytes, respectively, in circulation and have important consequences for the host's ability to mount an effective immune response. Some cytopenias are inherited and result from mutations to cytokine receptors that inhibit the proper association of cytokines to receptors. Others may be acquired through transitory events including the action of drugs, including cytotoxic agents.

Cytotoxic chemotherapy is designed to disrupt cellular division to control cancer growth. Unfortunately, this cytotoxicity can have significant deleterious effects on other dividing cells in the body, including the rapidly-renewing terminally-differentiated neutrophils (Dale and Mackey 2015; ten Berg et al. 2011; Vansteenkiste et al. 2013). Chemotherapy-induced neutropenia and ensuing infections are a common side effect of cytotoxic chemotherapy (Brooks et al. 2012; Craig et al. 2015; Friberg et al. 2002; Glisovic et al. 2018), necessitating dose size reductions or complete therapy cessation. These therapy modifications leave the patient particularly vulnerable to infection, a major cause of treatment-related mortality even in malignancies with high survival rates (Gatineau-Sailliant et al. 2019; Glisovic et al. 2018). Thus, there is intense interest in dampening the adverse hematological events associated with cytotoxic chemotherapy, through the design of less toxic chemotherapy combinations or the introduction of prophylactic agents. To that end, G-CSF is frequently administered as a rescue drug (prophylactically or as an adjuvant) during cytotoxic chemotherapy (Brooks et al. 2012; Craig et al. 2015). The optimal scheduling of exogenous G-CSF during chemotherapy is an active area of research, and recent mathematical modelling efforts suggest that delaying G-CSF after chemotherapy reduces the incidence and severity of chemotherapy-induced neutropenia (Craig et al. 2015; Krinner et al. 2013; Vainas et al. 2012).

Mathematical modelling has long played a role in understanding healthy and pathological hematopoietic dynamics. Early models aimed to delineate HSC and hematopoietic dynamics (Glass and Mackey 1979; Loeffler and Wichmann 1980; Mackey 1978, 1979; Wichmann and Loeffler 1985; Wichmann et al. 1988) and understand the so-called dynamic diseases (Mackey 2020) like cyclic neutropenia (Haurie et al. 1999; von Schulthess et al. 1983) and thrombocytopenia (von Schulthess and Gessner 1986). Since the discovery of cytokines and their control of hematopoiesis, there has been more recent attention to the role cytokines play for endogenous maintenance of basal cell concentrations (Craig et al. 2016) and on the use of exogenous administration of cytokine mimetics for treatment (Brooks et al. 2012; Craig et al. 2015; Foley and Mackey 2009; Quartino et al. 2012, 2014; Schmitz et al. 1996; Câmara De Souza et al. 2018). Extensive reviews of mathematical models of hematopoiesis are available in Pujo-Menjouet (2016) and Craig (2017).

Despite the role of monocyte-derived macrophages (MDMs) in the resolution of infection, monocyte production has not been extensively studied using mathematical models. To our knowledge, most mathematical modelling efforts that include MDMs have revolved around the innate immune response to infection (Álvarez et al. 2017; Day et al. 2009; Eftimie et al. 2016; Marino et al. 2015; Smith 2011; Smith et al. 2013). As a particular example, Smith (2011) developed a mathematical model for the resolution of pneumonia infection in mice that included three distinct phases of the immune response to pneumonia, including the immune response from tissue resident macrophages and cytokine driven recruitment of circulating neutrophils, and the role of MDMs in infection clearance. There, recruitment of MDMs was driven by the neutrophil concentration in the infected area. Their model shows good agreement with murine data, and has been used to model pneumonia co-infection with influenza and the action of antibiotics (Smith et al. 2013; Schirm et al. 2016). Other formalisms, including hybrid ODE and agent-based models, have also been used to study the role of MDMs in infection. For example, Marino et al. (2015) studied the role of macrophage polarization in *M.tuberculosis* infection. While these models underscore the importance of MDMs in the resolution of infection, their focus was primarily on the site of infection and therefore did not describe the regulation of the production of neutrophils and monocytes in the bone marrow, assuming cytokine- or neutrophil-driven recruitment instead.

Conversely, the role of MDMs in the immune response to malignant tumours has been extensively studied (Boemo and Byrne 2019; Eftimie and Eftimie 2018; Mahlbacher et al. 2018; Owen et al. 2011). Modelling work has focused on macrophage interactions in hypoxic tumours, with recent work also studying the effects of macrophage polarization in tumour progression (Eftimie and Eftimie 2018). Similar to the infection modelling discussed earlier, these models do not study the production of monocytes in the bone marrow, since they are principally focused on understanding tumour-immune dynamics in the tumour microenvironment and not descriptions of systemic monocytopoiesis.

Chemotherapy affects circulating monocyte concentrations and may influence tumour-macrophage interactions. Accordingly, understanding the dynamics of monocyte production during chemotherapy is crucial. Though chemotherapy-induced monocytopenia has previously been linked to an increased risk of developing neutropenia (Kondo et al. 1999), and GM-CSF can be incorporated as a rescue drug during cytotoxic chemotherapy (Tafuto et al. 1995), little attention has been paid in the mathematical literature to understanding monocytopenia during chemotherapy.

Here, we address the issue of monocyte production during chemotherapy by developing a novel model of monocytopoiesis based on our previous work modelling granulopoiesis (Craig et al. 2016). This model is derived directly from the mechanisms of monocyte production and parameterized from homeostatic relationships and a limited amount of data fitting. Leveraging data from a cohort of childhood acute lymphoblastic leukemia patients, we then applied our combined model to understand monocyte and neutrophil dynamics during cyclic chemotherapy. We found that there is an important relationship between the observation of monocytopenia and ensuing neutrophil nadir during cytotoxic chemotherapy. Our results have potentially significant clinical implications as they suggest that the arrival of the monocyte nadir can be used as an indicator to mitigate neutropenia through exogenous G-CSF administration.

2 Neutrophil and Monocyte Modelling

2.1 Model of Granulopoiesis

We have previously described a mathematical model for the physiology of neutrophil production that explicitly accounts for free and bound G-CSF concentrations ($G_1(t)$ and $G_2(t)$, respectively) and their pharmacodynamic effects (Craig et al. 2016). Briefly, the model accounts for the hematopoietic stem cell population $Q(t)$, neutrophils sequestered in the bone marrow reservoir after undergoing exponential proliferation and a period of maturation (denoted by $N_R(t)$), and neutrophils in circulation, $N(t)$. The full set of model equations is given by

$$\begin{aligned}\frac{\mathrm{d}}{\mathrm{d}t}Q(t) &= -\big(\kappa_N(G_1(t)) + \kappa_\delta + \beta(Q(t))\big)Q(t)\\ &\quad + A_Q(t)\beta\left(Q(t-\tau_Q)\right)Q(t-\tau_Q) \end{aligned} \tag{1}$$

$$\begin{aligned}\frac{\mathrm{d}}{\mathrm{d}t}N_R(t) &= A_N(t)\kappa_N(G_1(t-\tau_N(t)))Q(t-\tau_N(t))\frac{V_{N_M}(G_1(t))}{V_{N_M}(G_1(t-\tau_{N_M}(t)))}\\ &\quad - \big(\gamma_{N_R} + \phi_{N_R}(G_{BF}(t))\big)N_R(t) \end{aligned} \tag{2}$$

$$\frac{\mathrm{d}}{\mathrm{d}t}N(t) = \phi_{N_R}(G_{BF}(t))N_R(t) - \gamma_N N(t), \tag{3}$$

$$\begin{aligned}\frac{\mathrm{d}}{\mathrm{d}t}G_1(t) &= I_G(t) + G_{\text{prod}} - k_{\text{ren}}G_1(t)\\ &\quad - k_{12}([N_R(t)+N(t)]V - G_2(t))G_1(t)^{Pow} + k_{21}G_2(t) \end{aligned} \tag{4}$$

$$\frac{\mathrm{d}}{\mathrm{d}t}G_2(t) = -k_{\text{int}}G_2(t) + k_{12}\big([N_R(t)+N(t)]V - G_2(t)\big)G_1(t)^{Pow} - k_{21}G_2(t), \tag{5}$$

with parameter definitions and estimations as in Craig et al. (2016). Throughout, κ_i, A_i, and γ_i denote differentiation, amplification due to proliferation, and death rates, respectively, in the i-th lineage. Variations on this model have previously been applied to the study of cyclic neutropenia (Colijn and Mackey 2005) and to optimize G-CSF regimens during chemotherapy (Brooks et al. 2012; Craig et al. 2015; Foley et al. 2006).

2.2 Pharmacokinetic and Pharmacodynamic Models

Cytotoxic chemotherapy agents interrupt cell division (through disruptions to microtubule assembly or DNA synthesis) to kill malignant cells. Contrary to targeted therapies, this broad cytotoxicity also affects rapidly-dividing hematopoietic progenitor cells in the bone marrow. We model circulating drug concentrations as impacting on granulopoiesis and monocytopoiesis (described below) by accounting for the anti-cancer agent's pharmacokinetics (PK). These PKs then determine the chemotherapeutic drug's pharmacodynamics (PD).

The PKs of chemotherapy are described by a four-compartment model (central C_p, fast-exchange C_f, and two slow-exchange compartments C_{sl_1} and C_{sl_2}) by

$$\begin{aligned}
\frac{\mathrm{d}}{\mathrm{d}t}C_p(t) &= I_C(t) + k_{fp}C_f(t) + k_{sl_1p}C_{sl_1}(t) - (k_{pf} + k_{psl_1} + k_{el_C})C_p(t) \\
\frac{\mathrm{d}}{\mathrm{d}t}C_f(t) &= k_{pf}C_p(t) + k_{sl_2f}C_{sl_2}(t) - (k_{fp} + k_{fsl_2})C_f(t) \\
\frac{\mathrm{d}}{\mathrm{d}t}C_{sl_1}(t) &= k_{psl_1}C_p(t) - k_{sl_1p}C_{sl_1}(t), \\
\frac{\mathrm{d}}{\mathrm{d}t}C_{sl_2}(t) &= k_{fsl_2}C_f(t) - k_{sl_2f}C_{sl_2}(t),
\end{aligned} \tag{6}$$

where k_i indicate rate parameters with values as in Craig et al. (2016).

Since neutrophils no longer divide after exponential expansion, we consider only the HSCs and progenitor cells to be affected by plasmatic chemotherapy concentrations. Monocytes can differentiate from classic (CD14+CD16-) to non-classic (CD16+) types in the blood (described below), but we discount the effects of chemotherapy on this conversion, owing to the short half-life of the drug in the circulation and the relatively low evolution rate from classic to non-classic type. Accordingly, the PD effects of chemotherapy are modelled as

$$A_Q(t) = 2e^{-\gamma_Q\tau_Q - h_Q\int_{t-\tau_Q}^{t} C_p(s)\mathrm{d}s}, \tag{7}$$

i.e. a decrease in the effective amplification (difference between proliferation and death over the time for HSC self-renewal)

$$\frac{\mathrm{d}}{\mathrm{d}t}A_Q(t) = [h_Q(C_p(t-\tau_Q) - C_p(t))]A_Q(t), \tag{8}$$

i.e. a decrease in the proliferation rate for neutrophil and monocyte progenitor cells. In both Eqs. (7) and (8), τ_Q represents the time for HSC self-renewal and h_Q denotes the effect of chemotherapy on HSC amplification A_Q.

G-CSF is an endogenous cytokine that is administered as a drug during chemotherapy. Since the action of G-CSF is to bind to the surface of target cells, we model the pharmacokinetics of G-CSF using a two-compartment model accounting for free and bound concentrations (Eqs. (4) and (5), respectively) where $I_G(t)$ accounts for exogenous administration. As described in Craig et al. (2016), increasing G-CSF concentrations induce egress from the neutrophil bone marrow into circulation at rate ϕ_{N_R}, and act upstream to restock the mature reservoir by increasing the speed of maturation V_{N_M}, the rate of progenitor proliferation η_{N_P}, and the differentiation rate $\kappa(G_1)$ from the HSCs into the granulocyte lineage. These actions are modelled (from HSCs to circulation) as

$$\kappa_N(G_1) = \kappa_N^* + (\kappa_N^* - \kappa_N^{\min}) \left[\frac{G_1^{s_1} - (G_1^*)^{s_1}}{G_1^{s_1} + (G_1^*)^{s_1}} \right]$$

$$\eta_{N_P}(G_1(t)) = \eta_{N_P}^* + (\eta_{N_P}^* - \eta_{N_P}^{\min}) \frac{b_{N_P}}{G_1^*} \left(\frac{G_1(t) - G_1^*}{G_1(t) + b_{N_P}} \right)$$

$$V_{N_M}(G_1(t)) = 1 + (V_{\max} - 1) \frac{G_1(t) - G_1^*}{G_1(t) - G_1^* + b_V}$$

$$\phi_{N_R}(G_{BF}(t)) = \phi_{N_R}^* + (\phi_{N_R}^{\max} - \phi_{N_R}^*) \frac{G_{BF}(t) - G_{BF}^*}{G_{BF}(t) - G_{BF}^* + b_G},$$

where G_{BF} denotes the fraction of G-CSF bound to neutrophil receptors, and all parameters have been estimated in Craig et al. (2016). Similar to the HSCs, we included the effect of cytotoxic chemotherapy on the proliferation of neutrophils by

$$\eta_{N_P}^{\text{chemo}}(G_1(t), C_p(t)) = \frac{\eta_{N_P}(G_1(t))}{1 + \left(C_p(t)/EC_{50}\right)^{s_c}}.$$

2.3 Mathematical Model of Monocyte Production

We model monocytopoiesis similarly to our granulopoiesis model. A schematic of both granulopoiesis and monocytopoiesis is given in Fig. 1.

2.3.1 From Stem Cell to Monocyte

As a simplifying modelling assumption, circulating G-CSF concentrations are used as a cipher for GM-CSF's control of production and proliferation of monocyte precursors by all cytokines. The HSC differentiation rate into the monocyte lineage is given by $\kappa_M(G_1(t))$, implying that the flux of cells entering the monocyte lineage at a given time is $\kappa_M(G_1(t))Q(t)$. the G-CSF dependent differentiation rate is

$$\kappa_M(G_1(t)) = \kappa_M^{\min} + 2\left(\kappa_M^* - \kappa_M^{\min}\right) \frac{G_1(t)}{G_1(t) + G_1^*},$$

where the homeostatic concentration of unbound G-CSF is G_1^*.

Equation (1) models differentiation from HSCs into the monocyte, platelet and erythrocyte lineages via κ_δ. Therefore, we augment the differential equation for the HSCs to explicitly include differentiation into the monocyte lineage by

$$\frac{\mathrm{d}}{\mathrm{d}t} Q(t) = -\big(\kappa_N(G_1(t)) + \kappa_M(G_1(t)) + \hat{\kappa}_\delta + \beta(Q(t))\big) Q(t) + A_Q(t)\beta\left(Q(t - \tau_Q)\right) Q(t - \tau_Q),$$

where $\hat{\kappa}_\delta = \kappa_\delta - \kappa_M^*$.

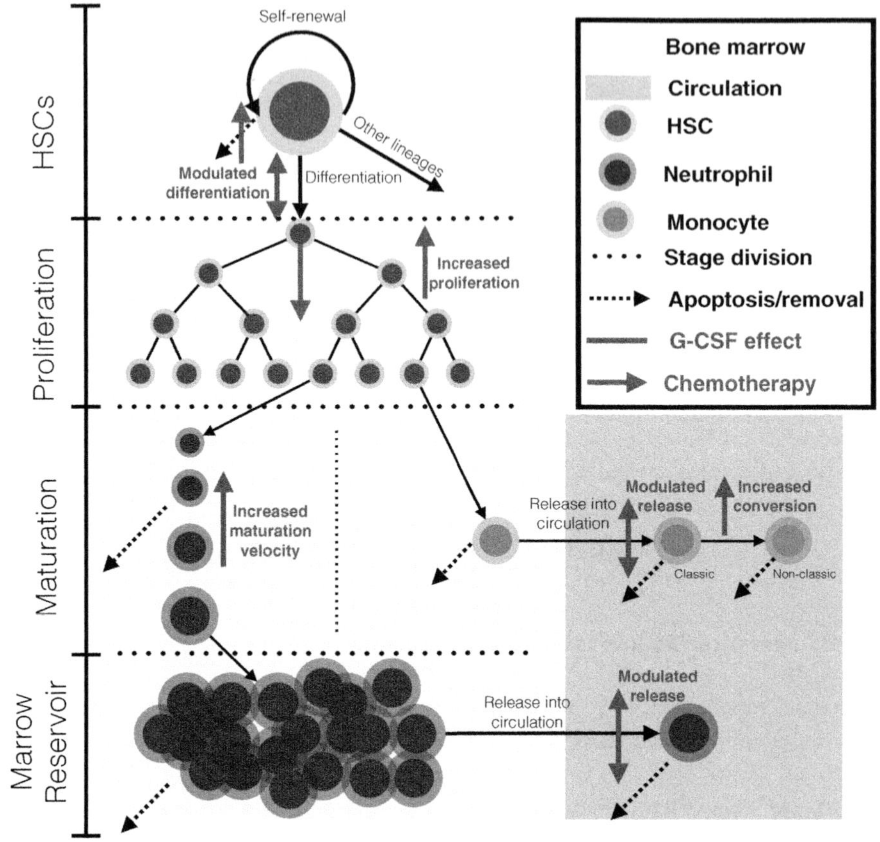

Fig. 1 Production of neutrophils and monocytes. Hematopoietic stem cells self-renew to maintain their population, die, or differentiate into the myeloid (or other) lineages. Progenitors then undergo a period of exponential expansion before committing to becoming neutrophils () or monocytes (). Circulating neutrophils leave the bone marrow after a period of maturation and sequestration in the mature marrow reservoir. Monocytes are released into the circulation, where they may convert from classic to non-class subtypes. Both circulating neutrophils and monocytes are removed from the circulation through apoptosis

After differentiation, monocyte precursors undergo both proliferation and death, with G-CSF acting to decrease the death rate of precursors. Rather than modelling these as separate processes, we combine them into an *effective* proliferation rate given by $\eta_{M_P}(G_1(t))$, similar to the granulopoiesis model (Craig et al. 2016). Monocyte proliferation has been shown to persist even in the absence of G-CSF and other stimulating factors (Boettcher and Manz 2017).

$$\eta_{M_P}(G_1(t)) = \eta_{M_P}^{\max} - (\eta_{M_P}^{\max} - \eta_{M_P}^{*})\frac{G_1^* + b_{MP}}{G_1(t) + b_{MP}},$$

with $\eta_{M_P}^{*}$ denoting the effective proliferation rate at homeostasis. This definition requires a constraint on b_{MP} to ensure that $\eta_{M_P}^{\min} = \eta_{M_P}(0) > 0$. Accordingly, we model the effect of chemotherapy on proliferating monocytes similarly to the neutrophils via

$$\eta_{M_P}^{\text{chemo}}(G_1(t), C_p(t)) = \frac{\eta_{M_P}(G_1(t))}{1 + \left(C_p(t)/EC_{50}\right)^{s_c}}.$$

To model the proliferative process, we use an age structured partial differential equation (Craig et al. 2016; Cassidy et al. 2019; Metz and Diekmann 1986) where the age variable models progression from HSC to mature monocyte. The density of monocyte precursors with age a at time t is denoted by $M_P(t, a)$. Then, as in the granulopoiesis model, the density of monocyte precursors satisfies the age structured partial differential equation (PDE)

$$\frac{\partial M_P(t,a)}{\partial t} + \frac{\partial M_P(t,a)}{\partial a} = \eta_{M_P}(G_1(t))M_P(t,a). \tag{9}$$

We assume that monocyte precursors leave the proliferative process in a deterministic process upon reaching the threshold age $a = a_M$. To define an initial value problem for the age structured PDE (9), appropriate initial and boundary conditions must be provided. As mentioned, HSC-derived monocyte precursors enter directly into proliferation at age $a = 0$, which gives the boundary condition

$$M_P(t, 0) = \kappa_M(G_1(t))Q(t). \tag{10}$$

Analogously, the initial condition of (9) defines the initial density of monocyte precursors throughout the ageing process, and is given by

$$M_P(0, a) = f(a).$$

In general, the function $f(a)$ is assumed to be integrable. However, as we will show, the biological interpretation of $M_P(t, a)$ provides a natural choice of initial density $f(a)$, while for general $f \in L_1(0, \infty)$, the corresponding initial value problem admits a solution $M_P(t, a)$ (Perthame 2007).

The density of monocyte precursors along the characteristic lines of (9) is

$$M_P(t,a) = M_P(t-a, 0)\exp\left[\int_{t-a}^{t} \eta_{M_P}(G_1(s))\mathrm{d}s\right],$$

where

$$M_P(t-a, 0) = \kappa_M(G_1(t-a))Q(t-a)$$

represents the flux of HSC into the monocyte lineage at time $t - a$, and the exponential term models population expansion due to proliferation.

Together, these relationships define the natural form for the initial condition $f(a)$. The density of monocyte precursors at time $t = 0$ given by $f(a)$ signifies that HSCs differentiated into the monocyte lineage some time in the past. Specifically, cells with age a at time 0 enter the monocyte lineage at time $t = -a$, when the flux of HSCs into the monocyte lineage is $\kappa_M(G_1(-a))Q(-a)$. These monocyte precursors proliferate from their entrance into the lineage until time $t = 0$. Thus, the initial density $f(a)$ is given by

$$f(a) = \kappa_M(G_1(-a))Q(-a)\exp\left[\int_{-a}^{0} \eta_{M_P}(G_1(s))\mathrm{d}s\right]. \tag{11}$$

Similar to the neutrophil lineage, we model the time delay between differentiation into the granulocyte-monocyte lineage and the production of a mature monocyte by τ_{M_P}. We assume for simplicity that the proliferative stage takes the same time for monocytes as neutrophils, so $\tau_{M_P} = \tau_{N_P}$. Moreover, we assume that monocyte precursors mature into classic monocytes after reaching the threshold age $a = a_m$ over τ_{M_P} time units. Accordingly, the flux of mature classic monocytes entering the bone marrow is

$$M_P(t, a_m) = \kappa_M(G_1(t - \tau_{M_P}))Q(t - \tau_{M_P})A_M(t),$$

where

$$A_M(t) = \exp\left[\int_{t-\tau_{M_P}}^{t} \eta_{M_P}(G_1(s))\mathrm{d}s\right].$$

These nascent classic monocytes reside in the bone marrow before entering circulation (Nguyen et al. 2013; Mitchell et al. 2014; Van Furth et al. 1973; Mandl et al. 2014). As for the neutrophil reservoir in Craig et al. (2016), we model the bone marrow concentration of classic monocytes as

$$\begin{aligned}\frac{\mathrm{d}}{\mathrm{d}t}M_R(t) &= \kappa_M(G_1(t - \tau_{M_P}))Q(t - \tau_{M_P})A_M(t)\\ &\quad - [\gamma_{M_R} + \phi_{M_R}(G_1(t))]M_R(t),\end{aligned} \tag{12}$$

where γ_{M_R} is the death rate of bone marrow monocytes and ϕ_{M_R} denotes the rate of release of classic monocytes into circulation.

Experimental data has shown an increase in circulating monocytes in response to G-CSF treatment (Hartung et al. 1995; Molineux et al. 1990), which we model by a G-CSF-dependent ϕ_{M_R}. However, during viral infections, circulating G-CSF levels increase without a noticeable increase in monocytes (Pauksen et al. 1994). Thus, slightly increased G-CSF concentrations do not lead to monocyte release from the bone marrow, while administration of large amounts of G-CSF does induce depletion of the bone marrow monocyte concentration and a corresponding increase in circulating monocyte concentrations. To account for this threshold G-CSF concentration, we introduce $G_1^{thres} = 2.5G_1^*$ to denote the minimal G-CSF concentration inducing increased monocyte recruitment, with ϕ_{M_R} given by

$$\phi_{M_R}(G_1(t)) = \phi_{M_R}^* + \Phi(G_1(t)), \tag{13}$$

where

$$\Phi(G_1(t)) = (\phi_{M_R}^{\max} - \phi_{M_R}^*)\left(\frac{\theta^{s_{M_R}}}{\theta^{s_{M_R}} + b_{M_R}^{s_{M_R}}}\right) \quad \text{and} \quad \theta = \max\left[0, G_1(t) - G_1^{thres}\right].$$

here b_{M_R} represents the G-CSF concentration for half maximal release of monocytes from the bone marrow. We assume that, if the G-CSF concentration is large enough to trigger increased monocyte egress from the bone marrow, then this release will be near maximal, rather than gradual, and thus set $s_{M_R} = 3$.

2.3.2 Monocyte Dynamics in Circulation

Egress of classic monocytes from the bone marrow into circulation is modulated by CCR2 (Patel et al. 2017; Shi and Pamer 2011). Circulating classic monocytes are then either cleared from the circulation at a constant rate (γ_{M_C}) or evolve into non-classic monocytes (Patel et al. 2017). Classic and non-classic monocytes are distinguished by their expression of CD14 and CD16 surface markers. While circulating monocytes express CD16 in a continuous spectrum (Shantsil et al. 2011), we separate the monocyte population into classic (low CD16) and non-classic (high CD16) compartments.

Increases to the population of CD16+ monocytes have been observed during inflammatory conditions (West et al. 2012; Patel et al. 2017; Strauss-Ayali et al. 2007; Wong et al. 2011). We therefore model the evolution of classic monocytes into non-classic monocytes at the G-CSF-dependent rate $\nu(G_1(t))$ (Ingersoll et al. 2011; Kratofil et al. 2017; Wong et al. 2012), and assume that the evolution from classic to non-classic monocyte takes place at a saturable rate given by

$$\nu(G_1(t)) = \nu^{\max} - (\nu^{\max} - \nu^*)\frac{G_1^* + c_{M_C}}{G_1(t) + c_{M_C}}. \tag{14}$$

The circulating classic monocyte population dynamics are then governed modelled by

$$\frac{\mathrm{d}}{\mathrm{d}t}M_C(t) = \phi_{M_R}(G_1(t))M_R(t) - \nu(G_1(t))M_C(t) - \gamma_{M_C}M_C(t). \tag{15}$$

Since we assume that non-classic monocytes are only produced via evolution from the classic compartment and are cleared at a constant rate γ_{M_N}, the dynamics of the non-classic monocytes are given by

$$\frac{\mathrm{d}}{\mathrm{d}t}M_N(t) = \nu(G_1(t))M_C(t) - \gamma_{M_N}M_N(t). \tag{16}$$

Taken together, the complete model of monocyte production is

$$\begin{cases} \frac{\mathrm{d}}{\mathrm{d}t}M_R(t) = \kappa_M(G_1(t - \tau_{M_P}))Q(t - \tau_{M_P}(t))A_M(t) \\ \qquad\qquad - [\gamma_{M_R} + \phi_{M_R}(G_1(t))]M_R(t) \\ \frac{\mathrm{d}}{\mathrm{d}t}M_C(t) = \phi_{M_R}(G_1(t))M_R(t) - \nu(G_1(t))M_C(t) - \gamma_{M_C}M_C(t) \\ \frac{\mathrm{d}}{\mathrm{d}t}M_N(t) = \nu(G_1(t))M_C(t) - \gamma_{M_N}M_N(t). \end{cases}$$

When coupled with Eqs. (1)–(5), the combined granulocytopoiesis-monocytopoiesis model is a system of state dependent discrete delay differential equations. To define the corresponding initial value problem, we must prescribe initial data defined over the delay interval. While we use homeostatic initial conditions throughout, this initial data typically is a continuous function defined over $[t_0 - \tau, t_0]$, where $\tau = \max[\tau_N(t), \tau_{M_P}, \tau_Q]$ is the maximal delay. Here, $\tau_N(t)$ is defined as in Craig et al. (2016) as the solution of the integral condition

$$\int_{t-\tau_N(t)}^{t} V_{N_M}(G_1(s))\mathrm{d}s = a_N,$$

where $V_{N_M}(G_1(s)) \geq V_{N_M}^{\min} > 0$, so $\tau_N(t) < a_N/V_{N_M}^{\min} < \infty$ and τ is well-defined and finite. Accordingly, the appropriate phase space for the mathematical model is the infinite dimensional space $C_0([-\tau, 0], \mathbb{R}^8)$. Further, the results from Câmara De Souza et al. (2018) and Cassidy et al. (2019) imply that solutions of the mathematical model evolving from non-negative initial data remain non-negative.

3 Results

3.1 Monocyte Parameter Estimation at Homeostasis

Monocytes and neutrophils share a significant portion of their early developmental pathway, and the point of divergence between the granulocyte and macrophage lineage is not well characterized. Accordingly, we assume that differentiation into the monocyte-macrophage lineage satisfies

$$\frac{\kappa_M}{\kappa_N} = \frac{M_C^* + M_N^*}{N^*}, \quad \text{which gives} \quad 6.0854 \times 10^5 \text{ cells/kg/day} = \kappa_M(G_1^*)Q^*.$$

The homeostatic circulating concentration of monocytes is $M^* = M_C + M_N = (0.034 - 0.062) \times 10^9$ cells/kg (Lichtman 2016; Meuret et al. 1974; Meuret and Hoffmann 1973; Whitelaw 1972). In the calculations that follow, we assume that $M^* = 0.060 \times 10^9$ cells/kg and a 90%:10% ratio of classic to non-classic monocytes (Patel et al. 2017; Strauss-Ayali et al. 2007; Wong et al. 2012, 2011; Ziegler-Heitbrock 2014; Zimmermann et al. 2010). Thus, $M_C^* = 5.4 \times 10^7$ cells/kg and $M_N^* = 6.0 \times 10^6$ cells/kg.

Patel et al. (2017) measured the appearance of deuterium label in circulating monocytes following a 3-h pulse labelling and concluded that monocytes spend between 1.5 and 1.7 days in the bone marrow following proliferation before entering circulation. setting

$$\phi_{M_R}^* = \frac{\log(2)}{1.7} \text{day}^{-1}. \tag{17}$$

To calculate the concentration of mature monocytes, we assume that influx and efflux of classic monocytes are balanced at homeostasis, i.e.

$$\left[\gamma_{M_C} + \nu(G_1^*)\right] M_C^* = \phi_{M_R}^* M_R^*.$$

Classic monocytes have a circulating half-life of roughly $t_{1/2} = 1$ day (Ginhoux and Jung 2014; Patel et al. 2017), which accounts for both clearance of classic monocytes from circulation and evolution into non-classic monocytes. Therefore,

$$\gamma_{M_C} + \nu(G_1^*) = \frac{\log(2)}{1.0} = 0.6931\ \text{day}^{-1},$$

The bone marrow concentration of monocytes at homeostasis is therefore

$$M_R^* = \frac{\gamma_{M_C} + \nu(G_1^*)}{\phi_{M_R}^*} M_C^* = 0.0918 \times 10^9 \text{cells/kg}. \tag{18}$$

The total bone marrow monocyte pool (TBMMP) consists of monocyte precursors and mature monocytes. The exponential nature during proliferation provides a natural upper bound on the number of monocyte lineage cells in the bone marrow. Since DNA synthesis in eukaryotic cells takes approximately 8 h, to compute this upper bound, we allow 1.5 h for the remainder of the cell cycle and assume that cells cannot complete more that 2.5 divisions. This corresponds to a division time of 9.6 h. Then, assuming that there is no death of proliferating cells at homeostasis, there are at most

$$\kappa_M^* Q^* \times 2^{\tau_{M_P} \times 2.5} = 0.4045 \times 10^9 \text{ cells/kg}$$

proliferating monocyte precursors. In our model, the number of monocyte precursors in the bone marrow at homeostasis is given by

$$\int_0^{\tau_{M_P}} \kappa_M(G_1^*) Q^* \exp(\eta_{M_P}^* t) \mathrm{d}t = \frac{\kappa_M(G_1^*) Q^*}{\eta_{M_P}^*} (\exp(\tau_{M_P} \eta_{M_P}^*) - 1).$$

Moreover, there are M_R^* mature monocytes in the bone marrow at homeostasis. Therefore, the TBMMP is

$$\text{TBMMP} = M_R^* + \frac{\kappa_M(G_1^*) Q^*}{\eta_{M_P}^*} [\exp(\tau_{M_P} \eta_{M_P}^*) - 1] \tag{19}$$

Re-arranging equation (19) gives

$$0 = \left[\text{TBMMP} - M_R^*\right] \eta_{M_P}^* - \kappa_M(G_1^*) Q^* \left[\exp(\tau_{M_P}^* \eta_{M_P}^*) - 1\right]. \tag{20}$$

Then, for a given TBMMP, the only unknown in (20) is the homeostatic proliferation rate $\eta_{M_P}^*$ that must satisfy $g(\eta_{M_P}^*) = 0$, where

$$g(x) = \left[\text{TBMMP} - M_R^*\right] x - \kappa_M(G_1^*) Q^* \left[\exp(\tau_{M_P}^* x) - 1\right]. \tag{21}$$

Since

$$g(0) = 0 \quad \text{and} \quad \lim_{x\to\infty} g(x) = -\infty, \tag{22}$$

and further, $g(x)$ is strictly increasing at $x = 0$ if

$$\text{TBMMP} - M_R^* > \tau_{M_P}^* \kappa_M(G_1^*)Q^*.$$

Accordingly, there can only be effective proliferation if the monocyte precursor pool is strictly larger than the accumulation of monocyte precursors solely due to input from the HSCs. With $\eta_{M_P}^*$ from Eq. (20), the death rate of bone marrow monocytes, γ_{M_R}, can then be calculated from

$$\kappa_M(G_1^*)Q^* \exp\left[\eta_{M_P}^* \tau_{M_P}^*\right] - \phi_{M_R}^* M_R^* = \gamma_{M_R} M_R^*. \tag{23}$$

Non-classic monocytes spend roughly 5–7 days in circulation (Ginhoux and Jung 2014; Patel et al. 2017), so

$$\gamma_{M_N} = \frac{\log(2)}{6} = 0.1155 \text{ 1/day}. \tag{24}$$

Therefore, at homeostasis,

$$\gamma_{M_N} M_N^* = \nu(G_1^*) M_C^*, \tag{25}$$

and

$$\gamma_{M_C} = \frac{\log(2)}{1.0} - \gamma_{M_N} \frac{M_N^*}{M_C^*}. \tag{26}$$

It has been observed during inflammatory conditions in trauma patients that non-classic monocytes comprise roughly 15% of the monocyte population on average (West et al. 2012). Thus, if M_N^{inf} and M_C^{inf} represent the inflammatory concentration of non-classic and classic monocytes, respectively,

$$\frac{15}{85} = \frac{M_N^{inf}}{M_C^{inf}} = \frac{\frac{\nu^{\max} M_C^{inf}}{\gamma_{M_N}}}{M_C^{inf}},$$

which gives a first approximation for $\nu^{\max} = \frac{3}{17}\gamma_{M_N}$. In the absence of any knowledge of the precise mechanism underlying cytokine driven evolution of classic to non-classic monocytes, we set $c_{M_C} = 100 \times G_1^*$.

Finally, TBMMP, $\phi_{M_R}^{\max}$, b_{M_R}, $\eta_{M_P}^{\max}$, and b_{M_P} remain to be estimated to fully parameterize the monocytopoiesis model. With the exception of TBMMP, these parameters represent the response of cells in the monocyte lineage to increased G-CSF concentrations. Accordingly, we integrated data for circulating monocyte concentration following the administration of exogenous G-CSF in healthy humans to estimate these parameters by minimizing the L_2 distance between our model's prediction and a spline interpolating the average data reported in Hartung et al. (1995) (Fig. 2). To validate our estimates, we compared model predictions to *a new* data set of two sequential

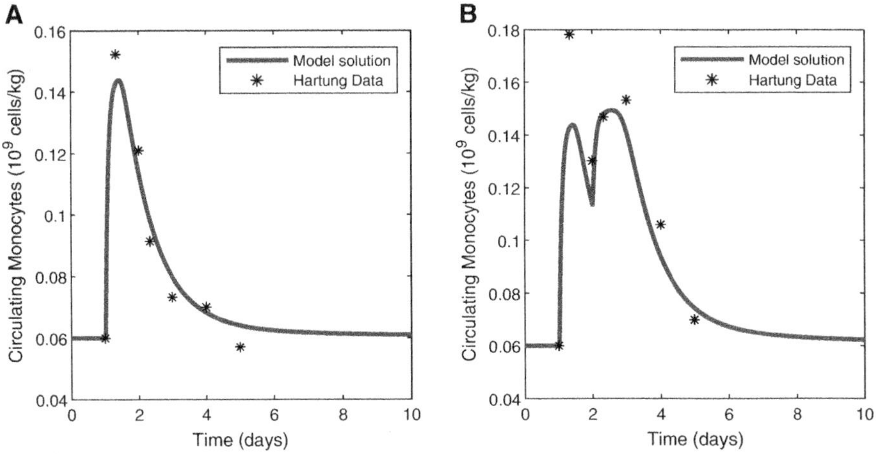

Fig. 2 Parameter fitting results. **a** Monocyte model fitting to the Hartung data for administration of a single dose of 480 000 ¯g/kg G-CSF to healthy human volunteers. **b** Monocyte prediction to Hartung data from administration of two doses of G-CSF to healthy volunteers

administrations of G-CSF (Hartung et al. 1995), and found that our model was able to accurately capture the monocyte dynamics in the second G-CSF administration cycle (Fig. 2).

Values for all parameters are listed in Table 1, and are comparable to previous measurements and estimates. For example, Van Furth et al. (1980) estimated that there are 0.1043×10^9 cells/kg mature monocytes in the bone marrow, while $M_R^* = 0.0918 \times 10^9$ cells/kg. Whitelaw (1972) predicted an egress rate of 1.34×10^7 cells/kg/day, while we calculate $\phi_{M_R}(G_1^*)M_R^* = 3.7410^7$ cells/kg/day. Our prediction that the evolution of classic into non-classic monocytes represents 1.8% of classic monocyte clearance is also close to the transition percentage of 1.4% calculated by Patel et al. (2017).

3.2 Parameter Sensitivity Analysis

To assess the dependency of model predictions to parameter values, we performed a local sensitivity analysis of the monocytopoiesis model (see Craig et al. (2016) for discussion of parameter values in the neutrophil model). As in Cassidy and Craig (2019); Crivelli et al. (2012), we quantify the sensitivity of the model output to changes of $\pm 10\%$ in each parameter value. We further evaluate the time between the monocyte and neutrophil nadirs and the duration of monocytopenia as they are clinically-relevant. The results of this sensitivity analysis show that model predictions are robust to parameter variations (Fig. 3).

3.3 Monocytopenia Precedes Neutropenia

Acute lymphoblastic leukemia (ALL) is characterized by an overabundance of immature lymphocytes and is the most common childhood malignancy. Treatment differs

Table 1 Summary of parameter values as calculated in the text

Parameter	Value	Biological interpretation	Source
TBMMP	0.1619×10^9 cells/kg	Total bone marrow monocyte	Fit
M_R^*	0.0918×10^9 cells/kg	Bone marrow monocyte concentration	(18)
γ_{M_R}	0.7654 /day	Bone marrow monocyte clearance rate	(23)
$\eta_{M_P}^*$	1.5518 /day	Proliferation Rate	(20)
$\eta_{M_P}^{\max}$	1.5620 /day	Proliferation Rate	Fit
b_{M_P}	1.1383 ng/mL	Proliferation half effect of G-CSF	Fit
b_{M_R}	3.9810 ng/mL	Proliferation half effect of G-CSF	Fit
$\phi_{M_R}(G_1^*)$	0.4077 /day	Release rate into circulation	(17)
$\phi_{M_R}^{\max}$	11.5451 /day	Maximal Release rate into circulation	Fit
γ_{M_C}	0.6803 /day	Classic monocyte clearance rate	(26)
$\nu(G_1^*)$	0.0128 /day	Classic to non-classic evolution rate	(14)
γ_{M_N}	0.1155 /day	Non-classic monocyte clearance rate	(24)

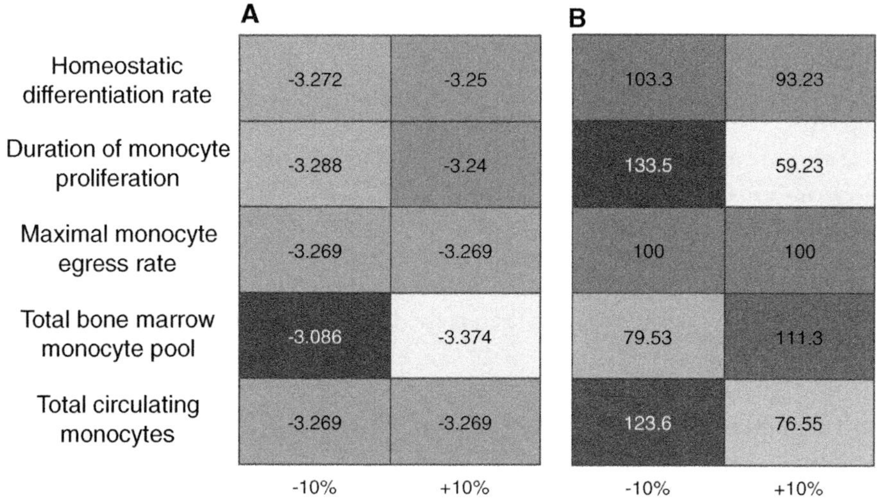

Fig. 3 Results of local parameter sensitivity analysis. **a** Time delay between the nadir of monocyte and neutrophil counts. The negative values indicates that the monocyte nadir occurs before the neutrophil nadir. **b** Parameter sensitivity in duration of monocytopenia when compared to the duration of monocytopenia with fitted parameters. In both cases, parameters were individually varied by 10% from their fitted values

defined based on a clinical risk-assessment, but childhood ALL is generally treated by chemotherapy administered in multiple phases depending on the protocol. Despite a high rate of recovery (5-year survival of 94% for children aged 0–14 in Canada (Canadian Cancer Statistics Advisory Committee 2019)), pediatric patients frequently suffer from neutropenia and monocytopenia during treatment.

Kondo et al. (1999) found that monocyte counts can be predictive for ensuing grade 3 or 4 neutropenia during 3- or 4-week cyclic chemotherapy, which could be explained by discordant production times between monocytes and neutrophils. To assess the impact of cyclic chemotherapy on monocytopoiesis and granulopoiesis, we analyzed neutrophil and monocyte counts from 286 pediatric ALL patients treated with the Dana-Farber Cancer Institute (DFCI) ALL Consortium protocols DFCI 87-01, 91-01, 95-01, or 00-0120-22 at the CHU Sainte-Justine (Montreal, Canada). Previous genomic analysis of this cohort has revealed loci (DARC, GSDM, and CXCL2) predictive of neutropenic complications (Gatineau-Sailliant et al. 2019; Glisovic et al. 2018), as measured by the absolute phagocyte count or APC. We have also identified resonance (chemotherapy induced oscillations) in the neutrophil counts of 26% of these patients during their treatment (Mackey et al. 2020). Here, monocyte and neutrophil numbers were collected at 3-week intervals after induction, prior to the beginning of the next chemotherapy cycle.

We begin by performing classical statistical analyses (Pearson coefficient analysis) on the childhood ALL cohort to investigate possible associations between monocyte counts and ensuing neutropenia during cyclic chemotherapy. Across all patients, we find a very low correlation (average Pearson correlation coefficient $r = 0.27$) between monocyte and neutrophil numbers (Fig. 4a). However, a positive correlative relationship between monocytes and neutrophils is detectable in a smaller subgroup (moderate positive Pearson correlation of $r = 0.53$; Fig. 4b).

The predictive relationship established by Kondo et al. (1999) was identified between monocytes with the onset of neutropenia counts 2–4 days prior to nadir. Unfortunately, due to the sampling rate in our data (monocytes and neutrophils measured once every 3 weeks), we were limited in our ability to discern a predictive relationship between monocytopenia and neutropenia. We therefore sought to leverage our model of granulo- and monocytopoiesis to understand the dynamics of monocyte and neutrophil production during cytotoxic chemotherapy with and without G-CSF support.

To predict monocyte and neutrophil dynamics during cyclic chemotherapy, we simulated chemotherapy administered every three weeks, similar to the DFCI ALL Consortium protocols described above. To quantify the effects of the treatment protocol, we calculated the time to nadir for both monocytes and neutrophils, as well as the beginning and duration of monocytopenic and neutropenic periods. Our results show that both the monocyte nadir and the beginning of monocytopenia precede the beginning of neutropenia, indicating that monocyte concentrations could be used to predict the onset of neutropenia (Fig. 5). We find that the monocyte nadir robustly preceded the neutrophil nadir by at least 3 days in all cases, as in Kondo et al. (1999). As previously mentioned, this observation is perhaps to be expected, since monocytes and neutrophils share much of their development pathway; the lack of a mature monocyte reservoir may account for this time delay.

3.4 G-CSF Support of the Granulocyte-Macrophage Lineage During Chemotherapy

G-CSF is used extensively during chemotherapy to avoid neutropenia. Accordingly, the dynamics of neutropenia during cyclic chemotherapy with and without G-CSF

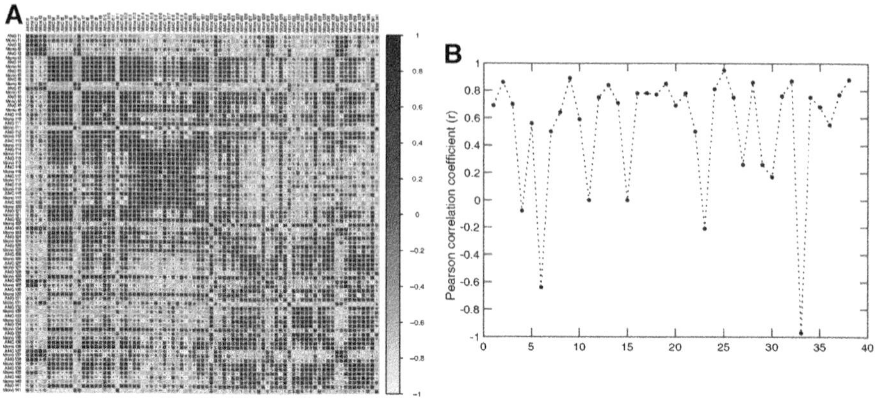

Fig. 4 Correlations between monocyte and neutrophil counts in the childhood ALL cohort. **a** Correlations between monocyte (Mono) and neutrophil (ANC) counts at each sampling point (t) in the full 286 patient cohort. Neutrophil and monocyte sampling points are indicated on horizontal and vertical axes. Deep purple circles represent perfect positive correlation, i.e. that monocyte and neutrophil counts at sampling point t_i and t_j are positively predictive for one another. Deep yellow indicates a strongly negative correlation. As indicated by the lack of clustering, no discernable correlative relationship was found within the complete cohort. **b** Restriction of **a** to comparison of monocytes to neutrophils at the same time point. At each sampling point, the Pearson correlation coefficient between monocytes and neutrophils is plotted. Though some points demonstrate a negative correlation, an average moderate positive correlation ($r = 0.53$) was found, indicating that monocyte and neutrophil counts are positively predictive for each other when measured on the same day during cyclic chemotherapy for this smaller patient group

support have been extensively investigated (Mackey et al. 2020; Craig et al. 2015, 2016; Friberg et al. 2002; Quartino et al. 2014). Previous studies have suggested that delaying the administration of G-CSF after chemotherapy may improve the neutrophil response and help to mitigate ensuing neutropenia (Craig et al. 2015; Vainas et al. 2012; Krinner et al. 2013). However, monocyte dynamics during chemotherapy with G-CSF support are not as well-described. As such, we investigate the impact of G-CSF support of the granulocyte-macrophage lineage during chemotherapy, with a particular emphasis on its effects on monocytopenia to inform the use of G-CSF during cyclic cytotoxic chemotherapy.

We studied 3-week treatment cycles that each began with chemotherapy administered on day 1, followed by subcutaneous G-CSF administrations (Craig et al. 2016) on the nth day following the beginning of monocytopenia. As shown in Fig. 3, neutropenia consistently occurred 3 days after the onset of monocytopenia in the absence of G-CSF support. We therefore restricted the range of possible G-CSF administration days to be $n = 0, 1, 2, 3$ after the monocyte nadir. G-CSF administrations were then continued for T days until the end of the treatment cycle, , and then discontinued until day n in the subsequent cycle. The duration of neutropenia was the metric with which we measured the effectiveness of each schedule; schedules that minimized the neutropenic period were selected as being the most effective.

Our results show that neutropenia is completely mitigated by administering G-CSF every 2 days immediately following the onset of monocytopenia (Fig. 5). The same is true for G-CSF cycles of lengths $T = 3, 4, 6, 7, 8$. In the more physiologically realistic

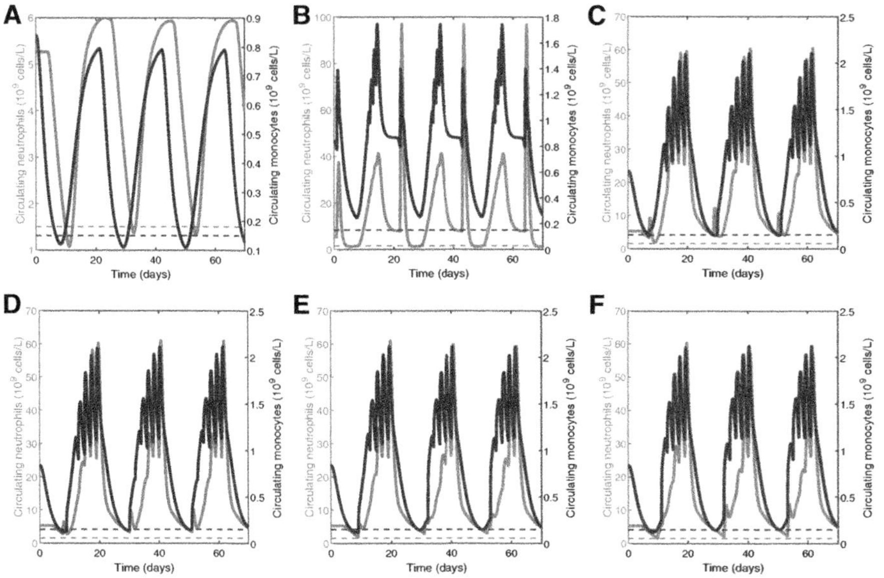

Fig. 5 Modelling predicts that monocytopenia precedes neutropenia **a** Circulating monocyte and neutrophil concentrations in response to chemotherapy administered every 21 days with the thresholds for monocytopenia and neutropenia. **b** Combination chemotherapy administered every 21 days with adjuvant G-CSF administered on 14 consecutive days beginning 1 day post chemotherapy. **c** The results for therapy beginning immediately following the beginning of monocytopenia. **d** Monocytopenia informed G-CSF schedule for therapy beginning 1 day following the beginning of monocytopenia. **e** Monocytopenia informed G-CSF schedule for therapy beginning 2 days following the beginning of monocytopenia. **f** Monocytopenia informed G-CSF schedule for therapy beginning 3 days following the beginning of monocytopenia. Solid dark maroon: monocytes concentrations, dashed dark maroon: monocytopenia threshold, solid orange: neutrophils, dashed orange: neutropenic threshold

cases of G-CSF support beginning at least 1 day after the onset of monocytopenia ($n = 1, 2, 3$), administering G-CSF with a period of $T = 2$ completely avoids neutropenia.

Crucially, we find that $T = 2$ is not the only treatment frequency that eliminates neutropenia. Letting T_i denote a treatment schedule with G-CSF administered every T days beginning on day i following monocytopoiesis, we found that schedules where $T_1 = \{3, 4, 6, 7, 8\}$, $T_2 = \{4, 7, 8\}$, and $T_3 = \{4, 8\}$ also all eliminate neutropenia. This further supports previous modelling work that found that delaying the administration of G-CSF following chemotherapy may decrease the duration of neutropenia by avoiding emptying the mature marrow reservoir (Craig et al. 2016; Vainas et al. 2012; Krinner et al. 2013). Finally, our results suggest that it may be beneficial to only administer G-CSF once the anti-proliferative effects of cytotoxic chemotherapy are observed in the circulating monocyte concentrations, as monocytopenia precedes neutropenia. Interestingly, our results do *not* support the idea that the daily administration of G-CSF is an optimal therapeutic strategy.

4 Discussion

Despite the rapid progression in the development of biologics, cytotoxic chemotherapy remains the standard treatment for most cancers. Given its broad effects on not only cancerous growths but also hematopoietic cells, particular interest should be paid to quantifying blood cell dynamics during cytotoxic chemotherapy. To this end, mathematical modelling has played a significant role, helping to delineate the mechanisms of observed responses to chemotherapy and suggesting improved treatment strategies to help mitigate neutropenia while reducing the therapeutic burden to patients.

Here, to further the quantitative effort to characterize the toxic hematopoietic effects of cytotoxic chemotherapeutic approaches, we extended our previous model of granulopoiesis to include monocytopoiesis. For this, we developed a model of monocyte production beginning from hematopoietic stem cells to their evolution into non-classic monocytes in circulation. Model parameters were then obtained from existing literature sources and through parameter estimation. Parameter estimations were also cross-validated to different data to assess the accuracy of predictions. We also confirmed the robustness of our estimations through local sensitivity analysis.

Importantly, we used our model to characterize monocyte and neutrophil dynamics during chemotherapy in a clinical context. We first assessed the relationship between monocytes and neutrophil concentrations in close to 300 childhood ALL patients treated with 3-week cyclic chemotherapy, and found positive correlations in a subset of the total cohort. Unfortunately, we were unable to validate this observation in all patients.

Based on the results of Kondo et al. (1999), we were interested in quantifying the temporal cascade of monocytopenia and neutropenia, but were unable to assess this relationship due to the frequency of sampling in our data. We therefore simulated chemotherapy administered in 3-week cycles and predicted monocyte and neutrophil responses and found that monocytopenia preceded neutropenia by 3 days. We then investigated the effects of G-CSF administration concurrent to chemotherapy on the granulocyte-macrophage lineage, predicting that delaying the first administration of G-CSF administration until the onset of monocytopenia can reduce or completely eliminate neutropenia. While this prediction differs from the optimal G-CSF administration from our previous work (Craig et al. 2015), the onset of monocytopenia is a clinically relevant measurement.

Despite validating our parameter estimations and model predictions with previous studies, our approach is not without limitations. Some of the parameters used to simulate the monocytopoiesis model were either determined from experimental studies with a limited number of participants, or fit to data following the administration of exogenous G-CSF, which is not the primary regulator of monocyte production. In particular, the fitted maximal proliferation rate $\eta_{M_P}^{\max}$ is very close to the (calculated) homeostatic proliferate rate $\eta_{M_P}^{*}$. This is not surprising, as monocyte proliferation is thought to be independent of G-CSF, but illustrates the need to integrate other relevant cytokines involved in the control of monocytopoiesis when translating our model to infection and other inflammatory conditions. Moreover, the mathematical model has not been parameterized to specifically represent physiological differences present in patients with ALL. Accordingly, our modelling should be interpreted as a qualitative description of the hematopoietic response to chemotherapy, rather than a specific quantitative model. Nonetheless, our model was able to capture dynamics from a broad spectrum of studies in a variety of scenarios, and represents an important development for understanding the relationship between monocyte and neutrophils in the bone marrow and in circulation after cyclic chemotherapy with or without G-CSF support. This work therefore more broadly underlines the contribution of mathematical modelling to rationalizing clinical lines of investigation.

Acknowledgements MCM would like to thank Jim Murray for over 40 years of collegial friendship, and Prof. Dr. Klaus Pawelzik, Universität Bremen, Germany for his hospitality during the time this was written. All authors wish to thank Sanja Glisovic, Drs. Jean-Marie Leclerc, Yves Pastore, and Maja Krajinovic, and the patients enrolled in the ALL study at the CHU Sainte-Justine.

Compliance with Ethical Standards

Conflict of interest The authors declare no conflict of interest.

References

Álvarez E, Toledano V, Morilla F, Hernández-Jiménez E, Cubillos-Zapata C, Varela-Serrano A, Casas-Martín J, Avendaño-Ortiz J, Aguirre LA, Arnalich F, Maroun-Eid C, Martín-Quirós A, Díaz MQ, López-Collazo E (2017) A system dynamics model to predict the human monocyte response to endotoxins. Front Immunol. https://doi.org/10.3389/fimmu.2017.00915

Athanassakis I, Iconomidou B (1995) Cytokine production in the serum and spleen of mice from day 6 to 14 of gestation: cytokines/placenta/spleen/serum. Dev Immunol 4:42412. https://doi.org/10.1155/1995/42412

Boemo MA, Byrne HM (2019) Mathematical modelling of a hypoxia-regulated oncolytic virus delivered by tumour-associated macrophages. J Theor Biol 461:102–116. https://doi.org/10.1016/j.jtbi.2018.10.044

Boettcher S, Manz MG (2017) Regulation of inflammation- and infection-driven hematopoiesis. Trends Immunol 38(5):345–357. https://doi.org/10.1016/j.it.2017.01.004

Brooks G, Provencher G, Lei J, Mackey M (2012) Neutrophil dynamics after chemotherapy and G-CSF: the role of pharmacokinetics in shaping the response. J Theor Biol 315:97–109. https://doi.org/10.1016/j.jtbi.2012.08.028

Câmara De Souza D, Craig M, Cassidy T, Li J, Nekka F, Bélair J, Humphries AR (2018) Transit and lifespan in neutrophil production: implications for drug intervention. J Pharmacokinet Pharmacodyn 45(1):59–77. https://doi.org/10.1007/s10928-017-9560-y. arXiv:1705.08396

Canadian Cancer Statistics Advisory Committee (2019) Canadian Cancer Statistics 2019

Cassidy T, Craig M (2019) Determinants of combination GM-CSF immunotherapy and oncolytic virotherapy success identified through in silico treatment personalization. PLOS Comput Biol 15(11):e1007495. https://doi.org/10.1371/journal.pcbi.1007495

Cassidy T, Craig M, Humphries AR (2019) Equivalences between age structured models and state dependent distributed delay differential equations. Math Biosci Eng 16(5):5419–5450. https://doi.org/10.3934/mbe.2019270. arXiv:1811.05930

Colijn C, Mackey M (2005) A mathematical model of hematopoiesis: II. Cyclical neutropenia. J Theor Biol 237:133–46. https://doi.org/10.1016/j.jtbi.2005.03.034

Craig M (2017) Towards quantitative systems pharmacology models of chemotherapy-induced neutropenia. CPT Pharmacomet Syst Pharmacol 6(5):293–304. https://doi.org/10.1002/psp4.12191

Craig M, Humphries AR, Nekka F, Blair J, Li J, Mackey MC (2015) Neutrophil dynamics during concurrent chemotherapy and G-CSF : mathematical modelling guides dose optimisation to minimize neutropenia. J Theor Biol 385:77–89. https://doi.org/10.1016/j.jtbi.2015.08.015

Craig M, Humphries AR, Mackey MC (2016) A mathematical model of granulopoiesis incorporating the negative feedback dynamics and kinetics of G-CSF/neutrophil binding and internalization. Bull Math Biol 78(12):2304–2357. https://doi.org/10.1007/s11538-016-0179-8

Crivelli JJ, Földes J, Kim PS, Wares JR (2012) A mathematical model for cell cycle-specific cancer virotherapy. J Biol Dyn 6(supl):104–120. https://doi.org/10.1080/17513758.2011.613486

Dale D, Hammond W (1988) Cyclic neutropenia: a clinical review. Blood Rev 2:178–185. https://doi.org/10.1016/0268-960X(88)90023-9

Dale D, Mackey M (2015) Understanding, treating and avoiding hematological disease: better medicine through mathematics? Bull Math Biol 77(5):739–757. https://doi.org/10.1007/s11538-014-9995-x

Dale D, Welte K (2011) Hematopoietic growth factors in oncology. Springer, Heidelberg

Day J, Friedman A, Schlesinger LS (2009) Modeling the immune rheostat of macrophages in the lung in response to infection. Proc Natl Acad Sci 106(27):11246–11251. https://doi.org/10.1073/pnas.0904846106

De Souza DC, Humphries AR (2019) Dynamics of a mathematical hematopoietic stem-cell population model. SIAM J Appl Dyn Syst 18(2):808–852. https://doi.org/10.1137/18M1165086

Eftimie R, Eftimie G (2018) Tumour-associated macrophages and oncolytic virotherapies: a mathematical investigation into a complex dynamics. Lett Biomath 5:S6–S35. https://doi.org/10.1080/23737867.2018.1430518

Eftimie R, Gillard JJ, Cantrell DA (2016) Mathematical models for immunology: current state of the art and future research directions. Bull Math Biol 78(10):2091–2134. https://doi.org/10.1007/s11538-016-0214-9

El Ouriaghli F, Fujiwara H, Metenhorst J, Sconocchia G, Hensel N, Barrett A (2003) Neutrophil elastase enzymatically antagonizes the in vitro action of G-CSF: implications for the regulation of granulopoiesis. Blood 101(5):1752–1758. https://doi.org/10.1182/blood-2002-06-1734

Foley C, Mackey M (2009) Mathematical model for G-CSF administration after chemotherapy. J Theor Biol 257:27–44. https://doi.org/10.1016/j.jtbi.2008.09.043

Foley C, Bernard S, Mackey MC (2006) Cost-effective G-CSF therapy strategies for cyclical neutropenia: mathematical modelling based hypotheses. J Theor Biol 238(4):754–763. https://doi.org/10.1016/j.jtbi.2005.06.021

Friberg LE, Henningsson A, Maas H, Nguyen L, Karlsson MO (2002) Model of chemotherapy-induced myelosuppression with parameter consistency across drugs. J Clin Oncol 20(24):4713–4721. https://doi.org/10.1200/JCO.2002.02.140

Gatineau-Sailliant S, Glisovic S, Gagné V, Laverdière C, Leclerc JM, Silverman LB, Sinnett D, Krajinovic M, Pastore Y (2019) Impact of DARC, GSDMA and CXCL2 polymorphisms on induction toxicity in children with acute lymphoblastic leukemia: a complementary study. Leuk Res 86(September):10–13. https://doi.org/10.1016/j.leukres.2019.106228

Ginhoux F, Jung S (2014) Monocytes and macrophages: developmental pathways and tissue homeostasis. Nat Rev Immunol 14(6):392–404. https://doi.org/10.1038/nri3671

Glass L, Mackey MC (1979) Pathological conditions resulting from instabilities in physiological control systems. Ann N Y Acad Sci 316:214–235

Glisovic SJ, Pastore YD, Gagne V, Plesa M, Laverdière C, Leclerc JM, Sinnett D, Krajinovic M (2018) Impact of genetic polymorphisms determining leukocyte/neutrophil count on chemotherapy toxicity. Pharmacogenomics J 18(2):270–274. https://doi.org/10.1038/tpj.2017.16

Hartung T, Docke W, Gantner F, Krieger G, Sauer A, Stevens P, Volk H, Wendel A (1995) Effect of granulocyte colony-stimulating factor treatment on ex vivo blood cytokine response in human volunteers. Blood 85(9):2482–9

Haurie C, Dale D, Mackey M (1999) Occurrence of periodic oscillations in the differential blood counts of congenital, idiopathic, and cyclical neutropenic patients before and during treatment with G-CSF. Exp Hematol 27:401–409. https://doi.org/10.1016/S0301-472X(98)00061-7

Ingersoll M, Platt A, Potteaux S, Randolph G (2011) Monocyte trafficking in acute and chronic inflammation. Trends Immunol 32(10):470–477. https://doi.org/10.1016/j.it.2011.05.001

Kaushansky K (2006) Lineage-specific hematopoietic growth factors. N Engl J Med 354(19):2034–2045

Kondo M, Oshita F, Kato Y, Yamada K, Nomura I, Noda K (1999) Early monocytopenia after chemotherapy as a risk factor for neutropenia. Am J Clin Oncol 22(1):103–105. https://doi.org/10.1097/00000421-199902000-00025

Kratofil RM, Kubes P, Deniset JF (2017) Monocyte conversion during inflammation and injury. Arterioscler Thromb Vasc Biol 37(1):35–42. https://doi.org/10.1161/ATVBAHA.116.308198

Krieger MS, Moreau JM, Zhang H, Chien M, Zehnder JL, Nowak MA, Craig M (2018) Novel cytokine interactions identified during perturbed hematopoiesis. bioRxiv

Krinner A, Roeder I, Loeffler M, Scholz M (2013) Merging concepts—coupling an agent-based model of hematopoietic stem cells with an ode model of granulopoiesis. BMC Syst Biol 7:117

Langlois GP, Craig M, Humphries AR, Mackey MC, Mahaffy JM, Bélair J, Moulin T, Sinclair SR, Wang L (2017) Normal and pathological dynamics of platelets in humans. J Math Biol 75(6–7):1411–1462. https://doi.org/10.1007/s00285-017-1125-6. arXiv:1608.02806

Lee-Six H, Abro NF, Shepherd MS, Grossmann S, Dawson K, Belmonte M, Osborne RJ, Huntly BJP, Martincorena I, Anderson E, ONeill L, Stratton MR, Laurenti E, Green AR, Kent DG, Campbell PJ (2018) Population dynamics of normal human blood inferred from somatic mutations. Nature 561(7724):473–478. https://doi.org/10.1038/s41586-018-0497-0

Lichtman M (2016) Monocytosis and monocytopenia chap 70. In: Kaushansky K, Lichtman M, Prchal J, Levi M, Press O, Burns L, Caligiuri M (eds) Williams hematology, 9th edn. McGraw-Hill, New York

Loeffler M, Wichmann H (1980) A comprehensive mathematical model of stem cell proliferation which reproduces most of the published experimental results. Cell Tissue Kinet 13:543–561

Mackey MC (1978) Unified hypothesis for the origin of aplastic anemia and periodic hematopoiesis. Blood 51(5):941–956

Mackey MC (1979) Dynamic haematological disorders of stem cell origin. In: Vassileva-Popova JG, Jensen VE (eds) Biophysical and biochemical information transfer in recognition. Plenum Publishing Corporation, New York

Mackey MC (2001) Cell kinetic status of haematopoietic stem cells. Cell Prolif 34(2):71–83. https://doi.org/10.1046/j.1365-2184.2001.00195.x

Mackey MC (2020-submitted) Periodic hematological disorders: quintessential examples of dynamical diseases

Mackey MC, Glisovic S, Leclerc JM, Pastore Y, Krajinovic M, Craig M (2020) The timing of cyclic cytotoxic chemotherapy can worsen neutropenia and neutrophilia. Br J Clin Pharmacol. https://doi.org/10.1111/bcp.14424 (In press)

Mahlbacher G, Curtis LT, Lowengrub J, Frieboes HB (2018) Mathematical modeling of tumor-associated macrophage interactions with the cancer microenvironment. J Immunother Cancer 6(1):10. https://doi.org/10.1186/s40425-017-0313-7

Mandl M, Schmitz S, Weber C, Hristov M (2014) Characterization of the CD14++CD16+ monocyte population in human bone marrow. PLoS ONE 9(11):e112140. https://doi.org/10.1371/journal.pone.0112140

Marino S, Cilfone NA, Mattila JT, Linderman JJ, Flynn JL, Kirschner DE (2015) Macrophage polarization drives granuloma outcome during Mycobacterium tuberculosis infection. Infect Immun 83(1):324–338. https://doi.org/10.1128/IAI.02494-14

Metcalf D (1990) The colony stimulating factors discovery, development, and clinical applications. Cancer 65(10):2185–2195

Metz J, Diekmann O (eds) (1986) The dynamics of physiologically structured populations, vol 68, 3rd edn. Lecture Notes in Biomathematics. Springer, Berlin. https://doi.org/10.1007/978-3-662-13159-6

Meuret G, Hoffmann G (1973) Monocyte kinetic studies in normal and disease states. Br J Haematol 24(3):275–285. https://doi.org/10.1111/j.1365-2141.1973.tb01652.x

Meuret G, Bammert J, Hoffmann G (1974) Kinetics of human monocytopoiesis. Blood 44(6):801–816

Mitchell A, Roediger B, Weninger W (2014) Monocyte homeostasis and the plasticity of inflammatory monocytes. Cell Immunol 291(1–2):22–31. https://doi.org/10.1016/j.cellimm.2014.05.010

Molineux G, Pojda Z, Dexter T (1990) A comparison of hematopoiesis in normal and splenectomized mice treated with granulocyte colony-stimulating factor. Blood 75:563–569

Nagata S, Tsuchiya M, Asano S, Kaziro Y, Yamazaki T, Yamamoto O, Hirata Y, Kubota N, Oheda M, Nomura H et al (1986) Molecular cloning and expression of cDNA for human granulocyte colony-stimulating factor. Nature 319(6052):415–418

Nahrendorf M, Pittet M, Swirski F (2010) Monocytes: protagonists of infarct inflammation and repair after myocardial infarction. Circulation 121(22):2437–2445. https://doi.org/10.1161/CIRCULATIONAHA.109.916346

Nguyen K, Fentress S, Qiu Y, Yun K, Cox J, Chawla A (2013) Circadian gene Bmal1 regulates diurnal oscillations of Ly6Chi inflammatory monocytes. Science 341(6153):1483–1488. https://doi.org/10.1126/science.1240636

Owen MR, Stamper IJ, Muthana M, Richardson GW, Dobson J, Lewis CE, Byrne HM (2011) Mathematical modeling predicts synergistic antitumor effects of combining a macrophage-based, hypoxia-targeted

gene therapy with chemotherapy. Cancer Res 71(8):2826–2837. https://doi.org/10.1158/0008-5472.CAN-10-2834
Patel A, Zhang Y, Fullerton J, Boelen L, Rongvaux A, Maini A, Bigley V, Flavell R, Gilroy D, Asquith B, Macallan D, Yona S (2017) The fate and lifespan of human monocyte subsets in steady state and systemic inflammation. J Exp Med 214(7):1913–1923. https://doi.org/10.1084/jem.20170355
Pauksen K, Elfman L, Ulfgren A, Venge P (1994) Serum levels of granulocyte-colony stimulating factor (G-CSF) in bacterial and viral infections, and in atypical pneumonia. Br J Haematol 88(2):256–260
Perthame B (2007) Transport equations in biology. Frontiers in Mathematics. Springer, Birkhäuser. https://doi.org/10.1007/978-3-7643-7842-4
Pittet MJ, Nahrendorf M, Swirski FK (2014) The journey from stem cell to macrophage. Ann N Y Acad Sci 1319(1):1–18. https://doi.org/10.1111/nyas.12393
Price T, Chatta G, Dale D (1996) Effect of recombinant granulocyte colony-stimulating factor on neutrophil kinetics in normal young and elderly humans. Blood 88:335–340
Pujo-Menjouet L (2016) Blood cell dynamics: half of a century of modelling. Math Model Nat Phenom 11(1):92–115. https://doi.org/10.1051/mmnp/201611106
Qian S, Fu F, Li W, Chen Q, de Sauvage FJ (1998) Primary role of the liver in thrombopoietin production shown by tissue-specific knockout. Blood 92(6):2189–2191. https://doi.org/10.1182/blood.V92.6.2189
Quartino AL, Friberg LE, Karlsson MO (2012) A simultaneous analysis of the time-course of leukocytes and neutrophils following docetaxel administration using a semi-mechanisitic myelosuppression model. Investig New Drugs 30:833–845
Quartino AL, Karlsson MO, Lindman H, Friberg LE (2014) Characterization of endogenous G-CSF and the inverse correlation to chemotherapy-induced neutropenia in patients with breast cancer using population modeling. Pharm Res 31(12):3390–3403. https://doi.org/10.1007/s11095-014-1429-9
Rapoport A, Abboud C, DiPersio J (1992) Granulocyte-macrophage colony-stimulating factor (GM-CSF) and granulocyte colony-stimulating factor (G-CSF): receptor biology, signal transduction, and neutrophil activation. Blood Rev 6(1):43–57. https://doi.org/10.1016/0268-960X(92)90007-D
Roberts A (2005) G-CSF: a key regulator of neutrophil production, but that's not all!. Growth Factors 23(1):33–41. https://doi.org/10.1080/08977190500055836
Schirm S, Ahnert P, Wienhold S, Mueller-Redetzky H, Nouailles-Kursar G, Loeffler M, Witzenrath M, Scholz M (2016) A biomathematical model of pneumococcal lung infection and antibiotic treatment in mice. PLoS ONE 11(5):1–22. https://doi.org/10.1371/journal.pone.0156047
Schmitz S, Franke H, Lffler M, Wichmann HE, Diehl V, Loeffler M, Wichmann HE, Diehl V (1996) Model analysis of the contrasting effects of GM-CSF and G-CSF treatment on peripheral blood neutrophils observed in three patients with childhood-onset cyclic neutropenia. BMC Syst Biol 95(4):616–625
Shantsil E, Wrigley B, Tapp L, Apostolakis S, Montoro-Garcia S, Drayson MT, Lip GYH (2011) Immunophenotypic characterization of human monocyte subsets: possible implications for cardiovascular disease pathophysiology. J Thromb Haemost 9(5):1056–1066. https://doi.org/10.1111/j.1538-7836.2011.04244.x
Shi C, Pamer E (2011) Monocyte recruitment during infection and inflammation. Nat Rev Immunol 11(11):762–774. https://doi.org/10.1038/nri3070
Smith AM, Adler FR, Ribeiro RM, Gutenkunst RN, McAuley JL, McCullers JA, Perelson AS (2013) Kinetics of coinfection with influenza A virus and Streptococcus pneumoniae. PLoS Pathog 9(3):e1003238. https://doi.org/10.1371/journal.ppat.1003238
Smith H (2011) An introduction to delay differential equations with applications to the life sciences, vol 57. Texts in Applied Mathematics. Springer, New York. https://doi.org/10.1007/978-1-4419-7646-8
Souza LM, Boone TC, Gabrilove J, Lai PH, Zsebo KM, Murdock DC, Chazin VR, Bruszewski J, Lu H, Chen KK et al (1986) Recombinant human granulocyte colony-stimulating factor: effects on normal and leukemic myeloid cells. Science 232(4746):61–65
Strauss-Ayali D, Conrad S, Mosser D (2007) Monocyte subpopulations and their differentiation patterns during infection. J Leukoc Biol 82(2):244–252. https://doi.org/10.1189/jlb.0307191
Swirski F, Hilgendorf I, Robbins C (2014) From proliferation to proliferation: monocyte lineage comes full circle. Semin Immunopathol 36(2):137–148. https://doi.org/10.1007/s00281-013-0409-1
Tafuto S, Abate G, D'Andrea P, Silvestri I, Marcelin P, Volta C, Monteverde A, Colombi S, Andorno S, Aglietta M (1995) A comparison of two GM-CSF schedules to counteract the granulo-monocytopenia of carboplatin-etoposide chemotherapy. Eur J Cancer 31(1):46–49. https://doi.org/10.1016/0959-8049(94)00270-F

ten Berg MJ, van den Bemt PM, Shantakumar S, Bennett D, Voest EE, Huisman A, van Solinge WW, Egberts TC (2011) Thrombocytopenia in adult cancer patients receiving cytotoxic chemotherapy. Drug Saf 34(12):1151

Vainas O, Ariad S, Amir O, Mermershtain W, Vainstein V, Kleiman M, Inbar O, Ben-Av R, Mukherjee A, Chan S, Agur Z (2012) Personalising docetaxel and G-CSF schedules in cancer patients by a clinically validated computational model. Br J Cancer 107:814–822. https://doi.org/10.1038/bjc.2012.316

Van Furth R, Diesselhoff-Den Dulk MMC, Mattie H (1973) Quantitative study on the production and kinetics of mononuclear phagocytes during an acute inflammatory reaction. J Exp Med 138:1314–1330

Van Furth R, Diesselhoff-DenDulk MMC, Raeburn JA, Van Zwet TL, Croften R, van Oud Blusse, Albas A (1980) Characteristics, origin, and kinetics of human and murine mononuclear phagocytes. Mononucl PhagocytesFunctional Asp I:279–316

Vansteenkiste J, Wauters I, Elliott S, Glaspy J, Hedenus M (2013) Chemotherapy-induced anemia: the story of darbepoetin alfa. Curr Med Res Opin 29(4):325–337

von Schulthess GK, Gessner U (1986) Oscillating platelet counts in healthy individuals: experimental investigation and quantitative evaluation of thrombocytopoietic feedback control. Scandanavian J Haematol 36(5):473–479

von Schulthess GK, Fehr J, Dahinden C (1983) Cyclic lithium neutropenia: and long-term. Blood 62(2):320–326

West SD, Goldberg D, Ziegler A, Krencicki M, Du Clos TW, Mold C (2012) Transforming growth factor-β, macrophage colony-stimulating factor and C-reactive protein levels correlate with CD14highCD16+ monocyte induction and activation in trauma patients. PLoS ONE 7(12):e52406. https://doi.org/10.1371/journal.pone.0052406

Whitelaw D (1972) Observations on human monocyte kinetics after pulse labeling. Cell Tissue Kinet 5(4):311–317

Wichmann H, Loeffler M (eds) (1985) Mathematical modeling of cell proliferation: stem cell regulation in hemopoiesis. CRC Press, Boca Raton

Wichmann H, Loeffler M, Schmitz S (1988) A concept of hemopoietic regulation and its biomathematical realization. Blood Cells 14:411–429

Wong K, Yeap W, Tai J, Ong S, Dang T, Wong S (2012) The three human monocyte subsets: implications for health and disease. Immunol Res 53(1–3):41–57. https://doi.org/10.1007/s12026-012-8297-3

Wong KL, Tai JJY, Wong WC, Han H, Sem X, Yeap WH, Kourilsky P, Wong SC (2011) Gene expression profiling reveals the defining features of the classical, intermediate, and nonclassical human monocyte subsets. Blood 118(5):e16–e31. https://doi.org/10.1182/blood-2010-12-326355

Ziegler-Heitbrock L (2014) Monocyte subsets in man and other species. Cell Immunol 291(1–2):11–15. https://doi.org/10.1016/j.cellimm.2014.06.008

Zimmermann H, Seidler S, Nattermann J, Gassler N, Hellerbrand C, Zernecke A, Tischendorf J, Luedde T, Weiskirchen R, Trautwein C, Tacke F (2010) Functional contribution of elevated circulating andhepatic non-classical CD14+CD16+ monocytes to inflammation and human liver fibrosis. PLoS ONE. https://doi.org/10.1371/journal.pone.0011049

Affiliations

Tyler Cassidy[1] · Antony R. Humphries[2,3] · Morgan Craig[4,5] · Michael C. Mackey[6,7,8]

✉ Morgan Craig
morgan.craig@umontreal.ca

Tyler Cassidy
tcassidy@lanl.gov

Antony R. Humphries
tony.humphries@mcgill.ca

Michael C. Mackey
michael.mackey@mcgill.ca

[1] Theoretical Biology and Biophysics, Los Alamos National Laboratory, Los Alamos, NM 87545, USA

[2] Department of Mathematics and Statistics, McGill University, Montréal, QC H3A 0B9, Canada

[3] Department of Physiology, McGill University, Montréal, QC H3A 0B9, Canada

[4] Department of Mathematics and Statistics, Université de Montréal, Montréal, Canada

[5] CHU Sainte-Justine Research Centre, University of Montreal, Montréal, Canada

[6] Department of Physiology, McGill University, 3655 Drummond, Montréal, QC H3G 1Y6, Canada

[7] Department of Mathematics and Statistics, McGill University, 3655 Drummond, Montréal, QC H3G 1Y6, Canada

[8] Department of Physics, McGill University, 3655 Drummond, Montréal, QC H3G 1Y6, Canada